The Encyclopedia of Ecotourism

The Encyclopedia of Ecotourism

Edited by

David B. Weaver

Department of Health, Fitness and Recreation Resources,
George Mason University, USA

Advisory editors

Kenneth F. Backman, *Department of Parks, Recreation and Tourism Management, Clemson University, USA*
Erlet Cater, *Department of Geography, University of Reading, UK*
Paul F.J. Eagles, *Department of Recreation and Leisure Studies, University of Waterloo, Canada*
Bob McKercher, *Department of Hotel and Tourism Management, Hong Kong Polytechnic University, Hong Kong, China*

CABI *Publishing*

CABI *Publishing* is a division of CAB *International*

CABI Publishing
CAB International
Wallingford
Oxon OX10 8DE
UK

Tel: +44 (0)1491 832111
Fax: +44 (0)1491 833508
Email: cabi@cabi.org
Web site: http://www.cabi.org

CABI Publishing
44 Brattle Street
4th Floor
Cambridge, MA 02138
USA

Tel: +1 617 395 4056
Fax: +1 617 354 6875
Email: cabi-nao@cabi.org

A catalogue record for this book is available from the British Library, London, UK.

Library of Congress Cataloging-in-Publication Data
The encyclopedia of ecotourism / edited by David B. Weaver.
p. cm.
Topical chapters arranged under a common theme.
Includes bibliographical references.
ISBN 0-815199-368-0 (alk. paper)
1. Ecotourism. I. Weaver, David B. (David Bruce)
G156.E26 E53 2000
338.4´7910–dc21

00-044459

ISBN 0 85199 368 0 (hb)
ISBN 0 85199 682 5 (pb)

First printed 2001
Reprinted 2003

Typeset by Columns Design Ltd, Reading
Printed and bound in the UK by Biddles Ltd, Guildford and King's Lynn

Contents

Contributors

Kenneth F. Backman Department of Parks, Recreation and Tourism Management, Clemson University, Clemson, SC 29634-1005, USA

Sheila Backman Department of Parks, Recreation and Tourism Management, Clemson University, Clemson, SC 29634-1005, USA

Russell K. Blamey Urban and Environmental Program, Research School of Social Sciences, Australian National University, ACT 0200, Australia

Sylvie Blangy Department of Tourism and Sustainable Development, SECA (Société d'Eco-Amènagement), Parc Scientifique Agropolis, 34 397 Montpelier Cedex 5, France

Katrina Brandon 4110 Gallatin Street, Hyattsville, MD 20781-2132, USA. Present address: Center for Applied Biodiversity Science, Conservation International, Washington, District of Columbia, USA

Ralf Buckley International Centre for Ecotourism Research, School of Environmental and Applied Science, Griffith University Gold Coast Campus, Parklands Drive, Southport, Queensland 4217, Australia

Richard W. Butler School of Management Studies for the Service Sector, University of Surrey, Guildford, GU2 5XH, UK

Carl Cater School of Geographical Sciences, University of Bristol, Bristol BS8 1SS, UK

Erlet Cater Department of Geography, University of Reading, Whiteknights, Reading RG6 2AB, UK

Hwan-Suk Choi Department of Recreation, Parks and Tourism Sciences, Texas A&M University, College Station, TX 77843-2261, USA

Judy Cohen Marketing Department, Rider University, 2083 Lawrenceville Road, Lawrenceville, NJ 08648-3099, USA

Peter U.C. Dieke University of Strathclyde, The Scottish Hotel School, 94 Cathedral Street, Glasgow G4 0LG, UK

Ross K. Dowling School of Marketing, Tourism and Leisure, Faculty of Business and Public Management, Edith Cowan University, Joondalup, WA 6027, Australia

Paul F.J. Eagles Department of Recreation and Leisure Studies, University of Waterloo, Waterloo, Ontario, Canada N2L 3G1

David A. Fennell Department of Recreation and Leisure Studies, Faculty of Physical Education and Recreation, Brock University, St Catherines, Ontario, Canada L2S 3A1

Warwick Frost Department of Management, Monash University, Clyde Road, Berwick 3806, Australia

John Gardner Cooper and Gardner Architects, 14 Par-la-Ville Road HM 08, PO Box HM 1376, Hamilton HM FX, Bermuda

Elizabeth A. Halpenny Nature Tourism Solutions, R.R.#2, Almonte, ON K0A 1A0, Canada

Sam H. Ham College of Forestry, Wildlife and Range Sciences, Department of Resource Recreation and Tourism, University of Idaho, Moscow, ID 83844-1139, USA

William E. Hammitt Department of Parks, Recreation and Tourism Management, Clemson University, Clemson, SC 29634-1005, USA

Donald E. Hawkins School of Business and Public Management, The George Washington University, 2121 Eye Street, NW, Washington, DC 20052, USA

Bryan R. Higgins Department of Geography and Planning, State University of New York at Plattsburgh, 101 Broad Street, Plattsburgh, NY 12901-2681, USA

Tom Hinch Faculty of Physical Education and Recreation, University of Alberta, Edmonton, Alberta, Canada T6G 2H9

Jean-Pierre Issaverdis Victoria University, Footscray Park Campus, K538, PO Box 14428, Melbourne, MC 8001, Australia

Tazim B. Jamal Department of Recreation, Parks and Tourism Sciences, Texas A&M University, College Station, TX 77843-2261, USA

Kristin Lamoureux School of Business and Public Management, The George Washington University, 2121 Eye Street, NW, Washington, DC 20052, USA

Jeff Langholz 312 Fernow Hall, Department of Natural Resources, Cornell University, Ithaca, NY 14853, USA. Present address: Program in International Environmental Policy, Monterey Institute of International Studies, Monterey, California, USA

Laura J. Lawton School of Business, Bond University, Gold Coast, Queensland 4229, Australia

Alan A. Lew Department of Geography and Public Planning, Northern Arizona University, PO Box 15016, Flagstaff, AZ 86011-5016, USA

Kreg Lindberg School of Tourism and Hotel Management, Griffith University Gold Coast Campus, PMB 50, Gold Coast Mail Centre, Queensland 9726, Australia

Neil Lipscombe School of Environmental and Information Science, Charles Sturt University, PO Box 789, Albury, NSW 2640, Australia

Bob McKercher Department of Hotel and Tourism Management, Hong Kong Polytechnic University, Hung Hom, Kowloon, Hong Kong, China

Duarte B. Morais School of Hotel, Restaurant and Recreation Management, 201 Mateer Building, Pennsylvania State University, University Park, PA 16802-1307, USA

Mark B. Orams Centre for Tourism Research, Massey University at Albany, Private Bag 102 904, North Shore MSC, New Zealand

Steven Parker Department of Political Science, University of Nevada, Las Vegas, NV 89154-5029, USA

James Petrick Department of Parks, Recreation and Tourism Management, Clemson University, Clemson, SC 29634-1005, USA

Regina Schlüter Centro de Investigaciones y Estudios Turisticos, Avenida del Libertador 774, Piso 6 – Of. 'W', 1001 Buenos Aires, Argentina

Tej Vir Singh Centre for Tourism Research and Development, A-965/6 Indira Nagar, Lucknow, India 226016

Ercan Sirakaya Department of Recreation, Parks and Tourism Sciences, 308 Francis Hall, Texas A&M University, College Station, TX 77843-2261, USA

Bernard Stonehouse Scott-Polar Research Institute, University of Cambridge, Lensfield Road, Cambridge CB2 1ER, UK

Mathew C. Symmonds Department of Parks, Recreation and Tourism Management, Clemson University, Clemson, SC 29634-1005, USA

Rik Thwaites School of Environmental and Information Science, Charles Sturt University, PO Box 789, Albury, NSW 2640, Australia

Sandrine Vautier Wittelsbacherallee 121, D-60385 Frankfurt am Main, Germany

Stephen Wearing School of Leisure and Tourism Studies, University of Technology, Sydney, PO Box 1, Lindfield, NSW 2070, Australia

David B. Weaver Department of Health, Fitness and Resources, George Mason University, Prince William 1, 10900 University Blvd., Manassas, VA 20115, USA

Betty Weiler Department of Management, Monash University, Berwick Campus, PO Box 1071, Narne Warren, VIC 3078, Australia

Pamela A. Wight Pam Wight and Associates, 14715-82 Avenue, Edmonton, Alberta, Canada T5R 3R7

Peter W. Williams School for Resource and Environmental Management, Simon Fraser University, Burnaby, BC, Canada V5A 1S6

Brett A. Wright Center for Recreation Resources Policy, George Mason University, Mail Stop 4E5, 10900 University Blvd, Manassas, VA 20110, USA

Preface

The beginning of the new century is a good time to take stock of the activities and topics that have been studied under the guise of 'ecotourism' for just over a decade. This *Encyclopedia of Ecotourism* is the first attempt to review the sector in a comprehensive way within a single volume. Its 41 chapters, whose authors include many leaders in the field, represent a diversity of perspectives and styles. This diversity, within reason, has been encouraged, since there is no one correct way to analyse the ecotourism sector or gain insight into its evolution. Topically, the chapters have been divided into eight sections, each representing a common theme. The first section establishes the context for the volume by considering fundamental issues of definition, categories, market development, growth and relationship to other forms of tourism. The next set of chapters reviews the status of ecotourism in all of the world's major regions, while the spatial theme is continued in Section 3 with its focus on major biomes, including rainforest, alpine and polar regions, savannahs, islands and marine environments. Generic ecotourism settings, such as protected areas, modified spaces and indigenous territories, are examined in the fourth section. Section 5 changes the focus by considering the environmental, socio-cultural and economic impacts of ecotourism on host destinations, including rural areas, and by reviewing sustainability indicators. This is followed by a series of chapters that considers aspects of planning, management and institutions, including those that are external to ecotourism itself. The theme of Section 7 is the business of ecotourism, which incorporates accommodations, tour operators, tour guiding and interpretation, planning and marketing, and quality control issues. The eighth and final section concentrates on the creation and dissemination of ecotourism knowledge by considering prevalent research methodologies, information sources, training and education, and research needs.

An ambitious volume such as this would not have been possible without the fine efforts and dedication of numerous individuals. My heartfelt thanks first of all go to the authors, who made time in their full schedules to prepare an impressive array of contributions that will mark this volume as an indispensable reference to those with any kind of interest in ecotourism. Special thanks go to authors such as Dave Fennell, Elizabeth Halpenny and Ralf Buckley who readily and willingly agreed to make additional chapter contributions when other potential contributors were forced to withdraw from the project.

As Chief Editor, I was fortunate to work with an excellent team of Advisory Editors, consisting of Ken Backman, Erlet Cater, Paul Eagles and Bob McKercher. This team of experts laboured away at the coalface, working directly and effectively with the authors through their first and second drafts, and thereby making my final editorial responsibilities a pleasant rather than onerous task. The team at CABI *Publishing* also deserves the highest praise. As my direct contact with the publisher, Development Editor Rebecca Stubbs performed a great number of essential tasks, and also provided encouragement and expert advice when required. Her professionalism is a major reason for the successful production of the Encyclopedia. As the originator of the project, Tim Hardwick, Publisher, must be congratulated for his inspired and timely book idea. I would also like to extend my gratitude to Production Editor Zoe Gipson for her excellent work in bringing the manuscript to final production. The reviewers of the original book proposal were also most supportive, by encouraging publication and providing excellent feedback on refining the scope and contents of the book. Finally, I extend my deep gratitude to Laura Lawton, who assisted with the Encyclopedia in a variety of ways, provided unwavering support, and kept me sane during the hectic final phases.

Section 1

Introduction to Ecotourism

D.B. Weaver
Department of Health, Fitness and Recreation Resources, George Mason University, Manassas, Virginia, USA

Authors who contribute to the introductory section of a volume such as this face the special responsibility of having to establish the context for the sections that follow. In the case of ecotourism, this is a particularly vexatious task since the knowledge base is incipient, and no consensus currently exists as to the meaning and interpretation of the term itself. Yet, even in this atmosphere of conceptual fuzziness, these introductory chapters perform the extremely useful function of pointing out the debates, disputes, shortfalls and ambiguities that characterize a field which is, after all, still only in its infancy. Moreover, where warranted, they suggest areas in which some degree of consensus or cohesion may be emerging; indicators, perhaps, that ecotourism is moving toward a higher level of maturity.

This duality between persisting ambiguity and emerging consensus is evident in Chapter 1, where Blamey puts forward criteria around which ecotourism seems to be coalescing, but also emphasizes the debates and uncertainties that continue to dog all of these criteria. For example, he acknowledges the widespread perception that ecotourism is 'nature-based', but then bursts the bubble of consensus by citing unresolved issues such as how proximate to nature the experience should be to qualify as ecotourism, and how disturbed a landscape can be yet still qualify as an ecotourism venue. In similar vein, Blamey cites the learning imperative of ecotourism, but asks whether product interpretation should merely satisfy consumer demand for information at a superficial level, or whether it should try to change consumer attitudes toward an enhanced sense of environmental responsibility.

The third criterion that he discusses, sustainability, is even more contentious and ambiguous. While most stakeholders agree that there is an onus on ecotourism to be environmentally and socio-culturally sustainable, stakeholders have dramatically variant perspectives on what this means and how it should be effected. For example, there are those who argue that ecotourism must contribute actively to the enhancement of the resource base, while others contend that it is sufficient for ecotourism to make things no worse than the status quo. But it is not a question of adopting just one or the other perspective. In Chapter 2, Orams demonstrates how ecotourism can adhere to its core criteria yet embrace a spectrum of motivations, levels of involvement and outcomes, ranging from 'hard' to 'soft' types of ecotourism, and from the 'active' to the 'passive'. In the hard, active pole of ecotourism, resource enhancement is generally regarded as an imperative. This is not the case in soft, passive ecotourism, although Orams supports the implementation of measures that would move the latter toward the more enhancive side of the continuum.

Some researchers and practitioners demonstrate a strong bias in favour of hard ecotourism, sometimes even to the point of excluding the soft perspective as a legitimate expression of the sector. This elitist approach, however, may be seen as misguided for a number of reasons. In reality, this type of ecotourism involves such a small number of participants as to render it almost irrelevant in terms of economic impacts on destinations, and in its capacity to foster adequate lobbying clout in the face of larger stakeholders such as the forestry and mining industries. The soft type of ecotourism, in contrast, is much more prevalent and therefore potentially more advantageous in both respects. As well, soft ecotourism is far more accessible to those who are not wealthy, young or healthy. Yet, even here there is risk, especially in the possibility that such a mode of ecotourism might mutate into the sort of conventional mass tourism that academics have been criticizing for so many years – and to which, ironically, ecotourism was originally conceived as a more appropriate alternative.

Despite this risk, the option of embracing both the hard and soft types of activity into a single ecotourism spectrum seems to be gaining support, and is reflected in the remaining three chapters in the section. In Chapter 3, Wight draws from a growing base of market knowledge to identify how the emerging 'ecotourist' differs from the tourist and consumer markets in general. The 'typical' ecotourist, it appears, tends to originate in a more developed country, is female, has higher-than-average income and education levels, and is somewhat older than the average tourist. However, Wight also stresses that the ecotourist market differs internally with respect to age, income, activity patterns, motivation, etc., differences that in large part reflect the hard–soft continuum. Soft ecotourists, for example, appear to be younger than hard ecotourists in some origin regions, though more research is required to determine whether such findings are more than just a regional trend.

The hard–soft spectrum is also implicitly embraced in Chapter 4 by Hawkins and Lamoureux, who show that ecotourism, perceived in this liberal way, constitutes a substantial portion of the overall tourist market. Furthermore, ecotourism is a rapidly expanding sector, as evidenced by the growth of indicators such as the provision of ecotourism-related educational opportunities, the formulation of strategic plans and policies, and the availability of funding from international agencies. But Hawkins and Lamoureux also point out the difficulties in trying to quantify the magnitude and growth of a sector that is often regarded interchangeably or included with other forms of tourism, such as 'nature-based', 'adventure' and 'sustainable'. This confusion of terminology is a hindrance to the systematic study of ecotourism as a discrete sector, and is another indication of the sector's immaturity.

In Chapter 5, Weaver helps to alleviate this situation by exploring the relationship between ecotourism and other relevant types of tourism. After reviewing much of the available literature, he concludes that ecotourism is a subset of both nature-based and sustainable tourism, and overlaps with adventure, cultural and 3S (sea, sand, sun) tourism. In many cases, as with trekking, ecotourism hybridizes with these other sectors, making it impossible to differentiate the constituent components. More contentiously, and in concert with his own support for the hard–soft continuum, Weaver asserts that ecotourism can be a subset of alternative tourism *or* mass tourism, as long as the basic criteria are met. Reflecting a point made in Chapter 1 by Blamey, he indicates that ecotourism has long been and is still widely regarded as a form of alternative tourism. This tendency owes to the origins of ecotourism in the 'adaptancy platform' of the 1980s (Jafari, 1989), which proposed small-scale alternatives to conventional mass tourism that were reputed to be inherently more benign. However, the 'knowledge-based platform' that appeared in the late 1980s, by adopting a more objective and less ideological approach to tourism, has served to erode the association between the scale of a tourism product and the perception that it is either good or bad as a result. Depending on the given circumstances, alternative tourism and mass tourism can both have either positive or negative consequences, and hence there is no inherent reason for making a categorical disassociation between ecotourism and mass tourism. Going even further, a good argument can be made that the majority of ecotourism *already occurs* in the guise of mass tourism, and that large scales of operation afford certain advantages in the provision of sustainable outcomes and quality education that are not available in small-scale alternative tourism operations. As stated earlier, this is a controversial line of reasoning, and the issue of scale remains one of the focal points of debate and dispute within the evolving area of ecotourism studies. Either implicitly or explicitly, each chapter in this section, and indeed in this volume, provides its own perspective on this point.

Reference

Jafari, J. (1989) An English language literature review. In: Bystrzanowski, J. (ed.) *Tourism as a Factor of Change: a Sociocultural Study.* Centre for Research and Documentation in Social Sciences, Vienna, pp. 17–60.

Chapter 1

Principles of Ecotourism

R.K. Blamey

Urban and Environmental Program, Research School of Social Sciences, Australian National University, Australian Capital Territory, Australia

> Around the world, ecotourism has been hailed as a panacea: a way to fund conservation and scientific research, protect fragile and pristine ecosystems, benefit rural communities, promote development in poor countries, enhance ecological and cultural sensitivity, instill environmental awareness and a social conscience in the travel industry, satisfy and educate the discriminating tourist, and, some claim, build world peace.
>
> (Honey, 1999, p. 4)

Introduction

Although the origins of the term 'ecotourism' are not entirely clear, one of the first to use it appears to have been Hetzer (1965), who identified four 'pillars' or principles of responsible tourism: minimizing environmental impacts, respecting host cultures, maximizing the benefits to local people, and maximizing tourist satisfaction. The first of these was held to be the most distinguishing characteristic of 'ecological tourism ("EcoTourism")' (Fennell, 1998). Other early references to ecotourism are found in Miller's (1978) work on national park planning for ecodevelopment in Latin America, and documentation produced by Environment Canada in relation to a set of road-based 'ecotours' they developed from the mid-1970s through to the early 1980s. Each tour focused on a different ecological zone found along the corridor of the Trans-Canada highway, with an information pack available to aid interpretation (Fennell, 1998).

Ecotourism developed 'within the womb' of the environmental movement in the 1970s and 1980s (Honey, 1999, p. 19). Growing environmental concern coupled with an emerging dissatisfaction with mass tourism led to increased demand for nature-based experiences of an alternative nature. At the same time, less developed countries began to realize that nature-based tourism offers a means of earning foreign exchange *and* providing a less destructive use of resources than alternatives such as logging and agriculture (Honey, 1999). By the mid 1980s, a number of such countries had identified ecotourism as a means of achieving both conservation and development goals.

The first formal definition of ecotourism is generally credited to Ceballos-Lascuráin (1987), who defined it as: 'travelling to relatively undisturbed or uncontaminated natural areas with the specific objective of studying, admiring, and enjoying the scenery and its wild plants and animals, as well as any existing cultural manifestations (both past and present) found in these areas'. While definitions such as that of Ceballos-Lascuráin (1987) and Boo (1990) tended to emphasize the nature-based experience sought by the tourist, more recent definitions have tended to highlight various principles associated with the concept of sustainable development. According to Wight (1993), sustainable ecotourism imposes an 'ethical overlay' on nature-based tourism that has an educative emphasis. Although this overlay has arguably been implicit, if not explicit, in earlier discussions of ecotourism, the concept does appear to have evolved into something explicitly normative over the past decade. This is in part a reflection of increasing recognition among industry and government that nature-based tourism can only be sustained in the long term if a principled and proactive supply-side management approach is adopted.

Some of the definitions of ecotourism that have proved popular in recent years, and which are consistent with the definition offered in the introduction to this volume, are listed in Table 1.1. Although any number of principles of ecotourism can be devised, an analysis of definitions such as these indicates that three dimensions can represent the main essence of the concept. According to this interpretation, ecotourism is:

- nature based,
- environmentally educated, and
- sustainably managed.

The last dimension is taken to encompass both the natural and cultural environments involved in supplying the ecotourism experience. Thus, where Ross and Wall (1999) outline five fundamental functions of ecotourism; namely: (i) protection of natural areas; (ii) education; (iii) generation of

Table 1.1. Selected definitions of ecotourism.

Source	Definition
Ceballos-Lascuráin (1987, p. 14)	Travelling to relatively undisturbed or uncontaminated natural areas with the specific objective of studying, admiring, and enjoying the scenery and its wild plants and animals, as well as any existing cultural manifestations (both past and present) found in these areas
The Ecotourism Society (1991a, b)	Responsible travel to natural areas which conserves the environment and improves the well-being of local people
Ecotourism Association of Australia (1992)	Ecologically sustainable tourism that fosters environmental and cultural understanding, appreciation and conservation
National Ecotourism Strategy of Australia (Allcock *et al.*, 1994)	Ecotourism is nature-based tourism that involves education and interpretation of the natural environment and is managed to be ecologically sustainable
	This definition recognizes that 'natural environment' includes cultural components and that 'ecologically sustainable' involves an appropriate return to the local community and long-term conservation of the resource
Tickell (1994, p. ix)	Travel to enjoy the world's amazing diversity of natural life and human culture without causing damage to either

money; (iv) quality tourism; and (v) local participation, the last three fall under the heading 'sustainably managed' in this chapter. The three-dimensional interpretation is also consistent with Buckley's (1994) restrictive notion of ecotourism in which ecotourism is nature based, environmentally educated, sustainably managed and conservation supporting.

One further dimension of ecotourism, not referred to in most definitions, but worthy of the status of at least a 'secondary principle', involves the small-scale, personalized and hence alternative nature of many classical ecotourism experiences. The above three principles, together with this fourth, provide the defining characteristics of classical ecotourism as shown in Fig. 1.1. Popular ecotourism is similar to classical ecotourism with the exception that it does not qualify as a form of alternative tourism. Each principle is now described in detail, beginning with the nature-based dimension.

Nature Based

The most obvious characteristic of ecotourism is that it is nature based. As noted above, it is this dimension that is emphasized in earlier definitions. Valentine (1992a, p. 108) defines nature-based tourism as tourism 'primarily concerned with the direct enjoyment of some relatively undisturbed phenomenon of nature'. A variety of motivations for nature-based tourism have been suggested, including the desire to get back in touch with nature, a desire to escape the pressures of everyday life, seeing wildlife before it is too late, and specific interests and activities such as trekking, birdwatching, canyoning and white-water rafting and kayaking (Whelan, 1991).

Valentine identified three main dimensions of nature-based tourism (NBT) pertaining to the experience, style and location. In terms of the type of experience involved, different NBT experiences vary in nature dependency, intensity of interaction, social context and duration. Different styles are associated with different levels of infrastructure support, group size and type, cultural interaction factor, willingness to pay and length of visit. Locations vary in terms of accessibility (remoteness), development contribution, ownership and fragility (Valentine, 1992a, b).

Questions arise as to what does and does not constitute a nature-based experience. Does a drive through a forested area

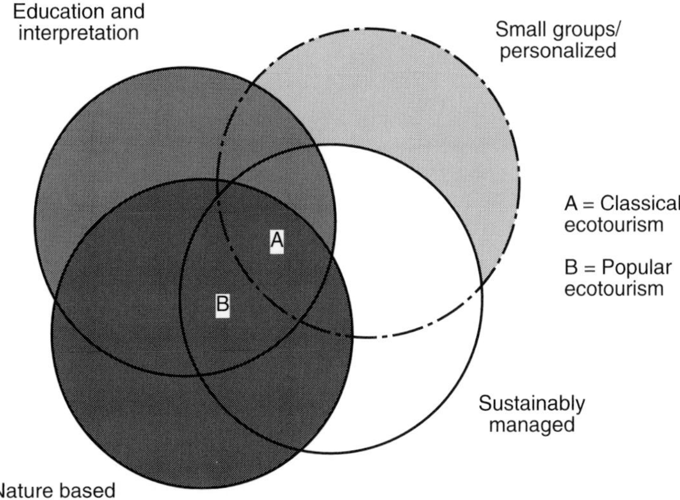

Fig. 1.1. Dimensions of ecotourism.

qualify as nature based, or must the driver actually pull over and go for a wander (the activity/experience component)? If he or she does wander, how long must this wander be for the individual to be considered a nature-based tourist (the duration component)? A further question relates to the natural environment itself (the attraction/experience component). Does walking through regenerated forest areas, or swimming in slightly polluted or littered lakes or streams, or for that matter any public beach, constitute a nature-based experience? (Blamey, 1997).

The issue of proximity is one that commonly arises when considering whether a tourism experience involving nature can be considered nature based. Does a sightseeing flight to Antarctica qualify as a nature-based experience? What about the tourist who travels to Nepal, arguably an 'ecotourism destination', but gets no closer to the Himalayas than sitting in a restaurant with great views of the Annapurna range. Alternatively, the tourist might take a short stroll up a nearby hill to get better views. Even the dedicated trekker may get little closer to true wilderness than the lower altitude areas dominated by subsistence farming. In all these cases, a guide can readily be hired to provide interpretation.

Requiring ecotourism to take place in protected areas (Kutay, 1989) does not resolve the question of what qualifies as nature based. While ecotourism often occurs within 'protected areas' such as the Annapurna Sanctuary, these areas may contain environments that have been quite disturbed by human activity (possibly leading to the protected area status). Furthermore, relatively undisturbed areas are also commonly found *outside* protected areas (Kusler, 1992). Indeed, it can be argued that ecotourism should occur outside protected areas since, by definition, protected areas are worthy of protection from development, and tourism represents a form of development. Unprotected areas are also most suited to another claimed benefit of ecotourism activity: promoting conservation of non-protected areas (Bottrill and Pearce, 1995).

Some authors have also questioned whether environments with significant evidence of human disturbance might qualify as ecotourism, particularly if they demonstrate adherence to other ecotourism principles. Whelan (1991), for example, asks whether farmstays might in some cases qualify as ecotourism, and Chirgwin and Hughes (1997, p. 97) suggest that modified areas such as wetlands associated with human-made watercourses can serve as an 'ecotourist venue if they are well presented and managed, aesthetically pleasing and provide the opportunity to observe wildlife'. While such areas are not natural, they may qualify as 'nature based', depending on how one interprets terms such as 'relatively undisturbed'. Tourism in urban environments can be sustainably managed (Hinch, 1996), but the 'natural' environments within them are often highly influenced by humans and do not generally satisfy the ecotourism criterion of being nature based. Acott *et al.* (1998) argue that it is possible for individuals to be ecotourists in 'non-ecotourist locations'.

Operational definitions of nature-based tourism, and hence also ecotourism will clearly require the line to be drawn somewhere. Subjective decisions cannot be avoided. Ultimately, any definition of these concepts will have an arbitrary component (Blamey, 1997). Nature-based tourism, and hence ecotourism, are indeed fuzzy concepts.

Environmentally and Culturally Educative

A feature of ecotourism experiences is education and interpretation about the natural environment and any associated 'cultural manifestations'. In contrast to learning, which is 'a natural process, occurring throughout life and mostly incidental', education involves 'a conscious, planned, sequential and systematic process, based on defined learning objectives and using specific learning procedures' (Kalinowski and Weiler, 1992). Interpretation is 'an educational activity which aims to reveal meanings and relationships through the use of original objects, by first hand experience, and by illustrative media, rather than simply to communicate factual information' (Tilden, 1977 cited in Moscardo, 1998). Although virtually all nature-based tourism involves some degree of learning, it is education and interpretation that serves as a key element and defining characteristic of ecotourism experiences. For the purposes of this chapter, the term education includes interpretation.

Two main purposes of environmental education can be distinguished in the ecotourism context. The first involves satisfying tourist demand for information regarding natural and cultural attractions, thereby providing a satisfying recreational experience. The second involves changing in a pro-environmental way, the knowledge, attitudes and/or behaviour of tourists, with a view to minimizing negative impacts and producing a more environmentally and culturally aware citizenry. The two purposes will often be related, for example, when information provided about the sensitive nature of the ecology serves both purposes. Education can, however, be environmentally informative without being environmentally supportive (Blamey, 1995).

The first function of environmental education in ecotourism coincides with what will often be the primary motivation for undertaking an ecotourism experience: learning about the plants, animals, landscapes and so on that are unique to an area. Interpretation may be static, as with self-guided walks involving information signs, displays and the like, or personalized, where guides provide information (Burgess, 1993). To varying degrees, individuals can tailor the educative experience to meet their own interests, for example by asking questions, moving closer, smelling, having eye contact with particular species and learning the mannerisms of species.

It is important to recognize that different tourists will have different needs for cognition in the form of formalized education and interpretation. A distinction can be drawn between more passive forms of learning and those with a heavier emphasis on active learning and formalized education. As Urry (1990, p.1) states, 'When we "go away" we look at the environment with interest and curiosity. It speaks to us in ways we appreciate, or at least we anticipate that it will do so. In other words, we gaze at what we encounter'. Most gazing involves learning. It is difficult to view an animal for the first time without learning something. This experiential form of learning is distinct from formal education regarding the biology, zoology or ecology of areas. Many tourists work hard for most of the year and primarily seek spiritual renewal and rejuvenation of the mind as a first priority for their holidays (Honey, 1999). Selectively attending to and processing the words of a tour guide is one thing, while studying information packs in detail before embarking on a tour may be quite another.

To the extent that 'ecotourists' are motivated by an interest in learning about nature, ecotourism can be considered a form of special interest tourism, or tourism in which 'the traveller's motivation and decision making are primarily determined by a particular special interest' (Hall and Weiler, 1992, p. 5). Adventure tourism is another case of special interest tourism. The first type of education does more to distinguish ecotourism from nature-based adventure tourism than it does to distinguish ecotourism from other nature-based sightseeing tours.

The second function of environmental education can be achieved in a variety of ways, some more overt than others. For example, tourists can be educated about how best to minimize their impacts while visiting a site, and presented with a code of ethics for tourist conduct. This recognizes that pursuit of the first educative function often impacts on local cultures and flora and fauna. It is often the desire for experiential learning that drives the tourist to move closer to the objects of their interest, thereby increasing the impact. Another approach is to simply provide information pertaining to ecological relationships and the sensitive and possibly threatened nature of such relationships and the species involved. The resultant knowledge changes will sometimes translate into attitudinal and behavioural change, particularly when presented in a caring and emotionally involving manner. The ethic of care implicit in some interpretation styles can readily flow over to the tourist. Another fairly covert way of influencing behaviour is to provide information about alternative sites, routes or activities with a view to moving visitors away from heavily used and ecologically and/or culturally sensitive sites (Moscardo, 1998).

The longer-term objective of the second function of environmental education is summarized well by Ceballos-Lascuráin (1988 cited in Ziffer, 1989, p. 5):

> the person who practices ecotourism has the opportunity of immersing him or herself in nature in a way most people cannot enjoy in their routine, urban existences. This person will eventually acquire a consciousness ... that will convert him into somebody keenly involved in conservation issues.

Individuals who do not normally consider in any detail their impacts on the environment may adopt a more reflective, sensitive and enlightened perspective once engaged in ecotourism experiences, which may last beyond the life of that experience. In Norton's terminology, the objective here is to instil transformative values in tourists and locals (Ross and Wall, 1999). Of course, the possibility exists that those who take most notice of tourist codes of behaviour and other information or material pertaining to the minimization of tourist impacts, may well be those who least need it (Hall, 1993).

In addition to being educated for the purposes of short- or long-term pro-environmental changes in knowledge, attitudes and/or behaviour, tourists can be educated about how to get the most out of their experiences in other ways. This involves providing information about the experiences available at a site, suggested walks and routes, the location of toilets and other facilities, safety and warning messages, how to cope with sea and road sickness, and so on (Moscardo, 1998).

Local communities and industry can also be the target of environmental communications, as the principles outlined by Wight (Table 1.2) demonstrate. For example, local communities can be educated regarding the sensitive nature of natural areas, and how best to protect these areas and maximize tourism-related revenues and benefits. Industry can also be educated about best environmental and/or business practice. More generally then, the ecotourism industry 'consists of a number of sectors which need to work together effectively for the industry to flourish. Each sector has specific needs for formal training and access to information. Each also has its own communication task to deliver information to other sectors' (Social Change Media, 1995). The sectors considered by the strategy include consumers, travel agents, operators and guides, natural resource managers, trainers, industry bodies, state tourism bodies and the media. Of course, most if not all industries are dependent for their success on communication within and between sectors. It is the two environmental education functions referred to above that can be considered as defining characteristics of ecotourism.

Sustainably Managed

The term sustainable tourism development is a derivative of the more general concept of sustainable development, brought to prominence with the publication of *Our Common Future*, the report of the World Commission on Environment and Development (WCED, 1987). The WCED defined sustainability as 'meeting the needs of the present without compromising the ability of future generations to meet their own needs' (WCED, 1987, p. 8). This report stimulated much discussion regarding definitions of sustainability and sustainable development, and the principles and practices held to be consistent with any one definition.

Responses to the sustainable development concept appear to take one of two main forms (Barbier, 1989). The first is a

Table 1.2. Ecotourism principles and guidelines.

Wight (1994)	The Ecotourism Society (Lindberg and Hawkins, 1993)	National Ecotourism Accreditation Program (NEAP), Australia. Eligibility principles
It should not degrade the resource and should be developed in an environmentally sound manner	Prepare travellers to minimize their negative impacts while visiting sensitive environments and cultures before departure	Focuses on personally experiencing natural areas in ways that lead to greater understanding and appreciation
It should provide long-term benefits to the resource, to the local community and industry	Prepare travellers for each encounter with local cultures and with native animals and plants	Integrates opportunities to understand natural areas into each experience
It should provide first-hand, participatory and enlightening experiences	Minimize visitor impacts on the environment by offering literature, briefings, leading by example, and taking corrective actions	Represents best practice for ecologically sustainable tourism
It should involve education among all parties: local communities, government, non-government organizations, industry and tourists (before, during and after the trip)	Minimize traveller impacts on cultures by offering literature, briefings, leading by example, and taking corrective actions	Positively contributes to the ongoing conservation of natural areas
It should encourage all-party recognition of the intrinsic values of the resource	Use adequate leadership, and maintain small enough groups to ensure minimum group impact on destinations. Avoid areas that are under-managed and over-visited	Provides constructive ongoing contributions to local communities
It should involve acceptance of the resource in its own terms, and in recognition of its limits, which involves supply-oriented management	Ensure managers, staff and contact employees know and participate in all aspects of company policy to prevent impacts on the environment and local cultures	Is sensitive to, interprets and involves different cultures, particularly indigenous cultures
It should promote understanding and involve partnerships between many players, which could involve government, non-governmental organizations, industry, scientists and locals (both before and during operations)	Give managers, staff and contact employees access to programmes that will upgrade their ability to communicate with and manage clients in sensitive natural and cultural settings	Consistently meets client expectations
It should promote moral and ethical responsibilities and behaviour towards the natural and cultural environment by all players	Be a contributor to the conservation of the region being visited	Marketing is accurate and leads to realistic expectations
	Provide competitive, local employment in all aspects of business operations	
	Offer site-sensitive accommodations that are not wasteful of local resources or destructive to the environment, which provide ample opportunity for learning about the environment and sensitive interchange with local communities	

generalized, normative and energized response associated with the pursuit of synergisms and balance among environmental impacts, economic development, participatory processes, intergenerational and intragenerational equity, sustainable livelihoods and so on. The second, while overlapping with the first, is narrower and involves the development of formal rules for sustainability. Different rules or models are associated with different assumptions regarding what it is that is to be sustained. This has become known as the constant capital perspective and is outlined later in this chapter.

The first type of response appears to have dominated discussions regarding sustainable tourism development and is reflected in a wide range of government and industry initiatives in the ecotourism context, including ecotourism strategies, sustainability indicators, accreditation and so on. Sustainable development provides an organizing concept for the development of such initiatives (de Kadt, 1992). This perspective tends to see the pursuit of sustainable tourism development as involving the balancing of social, economic and environmental goals (Wight, 1993). Discussions and initiatives are also commonly focused around lists of sustainability principles and guidelines. In the first edition of the *Journal of Sustainable Tourism*, for example, Bramwell and Lane (1993) outline four basic principles of sustainable development and sustainable tourism development: (i) holistic planning and strategy-making; (ii) preservation of essential ecological processes; (iii) protection of both human heritage and biodiversity; and (iv) development to ensure that productivity can be sustained over the long term for future generations.

Another well known list of principles and guidelines is that developed by Tourism Concern (1991) in association with the Worldwide Fund for Nature (WWF). Each of the ten sustainability principles listed in Box 1.1 is accompanied by a list of recommendations. Associated with the first general principle, for example, are recommendations that the tourism industry should: (i) prevent damage to environmental resources; (ii) act as a force for conservation; (iii) develop and implement sound environmental policies in all areas of tourism; (iv) install appropriate systems to minimize pollution from tourism developments; (v) develop and implement sustainable transport policies; (vi) adhere to the precautionary principle; (vii) research, establish and abide by the carrying capacity of a destination; (viii) respect the needs and rights of local people; (ix) protect and support the cultural and historical heritage of peoples worldwide; (x) carry out practices in a responsible and ethical manner; and (xi) actively discourage the growth of exploitative sex tourism[1].

Two sustainability principles that are commonly highlighted in the ecotourism context are that ecotourism should: (i) support local economies; and (ii) support conservation. The Australian National Ecotourism Strategy, for example, 'recognizes that "natural environment" includes cultural components and that "ecologically sustainable" involves an appropriate return to the local community and long-term conservation of the resource' (Table 1.1) (Allcock *et al.*, 1994, p. 3). The requirement that local communities and regions benefit from ecotourism and participate in decision making, or at least be no worse off, appears to be based on two main premises. The first draws on the principles of intragenerational equity and intergenerational equity underlying the concept of

[1] Although such wish-lists have value, several authors have argued that they are typically very general in nature and that this leaves a significant gap between policy endorsement and implementation (Berry and Ladkin, 1997; Hunter, 1997; Garrod and Fyall, 1998). Ashcroft (cited in Wheeller, 1995) argues that 'Reading each principle in turn I found myself increasingly asking the questions, Why? How? When? With what? It soon became very tiresome ploughing through so many platitudinous points'. Consequently, recent years have seen attention turning to the development of guidelines, codes of conduct, indicators of sustainable tourism, accreditation and so on.

Box 1.1. Principles for sustainable tourism (Tourism Concern, 1991).

Using resources sustainably
The conservation and sustainable use of resources – natural, social and cultural – is crucial and makes long-term business sense
Reducing over-consumption and waste
Reduction of over-consumption and waste avoids the costs of restoring long-term environmental damage and contributes to the quality of tourism
Maintaining biodiversity
Maintaining and promoting natural, social and cultural diversity is essential for long-term sustainable tourism, and creates a resilient base for the industry
Integrating tourism into planning
Tourism development which is integrated into a national and local strategic planning framework and which undertakes environmental impact assessments, increases the long-term viability of tourism
Supporting local economies
Tourism that supports a wide range of local economic activities and which takes environmental costs and values into account, both protects these economies and avoids environmental damage
Involving local communities
The full involvement of local communities in the tourism sector not only benefits them and the environment in general but also improves the quality of the tourism experience
Consulting stakeholders and the public
Consultation between the tourism industry and local communities, organizations and institutions is essential if they are to work alongside each other and resolve potential conflicts of interest
Training staff
Staff training which integrates sustainable tourism into work practices, along with recruitment of personnel at all levels, improves the quality of the tourism product
Marketing tourism responsibly
Marketing that provides tourists with full and responsible information increases respect for the natural, social and cultural environments of destination areas and enhances customer satisfaction
Undertaking research
Ongoing research and monitoring by the industry using effective data collection and analysis is essential to help solve problems and to bring benefits to destinations, the industry and consumers

sustainable development, and essentially holds that it is the socially responsible, or right, thing to do. The second is instrumental in nature and involves the assumption that local communities are most likely to protect or maintain a resource base in a form that is suitable for tourism if they stand to benefit from it. In this case, they have an incentive to protect the resource.

Support for local economies and conservation can take a variety of forms. Potential economic benefits include foreign exchange earnings, employment, infrastructure development, long-term economic stability and economic diversification (Lindberg, 1991; Wight, 1994). To the extent that the revenues obtained through entrance fees, donations and ancillary goods and services (accommodation, souvenirs, etc.) are sufficient in magnitude and earmarked for conservation, they can be used for conservation purposes (Ziffer, 1989). Other ways of supporting conservation include participating in rehabilitation projects, participating in scientific monitoring and removing litter from sites visited. The instrumental effect referred to above is additional. It is recognition of the potential of ecotourism to assist the twin goals of conservation and economic development that has resulted in its popularity as a core development strategy in numerous less developed countries (Honey, 1999).

The extent to which a region can supply locally made products for consumption by the tourism industry depends on the extent of development and the diversity of economic development (Ziffer, 1989). By integrating ecotourism development into broader regional development strategies, leakages can be minimized and the economic benefits from ecotourism maximized. The extent to which nature-based tourists seek out locally owned accommodation, tours and so on is still somewhat unclear. It is one thing to buy a locally made souvenir and quite another to seek out locally made products when they have no clear linkages to the local area. For example, wooden sculptures are likely to be made locally (local in supply) and tourists are likely to be looking for local product (local in demand). However, batteries for one's camera will often not be locally made, and partly in expectation of this, consumers are unlikely to search for local product.

As with the other principles discussed in this chapter, operational definitions of ecotourism will require some means of distinguishing between those tourism experiences that do and don't qualify. How does one decide when a nature-based tourism experience is sufficiently supporting of local communities and/or conservation to qualify as ecotourism? One approach is to assess these principles in terms of absolute levels of support. For example, ecotours might be required to make contributions to conservation organizations to the amount of 5% of the tour price. Alternatively, these benefits can be assessed in relation to the costs, with a requirement of no net loss in welfare. While cost–benefit analysis (CBA) offers a means of conducting such an analysis, it is both time consuming and costly, and of questionable use when dealing with matters of equity. How does one decide if benefits to local communities in the form of employment and income justify irreversible losses in cultural identity and sense of purpose? Compensating locals with tourist income may simply add fuel to their decline in identity[2].

Several authors have considered the cultural impacts of tourism. Pearce (1992), for example, refers to a number of studies that have looked at impacts on host communities, covering such areas as language changes, land tenure, desecration of community life, begging, prostitution and crime. Finucane (1992, p. 13) expressed a concern that 'heavy tourist exposure will result in a gradual erosion of indigenous language and culture or the creation of a commercialized culture'. Johnston and Edwards (1994) argue that codes of conduct and other strategies associated with responsible tourism may represent structural *adjustments*, but not necessarily the structural *transformations* required to make tourism sustainable. They argue (p. 475) that sustainability is a 'distracting, and arguably unobtainable notion. Perhaps what is needed is a more realistic articulation of what ecotourism can and cannot offer hosts and guests'. Cultural impacts can often be minimized by involving local communities in decisions that affect them, particularly regarding the kind and amount of tourism that should occur (Scheyvens, 1999; Wallace, 1999). The assumption here is that communities know what is good for them, and can put aside short-term interests in order to achieve the best long-term outcomes. A related tension exists between revenue raising and keeping the number of tourists below social and environmental carrying capacities (Honey, 1999) (see also Chapter 25 in this volume).

Twining-Ward (1999) argues that new types of development based on 'sound principles' may be a case of treating the symptoms rather than the cause and that care is required to ensure that ecotourism does not divert attention from issues of scale and intensity of tourism development. To the extent that ecotourism involves a growing means by which 'development' encroaches on to pristine natural areas, it may be operating at the margins of serious attempts to move towards sustainable tourism. As an ecotourism destination becomes more popular, it begins to lose its appeal, thereby prompting operators to move into new, pristine areas, with the

[2] To some extent, the cultural values at stake may have existence value among other members of society, which a comprehensive CBA might attempt to incorporate.

cycle repeating itself indefinitely if not controlled. Burton (1998) found that it is often the genuine ecotourism operators that seek to initiate the geographic spread of tourism by seeking new unspoiled environments. Lawrence *et al.* (1997) argue that the legitimacy of ecotourism is threatened by the tension between sustainability principles and the basic fact that growth in ecotourism involves more and more tourists moving into pristine areas. Even the most benign forms of ecotourism will still have some negative impact on the environment. Wheeller (1995) has questioned whether there can ever be a symbiotic relationship between tourism and the environment, arguing that the commitment of tour operators, tourists and host communities to principles of sustainability will tend to be conditional on self-interest: 'we rarely sacrifice so much as to cause any adverse effect on ourselves. The utility derived (by us) usually outweighs the cost of that sacrifice. So too ... with expressed support for sustainable tourism' (Wheeller, 1995, p. 128).

While such concerns are generally regarded as important, they have taken little of the gloss off the growing ecotourism movement. One way that the continued pursuit of ecotourism has been justified, in light of such concerns, is to argue that ecotourism can serve as a model for other forms of tourism, thereby facilitating the greening of tourism as a whole. The ultimate goal of the ecotourism 'movement' is thus to infuse the entire travel industry with sustainability principles (Honey, 1999). Clearly, there are substantial benefits to be gained by integrating environmental technologies and practices into mainstream tourism development, rather than restricting their application to a small niche market.

The constant capital perspective

As mentioned earlier, the constant capital perspective of sustainable development focuses on the development of formal rules for sustainability, with different rules or models being associated with different assumptions regarding what it is that is to be sustained. Sustainable development is interpreted to imply a requirement that human welfare does not decline with time. According to the *constant capital rule*, this can be achieved by leaving the next generation a stock of capital assets no less than the current stock. Intergenerational equity is achieved by acknowledging the right of future generations to 'expect an inheritance sufficient to allow them the capacity to generate for themselves a level of welfare no less than that enjoyed by the current generation' (Turner *et al.*, 1992, p. 2). Capital can take several forms, including man-made, human, natural, moral and cultural capital. Discussions most commonly focus on man-made and natural capital, with the latter being divided into categories such as non-renewable, renewable, semi-renewable and recyclable. The preservation of cultural capital involves the maintenance of sustainable livelihoods and hence a diversity of knowledge and perspectives on how to adapt to one's environment. Cultural diversity is hence related to biological diversity, and both have clear interrelations with moral capital.

A key issue that arises when implementing the constant capital rule is the extent to which the different types of capital can or should be substituted for one another, or in other words, what it is that is to be sustained. Several different schools of thought have emerged on this issue and these are summarized in Box 1.2. Stronger forms of sustainable development are associated with less optimistic views regarding the extent to which human-made capital can be substituted for natural capital in the long term. These two forms of capital are instead held to be complementary in the majority of cases. A precautionary and proactive environmental management approach is advocated, which involves taking a risk-averse perspective when considering impacts that are uncertain and potentially irreversible. Strong notions of sustainable development are also associated with *environmental (or ecological) sustainability.* This involves the maintenance of natural capital, consisting of

Box 1.2. The sustainability spectrum.

Very weak sustainability
Sustainability requires the total capital stock, consisting of the aggregate of natural, man-made, human, moral and cultural stocks, to remain constant over time. Any type of capital can hence be reduced as long as it is compensated for by the provision of other capital assets deemed to be of equal value to humans. Different types of capital are hence assumed to be perfectly substitutable. Renewable capital can hence be substituted for non-renewable capital and man-made capital can be substituted for any type of natural capital. This position is consistent with exploitation and an emphasis on economic growth. Economic growth increases consumer choice and satisfaction, and provides the human capital in the form of research and development expertise required to devise technical fixes and generally maintain human welfare at a constant level (Turner, 1991).

Weak sustainability
Very weak sustainability is modified here in response to the problematic nature of the assumption of perfect substitutability. Some types of natural capital are assumed to be complements rather than substitutes, and some key species and processes are not considered to be substitutable at all (Common and Perrings, 1992). The latter are referred to as critical natural capital. Ecological constraints are imposed on the use of natural assets such that stocks remain within the bounds thought to coincide with ecosystem stability and resilience (Turner *et al.*, 1992). A broad, systemic management perspective, drawing on the precautionary principle and safe minimum standards is likely to be involved. A set of environmental indicators may be derived to provide an indication of the state of ecosystems.

Strong sustainability
This rule requires that natural capital remains constant in aggregation, but one form of natural capital can be substituted for another, subject to certain ecological constraints. Physical indicators are again used to provide the monitoring necessary to inform managers. This approach assigns primary importance of maintaining ecosystem structure and function and responds to and takes a precautionary approach to uncertainties and irreversibilities. Unlike the safe minimum standard, constant natural capital must be maintained even when the opportunity costs (benefits forgone) are considered to be very high.

Very strong sustainability
This rule coincides with a steady-state economy, characterized by zero economic growth and zero population growth, and is motivated in part by a consideration of thermodynamic limits. This is likely to require constant stocks of individual natural assets in addition to a constant aggregate stock of natural capital. This position coincides with a more biocentric view in which the intrinsic rights of nature are acknowledged and given significant weight in decision making.

both source and sink functions, over a specified time and space. Box 1.3 offers a more rigorous definition of this natural science concept (Goodland, 1999).

The constant capital perspective is attracting increasing attention in the sustainable tourism literature (see, for example, Hunter, 1997; Garrod and Fyall, 1998). Hunter (1997) argues that there is a need to conceptually reconnect sustainable tourism to the sustainable development literature in general, and the constant capital literature in particular. According to Hunter (1997), the plethora of definitions of sustainable tourism that now exist most commonly reflect a weak sustainability position (Box 1.2), although little attention is typically given to such distinctions. One reason for the lack of consensus on a definition is that sustainability 'has been used by both industry and the conservation movement to legitimize and justify their existing activities and policies' (McKercher, 1993, p. 131).

Hunter (1997) argues that different sustainable development paths, corresponding to different positions in Box 1.2 and different manifestations of these positions, are

Box 1.3. A definition of environmental sustainability (adapted from Goodland, 1999, p. 716).

Environmental sustainability involves the maintenance of natural capital according to the following rules:

1. Output rule:
Waste emissions from a project or action should be kept within the assimilative capacity of the *local* environment without unacceptable degradation of its future waste absorptive capacity or other important services.

2. Input rule:
(a) Renewables: Harvest rates of renewable resource inputs must be kept within regenerative capacities of the natural system that generates them.
(b) Non-renewables: Depletion rates of non-renewable resource inputs should be set below the rate at which renewable substitutes are developed by human invention and investment. An easily calculable portion of the proceeds from liquidating non-renewables should be allocated to the attainment of sustainable substitutes.

suited to different circumstances. For example, a weak sustainability perspective with a strong emphasis on economic growth will most easily be justified when there is a strong link between poverty and environmental degradation and when tourism activity would result in a decline in more harmful activities such as uncontrolled logging. Stronger notions of sustainable tourism, on the other hand, may be better suited to circumstances that are less characterized by poverty relationships and where the alternative to tourism activity is complete protection. Hunter (1997) argues that sustainable tourism should be conceptualized as an adaptive paradigm.

The constant capital perspective is only workable if the relevant stocks and flows can be measured (Garrod and Fyall, 1998). Garrod and Fyall (1998) consider some of the issues arising. An interesting question is whether activities traditionally viewed as consumptive in orientation can qualify as ecotourism experiences. Holland *et al.* (1998) argue that billfish angling can often satisfy ecotourism criteria since it often involves, *inter alia*, a unique natural resource, an emphasis on catch and release, experiential learning and positive contributions to resource conservation and local economies. In terms of strong notions of sustainable development (Box 1.2) and the associated concept of environmental sustainability (Box 1.3), if the loss of species due to non-return and non-recovery following release is sufficiently low to keep the total loss, from all sources, below the maximum sustainable (or economic) yield, billfish angling can be considered environmentally sustainable. Applying the concept of environmental sustainability to tourism in natural areas thus involves adapting the input and output conditions specified in Box 1.3 to the tourism context in question and developing a holistic and multiple-use management plan to ensure that the conditions are met.

The first perspective on sustainable development tends to address ecological sustainability via the concepts of carrying capacity, indicators of sustainable development and so on. Along these lines, the Australian National Ecotourism Strategy states that 'Planning for ecotourism is based on resource constraints. Ecotourism opportunities will be lost if the resilience of an area and the ability of its community to absorb impact are exceeded, or if its biodiversity and physical appearance are altered significantly' (Allcock *et al.*, 1994, p. 17). The two perspectives on sustainable tourism development clearly have much in common, and both advocate a holistic approach to environmental management in order to avoid the tyranny of incrementalism.

An Alternative to Mass Tourism

Ecotourism is more than just sightseeing. It is an experience. For many ecotourists and others involved in the ecotourism industry, this experience differs fundamentally to the mass tourism experience. As noted earlier, disillusionment with mass tourism may have triggered the emergence of ecotourism. To the extent that ecotourism, as a form of alternative tourism, offers a less problematical form of tourism than mass tourism, it is potentially as diverse as the problems associated with mass tourism. Butler (1992, p. 33) observes that tourism can have detrimental impacts with regard to

> price rises (labor goods, taxes, land); changes in local attitudes and behavior; pressure on people (crowding, disturbance, alienation); loss of resources, access, rights, privacy; denigration or prostitution of local culture; reduction of aesthetics; pollution in various forms; lack of control over the destination's future; and specific problems such as vandalism, litter, traffic, and low-paid seasonal employment.

Many of these problems are addressed by the sustainability principles described in the previous section. However, ecotourism often offers an alternative to mass tourism in ways that do not necessarily fall within the principles described thus far in this chapter. And some of these attributes appear to be important from a demand perspective. In particular, tourists are increasingly seeking authenticity, immersion, self-discovery and quality rather than quantity (Hall and Weiler, 1992). They are seeking novel, adventuresome and personalized experiences in unique, remote and/or primitive locations (Wight, 1996b). The less popularized ecotourism developments thus tend to be small scale and low key and involve a high degree of participation by the local population. Crowded areas are to be avoided.

The fact that a significant proportion of the ecotourism market offers small-scale and personalized experiences suggests that such characteristics often represent an important element of the ecotourism experience. However, it is debatable whether these should be viewed as necessary characteristics of ecotourism, since this would rule out a large number of tours that satisfy the other criteria, some of which may excel on the sustainability and education dimensions. Hence, this dimension is perhaps best thought of as a secondary characteristic or principle that distinguishes classical ecotourism from popular ecotourism.

As with the other dimensions of ecotourism, this dimension is associated with certain tensions. In particular, a tension exists between the alternative tourist's desire for authentic, low key and intimate experiences and the need to be sensitive to community values. As Butler (1989, cited in Butler, 1990, p. 44) observed, it can be argued

> that tourism which places tourists in local homes, even when they are culturally sympathetic, and not desiring a change in local behavior, is much more likely to result in changes in local behavior in the long run than is a larger number of tourists in more conventional tourist ghettos, where contact with locals is limited.

More generally, small-scale tourism operations may or may not be more sustainable than larger-scale operations (Thomlinson and Getz, 1996). Destinations that are targeted towards the exclusive, quality rather than quantity, type of ecotourism experience also run the risk of being seen as elitist (Whelan, 1991).

Differing Perspectives

As with many other types of tourism, a good deal of confusion has surrounded the concept of ecotourism. To a large extent, this is a result of different stakeholders adopting different perspectives (Blamey, 1997). In particular, different perspectives have been adopted by: (i) scientific, conservation and non-governmental organizations; (ii) multilateral aid organizations; (iii) developing countries; and (iv) the travel industry and travelling public (Honey, 1999). Honey (1999, p. 11) observes that 'almost simultaneously but for different reasons, the principles and practices of ecotourism began taking shape within these four areas, and by the early 1990's, the concept had coalesced into the hottest new genre of environmentally and socially responsible travel'.

One of the key distinctions is between demand and supply-based perspectives on the ecotourism *product*. Those adopting the former tend to begin with an observation that for several decades now, a minority of tourists have sought to get away from the masses and seek experiences of a small scale, educative, personalized and unique nature. Ecotourism is a response to this demand. The ecotourism industry can only be profitable in the long term, however, if measures are taken to protect the natural-resource base upon which ecotourism depends. To this end, a holistic approach to environmental and tourism management is required, consisting primarily of supply-side initiatives such as restrictions of access, zoning systems, pricing mechanisms and monitoring. While ecotourism must be conducted in a sustainable way, the product is defined primarily from a demand perspective. While some tourists interested in ecotours actively seek out the environmentally friendly ones, this is very much the minority.

The second perspective tends to view the above indications of changing consumer demand as an opportunity to develop a new class of nature-based tourism product. The product is 'environmentally and culturally friendly nature-based experiences with an educative emphasis'. It is environmentally friendly in much the same way that toilet paper made from recycled and unbleached paper is seen as environmentally friendly. To the extent that consumers do not already demand the environmentally friendly product attribute, the demand for it needs to be created via advertising and other forms of promotion.

While the above demand perspective might be viewed as insufficiently proactive and less socially responsible, the supply-based perspective runs the risk of unnecessarily isolating a portion of the nature-based tourism market. Blamey (1997, p. 117) observes that if

> tourists are not currently demanding environmentally responsible tourism practices, there is no reason to focus discussions and initiatives regarding environmental responsibility around ecotourism in preference to other forms of tourism, unless ecotourism offers the greatest potential for improvement, or the best display of green tourism practices.

Preece *et al.* (1995) argue that a 'false distinction' is being made between tourism and ecotourism. Nature-based tourism and ecotourism 'should be considered as woven into the broad fabric of tourism, and should not be limited by artificially trying to categorise the phenomenon' (p. 10). They go on to state that 'the narrow focus on the term "ecotourism" has blinkered the view of planners and policy makers' (p. 13).

Although the above perspectives differ in how they define the ecotourism product, they both hold that a sustainable ecotourism industry requires supply-side management (Wight, 1993). The combination of supply-side initiatives that best serves the industry is itself subject to differing perspectives. Questions about how best to manage the environmental impacts of an industry, whether it be ecotourism or the production of automobiles, are typically distinct from the issue of how one defines the industry producing those impacts.

The supply-side management challenge for environmental and tourism management is to ensure 'that ecotourism doesn't occur willy-nilly wherever there is a demand for it, but that governments, tour operators, conservation groups, and local communities, among others, plan together where ecotourism sites should be established and how they should be managed' (Whelan, 1991, p. 20). Local communities may need to be involved in management decisions. As previously observed, supply-side initiatives pertaining to restrictions of access, use of permits and pricing mechanisms can be accused of being elitist or inequitable (Allcock *et al.*, 1994).

Ultimately, the complete integration of

environmental and developmental objectives requires a convergence among differing perspectives (Barbier, 1989). In the tourism context, this will require convergence in demand and supply perspectives and the different perspectives within each (particularly the latter) (Fig. 1.2). Wheeller (1995, p. 64) argues that people will interpret notions of ecotourism and sustainable tourism as suits them, and that 'no international decree will disperse the convenient clods of confusion' that have enveloped the terms. Ecotourism is, however, a fairly new concept in many countries, and the potential thus exists for the divergent current perspectives to merge into a coherent whole in the medium to long term (Blamey, 1995). Although ecotourism 'is indeed rare, often misdefined, and usually imperfect, it is still in its infancy, not on its deathbed' (Honey, 1999, p. 25).

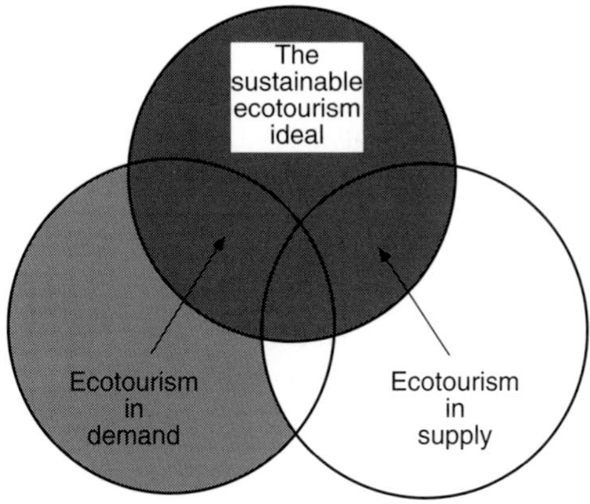

Fig. 1.2. One potential merging of perspectives.

References

Acott, T.G., La Trobe, H.L. and Howard, S.H. (1998) An evaluation of deep ecotourism and shallow ecotourism. *Journal of Sustainable Tourism* 6, 238–253.

Allcock, A., Jones, B., Lane, S. and Grant, J. (1994) *National Ecotourism Strategy*. Commonwealth Department of Tourism, Australian Government Publishing Service, Canberra.

Barbier, E. (1989) *Economics, Natural-Resource Scarcity and Development, Conventional and Alternative Views*. Earthscan, London.

Berry, S. and Ladkin, A. (1997) Sustainable tourism: a regional perspective. *Tourism Management* 18, 433–440.

Blamey, R.K. (1995) *The Nature of Ecotourism*, Occasional Paper No 21. Bureau of Tourism Research, Canberra.

Blamey, R.K. (1997) Ecotourism: the search for an operational definition. *Journal of Sustainable Tourism* 5, 109–130.

Boo, E. (1990) *Ecotourism: the Potentials and Pitfalls*, Vols 1 and 2. WorldWide Fund for Nature, Washington, DC.

Bottrill, C.G. and Pearce, D.G. (1995) Ecotourism: towards a key elements approach to operationalising the concept. *Journal of Sustainable Tourism* 3, 45–54.

Bramwell, B. and Lane, B. (1993) Sustainable tourism: an evolving global approach. *Journal of Sustainable Tourism* 1, 1–5.

Buckley, R. (1994) A framework for ecotourism. *Annals of Tourism Research* 21, 661–669.

Burgess, A.S. (1993) The environmental responsibilities of commercial ecotourism in Tasmania. Thesis, University of Tasmania, Hobart.

Burton, R. (1998) Maintaining the quality of ecotourism: ecotour operators' responses to tourism growth. *Journal of Sustainable Tourism* 6, 117–142.

Butler, R.W. (1990) Alternative tourism: pious hope or Trojan horse? *Journal of Travel Research* 28 (Winter), 40–45.

Butler, R.W. (1992) Alternative tourism: the thin edge of the wedge. In: Smith, V.L. and Eadington, W.R. (eds) *Tourism Alternatives: Potentials and Problems in the Development of Tourism*. University of Pennsylvania Press, Philadelphia, pp. 31–46.

Ceballos-Lascuráin, H. (1987) The future of ecotourism. *Mexico Journal* January, 13–14.

Chirgwin, S. and Hughes, K. (1997) Ecotourism: the participants' perceptions. *Journal of Tourism Studies* 8, 2–8.

Common, M.S. and Perrings, C. (1992) Towards an ecological economics of sustainability. *Ecological Economics* 6, 7–34.

Ecotourism Association of Australia (1992) *Newsletter* 1, 2.

The Ecotourism Society (1991a) *The Ecotourism Society Newsletter* Number 1, Spring.

The Ecotourism Society (1991b) *Ecotourism Guidelines for Nature-Based Tour Operators.* The Ecotourism Society, North Bennington, Vermont.

Fennell, D.A. (1998) Ecotourism in Canada. *Annals of Tourism Research* 25, 231–235.

Finucane, S.J. (1992) A report on the environmental impacts of ecotourism in Western Australia. Postgraduate report, Murdoch University, Western Australia.

Garrod, B. and Fyall, A. (1998) Beyond the rhetoric of sustainable tourism? *Tourism Management* 19, 199–212.

Goodland, R. (1999) The biophysical basis of environmental sustainability. In: Van den Bergh, J. (ed.) *Handbook of Environmental Economics.* Edward Elgar, Cheltenham, UK, pp. 709–721.

Hall, C.M. (1993) Submission on strategic direction for 'ecotourism'. Submission to National Ecotourism Strategy, Commonwealth Department of Tourism, Canberra.

Hall, C.M. and Weiler, B. (1992) Introduction: What's special about special interest tourism? In: Weiler, B. and Hall, C.M. (eds) *Special Interest Tourism.* Belhaven Press, London, pp. 1–14.

Hetzer, W. (1965) Environment, tourism, culture. *Links* July, 1–3.

Hinch, T.D. (1996) Urban tourism: perspectives on sustainability. *Journal of Sustainable Tourism* 4, 95–110.

Holland, S.M., Ditton, R.B. and Graefe, A.R. (1998) An ecotourism perspective on billfish fisheries. *Journal of Sustainable Tourism* 6, 97–115.

Honey, M. (1999) *Ecotourism and Sustainable Development: Who Owns Paradise?* Island Press, Washington, DC.

Hunter, C. (1997) Sustainable tourism as an adaptive paradigm. *Annals of Tourism Research* 21, 850–867.

Johnston, B.R. and Edwards, T. (1994) The commodification of mountaineering. *Annals of Tourism Research* 21, 459–478.

de Kadt, E. (1992) Making the alternative sustainable: lessons from development for tourism. In: Smith, V.L. and Eadington, W.R. (eds) *Tourism Alternatives: Potentials and Problems in the Development of Tourism.* University of Pennsylvania Press, Philadelphia, pp. 47–75.

Kalinowski, K.M. and Weiler, B. (1992) Educational travel. In: Weiler, B. and Hall, C.M. (eds) *Special Interest Tourism.* Belhaven Press, London, pp. 15–26.

Kusler, J. (1992) Ecotourism and resource conservation: introduction to issues. Paper presented at the World Congress on Adventure Travel and Ecotourism, August.

Kutay, K. (1989) The new ethic in environmental travel. In: *The Environmental Journal,* cited in Ziffer, K.A. (1989) *Ecotourism: the Uneasy Alliance,* Working Paper #1. Conservation International, Washington, DC.

Lawrence, T.B., Wickins, D. and Phillips, N. (1997) Managing legitimacy in ecotourism. *Tourism Management* 18, 307–316.

Lindberg, K. (1991) *Policies for Maximizing Nature Tourism's Ecological and Economic Benefits.* International Conservation Financing Project Working Paper, World Resources Institute, Washington, DC.

Lindberg, K. and Hawkins, D.E. (eds) (1993) *Ecotourism: a Guide for Planners and Managers.* The Ecotourism Society, Vermont, USA.

McKercher, B. (1993) Some fundamental truths about tourism: understanding tourism's social and environmental impacts. *Journal of Sustainable Tourism* 1, 6–16.

Miller, K. (1978) *Planning National Parks for Ecodevelopment: Methods and Cases from Latin America.* University of Michigan, Ann Arbor.

Moscardo, G. (1998) Interpretation and sustainable tourism: functions, examples and principles. *Journal of Tourism Studies* 9, 2–12.

Pearce, D.G. (1992) Alternative tourism: concepts, classifications and questions. In: Smith, V.L. and Eadington, W.R. (eds) *Tourism Alternatives: Potentials and Problems in the Development of Tourism.* University of Pennsylvania Press, Philadelphia, pp. 15–30.

Preece, N., Van Oosterzee, P. and James, D. (1995) *Two Way Track – Biodiversity Conservation and Ecotourism: an Investigation of Linkages, Mutual Benefits and Future Opportunities.* Biodiversity Series, Paper No. 5. Department of the Environment, Sport and the Territories, Canberra.

Ross, S. and Wall, G. (1999) Ecotourism: towards congruence between theory and practice. *Tourism Management* 20, 123–132.

Scheyvens, R. (1999) Ecotourism and the empowerment of local communities. *Tourism Management* 20, 245–249.

Social Change Media (1995) *A National Ecotourism Education Strategy.* Commonwealth Department of Tourism, Canberra.

Thomlinson, E. and Getz, D. (1996) The question of scale in ecotourism: case study of two small ecotour operators in the Mundo Maya region of Central America. *Journal of Sustainable Tourism* 4, 183–200.

Tickell, C. (1994) Foreword. In: Cater, E. and Lowman, G. (eds) *Ecotourism: a Sustainable Option?* John Wiley & Sons, Brisbane, pp. ix–x.

Tourism Concern (1991) *Beyond the Green Horizon.* Tourism Concern and WWF, Roehampton Institute, London.

Turner, R.K. (1991) Environment, economics and ethics. In: Pearce, D., Barbier, E., Markandya, A., Barrett, S., Turner, R.K. and Swanson, T. (eds) *Blueprint 2: Greening the World Economy.* Earthscan, London, pp. 208–224.

Turner, R.K., Doktor, P. and Adger, N. (1992) Sea level rise and coastal wetlands in the UK: mitigation strategies for sustainable management. Paper presented to the conference of the International Society for Ecological Economics, Stockholm, 6–8 August.

Twining-Ward, L. (1999) Towards sustainable tourism development: observations from a distance. *Tourism Management* 20, 187–188.

Urry, J. (1990) *The Tourist Gaze: Leisure and Travel in Contemporary Societies.* Sage Publications, London.

Valentine, P.S. (1992a) Review. Nature-based tourism. In: Weiler, B. and Hall, C.M. (eds) *Special Interest Tourism.* Belhaven Press, London, pp. 105–128.

Valentine, P.S. (1992b) Ecotourism and nature conservation: a definition with some recent developments in Micronesia. In: Weiler, B. (ed.) *Ecotourism: Incorporating the Global Classroom.* Bureau of Tourism Research, Canberra, pp. 4–9.

Wallace, G.N. (1999) Toward a principled evaluation of ecotourism ventures. Working paper located at http://www.ecotourism.org/textfiles/wallacea.txt

Wheeller, B. (1995) Egotourism, sustainable tourism and the environment – a symbiotic, symbolic or shambolic relationship? In: Seaton, A.V. (ed.) *Tourism: the State of the Art.* John Wiley and Sons, Brisbane, pp. 647–654.

Whelan, T. (1991) Ecotourism and its role in sustainable development. In: Whelan, T. (ed.) *Nature Tourism: Managing for the Environment.* Island Press, Washington, DC, pp. 3–22.

Wight, P.A. (1993) Sustainable tourism: balancing economic, environmental and social goals within an ethical framework. *Journal of Tourism Studies* 4, 54–66.

Wight, P.A. (1994) Environmentally responsible marketing of tourism. In: Cater, E. and Lowman, G. (eds) *Ecotourism: a Sustainable Option?* John Wiley & Sons, Brisbane, pp. 39–56.

Wight, P.A. (1996a) North American ecotourists: market profile and trip characteristics. *Journal of Travel Research* Spring, 2–10.

Wight, P.A. (1996b) North American ecotourism markets: motivations, preferences and destinations. *Journal of Travel Research* Summer, 3–10.

World Commission on Environment and Development (WCED) (1987) *Our Common Future.* Oxford University Press, Oxford.

Ziffer, K.A. (1989) *Ecotourism: the Uneasy Alliance*, Working Paper no. 1. Conservation International, Washington, DC.

Chapter 2

Types of Ecotourism

M.B. Orams

Centre for Tourism Research, Massey University at Albany, North Shore MSC, New Zealand

Introduction

The rise of the term 'ecotourism' has been relatively rapid. In 1980 the term did not exist and now, 20 years on, this Encyclopedia represents the thinking of many different authors from around the world on the topic. The term ecotourism has also become the subject of much debate; what it is, what it should be and how it can work are all questions that continue to dominate the literature. This chapter continues with the debate initiated in Chapter 1, for this kind of discussion is healthy. Complex problems, such as managing tourists' impacts on natural ecosystems, seldom have simple answers. Because of this complexity a concept such as ecotourism should not be expected to provide a universal simple answer. Consequently, ecotourism is not 'the' answer to the problems caused by nature-based tourism. It is, however, a concept that has value, and realizing that value is best achieved by constructive thinking and debate, but most of all by learning from practical application of the ideas contained within the ecotourism concept. A consideration of the types of ecotourism activity is one way to contribute to this debate, and this chapter attempts to do just that. In addition, this chapter also contains two case studies that are illustrative of the practical component that is now so important in the further evolution of the ecotourism concept.

Semantic Debates

The evolution of language is an interesting phenomenon. It seems that a natural part of this evolutionary process is the invention, adoption and eventual common usage of terms that describe valuable concepts or ideas. There are many examples throughout history. In the context of this discussion, namely the management of human influences on natural ecosystems, three terms provide examples of this process.

The terms 'conservation' and 'sustainability' have become popular terms for valuable approaches to managing natural resources. Each evolved in a similar way. First, the concepts themselves appear to have existed for many centuries. Second, the invention of the specific terms occurred and usage of them became more widespread and diverse in their application. Third, debates ensued regarding definitions, typologies and the value of the terms/concepts. For example, Fennell (1999, p. 72) outlines three differing perspectives that evolved in the early use of the term conservation:

> The first involved the view that conservation should entail the maintenance of harmony between humankind and nature, the second that conservation was related to the efficient use of resource. The final perception was that conservation – preservation could ideally be attained from the standpoint of religion and spirituality.

Finally, the terms became used in more formal contexts, such as in law, in international agreements and within government agencies.

The semantic evolutionary process illustrated by 'conservation' and 'sustainability' has also occurred with regard to the concept of 'ecotourism'. The idea of integrating tourism with conservation has probably been around since the early days of African safaris and the development of the national park concept in the 19th century. In the tourism literature, the concept was given wider and more explicit exposure by Budowski in a 1976 article entitled 'Tourism and conservation: conflict, coexistence or symbiosis?'. However, while Budowski's paper made a case for the ideals of ecotourism, as generally understood today, he did not explicitly use the term 'ecotourism' to describe his thinking. There is no general agreement on who invented or first used the term 'ecotourism' (Fennell, 1999), but it is clear that the term first appeared in the published material during the 1980s (for example, Romeril, 1985; Ceballos-Lascuráin, 1987; Laarman and Durst, 1987; Ziffer, 1989). Certainly, even if the term itself was not used, the concept has been around for centuries (Fennell, 1999). It seems likely that the invention of the term 'ecotourism' itself is related to the connotations associated with terms with the 'eco' prefix such as 'ecology' and 'ecosystem'. Thus combining this prefix with the term tourism provides a suitable label for the concept that authors like Hetzer (1965), Budowski (1976) and others have advocated.

The idea of visiting areas for the purposes of observing and experiencing elements of the natural environment pre-dates 'ecotourism'. Safaris to places such as Africa to view wildlife were popular among explorers and adventurers from Western Europe during the 1800s (Adler, 1989). Recreational activities such as hiking, birdwatching, climbing, cross-country skiing, fishing, canoeing and other boating are all based on the natural environment and have been popular for centuries. Therefore, many of the activities that can now be included under the label 'ecotourism' have existed long before the term itself was invented. The successful evolution of the term 'ecotourism' (in the sense that it is now widely used) is related to three main issues. First, it is a reaction against the negative impacts associated with 'mass tourism' (Weaver, 1998).' For example, Glasson *et al.* (1995, p. 27) stated: 'Tourism contains the seeds of its own destruction; tourism can kill tourism, destroying the very environmental attractions which visitors come to a location to experience'. Second, it has developed in response to the growth of tourism based on natural environmental attractions and third, as an outcome of the growing understanding and acceptance of the principles of environmental conservation and sustainability (Orams, 1995) (see Chapter 1).

Defining Ecotourism: Lines in the Sand

It is difficult to consider the subject 'types of ecotourism' without having a clear definition or understanding of what ecotourism actually is. Thus, a consideration of the topic occurs within the semantic debate that has dominated the literature on ecotourism to date. As with all such debates the difficulties arise when considering the 'margins'. That is, while there may be a general acceptance of what the basic concept includes and a corresponding general acceptance of what it does not include, there is little agreement regarding those activities or operations that don't clearly fit into either scenario. For example, does a safari that includes both an educational component and the hunting of wildlife constitute an 'ecotourism' experience? Similarly, what of a tourism operation that has the best conservation intentions but through ignorance and/or accident the natural attraction is harmed? An example of this is the September 1998 collisions between whale-watching vessels and whales off Stellwagen Bank north of Cape Cod, Massachusetts. These collisions resulted in injury for a humpback whale and the death of a minke whale (Associated Press, 1998).

There are now so many definitions of ecotourism and so many papers discussing these definitions that to review all of these again here would be repetitious and perhaps even annoying. Instead, the reader should consult useful publications such as those by Valentine (1990), Figgis (1993), Miller and Kaae (1993), Moore and Carter (1993), Hvenegaard (1994), Orams (1995), Higgins (1996), Weaver (1998), Fennell (1999) and Chapter 1 of this Encyclopedia. At the risk of inducing 'eco-nausea', however, it is useful to consider the range of definitions offered in terms of the conceptual approaches they represent. This analysis is important, for if we are to understand the range of 'types of ecotourism' it is helpful to clarify what this range includes.

Mass and alternative tourism

Fennell (1999) considers that ecotourism exists within the broader classification of tourism types which, at an initial level, can be divided into 'mass tourism' and 'alternative tourism'. Mass tourism is seen as the more traditional form of tourism development where short-term, free-market principles dominate and the maximization of income is paramount. The development of the tourism industry was originally seen as a desirable and relatively 'clean' industry for nations and regions to pursue. This was particularly true in terms of benefits in foreign exchange earnings, employment and infrastructural development such as transport networks (Warren and Taylor, 1994). However, in the past two decades the 'worm has turned'. 'These days we are more prone to vilify or characterise conventional mass tourism as a beast; a monstrosity which has few redeeming qualities for the destination region, their people and their natural resource base' (Fennell, 1999, p. 7). It should be remembered, however, that there are many locations where the mass tourism 'beast' has been an economic saviour and has, on balance, not been the destructive, exploitive industrial development that the modern tourism literature paints it. This is particularly true when one considers the potential alternative uses of those resources which formed the basis of the development of tourism, for example, as when a beach area is developed as a tourist resort as an alternative to sand mining.

This is not to deny that 'mass tourism' has caused problems, because it has. There has, quite justifiably, been a need to identify an alternative approach to tourism development that lessens the negative consequences of the mass tourism approach. Thus the 'alternative tourism' perspective has become a popular paradigm. This alternative approach has been described as a 'competing paradigm' to mass tourism (Fennell, 1999), but it can also be viewed as a complementary approach to tourism. That is, it is not possible to have 'alternative tourism' without something for it to be 'alternative' to. So, the discussion returns to a semantic debate ... perhaps it is best to accept that alternative tourism is a natural outcome of the maturing understanding of tourism development and its strengths and weaknesses. Fennell (1999, p. 9) states that:

> AT is a generic term that encompasses a whole range of tourism strategies (e.g. 'appropriate', 'eco-', 'soft', 'responsible', 'people to people', 'controlled', 'small scale', 'cottage', and 'green' tourism) all of which purport to offer a more benign alternative to conventional mass tourism in certain types of destinations.

However, Weaver (1998) quite rightly points out that there are also many criticisms of alternative tourism. It is clear that just because alternative tourism has developed as a reaction to the negative consequences of mass tourism it is not necessarily less harmful or better than its alternative. It does however, provide a useful means to conceptualize the relationship between differing views of tourism (Fig. 2.1). In Chapter 5 of this Encyclopedia, Weaver challenges the dichotomous perspective on alternative tourism and mass tourism in suggesting circumstances under which ecotourism can be considered as a form of mass tourism.

Nature-based tourism versus ecotourism

There appears to be a consensus in the literature on tourism that demand for opportunities to interact with nature has been increasing rapidly (Jenner and Smith, 1992). This general interest in nature and experiences based upon natural attractions is reflected in an increasing demand and value being placed on relatively undisturbed natural environments and, in particular, wild animals (Gauthier, 1993). Tourism of this type has been applauded by many as a suitable saviour for threatened wildlife populations (Davies, 1990; Borge *et al.*, 1991; Groom *et al.*, 1991; Barnes *et al.*, 1992; Burnie, 1994) and thus is widely viewed as 'ecotourism'. However, many authors point out that significant negative environmental impacts can result from nature-based tourism (Butler, 1990; Wheeller, 1991; Zell, 1992; Pleumarom, 1993; Wheeller, 1994; Glasson *et al.*, 1995) and question whether nature-based tourism is automatically 'ecotourism' (see also Chapter 5).

An additional issue to consider is the now commonplace inclusion of a sociocultural component within the discussion of ecotourism. For example, Wallace and Pierce (1996, p. 848) argue that ecotourism is travel that is based not only on nature but also on 'the people (caretakers) who live nearby, their needs, their culture, and their relationships to the land'. This inclusion of a human component within ecotourism is a significant extension of the concept. However, the inclusion of the needs of people does little to clarify the concept of ecotourism. For if humans, and human behaviour, are considered to be part of the 'natural' environment the lines in

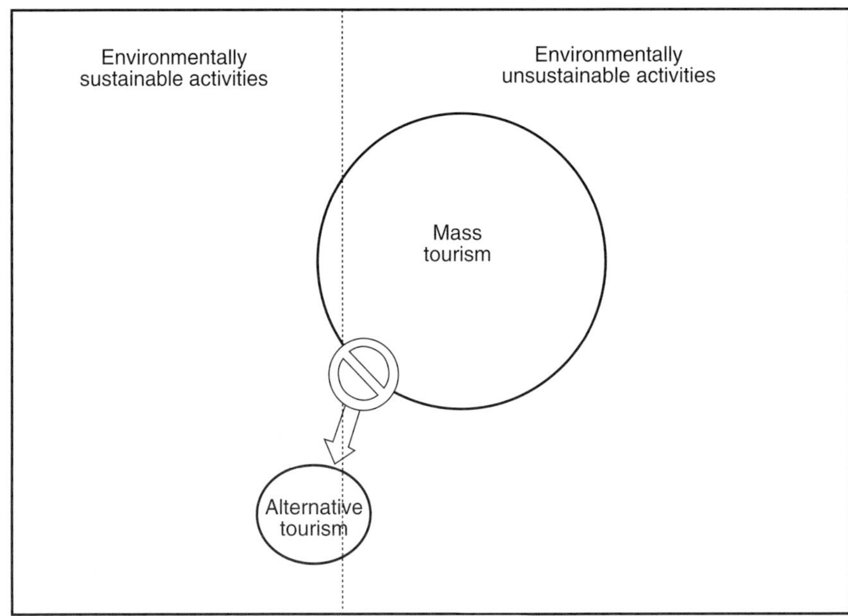

Fig. 2.1. Relationships between mass and alternative tourism (after Weaver, 1998).

the sand bounding the ecotourism concept are extended even further. If ecotourism is interpreted too broadly, it becomes meaningless. For example, if the viewing of a cultural dance performance by an indigenous group is considered to be ecotourism why not include other human activities, sport, art, music and so on?

If ecotourism is to have any meaning, lines bounding its application must be drawn. As discussed in Chapter 1, ecotourism is considered to share three basic characteristics.

1. The natural (non-human) environment or a feature of it is the prime attraction for the tourist.
2. The basis of that attraction for the tourist is an inherent appreciation/educational interest in that natural environment or natural environmental feature. Based on this and the first criterion, ecotourism therefore incorporates such related activities as birdwatching, nature observation, hiking and bushwalking, nature photography, outdoor education, stargazing and whale-watching. The status of diving, snorkelling and scuba-diving is a matter of some debate, while activities such as trekking and safaris are hybrids with adventure and other forms of tourism.
3. A management regime/effort directed at the conservation/sustainable use of that natural environment exists.

Thus, ecotourism is seen as a subset of nature-based tourism. One of the difficulties in applying such an interpretation of the concept is in assessing the application of the last criterion. Many tourism operators and their clients may satisfy each of the three criteria, however, despite the best intentions for their activities to be conservation-oriented and sustainable, the result of their activities may still be detrimental to the attraction. This is the essence of the problem in defining ecotourism. It is a concept with worthy intentions. However, its results are still often no better than those obtained from other forms of tourism. Consequently, ecotourism may not always be beneficial to the natural environment. There are, unfortunately, a significant number of relevant examples (Hanna and Wells, 1992; Burger and Gochfield, 1993; Griffiths and Van Schaik, 1993; Ingold *et al.*, 1993; Muir, 1993; Viskovic, 1993). As a result, many authors have expressed concern over the negative impacts that are being inflicted on natural ecosystems in the name of 'ecotourism' (Hegerl, 1984; Mellor, 1990; Ward, 1990; Laycock, 1991).

For the purposes of this chapter, the ecotourism concept is considered broadly. However, it is acknowledged that even though the philosophy of ecotourism is ethically sound – namely that it attempts at the very least to minimize its negative impacts – the application of that approach is not always successful. Because of this broad interpretation and application of the ecotourism approach it is useful to consider the types of ecotourism operations within a conceptual framework.

Conceptual Frameworks for Analysing Types of Ecotourism

It is clear that there is a great variety of definitions of ecotourism. Equally, there is a great variety of tourism operators and agencies that have adopted the label with different interpretations of what the label actually means. Thus, the types of ecotourism occur within this range of definitions and use of the term. While some may argue that certain types of operations should not use the term 'ecotourism' to describe their activities, the reality is they are using the label and they cannot be prevented from doing so. There is no copyright on the term, there is no patent on what the approach entails. As a consequence any debate over who has the right to call themselves an ecotourism operator is meaningless. What can be done, however, is to review the range of ecotourism types and to categorize them according to the nature of their operation or the definition of ecotourism that they subscribe to. A number of authors have attempted to do just that, as described below.

Soft–hard

Laarman and Durst (1987) describe 'hard' and 'soft' dimensions of ecotourism (Fig. 2.2). These terms refer to the level of dedication of the ecotourist to the experience in terms of the physical rigour/effort involved and the level of interest in the natural attraction. 'Hard-core' ecotourists have a deep level of interest and often expertise in the subject matter; for example, they may have a life-long passion for birdwatching or other forms of nature observation. In addition, ecotourists have differing dispositions regarding the level of physical challenge and comfort they wish to experience or are prepared to tolerate. A 'hard' ecotourist is prepared and may even desire to live basically, with few comforts, and to travel in difficult circumstances for long periods within a wilderness context in order to truly 'experience' nature. Conversely, the 'soft' ecotourist has casual interest in the natural attraction but wishes to experience that attraction on a more superficial and highly mediated level. Similarly, the soft ecotourist is less prepared to accept discomfort and physical hardship as part of the experience, and may be content to spend a considerable amount of their time in an interpretive centre surrounded by other tourists. Typically, hard ecotourists are engaged in specialized ecotourism travel, while soft ecotourists engage in ecotourism as one, usually short duration, element of a multi-purpose and multidimensional travel experience.

Laarman and Durst's (1987) discussion and subsequent work (1993) provide a useful context when considering the types of ecotourists themselves (see Chapter 3). However, at a more fundamental level types of ecotourism can be considered in terms of their relationships with nature.

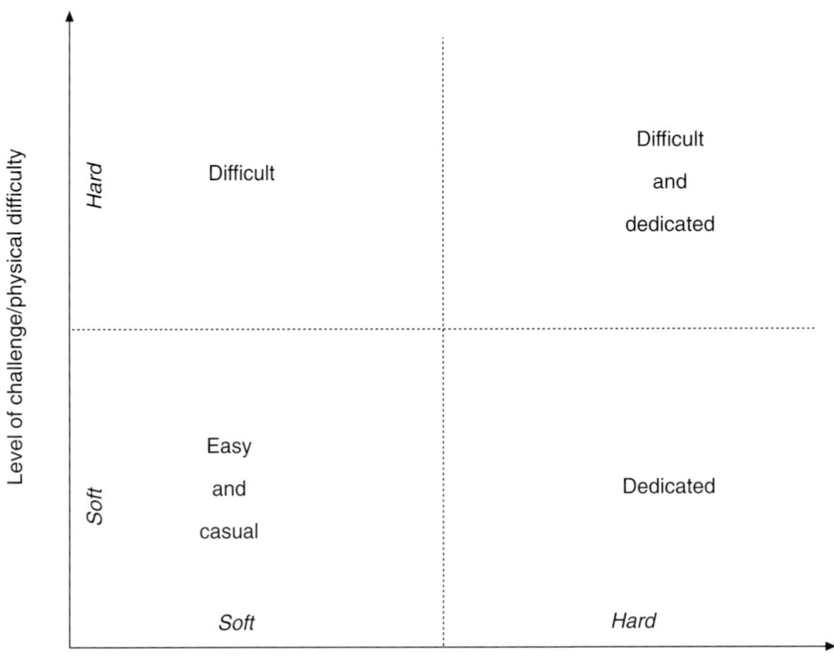

Fig. 2.2. Hard and soft ecotourism (after Laarman and Durst, 1987).

Natural–unnatural

Miller and Kaae (1993) have described the diverse number of definitions and applications of the concept of ecotourism as part of a continuum of relatedness to nature (Fig. 2.3). This continuum of ecotourism paradigms is bounded by polar extremes. At one pole is the view that all tourism (including ecotourism) has negative impacts on the natural world. That is, no matter what management strategies are in place, humans through their mere presence have an unnatural impact. Therefore, ecotourism, in this view, is impossible because any kind of tourism will have a negative effect. At the other extreme, humans are viewed as living organisms – fauna – whose behaviour is inevitably 'natural'. That is, humans are part of the natural world, just like all other living things and therefore human behaviour is 'natural behaviour' and contributes to the natural evolution of life. This view holds that because humans are part of 'nature' they are part of the 'natural process' and, as a result, they are literally unable to behave unnaturally. There is, therefore, no difference between ecotourism and other forms of tourism in terms of their 'naturalness' and thus, all ecotourism is tourism and vice versa. These two positions represent extreme and unrealistic views. In reality, types of ecotourism can be considered as lying somewhere between these polar extremes.

Exploitive–passive–active

Ecotourism types can also be classified according to their tendency to be consistent with their degree of impact on the natural environment. This classification is linked with a consideration of ethics in ecotourism. This consideration is seen by a number of authors as an integral part of any discussion of ecotourism (Kutay, 1989; Wight, 1993; Duenkel and Scott, 1994; Karwacki and Boyd, 1995; Orams, 1995; Fennell, 1999), and is pursued further in Chapter 41.

It is, in fact, difficult to view ecotourism in any other way; for inherent in almost all definitions of ecotourism is the suggestion that ecotourism is attempting to 'do the right thing'. The concepts of conservation, sustainability and alternative tourism discussed earlier have a similar ethical component and are concepts closely linked with that of ecotourism. The variation within the ecotourism realm surrounds what the 'right thing' actually is. Without delving too deeply into deep ecology and environmental ethics, the oft-quoted contention of Aldo Leopold (1949, p. 224) is a useful fundamental guideline with regard to ecotourism. 'A thing is right when it tends to preserve the integrity, stability and beauty of the biotic community. It is wrong when it tends otherwise.' Thus, ecotourism operations that actively contribute to the improvement of the natural environment can be viewed as 'better' (Orams, 1995), or

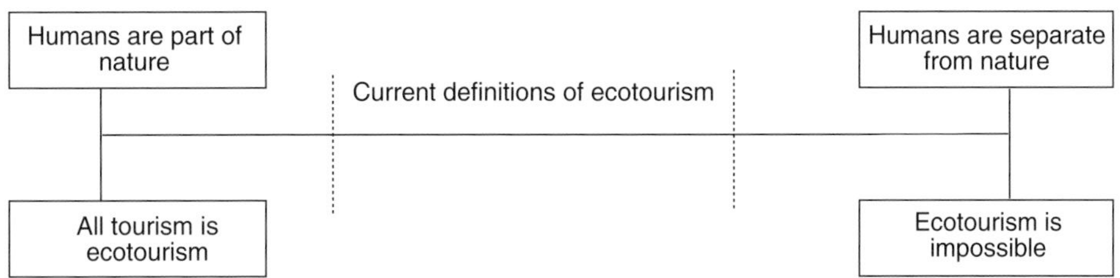

Fig. 2.3. Humans as natural and unnatural influences and ecotourism (after Miller and Kaae, 1993).

as more positive and responsible. Operations that detract from the quality of the natural environment can be viewed as 'worse' or more exploitive and irresponsible. Between these types of ecotourism are those types that can be viewed as more neutral and passive, operations that simply seek to minimize their impacts on the natural environment. This kind of conceptual approach can be represented diagrammatically (Fig. 2.4). It acknowledges that there are many different types of ecotourism but considers that some are better than others. Furthermore, this typology has been used as a basis for arguing that the role of ecotourism operators and agencies charged with managing ecotourism should be to prompt movement from less desirable to more desirable states along this continuum (Orams, 1995).

Case Studies of Differing Types of Ecotourism

The following two case studies provide examples of types of ecotourism activity. The first of these can be viewed as an ecotourism 'type' that is located toward the exploitive, negative end of the continuum, the second as a 'type' located toward the responsible, positive end.

Case study 1: Heron Island, Great Barrier Reef, Australia

Heron Island is located in the southern region of Australia's Great Barrier Reef among a group of around 13 islands known as the Capricorn–Bunker Group. Heron is a sand cay located at the western end of a coral reef lagoon and is relatively small – around 800 m long and 280 m wide at its widest point. The island is an ecologically important breeding site for a number of species of birds and sea turtles. Early human use of the island was based on the harvesting of turtles, and a turtle soup factory operated at Heron from 1925 to 1929 (Limpus *et al.*, 1983). This factory became a tourist resort in 1932 (Great Barrier Reef Committee, 1977) and it has been subsequently bought and developed by the P&O company. Additional development on the island includes a research station (established in 1951) now operated by the University of Queensland (Jones, 1967) and a park centre for Heron Island National Park (established in 1983) (Neil, 1993).

Heron Island Resort promotes itself as an 'environmentally friendly' ecotourism facility:

> Whilst the island is an international resort, great care is taken to make it a 'live and let live' situation with nature. The resort takes up only one corner of the island, with the remainder belonging to the seabirds and the turtles ... The golden rule of Heron is to enjoy nature, without disturbing it. This is carried out right through to details like garbage disposal. Resort waste that is not biodegradable such as bottles and cans, is actually shipped back to the mainland and properly disposed. Every precaution is taken so as not to upset the delicate balance of nature.
>
> (Heron Island Resort, 1986, p. 2)

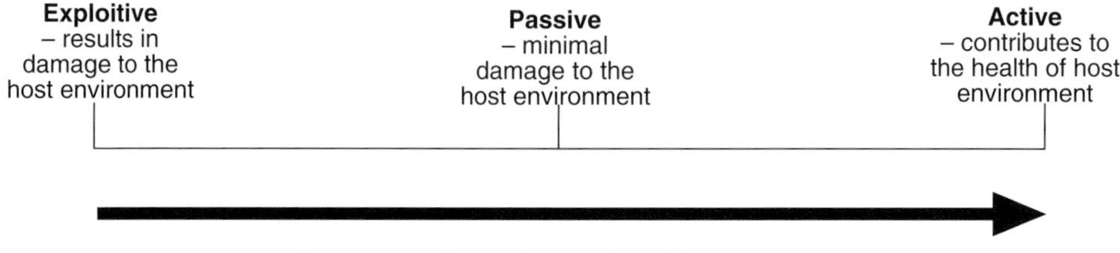

Fig. 2.4. The continuum of ecotourism types (after Orams, 1995).

However, despite these 'green' sentiments there is no doubt that the development and operation of the resort at Heron Island has caused significant detrimental environmental impacts. It should also be noted that the operation of the University of Queensland research station and the national park centre also contributes to these impacts.

Examples of these impacts include the building of structures resulting in a reduction in the area available for nesting for both seabirds and turtles. This is particularly significant for the green sea turtle and the burrow nesting wedge-tailed shearwater, both of which are thought to return to the same site each year for nesting. The lights from buildings are also thought to confuse hatchlings from both these species resulting in increased mortality rates (Gibson, 1976; Hulsman, 1983; Lane, 1991). Tourists who walk around the island have been observed to collapse the shearwater burrows, killing the nesting bird and chick (Dyer, 1992). In addition, disturbance of seabirds can cause them to abandon their nests and hatchlings can die if the parent does not return after as little 15 min (Hulsman, 1984).

Of more direct concern with regard to the marine environment, is the dredging and maintenance of the boat channel through the reef flat to the western side of the island. This channel has significantly altered the tidal flow around the island and the surrounding reef flat (Gourlay, 1991), altering the benthic fauna adjacent to the channel (Neil, 1988) and accelerating beach erosion on the northern and southern sides of the island (Neil, 1993). Tourists and their actions also have an impact on Heron's ecology. The common practice of walking over the coral reef flat at low tide at Heron can significantly reduce the health and abundance of coral (Woodland and Hooper, 1977). Litter is common on both the intertidal areas and the reef flat (personal observation). Recreational fishing is also likely to alter the natural composition of reef fish populations.

Heron Island provides an interesting example of an ecotourism destination that can be classified as being toward the exploitive end of the continuum. This is because, despite best intentions to minimize impacts, they still have occurred and have resulted in a significant deterioration of the natural environment. This pattern is common among many ecotourism destinations.

Case study 2: Tiritiri Matangi Island, Hauraki Gulf, New Zealand

Tiritiri Matangi Island is located east of New Zealand's Whangaparaoa Peninsula, approximately 30 km north of Auckland. The island was partially cleared of native forest by Maori, then almost completely cleared by European settlers near the turn of the century for conversion to pastoral farmland, primarily for sheep and cattle grazing (Moon, 1998). In 1971 the New Zealand government assumed responsibility for the island. As a result, public access to the island was granted (prior to this time, even though the island was publicly owned an exclusive grazing lease was in place and the public had restricted access). In 1982 a working plan was developed which adopted a strategy of replanting the island in native plants and to create an environment where natural bush, forest and native bird species could flourish (Drey, 1982). A critical component of this strategy was the involvement of visitors to the island and the promotion of the island as an easily accessible wildlife sanctuary for tourists (domestic and international) to visit (Hawley, 1997).

Throughout the 1980s and 1990s visitors to the island have been encouraged to participate in planting programmes, to become members of a community supporters' programme, to raise money, and to become aware of the conservation values of the island (Moon, 1998). In the mid-1990s the island had become so popular that a daily commercial ferry service began to operate from nearby Gulf Harbour on the mainland. In addition, a number of 'ecotour' operators began to take package tours to the island. With the assistance of the supporters' programme and a significant number of volunteers the gradual reintroduction of a number of endangered native bird species has begun. These birds have added to the attraction of the island and a remarkable success rate has been achieved in re-establishing these birds and their breeding on the island (Moon, 1998).

The support of tourists to the island has been fundamental in re-establishing the natural habitat that existed on Tiritiri before humans arrived in this part of the world. The continued involvement of these tourists in planting programmes, in financial support for management on the island, in providing educational programmes for other visitors and in ensuring that inappropriate behaviour (such as the bringing of pets) is minimized on the island, has resulted in a major improvement in the quality of the island's environment over the past decade (Cessford, 1995).

The development of ecotourism has been of great benefit to Tiritiri Matangi island. Quite simply, without tourism it would have remained an ecologically insignificant location. Fortunately, with the help of island visitors and supporters, it is fast becoming one of the most important bird sanctuaries in New Zealand – a nation with many of the world's most endangered bird species. Thus, Tiritiri Matangi is a location that conforms to the responsible, active end of the ecotourism continuum.

These two case studies may not necessarily reflect their status as more or less desirable ecotourism when considered in a temporal context. Rather they may simply represent their differing stages along the tourism development path. Tourism is a complex social and economic process that does not remain static over time. There are many examples of the development of tourism – or ecotourism – at a particular site that in its early stages had few negative impacts, appeared to be sustainable and was viewed as a positive outcome for local communities, both human and natural. However, over time many such locations have been transformed gradually from the positive end of the continuum to the exploitive.

This temporal transition is of particular concern with regard to ecotourism because much of this type of tourism is conducted in high-quality natural sites with few human influences. Ecotourism can, in some situations, simply be the catalyst for more widespread tourist development 'jump-starting' a pristine natural environment down the path of development. This risk is possibly one of the most significant with regard to ecotourism. The reality is for many ecotourism locations and operators with the best intentions, that once the economic development 'ball starts rolling' it is extremely difficult to control the outcomes. It is understandable then that many are cynical regarding the results of ecotourism, irrespective of whether the type of ecotourism is responsible and active or not (for example, see the writings of Wheeller, 1991, 1994). It can be argued that ecotourism, while a laudable concept in theory has little value in practice. However, cases such as that of Tiritiri Matangi island, provide hope that ecotourism can have positive practical outcomes.

Conclusions Regarding Types of Ecotourism

It is difficult when examining ecotourism not to become cynical regarding its application. The words of the Reverend Francis Kilvert, written in his diary in 1870 seem, in many situations, as apt today (Croall, 1995): 'of all the noxious animals, the most noxious is the tourist'. In fact many present day commentators may argue that because the ecotourist is 'dressed up' with a palatable label that 'of all the tourists, the most dangerous is the ecotourist'. Others, however, argue that ecotourism is the 'answer' to tourism development and claim 'that ecotourism is the only tourism development that is sustainable in the long term' (Warren and Taylor, 1994, p. 1). The argument that ecotourism should only fall towards the active and hard end of the continuum and contribute to the health and viability of the natural attractions upon which it is based is an appealing one. However, these lofty aspirations appear to be unattainable in many situations, and possibly incompatible with the dominant market behaviours and expectations. Perhaps this is because the value of the ecotourism concept has been diluted through its popularity. It is certainly an attractive label that has been rapidly adopted by many operators and nations because it expresses *ideals* that are attractive to the market place. Authors such as Wight (1993) recognize that the ecotourism label is being utilized to take advantage of a 'greening' of the market and to 'eco-sell' tourism and travel. In many cases ecotourism has become nothing more than a marketing gimmick which dresses up existing tourism attractions in an attempt to increase market share (Wight, 1993).

Irrespective of the marketing of the term, it is now clear that ecotourism is not a panacea that always both protects the environment and supports economic activity (Butler, 1990; Zell, 1992; Pearce, 1994; Wall, 1994; Wheeller, 1994; MacKinnon, 1995). However, it was always extremely naive to expect it to be so. The natural environment is complex and dynamic, and the invention of a concept does not provide an automatic answer to the problems caused by humans visiting it. The widespread adoption of the terms 'conservation' and 'sustainability' have not solved the planet's environmental problems. Neither will the concept of ecotourism. However, it is important to remember that the goal of ecotourism is a sound and worthy one. Just because it is difficult to achieve in practice does not mean that the concept is flawed. Rather, it may be a reflection of the relative immaturity of the tourism management field and the difficulty of the task. This does not negate the worth of the ecotourism concept. The next stage in the management of ecotourism is to move beyond the semantic evolutionary process discussed earlier and to further evolve the successful application of the concept. Part of this evolution involves the recognition of distinctive ecotourism typologies and types that utilize different habitats, attract different markets, and require distinctive planning and management measures. The battle that remains is in developing effective techniques and strategies to make these various types of ecotourism work in practice, in its active and responsible sense. We now know where we want to go, and that is the first important step in getting there.

References

Adler, J. (1989) Origins of sightseeing. *Annals of Tourism Research* 16, 7–29.

Associated Press (1998) *Conservationists Want Whale-Watching Boats to Slow Down.* Associated Press, Boston, 15 September.

Barnes, J., Burgess, J. and Pearce, D. (1992) Wildlife tourism. In: Swanson, T.M. and Barbier, E.B. (eds) *Economics for the Wilds: Wildlife, Wildlands, Diversity and Development.* Earthscan Publications, London, pp. 136–151.

Borge, L., Nelson, W.C., Leitch, J.A. and Lestritz, F.L. (1991) *Economic Impact of Wildlife Based Tourism in Northern Botswana.* Agricultural Economics Report No 262. Agricultural Experiment Station, North Dakota University.

Budowski, G. (1976) Tourism and environmental conservation: conflict, coexistence or symbiosis? *Environmental Conservation* 3(1), 27–31.

Burger, J. and Gochfield, M. (1993) Tourism and short term behavioural responses of nesting masked, red-footed, and blue-footed boobies in the Galapagos. *Environmental Conservation* 20(3), 255–259.

Burnie, D. (1994) Ecotourists to paradise. *New Scientist* April, 24–27.

Butler, R.W. (1990) Alternative tourism: pious hope or Trojan horse? *Journal of Travel Research* 28(3), 40–45.

Ceballos-Lascuráin, H. (1987) Estudio de prefactibilidaad socioeconomica del tourismo ecologico y anteproyecto asquitectonico y urbanistico del Centro de Turismo Ecologico de Sian Ka'an, Quintana Roo. Study completed for SEDUE, Mexico.

Cessford, G.R. (1995) *Conservation Benefits of Public Visits to Protected Islands.* Science and Research Series No. 95, Department of Conservation, Wellington.

Croall, J. (1995) *Preserve or Destroy. Tourism and the Environment.* Calouste Gulbenhin Foundation, London.

Davies, M. (1990) Wildlife as a tourism attraction. *Environments* 20(3), 74–77.

Drey, R. (1982) *Tiritiri Matangi Island Working Plan.* Department of Lands and Survey for the Hauraki Gulf Maritime Park Board, Auckland.

Duenkel, N. and Scott, H. (1994) Ecotourism's hidden potential – altering perceptions of reality. *Journal of Physical Education, Recreation and Dance* October, 40–44.

Dyer, P.K. (1992) Wedgetailed shearwater nesting patterns: a spatial and ecological perspective, Capricorn Group, Great Barrier Reef. PhD thesis, The University of Queensland, Brisbane, Australia.

Fennell, D.A. (1999) *Ecotourism: an Introduction.* Routledge, London.

Figgis, P. (1993) Ecotourism: special interest or major direction? *Habitat Australia* February, 8–11.

Gauthier, D.A. (1993) Sustainable development, tourism and wildlife. In: Nelson, J.G., Butler, R.W. and Walls, G. (eds) *Tourism and Sustainable Development: Monitoring, Planning, Managing.* Heritage Resources Centre Joint Publication No. 1, University of Waterloo, Ontario, pp. 98–109.

Gibson, J.D. (1976) Seabird islands no. 38, big island, five islands, New South Wales. *Australian Bird Bander* 14, 100–103.

Glasson, J., Godfrey, K. and Goodey, B. (1995) *Toward Visitor Impact Management.* Ashgate Publishing, Aldershot, UK.

Gourlay, M.R. (1991) Coastal observations for monitoring environmental conditions on a coral reef island. In: *Proceedings of the 10th Australasian Conference on Coastal and Ocean Engineering, Auckland.* The University of Auckland, Auckland, pp. 73–77.

Great Barrier Reef Committee (1977) *Conservation and Use of the Capricorn and Bunker Groups of Islands and Coral Reefs.* GBRC, Brisbane.

Griffiths, M. and Van Schaik, C.P. (1993) The impact of human traffic on the abundance and activity periods of Sumatran rain forest wildlife. *Conservation Biology* 7(3), 623–626.

Groom, M.J., Podolsky, R.O. and Munn, C.A. (1991) Tourism as a sustained use of wildlife: a case study of Madre de Dios, southeastern Peru. In: *Neotropical Wildlife Use and Conservation.* University of Chicago Press, Chicago, pp. 393–412.

Hanna, N. and Wells, S. (1992) Sea sickness. *In Focus (Tourism Concern)* 5, 4–6.

Hawley, J. (1997) *Tiritiri Matangi Working Plan.* Department of Conservation, Auckland.

Hegerl, E.J. (1984) An evaluation of the Great Barrier Reef Marine Park concept. In: Ward, W.T. and Saenger, P. (eds) *The Capricornia Section of the Great Barrier Reef. Past, Present and Future.* Royal Society of Queensland and Coral Reef Society, Brisbane, pp. 173–180.

Heron Island Resort (1986) Advertising brochure. Heron Island Resort, Queensland.

Hetzer, N.D. (1965) Environment, tourism, culture. *LINKS* (July). Reprinted in *Ecosphere* (1970) 1(2), 1–3.

Higgins, B.R. (1996) The global structure of the nature tourism industry: ecotourists, tour operators, and local businesses. *Journal of Travel Research* 35(2), 11–18.

Hulsman, K. (1983) Survey of seabird colonies in the Capricornia section of the Great Barrier Reef Marine Park ii: population parameters and management strategies. Unpublished research report to the Great Barrier Reef Marine Park Authority, Townsville, Queensland.

Hulsman, K. (1984) Seabirds of the Capricornia section of the Great Barrier Reef Marine Park. In: Ward, W.T. and Saenger, P. (eds) *The Capricornia Section of the Great Barrier Reef Marine Park: Past, Present and Future.* Royal Society of Queensland and Australian Coral Reef Society, Brisbane, pp. 159–168.

Hvenegaard, G.T. (1994) Ecotourism: a status report and conceptual framework. *Journal of Tourism Studies* 5(2), 24–35.

Ingold, P., Huber, B., Neuhaus, P., Mainini, B., Marbacher, H., Schnidrig-Petrig, R. and Zeller, R. (1993) Tourism and sport in the alps – a serious problem for wildlife? *Revue Suisse de Zoologie* 100(3), 529–545.

Jenner, P. and Smith, C. (1992) *The Tourism Industry and the Environment.* Special Report No. 2453. The Economist Intelligence Unit, London.

Jones, O.A. (1967) The Great Barrier Reef Committee – its work and achievements, 1922–1926. *Australian Natural History* 15, 315–318.

Karwacki, J. and Boyd, C. (1995) Ethics and ecotourism. *Business Ethics* 4(4), 225–232.
Kutay, K. (1989) The new ethic in adventure travel. *Buzzworm: the Environmental Journal* 1(4), 174–176.
Laarman, J.G. and Durst, P.B. (1987) Nature travel and tropical forests. FREI Working Paper Series. Southeastern Center for Forest Economics Research, North Carolina State University, Raleigh.
Laarman, J.G. and Durst, P.B. (1993) Nature tourism as a tool for economic development and conservation of natural resources. In: Nenon, J. and Durst, P.B. (eds) *Nature Tourism in Asia: Opportunities and Constraints for Conservation and Economic Development.* United States Forest Service, Washington, DC.
Lane, S.G. (1991) Some problems during the exodus of young shearwaters from Mutton Bird Island, New South Wales. *Corella* 15, 108.
Laycock, G. (1991) Good times are killing the Keys. *Audubon* 93(5), 38–41.
Leopold, A. (1949) *A Sand County Almanac.* Oxford University Press, Oxford.
Limpus, C.J., Fleay, A. and Guinea, M. (1983) Management and turtles. In: Ward, W.T. and Saenger, P. (eds) *Proceedings of The Inaugural Great Barrier Reef Conference.* James Cook University Press, Townsville, pp. 61–78.
MacKinnon, B. (1995) Beauty and beasts of ecotourism. *Business Mexico* 5(4), 44–47.
Mellor, B. (1990) Loving the reef to death? *Time* 45(Nov), 48–55.
Miller, M.L. and Kaae, B.C. (1993) Coastal and marine ecotourism: a formula for sustainable development? *Trends* 30(2), 35–41.
Moon, L. (1998) The Singing Island. *The Story of Tiritiri Matangi.* Godwit, Auckland.
Moore, S. and Carter, B. (1993) Ecotourism in the 21st century. *Tourism Management* 14(2), 123–130.
Muir, F. (1993) Managing tourism to a seabird nesting island. *Tourism Management* 14(2), 99–105.
Neil, D.T. (1988) Holothurian populations on Heron Reef, GBR: effect of the boat channel. Unpublished report to the Australian Marine Sciences Association meeting, 28 May.
Neil, D.T. (1993) Barrier Reef Studies. Unpublished course handbook. Department of Geographical Sciences and Planning, The University of Queensland, Brisbane.
Orams, M.B. (1995) Towards a more desirable form of ecotourism. *Tourism Management* 16(1), 3–8.
Pearce, D.G. (1994) Alternative tourism: concepts, classifications and questions. In: Smith, V.L. and Eadington, W.R. (eds) *Tourism Alternatives: Potentials and Problems in the Development of Tourism.* University of Pennsylvania Press, Philadelphia, pp. 15–30.
Pleumarom, A. (1993) What's wrong with mass tourism. *Contours* (Bangkok) 6(3/4), 15–21.
Romeril, M. (1985) Tourism and the environment – towards a symbiotic relationship. *International Journal of Environmental Studies* 25, 215–218.
Valentine, P.S. (1990) Nature-based tourism: a review of prospects and problems. In: Miller, M.L. and Auyong, J. (eds) *Proceedings of the 1990 Congress on Coastal And Marine Tourism,* Vol. 2. National Coastal Resources Research and Development Institute, Corvallis, Oregon, pp. 475–485.
Viskovic, N. (1993) Zootourism. *Turizam* 41(1–2), 23–25.
Wall, G. (1994) Ecotourism: old wine in new bottles? *Trends* 31(2), 4–9.
Wallace, G.N. and Pierce, S.M. (1996) An evaluation of ecotourism in Amazonas, Brazil. *Annals of Tourism Research* 23(4), 843–873.
Ward, F. (1990) Florida's coral reefs are imperilled. *National Geographic* 178(1), 115–132.
Warren, J.A.N. and Taylor, C.N. (1994) *Developing Ecotourism in New Zealand.* The New Zealand Institute for Social Research and Development Ltd, Wellington.
Weaver, D.B. (1998) *Ecotourism in the Less Developed World.* CAB International, Wallingford, UK.
Wheeller, B. (1991) Tourism's troubled times: responsible tourism is not the answer. *Tourism Management* 12(1), 91–96.
Wheeller, B. (1994) Ecotourism: a ruse by any other name. In: Cooper, C.P. and Lockwood, A. (eds) *Progress in Tourism, Recreation and Hospitality Management,* Vol. 7. Belhaven Press, London, pp. 3–11.
Wight, P.A. (1993) Ecotourism: ethics or ecosell? *Journal of Travel Research* 21(3), 3–9.
Woodland, D.J. and Hooper, J.N.A. (1977) The effect of human trampling on coral reefs. *Biological Conservation* 11, 1–4.

Zell, L. (1992) Ecotourism of the future – the vicarious experience. In: Weiler, B. (ed.) *Ecotourism Incorporating the Global Classroom.* International Conference Papers, The University of Auckland, Auckland, pp. 30–35.

Ziffer, K. (1989) *Ecotourism: the Uneasy Alliance.* Working Paper No. 1. Conservation International, Washington, DC.

Chapter 3

Ecotourists: Not a Homogeneous Market Segment

P.A. Wight
Pam Wight and Associates, Edmonton, Alberta, Canada

Who is the Ecotourist?

The purpose of this chapter is to consolidate reliable information about ecotourism markets, from as global a perspective as possible. Sections will examine the identity of ecotourists, trends, market and trip characteristics, origins and destinations, satisfaction and motivations, as well as information to assist in reaching the ecotourist market.

The oft-asked question, 'Who are ecotourists?' has no definitive answer for many reasons, including the limited studies of markets, poor definitional understanding, and the fact that ecotourist markets are not homogeneous. Despite the large body of literature on ecotourism, markets' studies are limited to destination area markets, to tour operator perceptions, or to more general studies of nature or adventure-based tourists. Studies tend to discuss general growth in interest, or markets to particular destinations, rather than identifying characteristics, preferences and motivations of broad 'origin' populations. Studies at the global scale do not exist.

There is no clear agreement on ecotourism

As Wylie (1994) points out, there are many dimensions of ecotourism. It can be seen as an activity, a business, a philosophy, a marketing device, a symbol, or a set of principles and goals. Since no universally accepted definition exists, there is considerable overlap with nature, adventure and culture markets (Center for Tourism Policy Studies, 1998; Klenosky *et al.*, 1998) (see also Chapter 5). For example, US Travel Data Center research (TIAA, 1994) found that 50% of travelling adult Americans (146.9 million) had taken an 'adventure vacation' in the past 5 years. However, many of the activities described could apply to ecotourists (46% had taken soft adventure vacations, including camping, bird-watching, animal watching, hiking, snorkelling and scuba-diving). Also, of those who had not taken any adventure trips, over one-quarter (28%) indicated they would probably do so in the next 5 years. This would mean that over 60% have taken or will take an adventure vacation, of which many activities are considered to overlap with ecotourism.

Agreement on a definition, however, does not provide a sufficient basis for reliable measurement of any concept (Blamey,

1995). A number of tourism products can incorporate elements of adventure, such as excitement or the outdoors, which may be with or without the components that together contribute to ecotourism. These include education, interpretation, and environmental and cultural protection. Further, even when there is an educational focus, this may not necessarily relate to the environment. Similarly, nature-based markets are not necessarily ecotourists (Wight, 1993; Ceballos-Lascuráin, 1998), because the mere desire to be in/see natural environments does not equate with being ecologically benign or beneficial. Confusion about the identity of ecotourists may help to explain the wide range of ecotourism growth projections.

Ecotourism markets are not homogeneous

Even if a definition is established, ecotourism markets are not a homogeneous group. In the same way that there is a spectrum of products/experiences which may be termed ecotourism, there is also great variation in the activities, motivations and characteristics of markets: a spectrum of demand (Wight, 1993). Indeed, individuals may be interested in a number of overlapping experiences and activities.

In Australia, Tourism Queensland surveyed residents and found that 'consumers' interest in ecotourism lies along a spectrum based on a number of elements' (*Ecotrends*, 1999). These elements of understanding were:

- taking vacations in natural locations;
- understanding the term ecotourism;
- attitudes towards nature and nature-based tourism;
- reasons for choosing where to take a vacation, in particular the role of nature and learning about nature;
- the extent of planning for the vacation;
- nature-based activities conducted while on vacation.

The survey used cluster analysis (Law, 1999, personal communication) to classify respondents into four major classes, as shown in Table 3.1. The conclusions are that 'nearly half the travelling public have an underlying disposition towards nature and learning as part of their vacation' (Charters, 1999). Overall, Tourism Queensland considers almost 30% of the travelling public as ecotourists. Only 1.6% of these respondents see the term ecotourism as a 'fad'. Over half (51.5%) see ecotourism as environmentally friendly tourism that is not harmful to the environment.

The market has previously been segmented many different ways (e.g. free independent travellers (FIT), group, business, pleasure, occasional, frequent, experienced, specialized, general, etc.). However, in any destination, one individual may belong to more than one segment. In addition, ecotourists can be viewed as soft through to hard (Wight, 1993; Weiler and Richins, 1995; Diamantis, 1999) with respect to degree of interest in nature, degree of physical challenge, difficulty or comfort, occasional or frequent, or portion of total trip desired to be ecotourism activities (see Chapter 2). This chapter attempts to examine the market as a whole.

Trends Affecting Ecotourism

Increased overall demand

Ecotourism has been considered the fastest growing area of tourism for many years (Cook *et al.*, 1992; Laarman and Durst, 1993; Parker, 1993; WTO News, 1997, 1998). One of the trends fuelling this

Table 3.1. Degree of commitment/understanding of ecotourism (*Ecotrends*, 1999).

Class of ecotourist	% of Respondents
Definite	27.2
Probable	20.6
Possible	18.2
No commitment/limited understanding	32.2
Total number of respondents	780

growth is the increasing propensity of travellers to take life-enriching vacations that involve education, the outdoors and nature (Wylie, 1997, citing Mass, 1995). The desire to learn and experience nature is influenced by changing attitudes to the environment (based on the recognition of interrelationships among species and ecosystems), development of environmental education in primary and secondary schools, and the emergence of environmental mass media (Eagles and Higgins, 1998).

One of the most important trends is the ageing of populations in those countries where the international market demand for ecotourism is centred: North America, Northern Europe and, to a lesser extent, Japan. As people age they are attracted less to active, dangerous, outdoor recreational activities, and more to appreciative and less strenuous activities. This change in demographics is creating more demand for both ecotourism trips and related soft adventure and culture trips (HLA/ARA, 1994; Eagles, 1995). Mature Americans (55+) account for 21% of the total US population, and by the year 2010 this is expected to rise to 25% (75 million). This rapid growth of mature travellers, combined with their financial ability and availability of leisure time, make them an excellent potential market. This trend is also evident in Canada, the EU and Japan (Center for Tourism Policy Studies, 1998).

Growth estimates of ecotourism vary considerably (7–30%), even within one organization. In October 1997, the World Tourism Organisation (WTO News, 1997) presented information to indicate that ecotourism 'now accounts for between 10 and 15 per cent of world tourism', but by December, they had revised their estimates upward to 20%, under the title *Ecotourism, Now One-Fifth of Market* (WTO News, 1998). To add a level of complexity, rates of growth vary by destination, and even by region. For example, estimates for Australia vary from 5 to 10% of domestic nature-based tourism experiences, but much higher percentages in some locations, such as the wet tropics. Additionally, growth varies by country of origin, and may depend on the activity (e.g. in Australia, outback safari tour average annual growth rates are 47% for Germans, 21% for Swiss and 44% for other Europeans, but only 5% for Scandinavians; Blamey, 1995).

Many estimates are probably too high, being based on an overly liberal definition of the soft ecotourist. Suffice it to say that demand is strong and growing. However, it is also changing, deepening and broadening. Further data that are specific to ecotourism and nature-based tourism are clearly required before its size, significance and growth can be estimated accurately.

Mainstreaming

There is a shift of interest by wider tourism markets toward experiencing the natural environment (WTO News, 1997), and to greater awareness about environmentally and socially benign travel (TIAA, 1994; Center for Tourism Policy Studies, 1998). A North American survey of travellers in middle to upper income households found that 77% have taken vacations which include a nature/adventure/culture component, and the remainder intend to do so in the future (HLA/ARA, 1994). In Australia, about a quarter of adults reported being likely to take a nature-based trip in the next 12 months that involves learning about nature (Blamey, 1995). These are significant numbers, and all the more credible because of their consistency across an array of origin regions.

Operators offering ecotourism in the Asia-Pacific region identified that a broadening of the clientele for ecotourism and a high growth in both the number of clients taking ecotours (23.4%), and the revenues that they generate (18.3%), were key trends in the Asia-Pacific region over the period 1993–1995 (Lew, 1998a, b). A number of studies in North America and Australia highlight the fact that the profile of potential ecotourists tends to be broader than the profile of actual ecotourists. Blamey and Hatch (1998) state that as ecotourism becomes more extreme or specialized, the profile of participants becomes narrower, and less representative of source populations. However, other opinions abound. Lew's view (1998b) is that while the trends appear to show that 'ecotourism is about to break out of the limited speciality travel market, they actually reflect broader patterns found throughout the tourism industry'. Epler Wood (1998) indicates that ecotourism operators say their clientele is both broadening and deepening to attract inexperienced travellers. In fact, since ecotourism is actually a subset of sustainable tourism, then whether ecotourism markets are mainstreaming (Wight, 1995) or whether the entire tourism industry is experiencing a shift, may be a moot point; what matters is that the trends exist.

Market Characteristics

There are few globally significant ecotourism studies. In Europe, a known source of ecotourism demand, there are no substantial studies. This chapter will therefore focus on Canada, the USA, UK and Australia, where rigorous studies have taken place, and are summarized in Tables 3.2, 3.9 and 3.12. In addition, information will be presented which may relate to nature-based, adventure, or nature/adventure/culture markets. Table 3.2 summarizes ecotourism and related markets' characteristics. The text expands, rather than repeats the information.

Income

In general, household incomes are higher for ecotourists than for travellers overall (Backman and Potts, 1993; Liu, 1994; Eagles and Cascagnette, 1995). However, this is not supported by a US adventure/outdoors travel survey (TIAA, 1994), nor by a previous survey of US ecotourists, which found little divergence from the income profiles of the average traveller (Cook *et al.*, 1992). Diamantis (1999) found that while age affected incomes, the majority of UK ecotourists had mid- to high-range incomes.

The relationship between income and activities varies. In Australia, there were relatively high intentions to visit a natural attraction or national park in the upcoming 12 months, whether or not respondents had a full-time or part-time job. Potential ecotourists are only slightly more likely to have higher gross incomes than non-ecotourists (Blamey, 1995). Yet, participation in various activities *may* vary by income: the most affluent US travellers (household income > US$50,000) participated more in snorkelling/scuba-diving (39%) and sailing (29%); those with US$20,000–$30,000 hiked more. The least affluent (< US$20,000) camped more than the most affluent (95% vs. 77%) (TIAA, 1994).

Occupation

Little information exists on occupation, but in the USA, 35% of ecotourists are professional/managerial (TIAA, 1994). In Australia, those with the highest propensity to participate in nature-based activities were professional/technical; this group constitutes a larger percentage of visitors overall (Blamey and Hatch, 1998). In fact, blue collar, skilled trades, students and clerical/sales generally participate more than their percentage of visitation overall, although this percentage varies by activity.

Education

There is a finding in all the sampled markets that ecotourists are generally well educated. In North America 75% of general and 96% of experienced ecotourists had degrees or at least some college education. In the UK, 61% of frequent ecotourists were educated to degree or postgraduate level (Diamantis, 1998). This was related to income; the higher the educational status of frequent ecotourists, the higher their total income earnings. However, studies show interest in ecotourism at all levels of education.

Table 3.2. Profiles of ecotourists and related travellers.

Market characteristics	US adventure and outdoor traveller[a] *N* = 1172	North American general ecotourist[b] *N* = 1384	North American experienced ecotourists[b] *N* = 424	Canadian ecotourists[c]	Australian ecotourists[d]	UK group frequent ecotourists[e] *N* = 379	UK group occasional ecotourists[f] *N* = 247
Household income	76% > US$30,000 (CDN$40,000) 24% < US$30,000 (CDN$40,000) 18% > US$75,000 (CDN$100,000)	Live in neighbourhoods with > US$35,000 (CDN$45,000)	No information	57% > CDN$50,000 36% > CDN$70,000 average CDN$64,000	Higher incomes	13% < £10,000 15% £10–15,000 22% £15–20,000 17% £20–25,000 12% £25–30,000 21% > £30,000	33% < £10,000 12% £10–15,000 14% £15–20,000 14% £20–25,000 9% £25–30,000 18% > £30,000
Age	51% 25–40 25% 45–64 10% 65+ Average is 40	10% 18–24 24% 25–34 25% 35–44 18% 45–54 23% 55+	2% 18–24 20% 25–34 28% 35–44 28% 45–54 23% 55+	46% 45–64 22% 35–44 11% 25–34 11% 70+	36% 20–29 23% 30–39 27% 50+	7% 17–24 15% 25–34 27% 35–44 24% 45–54 18% 55+	28% 17–24 28% 25–34 16% 35–44 17% 45–54 11% 55+
Gender	51% male (vs. 60% travellers overall), varies by activity	Males and females, varies by activity	Males and females, varies by activity	50 : 50	55% female 45% male, varies by activity	54% females 46% males	57% females 43% males
Household	66% married (vs. 56% travellers overall) 50% have children (vs. 37% travellers overall)	44% couples $\frac{1}{3}$ families	47% couples $\frac{1}{4}$ families	No information	No information	58% married 34% single 8% divorced	54% single 39% married 7% divorced
Education	41% completed college	45% college grads 30% some college 21% high school 5% some HS	82% college graduates 14% some college 4% high school 1% some HS	> 64% university education 24% some post secondary	All levels, but potential ecotourists tend to be more highly educated	38% first degree 25% secondary education 23% postgrad. 15% high school	46% first degree 22% secondary education 21% postgrad. 11% high school

Continued

Table 3.2. *Continued*

Market characteristics	US adventure and outdoor traveller[a] N = 1172	North American general ecotourist[b] N = 1384	North American experienced ecotourists[b] N = 424	Canadian ecotourists[c]	Australian ecotourists[d]	UK group frequent ecotourists[e] N = 379	UK group occasional ecotourists[f] N = 247
Party composition/ travelling companions	58% couple 36% with (grand) children 34% with other adults 11% with (grand) parents 4% alone	59% couples 26% families with children 7% alone	61% couples 15% family 13% alone	No information	30% couples 14% family/friends 45% alone (visitors less likely to be unaccompanied)	66% one 18% two 9% three 4% four 3% other	63% one 15% two 9% three 5% four 9% other
Occupation	35% professional/ manager 14% blue collar 12% retired 10% clerical	No information	No information	No information	Professionals greatest number	No information	No information

[a] TIAA, 1994; [b] HLA/ARA, 1994; [c] Eagles and Cascagnette, 1995; [d] Blamey and Hatch, 1998; [e] Diamantis, 1998; [f] Diamantis, 1999.

Age

Although it cannot be assumed that all park visitors are ecotourists, Australian national parks have broad appeal, and visitation by each age group is roughly proportional to shares of total visitors in these groups. This is unlike most other nature-based activities, where participation is greatest among those aged 20–29 years (Blamey and Hatch, 1998). Diamantis (1998) found that two-thirds of UK ecotourists were aged 25–54. Similarly, in North America, most ecotourists were aged 25–54 (67% general ecotourism, 76% experienced; HLA/ARA, 1994). However, all ages are interested in ecotourism with a tendency for general or occasional ecotourists to be younger, and more frequent or experienced ecotourists to be older.

Most market surveys use adults as their base, so it is difficult to obtain participation rates for children. Previous studies have indicated that a large proportion of ecotourists appear to be childless or 'empty-nesters' (Reingold, 1993; WTO, 1994), but there is evidence that ecotourism and nature-based visitation has broad appeal. Family vacations became a major trend in the 1990s (49% increase in a decade; The Center for Tourism Policy Studies, 1998). In the USA, 36% of adventure/outdoor parties take children or grandchildren (TIAA, 1994). In Australia, 61% of respondents with children intended to visit a natural area in the next 12 months as opposed to 48% of those without children (Blamey, 1995, quoting a Newspoll study).

Age may influence activity participation rates. In the USA, those under 24 years have a higher than average participation rate for physically demanding activities such as hiking, kayaking/white-water rafting, biking, rock-climbing and sailing (TIAA, 1994). In Australia, those under 24 were more interested in bushwalking/outback/safari tours; scuba-divers/snorkellers were mainly 25–34; and most of those visiting a national/state park were 55 and over (Blamey, 1995).

Gender

The tradition of male dominance in nature-based activities has been superseded by female dominance. In North America, both males and females are equally interested, but participation varies by activity. In Australia, 55% of ecotourists are women, and women aged 20–29 are the most active of all nature-based participants (Blamey and Hatch, 1998). Similarly, a higher proportion of US adventure/outdoor travellers are women, compared with the profile of average US travellers, with popularity of various activities varying by gender (e.g. snorkel/scuba 35% of males: 26% of females; kayak/white-water rafting 29% of males: 19% of females; sailing 30% of females: 22% of males) (TIAA, 1994). In the UK, Diamantis (1998, 1999) found more female ecotourists (both frequent and occasional), and a relationship between gender and age: young frequent ecotourists were predominantly female, while males were equally distributed between the young and older age groups.

The reason for this increase in female participation is not clear, but could be related to women's increasing independence and incomes, the higher population of older women, their growing majority in universities (and thus their higher education levels), and the desire to socialize with like-minded women. Other factors consider that ecotourism may be an acceptable way for women to travel on their own, and that greater safety and security are available in guided tours. Whatever the reasons, ecotourism is satisfying women's needs and desires. This may be inadvertent, although there are a number of commercial operators specializing in women's groups.

Trip Characteristics

In North America, ecotourists are more frequent travellers, with 41% having travelled out of their state/province six or more times in the previous 3 years, versus 24% of travellers in general. This amounts to an average of two or more trips per year.

Season

There has been little information in the literature on what times of the year ecotourism travellers prefer to travel, and that which exists is often contradictory or ambiguous. The majority of North American ecotourists prefer to travel in summer (23% June, 40% July, 40% August). However, there is winter interest, and strong shoulder season interest (16% May, to 29% September). Experienced ecotourists, who tend to be more frequent travellers, are more interested in all seasons of travel, particularly shoulder seasons. This finding presents useful opportunities for destinations to extend their season well into shoulder seasons and beyond. In any case, both the destination and reason for travel have some bearing on season of travel.

Trip length

Ecotourists' trip length varies tremendously, and may vary by *activity*. The general tourist or vacationer tends to stay for a shorter time than the nature-based traveller. Yuan and Moisey (1992) found that Montana visitors interested in wildland-based activities spent more time in the state than non-wildland-based visitors (8.22 days backpacking and 5.99 in nature study, vs. non-wildland-based visitors, who spent only 3.66 days). In addition, the ecotourism portion of the trip may be quite different from the total trip duration. One half of North Americans preferred ecotourism portions of trips to be over 1 week (this portion is not known for experienced ecotourists) (HLA/ARA, 1994; Wight, 1996) (Fig. 3.1).

In Australia, nature-based visitors spent a longer time (average 33 nights) in the country than the traveller average of 24 nights. But most nature-based visitors on an organized tour went for a day, a half-day or less (not overnight). Only 20% took tours of four or more nights; the average length was 2.9 nights, although for males it was 3.4 nights vs. 2.3 for females (Blamey and Hatch, 1998).

Thus, trip length may vary by *destina-*

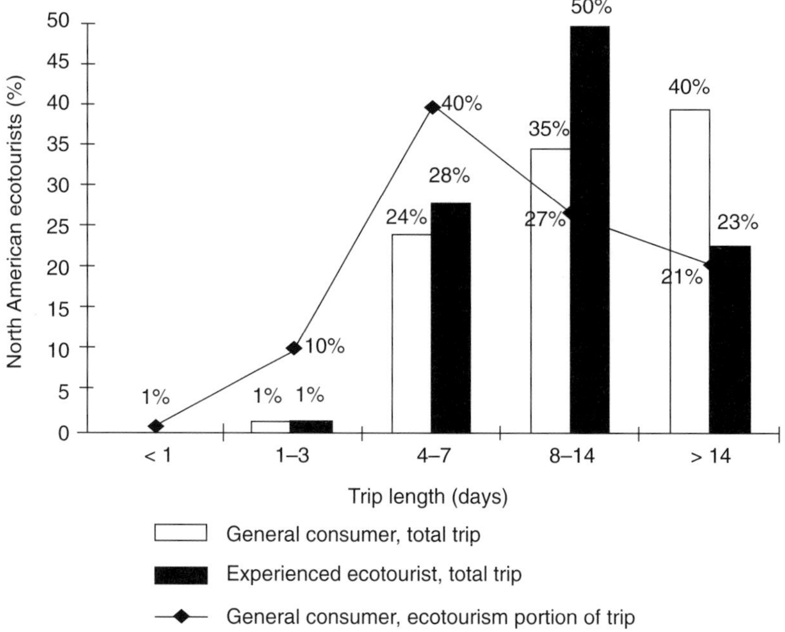

Fig. 3.1. Preferred trip length for North American ecotourists (Wight, 1996).

tion. Very often, for more distant locations, the length of stay is understandably greater. For example, nature-based travellers stay an average of five nights at their destination in the south-east USA (Backman and Potts, 1993), yet in Alaska, conservation group trips averaged 12 days (ARA *et al.*, 1991). Trip length may also vary by origin. National park visitors from all countries stayed longer in Australia than non-national park visitors, and some origins stayed more than twice the average of 23 nights (UK and Ireland, Scandinavia, Switzerland). Trip length may also vary by activities, as in Australia (Table 3.3).

Expenditures and willingness to spend

One quarter of North American ecotourists were prepared to spend US$2000 per person per trip that involved ecotourism experiences and, on average, would be willing to spend US$238 per person per day excluding travel (HLA/ARA, 1994). The more specialized ecotourists had a higher willingness to spend.

Laarman and Gregersen (1996) point out that certain destinations or sites ('jewels of nature') have high scarcity value, and so markets are more willing to pay for these products, compared with sites for which several alternatives provide roughly similar experiences. However, they point out that willingness to pay also reflects such experience-enhancing features as presence and quality of accommodation, guide service, ground transportation and cooperative governments. UK ecotourists claim that ecotourism holidays are not expensive (59.9% agreement overall) which 'suggests that frequent ecotourists were not concerned about the expensive price, hence it was not a determinant during the decision making process' (Diamantis, 1998).

Willingness to pay is not the same thing as actual expenditures, but expenditure studies support willingness-to-pay studies. Montana tourists interested in wildland-based activities spent 25–50% more per day than non-wildland tourists (Yuan and Moisey, 1992). Canadian ecotourists spent considerably more per day than general Canadian travellers (Eagles and Cascagnette, 1995). US outdoor/adventure travellers spent US$871 on average. Those who spent the most had the highest incomes (US$1275) and were older (US$1461 over 65s); those who spent least were aged 25–34 (US$589) and had the lowest household incomes (US$326). Also, men spent more than women (US$908 vs. US$831) (TIAA, 1994).

Ecotourists have been found to have differences in expenditure characteristics from other visitors going to the same natural areas. For example, Leones *et al.*, (1998) found that nature tourists spend more per trip in Sierra Vista, Arizona (US$177 per party), than other visitors (US$111) to the same natural areas, but they stay longer; thus per diem expenditures are lower. Similarly, in Australia, nature-based visitors spend considerably more while in Australia than other visitors, with average expenditure amounting to AU$2256 (13% more than that spent by other visitors, in part because of their longer duration of stay). But their average expenditure per night is only AU$78 excluding tour costs (vs. AU$90 for all visitors) (Blamey and Hatch, 1998).

Table 3.3. Trip length varies by activity (Blamey, 1995).

Nights by international visitors engaged in selected activities in Australia						
Bush walking	Scuba-diving/ snorkelling	Rock-climbing/ mountaineering	Horse riding/ trail riding	Outback safari tours	Wildflower viewing	All visitors
49	37	58	74	64	44	23

Those strongly motivated to visit Australia's natural areas had the greatest expenditures, with AU$3222 per person spent per trip (Table 3.4). Also, as in North America, visitor expenditure varies by activity, as well as by country of origin. Expenditures were greatest, on average, for horse/trail riding, followed by outback safari tours, and rock-climbing/mountaineering. Those markets who spend the most, on average, are Swiss, German, Scandinavian, other European, Canadian and 'other Asian' (Blamey and Hatch, 1998).

The essence of these studies is similar: ecotourists tend to spend more than the average traveller. However, while ecotourists in general are willing to spend substantially more than most tourists, they expect *value* for their expenditures. They tend to be experienced, discerning travellers who are willing to pay for *quality* experiences. The price that an ecotourism operator charges may have little bearing on the value of the product. As McKercher and Robbins (1998) and Laarman and Gregerson (1996) have stated, it appears than many new operators base prices on the component parts of the trip (cost plus pricing), whereas the level and quality of personalized service offered is the true value, not those of the component parts (see Chapter 36).

Table 3.4. Influence on trip expenditure of desire to visit natural areas (Blamey and Hatch, 1998).

Importance of 'desire to visit natural areas', as motivator to visit Australia	Expenditure per trip (AU$)
Most important motivator	3222
Major factor	2778
Influence, but not major	1946
Little or no influence	2202
Total nature-based trip expenditure (excluding tour costs)	2256

Table 3.5. Most popular Asia-Pacific destinations (Lew, 1998b).

Destination country or region	Tour companies (%)
Indonesia	40.0
India	32.5
Australia	30.0
Nepal	30.0
Bhutan	25.0
New Zealand	20.0

Origins and Destinations

Destinations of interest

Popular belief, and much key research, has discussed ecotourism in the context of tropical and developing countries as prime ecotourism destinations. However, market research on ecotourists draws a different picture (Wight, 1996; Blamey and Hatch, 1998). Lew's study (1998b) of 44 ecotour companies operating in the Asia-Pacific region found that the top six destinations in terms of popularity included Australia and New Zealand (Table 3.5). In addition, this does not account for the free independent travellers' (FIT) ecotourism destination preferences in the region (e.g. 61% of Australia's total visitors were FIT). In the UK, Europe was the top continent of interest for ecotourists' next ecotourism holiday (58%), related in part to air travel and expenses (Diamantis, 1999). Ranked preferences are shown in Table 3.6. The preferred location for one-third of North Americans' next ecotourism vacation was Canada (HLA/ARA, 1994). There are, in fact, a huge range of current and potential destinations of interest to ecotourists. Attractiveness also varies with specific destination attributes and services.

Origins of ecotourists

There is no definitive information on ecotourism market origins. However, based on demand data, it is clear that the international market demands are centred in North America and Europe. Eagles and Higgins (1998) estimate from anecdotal sources that the most prominent countries

Table 3.6. Preference, by continent, of UK ecotourists (Diamantis, 1999).

Destination continent	UK frequent ecotourists (%)	UK occasional ecotourists (%)
Europe	58.3	30.0
Asia	49.6	24.3
Americas	36.1	23.1
Australia/New Zealand/Pacific	33.8	10.9
Africa	31.9	11.7

supplying ecotourists, in order of market size, are the USA, the UK, Germany, Canada, France, Australia, The Netherlands, Sweden, Austria, New Zealand, Norway and Denmark. In addition, Japan, southern Europe and the newly industrialized Asian countries are also generating increasing numbers of ecotourists.

In Australia, a study of nature-based tourists asked about the importance of nature-based outdoor activities in destination selection. Germans, Scandinavians and Canadians were much more influenced by opportunities for nature activities than other origin areas. The percentage of national park visitors was analysed as a proportion of total visitation from each country, to determine those origins with a higher than average propensity to visit. Although the largest *numbers* of visitors to Australian national parks were from Asia, a smaller than average proportion from this market visited these sites. Japanese visitors account for the largest percentage of participants in most nature-based activities, because they account for a very large share of total visitors (23%). They do not, however, reflect a particularly high propensity to participate in such activities (Blamey, 1995). British, US, German and Scandinavian visitors appear to be more active participants in nature-based activities, relative to visitors from other origins. Additionally, Blamey (1995) suggests that outback safari tours are most similar to ecotourism experiences. Table 3.7 shows origin interests by select experience.

Origin markets may vary with location of ecotour companies. Lew (1998b) found that for Asia-Pacific ecotour companies *based in the USA* (22 of 44), the largest proportion of clients originated from the USA (plus some from Australia and Canada). For companies based in Asia-Pacific (16), the main sources of clients were more varied, and included Australia, the USA and Europe. Asian sources (Japan, Indonesia and the Philippines) were less important.

Table 3.7. Origins of Australian visitors by experience interest (Blamey, 1995).

Country of residence	% All visitors	% Non-national park visitors	% National park visitors	% Outback safari visitors	Degree of interest
UK and Ireland	11	8	14	22	More
Germany	4	2	6	17	interested
USA	10	7	12	13	↑
Other Europe	5	3	7	13	
Scandinavia	1	1	2	4	
Switzerland	2	1	3	7	
Canada	1	0	2	7	
Japan	23	25	21	8	↓
Other Asia	22	24	19	5	Less
New Zealand	17	22	10	2	interested

Domestic visitors

A neglected group of ecotourists are those who originate in the same country, domestic visitors. For example, in Australia, recent research found ecotourists to represent nearly 30% of domestic travellers (*Ecotrends*, 1999). Similarly, in North America as a whole, there is a high propensity to select domestic destinations for ecotourism. This varied by country, with USA respondents regarding the USA and other destinations as almost equally attractive, while Canadian respondents by far prefer to visit their own country for their next ecotourism trip (HLA/ARA, 1994) (Table 3.8). Thus domestic ecotourism visitors represent an important and hitherto neglected market.

Preferences

A survey of US travellers has found that adventure and outdoor travellers are similar to the general population of travellers with respect to income, number of wage earners, education, occupation, household size and region of origin (TIAA, 1994). It may therefore be preference and motivational features that mainly differentiate ecotourists from other markets. Table 3.9 shows preferences of ecotourists from global survey information.

Activities

Typically, the most popular ecotourism activities in all surveys are visiting national parks, hiking, water-based activities (especially rafting), admiring nature, camping and touring. Additionally, cultural/aboriginal experiences may be of interest. Preferences also vary by site type. McNeely (1988), Dixon and Sherman (1990), and Laarman and Gregersen (1996) feel the main attraction for nature-based tourism is publicly owned national parks, wildlife reserves and other protected areas (see Chapters 18 and 19). This is the main activity by international visitors to Australia (50%). For Australian domestic markets, the most strongly supported recreational activity is bushwalking for all ages (< 54 years: 72.3% average; 55–64: 57.1%; > 65 years: 45.1%). Activity preferences may also vary by origin, age or gender. Interest may even vary within one destination country, as shown in Table 3.10.

The HLA/ARA (1994) study of North American ecotourists took into account the dynamic nature of markets, and asked for last trip *and* next trip preferences. Ecotourism market interests changed. Activities which increased in attractiveness for the *next* trip include: hiking/backpacking, boating, fishing, cycling and camping. Of note is the fact that more general interest ecotourist preferences moved towards preferences of the experienced ecotourist, supporting earlier suggestions about the mainstreaming of ecotourism.

Accommodation

Recent ecotourism research has found that ecotourism markets prefer more than the conventional hotel/motel options, or camping. They desire more rustic, intimate, adventure-type roofed accommodation (such as bed and breakfasts, cabins, lodges, inns) which is a growing market trend

Table 3.8. Preferred country for ecotourism vacation (HLA/ARA, 1994).

	Preferred destination (%)			
Origin	Canada (451)	USA (417)	All other (486)	Total number (1354)
USA (5 cities)	21.5	38.5	40	970
Canada (2 cities)	64	12	24	384

Table 3.9. Preferences of ecotourism and related travellers.

Market characteristic	US adventure and outdoor travellers[a] $N = 1172$	North American general ecotourist (next trip)[b] $N = 1384$	North American experienced ecotourist[b] (next trip) $N = 424$	Australia nature-based tourists[c]	UK group frequent ecotourists[d] $N = 379$	UK occasional ecotourists[d] $N = 247$
Activity preferences	Camping (85%) Hiking (74%) Skiing (51%) Snorkel/scuba (30%) Sailing (26%) Kayaking/white-water (24%) Biking (24%) Rock-climbing (18%) Cattle/dude ranch (14%) Hang-glide/parasail (8%)	Hiking (37%) Touring (20%) Camping (19%) Boating (17%) Walking (17%) Fishing (16%) Scenery, other than mountain/ocean (14%) Swimming (12%) Other water (9%) Local cultures (8%) Cycling (8%)	Hiking (60%) Rafting (25%) and other boating (13%) Cycling (25%) Camping (21%) Wildlife viewing (15%) Scenery other than mountain/ocean (13%) Skiing (13%) Canoeing (13%) Kayaking (13%) Fishing (12%) Local cultures (12%)	All international visitors National parks (50%) Bushwalking (19%) Scuba/snorkelling (13%) Aboriginal sites (11%) Outback safari tours (3%) Rafting (2%) Horse-riding (2%) Rock-climbing/ mountaineering (2%)	Educational guided tours (72%) Admiring nature (72%) Observing animals (68%) Bushwalking (54%) Adventure tours (46%) Nature photography (45%) Observing flowers (40%) Snorkelling (38%) Birdwatching (35%) Whale-watching (31%) Horse-riding (22%) White-water rafting (22%) Scuba-diving (22%) Rock-climbing (19%)	Admiring nature (76%) Observing animals (71%) Snorkelling (62%) Educational guided tours (60%) Natural photography (51%) Camping (49%) Scuba-diving (48%) Bushwalking (47%) Adventure tours (37%) Observing flowers (37%) Whale-watching (33%) White-water rafting (28%) Turtle-watching (26%) Birdwatching (25%) Horse-riding (21%)
Accommodation	No information	56% hotel/motel 17% camping 14% lodge/inn 14% cabins 10% bed and breakfasts 6% friends/relatives 5% RVs 60% mid-range luxury 31% basic/budget	66% cabin/cottage 60% lodge/inn 58% camping 55% bed and breakfasts 41% hotel/motel 40% ranch 56% mid-range luxury 38% basic/budget	National parks visitors: 70% hotel/motel 42% friend/relative 8% camp 8% backpacker hostel 6% rented house/flat 4% youth hostel	60% hotels/motels 41% tent 32% cabins 30% bed and breakfasts 29% ecolodges 20% Inns 6% ranches 58% mid-range luxury 40% basic/budget	53% tent 41% hotels/motels 25% bed and breakfasts 23% cabins 19% ecolodges 10% inns 3% ranches 48% mid-range luxury 47% basic/budget

[a] TIAA, 1994; [b] HLA/ARA, 1994; [c] Blamey and Hatch, 1998; [d] Diamantis, 1998.

Table 3.10. Interest areas by market origins to Canada (*Tourism Canada,* 1995).

Market	Principal activities of interest in Canada
USA	• sea kayaking (37%) • nature observation (36%) • whale-watching (30%) • scuba-diving (29%) • other wildlife viewing (24%) • birdwatching (19%) • canoeing (14%) • rock and ice climbing (13%) • hiking (12%)
Germans	• canoeing (31%) and trail riding • other wildlife viewing (11%) • nature observation (6%) • scenery, national parks, forests and wildlife
French	• culture and nature
British	• canoeing
Japanese	• nature observation, soft adventure/ecotourism • locations 'abundant in nature', with Canada's attraction being: – natural attractions (96%) – national parks (88%) – mountainous areas (88%) – seeing wildlife (83%)
Ontario Québécois British Columbia	• birdwatching, sea kayaking and canoeing • scuba-diving (30%) • hiking (39%) • sea kayaking (17%), • climbing, hiking and scuba-diving to a lesser degree • whale-watching (15%) • nature observation (14%)

(Hawkins *et al.*, 1995; Selengut, 1995; HLA Consultants, 1996; Wight, 1997; Diamantis, 1998). The overall vacation experience seems to determine the accommodation choice.

In addition, preferences vary by activity, and by destination. In North America, those who prefer activities considered more as 'ecotourism' in nature (hiking, boating, camping or fishing) and experienced in relatively wild settings, were more likely to select accommodation that tends to be found in those settings (tent, cabin/cottage) (HLA Consultants, 1996). The opposite was found for activities less associated with ecotourism. Similarly, the most common type of accommodation used by all international visitors in Australia, including nature-based visitors, is hotel/motels (69%), followed by the home of a friend or relative (36%). However, nature-based visitors are significantly more likely than other visitors to use backpacker accommodation or to go camping, and show at least a four times greater than average preference for these accommodation types, as well as youth hostels and rented campervans. A recent study of UK ecotourists also found that ecotourists prefer hotels, but that alternative forms of accommodation such as tents, cabins, bed and breakfasts, ecolodges, etc. are also popular (Diamantis, 1998). It is probable that the lack of alternative more rustic and intimate accommodation explains why hotels/motels rank quite highly. In terms of comfort level, ecotourists appear to prefer mid-range luxury levels (60% in North America, 58% in UK), followed by basic/budget (31% in North America, 40% in UK) (HLA, 1996; Diamantis, 1998).

FIT vs. group ecotourists

There is an important conceptual distinction between ecotourists who are FITs (free and independent travellers) and those who go on group tours. Studies tend to examine either speciality tour operators and clients, or to examine all visitors to a destination. Exceptions include the North American market research which included separate ecotourist and tour operator surveys (HLA/ARA, 1994), and the Australian research, where group tour travellers were analysed separately, both for volume and for characteristics.

In Australia, 39% of all visitors came into the country on a fully pre-paid package tour, although this varied by activity (Table 3.11). In addition, FIT travellers may take tours once at their destination. Australian visitors, on average, participated in three nature-based tours, but this ranged from one to four tours. Also, the origin of visitors influenced the number of tours (Germans participated in 6.7, whereas Asian visitors participated in the fewest) (Blamey and Hatch, 1998). All activities except non-guided walks tended to be part of an organized tour.

Lew (1998b) reports a large increase in FITs in the Asia-Pacific region. It is possible that as ecotourism matures, this unorganized dimension will increase. Clearly, changes in ecotour markets are occurring. Asia-Pacific operators are seeing more people of varied ages and incomes taking ecotours, leading to a softening of the adventure aspect of the tours, as well as an increased market sensitivity to pricing. By product modification, operators may attract or generate new markets, and operators predict the future broadening of market distribution channels, expansion of ecotourism into new areas, and new products for FITs (Lew, 1998b).

Purpose, Satisfaction and Motivation

Reasons for ecotourism trip

Table 3.12 summarizes the reasons and motivations for taking an ecotourism trip from various surveys. In all surveys, the primary reasons largely revolve around experiencing various elements of nature and scenery. Also prominent are new experiences, learning and exposure to local cultures. Even adventure/outdoors travellers had a strong interest in nature, with the 55–66-year-olds more interested than average in taking the trip because of an interest in the environment.

The location of residence, age and other factors may have considerable influence on

Table 3.11. Fully inclusive prepaid package tour use.

Activity	% Not on inclusive package	% Package tours
Snorkelling	49	51
Scuba	68	32
White-water rafting	49	50
Horse-racing/riding	60	40
Rock-climbing/mountaineering	65	35
National parks	57	43
Aboriginal sites	62	38
Bushwalking	67	33
Outback safari tours	65	35
Whale-watching	80	20
All visitors 3,422,000	61% (2,097,300)	39% (1,324,700)

Table 3.12. Trip reason and motivations of ecotourism and related travellers.

Market characteristic	US adventure and outdoor traveller[a] $N = 1172$	North American general ecotourist (next trip)[b] $N = 1384$	North American experienced ecotourist[b] (next trip) $N = 424$	Canadian ecotourists[c]	Australia nature-based tourists[d]	UK group frequent ecotourists[e] $N = 379$	UK group occasional ecotourists[e] $N = 247$
Reasons, motivations	71% fun and entertainment 51% get away from it all 21% thrill 15% try/learn something new 14% interest in environment 7% learn/test something about selves 7% health	45% scenery and nature 28% new experiences/places 16% been and want to return 15% cultural attraction 15% see mountains 14% study/learn nature and cultures 13% relax and get away from it all	45% scenery and nature 22% new experiences/places 16% land activities 15% wildlife viewing 14% see mountains 11% wilderness 11% not crowded 11% water activities 10% cultural attraction 10% study/learn nature and culture	1 wilderness/ undisturbed nature 2 learn about nature 3 tropical forests 4 birds 5 photography 6 trees and wildflowers 7 mammals 8 national and provincial parks 9 lakes and streams 10 see maximum in time available 11 mountains 12 oceanside	1 natural beauty of sites 2 new experience 3 wildlife 4 close to nature 5 different way of experiencing nature 6 exciting experiences 7 something to tell friends 8 educational/ learning experience 9 being physically active 10 chance to escape crowds 11 escape towns and cities	93% see natural environment 85% experience local culture 78% experience traditional and natural lifestyles 74% travel to wild places on earth 70% survey/study natural habitats 61% historical attractions 59% experience unique exclusive place 45% outdoor/ recreational activities 34% third world countries 14% expensive holiday	1 experience new and different lifestyle 2 explore area and be educated 3 increase knowledge 4 meeting new people 5 outdoor activities 6 undisturbed natural area 7 enjoy weather 8 study/admire/ understand area 9 interesting countryside 10 cultural attractions 11 experience tranquillity 12 visit national parks

[a] TIAA, 1994; [b] HLA/ARA, 1994; [c] Eagles, 1992; [d] Blamey and Hatch, 1998; [e] Diamantis, 1998.

motivations and reasons for the trip. In Australia, survey research found that people living in the city have a greater desire to see somewhere different from home (55.9%) than people living in rural areas (41.9%) (*Ecotrends*, 1999). The opportunity to experience nature-based outdoor activities is a significant factor influencing visitors to Australia from Germany, the UK and Ireland.

Motivations

Reasons for the trip are different from motivations. Motivations are associated with the needs of the individual. Reasons for taking a trip may be fairly broad, whereas motivational information is more helpful in differentiating ecotourists. In the USA, motivations varied by age among outdoor/adventure travellers (TIAA, 1994):

- the younger age group (39% of 18–24-year-olds vs. 21% of the total group) tended to look for the thrill;
- the older age group (30% of 55–64-year-olds, vs. 14% of the total group) tended to have an interest in the environment;
- the middle ages (59% of 45–54-year-olds vs. 51% of total group) tended to want to get away from it all.

Eagles (1992) examined types of motivation for group tour ecotourists, including attractions (related to desired features/attractions of the destination) and social factors (related to opinions on personal goals and interaction with others). The motivations that are significantly more important to group ecotourists are shown in Table 3.13.

According to several sources (Crossley and Lee, 1994; Wight, 1996), motivations that differentiate ecotourists from more mass travellers include:

- uncrowded locations,
- remote, wilderness areas,
- learning about wildlife, nature,
- learning about natives, cultures,
- community benefits,
- viewing plants and animals,
- physical challenge.

Table 3.13. Motivations of group tour ecotourists (Eagles, 1992).

Significant motivations
• wilderness and undisturbed nature • lakes and streams • being physically active • mountains • national or provincial parks • experiencing new lifestyles • rural areas • oceanside • meet people with similar interests

Getting close to nature is a motivation for ecotourism in Australia (*Ecotrends*, 1999). Recent research on ecotourists and potential ecotourists has revealed three broad market segments (Bureau of Tourism Research, 1998). Each has different motivations and different determinants of satisfaction (Table 3.14). What this research stresses is the link between the motivations of the segments, and the aspects which determine satisfaction.

Satisfaction

Satisfaction is strongly related to meeting visitor expectations, which are largely built on destination image. Images are partly connected with the landscape, and partly with many other elements of the experience. In this respect, it is important to note the relative importance which ecotourists place on various experience elements. Service/activity importance ratings for generalist and specialist North American ecotourists are shown in Table 3.15. Walking and wildlife viewing are clearly top priorities, as is visiting a park or protected area. This relates to the importance of setting and landscape to the ecotourism experience. Important services appear to be those that are related to learning and cultures, guides, and interpretative education programmes. Knowledgeable guides and good education programmes or interpretive materials are critical.

Table 3.14. Australian ecotourism segments, motivations and satisfaction (Bureau of Tourism Research, 1998).

Impulse markets	Active market	Personalized market
Characteristics		
• nature-based day trips away from main tourist destinations • domestic and international	• young-mid aged professionals • usually book in advance • mainly domestic	• older professionals • expecting more comfort • international, overnight bookings prior to arrival
Motivations		
• getaway from masses, small group • relax, fun, enjoyment • nature-based tour • best possible experience • convenient transport, no planning • realistic brochure information	• enjoy nature and scenic wonder • challenge and achievement • no high comfort expectations • clear, good pre-trip information • social interaction • spontaneity and flexibility to individual needs	• interaction with environment • education and learning • quality accommodation and food with local produce • details associated with well-scheduled/organized tour at premium price
Satisfaction determinants		
• see/experience as advertised • relax, fun, enjoyment • hassle-free day, with pick-up and drop-off	• accomplishment more significant the better known the attraction • escape from daily stresses • enhanced by environmental knowledge • learn how to contribute to ecological sustainability	• desire to see/learn about environment, local history, sense of special experience • maximum enjoyment for time available

Table 3.15. Relative importance ratings, North American ecotourists (HLA/ARA, 1994).

Experienced ecotourist	General interest ecotourist	Travel trade
Wilderness setting	Casual walking	Wilderness setting
Wildlife viewing	Wildlife viewing	Guides
Hiking/trekking	Learn about other cultures	Outdoor activities
Visit national park/other protected area	Visit national park/other protected area	All inclusive packages
Rafting/canoeing/kayaking on river/lake	Wilderness setting	Parks/protected areas
Casual walking	Hiking/trekking	Interpretive/educational programmes
Learn about other cultures	The importance of guides	Cultural experiences
Participate in physically challenging programmes	Interpretive education programmes	Communicate in client's language
The importance of guides	Cycling	
Interpretive education programmes	Participate in physically challenging programmes	

Klenosky *et al.* (1998) put forward the means–end theory to develop a better understanding of factors influencing park visitors' use of specific interpretive services (i.e. individuals select products or services that produce desired consequences or benefits, which are a function of personal values). Thus the tourism product attribute is related to the benefits (or consequences) it provides, which are important to satisfying personal values (the 'ends'). They found that whatever the park product (self-guided trails, naturalist-led hikes or walks, or night/overnight programmes), learning was the greatest benefit sought. Even in canoe or fishing programmes, learning was highest, together with doing something different. They found that environmental ethics were part of the product attributes that contributed to obtaining personal benefits, for both the canoe/fishing programmes, and the self-guided trails.

Diamantis (1998) found that UK frequent ecotourists aim to be educated by the ecotourism holiday (63.3%). In addition, 91.3% agreed/strongly agreed with the statement that the value of an ecotourism vacation is to 'become more knowledgeable' (the second highest score of all values examined). However, he also found that whether or not frequent ecotourists were seeking to increase their knowledge depended on how concerned they were about the state of the natural environment. A recent Australian survey found that 62% of respondents will pay extra to go on a tour with an expert guide (*Ecotrends*, 1999). Sixty-nine per cent of all nature-based visitors report educational or learning experiences were important or very important to them. Table 3.16 shows their learning preferences. When nature-based visitors were asked about satisfaction after tours, the friendliness/helpfulness of the staff (86% very/somewhat satisfied) was the individual element that gave them the most satisfaction (Table 3.17). That which gave highest dissatisfaction was the overall size of tour numbers (6%).

Table 3.16. Learning preferences of nature-based visitors (Blamey and Hatch, 1998).

Learning preferences		Importance (%)
1	seeing and observing animals, plants, landscapes	97
2	being provided with information about the biology/ecology of species	84
3	cultural and/or historical aspects of the area	74
4	information about geology/ landscapes	70

Table 3.17. Satisfaction with aspects of nature-based tours[a] (Blamey and Hatch, 1998).

Element	Very/somewhat satisfied (%)	Very/somewhat dissatisfied (%)
Information about the natural environment (plants, animals, geology, etc.)	84[a]	1
Overall size of tour (numbers of participants)	74	6
No. of guides on tour	84	1
Value for money	80[a]	1
Time spent at sites	76	4
Quality of sites	82	1
Friendliness/helpfulness of staff	86	—
Food	65	5
Measures to minimize environmental impacts of tours	75	1
The whole tour	87	1

NB Table excludes those who didn't know or felt neutral about evaluations.

[a] Statistically significant aspects for 2 of 3 clusters of ecotourists examined.

One of the elements of ecotourism that also distinguishes it from many other tourism experiences is interpretation. Convincing arguments have been made that quality interpretation helps to minimize the negative environmental impacts of tourism (Moscardo, 1996, 1998). It has been noted that ecotourists tend to be experienced/frequent travellers who demand more from the experience, and are more likely to seek learning and educational components in the tourism experience (Urry, 1990; Cleverdon, 1993; Poon, 1993; Moscardo, 1996; Aiello, 1998). Hence, it is not surprising that the major source of dissatisfaction in the ecotourism experience is lack of interpretation, education or good guiding (Almagor, 1985; Blamey and Hatch, 1998). Roggenbuck and Williams (1991) and Aiello (1998) show that giving commercial guides interpretive training and area knowledge improves tourist satisfaction. Also, satisfaction improves experiences for staff, company owners and governments/destinations (see Chapter 35).

Diamantis (1998) found that UK ecotourists mainly expressed social intentions when participating in ecotourism holidays. In terms of values, he found them to be motivated by maturity and internal values, such as 'appreciate and respect the world we live in'. In Australia, mature travellers strongly believe that learning about nature enriches life (80.1% for > 45 years), while this is a less important attitude for those under 45 years (65.3%).

Social and environmental values

Ecotourists can be considered as a growing group of tourists who are shifting away from the consumption of *things*, toward the consumption of meaningful, learning and experiential vacations. More are travelling for self-improvement or self-enrichment, to learn and acquire new interests and friends, or to improve their physical and mental well-being. Part of the reason for the surge in ecotourism has been increased levels of environmental awareness among travellers. The Center for Tourism Policy Studies (1998) states that 'there is likely to be more development of eco-resorts as environmentally-conscious travel segments are expected to grow in the US as well as globally'. Operators in the Asia-Pacific region predict an increased awareness about environmentally friendly tourism overall in the future. They also observe that clients are more aware of social and environmental issues in the destination, and are generally interested in financially supporting local environmental conservation and social development projects, as shown in Table 3.18 (Lew, 1998b).

US 'green' travellers are willing to spend, on average, 8.5% more for travel services and accommodation provided by environmentally responsible operators (Cook *et al.*, 1992). In an analysis of the data tables, it is found that those who took an ecotourism trip are more likely (92%) to support environmentally responsible companies than non-ecotourists (86%). Results were similar for potential ecotourists (Table 3.19). Similarly, ecotourists were more willing to pay extra for an eco-aware company, than the average traveller or non-ecotourist, and 46% would pay 6–20% more, whereas only 29% of non-ecotourists would pay 6–20% more (Table 3.20). Table 3.21 shows that frequent UK ecotourists were mainly interested in social and conservation-oriented elements in terms of the *consequences* of their trip (Diamantis, 1998). The realization that their visits could disturb nature (85%) suggested that frequent UK ecotourists were a group of travellers who understood the fragility of such landscapes. Diamantis (1998) also found that while males and females were

Table 3.18. Willingness of clients to donate money to local environmental/social causes (Lew, 1998b).

Ecotour clients' willingness to contribute $	Number of operators	%
Very willing	14	38.9
Somewhat willing	20	55.6
Not interested or willing	2	5.6

Table 3.19. Ecotourists, potential ecotourists and non-potential ecotourists: likelihood to support eco-aware tourism companies (US Travel Data Center Data Tables, 1991).

Support for eco-aware companies	Took ecotourism trip	Did not take ecotourism trip	Did not take ecotourism trip: Interest in future ecotourism trip	Did not take ecotourism trip: No potential for ecotourism trip
Somewhat/very likely to support companies	92%	86%	94%	83%
Not very/at all likely to support companies	8%	14%	6%	17%
Totals	63	886	262	624

Table 3.20. Willingness to pay extra for sightseeing tours by environmentally responsible travel suppliers (US Travel Data Center Data Tables, 1991).

% Extra willing to pay for eco-aware companies	Total travellers (%)	Likely to take ecotourism trip (%)	Not likely to take ecotourism trip (%)
0	13.5	6.2	17.2
1–5	43.1	40.4	44.8
6–10	21.1	25.7	18.9
11–20	13.3	20.0	9.7
21–40	1.4	2.9	1.0
41–60	0.9	0.5	0.6
61–99	0.4	0.2	0.2
100	0.1	—	0.2
DK	6.3	4.1	7.4
Mean	8.4	9.7	7.4
Total No. US travellers	963	271	634

both more concerned with social benefits of ecotourism, females had more environmental and educational interests/concerns than males.

Reaching Ecotourists

Publications read by ecotourists

Ecotourists are extremely well educated. Significant numbers read nature-related magazines (61%) and experienced ecotourists even more (72%), with the most popular being the *National Geographic Magazine* (Table 3.22). Other publications were related to fishing/hunting, clubs, activities and travel, and such speciality publications as *The Educated Traveler: Directory of Special Interest Travel*. Double the percentage of experienced ecotourists read club publications than general consumers, since many more experienced ecotourists (50%) are members of organizations than general consumers (11%) (HLA/ARA, 1994)

Membership in clubs/organizations

A potential avenue for reaching the ecotourist is through nature-related organizations or clubs. The experienced ecotourist in North America exhibits a much greater

Table 3.21. Trip consequence interests (Diamantis, 1998).

%	Consequences of ecotourism trip
96	• respect the local population and indigenous people
93	• have awareness of the world's natural environment
85	• be concerned that your presence there may damage the natural environment
82	• go again when possible
75	• maintain environmental standards for future holiday makers
63	• create a memory that normal holidays could not give
61	• contribute actively in conservation of these areas
56	• be more energetic and adventurous
54	• feel calm and relaxed
51	• feel travel companies just use the word 'eco-holidays' to attract more people

Table 3.22. Publications read (HLA/ARA, 1994).

	General consumer	Experienced ecotourist
	% of total sample which reads such publications	
Publication	61 % (540 respondents)	72 % (271 respondents)
National Geographic	35	17
Outdoor Life[a]	10	36
Club publications	7	15
Fishing/hunting related	6	2
General nature[b]	5	3
Field and Stream	5	—
General travel	5	6
General activity/sports	5	14
Wildlife related	4	3

NB Multiple responses were permitted.
[a] Refers to *Outdoor Life/Outdoors/Outside/Outdoor Canada.*
[b] Refers to nature/natural history publications.

propensity to belong to a nature-oriented club or organization (50%) than the general interest ecotourist (11%). However, general interest ecotourists represent a large target population, so the actual numbers who are members of nature-oriented organizations are by no means small (Table 3.23). It is interesting to note that the experienced ecotourism traveller, while tending to prefer activity-related magazines, belongs to more nature/wilderness-related organizations. In the UK, Diamantis (1998) found that 22% of frequent ecotourists are members of an environmental group or society. Most of these (67%) are highly involved ecotourists. For occasional UK ecotourists, Diamantis found that 37% belong to at least one group, society or organization. Singles and those with higher than average education are more likely to be members.

The information about reaching ecotourists is helpful, but not critical, in terms of marketing. Packaging the right product for the right market is very important, as are customer service, quality and value for money. In addition, the ability to provide quality interpretive guides, and to meet and exceed customer expectations is critical (Pam Wight and Associates, 1999).

Table 3.23. Membership in organizations/clubs (HLA/ARA, 1994).

	General consumer	Experienced ecotourist
	% of total sample which belongs to club/organization	
Club/organization	11 % (153 respondents)	50 % (189 respondents)
Sierra Club	18	34
Outdoor activity club	13	11
Nature organization[a]	10	37
Audubon Society	5	17
Other wildlife organizations[b]	10	8
Fishing and hunting	6	—
Greenpeace	5	2
Worldwide Fund for Nature	4	3
National Wildlife Federation	4	3
Boy/Girl Scouts	4	—

NB Multiple responses were permitted, therefore total percentage may exceed 100%.
[a] Refers to nature/naturalist/conservation/park organizations.
[b] Refers to organizations such as Ducks Unlimited and wilderness societies.

Conclusions

Ecotourism is a complicated subject, involving specialized niche markets that may share many characteristics, preferences and motivations, or vary by these same attributes. Markets today reflect greater sophistication, as well as changing lifestyles, attitudes, values and interests. They exhibit well-defined expectations, and seek new experiences and purposes for travel based on these diverse interests and preferences. Destinations and operators need to be able to provide convenience and customization for these diverse markets.

Ecotourism markets are not homogeneous. However, it may be that ecotourism markets cannot be segmented well at the global level. For example, Diamantis (1998) found in the UK that ecotourists were frequent or occasional, in North America, HLA/ARA found ecotourists were more generally interested or experienced. It may be that particular destinations attract certain ecotourist segments. In any event, previous studies have revealed a variety of segments with distinctive differences and similarities, as summarized in Tables 3.2, 3.9 and 3.12.

While this chapter has attempted the challenge of analysing and summarizing globally significant market studies, it should be pointed out that the application of these findings requires care. Like any other travellers, ecotourists at the aggregate level tend to be educated, time-poor, and desire value for money. They are interested not simply in a menu of choices, but in quality customer services, customization, interpretation by knowledgeable guides, a sense of authenticity, and opportunities to experience a number of destination areas, all conveniently packaged or available to the FIT.

Ecotourism markets are dynamic. They have changed somewhat in the last decade, and are likely to continue to refine their preferences and seek benefits related to their motivations, as well as to support companies which provide experiences which support their social and environmental value systems. Operators (communities and destinations) must increase the value of their products, and respond to the needs and preferences of ecotourists. It is how the market findings provided in this chapter are *applied*, which is relevant, and this will vary by each operation/destination.

References

Aiello, R. (1998) Interpretation and the marine tourism industry, who needs it?: a case study of Great Adventures, Australia. *Journal of Tourism Studies* 9(1), 51–61.

Almagor, U. (1985) A tourist's 'Vision Quest' in an African game reserve. *Annals of Tourism Research* 12, 31–48.

ARA Consulting Group, Eureka Tourism and Hospitality Management Consultants and the Tourism Research Group (1991) *Yukon Wilderness Adventure Travel Market Awareness Study.* Yukon Department of Tourism, Whitehorse, Yukon.

Backman, K.F. and Potts, T.D. (1993) *Profiling Nature-Based Travelers: Southeastern Market Segments.* Strom Thurmond Institute, South Carolina.

Blamey, R. (1995) *The Nature of Ecotourism*, Occasional Paper No. 21. Bureau of Tourism Research, Canberra.

Blamey, R. and Hatch, D. (1998) *Profiles and Motivations of Nature-Based Tourists Visiting Australia*, Occasional Paper No. 25. Bureau of Tourism Research, Canberra.

Bureau of Tourism Research (1998) *Ecotourism Snapshot: a Focus on Recent Market Research.* Office of National Tourism, Australia.

Ceballos-Lascuráin, H. (1998) Introduction. In: Lindberg, K., Epler-Wood, M. and Engeldrum, D. (eds) *Ecotourism: a guide for Planners and Managers*, Vol. 2. The Ecotourism Society, North Bennington, Vermont, pp. 7–10.

Center for Tourism Policy Studies, University of Hawaii at Manoa (1998) Repositioning Hawaii's Visitor Industry Products. Prepared for the Department of Business, Economic Development and Tourism, Hawaii Tourism Office.

Charters, T. (1999) Environmental Tourism Manager for Tourism Queensland, quoted in *Ecotrends*, March.

Cleverdon, R. (1993) Global tourism trends: influences, determinants and directional flows. *World Travel and Tourism Review* 3, 81–89.

Cook, S.D., Stewart, E. and Repass, K., US Travel Data Centre (1992) *Discover America: Tourism and the Environment.* Travel Industry Association of America, Washington, DC.

Crossley, J. and Lee, B. (1994) Ecotourists and mass tourists: a difference in 'benefits sought'. Proceedings of the Travel and Tourism Research Association Conference, Bal Harbour, Florida.

Diamantis, D. (1998) Ecotourism: characteristics and involvement patterns of its consumers in the United Kingdom. PhD dissertation, Bournemouth University, UK.

Diamantis, D. (1999) The characteristics of UK's ecotourists. *Tourism Recreation Research* 24(2), 99–102.

Dixon, J.A. and Sherman, P.B. (1990) *Economics of Protected Areas.* Island Press, Washington, DC.

Eagles, P. (1992) The travel motivations of Canadian ecotourists. *Journal of Travel Research* 31(2), 3–7.

Eagles, P. (1995) Understanding the market for sustainable tourism. In: McCool, S. and Watson, A.E. (eds) *Linking Tourism, the Environment, and Sustainability.* Special session of the Annual meeting of the National Recreation and Park Association, Minneapolis, Minnesota, 12–14 October. General Technical Report INT-GTR-323, USDA Intermountain Research Station, Ogden, Utah.

Eagles, P.F.J. and Cascagnette, J.W. (1995) Canadian ecotourists: who are they*? Tourism Recreation Research* 20(1), 22–28.

Eagles, P.F.J. and Higgins, B.R. (1998) Ecotourism market and industry structure. In: Lindberg, K., Epler Wood, M. and Engeldrum, D. (eds) *Ecotourism: a Guide for Planners and Managers*, Vol. 2. The Ecotourism Society, North Bennington, Vermont.

Epler Wood, M. (1998) New directions in the ecotourism industry. In: Lindberg, K., Epler Wood, M. and Engeldrum, D. (eds) *Ecotourism: a Guide for Planners and Managers*, Vol. 2. The Ecotourism Society, North Bennington, Vermont.

Hawkins, D., Epler Wood, M. and Bittman, S. (1995) *The Ecolodge Source book for Planners and Developers.* The Ecotourism Society, North Bennington, Vermont.

HLA Consultants (1996) *Ecotourism Accommodation: an Alberta Profile.* Alberta Economic Development and Tourism, Edmonton, Alberta.

HLA Consultants and The ARA Consulting Group (1994) *Ecotourism – Nature/Adventure/Culture: Alberta and British Columbia Market Demand Assessment.* Canadian Heritage, Industry Canada, British Columbia Ministry of Small Business, Tourism and Culture, Alberta Economic Development and Tourism, and the Outdoor Recreation Council of British Columbia.

Klenosky, D.B., Frauman, E., Norman, W.C. and Gengler, C.E. (1998) Nature-based tourists' use of interpretive services: a means-end investigation. *Journal of Tourism Studies* 9(2), 26–36.

Laarman, J.G. and Durst, P.B. (1993) Nature tourism as a tool for economic development and conservation of natural resources. In: Nenon, J. and Durst, P.B. (eds) *Nature Tourism and Asia: Opportunities and Constraints for Conservation and Economic Development.* USDA, Forest Service, USAID, USDA, Office of International Cooperation and Development, Washington, DC, pp. 1–19.

Laarman, J.G. and Gregersen, H.M. (1996) Pricing policy in nature-based tourism. *Tourism Management* 17(4), 247–254.

Leones, J., Colby, B. and Crandall, K. (1998) Tracking expenditures of the elusive nature tourists of Southeastern Arizona. *Journal of Travel Research* 36(3), 56–64.

Lew, A. (1998a) Ecotourism Trends. *Annals of Tourism Research* 25(3), 742–746.

Lew, A. (1998b) The Asia-Pacific ecotourism industry: putting sustainable tourism into practice. In: Hall, C.M. and Lew, A.A. (eds) *Sustainable Tourism: a Geographical Perspective.* Addison Wesley Longman Ltd, Harlow, UK.

Liu, J.C. (1994) *Pacific Islands Ecotourism: a Public Policy and Planning Guide.* University of Hawaii, The Pacific Business Center Program, Honolulu.

McKercher, B. and Robbins, B. (1998) Business development issues affecting nature-based tourism operators in Australia. *Journal of Sustainable Tourism* 6(2), 173–188.

McNeely, J.A. (1988) *Economics and Biological Diversity.* International Union for the Conservation of Nature and Natural Resources, Gland, Switzerland.

Moscardo, G. (1996) Mindful visitors. *Annals of Tourism Research* 23(2), 376–397.

Moscardo, G. (1998) Interpretation and sustainable tourism: functions, examples and principles. *Journal of Tourism Studies* 9(1), 2–13.

Pam Wight and Associates (1999) *Catalogue of Exemplary Practices in Adventure Travel and Ecotourism.* Prepared for the Canadian Tourism Commission, Ottawa.

Parker, T. (1993) Nature tourism in Nepal. In: Nenon, J. and Durst, P.B. (eds) *Nature Tourism and Asia: Opportunities and Constraints for Conservation and Economic Development.* USDA, Forest Service, USAID, USDA, Office of International Cooperation and Development. Washington, DC, pp. 21–30.

Poon, A. (1993) *Tourism, Technology and Competitive Strategies.* CAB International, Wallingford, UK.

Reingold, L. (1993) Identifying the elusive ecotourist. In: *Going Green,* a supplement to *Tour and Travel News* (October 25), 36–39.

Roggenbuck, J.W. and Williams, D.R. (1991) Commercial tour guides's effectiveness as nature educators. Paper presented at the World Congress on Leisure and Tourism, Sydney.

Selengut, A. (1995) 'Foreword'. In: Hawkins, D., Epler Wood, M. and Bittman, S. (eds) *The Ecolodge Source Book.* The Ecotourism Society, North Bennington, Vermont, pp. v-vi.

Tourism Canada (1995) *Adventure Travel in Canada: an Overview of Product, Market and Business Potential.* Tourism Canada, Canada Directorate, Ottawa, February.

Tourism Queensland (1999) *Ecotrends* March, quoting a survey commissioned by the Environmental Tourism Department of Tourism Queensland.

Travel Industry Association of America (TIAA) (1994) *Adventure Travel: Profile of a Growing Market,* conducted by US Travel Data Center, Washington, DC.

Urry, J. (1990) *The Tourist Gaze.* Sage, London, UK.

US Travel Data Center (1991) *Tourism & The Environment Study Tables.* TIAA, Washington, DC.

Weiler, B. and Richins, H. (1995) Extreme, extravagant and elite: a profile of ecotourists on Earthwatch expeditions. *Tourism Recreation Research* 20(1), 29–36.

Wight, P.A. (1993) Sustainable ecotourism: balancing economic, environmental and social goals within an ethical framework. *Journal of Tourism Studies* 4(2), 54–66.

Wight, P.A. (1995) Planning for success in sustainable tourism. Invited presentation to 'Plan for Success', National Conference of the Canadian Institute of Planners, Saskatoon, Saskatchewan, 2–5 June.

Wight, P.A. (1996) North American ecotourists: market profile and trip characteristics. Second in a two-part series for *Journal of Travel Research* 34(4), 2–10.

Wight, P.A. (1997) Ecotourism accommodation spectrum: does supply match the demand? *Tourism Management* 18(4), 209–220.

World Tourism Organization (WTO) (1994) *Global Tourism Forecasts to the Year 2000 and Beyond: East Asia and the Pacific*, 4. WTO, Madrid.

WTO News (1997) *Eco-tourism: a Rapidly Growing Niche Market* December/January. http://www.world-tourism.org/newslett/decjan97/decjan6.htm

WTO News (1998) *Ecotourism, Now One-Fifth of Market* January/February. http://www.world-tourism.org/newslett/janfeb98/ecotour.htm

Wylie, J. (1994) *Journey Through a Sea of Islands: a Review of Forest Tourism in Micronesia.* USDA Forest Service Institute of Pacific Islands Forestry, Honolulu, Hawaii.

Wylie, J. (1997) *Tourism and the Environment.* Executive Development Institute for Tourism, University of Hawaii at Manoa, 23–24 June. Quoting J. Mass (1995) Ten travel trends and what you can do about them. Travel Industry Association of America National Conference.

Yuan, M.S. and Moisey, N. (1992) The characteristics and economic significance of visitors attracted to Montana wildlands. *Western Wildlands* Fall 18(3), 20–24.

Chapter 4

Global Growth and Magnitude of Ecotourism

D.E. Hawkins and K. Lamoureux

School of Business and Public Management, The George Washington University, Washington, District of Columbia, USA

Even before Thomas Cook began the world's first travel agency in 1841 (Gartner, 1996) or before the first group of young European men took their once in a lifetime 'Grand Tour', humans have been inclined to travel or participate in tourism activities. Regardless of the reasons why we travel – religious practice, social interaction, leisure activities, cultural exchange or just plain curiosity – people like to travel. Proof of this can be found in the fact that the demand for travel and tourism services is growing at a faster rate than ever before.

As tourist numbers increase around the world, so do the types of activities they choose to undertake during their trip. While 'traditional' tourism still exists and continues to grow, 'new' types of tourism, or alternative tourism, such as ecotourism, cultural/heritage, educational or health tourism have emerged, as well. Not only does the market for these new types of tourism exist, but trends indicate that the market for alternative tourism is growing faster than could be imagined. Tourists are seeking more out of their precious vacation time than just relaxation. Advances in both transportation and communication have opened new doors, allowing us to travel further and experience more of the world, than ever before.

The focus of this chapter is the growth and magnitude of ecotourism. However, in order to better understand how ecotourism is growing, it is important to look briefly at the tourism industry in general. Just like other forms of tourism, ecotourism is dependent upon a series of global and regional trends that dictate the ability and desire people will have to travel in the future. Past trends have shown that factors such as available leisure time, disposable income, and education, among others, influence the amount and type of tourism an individual or group will partake in. By analysing both past and projected data for the general tourism industry, we can not only forecast upcoming tourism trends, but also trends in specific types of tourism.

The Tourism Industry

The World Tourism Organization (WTO) estimates that by the year 2020 there will be 1.6 billion international tourist arrivals worldwide, spending over US$2 trillion. This means that globally, arrivals will continue to grow at an average of 4.3% and spending at 6.7% per year. This surpasses the maximum probable expansion in the world's wealth estimated at a 3% increase per year (WTO, 1988a). Regardless of whether these estimates are realistic or not, it is obvious that tourism, growing at an exorbitant rate, has yet to reach its full potential.

Travel and tourism is also one of the world's fastest growing economic activities. According to the World Travel and Tourism Council (WTTC), tourism is 'expected to contribute 10.8% to the global gross domestic product in 2000 (US$3.6 trillion), rising to 11.6% or US$6.6 trillion by 2010' (World Travel and Tourism Council, 2000). They also estimate that capital investment for tourism will reach US$701 billion or 9.4% of total investment in 2000 and reach US$1.4 trillion or 10.6% of total by 2010 (WTTC, 2000). The WTTC also estimates that travel and tourism-related jobs in 2000 will reach 192 million, accounting for over 8% of global employment. Over 59 million new jobs will be created over the next 10 years (WTTC, 2000).

Tourism arrivals and receipts

According to the WTO (1998b, 1999) between 1989 and 1998, international tourism arrivals grew at an average annual rate of 10% and international tourism receipts (excluding transport) at 9% (Fig. 4.1). Annual percentage increases declined significantly in all markets during the 1990–1995 period due to the Gulf War and a poor global economy. When comparing 1997 with 1998, market increases occurred in all major outbound markets with the exception of the East Asia/Pacific region, primarily due to the Asian economic crisis of 1997. International arrivals are projected to increase from 673 million in 2000 to 1.05 billion in 2010, and 1.6 billion in 2020. These forecasts are based on annual growth rates of 4.2% to the end of the 1990s and between 4 and 5% during the first decade of the 21st century.

Regional Breakdown

Given the information above, it is difficult not to believe, barring a huge disaster, that tourism will demonstrate continued growth throughout the world. Tourism destinations, however, are likely to change. According to the WTO, Europe will continue to be the largest receiving region well into 2020 but will decline from 59% of the market share to 49%. As Table 4.1 indicates, East Asia and the Pacific region will pass the Americas, as the second largest receiving region. Africa, the Middle East and South Asia are projected to increase at a slower rate of 5%, 4% and 1%, respectively, by 2020 (WTO, 1998b).

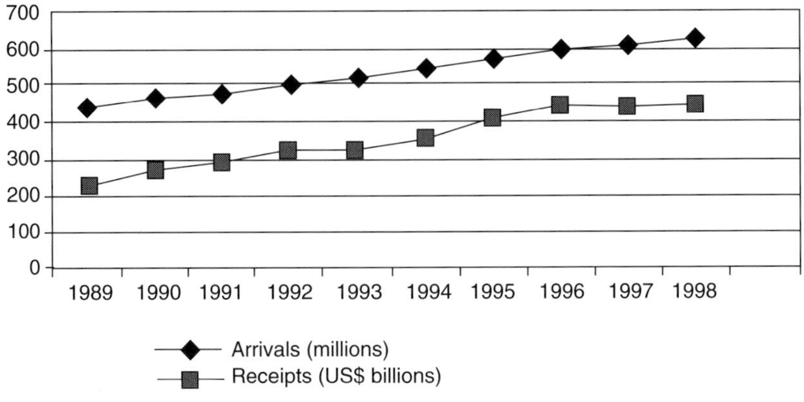

Fig. 4.1. International tourist arrivals and receipts: 1989 to 1998 (WTO, 1999).

Table 4.1. Forecast of international tourist arrivals 1995–2020 (World Tourism Organization, 1998b).

	Tourist arrivals (millions)			
Regions	1995	2000	2010	2020
Europe	335	390	527	717
East Asia/Pacific	80	116	231	438
Americas	110	134	195	284
Africa	20	27	46	75
Middle East	14	19	37	69
South Asia	4	6	11	19
Total	563	692	1047	1602

Speciality Market Forces

Travel motivations are major push factors in determining what kind of tourism people will seek. Overall the tourism market is becoming increasingly segmented with many new tourism alternatives. 'The new consumers want to be involved – to discover new experiences, to interact with the community, and to learn about and appreciate the destination at more than a superficial level' (Jones, 1998). In response to this trend, destinations are increasingly targeting their tourism product at specific markets or 'speciality travel markets'. Below is a list (International Institute of Tourism Studies, 1999) of some of these specific types of tourism; however, in many cases a destination is able to develop its tourism product in such a way as to satisfy various different markets simultaneously.

- *Ecotourism.* Increasingly, parks, nature reserves and natural settings are becoming popular tourist destinations. The development of environmentally sensitive hotels and resorts responds to the fast growing ecotourism markets and the general public's awareness of environmental preservation and sustainability. Sustainability is to improve the quality of life today without destroying it for future generations.
- *Cultural and heritage tourism.* Culture and history are the most popular of tourist activities. Experiencing different customs and lifestyles, learning about historical cultures, visiting historic sites, folklore and theatre characterizes this type of tourism.
- *Adventure tourism* can range from high to soft adventure and, in that order, includes activities such as mountain climbing, scuba-diving or walking along park trails.
- *Health tourism.* Natural hot springs, weight reduction.
- *Sports tourism.* Tennis, golfing, World Cup football, Olympic events.
- *Cruise ship.* 81% sail from North American ports. US$18 billion in 1997, 205 ships. 8.5% growth since 1990 in North America, Europe growing rapidly.

In 1997, WTO Secretary General Francesco Franigialli confirmed that even as tourism in general continues to grow dramatically, the area of greatest expansion is that of speciality travel. Specifically, he stated that 'trends show a higher than average increase in new types of tourism including ecotourism and all nature-related forms of tourism, which today account for approximately 20% of total international travel' (WTO, 1998c). This is not to say that traditional tourism, such as mass tourism to amusement parks or beaches will cease to exist. The fact remains that there will always be a market for traditional tourism; however, consumer preferences are expanding to alternative forms of vacationing, often as an 'add-on' to more 'traditional' tourism activities.

According to the Adventure Travel Society, 'Special interest travel is now the fastest growing segment of the travel market. It has grown 12% annually over the past 5 years. The last decade brought increased levels of physical fitness and a growing respect and concern for the environment' (Adventure Travel Society, 1998). It has been largely acknowledged that tourists are looking for more out of their vacation. Just as people's tastes vary, so does their willingness to experiment, whether that is by trekking through an Amazon jungle, taking a class about Australian aboriginal culture or observing sea turtles in Costa Rica.

Ecotourism and Nature-based Tourism

Many have said that ecotourism is one of the fastest growing, if not 'the' fastest growing type of 'new tourism'. Trends indicate that the growth of ecotourism coupled with the larger market segment of nature tourism far surpasses that of tourism in general. While the lack of clear differentiation between ecotourism and other forms of nature tourism makes tracking ecotourism development difficult, it is obvious that travel to natural areas is increasing at a tremendous rate. In 1996, the WTO predicted that there would be an 86% increase in tourism receipts, of which the majority would come from 'active, adventurous, nature and culture-related travel' (Honey, 1999).

In economic terms, the WTO announced in 1997 that 'ecotourism is worth US$20 billion a year' at the World Ecotour Conference. Although this amount pales in comparison to tourism's estimated US$3.6 trillion contribution to the global gross domestic product, it still represents a substantial portion of total tourism receipts (WTO, 1998c). Some examples (mainly in North America) of the magnitude of ecotourism or nature tourism follow.

At the same time, travellers seem to be willing to pay for an 'eco' experience. In a recent study, Travel Industries of America (TIA) found that '83% of all American travellers are inclined to support "green" companies and are willing to spend, on average, 6.2% more in travel services and products provided by environmentally responsible travel suppliers' (TIA, 2000). Another study of North American travel consumers in 1994 showed that '77% had already taken a vacation involving activities related to nature, outdoor adventure, or learning about another culture in the countryside or wilderness. Of the 23% remaining who had not, all but one respondent stated that they were interested in doing so' (Wight, 1996).

Ecotourism Indicators

As previously mentioned, ecotourism is often 'clumped' together with other forms of nature tourism making it difficult to grasp its real magnitude and growth rate. In the absence of some of the traditional measurement tools often used to calculate general tourism expansion, those wishing to measure ecotourism growth must employ other indicators, which do provide insight into the current and future magnitude of the sub-industry. The following indicators include growth in ecotourism education, international recognition and regional support, international funding opportunities, and growth in tourism eco-certification and eco-labelling programmes. Each of these indicators provides a glimpse beyond the traditional statistical analysis, into the current position of ecotourism and its expected future growth.

Increase in 'nature-motivated' tourism visitation worldwide

Visitation to ecotourism or nature tourism destinations is on the increase worldwide. More and more destinations are trying to 'jump' on the nature tourism 'bandwagon'. Many destinations are experiencing an increased number of visitors looking to include eco-activities in their vacation. Demand for destinations that include natural elements such as national parks and local parks, forests, waterways and others continues to increase. For example, according to a 1998 Travel Poll by TIA, 'National parks are one of America's biggest attractions. Nearly 30 million US adults (20% of travellers or 15% of all US adults) took a trip of 100 miles or more, one-way, to visit a national park during 1997–1998. Residents of the Rocky Mountain region of the US were the most likely to visit a national park with 37% saying they included a park visit while travelling (during 1997–1998)' (TIA, 2000). Although not all of the activities undertaken at these parks can be considered ecotourism, according to TIA 'a large share of these travellers (70%) participated in outdoor activities while visiting the national parks. Among these outdoor activities, hiking (53%) was the most popular, followed by camping (33%) and fishing (19%)' (TIA, 2000). TIA also found that 'one-half of US adults, or 98 million people, have taken an adventure trip in the past 5 years. This includes 31 million adults who engaged in hard adventure activities like white-water rafting, scuba-diving and mountain biking. Adventure travellers are more likely to be young, single and employed compared to all US adults' (TIA, 2000).

As visitation to natural areas increases, including ecotourism visitation, so does the demand for travel professionals to accommodate these tourists. Over the last few years there has also been an increase in tourism professionals, such as travel agents, tour operators, tour guides, etc., that focus either on the ecotourism or at least, the nature tourism markets. Today, it is not difficult to find a tour operator that specializes in an 'eco-niche' such as birdwatching or hiking, among others.

At the same time, travel industry organizations have also developed around this trend. Some of these groups include Partners in Responsible Tourism (PIRT), The Ecotourism Society (TES), Business Enterprises for Sustainable Travel (BEST), along with other regional ecotourism societies, such as in Australia, Pakistan, etc. These are just examples of the growing number of travel industry affiliated organizations developed to address different aspects and issues related to this growing trend for tourism to natural areas.

Growth in ecotourism education

Tourism, in general, is a relatively new field of study. Only in the last half a century or so, has tourism studies gained recognition as a social science. Twenty-five years ago, the George Washington University was one of the first colleges in the world to offer a graduate level degree in tourism administration. Today, there are numerous institutions around the world, offering everything from undergraduate and graduate degrees to tourism certificate or diploma programmes. In addition to general tourism education, training specific to speciality tourism such as ecotourism is also expanding. In 1999, the Ecotourism Society put together a sample list of universities in the US, Canada and the UK that offer programmes or courses in ecotourism. In these three countries alone, there are over 25 institutions of higher learning involved in ecotourism training (Ecotourism Society, 1999) Of course, ecotourism training is not limited to these few countries. Ecotourism-specific education is gaining recognition throughout the world in both developed and developing countries. The Commonwealth Department of Tourism in Australia also offers a list of programmes in Australia. In addition, universities in Costa Rica, Ecuador and Ghana are just some of the many institutions worldwide delivering post-secondary ecotourism education.

International recognition and regional support

Internationally and regionally, nature tourism and ecotourism are gaining public awareness. Internationally, one of the greatest 'achievements' of the ecotourism industry is to have 2002 declared the 'International Year of Ecotourism' by the United Nations (UN). The UN also declared 1999 to be the 'Year of Sustainable Tourism, Small Island States and the Ocean'. The choice of ecotourism specifically only 3 years later, is not only a great success for industry, but it also signifies the gradually accepted distinction between sustainable tourism and ecotourism on the international development level.

Regionally, many governments have adopted either ecotourism strategies within their national tourism plan or implemented ecotourism-specific development projects. The Vietnamese government planned to adopt a 'National Ecotourism Strategy' in 1999, designed to highlight ecotourism as a national development priority. The government of Nepal implemented a 5-year Manaslu Ecotourism Project designed to build infrastructure, help local communities benefit from tourism and at the same time help to conserve the environment (Shrestha, 1997). In the US, the Environmental Protection Agency has begun a tourism industry-wide discussion on sustainable tourism development. The intention of this project is to create a working dialogue aimed at promoting sustainable tourism development and management by identifying what impediments and opportunities for sustainability the tourism industry is currently faced with.

According to the WTO, Brazil 'launched a US$200 million program to develop ecological tourism in the Amazon' in 1997. The programme was funded by the Inter-American Development Bank, with a focus on private sector investment in Amazon ecotourism, and the expansion and improvement of conditions for ecotourism within this area (WTO, 1998c). Ecotourism investment made by the Brazilian government in 1996 totalled US$3 billion. However, in terms of gross domestic product, this was less than Costa Rica, the Dominican Republic, Ecuador, Peru, Chile and Argentina, for example. Brazil expects that investment levels in ecotourism will reach US$12 billion in 1999 (WTO, 1998c).

In addition to international agreements and discussions, such as the UN's Council for Sustainable Development, there are a great number of smaller strategies taking place on a national or regional level to stimulate tourism at the micro-enterprise level and promote business linkages intended upon developing alternative tourism, such as ecotourism, that is planned from the beginning with the local community and the sustainability of the industry, as the prime benefactors. Some of the strategies that are already being implemented include the following (DFID, 1999):

- The Fiji Tourism Development Plan has begun a programme of removal of red tape and regulations that suppress the informal sector.
- Some South African national and provincial parks have facilitated access to markets for local entrepreneurs by providing opportunities to initiate suitable lodging outside the park or small business endeavours or advertising inside the park. The South African government asks potential investors to submit their plans for boosting local development when they bid for a tourism lease.
- Enhanced community participation in areas where tourism is being used as a tool for economic development, such as the north coast of Honduras.

At the same time, strategies currently being employed throughout the world at the non-governmental level include the following (DFID, 1999):

- Provide credit and non-financial services for micro-enterprise.
- Build the capacity of poor people to assess tourism options.
- Facilitate communication and negotiation between the tourism industry and local people.
- Understand tourism businesses so they are well positioned so that all stakeholders are satisfied.

Some examples of the initiatives being taken by business to enhance linkages include (DFID, 1999):

- out-source contracting (e.g. laundry, transportation, food service, water sports concessions);
- support local enterprise;
- encourage tourists to visit local sellers;
- develop partnerships with communities;

- enhance partnerships with donors, NGOs and governments.

International funding or aid

One of the strongest indicators that ecotourism, and more generally, sustainable tourism, are gaining worldwide recognition, is the increasing amount of funding available for these endeavours. Official development financing or aid is usually mobilized through multilateral (representative of several governments) institutions, bilateral (one government) agencies, or regional development banks. Many of these institutions have begun to incorporate sustainable tourism development into larger regional or national sustainable development projects. Some of the organizations that have accepted sustainable tourism as a viable economic device for development include the World Bank, UN, USAID, and the European Union.

In the past, tourism has received 'mixed reviews' from various development organizations, largely due to the bad reputation the industry has acquired from poorly planned, non-sustainable tourism development. This problem occurs when tourism is developed without stakeholder participation and planning. Often, tourism projects that are deemed 'economically beneficial' are not necessarily socially or environmentally beneficial, nor do the economic benefits fall to those most closely affected by the development. While it is true that this type of non-sustainable tourism development still occurs throughout the world, the popularization of alternative tourism development such as ecotourism or cultural heritage tourism, is once again allowing tourism to be accepted as an economic tool for sustainable development (International Institute of Tourism Studies, 1999).

Increased participation in tourism development on the part of international funding is particularly important to the world's poorest countries. According to the UK's Department for International Development (DFID), 'while poor countries command a minority share of the international tourism market, tourism can still make a significant contribution to their economy'. Of the earth's poor '80% live in 12 countries (under \$1 per day). In 11 of these, tourism is significant and/or growing'. Furthermore, for the world's 100 poorest countries, 'tourism is significant in almost half the low income countries and virtually all the lower-middle income countries (accounting for over 2% of GDP or 5% of exports)' (DFID, 1999).

Some of the key projects developed or being developed by major multi-lateral or national funding organizations are explained below. It is important to note that the most significant aspect of this particular indicator is the recent acceptance and continued growth of tourism as a means for stimulating economic growth, reducing poverty, and in the case of ecotourism, protecting the environment.

In 1995, the World Bank boasted that since the 1992 Earth Summit, they had become the 'world's leading financier of environmental projects in the developing world' (Honey, 1999). Recently, the World Bank Group, which includes the International Finance Corporation (IFC) and the Multilateral Investment Guarantee Agency (MIGA), has begun to formally reintroduce sustainable tourism as a tool for economic, social and environmental development. It is important to note that the IFC never stopped funding tourism projects, as did the World Bank, although ecotourism does not account for a large portion of their agenda (Honey, 1999). The IFC has devoted a special unit to general tourism development, which has invested over US\$600 million in loans and equity investments for general tourism development projects (Honey, 1999). MIGA only began its involvement with ecotourism in 1994 with the Rain Forest Aerial Tram in Costa Rica (Honey, 1999). Below is a list of some examples of World Bank projects that include ecotourism development:

- Lesotho-Maloti-Drakensberg Transfrontier Conservation and Development Area Project

- Uganda-Protected Areas Management and Sustainable Use Project
- Madagascar-Second Environment Program Support Project
- Panama Atlantic Biological Corridor
- Costa Rica Development of Ecomarkets
- Georgia-Integrated Black Sea Environmental Project
- Indonesia-Maluku Conservation & Natural Resources Management Project.

The UN has also become involved with tourism projects through the United Nations Development Program (UNDP), which is the world's largest multilateral source for development cooperation (UNDP, 1999) and the Global Environment Facility (GEF). GEF was established as a joint international effort to help solve global environmental problems. The GEF Trust Fund was established by a World Bank resolution as a joint programme between the United Nations Development Program, UNEP and the World Bank (UNDP, 1999).

In 1992, UNDP launched the GEF Small Grants Program (GEF/SGP). The GEF/SGP provides grants of up to US$50,000 and other support to community-based groups (CBOs) and non-governmental organizations (NGOs) for activities that address local problems related to the GEF areas of concern. Since its inception, the GEF/SGP has funded over 750 projects in Africa, North America and the Middle East, Asia and the Pacific, Europe and Latin America and the Caribbean. Today, the programme is operational in 46 countries (Christopher Holtz, Washington 1999, personal communication). Here are a few tourism projects from their small-grants portfolio:

- Belize: Red Band Scarlet Macaw Conservation and Tourism Development Project. Project dates: June 1997–May 1998
- Brazil: Training Program for Income Generation and Environmental Education for Communities near the Chapada dos Veadeiros National Park. Project dates: September 1997–February 1998
- Chile: Training and Capacity Building in Tourism
- Costa Rica: Community-Based Ecotourism for Conservation of the Marine Turtle and Marine Resources in Gandoca. Project dates: February 1997–February 1998
- Dominican Republic: Ecotouristic Promotion with Gender Participation on Ecological Transept Los Calabozos-La Guazara in High Watershed Yaque del Norte River. Project dates: September 1997–February 1998
- Dominican Republic: Sustainable Ecotourism and Environmental Education in the Surrounding of Los Haitises National Park. Project dates: September 1997–September 1998
- Zimbabwe: Umzingwane Ecotourism Project. Project dates: October 1997–October 1998.

The two groups discussed above, the World Bank and the UN, are arguably the largest international funding organizations on Earth. However, there are a number of national or governmental aid organizations such as USAID in the US, JICA in Japan, DFID in the UK and many others. Each of these programmes also provides developmental funding to other countries around the world, including the funding of sustainable tourism projects.

In addition to multilateral or governmental aid, many not-for-profit organizations, particularly those involved in conservation and wildlife protection, have also begun to incorporate tourism and ecotourism into their strategic planning. Later in this book the role of not-for-profit organizations will be discussed at length, therefore it is not necessary to go into too much detail here on the efforts of the NGO community. However, it is worth mentioning that many governmental or internationally funded tourism projects are generally developed in conjunction with or at least under the guidance of a not-for-profit organization, either locally or internationally.

Green certification and eco-labels for tourism

In general, the tourism industry is dependent on the environment for its sustainability and makes extensive use of the natural and cultural resources in its area of operation. The industry's prosperity is thus somewhat dependent on the conservation and responsible use of the environment. Several organizations including government organizations, not-for-profit industry organizations and NGOs have addressed the issue pertaining to environmental conservation and best practices within the tourism industry by introducing ecolabelling and green certification schemes. Each certification programme defines criteria and standards that enhance efficiency and reduce overuse and wastage. Each scheme is unique in that the certification period varies and may range from 1 to 3 years. Evaluation methods also vary from scheme to scheme. Although there is abundant information on the criteria required to participate in these schemes, there is a shortage of information on the evaluation mechanisms and duration period of each scheme. The individual certification programmes are discussed in greater detail below (United Nations Environment Program, 1998).

- *PATA Green Leaf* (Asia-Pacific). This is a green certification scheme developed by the Pacific Asia Travel Association, which is an industry association.
- *Tyrolean Environmental Seal of Quality* (Austria and Italy). This scheme is promoted by Tirol Werbung and Sudtirol Werbung, which are public authorities operating in the area of accommodation and catering.
- *Green Globe* (International). This programme was developed by the WTTC, which is an industry association and focuses on all industries within the tourism sector.
- *The Audubon Cooperative Sanctuary System* (International). This scheme is promoted by Audubon International, which is a non-governmental association.
- *Blue Flag* (Europe). This certification scheme is promoted by the Foundation For Environmental Education in Europe, which is a non-governmental organization and is aimed primarily at the preservation and responsible use of beaches in Europe.
- *We Are an Environmentally-friendly Operation* (Germany). This certification scheme is promoted by the Deutscher Hotel and Gaststatten Verbans DEHOGA (Hotel and Restaurant Association of Germany), which is an industry association operating in the field of lodging and catering.
- *Committed to Green* (Europe). This certification scheme has been developed by the European Golf Association's Ecology Unit.
- *Ecotel* (International). This scheme has been developed by HVS Eco Services, the environmental consulting division of HVS dedicated exclusively to the hospitality industry.
- *British Airways Tourism for Tomorrow Awards* (International). These awards have been developed by British Airways and are directed at tour operators, individual hotels and chains, national parks and heritage sites and other activities associated with tourism in order to promote the responsible use of the environment by these agencies.
- *Code of Practice for Ecotourism Operators* (Regional). This code was developed in 1991 by the Ecotourism Association of Australia, a not-for-profit organization, to develop ethics and standards for ecotourism and to facilitate understanding and interaction between the tourist, host communities, the tourism industries and government and conservation groups.

Conclusion

In its early stage of development, ecotourism was regarded as a completely new concept. However, today, as we can see from the indicators highlighted here, the areas of both ecotourism and nature tourism have become a significant portion of the tourism industry in general. Growing environmental awareness worldwide, paired with advances in transportation and communication, will only help to foster future ecotourism growth. The markets for these types of sustainable tourism are likely to expand as more people in the world achieve the financial resources needed to travel. In addition, ecotourism development will continue to expand and increase in importance as more communities around the world begin to accept it as an essential strategy for their overall sustainable development plan.

Thirty years ago, the term 'ecotourism' did not exist. In 2000, ecotourism was practised on every one of the Earth's continents. There is no doubt that the industry made tremendous strides in the last part of the 20th century. However, there is still much to be done. 'Green cloaking', mismanaged or poorly planned projects and lack of education, are just some of the obstacles 'real' ecotourism must overcome in order to continue to serve its purpose. Strong leadership, solid foundations and idealistic principles have brought ecotourism to where it is today. We need to recommit to the promise of ecotourism so that future generations can enjoy sustainable tourism experiences which produce sound economic, social and environmental outcomes.

References

Adventure Travel Society, Inc. (1998) Adventure Travel Society Information Packet. The Adventure Travel Society, Englewood Cliffs, New Jersey.

Department for International Development (DFID) (1999) *Tourism and Poverty Elimination: Untapped Potential.* DFID, London.

The Ecotourism Society (1999) Ecotourism Education University and College Fact Sheet. Bennington, Vermont, pp. 2–6.

Gartner, W.C. (1996) *Tourism Development: Principles, Processes and Policies.* Van Nostrand Reinhold, New York.

Honey, M. (1999) *Ecotourism and Sustainable Development: Who Owns Paradise?* Island Press, Washington, DC.

International Institute of Tourism Studies (1999) *Tourism Investment Promotion and Facilitation.* Draft. The George Washington University, Washington.

Jones, C.B. (1998) The new tourism and leisure environment: a discussion paper. Economics Research Associates, San Francisco.

Shrestha, M. (1997) Can Trekkers Help Manaslu? *People and Planet* 6 (4).

Travel Industries of America (TIA) (1999) Fast Facts. TIA web site: http://www.tia.org/press/fastfacts8.stm Washington.

Travel Industries of America (TIA) (2000) Travel Trends. TIA web site: http://www.tia.org/Travel/TravelTrends.asp Washington.

United Nations Development Program (UNDP) (1999) About UNDP. http://www.undp.org.br/pnudmund ING.htm?BI=UNDP New York.

United Nations Environment Program (1998) *Ecolabels in the Tourism Industry.* United Nations Environment Program Industry and Environment, Paris.

Wight, P. (1996) *North American Ecotourists: Market Profile and Trip Characteristics.* Sage Publications, California.

World Tourism Organization (WTO) (1998a) *Global Tourism Forecasts.* World Tourism Organization, Madrid.

World Tourism Organization (WTO) (1998b) *Tourism 2020 Vision.* World Tourism Organization, Madrid, pp. 3–11.

World Tourism Organization (1998c) Ecotourism. In: *WTO News* Jan/Feb Issue I. World Tourism Organization, Madrid.

World Tourism Organization (WTO) (1999) *Tourism Highlights 1999.* World Tourism Organization, Madrid.

World Travel and Tourism Council (WTTC) (2000) *World Travel and Tourism Council Year 2000 TSA Research Tables.* World Travel and Tourism Council, London.

Chapter 5

Ecotourism in the Context of Other Tourism Types

D.B. Weaver
Department of Health, Fitness and Recreation Resources, George Mason University, Manassas, Virginia, USA

Introduction

The term 'ecotourism' has been used in the literature and by the tourism industry since the mid-1980s, and during this time has co-evolved along with a number of related terms, including 'nature-based tourism', 'adventure tourism', 'alternative tourism', 'trekking', 'non-consumptive tourism' and 'sustainable tourism'. Such terms are often used synonymously with ecotourism, leading to a confusion in semantics that hinders the development of tourism as a coherent field of studies with its own logical network of distinct tourism categories. At the same time, ecotourism and such terms as 'mass tourism' and '3S' (sea, sand, sun) tourism are typically seen as being mutually exclusive. The purpose of this chapter is, firstly, to consider the various perspectives on the relationship between ecotourism and the other terms cited above that have emerged since the 1980s. Secondly, as a basis for future discussion, Venn diagrams are used to suggest appropriate relationships between ecotourism and each of these terms.

The chapter begins with the terms that are essentially descriptive, in that they indicate the relevant resource base, attractions and activities used by that sector (i.e. 'nature-based tourism', 'adventure tourism', 'trekking' and '3S tourism'). These are followed by the terms that connote certain values or end results (i.e. 'alternative' or 'mass' tourism, 'sustainable tourism', and 'non-consumptive' or 'consumptive tourism'). 'Mass tourism' has both a descriptive and a value component, but is considered under the second category because of its status as a counterpoint to alternative tourism. The net result of such a structure is to clarify both the descriptive and evaluative dimensions of ecotourism within the context of this constellation of tourism categories.

Nature-based Tourism

Ecotourism was often portrayed in the early literature (e.g. Boo, 1990; Sherman and Dixon, 1991; Whelan, 1991; WTTERC, 1993) as being indistinguishable from 'nature-based', 'nature-oriented' or 'nature' tourism. This tendency was no doubt fostered by the equation of 'nature' with a relatively unspoiled natural environment, and with the close association between ecotourism and that same sort of environment. However, even at that early stage, some analysts such as Ziffer (1989) argued for the differentiation between ecotourism and nature-based tourism on the grounds that the former implied adherence to a particular set of sustainability values. In contrast, nature-based tourism according to Ingram and Durst (1987) is simply leisure travel that involves utilization of the natural resources of an area.

This early recognition of a distinction between ecotourism and nature-based tourism is now more normative in the literature and among practitioners (e.g. Goodwin, 1996; Ceballos-Lascuráin, 1998). Fennell (1999) suggests a growing consensus around the view that ecotourism is but one component within the latter broad category. The breadth of options that is accommodated within the 'nature-based tourism' category is apparent in Goodwin (1996), who includes 3S-type mass tourism, adventure tourism and trekking, as well as ecotourism. To this array could be added hunting and fishing, which are seldom described as forms of ecotourism. More open to debate is the specification of ecotourism's 'territory' within the nature-based tourism realm, which depends upon the extent to which one accepts interactions with nature that are 'soft' and high-volume as a legitimate component of ecotourism (see discussion on mass tourism below). Beyond this question of scale and intensity of interaction, ecotourism is differentiated from other forms of nature-based tourism by such factors as sustainability (that is, nature-based tourism is not necessarily sustainable) and the nature of the interaction between the tourist and the attraction, as discussed below.

The Venn diagram depicted in Fig. 5.1 puts forward the view, with one major qualification, that ecotourism is a subset of nature-based tourism. The fact that ecotourism is not subsumed entirely under this category recognizes that certain past and present cultural attractions may constitute a secondary component of ecotourism. Such a view, for example, is contained in the original definition of ecotourism provided by Ceballos-Lascuráin (in Boo, 1990). The logic of incorporating this cultural component is best demonstrated with respect to the influence and presence of indigenous cultures, wherein the boundary between culture and nature may be perceived as cloudy at best, and arguably even as an entirely artificial construct.

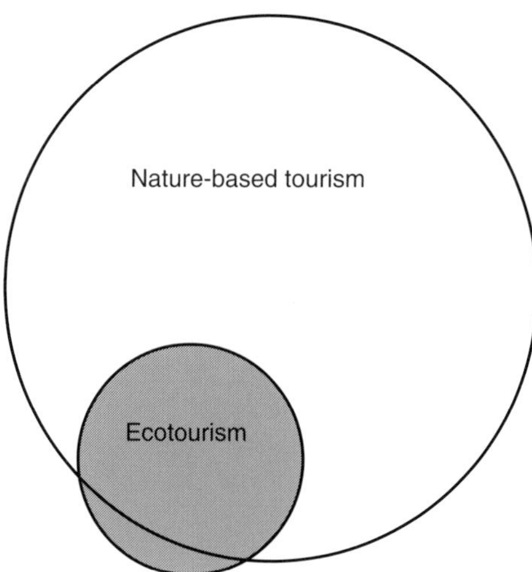

Fig. 5.1. Nature-based tourism and ecotourism.

Adventure Tourism

As with nature-based tourism, the term 'adventure tourism' has sometimes been used interchangeably with ecotourism (e.g. Kutay, 1989). Others, such as the Canadian Tourism Commission, have included 'nature observation' and 'wildlife viewing' under the adventure tourism umbrella (Fennell, 1999). However, more usually, adventure tourism is differentiated by its emphasis on three factors:

- an element of *risk* in the tourism experience (Ewart, 1989; Hall, 1992; Fennell, 1999);
- higher levels of *physical exertion* by the participant (Ewart, 1989); and
- the need for specialized *skills* to facilitate successful participation.

Although there is a tendency in the literature to associate adventure tourism with natural settings (Sung *et al.*, 1996/97), this form of tourism also has a connection with non-nature-based venues. This is demonstrated by tourist activities that involve interaction with remote cultures, or with situations involving conflict (as in *Fielding's the World's Most Dangerous Places* guidebook – Pelton, 1999).

On the nature side, activities typically associated with adventure tourism include white-water rafting, wilderness hiking, skydiving, sea-kayaking, caving, orienteering, mountain climbing, diving and hang-gliding (Sung *et al.*, 1996/97). Aside from the characteristics listed above, the essential factor that tends to place such activities within adventure tourism, and not ecotourism, is the nature of interaction with the surrounding natural environment. Whereas ecotourism places the stress on an educative or appreciative interaction with that environment or some element thereof, adventure tourists are primarily interested in accessing settings that facilitate the desired level of risk and physical exertion. Steep mountain slopes, wilderness settings and white-water rapids, in this view, are valued primarily for the personal thrills and challenges that they offer, not for their scientific interest or associated species of wildlife.

However, there are of course many situations where the 'adventure tourist' *is* equally interested in the above qualities. Similarly, there are many 'ecotourists' who are willing to incur an element of risk in order to access or experience a particular natural attraction. Examples include the wilderness hiker who seeks to find some undisturbed habitat, or the birdwatcher who takes physical risks in order to observe a rare bird of prey in its high mountain habitat. Accordingly, the relationship between ecotourism and adventure tourism is one of partial overlap, as depicted in Fig. 5.2. The reason for providing only a limited scope for overlap is the probability that only the 'harder' and more dedicated forms of ecotourism, which account for only a very small proportion of all ecotourism activity (see Chapter 2), will entail a significant element of risk. If minimal-risk 'soft adventure' activities are allowed, which encompasses just about any interaction with the natural environment, then the overlap will of course be much greater (Canadian Tourism Commission, 1997).

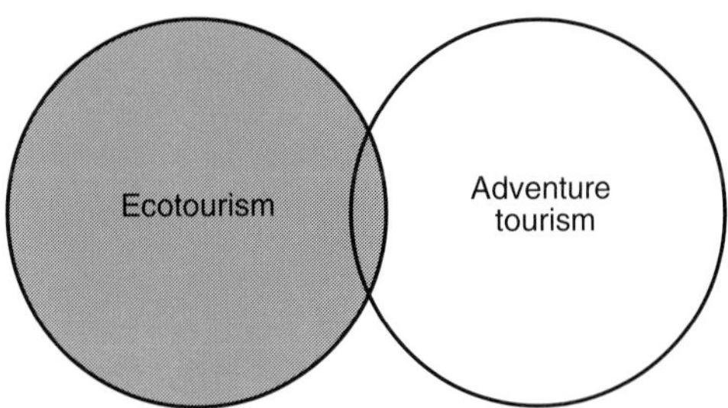

Fig. 5.2. Adventure tourism and ecotourism.

Trekking

'Trekking' is a form of tourism activity usually associated with mountainous venues such as the Himalayas and northern Thailand (Cohen, 1989; Rai and Sundriyal, 1997; Weaver, 1998). A trek typically entails some combination of distance hiking, visits to local villages, adventure experiences (such as using rope bridges to cross streams) and nature/scenery appreciation (Dearden and Harron, 1994). Hence, trekking can legitimately be portrayed as an amalgam that incorporates, in varying degrees, elements of cultural tourism, ecotourism and adventure tourism (see Fig. 5.3). While advocating that ecotourism should maintain its focus as a distinct tourism type, Fennell (1999) acknowledges that such combinations, which he describes as ACE tourism (adventure, cultural, ecotourism), are preferred by some practitioners and marketers over terms such as ecotourism. At least two related factors account for this popularity. Firstly, there are many circumstances where it is virtually impossible to differentiate in any meaningful way among the three components; the distinctions, for example, are not likely to be made by the tourists themselves as they simultaneously engage in cultural, nature-based and adventurous activities. Secondly, such a

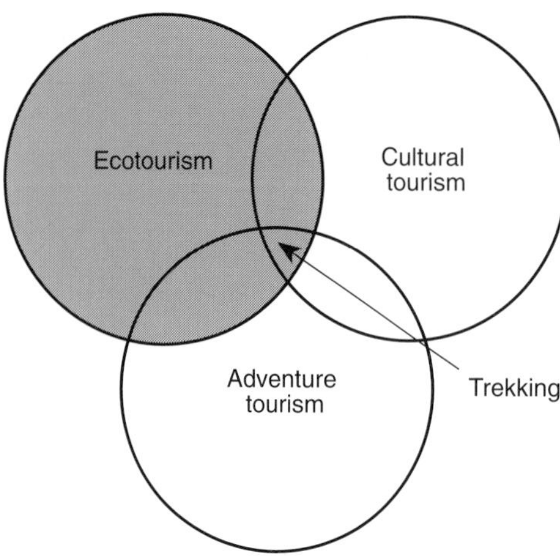

Fig. 5.3. Trekking and ecotourism.

synthesized tourism product may be popular among consumers seeking a diversified and more holistic tourism experience, as opposed to one that is perceived to be mono-directional and overly specialized. Another example of such an acronym, coined as a result of similar concerns, and also applicable to trekking, is the concept of NEAT tourism (nature, ecotourism and adventure tourism) (Ralf Buckley, 1999, personal communication).

3S Tourism

With its resource base of 'sea, sand and sun', 3S tourism clearly fits within the category of nature-based tourism. However, given its affiliation with mass tourism and its emphasis on hedonism, a link is rarely made with ecotourism, which is usually positioned at the opposite end of the tourism spectrum in terms of motivation, impact and scale (see below). Despite this apparent incompatibility, there is one major cluster of activities that account for an area of overlap. This group includes scuba-diving, skin diving, snorkelling, submarine tours and other types of marine observation. Such activities are commonly associated with 3S-oriented vacations, yet can be entirely consistent with widely accepted definitions of ecotourism if they focus on the observation of the marine environment and are pursued in a sustainable manner. This is not to say that marine observation is *necessarily* a form of ecotourism, but rather that there is no inherent grounds for exclusion on the basis of a close association with 3S tourism. The case for a linkage is strengthened by historical trends in the diving sector that have seen a movement away from consumptive activities such as spear-fishing (which is now illegal in many recreational diving venues) toward the passive viewing of marine fauna. As well, marine protected areas are growing in importance as a preferred non-consumptive diving venue (Tabata, 1991; Davis *et al.*, 1996).

The amount of overlap provided in Fig. 5.4 attests to the importance of diving as a rapidly growing component of tourism, and hence of ecotourism (Tabata, 1991). This linkage between the two sectors is not merely a matter of incidental interest, since it leads to the need for a major reassessment of the overall magnitude of ecotourism, its composition, and of its importance within destinations that are not usually associated with the sector. The latter include stereotypical 3S locales such as the Cayman Islands, the British Virgin Islands and the Bahamas (Weaver, 1998), as well as the Egyptian coastal resort of Sharm-el-Sheikh, where divers in the mid-1990s accounted for about 50,000 of 200,000 visitors in total (Hawkins and

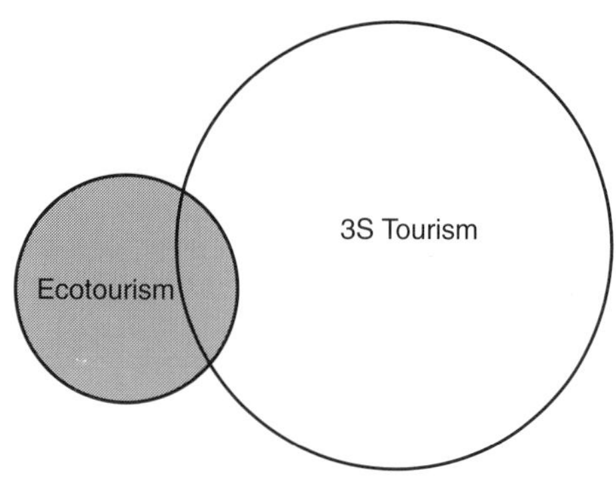

Fig. 5.4. 3S Tourism and ecotourism.

Roberts, 1994). In addition, specialized diving venues such as Saba (Netherlands Antilles) and Palau would need to be more firmly repositioned as specialized ecotourism destinations if the diving that occurs there is recognized as a form of ecotourism.

Alternative Tourism and Mass Tourism

The discussion of ecotourism within the context of alternative tourism cannot be divorced from the context of mass tourism, and hence the two are discussed together in this section. Within the tourism literature, ecotourism is commonly regarded as a form of 'alternative tourism' that places its emphasis on natural rather than cultural attractions (e.g. Goodwin, 1996; Weaver, 1998). To appreciate this value-based linkage, it is necessary to review some of the major conceptual developments that have occurred in the field of tourism studies since the 1960s. In brief, the post-war period was dominated by an 'advocacy platform' that tended to view tourism uncritically as an economic benefit to most destinations. This perspective both condoned and encouraged the emergence of the mass tourism sector (Jafari, 1989). During the 1970s, the advocacy platform was challenged by a 'cautionary platform' that regarded mass tourism in a far more critical light because of its purported negative impacts.

The next logical step in this evolution was the emergence in the early 1980s of the 'adaptancy platform', which introduced the concept of 'alternative tourism' as a more benign alternative to mass tourism (Dernoi, 1981; Holden, 1984; Gonsalves, 1987). Typically, the relationship between the two forms of tourism was depicted in dialectical and dichotomous terms, with alternative tourism clearly being the 'good' option, and mass tourism the 'bad' option (Lanfant and Graburn, 1992; Clarke, 1997) (Table 5.1). With the two types of tourism conceived in this way, ecotourism is logically subsumed under alternative tourism, with the remainder of the latter category being accounted for by mainly socio-cultural forms of alternative tourism such as vacation farms, homestays, feminist travel, etc. Mass tourism stands in relationship to this model as a mutually exclusive category of tourism, separated by what Clarke (1997) describes as a conceptual barrier.

As stated above, this structure, depicted in Fig. 5.5, is the one that has been prevalent in the literature. However, this perspective is now being increasingly challenged in conjunction with changing views about the nature of the relationship between alternative tourism and mass tourism. In effect, this view suggests that the alternative tourism and mass tourism ideal types are merely the poles of a continuum, and that the movement from one to the other is therefore a matter of subtle transition rather than abrupt boundary. This rethinking, which also entails a reassessment of the 'good' and 'bad' values assigned to each of the ideal types (see the section on sustainable tourism below), is coherent with yet another shift in philosophy within tourism studies, to what Jafari (1989) describes as a more objective 'knowledge-based platform'.

As the line between alternative and mass tourism becomes increasingly vague, so too does the boundary between ecotourism and mass tourism. For many academics and practitioners, the concept of 'mass' or 'large-scale ecotourism' is an oxymoron or a betrayal of principle. Yet, while it is logical to assume that a small-scale ecotourism enterprise is more likely to meet the requirements of sustainability in most circumstances, there is no inherent reason why a large-scale product cannot be sustainable, while simultaneously meeting the other criteria assigned to ecotourism. Upon closer examination, one can identify additional grounds for including some portion of ecotourism activity under the category of mass or large-scale tourism:

- 'Soft' ecotourism activity in some destinations is in practice already large enough in volume to warrant the label of mass tourism (without its negative

Table 5.1. Mass tourism and alternative tourism: ideal types as portrayed by the advocacy platform (Weaver, 1998).

Characteristic	Mass tourism	Alternative tourism
Markets		
Segment	Psychocentric–midcentric	Allocentric–midcentric
Volume and mode	High; package tours	Low; individual arrangements
Seasonality	Distinct high and low seasons	No distinct seasonality
Origins	A few dominant markets	No dominant markets
Attractions		
Emphasis	Highly commercialized	Moderately commercialized
Character	Generic, 'contrived'	Area specific, 'authentic'
Orientation	Tourists only or mainly	Tourists and locals
Accommodation		
Size	Large scale	Small scale
Spatial pattern	Concentrated in 'tourist areas'	Dispersed throughout area
Density	High density	Low density
Architecture	'International' style; obtrusive, non-sympathetic	Vernacular style, unobtrusive, complementary
Ownership	Non-local, large corporations	Local, small businesses
Economic status		
Role of tourism	Dominates local economy	Complements existing activity
Linkages	Mainly external	Mainly internal
Leakages	Extensive	Minimal
Multiplier effect	Low	High
Regulation		
Control	Non-local private sector	Local 'community'
Amount	Minimal; to facilitate private sector	Extensive; to minimize local negative impacts
Ideology	Free market forces	Public intervention
Emphasis	Economic growth, profits; sector-specific	Community stability and well-being; integrated, holistic
Timeframe	Short term	Long term

value connotations), while remaining coherent with the principles of sustainability. This situation is evident in many popular protected areas, where stringent zoning regulations and site-hardening tactics facilitate a high but apparently sustainable volume of visitation (see Chapter 18). The argument has even been made that higher volumes of visitation create the economies of scale and revenue flow that justify the implementation of sophisticated management techniques that facilitate sustainable outcomes. In addition, they help to position ecotourism as a resource stakeholder capable of lobbying government on a more equal foothold with traditional resource users such as agriculture, mining and logging.

- Many if not most 'soft' ecotourism participants *are* mass tourists engaged in such activities as part of a broader, multi-purpose vacation that often places the emphasis in the 3S realm. Most ecotourism activity in the high profile ecotourism destinations of Costa Rica and Kenya, for example, fits into this category (Weaver, 1999). Moreover, it appears that the possibility of accessing both the well-serviced beach-based

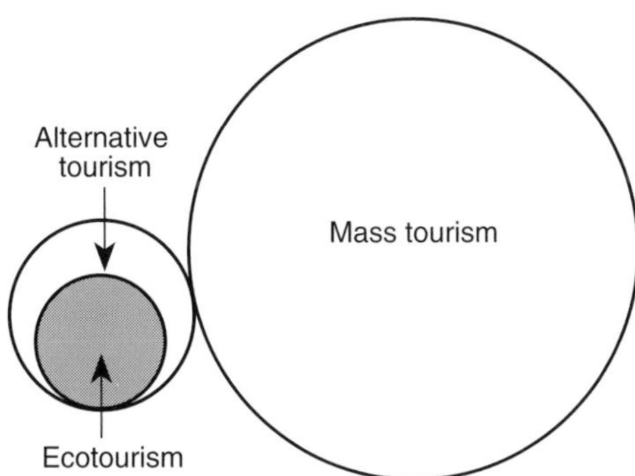

Fig. 5.5. Alternative tourism and ecotourism: conventional approach.

> resorts and the natural attractions of well-known protected areas is a primary motivation for these tourists to visit those two countries, rather than countries that are wildlife-rich but service-poor.

This evidence suggests that the emerging relationship between mass tourism and ecotourism may be moving in the direction of synthesis, convergence and symbiosis. As shown in Fig. 5.6, ecotourism can serve to strengthen the mass tourism product by offering opportunities for diversification that are especially attractive in the light of the increased 'greening' of the tourist market. As well, ecotourism, with its roots in the cautionary and adaptancy platforms, can help to impart an ethos of sustainability and environmental awareness to the mainstream sector, thus assisting its movement in the direction of sustainability (Ayala, 1996; Clarke, 1997). In turn, mass tourism supplies a large market of soft ecotourists that helps to position ecotourism as an important stakeholder capable of lobbying on an equal footing with stakeholders in other sectors such as agriculture and logging. Furthermore, as mentioned above, the mass tourism industry can introduce sophisticated environmental management strategies to ecotourism that Clarke (1997) suggests are beyond the capability of most traditional small-scale ecotourism operations.

The relationships fostered by this less conventional perspective are depicted in Fig. 5.7. All tourism is depicted by a single circle, within which smaller-scale, alternative tourism-type products constitute one relatively small component that gradually gives way to large-scale tourism (the dotted line represents the presence of a transition zone rather than a sharp boundary). Ecotourism is positioned as a diverse activity that overlaps both the alternative and mass tourism components of the circle, thereby encompassing all options from the lone wilderness hiker (hard ecotourism) to the busload of resort patrons engaged in a half-day excursion to a local wildlife interpretation centre (soft ecotourism).

Without doubt, this association between mass tourism and ecotourism is controversial, and is a linkage not likely to be universally embraced by ecotourism stakeholders. One counter-argument holds that the disparity in power between the two sectors will mean that the influence of mass tourism over ecotourism is likely to

Fig. 5.6. Converging and symbiotic relationship between ecotourism and mass tourism.

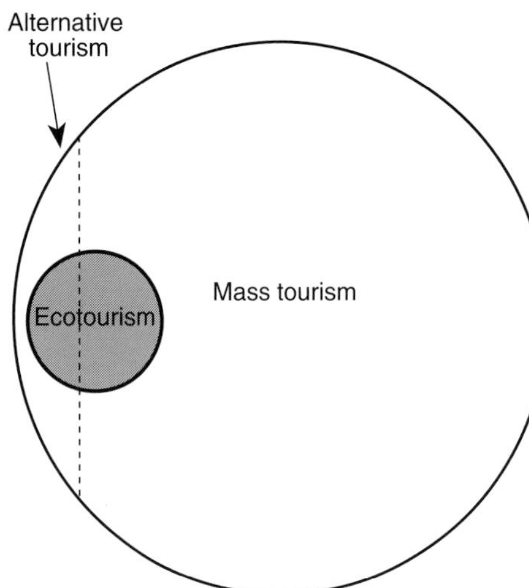

Fig. 5.7. Alternative tourism, mass tourism and ecotourism: emergent approach.

be much greater than the reverse situation, and that mass tourism will therefore effectively appropriate ecotourism for its own purposes. This, however, assumes a cautionary platform model of large-scale tourism as a sector guided by sinister intentions and essentially unaffected by the positive dimensions of the sustainability paradigm. It would follow from this assumption that the outcomes of such an annexation would be unsustainable, and therefore ecotourism could not become established on the mass tourism side of the tourism spectrum.

Sustainable Tourism

The concept of 'sustainable tourism' has proven to be just as or even more contentious than ecotourism or alternative tourism since its introduction in the late 1980s. While the idea of sustainability had been alluded to much earlier in the tourism literature, the appearance of the term 'sustainable tourism' itself followed the release of the so-called 'Brundtland Report' in 1987, which popularized the concept of 'sustainable development' (WCED, 1987). In emulation of its namesake, sustainable tourism was and still is broadly conceived as tourism that does not threaten the economic, social, cultural or environmental integrity of the tourist destination over the long term (Butler, 1993). Since no one is likely to disagree in principle with this goal, it is not surprising that sustainable tourism has emerged as the 'great imperative' of the global tourism sector.

In its initial conception, sustainable tourism was commonly perceived as being synonymous with alternative tourism, in keeping with the philosophy of the cautionary and adaptancy platforms (Clarke, 1997). Accordingly, ecotourism, as a subset of alternative tourism, was also regarded as a subset of sustainable tourism. However, how is this set of relationships affected once ecotourism is extended into the realm of large-scale or mass tourism, as depicted in Fig. 5.7? Under the cautionary and adaptancy perspectives, such an extension is a contradiction, as such activity would cease to meet the criterion of sustainability. The knowledge-based platform, in contrast, de-emphasizes the relationship between scale and sustainability; small-scale operations can be bad or good, depending on the way that they are managed, while the same can be said for large-scale operations (Weaver and Lawton, 1999). The core criterion of sustainability can therefore be retained when ecotourism is extended into the mass tourism arena, and ecotourism remains a subset of sustainable tourism (Fig. 5.8), since sustainability is one of the core ecotourism criteria. The actual positioning of the sustainable tourism circle in this figure, however, concedes that alternative tourism is at present more likely than mass tourism to be sustainable, with the qualification that the entire issue of sustainability is fraught with uncertainty.

Consumptive and Non-consumptive Tourism

The distinction between 'consumptive' and 'non-consumptive' activity has long been recognized in the tourism and outdoor recreation literature. Non-consumptive activi-

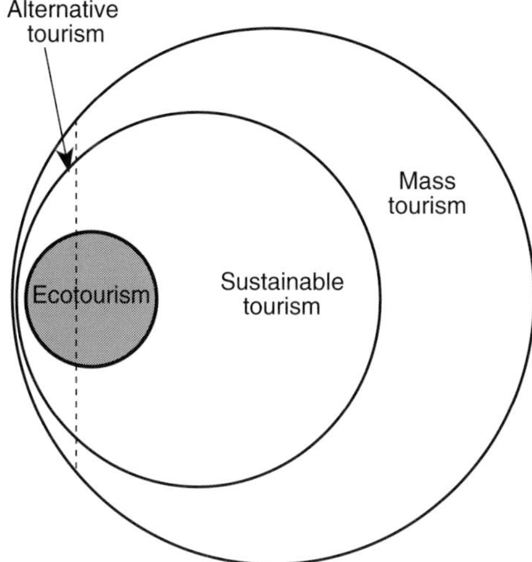

Fig. 5.8. Sustainable tourism and ecotourism.

ties are commonly perceived as those that offer *experiences* to the market, while consumptive activities offer tangible *products* (Applegate and Clark, 1987). Activities such as hunting and fishing (except perhaps for 'catch and release' fishing) are commonly identified as being consumptive, while ecotourism-related activities such as birdwatching are usually perceived as experiential, and hence non-consumptive.

Vaske *et al.* (1982), among others, have questioned the merit of this dichotomous approach, and have suggested instead that activities fall along a consumptive–non-consumptive continuum, and that all activities actually incorporate elements of both. This is illustrated by the observation that most hunting excursions actually end without any kills being achieved, and that the aesthetic experiences of being in a rural or semi-wild environment are rated just as highly by many hunters as the hunting/killing element itself. Conversely, ecotourism 'experiences' potentially involve several forms of 'consumption', as follows:

- the consumption of fossil fuels in the process of transit, and when using vehicles or boats in the process of wildlife observation; also the consumption of food and other products for the duration of the ecotourism experience;
- the purchase of material souvenirs, which require at least some degree of resource consumption;
- the gradual and imperceptible deterioration of the environment through erosion, trampling of vegetation and other stresses incurred during the process of wildlife observation or in the establishment of facilities; these can be described as a form of unintended resource consumption;
- the keeping of checklists of wildlife species as a type of score-keeping or consumption; that is, once a species has been sighted, it is checked off and is no longer sought, and is thus 'consumed'.

On the other hand, Duffus and Dearden (1990) suggest that the consumptive–non-consumptive distinction is useful, since there is a fundamental difference between activities that deliberately seek to destroy and remove an organism and those that do not. In this view, it would still be legitimate to describe ecotourism as an essentially non-consumptive activity, despite the four facets of 'consumption' bulleted above. Figure 5.9 positions ecotourism in a

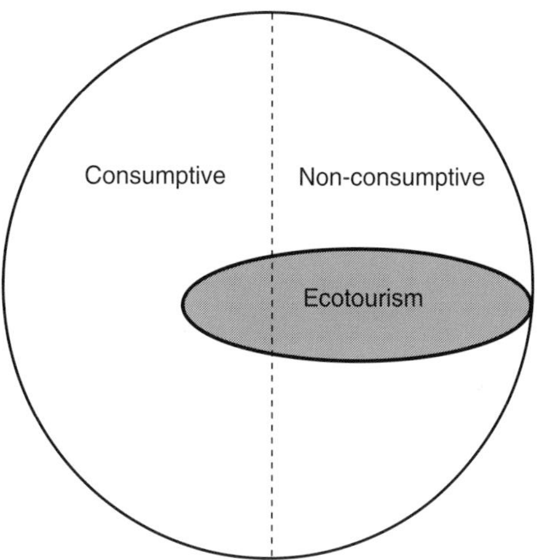

Fig. 5.9. Consumptive/non-consumptive tourism and ecotourism.

way that takes into account both of these perspectives. The larger circle represents all tourism, and incorporates the continuum by positioning consumptive and non-consumptive tourism at either side. Ecotourism, which is only a relatively small proportion of all tourism, is primarily non-consumptive, but is extended over into the consumptive side to recognize these consumptive aspects as well. However, it does not extend as far into this side as the same territory occupied by hunting or fishing.

Conclusions

This chapter has attempted to clarify the relationship between ecotourism and other types of tourism that are commonly associated or disassociated with ecotourism. In summing up these relationships, two distinct patterns are apparent. With all of the descriptive terms (i.e. 'nature-based', 'adventure', 'trekking' and '3S'), the association with ecotourism as depicted in the Venn diagrams is one of overlap. The overlaps allowed for in this chapter depend upon the specific way in which each term, including ecotourism, is defined. Obviously, a different conception of ecotourism, such as more of an emphasis on its 'hard' component or more accommodation of cultural attractions, would yield a different measure of overlap in the relevant case.

For the value-based terms, the relationships are more complex. During the era when the cautionary and adaptancy platforms were dominant, ecotourism was unambiguously affiliated with alternative and non-consumptive tourism. These affiliations survive under the knowledge-based platform. However, it is argued here that ecotourism can now also straddle the boundary with mass tourism and consumptive tourism, respectively, while still retaining its core defining characteristics. This is because of the new platform's tendency to position tourism activities along continuums rather than into dichotomies, and to remove the negative connotations from mass tourism in particular. The only situation in which ecotourism is still considered to be subsumed entirely under another term is with respect to sustainable tourism. Coherent with the disassociation between scale and value, both the alternative and mass tourism components of ecotourism fall under the sustainable tourism umbrella.

References

Applegate, J. and Clark, K. (1987) Satisfaction levels of birdwatchers: an observation on the consumptive–nonconsumptive continuum. *Leisure Sciences* 9, 129–134.

Ayala, H. (1996) Resort ecotourism: a paradigm for the 21st century. *Cornell Hotel and Restaurant Administration Quarterly* 37(5), 46–51.

Boo, E. (1990) *Ecotourism: the Potentials and Pitfalls*, Vol. 1. World Wildlife Fund, Washington, DC.

Butler, R.W. (1993) Tourism – an evolutionary perspective. In: Nelson, J.G., Butler, R.W. and Wall, G. (eds) *Tourism and Sustainable Development: Monitoring, Planning, Managing*. Department of Geography Publication Series 37, University of Waterloo, Waterloo, Canada, pp. 29–43.

Canadian Tourism Commission (1997) *Adventure Travel and Ecotourism: the Challenge Ahead*. Canadian Tourism Commission, Ottawa, Canada.

Ceballos-Lascuráin, H. (1998) Introduction. In: Lindberg, K., Epler Wood, M. and Engeldrum, D. (eds) *Ecotourism: a Guide for Planners and Managers*, Vol. 2. Ecotourism Society, North Bennington, Vermont, pp. 7–10.

Cohen, E. (1989) 'Primitive and remote' hill tribe trekking in Thailand. *Annals of Tourism Research* 16, 30–61.

Clarke, J. (1997) A framework of approaches to sustainable tourism. *Journal of Sustainable Tourism* 5, 224–233.

Davis, D., Banks, S. and Davey, G. (1996) Aspects of recreational scuba diving in Australia. In: Prosser, G. (ed.) *Tourism and Hospitality Research: Australian and International Perspectives.*

Proceedings from the Australian Tourism and Hospitality Research Conference, 1996. Bureau of Tourism Research, Canberra, pp. 455–465.

Dearden, P. and Harron, S. (1994) Alternative tourism and adaptive change. *Annals of Tourism Research* 21, 81–102.

Dernoi, L. (1981) Alternative tourism: towards a new style in North–South relations. *International Journal of Tourism Management* 2, 253–264.

Duffus, D. and Dearden, P. (1990) Non-consumptive wildlife-oriented recreation: a conceptual framework. *Biological Conservation* 53, 213–231.

Ewart, A. (1989) *Outdoor Adventure Pursuits: Foundations, Models, and Theories.* Publishing Horizons, Columbus, Ohio, USA.

Fennell, D.A. (1999) *Ecotourism: an Introduction.* Routledge, London.

Gonsalves, P. (1987) Alternative tourism – the evolution of a concept and establishment of a network. *Tourism Recreation Review* 12(2), 9–12.

Goodwin, H. (1996) In pursuit of ecotourism. *Biodiversity and Conservation* 5, 277–291.

Hall, C.M. (1992) Adventure, sport and health tourism. In: Weiler, B. and Hall, C.M. (eds) *Special Interest Tourism.* Belhaven Press, London, pp. 141–158.

Hawkins, J. and Roberts, C. (1994) The growth of coastal tourism in the Red Sea: present and future effects on coral reefs. *Ambio* 23, 503–508.

Holden, P. (ed.) (1984) *Alternative Tourism With a Focus on Asia.* Ecumenical Council on Third World Tourism, Bangkok.

Ingram, D. and Durst, P. (1987) *Nature Oriented Travel to Developing Countries.* FPEI Working Paper No. 28. Southeastern Center for Forest Economics Research, Research Triangle Park, North Carolina, USA.

Jafari, J. (1989) An English language literature review. In: Bystrzanowski, J. (ed.) *Tourism as a Factor of Change: a Sociocultural Study.* Centre for Research and Documentation in Social Sciences, Vienna, pp. 17–60.

Kutay, K. (1989) The new ethic in adventure travel. *Buzzworm: the Environmental Journal* 1(4), 31–36.

Lanfant, M.-F. and Graburn, N. (1992) International tourism reconsidered: the principle of the alternative. In: Smith, V.L. and Eadington, W. (eds) *Tourism Alternatives: Potentials and Problems in the Development of Tourism.* University of Pennsylvania Press, Philadelphia, pp. 88–112.

Pelton, R. (1999) *Fielding's the World's Most Dangerous Places*, 4th edn. Fielding Worldwide, Redondo Beach, California, USA.

Rai, S. and Sundriyal, R. (1997) Tourism and biodiversity conservation: the Sikkim Himalaya. *Ambio* 26, 235–242.

Sherman, P. and Dixon, J. (1991) The economics of nature tourism: determining if it pays. In: Whelan, T. (ed.) *Nature Tourism: Managing for the Environment.* Island Press, Washington, DC, pp. 89–131.

Sung, H., Morrison, A. and O'Leary, J. (1996/97) Definition of adventure travel: conceptual framework for empirical application from the providers' perspective. *Asia Pacific Journal of Tourism Research* 1(2), 47–67.

Tabata, R. (1991) Dive travel: implications for resource management and tourism development. In: Kusler, J. (ed.) *Ecotourism and Resource Conservation: a Collection of Papers.* Omnipress, Madison, Wisconsin, pp. 677–693.

Vaske, J., Donnelly, M., Heberlein, T. and Shelby, B. (1982) Differences in reported satisfaction ratings by consumptive and nonconsumptive recreationists. *Journal of Leisure Research* 14, 195–206.

WCED (1987) *Our Common Future.* Oxford University Press, Oxford.

Weaver, D. (1998) *Ecotourism in the Less Developed World.* CAB International, Wallingford, UK.

Weaver, D. (1999) Magnitude of ecotourism in Costa Rica and Kenya. *Annals of Tourism Research* 26, 792–816.

Weaver, D. and Lawton, L. (1999) *Sustainable Tourism: a Critical Analysis.* Research Report 1. CRC for Sustainable Tourism, Gold Coast, Australia.

Whelan, T. (ed.) (1991) *Nature Tourism: Managing for the Environment.* Island Press, Washington, DC.

WTTERC (1993) *World Travel and Tourism Environment Review 1993.* World Travel and Tourism Environment Research Centre, Oxford.

Ziffer, K. (1989) *Ecotourism: the Uneasy Alliance.* Conservation International, Washington, DC.

Section 2

A Regional Survey by Continent

E. Cater
Department of Geography, University of Reading, Reading, UK

Scale and Circumstance

A global overview of the state of the art with regard to ecotourism in the world's regions could not only result in overgeneralization, but also prove to be a counter-productive exercise. This is because the regions, as can be seen from the following chapters, are both products of, and produce, considerable difference and diversity. This is evident at varying spatial levels.

First, there are marked differences in approaches to ecotourism between continents. With relatively high population densities, a prevalence of humanized landscapes and high tourist visitation levels in Europe, described by Blangy and Vautier in Chapter 10, the focus on sustainable tourism rather than on ecotourism is understandable. In contrast, Dowling, in Chapter 9, describes the solid emphasis on ecotourism *per se* in Australia, where a strong market demand, based on rich and diverse natural landscapes, is coupled with considerable impetus from the Federal Government. There is also a solid foundation of demand and supply for ecotourism in North America, as described by Fennell in Chapter 7. Lew describes, in Chapter 8, a strong emphasis on ecotourism development in Southeast Asia, where countries capitalize on their comparative advantage of natural ecosystems in a markedly different way from those of temperate latitudes to attract Western ecotourists.

Second, within individual continents, significant differences exist between countries and regions. Witness the divide between the north and south of South America, described by Weaver and Schlüter in Chapter 11; in insular Southeast Asia compared with central Asia (Chapter 8); and the relative paucity of ecotourism in North and West Africa in contrast to the eastern and southern part of the continent, as described by Dieke in Chapter 6. This last scenario occurs despite the fact that one of the first sustainable ventures on the African continent, the Lower Casamance 'Tourism for Discovery' project is in Senegal, West Africa (Eber, 1992).

Third, significant differences also occur within individual countries. In coastal locations and near international gateways, soft ecotourism frequently acts as an adjunct to conventional mass tourism. At the opposite end of the spectrum, interior mountains and other peripheral locations often give rise to a harder form of ecotourism product, as illustrated by Dominica in Chapter 11.

Thus, it is important to avoid generalization, and to consider the specific attributes of a locality. As Dowling suggests in the case of Oceania, circumstances dictate that not all natural sites lend themselves to ecotourism. The reasons for this can be ascribed to contingencies of place (Williams and Shaw, 1998). These contingencies are shaped by economic, sociocultural, political, ecological, institutional, and technical forces that are exogenous and endogenous, as well as dynamic. A consideration of the various macro-regions covered in this section under these categories provides a useful framework for highlighting similarities and differences with regard to ecotourism experience across the globe.

Economic Forces

It is vital that ecotourism is set into context with regard to different levels, sectors and interests that will condition sustainability. With respect to the spatial hierarchy, it is interesting to note how the international community has become increasingly engaged with ecotourism, not only as a form of sustainable tourism, but also for its role in contributing towards sustainable development in general. The International Resources Group, for example, prepared a report for USAID on the potential role of ecotourism as a viable alternative for the sustainable management of natural resources in Africa. They suggest that, while it is probably not appropriate for overseas development assistance to support tourism in general, ecotourism is different (IRG, 1992). The UK Department for International Development (DFID) also adopts this train of thought, and developed an agenda for action in 1999, *Changing the Nature of Tourism* (DFID, 1999). Similarly, Chapter 10 describes how LEADER funds for economic restructuring in Central and Eastern Europe are targeted at rural tourism development in a strategy to restore balance between different levels of development among the regions of Europe.

At the national level, most tourism authorities continue to attach a high priority to conventional tourism development. As outlined in Chapter 11, Costa Rica persists in giving incentives to mass tourism despite its high profile as an ecotourism destination. However, an emerging and contrasting tendency sees some 3S destinations, such as Turkey, Spain and Cyprus, attempting to devolve tourism away from the coastal magnets to more remote, rural, locations.

Ecotourism's contribution to rural livelihoods is an important consideration at the local level. Dowling describes the integration of ecotourism and village-based community tourism in Fiji, but stresses that ecotourism can only complement, not replace, other forms of tourism. Equally, it is imperative that it should be regarded as a complementary and supplementary, not alternative, activity to agriculture.

Sectoral conflicts may compromise the success, if not the very existence of ecotourism. The impact of the oil and fishing industries on whale-watching in Patagonia and destructive logging practices in Indonesia, Haiti and the Dominican Republic are all illustrative examples. In addition, the inclusion of various stakeholders' interests is a much-vaunted principle of sustainable ecotourism, but Fennell suggests that even this has problems, since interests frequently diverge. In Saskatchewan, for example, the inclusion of several stakeholder groups has created friction between economic and environmental interests.

Socio-cultural Factors

The following Chapters point to sociocultural resources as an important component of ecotourism. In many locations so-called natural landscapes are often cultural: the product of many generations of traditional land management. Much of Europe, the terraces of the Himalayan foothills and the rice terraces of Luzon, Philippines all substantiate this fact. The indivisibility of nature and culture is also illustrated by the principle of free access to nature enshrined in northern European cultures. Blangy and Vautier describe how the Danish government buys up land expressly for that purpose.

Ecotourism, by definition, should be socio-culturally responsible, but there is a vexed issue of ethnocentricity. Not only do views of the environment differ between hosts and guests (Dowling cites the Maori of New Zealand), but also between domestic and foreign tourists. As the developing world modernizes, intraregional and domestic tourism proliferates. The agenda of these tourists, as Lew describes in the case of Southeast Asia, may be markedly different from those of visiting ecotourists. Wherever these two groups converge in natural areas, the needs of one may prejudice those of the other.

Political Forces

While stable political regimes, such as that of Belize, are undoubted assets for ecotourism development, the corollary is that instability militates against tourism in general, and ecotourism in particular. This is because attractions are often in remote, relatively unpoliced locations where foreign tourists become pawns in the power struggles of factional groups headquartered away from the seat of government. Examples of this ilk are the threats posed by the *Sendero Luminoso* in Peru and, more recently, by the massacre of foreign tourists in 1999 at Bwindi Impenetrable Forest gorilla reserve in western Uganda. Another important political criterion is that of will. Unless protected areas receive appropriate legislative backing, committed enforcement, and adequate funding, they remain little more than 'paper parks', as described by Weaver and Schlüter in the context of Latin America, and by Lew in the context of Asia.

Ecological Considerations

The consideration of protected areas raises the issue of environmental protection and conservation. Throughout the world's regions there is a diversity of natural and cultural resource conservation policies. In Asia, Lew describes how these range from economic to environmental priorities. In China conservation efforts often have significant economic goals, local governments being involved in resource development as much as resource protection. India, however, has a long history of conservation for the sake of the environment. Where protected areas do exist, the chapters describe a considerable range first in their percentage of a country's surface area, and second, in the level of protection afforded. Blangy and Vautier discuss this with regard to Europe. In the UK, for example, most protected areas fall in IUCN Category V.

Another debate raised is that of consumptive versus non-consumptive use. Fennell describes the inclusion of fishing as a form of ecotourism in Canada as eco-opportunism, while Dieke suggests that, although the licensed hunting of wildlife under the CAMPFIRE initiative in Zimbabwe benefits local communities, it can only be partially related to ecotourism. The financing of conservation also continues to be of concern, particularly in countries with beleagered, developing economies. The need for enhanced environmental stewardship, illustrated through underfunding of conservation, is described in several of the chapters. Not only is ecotourism an alternative to environmentally destructive activities such as logging or mining in Southeast Asia, Fiji, and Central and South America, but it can also provide revenue for conservation.

The relationship between ecotourism and the environment is two-way. Weaver and Schlüter point to the impact of environmental disasters in the Caribbean, and Dowling describes their omnipresent threat in the Pacific islands. The damage done to Mozambique's fledgling tourism industry by the massive flooding during 2000, although trivial compared with humanitarian concerns, must be considerable.

Institutional Factors

The myriad of institutions concerned with ecotourism at different levels, representing varied interests, are described in the chapters. A related factor of interest is expatriate involvement in ecotourism, particularly in the developing countries. Critics might point to eco-imperialism, but private capital fills a gap in capital-scarce situations. There are an increasing number of private ecotourism initiatives, described in Central America by Weaver and Schlüter and in southern and East Africa by Dieke. Furthermore, it is not simply a Western core versus under-developed periphery scenario. The South African based Conservation Corporation, for example, now has ecotourism partnerships throughout east and southern Africa (Christ, 1998).

Technical Forces

Eco-architecture and more environmentally sound infrastructure, described in Chapters 20 and 23, should be a feature of ecotourism. However, in many parts of the world ecolodges are few and far between, even in the much lauded ecotourism destination of Costa Rica. Ecotourism is the ideal candidate for appropriate technology. Unfortunately, this does not always come cheap. Dowling describes the upmarket ecolodge on Vanua Levu as one example of elite or luxury ecotourism.

Accessibility is another technical factor to consider. It is no accident that prime ecotourism sites develop close to existing urban nodes, such as Belém and Manaus in Brazil, and in countries readily accessible to the main tourist generating countries, such as Belize and Costa Rica relative to the USA. Other popular ecotourism destinations are accessed from popular stopover points such as the natural areas of Thailand and Malaysia. Conversely, a challenge is posed by the remoteness of the Pacific islands and the high cost of accessing these destinations.

Conclusion

It is evident from this discussion, more amply illustrated in the following chapters, that ecotourism must be set into context, and that the context is regional as well as site-specific. As well as common interests, there are likely to be conflicts. A better understanding of areas of discord, as well as concord, is essential, so that negative links can be broken, and positive links built upon (World Bank, 1992; Cater, 1995). Only then will ecotourism across the globe begin to live up to the reputation of sustainability that precedes it.

References

Cater, E. (1995) Environmental contradictions in sustainable tourism. *The Geographical Journal* 161(1), 21–28.

Christ, C. (1998) Taking ecotourism to the next level: a look at private sector involvement with local communities. In: Lindberg, K., Epler-Wood, M. and Engeldrum, D. (eds) *Ecotourism: a Guide for Planners and Managers*, Vol. 2. The Ecotourism Society, North Bennington, Vermont, pp. 183–195.

DFID (1999) *Changing the Nature of Tourism*. DFID, London.

Eber, S. (ed.) (1992) *Beyond the Green Horizon: Principles for sustainable tourism*. Godalming, WWF, UK.

IRG (1992) *Ecotourism: a Viable Alternative for Sustainable Management of Natural Resources in Africa*. Submitted to Agency for International Development Bureau for Africa, Washington, DC.

Williams, A. and Shaw, G. (1994) *Critical Issues in Tourism: a Geographical Perspective*. Blackwell, Oxford.

Williams, A.M. and Shaw, G. (1998) Tourism and the environment: sustainability and economic restructuring. In: Hall, C.M. and Lew, A.A. (eds) *Sustainable Tourism*. Addison Wesley Longman, Harlow.

World Bank (1992) *World Development Report*. World Bank, Washington, DC.

Chapter 6

Kenya and South Africa

P.U.C. Dieke
University of Strathclyde, The Scottish Hotel School, Glasgow, UK

Introduction

The purpose of this chapter is to provide a comparative analysis of ecotourism as a form of sustainable development in the context of Kenya and South Africa. These two African countries exemplify characteristics of well-established tourism destinations generally but they are also probably the highest profile ecotourism destination countries in the region. The chapter first briefly reviews Africa's place in the broader tourism perspective, then considers the situation in the two main case studies. An insight into wider continental ecotourism development experiences is then provided, by examining, in less detail, other African countries, and especially those in the same sub-region as Kenya and South Africa. This chapter also considers various management and policy-related issues that affect the status of ecotourism in the region.

Africa and International Tourism

Africa comprises 53 countries categorized into five geographical sub-regions: central, eastern, northern, southern and western. Like other developing regions, Africa is a continent of considerable cultural, economic, geographic, political and social diversity. Perhaps part of the variation has to do with differences in their colonial experience, following the balkanization of the continent into arbitrary nation-states to meet the needs of the European governing powers (see Afigbo *et al.*, 1992).

A recent study (Dieke, 2000) has shown that there is clearly a wide variety of different types of tourism available in the region: from safari tourism (e.g. wildlife, desert), beach tourism and 'roots' tourism to marine tourism. Some others include cultural/heritage and archaeological tourism, ethnic tourism, 'overland' tourism (or desert, as noted) and, perhaps to a lesser degree, sex tourism. In terms of global context, Africa received about 3.6% of all international stayover arrivals in 1998 (WTO, 1999). Within Africa, the northern sub-region had the highest share of traffic (34.6%) and revenue (33%), followed in descending order by southern Africa, eastern Africa, western Africa, and middle Africa (Table 6.1). Almost 40% of all visits to Africa originate in the region, with Europe accounting for more than one-third of total arrivals.

Table 6.2a indicates the most visited destinations in Africa in 1998. Of the 20 countries profiled here, South Africa was the most favoured destination, taking 24% of total traffic, followed by two northern

Table 6.1. Tourism trends by sub-regions, 1995–1998 (WTO, 1999, pp. 4–29).

Sub-regions	Tourist arrivals (000s) 1998	% change-over 1997	Market share of total Africa (%)		Tourist receipts (US$ million) 1998	% change-over 1997	Market share of total Africa (%)	
			1995	1998			1995	1998
Eastern	5,761	7.70	21.7	23.1	2,426	5.75	23.4	25.4
Middle	483	7.81	1.4	1.9	82	5.13	1.7	0.9
Northern	8,623	7.79	38.7	34.6	3,176	9.90	38.1	33.3
Southern	7,671	7.94	29.9	30.8	2,950	2.54	28.1	30.9
Western	2,365	4.97	8.3	9.5	917	4.23	8.5	9.6
Total Africa	24,903	7.50	100.0	100.0	9,551	5.90	100.0	100.0

Countries of the sub-regions:
Eastern: Burundi, The Comoros, Djibouti, Eritrea, Ethiopia, Kenya, Madagascar, Malawi, Mauritius, Mozambique, Réunion, Seychelles, Somalia, Uganda, Tanzania, Zambia, Zimbabwe.
Middle: Angola, Cameroon, Central Africa Republic, Chad, Congo (Brazzaville), Equatorial Guinea, Gabon, Sao Tomé and Principé, Democratic Republic of Congo (Kinshasa).
Northern: Algeria, Morocco, Sudan, Tunisia.
Southern: Botswana, Lesotho, Namibia, South Africa, Swaziland.
Western: Benin, Burkina Faso, Cape Verde, Côte d'Ivoire, The Gambia, Ghana, Guinea, Guinea Bissau, Liberia, Mauritania, Niger, Nigeria, Senegal, Sierra Leone, Togo.

Table 6.2a. Top 20 tourism destinations in Africa, 1998 (International tourist arrivals, 000s) (WTO, 1999, p. 31).

Rank				Arrivals (000)			
1990	1995	1998	Country	Estimated figures up to 11 Jan 1999	Figures received after 11 Jan 1999	% change 1998/97	% of total 1998
4	1	1	South Africa	5,981	5,981	10.0	24.0
2	2	2	Tunisia	4,700	4,471	10.7	18.9
1	3	3	Morocco	3,241	3,243	5.6	13.0
6	4	4	Zimbabwe	1,600	1,600	7.0	6.4
5	5	5	Kenya	1,062	951	−5.0	4.3
7	7	6	Botswana	740	740	0.8	3.0
3	8	7	Algeria	648	678	6.8	2.6
13	6	8	Nigeria	640	640	4.7	2.6
8	9	9	Mauritius	570	558	4.1	2.3
—	10	10	Namibia	510	560	11.6	2.0
15	15	11	Tanzania	447	447	28.8	1.8
—	11	12	Eritrea	414	414	1.0	1.7
17	20	13	Zambia	382	362	6.2	1.5
11	12	14	Réunion	377	377	1.9	1.5
16	14	15	Ghana	335	335	3.1	1.3
9	13	16	Swaziland	325	325	0.9	1.3
10	16	17	Senegal	309	332	6.1	1.2
12	18	18	Côte d'Ivoire	302	301	9.9	1.2
30	18	19	Uganda	238	238	4.8	1.0
18	17	20	Malawi	215	205	5.7	0.9
Total				23,036	—	7.6	92.5
Total Africa				24,903	—	7.5	100.0

Table 6.2b. Top 20 tourism earners in Africa, 1998 (International tourist receipts, US$ million) (WTO, 1999, p. 33).

Rank				Receipts (US$ million)			
1990	1995	1998	Country	Estimated figures up to 11 Jan 1999	Figures received after 11 Jan 1999	% change 1998/97	% of total 1998
2	1	1	South Africa	2366	2366	3.0	24.8
1	3	2	Morocco	1600	1600	10.9	16.8
3	2	3	Tunisia	1550	1550	8.9	16.2
5	4	4	Mauritius	502	503	3.7	5.3
11	7	5	Tanzania	431	431	9.9	4.5
4	5	6	Kenya	400	358	−5.0	4.2
9	6	7	Namibia	339	339	0.9	3.5
10	8	8	Ghana	274	274	3.0	2.9
—	9	9	Réunion	250	250	0.4	2.6
13	11	10	Zimbabwe	246	246	7.0	2.6
8	10	11	Botswana	185	185	0.5	1.9
6	12	12	Senegal	165	161	5.2	1.7
33	14	13	Uganda	142	142	5.2	1.5
25	19	14	Nigeria	124	124	5.1	1.3
7	13	15	Seychelles	120	111	−9.0	1.3
16	15	16	Côte d'Ivoire	97	97	10.2	1.0
18	21	17	Zambia	90	75	0.0	0.9
—	17	18	Eritrea	75	75	0.0	0.8
19	16	19	Madagascar	74	74	1.4	0.8
20	18	20	Sierra Leone	57	57	0.0	0.6
Total				9087	—	5.9	95.1
Total Africa				9551	—	5.9	100.1

countries, Tunisia (18%) and Morocco (13%), and two eastern countries, Zimbabwe (6.4%) and Kenya (4.3%). The pattern of receipts is similar (as shown in Table 6.2b), with South Africa the leading earner (24.8%), followed by Tunisia and Morocco. However, although Zimbabwe and Kenya attracted considerable numbers of tourists, Mauritius and Tanzania were able to gain and earn more from tourism.

This brief background illustrates the nature and scope of international tourism in Africa and the significance of tourism in some countries, which is clearly influenced by the wider nature of economic development. For the purposes of this chapter, this profile provides a framework within which to examine the ecotourism activity in the region.

General Tourism in Kenya and South Africa

The regional perspective on tourism has been summarized above. In this section we consider the case of Kenya and South Africa, as a prelude to the consideration of ecotourism. This section begins by presenting a general background to the countries, and then examines the current situation with regard to their overall tourism sectors.

Kenya

Kenya occupies a mainly arid to semiarid area of 560,367 km^2, and has an estimated population in excess of 29 million (ECA, 1999; Kenya, 1999). Between 1979 and 1989, Kenya experienced an annual net population increase of 4% (Kenya, 1995, p. 19). Although the indigenous people account for the vast majority of the population, other non-Africans, especially those from Asia, exercise a considerable influence over the economy. The economy is based on the export of agricultural products mainly related to coffee (7% of GDP) and tea (6%), with tourism (10%) also playing a leading role. GDP at factor cost in real terms declined at an annual average rate of 2% between 1996 and 1998, itself the result of many factors including poor infrastructure, depressed investments, labour unrest, etc. (Kenya, 1999, p. 17). Given this context, three broad conclusions may be drawn. First, there is a need to enhance and diversify export earnings. Second, there is a need to increase contribution to government revenues and the balance of payments. Third, there is a need to increase employment opportunity. Ecotourism can play a major role in all of these areas.

South Africa

South Africa is the most complex country in sub-Saharan Africa. First, the country has the most diversified natural resource base, from oceans to snow-capped mountains (in winter), from subtropical deserts to montane forests. Its landscape is varied, mainly related not only to the great mountains, as noted, but also to the plateaux of the high veld, or low veld steppe lands. Although the plateaux extend from east to west and from south to north, there are two distinct climatic features that can be discerned. The country is wet and green on the east coast, in the eastern mountains, and on the high plateaux, but dry and highly desertified on the lower plateaux and western dry lands. The country also has the most diverse wildlife sanctuaries in Africa. All these factors together indicate a high potential for ecotourism development.

The country of 44 million inhabitants has also had a complex history of human relations in the last 400 years. Its state of development and real economic power has for historical reasons been deliberately designed and implemented to favour white-dominated areas. The black African reservations are poor and over-populated and dominated by poverty-related crimes. But the elections of 1994 that ushered in the universal adult suffrage and democratic government opened the door for revolutionary changes since then. The economy is diversified; agriculture, industry, manufacturing or mining, and services (including tourism) are the main sectors.

As noted in the introduction, Kenya and South Africa are both considered well established and 'successful' as tourism destinations. Table 6.2 shows their relative volume and value significance comparative to selected main tourism players in the region. Both countries accord tourism a high priority in their national development plans. Their success in this field has demonstrated how other countries in the region should be able to use tourism as part of their economic development strategies. The importance of tourism in the process of national development is reflected in the current national development plan of Kenya:

> A sustained flow of tourists will contribute to industrial development through generation of foreign exchange, creation of income earnings opportunities ... The constraints to growth for the tourism industry include inadequate tourism promotion and marketing efforts ... weak institutional and regulatory support framework. The strategies that will be used to address the above issues with a view to improving the tourism sector include strengthening the Kenya Tourism Board (KTB) to become fully operational ... In this regard, the National Tourism Master Plan will be implemented fully to establish a sustainable tourism base.
>
> (Kenya, 1997a, pp. 201–202)

In South Africa, the government recently published a White Paper on Tourism (SATOUR, 1996) that was later approved by the Cabinet in June 1996. The main purpose was to set out the necessary development parameters and how these might be realized. In particular, emphasis was given to the key changes required in the organizational structures to permit the management of the tourism sector, including the regulatory and legislative systems.

One other important component of the White Paper and related initiatives (see SATOUR, 1998, 1999a, b, c) was to stimulate the dialogue between the private and public sectors in a partnership arrangement, thereby broadening participation in the sector. These developments are part of the major economic and political reforms in the country, catalysed and actuated by the events of 1994. They need to be seen against a background where South Africa, as previously noted, was regarded as an economic and a political pariah state as a consequence of its apartheid policy. Now the country has been re-incorporated into the international community. The policy now is for development of 'responsible tourism' with its emphasis on:

> The right and appropriate vision, structure and texture of the industry to facilitate sustainable tourism; the evolution of an economically active and integrated black majority into tourism; optimisation of the socio-economic benefits to the widest possible spectrum of society in all provinces; and evolution, therefore, of a long-term base for growth through vision, promotion and integration of environmental management into the different phases of project management.
>
> (SATOUR, 1996, p. 19)

Indeed, these are difficult objectives to achieve in a country where the apartheid policy of the government had been the focal point of police statism. But the country is anxious to make up for lost opportunities of the apartheid era, researching and planning for unique packages such as ecotourism, 'afro-tourism' and cultural tourism.

Features of tourism in Kenya and South Africa

Three characteristics define the current tourism sector: seasonality; concentration of generating countries; and the tourism product. In relation to seasonality, in 1998 1,062,000 tourists arrived in Kenya, 5,981,000 in South Africa (Table 6.2a). In the focus period, over 70% of the visitors arriving in Kenya (see Kenya, 1999) and 50% in South Africa (see South Africa, 1999a) came during the months of November 1998 to March 1999. These figures indicate a very high concentration of arrivals, a pattern that has consequent economic implications.

Europe was the major 'trigger' market for Kenya in 1998, with Germany and the

UK between them accounting for over 60% of all bed-nights occupied by European visitors. The next most important source of tourists came from regional countries, e.g. Tanzania and Uganda with 216,800 bed-nights in 1998. For South Africa 60% of arrivals in 1998 were from neighbouring countries. There was relative dependence on charter markets from the UK, Germany, US and France during the period in focus.

Finally, the countries' tourism product is defined here as all those facilities, amenities, and services, including the natural environment, which attract visitors. In particular the product has a number of distinct features: game viewing, beach, conferences and seminars, activity/adventure pursuits, shopping, cultural events and, for South Africa especially, hotel-based gambling and sports events (see Kenya, 1999; SATOUR, n.d., p. 11). In the context of this chapter, prime emphasis will be on the natural endowments of the countries and, in particular, those that are based on wildlife and its natural habitat. These are mainly available within the national parks and reserves. As will be discussed below, Kenya has 60 such parks and reserves and South Africa 212, out of which 17 are major ones, such as the 2 M ha Kruger National Park.

But the key question is whether these features are in themselves unique to justify the countries' successes, given that the features are also available in many destinations in neighbouring countries in both eastern and southern sub-regions. It should be noted, as tourism marketers always do, that no one activity is adequately attractive to motivate visitors. However, taken together, they provide a basket of options available to tourists (see Jefferson and Lickorish, 1988; Richards, 1997). In essence, the implication must be that very few countries have attractions that constitute unique selling propositions (USP). In the case of Kenya and South Africa one would have to look therefore at other factors, e.g. marketing, image and others, all considerations that can be found in 'successful' tourism destinations.

In specific terms, the World Tourism Organization (WTO) has identified a number of factors that shaped tourism development in Kenya and South Africa in 1998, and especially tourist product, transport, and marketing and promotion activities (WTO, 1999, p. 100). Each area has both positive and negative aspects. On Kenya's tourist product, there were positive developments in the liberalization of foreign exchange regimes, divestiture of government's interests in the sector and diversification of the product. Of course, whatever gains were made as a consequence of these measures, were off-set by adverse publicity in the international media labelling Kenya as an unsafe and insecure destination. Similar observations could be made of South Africa. On the positive side, the country upgraded its international airports, awarded mega-casino licences, and established provincial tourism and marketing development agencies. However, there was a lack of investment incentives and, in particular, local banks were unwilling to provide loans for tourism development projects.

On the transport front, Kenya widened its charter market networks to allow Hungary and France to increase their charters. Kenya Airways, the national carrier, struck strategic alliances with KLM and North West Airlines that gave it more routes for the benefit of international tourists. The down side was in the area of higher fuel prices, which led to some airlines having to over-fly Kenya and to cut down on frequencies. South Africa also allowed more air charter operators to enter the market. For instance, the airline Iberia re-established links with South Africa, even though South African Airways were still running at a loss.

In relation to the marketing and promotion activities, the formation of the Kenya Tourist Board to handle these tasks, the completion of a Tourism Master Plan and the relaunching of the East Africa Community were seen as positive developments. Negative factors included a lack of diversification of the source markets, relying heavily on the traditional markets; inadequate funding; and increasing competition by rival destinations (e.g. Zimbabwe).

South Africa transformed the institutional structures for tourism, e.g. SATOUR, but extensive media coverage of South Africa's crime was unhelpful.

Ecotourism in Kenya and South Africa

In considering ecotourism activity in Kenya and South Africa, there are three areas that will be emphasized: its early growth and magnitude; spatial distribution; and a number of development issues, albeit in a comparative setting.

Growth and magnitude

To consider the early growth and magnitude of ecotourism in Kenya and South Africa, it is helpful, even briefly, to distinguish between 'consumptive' and 'non-consumptive' tourist attractions and holiday opportunities (Gibson, 1999; Honey, 1999). This is because both concepts always recur in any discussion of wildlife-based tourism development in these countries, given the comparative use of the resource relative to other development sectors (see Olindo, 1991; ECA, 1997).

Simply put, consumptive uses of wildlife such as big game sport hunting expeditions, bird shooting, etc., mirror a master–servant relationship situation in which white settlers were the masters and indigenous Africans the servants. Colonial wildlife policy sought to advance the needs of the settlers at the expense of the native population relating to ownership, use and even the conservation of resources that subsequently followed (Anderson and Grove, 1987; Gibson, 1999; Honey, 1999). This may be tantamount, in the words of Akama (1996, p. 572), to 'the taking away of wildlife resource user rights from the rural peasants'. Post-colonial policy, with its emphasis on non-consumptive or ecotourism use of wildlife resources sought, conversely, to redress the injustice seen during the consumptive era. It has sought to empower the locals by giving them a voice in decisions regarding benefits-sharing arising from, and therefore ownership of, wildlife resources. Thus in essence it can be suggested that the consumptive activity is a necessary precursor to understanding ecotourism developments that were to follow.

The development of ecotourism in Kenya dates back to 1977 and 1978 when the country's government imposed a total ban on sport hunting and on the trade in game trophies. This apparent U-turn in wildlife policy was prompted by several considerations, not least being how to ensure that best use was made of wildlife resources (Dieke, 1991). This measure had an effect at three levels. First, it helped to concentrate people's minds on alternative uses of wildlife; second, it had a disastrous effect on earnings and employment in the country; and third, it ensured the adaptation of existing hunting structures (e.g. lodges, game parks, national reserves, etc.) to the cause of non-consumptive ecotourism (Olindo, 1991). The response to the new dispensation was swift: shooting wildlife with the camera took centre-stage; promotional activities highlighted the natural landscapes of the country, its biodiversity, unique ecosystems, beautiful scenery (e.g. the Rift Valley) and volcanic mountains (see The Ecotourism Society, 1998). Tour organizers developed ornithological trips and botanical study tours.

The extent of ecotourism in Kenya raises definitional and motivational issues. The first is a continuing example of the dearth of reliable tourism statistics in the country and of the fact that those which are available need to be interpreted with caution. This is not an unusual situation in a country where, as was observed some 10 years ago (Dieke, 1991), tourism statistics are generally under- or overestimated. The problems stem from variations in collection methods, processing of data, and definitions – problems which can be found in other developing countries. As the role of ecotourism increases, so too does the need for more reliable time series tourism data, as a basis for policy formulation.

Table 6.2a indicates that there were

about 1 million international visitors arriving in Kenya in 1998. However, on the basis of provisional figures from government sources of park- and reserve-visitations between 1995 and 1998 (Table 6.3a, b), it seems that many of these visitors were ecotourists, as they were engaged in safaris within parks and reserves. In 1998, 1,079,400 protected area visitors were reported (Table 6.3a), a figure that exceeds the 1 million tourists noted above. Not all of these visitors were international tourists, as indicated in Table 6.3c with respect to lodge bed-occupancy rates.

It should be stressed that the continued contraction in the number of visitors to parks and game reserves contributed to low bed-occupancy rates in game lodges. The number of bed-nights fell dramatically from 351,200 in 1997 to 167,000 in 1998, while the proportion accounted for by East African visitors increased from 9% of the total to 17%. This decline mirrored trends in other sectors of Kenyan tourism such as Nairobi and the coast (Kenya, 1999).

In the field of ecotourism, South Africa has been described as having 'a reputation as one of the world's leading countries, with its well-managed system of public protected areas, extensive private sector involvement and conservation-linked community development initiatives' ('t Sas-Rolfes, 1996). In this respect the country is similar to Kenya, becoming well known and 'successful' in this sector. But in contrast to Kenya, ecotourism in South Africa has a chequered history and has been controversial. It is worth examining, albeit briefly, a short historical context within which ecotourism activity exists in this

Table 6.3a. Number of visitors to parks and game reserves, 1995–1998 (000s) (Kenya, 1999, p. 164; Kenya, 1997b, p. 184).

Area	1995	1996	1997	1998
Nairobi	113.5	158.3	149.6	122.3
Animal Orphanage	212.1	210.6	193.7	164.8
Amboseli	114.8	109.1	117.2	62.9
Tsavo (West)	93.1	93.6	88.6	54.9
Tsavo (East)	228.8	137.5	123.2	66.9
Aberdare	70.1	60.2	59.0	47.9
Lake Nakuru	166.8	156.9	132.1	111.0
Maasai Mara	133.2	130.3	118.3	100.4
Bamburi Nature Park	109.2	107.0	86.8	77.9
Malindi Marine	38.8	39.3	27.0	13.7
Lake Bogoria	14.2	14.2	24.5	20.6
Meru	7.3	7.8	4.1	1.8
Shimba Hills	20.0	23.4	22.5	16.8
Mount Kenya	17.2	17.1	14.8	10.2
Samburu	9.1	9.1	8.3	7.0
Kisite/Mpunguti	32.4	39.9	35.1	29.2
Mombasa Marine	23.9	21.7	15.2	16.2
Watamu Marine	16.1	20.2	19.4	18.3
Hell's Gate	50.1	52.1	47.2	57.1
Impala Sanctuary (Kisumu)	3.5	65.6	62.4	65.6
Other[a]	18.9	14.8	15.5	13.9
Total	1493.1	1488.7	1364.5	1079.4

[a] Other includes Mount Elgon, Ol-Donyo Sabuk, Marsabit, Saiwa Swamp, Sibiloi, Ruma National Park, Mwea National Reserve, Central Island National Park, Nasolot National Reserve and Kakamega National Reserve.

Table 6.3b. Visitors to museums, snake park and sites, 1995–1998 (Kenya, 1999, p. 165; Kenya, 1997b, p. 184).

	1995	1996	1997	1998
National Museum (Main Gate)	215.4	218.0	184.5	173.4
National Museum (Snake Park)	181.6	170.6	148.6	75.9
Forth Jesus	245.3	180.2	124.4	88.9
Kisumu Museum	36.1	49.5	18.2	34.7
Kitale Museum	27.5	29.0	16.1	27.3
Gedi	43.7	29.6	29.7	14.8
Meru Museum	21.0	12.4	9.4	15.8
Lamu	10.7	12.2	8.6	6.2
Jumba la Mtwala	11.3	8.5	4.9	4.0
Ologesailie	—	—	2.2	1.9
Kariandusi	3.0	2.3	0.7	4.5
Hyrax Hills	—	1.9	1.5	2.8
Karen Blixen	46.1	43.7	38.6	41.1
Kilifi Mwarani	0.8	0.9	0.7	2.9
Total	842.5	758.8	588.1	494.2

Table 6.3c. Game lodges occupancy, 1995–1998 (000s) (Kenya, 1999, p. 163; Kenya, 1997b, p. 184).

	Bed-nights occupied							
	Foreign residents				East African residents			
Lodge locality/type	1995	1996	1997	1998	1995	1996	1997	1998
Game reserves	218.6	255.5	178.9	77.7	21.0	20.9	18.0	16.2
National parks	172.5	201.7	141.2	61.3	15.1	15.1	13.1	11.8
Total	391.1	457.2	320.1	139.0	36.1	36.0	31.1	28.0
Of which full catering	341.2	398.9	279.3	121.3	28.3	28.2	24.1	22.0
Self-service	49.9	58.3	40.8	17.7	7.8	7.8	7.0	6.0

country and considering current changes of thinking in and extent of the sector.

Early utilization of wildlife centred on hunting involving three different groups of people: sport hunters (mostly English-speaking), commercial hunters (mostly Afrikaners) and subsistence hunters (native Africans). As wildlife became scarce, these groups started to compete for the rights to hunt. The first people to lose their rights were the subsistence hunters. There was a philosophical divide between the other two groups over the justification for hunting. Commercial hunters earned a living from wildlife, and could not understand the rationale for sport hunting, which they regarded as wasteful. Conversely, sport hunters, mostly from wealthy landowners' and urban dwellers' backgrounds, saw no need to gain commercially from wildlife. They justified hunting as a glamorous recreational outlet and an indicator of social status, and regarded the killing of wildlife for commercial and subsistence purposes as cruel and unnecessary. These contrasting positions between people who live off the land and a more wealthy elite persist today in this country, although in a somewhat different way, thanks to the evolution of new forms of international tourism, and especially ecotourism. Thus, given the revolutionary reforms within the national, political, philosophical, economic and social attitudes, relationships and participatory integration of the black majority, the South African ecotourism is entering a phase of revolutionary growth.

The current South African tourism development strategy, *Tourism in Gear*, for the period 1998–2000 identified ecotourism – including safari, game-watching and birdwatching – as one of the seven core activities (SATOUR, 1998). Others include culture, adventure, sport, business, special interest and, finally, the MICE (meetings, incentives, conventions and exhibitions) sector. Clearly, scenic interest and wildlife (in the widest sense of ecotourism) emerge as the prime focus of attention. The importance of ecotourism and others is understandable considering:

> the slowly diminishing importance of South Africa's cities as tourist attractions ... There are signs that city dwellers are choosing rural locations for holidays in preference to cities ... At the same time, locations in the interior which are attractive either for their scenery, wildlife, or just for their rustic setting, are becoming popular among visitors, not only from South Africa but also from abroad.
>
> (TTI, 1999, p. 89)

It could be argued that this shift away from cities has in large part been caused by the perceived deterioration in the condition of many, if not most, of South Africa's urban centres over the past decade. By implication, the ending of apartheid has inevitably brought about an invasion of places that were in effect previously forbidden, with consequences for accommodation and occupancy.

The evidence shows that annual room occupancy in Johannesburg, for example, fell from around 45% in the mid-1990s to 41% in 1997 – and probably to below 40% in 1998 – and the Durban area registered a fall from 66% to well below 60% (TTI, 1999). Furthermore the estimates indicate that in 1996 the 8500 registered properties offered the equivalent of 600,000 beds, the bulk of which were in campsites (or 'ecolodge' type accommodation outlets). These figures may be underestimates because precise statistics are unavailable, particularly in relation to unregistered self-catering accommodation that may be better suited to the low-income levels of most South Africans, and also the foreign budget travellers.

Of the 212 parks in this country, the 2 million ha Kruger National Park is an icon in the South African ecotourism development experience. The Park is the largest in the country and offers an unrivalled variety of game animals: amphibians, reptiles, birds and 147 mammal species including the Big Five (buffalo, cheetah, elephant, giraffe and leopard). It also offers game or nature reserve type accommodation, covering the full spectrum from camping to luxury cottages. There are numerous camps to cater for every taste and budget ranging from rest-camps and bush-camps to bush-lodges. Latest estimates (TTI, 1999) put the visitation level at about 1 million a year (almost exactly half of the visitors stay at least one night), despite mounting criticism over the standard of service in its camps. Kruger is particularly popular with foreign visitors who, it is claimed, now number almost 200,000 a year, accounting for almost 50% of all foreign overnights in game/nature parks (SATOUR, 1999a).

The other parks operated by South African National Parks do not fare well in visitation terms, probably because they tend to lie off the foreigners' beaten track, and partly because they are not well known. As a consequence of tax incentives given in recent times, there is now a considerable increase in private game lodges, currently estimated at around 300, with total bed capacity of 18,000 (TTI, 1999). Some of them do not score well, perhaps because they are considered expensive, but they reflect a trend of over-supply.

Spatial distribution

From a spatial framework, the wildlife resource and the activity associated with it are largely confined within parks and reserves, although the activity is carried out in adjacent areas as well. In the case of Kenya, the distinction between the parks and reserves is important for two reasons.

The first is to clarify the issue of their ownership, management and financial arrangements, and the second to determine the relationship of local people and the protected areas vis-à-vis benefits sharing (Sindiga, 2000). In particular (see Yeager and Miller, 1986; KWS, 1990), the 'parks' refer to parcels of land belonging to, and fully administered and financed by, the central government. 'Reserves', in contrast, are areas set aside by local authorities (counties) for conservation of wildlife but which are managed and partly financed by the central government. County councils operate game reserves on trust lands for which they are responsible, and also participate in safari lodges or self-catering accommodations.

At present there are 60 such protected areas in Kenya. Estimates of the proportion of the country's land area occupied by the parks and reserves are imprecise, but range between 6 and 12% (Yeager and Miller, 1986; Akama, 1996; Sindiga, 2000). The imprecision is understandable, especially as more areas are usually incorporated into the 'protected' network as and when the need arises, given that there is no set numerical limit. However, as evident in Table 6.3a, visitation is spatially concentrated within or skewed towards small core 'protected' areas, partly because of easy access, and partly because of their proximity to international gateways. It may also be that these areas are popular with ecotourists. It is therefore not surprising that Nairobi is significant as both capital city and international gateway, as a centre for incoming tours, and as a base for game viewing tours. In addition, its proximity to the beach area provides the ancillary opportunities to combine beach and safari holidays.

In South Africa, as in Kenya, the extent of area occupied by the parks is unclear, but one source (Weaver, 1998), quoting World Resources Institute 1994, gives a figure of 6.1%. However, one member of a South African NGO adds a perspective to the debate by declaring that 'there are also something like 17 million ha of privately owned land that have been converted to game farming' (personal communication, April 2000). This statement is significant in three respects. First, it underscores the new opening-up of South Africa's development landscape to accommodate all shades of opinion on enterprise culture. Second, implicitly there are basic differences between government-owned and privately owned parks in respect to their relative roles in ecotourism. As noted above, it is not surprising that private parks, encouraged by generous incentives, are numerically significant. For government parks, they highlight the process that gives provincial governments responsibility for tourism promotion and development, leaving the central government with a restricted responsibility (see SATOUR, 1999a). Third, whether the parks/game farms fall within the jurisdiction of national or provincial government, or whether they are owned and managed by the government or private sector, it is pertinent that the parks are attractive to both foreign and domestic tourists. The Strategy document (SATOUR, 1998) proves the point: R15 billion were generated by domestic tourism, and R12 billion by inbound visitors.

Development issues

Various development issues, such as carrying capacity, marketing and image, pertain specifically to the ecotourism sector in Kenya and South Africa. Articulating the carrying capacity problems in Kenya (see also Table 6.3a, b, c), Weaver has succinctly described the situation thus:

> In Kenya, much attention has been focused on Amboseli National Park and Maasai Mara National Reserve. Visitor crowding and mismanagement in the former has long been associated with disruption of sensitive species such as cheetahs. Commonly, large numbers of safari vehicles would concentrate around a single predator group, as nearby safari vehicles would be alerted to the presence of their activity. Other problems have included scavenging by local wildlife in garbage dumps, and landscape degeneration as a result of extensive off-road vehicular traffic.
>
> (Weaver, 1999, p. 806)

Ecotourism in Other African Countries

In this section a small selection of some countries in the sub-regions of Africa are presented, in essence, to provide a broader continental context. Perhaps these area studies will also emphasize or demonstrate why there is relatively little ecotourism in some countries and more in others. The countries selected are not only significant because they exhibit in relatively varying degrees some of the features of ecotourism as discussed, but they also highlight problems that can be found in much of the region, considerations that might potentially threaten or enhance ecotourism activities. Emphasis is placed on other 'Safari Corridor' countries located between Kenya and South Africa (see Chapter 16).

Tanzania

Tanzania, once a tourism rival of Kenya, has the potential to re-launch itself as an ecotourism destination following the repeal of socialist policies that were inaugurated during the 1970s. Private enterprise in tourism is now encouraged and growing, leading to the rapid growth of tourist arrivals and receipts since the mid-1990s. In terms of its competitive ecotourism advantage, Tanzania has a wide variety of environmental resources, and its wildlife resources are unmatched in Africa beyond South Africa. Included in this inventory are ecotourism icons such as Kilimanjaro, the Ngorongoro, and Serengeti, the last two located within the so-called Northern Circuit wildlife area. Historically, ecotourism in Tanzania has focused on this area, though considerable potential is found along the coast and within the undeveloped southern wildlife sanctuaries. Such extensions would be logical from both a managerial and marketing perspective, given that 80% of all tourist bed-nights are concentrated in the Northern Circuit: a skewed pattern that has resulted in local pressures on environmental resources during the peak season, and a stereotypical 'safari' image of the Tanzanian tourism product.

Zambia

Zambian ecotourism potential is based on largely intact wildlife resources and the singular Victoria Falls. However, at the present time, this potential is not being realized due to the poor quality of its lodges, food, infrastructure, vehicles and guides (Zambia, 1995a). This situation does not bode well for Zambia, which appears to have no discernable competitive advantage against emerging sub-regional ecotourism destinations such as Tanzania, Zimbabwe and South Africa, despite the quality of its wildlife. Weaver also shares some of the above concerns as well as offering some perspectives, maintaining that:

> If the situation in Zimbabwe suggests overdevelopment, its Zambian counterpart reveals underutilization, with visitation averaging only 50 visitors each day. A similar situation pertains to Zambia's protected areas, which accommodate modest tourist numbers, despite outstanding natural qualities. Without associated revenues, there is little justification for maintaining such areas in a state conducive to the fulfilment of their conservation mandate. Because of inadequate funding, infrastructure for reaching the parks is poor (thus discouraging tourist traffic) and the number of rangers (600 as of the late 1980s) has been far less than 4000–5000 required to address serious poaching-related wildlife depletion.
>
> (Weaver, 1998, p. 132)

In response, Zambia has taken a number of sustainable development measures. Notable among these is the Administrative Management Design for Game Management (ADMADE), introduced in the Lupande Game Management Area of the South Luangwa National Park, and including the Luangwa Integrated Rural Development Project (LIRDP) (Inskeep, 1991). Both initiatives are intended to advance the cause of ecotourism, as broadly defined, but may be only superficial. More recently, the

Medium-Term Strategy for Tourism (see Zambia, 1995a) and the *White Paper on Tourism* (see Zambia, 1995b) have attempted to address the problems of tourism in the country, though neither constitutes a comprehensive tourism development plan. Basically the strategy identifies the principal problems and constraints to further tourism development in Zambia and prescribes the direction for future development as well as identifiying the necessary policy, legislative and development issues that need to be attended to in the short-to-medium term.

In terms of ecotourism, to implement the strategies will require investment in physical plant, facilities and associated services. The government has recently put accommodation facilities in the national parks out to tender, and this should result in an immediate inflow of both investment and expertise to revitalize the existing facilities through upgrading and development. This in turn should have a positive effect on the inflow of tourists. Revenue generated will benefit the National Parks and Wildlife Service (NPWS) through contributions to its Wildlife and Conservation Revolving Fund (WCRF) and to communities through the ADMADE programme.

Zimbabwe

Zimbabwe is endowed with considerable natural landscapes, having 29 wildlife protected areas that account for between 8% (Weaver, 1998) and 12% (ECA, 1997) of its territory. What is striking about this country is the quality of its infrastructure and support services that, by any comparative African standards, might be considered good. It has managed to maintain relative peace since independence in 1980, although current social unrest in the country might undermine future prospects.

Zimbabwe suffers from two interrelated problems, one of which is an upsurge in visitation levels, and the other a dramatic reduction in some key animal species on which ecotourism depends. It has been suggested that, although 'most protected areas have yet to experience intensive levels of visitation, an impact is being made on certain areas, by what resembles a form of incipient mass tourism' (Weaver, 1998, p. 131). Part of the problem is that preferred holiday areas outside Harare, the capital, are concentrated in the northern region of the country, notably Lake Kariba, Hwange National Park and the geographically proximate Victoria Falls.

Given, therefore, the huge popularity of the Victoria Falls (a World Heritage resource), with about 20% of total overnight stays in 1995, it is not surprising that this poses an increasing threat to the environmental sustainability of the country's most important tourist attractions (ECA, 1997). The second, and related, problem concerns the actual wildlife, and the problem is twofold. First there is the claim that over-crowding and over-utilization of the Kariba and Zambezi have changed some wildlife habits (ECA, 1997). For example, the water-skinner, which normally nests on the riverbank, has been disturbed to the extent that it has had to change its nesting habits. There is a likelihood of other birds and animals being affected in one way or another. The second problem is the decimation of wildlife, e.g. rhinoceros, by poaching, which was said to be serious before the introduction of the Communal Areas Management Programme for Indigenous Resources (CAMPFIRE) in 1989. The latter movement integrates local communities in tourism development by providing these communities with tangible benefits from the use of wildlife resources for hunting and other forms of tourism, hence the initiative is only partially related to ecotourism. However, this does not mean that it is unsustainable, since the movement has brought about a fundamental and positive change in community attitudes towards wildlife management.

The Gambia

Some mention should be made of African ecotourism beyond its East and South African strongholds. Ecotourism in The

Gambia is perhaps representative of the general trend in West Africa. Here, the sector is embryonic and small scale. In addition, the product is different due to, *inter alia*, limited land provision, harsh climatic conditions, desertification, destruction of rainforest and savannahs and, arguably, civil and military conflicts. The first major effort to develop ecotourism in The Gambia began in 1977 with the Banjul Declaration (The Gambia, n.d., p. 2), which was spawned by government's awareness of the loss of wildlife and biodiversity. The sole purpose of the Declaration, in the words of the government was to 'ensure the survival of the wildlife still remaining with us' through taking 'untiring efforts to conserve for now and posterity as wide a spectrum as possible of our remaining fauna and flora'. In one sense ecotourism policy in this country can therefore be seen as an outgrowth of the broad environmental master plan with its focus on arresting the degradation of vegetation, forest cover, biodiversity and wildlife, much of it attributable to other forms of tourism (see Dieke, 1993).

A 1967 study listed 67 species of mammals known or expected to have existed in The Gambia in the 20th century. Of this number, however, 13 are now locally extinct, and the remaining species are under the growing threat of high population growth and ever-increasing demands on land resources. For example, the aquatic antelope or sitatunga and the African manatee are teetering on the edge of extinction. Against this background, Gambian ecotourism is not to be sought in recent plans to develop ecotourism *sui generis*. These were foreshadowed by a number of government initiatives that followed in the wake of the Banjul Declaration or even preceded it. The initiatives have centred largely on attempts to bring specific areas under protection. There are already a number of nature reserves and forest parks that are proving effective in terms of preserving endangered animal species. For instance, Kiang West National Park is something of a trailblazer in that it seeks to conserve plant and animal life through close and active collaboration with the resident population, which derives significant economic benefits from the symbiotic relationship. To protect bird life, for which The Gambia enjoys a good reputation, a number of areas have recently been reserved as 'bird sanctuaries'.

Future development of ecotourism in The Gambia, in summary, will undoubtedly be hindered by the environmental problems of the past. Furthermore, as in many West African destinations, the marketing of this product will be hindered by the absence of 'dramatic' attractions such as lions, giraffes and elephants. However, by allying ecotourism with product strengths such as cultural tourism or even sports tourism, and by intelligently and responsibly accessing its existing natural attractions, The Gambia can succeed as an ecotourism destination. Another possibility would be to establish a more diversified tourism product by engaging in multilateral product development and marketing (The Gambia, n.d., p. 5).

Conclusion

The importance of developing sustainable ecotourism in Africa cannot be overemphasized because of its potential for diversifying the economy while protecting its still formidable environmental heritage. The case is made that its development can be based on using the many and varied wildlife and environmental assets of the region sensitively to stimulate economic development. This is particularly important in eastern and southern regional countries where a strong ecotourism tradition is already evident. Yet, despite this tradition and the richness and variety of the natural assets in Africa, these natural resources are very much under-utilized for ecotourism. Areas that appear promising for the effective and sustainable development of the sector include the fostering of domestic markets, community initiatives along the lines of CAMPFIRE, the return of funds by the industry into protected areas and wildlife management, and the formation of diversified, multilateral tourism circuits.

Regarding the last point, ecotourism should not be isolated from regional African trends, where there has been the move towards economic integration and cooperation, as advocated in the Abuja Treaty (OAU, 1991; Dieke, 1998). Cooperation in developing ecotourism between those adjoining African countries where this activity is meaningful is advocated, as the cooperating countries will be able to benefit from economies of scale. Such moves could be facilitated through existing regional and sub-regional organizations such as the Southern African Development Community (SADC), Common Market for Eastern and Southern Africa (COMESA), Regional Tourism Organisation of Southern Africa (RETOSA) and others.

References

Afigbo, A.E., Ayandele, E.A., Gavin, R.J., Omer-Cooper, J.D. and Palmer, R. (1992) *The Making of Modern Africa: the 19th Century,* Vol. 1. Longman, London.

Akama, J.S. (1996) Western environmental values and nature-based tourism in Kenya. *Tourism Management* 17(8), 567–574.

Anderson, D. and Grover, R. (eds) (1987) *Conservation in Africa: People, Policies and Practice,* University Press. Cambridge.

Dieke, P.U.C. (1991) Policies for tourism development in Kenya. *Annals of Tourism Research* 18(3), 269–294.

Dieke, P.U.C. (1993) Tourism and development policy in The Gambia. *Annals of Tourism Research* 20(3), 423–449.

Dieke, P.U.C. (1998) Regional tourism in Africa: scope and critical issues. In: Laws, E., Faulkner, B. and Moscardo, G. (eds) *Embracing and Managing Change in Tourism: International Case Studies.* Routledge, London, pp. 29–48.

Dieke, P.U.C. (ed.) (2000) *The Political Economy of Tourism Development in Africa.* Cognizant, Elmsford, New York.

ECA: Economic Commission for Africa (1997) *Environmental Issues in Transport and Tourism Sectors in Africa,* Vol. V. TRANSCOM/1067/REV. 1. ECA, Addis Ababa, Ethiopia.

ECA: Economic Commission for Africa (1999) *Economic Report on Africa 1999.* ECA, Addis Ababa, Ethiopia.

The Ecotourism Society (1998) *Ecotourism at a Crossroads: Charting the Way Forward* (A Summary of Conference Proceedings held in Nairobi, Kenya, October 1997). The Ecotourism Society, North Bennington, Vermont.

The Gambia (n.d.) *Ecotourism in The Gambia: Possible Strategies and Projections,* (Unpublished manuscript). Ministry of Information and Tourism, Banjul, The Gambia.

Gibson, C.C. (1999) *Politicians and Poachers: the Political Economy of Wildlife Policy in Africa.* Cambridge University Press, Cambridge.

Honey, M. (1999) *Ecotourism and Sustainable Development: Who Owns Paradise?* Island Press, Washington, DC.

Inskeep, E. (1991) *Tourism Planning.* Van Nostrand Reinhold, New York.

Jefferson, A. and Lickorish, L. (1988) *Marketing Tourism: a Practical Guide.* Longman, Harlow, UK.

Kenya, Republic of (1995) *Statistical Abstract.* Central Bureau of Statistics, Nairobi.

Kenya, Republic of (1997a) *National Development Plan 1997–2001.* Government Printer, Nairobi.

Kenya, Republic of (1997b) *Economic Survey 1997.* Central Bureau of Statistics, Nairobi.

Kenya, Republic of (1999) *Economic Survey 1999.* Central Bureau of Statistics, Nairobi.

KWS (1990) *A Policy Framework and Development Programme 1991–1996.* Kenya Wildlife Service, Nairobi.

Olindo, P. (1991) The old man of nature tourism in Kenya. In: Whelan, T. (ed.) *Nature Tourism: Managing for the Environment.* Island Press, Washington, DC, pp. 23–38.

OAU: Organisation of African Unity (1991) *Treaty Establishing the African Economic Community.* OAU, Abuja, Nigeria.

Richards, B. (1997) Tourism trades. In: Lickorish, L.J. and Jenkins, C.L. (eds) *An Introduction to Tourism.* Butterworth-Heinemann, Oxford, pp. 98–134.

Sindiga, I. (2000) Tourism development in Kenya. In: Dieke, P.U.C. (ed.) *The Political Economy of Tourism Development in Africa.* Cognizant, Elmsford, New York, pp. 129–153.
South African Tourism (SATOUR) (1996) *Development and Promotion of Tourism in South Africa* (Government White Paper). Department of Environmental Affairs and Tourism, Pretoria.
South African Tourism (SATOUR) (1998) *Tourism in Gear: Tourism Development Strategy 1998–2000.* Department of Environmental Affairs and Tourism, Pretoria.
South African Tourism (SATOUR) (1999a) *Annual Report 1997–1998.* Department of Environmental Affairs and Tourism, Pretoria.
South African Tourism (SATOUR) (1999b) *Institutional Guidelines for Public Sector Tourism Development and Promotion in South Africa.* Department of Environmental Affairs and Tourism, Pretoria.
South African Tourism (SATOUR) (1999c) *Tourism Action Plan.* Department of Environmental Affairs and Tourism, Pretoria.
South African Tourism (SATOUR) (n.d.) *Investing in Tourism: South Africa.* Department of Environmental Affairs and Tourism, Pretoria.
't Sas-Rolfes, M. (1996) *The Kruger National Park: a Heritage for All South Africans?* Pretoria: Africa Resources Trust. http://www.wildnetafrica.com/bushcraft/articles/article-kruger-start.html
Travel and Tourism Intelligence (TTI) (1999) *South Africa*, Country Reports No. 2, Travel and Tourism Intelligence, London, pp. 78–100.
Weaver, D.B. (1998) *Ecotourism in the Less Developed World.* CAB International, Wallingford, UK.
Weaver, D.B. (1999) Magnitude of ecotourism in Costa Rica and Kenya. *Annals of Tourism Research* 26, 792–816.
World Tourism Organization (WTO) (1999) *Tourism Market Trends: Africa.* WTO, Madrid.
Yeager, R. and Miller, N.N. (1986) *Wildlife, Wild Death: Land Use and Survival in Eastern Africa.* State University of New York, Albany, New York.
Zambia, Republic of (1995a) *Medium-Term National Tourism Strategy and Action Plan for Zambia.* Government of Zambia, Lusaka.
Zambia, Republic of (1995b) *A Tourism Policy for Zambia.* Government of Zambia, Lusaka.

Chapter 7
Anglo-America

D.A. Fennell
Department of Recreation and Leisure Studies, Faculty of Physical Education and Recreation, Brock University, St Catharines, Ontario, Canada

Introduction

> In looking over the accounts given of the Ivory-billed Woodpecker by the naturalists of Europe, I find it asserted, that it inhabits from New Jersey to Mexico. I believe, however, that few of them are ever seen to the north of Virginia, and very few of them even in that state.
>
> Alexander Wilson (1808), see Finch and Elder, 1990, pp. 79–80

In Anglo-America, particularly the US, the natural history accounts of intrepid travellers are well documented. Naturalists and explorers the likes of William Bartram, Alexander Wilson, John James Audubon, George Catlin, John Burroughs, Robert Service, Henry David Thoreau and John Muir, contributed greatly to the understanding of a region's nature and natural resources. In an age of industrialization, such accounts were instrumental in helping to change the population's perception that the forests of the land were more than just, in the words of Michael Wigglesworth (1662), 'A waste and howling wilderness, where none inhabited but hellish fiends, and brutish men' (quoted in Nash, 1982, p. 36). It was their words and actions that paved the way for the development of the world's first large preserves and parks, designed for the protection of habitat and the enjoyment of the public.

Americans and Canadians continue to use their parks systems by the millions, and governments in turn have responded by creating more protected areas to accommodate increasing outdoor recreational demands. In the USA, for example, 287 million people visited the national parks (natural and cultural) in 1997, representing an increase of 4.2% from the previous year (US Department of the Interior, 1998). International visitors too are drawn to the parks and natural resources of the continent. Based on research conducted by Filion *et al.* (1992), 69–88% of Europeans and Japanese reported that birds and wildlife were important factors in travelling to North America. Seventy per cent of these travellers visited national parks and 30–64% observed birds and other wildlife.

One of the most interesting outdoor recreational trends in Anglo-America is the changing interest and emphasis away from consumptive activities like hunting, to those which are more non-consumptive in their orientation. In projections of annual growth rates on various outdoor activities between 1996 and 2011, birding was reported to be the fastest-growing of all at

6%, compared with 3% for golf and 4.5% for fishing (Foot and Stoffman, 1996). In a recent study by Statistics Canada (1998) on the importance of nature to Canadians, 18% of Canadians in 1996 said that they had fished for recreational purposes compared with 26% 5 years earlier. The statistics for hunting in the survey show a similar decline, with 5% of the population hunting in 1996 compared to 7% in 1991. This survey also found that 85% of Canada's population, aged 15 years and over, participated in one or more nature-related activities in 1996 (e.g. camping, canoeing, hunting), and about one-third of these individuals (6.7 million) visited a provincial park, national park, or other protected area. In addition, Canadians spent an estimated CDN$11 billion (CDN$550 per person) on nature-related activities, e.g. outdoor clothes, binoculars, camera gear, hotels and transportation; while 1.3 million, or just over 5% of the population, joined or contributed to nature-related organizations such as naturalist, conservation, or sportsman's clubs.

These results mirror many of the trends in the USA. In the 1994–1995 US National Survey on Recreation and the Environment (US Federal Government, n.d.), birdwatching experienced the greatest positive percentage change from 1982–1983 to 1994–1995, of 30 outdoor recreation activities reported. Birding increased by 155% over the time period, followed by hiking (93.5%) and backpacking (72.7%). Conversely, fishing declined by 3.8% and hunting declined by 12.3%. Foot and Stoffman (1996) illustrate that in the USA, 65 million birders are spending US$5.2 billion annually on bird-related products, with a total economic output of US$15.9 billion.

In the latter part of the 1980s, ecotourism emerged to cater to a rather small market of travellers primarily interested in rainforests and exotic natural attractions. Anglo-Americans felt they had to go abroad to places like Costa Rica, the Galapagos Islands and Africa in order to be ecotourists. However, as the discussion above demonstrates, ecotourism in Anglo-America – or activities that adhere to some of the principles of ecotourism – has been thriving for some time, and rests on a solid foundation of supply and demand (see also Anderson, 1996). While this may be true from a statistical standpoint, further analysis points to the fact that the ecotourism industry in Anglo-America is rather loosely defined and encompasses a number of different products that may or may not be classified as ecotourism, including adventure and outdoor pursuits, wilderness, aboriginal culture and wildlife viewing. In this chapter the ecotourism industries of the USA and Canada are discussed, in addition to a number of key issues related to the development of ecotourism in these regions, including definitions of the term, policy, and the physical characteristics of Anglo-American regions where ecotourism is said to exist.

Ecotourism in the USA

In the USA, federal government involvement in ecotourism is virtually non-existent. Linda Harbaugh in the tourism policy unit of the US Department of Commerce (personal communication, June 22, 1999) acknowledges that while her department is concerned with the generation of tourism statistics, it has no direct concern for the development of ecotourism programme areas. The USA, therefore, has no federal laws dictating protocol for regional ecotourism development. This responsibility lies with the individual states which have their own mandates, budgets and products, and who help those working in the field, e.g. operators, with marketing and promotion. In a review of governmental agencies charged with the responsibility of administering the ecotourism industry, Edwards *et al.* (1998) found little consistency among states and provinces in Anglo-America. Tourism and ecotourism are administered by a wide variety of agencies of government, including policy, marketing, economic development, planning and environment. The implication is that tourism is not yet perceived as having enough importance to stand alone as a distinct department.

The definitions of ecotourism that are currently used by the US states are just as variable as the governmental bodies that administer tourism and ecotourism. The principle difference lies in the usage of the 'nature', 'nature-based' and 'ecotourism' labels. The examples below serve to illustrate this point.

- In Alaska, 'ecotourism' is defined as 'Environmentally responsible travel to experience the natural areas and culture of the region while promoting conservation and economically contributing to local communities' (Alaska Wilderness Recreation and Tourism Association, 1999, p. 1).
- In South Carolina, 'nature-based tourism' is defined as 'responsible travel to natural areas that conserves the environment and improves the welfare of local people' (South Carolina Nature-Based Tourism Association, 1997, p. 9).
- In Texas, 'nature tourism' is defined as 'discretionary travel to natural areas that conserves the environmental, social and cultural values while generating an economic benefit to the local community' (Texas Parks and Wildlife Department, 1995, p. 2).
- In Hawaii, 'ecotourism' is defined as 'nature-based travel to Hawaii's natural attractions to experience and study Hawaii's unique flora, fauna, and culture in a manner which is ecologically responsible, sustains the well-being of the local community, and is infused with the spirit of aloha aina (love of the land) (Centre for Tourism Policy Studies, 1994, p. i).
- In Florida, 'ecotourism' is defined as 'responsible travel to natural areas which conserves the environment and sustains the well-being of local people while providing a quality experience that connects the visitor to nature' (Florida Ecotourism/Heritage Tourism Advisory Committee, 1997, p. C-3).

South Carolina, for example, uses the nature-based tourism label to act as an umbrella for a variety of tourism activities, including backpacking and hiking, birding, boat tours, camping outfitters, canoe/kayak outfitters, cycling tours, environmental education, farms, fishing operators, gardens, horseback riding, hunting, lodging packages, retail outfitting businesses, scuba-diving, state parks, white-water rafting and zoos. Texas does the same, and acknowledges that although hunting and fishing are the mainstays of nature tourism in the state, other non-consumptive activities are emerging with the most significant market growth. The activities classified as nature tourism above, however, are classified as ecotourism in Alaska, Florida and Hawaii. Although many states use terms like ecotourism, nature tourism and nature-based tourism interchangeably (South Carolina views ecotourism and nature-based as being synonymous), some theorists argue that ecotourism and nature tourism are quite different in meaning (see Chapter 5).

In the USA a number of not-for-profit organizations have been instrumental in helping to support the industry from both policy and programme perspectives. Although they are centred in the USA and have been active in national ecotourism issues, these organizations (The Ecotourism Society, The Adventure Travel Society, and Conservation International) also have broader international mandates.

The Ecotourism Society (TES)

TES (founded in 1990) works with the mission of fostering a true sense of synergy between tourism, research and conservation (The Ecotourism Society, 1999). This membership-based society sponsors events, publishes ecotourism documents, undertakes research, is involved in many international endeavours related to ecolodge development, and has been a resource for governments on a number of policy-related issues.

The Adventure Travel Society

This organization, which runs an adventure tourism and ecotourism conference every year, is strongly tied to the adventure industry. Its Adventure Travel Business Trade Association (ATBTA) exclusively serves the business and industry-related needs of adventure travel professionals and businesses like guides, outfitters, tour operators and adventure resorts.

Conservation International (CI)

CI approaches ecotourism less from a business perspective, and more from a grassroots vantage point. Their mission is to 'conserve the earth's living natural heritage, biodiversity, and to demonstrate that human societies are able to live harmoniously with nature' (Sweeting *et al.*, 1999). Accordingly, this Washington-based organization is active in a number of development issues in less developed countries around the world.

Ecotourism in Canada

The importance of Canada's nature tourism product is underscored in the Canadian Tourism Commission's (CTC) vision statement, 'Canada will be the premier four-season destination to connect with nature and to experience diverse cultures and communities' (Canadian Tourism Commission, 1998). The vision points to the fact that nature and culture are central to the development of tourism in the country. Not surprisingly, the CTC, which is a crown corporation of the federal government, has spent a great deal of time positioning the adventure and ecotourism sector as a key component of the overall industry. The CTC operates on the basis of core funding from the federal government, but also through unique public and private sector partnerships. However, it also relies on territorial, provincial and First Nations governmental departments and community groups as stakeholders to further develop ecotourism products. As of 1997, Canada was ranked ninth in terms of international tourist arrivals, with 17.6 million international tourists visiting the country (Canadian Tourism Commission, 1998). Tourism spending by foreigners rose by more than 8% per year during the last decade, reaching CDN$12.7 billion in 1997. However, the domestic visitor is still the mainstay of the industry with Canadians spending more than CDN$31 billion of the CDN$44 billion injected into the economy (Canadian Tourism Commission, 1998).

While the CTC failed to make a distinction between ecotourism and adventure tourism (the nature-based tourism label does not appear to be as strongly supported in Canada) in some of their publications up to 1995, by 1997 the two types of tourism were deemed separate. However, in Canada the focus is still predominantly on the adventure tourism product, as evident in the following organization of categories of adventure tours across the country:

- air sport/activity (e.g. hang-gliding, helitours);
- land sport/activity (e.g. hiking, mountain biking);
- water sport/activity (e.g. fishing, waterskiing);
- winter sport/activity (e.g. dog sledding, snowshoeing);
- nature/wildlife (e.g. ecotourism, naturalist tours).

Almost 60 distinct adventure activities are listed by the CTC in the the first four categories above, compared with just six activities listed under the nature/wildlife category.

The value placed on adventure and ecotourism in Canada is further emphasized in a recently published report entitled *Adventure Travel and Ecotourism: the Challenge Ahead*, by the CTC (1997). The document examines the adventure tourism and ecotourism opportunities and constraints for each province and territory on the basis of seven criteria: product development, packaging, resource protection/ sustainability, business development/ management, marketing/promotion, training/

human resources, and industry organization. The strategic priorities identified for Ontario are outlined in Table 7.1.

Like the USA, Canadian provinces have initiated their own policies and definitions of ecotourism based on their recognized needs. In the remote Yukon Territory, for example, ecotourism falls under the umbrella of wilderness tourism. According to the Wilderness Tourism Licensing Act (Government of Yukon, 1997), 'wilderness' is any area of the Yukon in a largely natural condition in which ecosystem processes are generally unaltered by human activity, and may include areas of visible human activity that do not detract from wilderness tourism. Wilderness tourism includes all types of activities occurring in the wilderness, including canoeing, kayaking, river rafting, snowmobiling, photographic safaris, and First Nation cultural interpretive tours. The definitions and policies developed for the wilderness tourism industry in the Yukon are the result of a series of meetings and interviews involving over 70 key members of the tourism community, including tourism and recreation associations, commercial outfitters (native and non-native) and government personnel. In an era of public accountability, open forum meetings and debates are needed to satisfy the needs of industry stakeholders.

The inclusion of several stakeholder groups, however, may often create friction

Table 7.1. Ontario strategic priorities (CTC, 1997). Ontario offers a diverse range of adventure and ecotourism activities, but competition from mass tourism has prevented it from fully developing its potential. Ontario's priorities should be product upgrading, resource protection/sustainability, marketing/promotion, training/organization.

Strategic thrust	Highest priority strategies
1 Product development	Diversify off-season products Upgrade product quality Facilitate access to funding
2 Packaging	Establish programme to link operators with external partners Provide 'how-to' advice Improve access to distribution and markets
3 Resource protection/sustainability	Coordinate resource access Implement environmentally sensitive/sustainable practices Advocate and coordinate resource management Improve local land use management
4 Business development/management	Facilitate access to development financing Make management training accessible Develop risk management and appropriate insurance Establish business mentoring programmes
5 Marketing/promotion	Select distribution channels and define tactics Upgrade promotional materials Develop the US market Broker cooperative marketing/promotion initiatives Provide 'how-to' advice
6 Training/human resources	Identify training needs and priorities Coordinate access to training opportunities Facilitate the development of training programmes and resources Develop 'how-to' manuals Facilitate access to opportunities
7 Industry organization	Communicate the benefits of organizing Support organizational development Enhance the development of provincial organizations

in the development of ecotourism policy. Such is the case in Saskatchewan where recent deliberations over ecotourism definition and policy have served to accent the fundamental differences between groups representing the environment and those representing economic development. Environmentalists in Saskatchewan (government and non-government organizations) are in favour of firmly structured principles defining and guiding the industry. Conversely, those in economic development are in favour of broadening ecotourism to encompass many non-consumptive (birding) and consumptive activities (fishing). At present deliberations are on hold in an attempt to reconcile the disparate stances of those involved. One of the key issues being discussed is the writing and interpretation of the proposed accreditation application document (Tourism Saskatchewan, 1999). Under the section entitled 'Primary Activities' are three standards-based questions (operators are asked to respond 'yes' or 'no') that deal with on-site activities. These read as follows:

1. Where wild plant materials are gathered in a sustainable fashion from the local ecosystem for on-site food preparation and consumption, are they gathered under the authority of a permit or license where required?
2. Where wildlife is captured in a sustainable and culturally acceptable fashion from the local ecosystem for on-site food preparation and consumption, is it taken under the authority of a permit or license where required?
3. Where the sustainable local gathering of wild plants or animals, birds, or fish for on-site food preparation is part of a package, is this component of the package specifically mentioned in advertising and promotional material?

The Saskatchewan case study is indicative of what is felt to be the main problem constraining the industry in Anglo-America (and I suspect elsewhere): definition and policy. While it is important to realize that no absolute definition exists for ecotourism, there is consensus among governments and industry (less consensus among academics) that ecotourism should be responsible, contribute to local livelihoods and contribute to conservation. Leadership in the development of state and provincial definitions has no doubt come from TES which has defined ecotourism as 'Responsible travel to areas which conserves the environment and improves the welfare of local people' (Western, 1993). Saskatchewan is no exception, and defines ecotourism as: 'an enlightened nature travel experience that contributes to conservation of ecosystems and the cultural and economic resources of the host communities' (Tourism Saskatchewan, 1999). While the two definitions differ only marginally, Saskatchewan's interpretation of the term may or may not be consistent with the ideals of TES. Point 2 (using fishing as an example) of the Saskatchewan strategy, above, suffices to argue this position:

1. Fishing can be responsible/enlightened (e.g. the institution of regulations and catch limits);
2. It can conserve the environment (e.g. in much the same way that 'Ducks Unlimited' conserves habitat for duck hunting); and
3. It can contribute to or improve the welfare of local people (e.g. soliciting the use of an aboriginal guide).

Using this rationale, a case may be made to argue that fishing (the catching and killing of fish for consumption) is indeed ecotourism. This is what the many Saskatchewan operators are probably striving to emphasize in protecting, or rather substantiating, their product under the current interpretation of the term. Consequently, such middle-of-the-road definitions of ecotourism leave much to the interpretation of the individual or agency, and say nothing about how ecotourism relates to the philosophy of ecocentrism, specifically in regards to sustainability, ethics, learning about nature, low impact, non-consumptiveness, and appropriate management which, according to some, are hallmark principles of ecotourism (see Wallace and Pierce, 1996; Fennell, 1999). The importance of definition, therefore, cannot be underestimated as a starting point in the development of a provincial or state ecotourism industry.

Issues in Anglo-American Ecotourism

Policy

Liu (1994) writes that policy is particularly important for ecotourism as a means by which to balance economics (the viability of ecotour operators) and environmental protection. Effective policy involves all stakeholders who stand to be influenced by the development of the ecotourism industry. Recently Edwards *et al.* (1998) undertook a comprehensive overview of policy in the Americas. Their work, as it relates to Anglo-America, has been adopted and presented here (Fig. 7.1). The map illustrates Canadian provinces and US states: (i) which have developed policy; (ii) which have not; and (iii) whose status is uncertain. The map shows that 7 of the 11 Canadian political jurisdictions (including

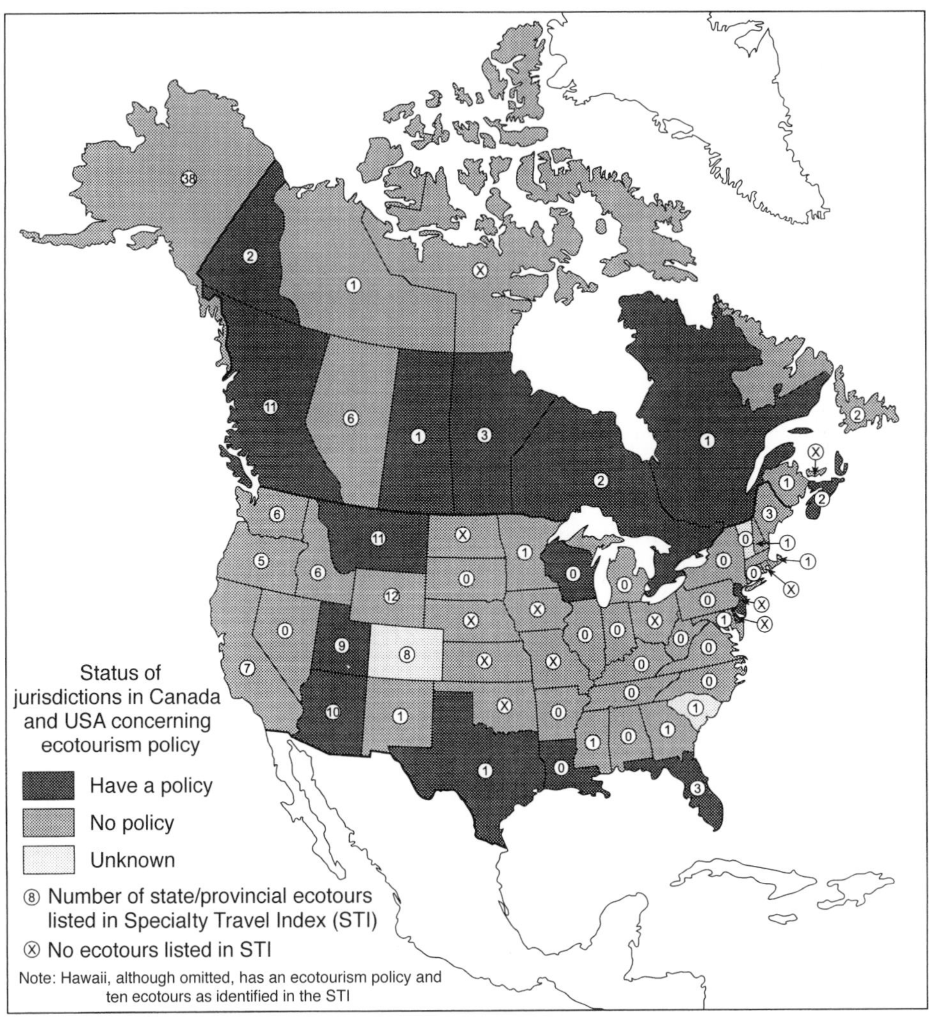

Fig. 7.1. Anglo-American ecotourism: policy and operator comparisons (adapted from Edwards *et al.*, 1998).

the newly established Nunavut Territory) have developed ecotourism policy, whereas this applies to only 10 of the 50 US states, most of which are located in the US south (Arizona, Texas, Louisiana and Florida).

Figure 7.1 also contains data on numbers of ecotours that occur in each of these political jurisdictions, as listed in the STI or Specialty Travel Index (1999). The STI is widely used as an advertisement medium by special interest tour operators. Although this list is certainly not representative of the numbers of ecotour operators that currently exist in states and provinces, it does provide an objective overview of numbers of ecotour operators from the perspective of one publication. Moreover, the STI was used because it lists tour operators by region and not just by type of tour, thereby allowing the researcher to determine the types of tours found within each state and province. (In gathering information on operators it became obvious that many states and provinces have only recently endeavoured to piece together lists of ecotour operators in their respective regions.) The methodology for identifying an ecotour operator in the STI involved simply scanning the special interest tours under each state and province, and identifying those that were ecotourism-oriented. The following STI listings were identified as ecotourism: birding, botany, butterfly tours, conservation, ecology, ecotourism, environmental education, marine biology, national parks, natural history, nature reserve, nature trips, rainforest, safari/game, whale-watching, wildflower viewing, wildlife viewing and zoology. As some tours could be found listed under six or seven of these names, the researcher had to record and control for any duplication of tours.

Figure 7.1 shows that there are very few ecotours listed in the STI in states east of the Rocky Mountains, with the biggest shortfall in the Midwest. Almost all of the western states have some degree of ecotourism, and Alaska, by far, has the greatest number of ecotours (38) in the STI, followed by Wyoming (12), Montana (11) and Hawaii and Arizona with 10 apiece. This western representation also holds true for Canada, with British Columbia and Alberta having the highest number of ecotours as listed in the STI.

Ecotourism operators

The overlap that exists between adventure and ecotourism products in Canada is well summarized in a recent magazine article entitled: 'Risky business: Canada's ecotourism outfitters will give you anything you want – from paddling wild rivers to hiking in the arctic' (Rapsey, 1995). In the article the author states that 'While different operators take different approaches to packaging and pitching their brand of wilderness experience – one man's adventure is another man's eco-tour – they generally agree on one thing: what they provide begins and ends with wilderness' (Rapsey, 1995, p. 30). Rapsey provides a diary of a trip down the Dumoine River in Quebec, which is easily representative of the types of adventure/ecotour trips outlined in the article and in certain parts of Canada as a whole. His description of the trip made reference to little other than rocks, white-water and instruction in canoeing. Based on descriptions like this, it appears as though it is 'open season' on ecotourism in Canada, especially in the north. Is it really ecotourism? To the uninitiated and those whose job it is to market ecotourism products, yes. Call it what you want, if it pays the bills it works. These days ecotourism is paying the bills.

The apparent confusion over adventure and ecotourism is alluded to in the article by Carolyn Wilde (cited in Rapsey), an Ottawa-based ecotourism consultant. She feels that Canada is not doing a good enough job of selling itself, and that no one really knows who is offering what to whom, or how reliable the outfitters are. Even more striking are Anderson's (1996) conclusions on the ecotourism industries in Canada and Alaska. Based on his discussions with a number of key industry stakeholders, Anderson found that there was a significant difference between the marketed experience and the actual product delivered. Operators had concerns about the overuse of the term ecotourism, and that they may not be able to live up to the strict definition of the concept. Anderson concluded by suggesting that, although there is considerable potential for the development of ecotourism in Canada and Alaska, the industry is negating this potential through a strong consumptive philosophy, anti-regulatory ideals, a lack of consistent standards, poorly developed ecological management standards, declining emphasis of private sector interpretation, and operator seasonality (a narrow window of opportunity).

In many respects the commercial enterprises in Alaska and Canada's north, often with government encouragement and support, have been left to re-package and retool their product from one initially based on consumptive activities (e.g. bear hunting), to one that is more non-consumptive (catch-and-release fishing or bear viewing, with some in-season hunting). Although many have not successfully made the transition, they see no reason not to use the conceptually elusive ecotourism label.

The same reasoning can be used by cultural tourism and adventure pursuits operators who specialize in, for example, week-long cycling or canoeing excursions. These operators encourage people to view wildlife on such trips in an effort to offer their clients as much of an on-site experience as possible, despite their lack of skills in natural history and environmental education. The interpretation, therefore, is often the responsibility of the client. The merging of adventure tourism, ecotourism and culture tourism that is increasingly apparent in Anglo-America and elsewhere is described by Fennell (1999) as ACE tourism. He suggests that this phenomenon, which has had a dilution effect on ecotourism, has grown stronger in recent years (Fig. 7.2). Depending on the activity(ies) and setting or region in which it occurs, ACE expands or contracts to represent the different focus of the product. While many tourists are in fact looking for combined nature, adventure and/or cultural experiences, the potential problem

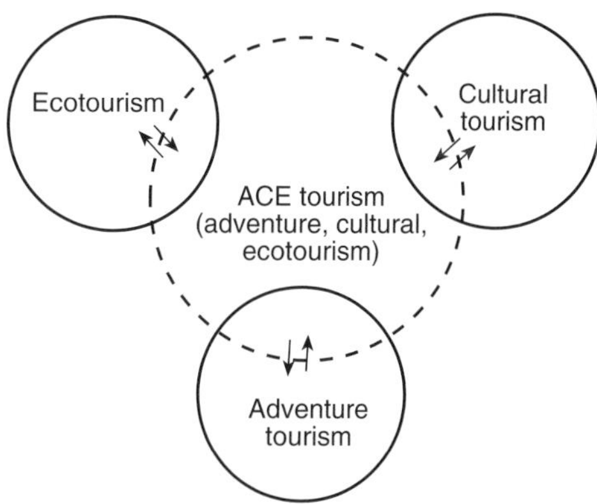

Fig. 7.2. The changing face of ecotourism (adapted from Fennell, 1999).

exists for those travellers looking for unique, hard path nature or adventure or cultural experiences in their travels.

The problem of ecotourism in name, but not necessarily content, also ventures further into the realm of environmental ethics. In conducting a content analysis of outdoor recreation magazines, Lenton (1993) found a vast number of Canadian tripping outfitters using terms like 'environmentally aware', 'minimum impact travel' and 'ecotourism' in their advertising. In fact, he found that there were enough tour operators waving environmentally friendly flags to sink the ecotourism boat. In a world that is hungry for ecotourism products, the competitiveness of the industry has tested the values of operators who are trying to stay afloat. This eco-opportunism is not solely an Anglo-American phenomenon but has been reported in destinations around the world. In some cases the problem has become so bad that many legitimate ecotour operators are refusing to use the label because of its poor image (Preece *et al.*, 1995).

While it has been necessary to highlight some of the dysfunctions of the ecotourism industry in Anglo-America, there are a number of operators who practise environmentally and culturally sound ecotourism. In his paper, Lenton (1993) identifies ten such Canadian ecotour operators, one of which is the Mingan Island Cetacean Study, and Lenton's description of this operator is as follows:

> The Mingan Island Cetacean Study (MICS) offers marine mammal educational programs to the public 'to help finance their research and enhance the public awareness of marine mammals'. This is a direct example of how ecotourism dollars are invested in conservation efforts. People pay to go out and spend anywhere from one to ten days on the ocean to observe the whales, dolphins, seals, and seabirds of the Mingan Island area in the Gulf of St. Lawrence ... MICS is a quintessential stop for the environmentally minded tourist travelling east of Quebec City. This group offers proof that conservation and ecotourism can be blended to yield many benefits. These include: boosting the local economy, initiating research, and educating the public about the plight of marine mammals – probably one of the best indicators of the quality of our environment.
>
> (Lenton, 1993, p. 13)

Ecotourism and ecoregions

In the 1980s, a paradigm shift occurred in the planning and management of parks and protected areas. Ecosystem management – the integrated management of human activities and the broad environments in which these take place – took the place of older models which viewed parks as discrete geopolitical entities. This new approach was based on the realization that in safeguarding natural areas, one must scientifically understand the human and biophysical processes that exist within these dynamic settings. Geology, landforms, soils, vegetation, climate, wildlife and water were all elements that had to be considered in park planning and management (Bailey, 1998). Map makers too were challenged with the task of trying to represent these dynamic systems at various scales. The results of the amalgamation of these elements, however, were important in enabling policy makers to make decisions on the basis of ecosystems (or ecoregions at various scales) instead of discrete sectors, and assisting in the setting of priorities and standards for resource management (Wilken and Gauthier, 1998).

Macroclimatic data have been especially valuable in the design of ecoregion maps, on the basis of the fact that climate is one of the most significant factors affecting soil composition, surface topography, vegetation and the distribution of life (Bailey, 1998). Figure 7.3 illustrates the main macroclimatic ecodivisions of Anglo-America, as developed by Bailey (the level of detail of the map represents the second of three ecoregion levels). In addition, a number of ecotour destinations/attractions are identified on the map as examples of the types of ecotourism activities that occur in these ecodivisions. Each of these regions is explained in greater detail below (climate descriptions are taken primarily from Bailey, 1998).

Tundra

Lying beyond the alpine tree line, this region is marked by slow-growing, low-formation, and mainly closed vegetation of dwarf-shrubs, moss and lichens. Normal January temperatures range from about −20 to −30°C, while normal July temperatures range from 10 to 15°C. In Churchill, Manitoba, a thriving tourist industry has developed around polar bears, which are forced to spend 3 or 4 months ashore due to the sea-ice melts. In Canada, there are 13 populations totalling some 15,000 bears (Churchill Northern Studies Centre, 1999), which feed typically on seals, walruses, beluga whales and narwhals. Late October is an excellent time to view polar bears in Churchill, as the bears congregate along the shores of Hudson Bay in anticipation of the formation of ice. Tundra buggies are used to help tourists view the bears, in addition to arctic foxes, ptarmigan and caribou. In the USA, Katmai National Park, 300 miles from Anchorage, Alaska, is one of world's most accessible locations from which to view brown bears. In July the Brook's River in the heart of the park attracts an abundance of bears which feed on sockeye

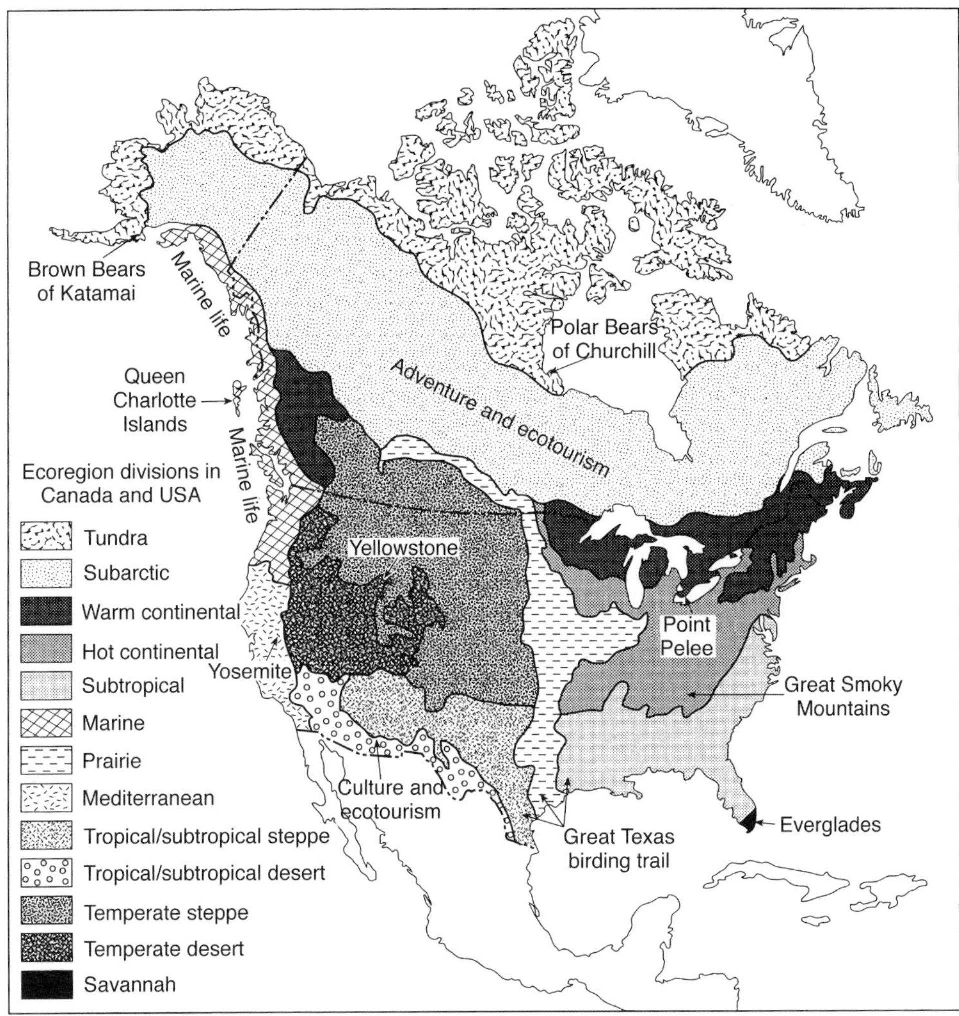

Fig. 7.3. Ecoregion divisions in Anglo-America (adapted from Bailey, 1998).

salmon. Although bears are the principal attraction, the surrounding lakes, forests, mountains and marshland, and marine environments are habitat to a diversity of birds, flowering plants and whales (USA National Parks Net, 1999d).

Subarctic

Temperatures in the Subarctic typically range from −25°C in the winter to +18°C in the summer, with moderate amounts (approximately 300 mm) of precipitation throughout the year. This vast region is dominated by boreal forests and the Canadian Shield (dominated by granite rocks with sedimentary and volcanic rocks), and contains a myriad of lakes and rivers which are ideal settings for those interested in water-based recreation (e.g. canoeing and white-water rafting). Consequently a number of adventure-ecotourism operations have developed in the region, and these tend to emphasize outdoor pursuits over wildlife viewing and environmental education (see above).

Warm continental

Temperatures in the Warm Continental zone typically average −5 to −10°C in the winter and +20 to 25°C in the summer, with moderate amounts of precipitation (600–1000 mm). Point Pelee National Park, despite its small size (20 km^2), is well noted for its diverse marsh, beaches, fields and forest habitats. As Canada's most southerly land mass, Point Pelee lies at the northern end of the broad belt that extends from the coastal zone of the Carolinas known as the eastern deciduous forest. Temperatures, which are moderated by the Great Lakes, are a major factor in limiting the northern expansion of this region. The park contains an amazing variety of plant and animal species, including 70 tree species, 47 species of reptiles and amphibians, and 50 species of spiders and insects found nowhere else in Canada (Parks Canada, 1999). However, the main attraction is the abundance of birds which frequent the park, especially in May, as a stopover on their annual migrations.

Hot continental

This region of broadleaved forests (mixed forest and meadow at higher elevations) contains Great Smoky Mountains National Park, which is located on the border of Tennessee and North Carolina. It is both a World Heritage Site and an International Biosphere Reserve, and hosts over 9 million visits per year. Five different forest types dominate the Great Smoky Mountains, and together they support 130 species of trees and 4000 other plant species (US National Parks Net, 1999c). Wildlife is also abundant in the park, including black bear, the reintroduced red wolf, coyote, bobcat, bats, European boar, the river otter, timber rattlesnake, copperhead, juncos, cardinals, blue jays, pileated woodpecker and wild turkeys.

Subtropical

The Subtropical realm is defined by broadleaved coniferous evergreen forests, significant amounts of rain (about 1300 mm per year) and mean temperatures that typically range from 8°C to 26°C. The eastern leg of the recently developed Great Texas Coastal Birding Trail is located in the south-western corner of the Subtropical region. This trail is an excellent example of how organizations and agencies in Texas (private citizens and landholders, conservation groups, businesses, government and communities) have come together with the common goal of enabling birders to gain access to the great diversity of avian resources found in the state. Over 450 of the state's 600 bird species are found along the coast, and the trail contains three sections spanning over 600 miles of coastline. In addition, the trail maintains 300 bird viewing areas in nine wildlife refuges, 11 state parks, one national seashore, and numerous city and county preserves (DiStefano and Raimi, n.d.).

Marine

Average annual temperatures in the Marine region typically range from 0°C to 18°C over the course of the year, with an abundance of precipitation (1000–2000 mm). This ecologically rich ecodivision is characterized by mixed forest, coniferous forests and meadow. Tour companies from Alaska down to Oregon offer a variety of ecotourism packages, including trips to view ice formations and glaciers, sea otters, Dall's porpoise, harbour seals, killer whales, grey whales, blue whales, mountain goats, Steller's sea lions, horned puffins, cormorants, black-legged kittiwakes, common murres, bald eagles, arctic terns and black oystercatchers (Wildlife Quest, 1999).

Prairie

The Prairies are characterized by tall grasses, most exceeding 1 m in height, and other broad-leaved herbs. Given the range of this region throughout Canada and the USA, temperatures vary significantly from the coastal areas of Texas to the interior of Saskatchewan. One of the cornerstones of a emerging ecotourism industry in Texas and Saskatchewan is birding. In fact many birds have Texas and Saskatchewan as their southern and northern ranges. In other areas through the Midwest, however, ecotourism is not well developed, as illustrated in Fig. 7.1.

Mediterranean

The Mediterranean region includes a combination of dry steppe, hard-leaved evergreen forests, open woodlands and scrub, and redwood forests. Yosemite National Park (mixed forest, coniferous forest and alpine meadows), in the mountainous part of this region is one of the most frequently visited parks in the USA, hosting upwards of 4 million visits per year. The popularity of the park is one of its chief concerns, however, as increasing numbers of vehicles in the summer months clog its roadways (Lovejoy, 1992). The park is home to 80 mammal species (e.g. coyote, mule deer, California bighorn sheep), 247 bird species (e.g. peregrine falcon, golden eagle, great grey owl), 40 reptile species, and hundreds of tree and wildflowers species, including the ponderosa pine and giant sequoias, which are the largest and oldest trees on earth (Delaware North Companies, 1999).

Tropical/subtropical steppe

This ecodivision is marked by steppes, shortgrass steppes and shrubs, and coniferous open woodland and semideserts. Mean temperatures range from approximately 10°C in the winter months to about 30°C in the summer, with ample precipitation throughout the year (600–700 mm). One of the main attractions in the north-west of this region is Grand Canyon National Park, and in the south-east the Great Texas Coastal Birding Trail (shortgrass steppes), which is home to a tremendous variety of birds (see Subtropical ecodivision, above).

Tropical/subtropical desert

Characterized by semideserts and deserts on sand, temperatures in this region typically range from about 12°C to 33°C, with very little rainfall (about 50 mm per year). The ecodivision contains Death Valley National Park, the Mojave Desert, Joshua Tree National Park, and a number of Native American settlements. The ecotours offered by the Indians of the region combine myth, archaeology and natural history. The desert is home to a tremendous diversity of life, including wild burros, roadrunners, turkey vultures, sidewinders, black-collared lizards, black widow spiders, scorpions, tarantulas, desert tortoises and desert iguanas (US National Parks Net, 1999a). Death Valley attracts scientists and nature enthusiasts from around the world and the increased demands placed upon this environment are of concern for organizations like The Death Valley Natural History Association, which is involved in preservation and interpretation of the natural and cultural history of the area.

Temperate steppe and desert

These two divisions, Temperate steppe and Temperate desert, have moderate temperatures, collectively ranging from 0°C to 20°C, with precipitation ranging from about 200 to 400 mm per year. The Temperate steppe is marked by steppes and dry steppes at lower elevations, and coniferous forests, open woodland and alpine meadows at higher elevations. The Temperate desert region consists of semideserts and deserts. The famed geothermal activity and fossil forests of Yellowstone National Park (the world's first national park, which contains open woodland, coniferous forest and alpine meadows), make this a principal attraction in the ecodivision. Yellowstone supports one of the continent's largest mammal populations, including grizzly bears, wolves, bison, elk, bighorn sheep, moose, pronghorn antelope and mule deer. Other animals include the marmot, bald eagles, osprey, sandhill cranes and trumpeter swans (US National Parks Net, 1999d).

Savannah

The Savannah ecodivision comprises open woodlands, and shrubs and savannahs (herbaceous vegetation with scattered woody plants, including low trees), with uniform temperatures averaging about 22°C throughout the year. In Everglades National Park, freshwater meets saltwater and the associated edge effect creates an area rich in biodiversity. Wildlife viewing (alligators, manatees, a variety of reptile species, roseate spoonbills, herons, egrets, eagles and osprey) is one of the principal activities of visitors. Other activities include biking, camping, diving, fishing, hiking, photography, ranger-led activities, snorkelling and swimming (US National Parks Net, 1999b). There is an abundance of trails, visitor centres and accommodation both inside and outside the park. In recent years a number of private ecotour operators have emerged to accommodate the needs of the area's growing tourist industry.

Conclusion

This chapter has examined the ecotourism industry in Anglo-America chiefly from definitional, policy, operator and ecoregional perspectives. Although it appears as though the industry will continue to grow into the 21st century, at least some of this potential will be compromised by the strong consumptive philosophy, antiregulatory ideals, and a lack of focus and consistent standards (Anderson, 1996) supported by those working in the field. The lack of consensus in defining ecotourism is partly to blame, especially in relation to what qualifies as ecotourism along the consumptive/non-consumptive continuum. As such, Anglo-America would benefit from future research which examines the region as one spatial unit, rather than 63 individual political jurisdictions.

The policy, operator and ecoregion data used in the maps of this chapter proved effective, as a starting point, in examining Anglo-American ecotourism from a broader continental and ecological perspective. While an attempt was made to compare regions on the basis of policy and numbers of operators, it soon became clear that much more work needs to be done in these two areas. The link between ecotourism and ecoregions was found to be a natural one. However, although ecotourism is reported to be an ecologically conscious form of tourism, there is little information available which documents its successes in preserving habitat and biodiversity in the ecoregions of Anglo-America (e.g. how can ecotourism in the prairie ecodivision be developed as an effective mechanism to preserve species like the black-footed ferret, whooping crane or burrowing owl?). As such, the union of ecoregions and ecotourism may prove to be a fruitful means by which to identify 'hot' and 'cold' spots of ecotourism development across the continent.

Anglo-America is often characterized as being economically and socially well developed. Although not a central theme in the chapter, there are often core–periphery relationships that exist between urban and remote communities in both Canada and the USA. To what extent ecotourism is able to help alleviate some of the economic and social disparities that exist in these remote regions is open to debate. Accordingly, future research should endeavour to examine if ecotourism contributes to local communities, and how such benefits are distributed. This information has not been adequately addressed in the research on ecotourism in Anglo-America, or elsewhere. Finally, a related concern surrounds the notion of estimating the economic worth of ecotourism in Anglo-America. If ecotourism is to subsume adventure and cultural tourism, in their various forms, economic impact studies will understandably be optimistic. Studies should be specific and cautious in how they define ecotourism, therefore, and the implications of over-visitation.

References

Alaska Wilderness Recreation and Tourism Association (1999) http://www.alaska.net/~awrta/

Anderson, M. (1996) *From Sea to Sea: Ecotourism Trends in Alaska and Canada.* Winston Churchill Memorial Trust Board, Wellington, New Zealand.

Bailey, R.G. (1998) *Ecoregions Map of North America.* USDA Forest Service, Fort Collins, Colorado.

Canadian Tourism Commission (1997) *Adventure Travel and Ecotourism: the Challenge Ahead.* Canadian Tourism Commission, Ottawa.

Canadian Tourism Commission (1998) *Annual Report: Achieving Critical Mass, 1997–1998.* Canadian Tourism Commission, Ottawa.

Centre for Tourism Policy Studies (1994) *Ecotourism Opportunities for Hawaii's Visitor Industry.* University of Hawaii, Manoa.

Churchill Northern Studies Centre (1999) http://www.brandonu.ca/CNSC/polar.htm

Delaware North Companies (1999) Yosemite National Park. http://www.yosemite.ca.us/cgi-bin/mirrory?www.yosemitepark.com/overview/index.htm

DiStefano and Raimi (n.d.) Great Texas Coastal Birding Trail. In: *Five Years of Progress: 110 Communities Where ISTEA is Making a Difference.* Texas Department of Transportation's Surface Transportation Policy Project.

The Ecotourism Society (1999) Information About TES. http://www.ecotourism.org/tesinfo.html

Edwards, S.N., McLaughlin, W.J. and Ham, S.H. (1998) *Comparative Study of Ecotourism Policy in the Americas – 1998: USA and Canada.* Department of Resource Recreation and Tourism, University of Idaho, Moscow, Idaho.

Fennell, D.A. (1999) *Ecotourism: an Introduction.* Routledge, London.

Filion, F.L., Foley, J.P. and Jacquemot, A.J. (1992) The economics of global ecotourism. Paper presented at the Fourth World Congress on Parks and Protected Areas, Caracas, Venezuela, 10–21 February.

Finch, R. and Elder, J. (eds) (1990) *The Norton Book of Nature Writing.* W.W. Norton, New York.

Florida Ecotourism/Heritage Tourism Advisory Committee (1997) *Planning for the Florida of the Future.* Government Document, Tallahassee, Florida.

Foot, D.K. and Stoffman, D. (1996) *Boom, Bust & Echo.* Macfarlane Walter & Ross, Toronto.

Government of Yukon (1997) *Wilderness Tourism Licensing Act (Draft).* Government Policy Document, Whitehorse, Yukon.

Lenton, P. (1993) Ecotourism: Canadian destinations for the conscientious tripper. *Explore* 58, 11–15.

Liu, J.C. (1994) *Pacific Island Ecotourism: a Public Policy and Planning Guide.* Office of Territorial and International Affairs, Hawaii.

Lovejoy, T.E. (1992) Looking to the next millennium. *National Parks* January/February, 41–44.

Nash, R. (1982) *Wilderness and the American Mind.* Yale University Press, New Haven.

Parks Canada (1999) Point Pelee National Park. http://parkscanada.pch.gc.ca/parks/ontario/point_pelee.htm

Preece, N., van Oosterzee, P. and James, D. (1995) *Biodiversity Conservation and Ecotourism: an Investigation of Linkages, Mutual Benefits and Future Opportunities.* Department of the Environment, Sport, and Territories, Canberra.

Rapsey, M. (1995) Risky business: Canada's ecotourism outfitters will give you anything you want – from paddling wild rivers to hiking in the arctic. *Outdoor Canada* April, 28–34.

South Carolina Nature-Based Tourism Association (1997) *The South Carolina Nature-Based Tourism Directory and Listing of the South Carolina Nature-Based Tourism Association Members.* SCNBTA, Greenville, South Carolina.

Specialty Travel Index (1999) *Specialty Travel Index: the Adventure Travel Directory,* Issue 38. Alpine Hansen Publishers, San Anselmo, California.

Statistics Canada (1998) *The Importance of Nature to Canadians.* Minister of Supply and Services, Ottawa.

Sweeting, J.E.N., Bruner, A.G. and Rosenfeld, A.B. (1999) *The Green Host Effect: an Integrated Approach to Sustainable Tourism and Resort Development.* Conservation International, Washington, DC.

Texas Parks and Wildlife Department (1995) *Nature tourism in the Lone Star State: Economic Opportunities in Nature.* State Task Force on Texas Nature Tourism, Austin, Texas.

Tourism Saskatchewan (1999) *Application Document to Accredit an Ecotourism Package (Draft).* Tourism Saskatchewan Publication, Regina, Saskatchewan.

US Department of the Interior (1998) *National Park Service Statistical Abstract.* Public Use Statistics Office, Denver, Colorado.

US Federal Government (n.d.) *The National Survey on Recreation and the Environment.* US Federal Government, Washington, DC.

US National Parks Net (1999a) Death Valley National Park. http://www.nps.gov/deva/

US National Parks Net (1999b) Everglades National Park. http://www.everglades.national-park.com/location.htm

US National Parks Net (1999c) Great Smoky Mountains National Park. http://www.great.smoky.mountains.national-park.com/info.htm

US National Parks Net (1999d) Katmai National Park. http://www.katmai.national-park.com/sights.htm

Wallace, G.N. and Pierce, S.M. (1996) An evaluation of ecotourism in Amazonas, Brazil. *Annals of Tourism Research* 23(4), 843–873.

Western, D. (1993) Defining ecotourism. In: Lindberg, K. and Hawkins, D.E. (eds) *Ecotourism: a Guide for Planners and Managers.* The Ecotourism Society, North Bennington, Vermont.

Wildlife Quest (1999) The Wildlife. http://wildlifequest.com/wildlife.htm

Wilken, E. and Gauthier, D. (1998) Ecological regions of North America. In: Munroe, N. and Willison, J.H. (eds) *Linking Protected Areas with Working Landscapes Conserving Biodiversity.* Science and Management of Protected Areas Association, Wolfville, Nova Scotia.

Chapter 8

Asia

A.A. Lew
Department of Geography and Public Planning, Northern Arizona University, Flagstaff, Arizona, USA

Introduction

Asia is, by far, the world's largest continent. It ranges in elevation from Mount Everest (8848 m; 29,028 ft) and the Tibetan Plateau (average 4500 m; 15,000 ft), to the earth's lowest point on the surface of the Dead Sea (over 400 m or 1312 ft below sea level). Not far away is the world's largest sand desert, the Rub' al Khali, on the Arabian Peninsula. In the south-east of the continent lies a large tropical archipelago spanning the Equator over an area larger than the USA, while in Siberia the largest forest in the world (the taiga) gives way northward to a landscape of desolate tundra and permafrost. Not surprisingly, each of these environments has its own potential for ecotourism development, and most have been used for this purpose as the tourism industry seeks ever more remote and different destinations.

Asia is also the world's most populous continent, with some 3.64 billion people (Table 8.1). East Asia, South Asia and Southeast Asia together are home to over half of the world's human inhabitants. With such a large population, Asia's ethnic diversity, often associated with ecotourism, is also great. Using language as a simple and conservative measure of ethnicity, Asia is home to 2165 living languages, about a third of an estimated world total of 6703 language (Grimes, 1999). However, this can be an unfair comparison, as Asia really consists of five subcontinents (listed in Table 8.1), in addition to Europe – the sixth subcontinent on the Eurasian landmass. Each of these subcontinents has its own internal wealth of environmental and cultural diversity and uniqueness. For example, some 500 languages and local dialects have been identified in Indonesia in Southeast Asia, and 300 different languages are spoken daily in India on the South Asian subcontinent, along with more than 1000 local dialects. Today, even the most remote of these cultures has been touched by organized groups of ecotourists and adventure tourists.

According to the World Tourism Organization (WTO), tourism in the Asia (East and Southeast Asia) and Pacific region grew in 1996 by 9.3% (arrivals) and 9.8% (receipts) over 1995 (WTO, 1997). These rates of growth were nearly twice that for the world overall (4.5% arrivals, 7.6% receipts in 1996). Arrival growth rates for South Asia (4.3%) and Southwest Asia (10.5%) were also significant in 1996, though South Asia was affected by internal political problems that year. Unfortunately, the 1997–1998 Asian economic crisis resulted in a 1.2% decline in international

Table 8.1. World regional populations, mid-1999 estimate and World Heritage Sites (PRB, 1999; UNESCO, 1999).

	Population		World Heritage Sites	
Region	(1,000,000)	%	Total	Endangered
World	5981	100.0	582	23
Asia total	3637	60.8	125	4
East Asia	1481	24.8	35	0
South Asia	1301	21.8	41	1
Southeast Asia	520	8.7	16	1
Southwest Asia	252	4.2	31	2
Central Asia	87	1.5	2	0
Africa	771	12.9	79	10
Europe	728	12.2	241	3
Central and South America	512	8.6	76	4
North America (US and Canada)	303	5.1	31	2
Pacific	30	0.5	18	0

Notes: Figures are adjusted from PRB (1999), which originally combined Central Asia with South Asia. Other differences in this table from the original data are: Central Asia includes Afghanistan and excludes Mongolia and Siberian Russia, both of which are in East Asia in the table above, and Iran and the Caucasus are included in Southwest Asia above.

arrivals for East and Southeast Asia (including the Pacific) for the second year in a row in 1998, and a 3.8% decline in receipts (WTO, 1999b) (Table 8.2). Some countries were hurt far more than others, with Indonesia down 5.5% while Thailand was up 6.9%. Furthermore, South and Southwest Asia both continued the strong percentage growth that they experienced through most of the late 1990s.

Once the East and Southeast Asian economies recover, however, tourism growth will probably rebound, bringing with it both good and bad impacts. While welcome economic news to many, Asia's robust growth in tourism has not been without its social and environmental costs. A survey of member countries of the Asia-Pacific Economic Cooperation group found that environmental pollution, air traffic congestion, and overcrowding at major attractions were the three leading constraints to the expansion of tourism in the region (Muqbil, 1996). Private sector respondents to the same survey, however, identified excessive governmental controls over the use of sensitive natural areas and conflicts between tour operators and natural resource managers as their major difficulties. The issue of exploitation versus the conservation of tourism resources has been a major concern throughout the Asia-Pacific region since the 1980s, when ecotourism (which as 'nature tourism' has been an important part of Asian tourism for decades) emerged as a visible and potentially major segment of the region's tourism industry.

A survey of mostly North American tour companies that offered ecotours to the Asia-Pacific region found that, as an economic activity, ecotourism to Asia has been growing at about 20% a year through most of the 1990s, at least prior to the Asian economic crisis (Lew, 1998). In this study, Indonesia was the most cited destination, followed by South Asian countries bordering the Himalayas (Table 8.3). Many secondary ecotour destinations are also shown in Table 8.3. Some of these, like China and Thailand are already major tourist destinations, although their role as ecotourist destinations may still be developing (Studley, 1999). Other countries, such as those in Indochina and Central Asia, are emerging destinations where all forms of tourism,

Table 8.2. 1998 tourist arrivals (estimates) (WTO, 1999b).

	Total arrivals (1,000,000)	% Change 1997–1998	Arrivals as % of world
World	625.2	2.4	100.0
East Asia/Pacific	86.9	−1.2	13.9
Middle East	15.6	5.3	2.5
South Asia	5.1	5.0	0.8
Europe	372.5	3.0	59.6
Americas	120.2	1.4	19.2
Africa	24.9	7.5	4.0

not just ecotourism, are as yet poorly developed. (Note that for the study cited in Table 8.3, Southwest Asia was excluded and the Pacific was included.)

A 1992 study by the Pacific Asia Travel Association (PATA), also of North American ecotour providers, reported that clients were primarily interested in rainforest destinations (62%), followed by islands (17%) and mountains (17%) (Yee, 1992). These are all features with which Asia is strongly endowed. In Lew's 1998 study, ecotour providers to the Asia-Pacific similarly focused on nature, although culture and educational activities were also significant elements (Table 8.4).

Ecotourism opportunities clearly abound in every corner of the Asian continent (About.com, 1999). The range of natural and cultural attractions in each of the major subregions (East Asia, Southeast Asia, South Asia, and Southwest and Central Asia) of this continent are discussed below.

East Asia

East Asia consists of Japan, North and South Korea, China, Taiwan, Mongolia and the Siberia region of Russia. Along the eastern edge of the continent the Pacific Ocean floor is colliding with the Eurasian plate to create a string of mostly offshore islands, from Japan to the Philippines. The Kamchatka Peninsula is also part of this 'Pacific Ring of Fire'. Volcanic activity in many of these coastal mountain areas provides opportunities for scientific exploration and mineral bathing from the 'Valley of the Geysers' on the Kamchatka Peninsula, through the islands of Japan and Taiwan and into Southeast Asia. Further inland the East Asian landscape gradually rises to the Tibetan Plateau and Himalayan mountain range.

East Asia is the most populated region of Asia and the most culturally homogeneous. The most heavily inhabited areas are in eastern China, on the Korean Peninsula and in Japan, where humanity has often overwhelmed the natural environment. The opposite is true of Siberia, Mongolia and western China (including Chinese occupied Tibet) which have very low population densities and expanses of relatively unexploited deserts, plains and forests.

Mountain peaks have been among the most cherished natural environments for the cultures of East Asia; monasteries are one of the few proper forms of development allowed. Pristine wilderness, however, is rare, though backcountry trekking areas do exist in Taiwan's high mountains. Following the Second World War, growing populations and an expanding middle class made the high mountain areas of coastal East Asia more accessible to the masses and starting in the 1950s national and local park systems were introduced to protect these resources. Management and enforcement, however, have been lax across the region and the excessive trampling of mountain soil and vegetation is a widespread problem.

Table 8.3. Asia-Pacific countries and regions in which North American ecotours operated (Lew, 1998).

Country or region	No. tour companies	% of all tour companies
Indonesia	16	40.0
India	13	32.5
Australia	12	30.0
Nepal	12	30.0
Bhutan	10	25.0
New Zealand	8	20.0
Tibet (region of China)	8	20.0
China	7	17.5
Thailand	7	17.5
Myanmar (Burma)	5	12.5
Cambodia	5	12.5
Laos	5	12.5
Pakistan	5	12.5
Malaysia	4	10.0
Papua New Guinea	4	10.0
Russian Far East	4	10.0
Vietnam	4	10.0
Central Asia[a]	3	7.5
Japan	3	7.5
Mongolia	3	7.5
Sikkim (State of India)	3	7.5
Philippines	2	5.0

[a] Uzbekistan and Kyrgyzstan.

Lowland areas, on the other hand, have been intensely settled, cultivated, and industrialized. The close relationship between human agricultural activity and the land has created distinctive ecosystems that have been of interest to alternative tour groups since the 1970s. Rice paddies that combine the growing of wet rice with the cultivation of bottom-feeding fish, mostly carp, are an example.

Away from the more densely settled portions of East Asia are areas that contain some of the greatest potential for ecotourism on the continent. Starting in the north, Siberia holds considerable opportunity for ecotourism development, although the costs of these experiences tend to be high due to the region's remoteness and limited infrastructure. Lake Baikal, Eurasia's largest freshwater lake and the world's deepest lake, is located in southern Siberia, and serves as a focus for the region's ecotourism, due both to its wildlife and accessible location near the Trans-Siberia railway. Wilderness (hiking and wildlife viewing), sporting activities (fishing and hunting) and arctic activities (dog and reindeer sledding and skiing) are the mainstay of the fledgling tourism industry in interior Siberia. Whale and other sea mammals, birds, and volcanoes are attractions along coastal and island areas, and especially on the Kamchatka Peninsula. Inuit and other nomadic peoples of the north compose the cultural attraction of the area. Unfortunately, it is estimated that 10% of the wildlife in northern Eurasia (including Central Asia) is seriously in danger of extinction (SEN, 1997). Similar to Central Asia (see below), Siberia is in need of more concerted conservation efforts, which ecotourism might be able to facilitate.

Moving further south, the steppes of

Table 8.4. Ecotour types and activities in the Asia-Pacific region (Lew, 1998).

	%	Tour descriptors
Tour type		
Nature	71	Wildlife, natural history, jungles/rainforests, science-based nature tours, fossil expeditions, national parks, nature reserves, orang-utans, ornithology, village wildlife conservation, zoos
Culture	45	Ethnic culture, agriculture, anthropology, countryside tours, culture exchanges, ethnic area lodge, food, local guides, sustainable technology
Adventure	13	Soft adventure, adventure, hard adventure, outdoor adventure
Tour activities		
Land	48	Trekking, walking, cycling/mountain biking, backpacking, bushwalking, day hiking, physical activity
Education	35	Educational, guest scholar/teachers/experts, animal riding safaris, birdwatching, local educational programmes, photo-taking safaris, study tours
Water	19	Boat rides, diving, rafting, sailing, sea kayaking, white-water

N = 31 respondents.

Mongolia and north-west China became an increasingly recognized ecotour destination in the late 1990s. Ecotours have emphasized visiting and living among the country's traditionally nomadic ethnic groups. Activities include camel treks, horse and yak riding, trekking, fishing, and wildlife viewing. This was also the path of the Silk Route from China to Europe, which has become a major theme for promoting tourism in north-west China. Many of north-west China's deserts are surrounded by high mountains, which are also emerging ecotour destinations, in addition to the cultural and archaeological history of oasis settlements. The Muslim Uyghur people in China's Xinjiang Autonomous Region are a major cultural attraction, although Buddhist archaeological sites (often found in caves) are more frequented by tour groups. Most tours are organized due to the political sensitivity of this region, where terrorist attacks by Uyghur separatists occasionally occur.

Even more politically sensitive for the Chinese government is Tibet (Xizang Autonomous Region). The Tibetan Plateau is a largely arid region with a base elevation of 3000 to over 4500 m (10,000 to 15,000 ft), bordered on the south by the Himalayan mountain range. Tibetan culture is largely pastoral and deeply religious, following the Lamaism branch of Buddhism. China forcibly reasserted its domination over Tibet in 1950, after which Tibetan Lamaism was severely suppressed. Many of the harsher measures were lifted in the 1980s and 1990s, but the political situation remains tense. Travel to Tibet has been expensive, mostly limited to groups, and rarely extended beyond the capital Lhasa and nearby mountains that have restored monasteries and mountain trekking opportunities. However, as part of China's officially designated 'Year of Ecotourism' in 1999, several new areas of Tibet have been opened to international tourism for the first time (Studley, 1999).

To the north-east and east of the Tibetan Plateau are areas of China that hold considerable opportunity for ecotourism development, both for nature and culture. As yet, these areas are not developed and many are environmentally quite sensitive, such as the panda regions below the Plateau's eastern slopes. South-east of the Tibetan Plateau, however, is the location of what is possibly China's premier ecotourism region. Yunnan Province, with an average elevation of about 2000 m (6500 ft), is famous for its year-round temperate-tropical climate, its biological diversity and for the cultural diversity of its many ethnic groups, which are related to the hill people of northern Southeast Asia. With the Tibetan Plateau on its western border, Yunnan has a wide variety of ecosystems giving it over half of China's plant diversity, 64% of its bird species, 42% of the country's reptiles, and many rare and endangered animals. The Nature Conservancy is working with the Chinese government to create the Yunnan Great Rivers National Parks System, which would encompass a region of 3000 m (10,000 ft)-deep gorges through which three of the world's great rivers pass. The goals of this project include biodiversity and cultural heritage protection, and sustainable economic development.

Ecotourism is an important part of the tourism industry in many parts of East Asia. However, the areas where ecotourism has the greatest potential tend to be very remote, making the cost and experience of getting there difficult. Because of this, ecotourism plays a smaller part in the overall tourism industry of East Asia than it does in most of the other parts of the Asian continent. In most people's minds, travel to East Asia remains an urban experience.

Southeast Asia

In contrast to East Asia, ecotourism is a pervasive aspect of the rapidly growing tourism industry in Southeast Asia. Geographically, Southeast Asia can be divided into two parts: Peninsular, consisting of Myanmar (Burma), Thailand, Laos, Cambodia and Vietnam; and Insular, consisting of Malaysia (including its mainland component), Singapore, Brunei, Indonesia and the Philippines. Peninsular Southeast Asia comprises a series of mountain ranges and rivers extending out of the Tibetan Plateau and south-west China. Trekking to visit the hill tribe people of northern Thailand is the most popular eco-adventure experience in the northernmost portion of Southeast Asia. Similar rainforest experiences may be possible in Myanmar once the political situation in that country becomes more stable, and in Laos as that country gradually becomes more open to foreigners. Further south, cultural and archaeological treasures, such as Cambodia's Angkor Wat, become more important. Much of Vietnam's environmental beauty lies in the rich diversity along its long coastline, although remote rainforests and hill villages are also of potential interest. Still further south, the tropical beaches of this region offer a wide variety of ecotour opportunities, including sea kayaking through limestone caves, underwater coral diving and coastal wildlife viewing. Most of the people of Peninsular Southeast Asia follow Theravadin Buddhism, with strong undercurrents of Hinduism and local animist beliefs that make for a very colourful ethnic landscape.

Thailand has had a well-developed tour industry for many years and Bangkok is one of the most internationally accessible cities in Southeast Asia. A larger proportion of the country's land area is under national park and conservation area status than any other country in Asia, outside of Bhutan, though management resources are a continuing problem (Weaver, 1998). The other countries of Peninsular Southeast Asia have only been opening to tourism since the 1990s and none has a well-developed travel sector, let alone an ecotourism industry. National parks are being created, but are not adequately financed, and major cultural heritage sites continue to be looted. Vietnam and Cambodia experienced rapid increases in tourist arrivals at the end of the 1990s, and the more insulated Laos and Myanmar are hoping to do the same. Their longer period of isolation, great environmental and cultural richness, and international accessibility via Bangkok, give these emerging countries the potential to be among the great ecotourism destinations in the next century, rivalling Indonesia before its economic and social turmoil (Lilley, 1998).

Insular Southeast Asia is rich in its variety of landscapes and biology due to its tropical location (astride the Equator), the diverse size and shape of its many thousands of islands (many of which are volcanic in origin), and the region's ethnic diversity. Malaysia is one of the more economically developed of the Southeast Asian countries and has many well-developed and protected nature areas. Mountain peaks, including traditional 'hill station' resorts and the highest peak in Southeast Asia, Mount Kinabalu (4101 m; 13,455 ft) on Malaysia's portion of Borneo, offer distinctive ecological transition zones. Taman Negara National Park, a highland rainforest in West Malaysia, has been a case study in efforts to balance ecosystem conservation with ecotourism development.

More than anything else, orang-utans have come to symbolize ecotourism in Malaysia and Indonesia. The natural habitat of these highly endangered animals is the dwindling rainforests of Sumatra and Borneo, both of which are popular ecotour trekking destinations, but they are most often seen at centres that specialize in reintroducing former pet orang-utans to the wild. While hiking the rainforest interiors of Borneo, ecotrekkers also come into contact with the culture of the traditional Dayak headhunters of the island and some of the 450 species of birds found there. The West Papua (formerly Irian Jaya) province of Indonesia, located on the island of New Guinea, also offers contact with pre-modern tribal groups in dense rainforests, while rare animals such as the Sumatran tiger, Javan rhinoceros and the sun bear, the world's smallest bear, are also found in the Indonesian archipelago. Komodo Island, in south-eastern Indonesia is home to the Komodo dragon, the world's largest land lizard, the protection of which has been described as another model of ecotourism development (Hitchcock, 1993).

The 17,000-plus islands of Indonesia have coastlines that stretch over 80,000 km (50,000 miles) and contain 15% of the world's coral reefs, making the country one of the greatest scuba-diving destinations on Earth (DTPT, 1997). The Philippines does not have as many volcanic attractions as Indonesia, but does have similar coastal coral diving opportunities. Another major ecotour attraction of the Philippines is the rice-terraced mountain slopes of northern Luzon Island.

Indonesia has benefited from its close proximity to Australia, for which it serves as an inexpensive and exotic vacation destination. In addition to significant tourist arrivals from their former colonial rulers – for example the Dutch are frequent travellers in Indonesia and the British in Malaysia – there has been a rapid increase in intra-regional and domestic ecotourism throughout Southeast Asia, especially in Thailand, Malaysia and Indonesia. For example, before the Asian economic crisis of 1997, increasing numbers of ecotourists from Jakarta and other major cities on Java were trekking into interior Sumatra and Borneo to see the wildlife and traditional cultures. Similarly, Thai nationals comprise the great majority of visitors to Thailand's national parks (Weaver, 1998).

The population densities of Southeast Asia have historically been well below those of East and South Asia, allowing for a greater expression of the region's natural environment. As such, Southeast Asia has become the Asian continent's premier ecotourism destination. With its tropical climate, rainforests, coastal coral reefs, mountain trekking, great variety of flora, fauna and ethnic cultures, developed and emerging destinations, and easy international access through the major international air hubs of Bangkok and Singapore, Southeast Asia has all the ingredients to keep it at the forefront of world ecotourism destinations.

South Asia

South Asia, often referred to as the 'Indian Subcontinent', is separated from the rest of Asia by an arc of mountains, including the Himalayas. Most of these mountain ranges are active earthquake areas, although volcanic activity is rare. The land within consists of a less mountainous peninsula dominated by India, with Pakistan to the west and Bangladesh to the east. India is large and diverse, with 14 official languages, each associated with a major province, and well over 1000 smaller ethnic groups. Pakistan, Nepal and Bhutan, along with India, share portions of the great mountain ranges to the north, while Sri Lanka and the Maldives are tropical island countries at the southern tip of India. The summer monsoon (winds from the south-west) bring some relief to the otherwise year-round hot and humid weather of lowland South Asia. Drier shrub and grasslands are found in much of Pakistan and adjacent areas of India, as well as on the Deccan Plateau in central India.

Next to Southeast Asia, South Asia is the most popular ecotour destination on the Asian continent (Table 8.3). Most of this interest lies in the Himalaya region; mountain trekking in this area is among the ultimate ecotour experiences. Large numbers of foreign mountain trekkers have had major impacts on the local culture and environment in the largely impoverished Himalayan region, particularly in Nepal and Kashmir (Weaver, 1998). Nepal's Annapurna Conservation Area Project has been a major conservation effort to address some of these impacts in one of the most popular trekking regions, although several other smaller efforts are also under way (Pobocik and Butalla, 1998). All, however, are seriously underfunded and understaffed.

Bhutan has created a particularly unique experience designed to sustainably maintain its strong traditional culture and religion. Long closed to the rest of the world, Bhutan limits its arrivals by requiring a fully planned itinerary supplied by Bhutanese tour operators and a minimum expenditure of US$200 per day, well beyond the means of a typical budget traveller. This has helped to keep Bhutan less commercialized than Nepal, and among the most traditional cultures on the planet. Kashmir (in the Himalayas) and Assam (south and east of Bhutan, and bordering Myanmar) are two corners of India that contain a diversity of cultural (non-Hindu) and natural attractions, but have been disrupted by separatist movements that have kept tourism to a minimum. Pakistan's Kashmir region is similar to that of India, both in its trekking opportunities and political sensitivity. The Karakoram Highway passes through this region, allowing a backdoor entry into China for tourists. Periodic military tensions with India on one side and Afghanistan on the other can make this journey through the Himalayas a risky prospect. More severe, however, has been ethnic and political conflict among Pakistan's major ethnic groups (Baluchis, Pushtun and Sindhi), which has limited all forms of tourism to Pakistan in the 1990s.

Below the northern mountains are the densely settled agricultural lands of the Indo-Gangetic Plain (Indus River in Pakistan and Ganges River in India and Bangladesh). Like East Asia, the many people in this region have caused considerable environmental degradation. Native ecosystems have been replaced with agricultural ecosystems that may have functioned well in earlier times, but are often environmentally dysfunctional under current population pressures. Volunteer tourism that supports efforts to make traditional agriculture more sustainable is one form of ecotourism experience that is potentially available in most South Asian countries, though the market is rather modest. Slightly lower populations reside in drier areas to the south of the Indo-Gangetic Plain, where several ecotour destinations have been developed in national parks and wildlife refuges, as well as on the Thar Desert with its popular camel treks.

Much of the ecotourism in India is centred on the country's many national parks and wildlife refuges. Jim Corbett National Park, for example, was created in 1936 from a popular hunting ground for the British (Rao, 1998). Ranthambor National Park's (390 km^2; 150 mile^2) dry forests in south-western Rajasthan preserve 300 tree species and 50 aquatic plant species. These forests, and many others on the Deccan Plateau, are havens for the Indian tiger, although poaching is a constant problem. Three hundred species of birds are protected in Keoladeo Ghana National Park, and Bandipur National Park, in south-western India, is a habitat for wild Asian elephants. Outside Kashmir in the north, Pakistan has far fewer floral and faunal resources, although its Arabian Sea coast is rich in shellfish and sea turtles, as well as sharks.

The two Indian Ocean countries of Sri Lanka and the Maldives are both significant ecotour destinations, although political unrest in northern Sri Lanka has seriously reduced the country's tourist arrivals. Scuba-diving and snorkelling are major activities in both countries, with dive safaris and reef tours a speciality on the coral atolls of the Maldives. Sri Lanka also has varied and lush jungle vegetation extending into its cooler highland areas where animal and birdwatching and trekking are popular activities. The island's diverse wildlife resources are complemented by ancient and contemporary Buddhist sites.

South Asia accommodates many people, cultures, landscapes and ecotour opportunities into a relatively confined area. Human impacts are significant throughout and efforts to protect sensitive environments present major challenges. Trampled earth and litter are common in remote trekking areas, while poaching and loss of habitat are problems in nature preserves. The competition for survival between humans and the environment is greater here than in any other region of Asia, making it perhaps the most in need of a vibrant and influential ecotourism industry.

Southwest Asia and Central Asia

Southwest Asia and Central Asia together cover a diverse land area second in size only to East Asia among Asia's subcontinental regions. They are combined here because of a high degree of shared cultural and environmental characteristics, including an overall low population density, mostly arid grassland and desert climates, mostly Islamic religious beliefs, and historical ties that involve the Arab and Ottoman empires. The major differences between Southwest Asia and Central Asia are colder winter temperatures and a history of Soviet control in the north.

The cultural and geological diversity of Southwest Asia and Central Asia has created more countries in this region than in any other part of Asia. Asia Minor and the Caucasus include: Turkey, Georgia, Armenia, Azerbaijan, and sometimes Cyprus. The Arabian Peninsula includes: Syria, Lebanon, Israel, Jordan, Iraq, Kuwait, Saudi Arabia, Bahrain, Qatar, United Arab Emirates, Oman and Yemen. Iran and Afghanistan comprise most of the highland areas just south of the Caspian Sea and the countries of Central Asia. Turkmenistan, Uzbekistan, and Kazakhstan cover the internal drainage basins of Central Asia, while Tajikistan and Kyrgyzstan are situated in the high mountains of the Pamir Range and Tian Shan, respectively.

The arid conditions throughout the region create highly sensitive and endangered aquatic ecosystems, the largest of which are the Red Sea and the Persian Gulf in Southwest Asia, and the Black Sea, the Caspian Sea and the Aral Sea in Central Asia. These water bodies concentrate drainage from their surrounding watershed and serve as sensitive measures of regional environmental degradation. Most are severely polluted, though concerted international efforts in the 1990s to address these problems have helped to some degree. Ecotourism offers an alternative, especially in Central Asia, to the large-scale and heavy industries that have caused environmental widespread damage.

The potential for ecotourism in the many countries that span Central and Southwest Asia is considerable. Unfortunately, political and civil unrest and conflicts in many parts of this region make it the most dangerous of all the regions in Asia in which to travel. Yemen, Israel, Lebanon, Turkey, Cyprus, Iraq, Kuwait, Iran, Afghanistan, the Caucasus and Tajikstan have all experienced major civil wars, military incursions and terrorist conflicts that have closed major portions of their territory to international travel for varying periods of time during the 1980s and 1990s. The Central Asian republics have had additional growing pains in establishing legitimate rule and economic stability following their independence from the Soviet Union in the early 1990s, though their remote locations and poor transportation infrastructure may be even greater barriers for international visitors. Every country in the region has its politically sensitive areas and these issues need to be kept in mind in any discussion of ecotourism resources in Central and Southwest Asia.

Throughout Southwest Asia, nature tourism is poorly developed, with most tours focusing on archaeology, ancient cities, churches and monasteries, traditional markets, and village and urban life (Ady, 1997). On the Arabian Peninsula, day and overnight trips to Bedouin villages offer the most intimate desert experience for most foreign visitors. Modern technology has been developed in efforts to make the most efficient use of water in this arid land, and this has also created a type of educational ecotourism in the more developed areas of the region. In Israel, for example, people from other arid environments come to learn new ways of addressing shared problems.

Ecotourism opportunities are more plentiful on the outer edges of the Arabian Peninsula, where civilizations have existed for thousands of years. Major ecotourism activities include:

- diving in coral areas and viewing sea turtles, usually away from major cities where pollution has caused considerable coral damage;
- desert treks, usually by four-wheel drive, to see and photograph flora and fauna; these often focus on oases and coastal wetlands that attract large numbers of birds;
- hill trips to see wildflowers, wildlife, rock and cave art, and to collect fossils.

There are also areas of considerable ecotourism potential. These include the Farasan Islands, which support the largest wild gazelle population in Saudi Arabia, in addition to a great variety of birds, dolphins, turtles and whales in its coral reefs and mangroves (Ady, 1997; see also *Arabian Wildlife* online magazine). Israel has some 20 nature reserves protecting a diverse array of wetland areas and the plant and animal life that depend on them. In western Jordan runs the Great Rift Valley, a deep depression which includes the Jordan Valley, the Wadi ('oasis') Araba, Lake Tiberias (Sea of Galilee) and the Dead Sea, the lowest point on earth. In eastern Jordan there are wetland parks managed by the Worldwide Fund for Nature in the oases of the Shaumari and Azraq.

North of the Arabian Peninsula area lies a large area of highland plateaux and mountains covering the non-Arabic speaking countries of Turkey, Iran, Afghanistan and those in the Caucasus. Turkey has the best developed tourism industry in this sub-region, and while the others have considerable potential, political unrest has caused most of them to be largely closed to travel for most of the 1980s and 1990s. As in the Arabian Peninsula, archaeological and historic sites are the mainstay of tourism throughout highland Southwest Asia, which has historically served as the 'crossroads' of the world. The higher elevations are considerably cooler than the Arabian Peninsula, and many distinct forested regions are protected in Turkey's nature reserves and 21 national parks. Some of these parks, created as early as 1958, were initially established for archaeological and historical purposes but have since become rich habitats of protected biological diversity. Wetlands are also more plentiful in Asia Minor and the Caucasus. Trekking in the mountains (including ski-mountaineering in Georgia), bicycle touring and sea kayaking along the Mediterranean coast have become popular eco-adventure travel experiences.

Further to the east, tourism returned to Iran by the late 1990s, though visas for independent travel are very difficult to obtain. Rampant urban and industrial development, combined with devastation from the Iran–Iraq War in the 1980s, have caused widespread environmental damage, especially along coastal areas of the Persian Gulf and Caspian Sea. The country has established a few national parks, mostly in forested areas of the Alborz mountain range where the rare black-bearded and spiral-horned Alborz sheep reside. However, these parks are almost entirely unmanaged and unprotected. The situation is even worse in civil war-torn Afghanistan.

Central Asia's northern location deep within the continent of Asia results in extremes of summer heat and winter cold. Arid conditions are most prominent in the southern lowland areas, with higher precipitation in the mountains in northern Kazakhstan, which borders Siberia. Desert-based ecotourism resources are quite similar in these drier, sand-covered areas to those of the Arabian Peninsula; migratory birdwatching on wetland areas is the principal attraction. However, the greater difference between summer and winter temperatures in Central Asia results in large numbers of summer insects, which become a major distraction. Because of their location on the former edge of a large Jurassic sea, the south-eastern mountains of Turkmenistan and Uzbekistan have some outstanding dinosaur sites, including the Kugitang Reserve, located on the Turkmen side of their shared border. These mountains are also rich in mineral resources. The intersection of desert, grasslands and mountains make for a rich diversity of ecosystems that have already attracted hunters and fishermen from around the world who are starting to threaten some of the region's endangered large game (Sievers, 1998).

Kazakhstan is the ninth largest country in the world in land area and, in addition to its southern sand dune areas, consists mostly of rolling plains rising to older mountain areas in its eastern portions. It is also one of six countries that border the Caspian and Aral Seas. Despite serious environmental degradation due to the diversion of tributaries and fertilizer runoff, there is still considerable wildlife to be seen along the shores of these landlocked seas, and they have become the focus of major international ecological restoration efforts.

Highland Central Asia consists of the much smaller countries of Takjikistan and Kyrgyzstan, each of which has lowland plains in its western portions and high mountains in the eastern parts. For most of the 1990s, Tajikistan has suffered through civil war, resulting in the deaths of tens of thousands and in hundreds of thousands of refugees. Such circumstances make travel close to impossible to Central Asia's highest mountain peaks (over 7000 m or 23,000 ft high) where intense winter blizzards can last for several days. Kyrgyzstan, on the other hand, is possibly the most democratic and stable of the Central Asian countries. Most of the country is located in the Tian Shan mountain range, which is a major potential ecotourism area for China, as well. Ala-Archa Canyon is a national nature park not far from the Kyrgyz capital of Bishkek where numerous trekking trails lead visitors to glaciers on the country's highest peaks.

Southwest Asia and Central Asia cover a large area with many shared characteristics, including history, religion, politics and sensitive environmental resources. Ecotourism, combined with archaeological tourism, has great potential in almost every corner of this region. However, there remains a great need throughout to increase environmental awareness and to address major environmental problems. Political turmoil is also a major problem that stands in the way of tourism development. Once these issues are addressed, Southwest Asia and Central Asia could both blossom as major international ecotourism destinations.

Ecotourism Planning and Development

Although growing in popularity, ecotourism remains a relatively small, niche market. In some cases ecotourism products are offered as a means of education, public relations and financial support for organizations whose primary interests are nature and cultural conservation. In other instances, ecotourism is used as a marketing tool to entice conservation-oriented consumers to purchase tourism products that support biodiversity research efforts.

Both of these uses of ecotourism have been approached through regional, national and international organizations, many of which are focused on Asia. The activities of some of these major international groups are discussed below.

Tourism industry organizations: Pacific Asia Travel Association and national organizations

The PATA is a major international association of tourism industry providers, including travel agents, hoteliers, tour providers and government tourism organizations. PATA covers all of Asia, with the exception of Southwest Asia and Central Asia. Its member countries also include Oceania (Australia and the Pacific island nations) and North America, which is considered part of the Pacific Rim.

Even though environmental ethics were written into its original charter in 1952, it was not until the 1990s that PATA became involved in developing and promoting ecotourism destinations in a major way (PATA, 1992). In 1992 PATA members adopted the 'PATA Code for Environmentally Responsible Tourism' (PATA, 1999a). A few years late PATA started its Green Leaf programme, through which tourism-related companies reaffirm their support of the Code by paying an annual membership fee and, in return, receive the right to promote themselves through the PATA Green Leaf symbol. This is a membership programme only, not a certification programme.

PATA held its first Adventure Travel Conference in 1988 in Kathmandu, Nepal. The following year, the conference was expanded to add a travel mart (bringing together ground tour providers, mostly in Asia, and tour sellers, mostly in the developed world), and the year after that the title of the conference was expanded to include 'ecotourism'. Every year attendees at the PATA Adventure Travel and Ecotourism Conference discuss how they are promoting ecotourism and debate issues such as certifying ecotour products and industry responsibility (annual proceedings are available at PATA, 1999b). The success of these conferences, combined with the growing importance of ecotourism to most PATA member countries, has resulted in the establishment of the PATA Office of Environment and Culture, centred in PATA's European headquarters in Monaco. This office serves as a clearing house for news and publication on ecotourism and sustainable tourism, as well as promoting these among PATA members. It is also responsible for administering PATA's Green Leaf programme.

In addition to PATA, several national-level ecotourism associations have formed, primarily to promote the interests of ecotour operators, including environmental conservation organizations that provide ecotourism services. The Indonesia Ecotourism Network (formed in 1995) and Ecotourism Society Pakistan (formed in 1998; ESP, 2000) are among the more active national-level ecotourism industry groups. Efforts to establish similar groups in other countries, including China and India, have been less successful.

Intergovernmental organizations: World Tourism Organization and UNESCO

While PATA is the primary international tourism industry association operating in Asia, the WTO serves as an intergovernmental association of national tourism agencies. Most of the world's countries are members, in addition to several hundred affiliated private groups. In addition to gathering and standardizing tourism statistics among member countries, the WTO sponsors meetings that address contemporary tourism issues. It has focused on sustainable tourism issues for many years, starting with the Manila Declaration in 1980 (WTO, 1999a). Several of its conferences (and resulting proceedings) focus on tourism in Asia. Of particular interest for the ecotourism industry have been several conferences in the late 1990s on developing the Silk Road for tourism, affecting China, Central Asia and Southwest Asia, and the Asia-Pacific Ministers Conference on Tourism and Environment (PATA,

1997). While such ministerial meetings can often seem stilted and hyperbolic, they are worthwhile in identifying issues that have reached the attention of national and international political bodies. Three major political barriers to the sustainable development of tourism (and ecotourism) in Asia were identified at the 1997 meeting:

1. The dilemma of balancing economic development and ecosystem management.
2. Rapid population growth and its impact on the environment and travel demand.
3. A lack of political and lobbying strength in the tourism industry to promote a sustainable tourism agenda.

The United Nations Educational and Scientific Commission (UNESCO) has come to play a major role in ecotourism development through its World Heritage Centre and its designation of World Heritage Sites (UNESCO, 1999). While many of these sites are cultural (including almost all of those in Europe), a large number in Asia are nature related (Table 8.1). Their designation often leads to a considerable increase in tourist arrivals and UNESCO works with other organizations to promote sustainable tourism in the conservation of designated World Heritage Sites.

Non-governmental organizations

Most non-governmental organizations (NGOs) are primarily concerned with sustainable development issues, and secondarily with tourism. Examples of NGOs that are particularly active in Asia include the Sacred Earth Network, the Mountain Institute and the Hong Kong-based Ecumenical Coalition on Third World Tourism, which focuses on the social impacts of tourism, including prostitution, child labour and community empowerment. The Mountain Institute (1999) sponsors environmental and cultural conservation and development projects in alpine regions in the developing world, including the Himalayas. 'Volunteers' are able to participate in programmes, including the development of a community-based ecotourism programme in the Sikkim region of India. The Earthwatch Institute (1999) is a similar organization offering ecotourists work on educational and scientific tours, for example, with wolves in India and on preserving Angkor Wat in Cambodia.

The Sacred Earth Network (1999a) facilitates communications for environmental NGOs in 'northern Eurasia' (defined as the states of the former Soviet Union). A search of their online database (SEN, 1999b) using the word 'ecotourism' resulted in 60 entries, all of which are involved (or hope to be involved) in ecotourism programmes. Some of these are:

- The Sustainable Tourism Center in Georgia;
- Kamchatka Institute of Ecology and Nature Protection;
- Dashkhovuz Ecological Club in Turkmenistan;
- Tabiat-Environmental Movement of Kyrgyzstan;
- ARMECAS/Armenian Ecotourism Association; and
- International Public Center of Study of Local Lore and Ecotourism, 'Caucasus' in Azerbaijan.

Conclusions

Asia is an incredibly vast and diverse continent, both in terms of human settlements and physical attributes. Each of its large subcontinental regions has much to offer the ecotourism industry. Some areas are already well developed, most notably insular Southeast Asia and the Himalayas. Others are just now emerging as ecotourism destinations, including Siberia, Mongolia, western China and parts of peninsular Southeast Asia. Great potential exists elsewhere, though major barriers to immediate development also exist.

Similarly diverse are the natural and cultural resource conservation policies in Asia, which span the entire spectrum from economic to environmental priorities. In China, for example, conservation efforts often have significant economic goals; local governments are involved in resource development as much as resource protection (Lindberg *et al.*, 1997). India, on the other hand, has had a long history of conservation for the sake of preserving the natural environment. In Southeast Asia and Central Asia ecotourism is being viewed as an alternative form of development to more destructive industries, such as mining, timber and large-scale agriculture.

Throughout Asia, however, there is a critical need for enhanced environmental stewardship, a need which ecotourism can help to realize. This is primarily seen in the underfunding of conservation efforts, despite the growing economic success and globalization of Asian countries. In South Asia, most of the economic growth has been in urban areas, while continuing rural poverty limits the extent of conservation efforts in India and the Himalayan countries. On the other hand, in Southwest and Central Asia, along with bordering areas in China, cultural divisions and ethnic conflicts have made travel difficult and, like the economic imperative, have jeopardized environmental and cultural conservation efforts and the ecotourism potential that these regions can offer.

A potentially positive trend in the 1990s has been a growth in domestic and intra-Asian tourism. For many years tourism to developing countries has been a one-way venture, seen by some as a form of exploitation. Ecotourism comes out of a tradition of less exploitative alternative tourism and it has had some significant impacts. With the growing Asian middle class, there has been a significant increase in domestic and intra-Asian ecotourism across the continent. These better educated and travelled citizens can provide a voice for increasing conservation efforts in their home countries. At the same time the varying cultural behaviour and expectations of domestic tourists, Asian tourists, and tourists from Europe and North America in sensitive ecotourism destinations further complicates the form of environmental conservation and tourism development that is emerging (Lew, 1995; Lindberg *et al.*, 1997). Ecotourism has a good future in Asia, but one that will take creativity, patience and an understanding of the social changes that every corner of this vast continent is experiencing.

References

About.com (1999) *Asia – Ecotourism* (online). http://ecotourism.about.com/msub10.htm

Ady, J. (1997) Nature-oriented tourism in Saudi Arabia. *Landscape Issues* 3 (online). http://www.chelt.ac.uk/cwis/pubs/landiss/vol13/page3.htm

Arabian Wildlife online magazine. http://www.arabianwildlife.com/homefrm.html

Cater, E. (2000) Tourism in the Yunnan Great Rivers National Parks System Project: prospects for sustainability. *Tourism Geographics* 2(4), 472–484.

Department of Tourism, Post and Telecommunications of Indonesia (DTPT) (1997) *Diving* (online). http://www.tourismindonesia.com/diving.htm

Earthwatch Institute (1999) *Earthwatch Institute* (online). http://www.earthwatch.org/

Ecotourism Society Pakistan (ESP) (2000) *Ecotourism Society Pakistan.* http://www.ecotourism.org.pk/

Grimes, B.F. (ed.) (1999) Geographic distribution of living languages. In: *Ethnologue: Languages of the World*, 13th edn, Internet version, SLI International (online). http://www.sil.org/ethnologue/distribution.html

Hitchcock, M. (1993) Dragon tourism in Komodo, eastern Indonesia. In: Hitchcock, M., King, V.T. and Parnwell, M.L.G. (eds) *Tourism in South-East Asia.* Routledge, London, pp. 303–316.

Lew, A.A. (1995) Overseas Chinese and compatriots in China's tourism development. In: Lew, A.A. and Yu, L. (eds) *Tourism in China: Geographical, Political, and Economic Perspectives.* Westview Press, Boulder, Colorado, pp. 155–175.

Lew, A.A. (1998) The Asia-Pacific ecotourism industry: putting sustainable tourism into practice. In: Hall, C.M. and Lew, A.A. (eds) *Sustainable Tourism: a Geographical Approach.* Routledge, London, pp. 92–106.

Lilley, P. (1998) Pacific and Asia report. *Travel Trade Gazette, UK & Ireland* 4 March, 37.
Lindberg, K., Goulding, C., Huang, Z., Mo, J., Wei, P. and Kong, G. (1997) Ecotourism in China: selected issues and challenges. In: Oppermann, M. (ed.) *Pacific Rim Tourism.* CAB International, New York, pp. 128–143.
The Mountain Institute (1999) *The Mountain Institute* (online). http://www.mountain.org/index.html
Muqbil, I. (1996) Growth highlights tourism quandary. *Travel News Asia* (online). http://web3.asia1.com.sg/timesnet/data/tna/docs/tna3537.html
Pacific Asia Travel Association (PATA) (1992) *PATA Chapters Environment Day Resource Manual.* PATA, San Francisco.
Pacific Asia Travel Association (PATA) (1997) Tourism 2000: Building a Sustainable Future for Asia-Pacific. *Final Report of the Asia Pacific Ministers' Conference on Tourism and the Environment, Maldives, 16–17 February 1997.* World Tourism Organization, Madrid.
Pacific Asia Travel Association (PATA) (1999a) *PATA Code for Environmentally Responsible Tourism* (online). http://www.pata.org/frame3.cfm?pageid=55
Pacific Asia Travel Association (PATA) (1999b) *PATA Online Publication Catalog* (online). http://www.pata.org/sub6.cfm?pageid=6
Pobocik, M. and Butalla, C. (1998) Development in Nepal: the Annapurna Conservation Area Project. In: Hall, C.M. and Lew, A.A. (eds) *Sustainable Tourism: a Geographical Approach.* Routledge, London, pp. 159–172.
Population Reference Bureau (PRB) (1999) *Mid-1999 World Population* (online). http://www.prb.org/pubs/wpds99/wpds99a.htm
Rao, N. (1998) India's National Parks. *Great Outdoor Recreation Pages (GORP)* (online). http://www.gorp.com/gorp/location/asia/india/np_intro.htm
The Sacred Earth Network (SEN) (1997) *The Sacred Earth Network Newsletter* Spring 97, No. 11 (online). http://www.igc.apc.org/sen/ENews11.html
SEN (1999a) *The Sacred Earth Network* (online). http://www.igc.org/sen/
SEN (1999b) *Sacred Earth Network Eurasian E-mail Users Database.* http://ecologia.org/SENdb/
Sievers, E. (1998) The potential for ecotourism in Uzbekistan. *Arid Lands Newsletter* 43, Spring/Summer (online). http://ag.arizona.edu/OALS/ALN/aln43/uzbekistan.html
Studley, J. (1999) *Ecotourism in China: Endogenous Paradigms for SW China's Indigenous Minority Peoples.* May 1999 (online). http://ourworld.compuserve.com/homepages/John_Studley/Ecotours.HTM
UNESCO World Heritage Committee (UNESCO) (1999) *World Heritage Sites* (online). http://www.unesco.org/whc/nwhc/pages/sites/main.htm
Weaver, D.B. (1998) *Ecotourism in the Less Developed World.* CAB International, Wallingford, UK.
World Tourism Organization (WTO) (1997) *Tourism to East Asia and the Pacific Continues to Climb.* Press release, 15 May (online). http://www.world-tourism.org/pressrel/eapclimb.htm
World Tourism Organization (WTO) (1999a) *Sustainable Tourism Development – Declarations* (online). http://www.world-tourism.org/Sustainb/SustHom.htm
World Tourism Organization (WTO) (1999b) *Tourism Grows Steadily Despite Asian Financial Crisis.* Press release, 26 January (online). http://www.world-tourism.org/pressrel/26_01_99.htm
Yee, J.G. (1992) *Ecotourism Market Survey: A Survey of North American Tour Operators.* The Intelligence Centre, Pacific Asia Travel Association, San Francisco.

Chapter 9

Oceania (Australia, New Zealand and South Pacific)

R.K. Dowling

School of Marketing, Tourism and Leisure, Faculty of Business and Public Management, Edith Cowan University, Joondalup, Western Australia, Australia

Introduction

There is a considerable divergence in style and scale of tourism in the island nations of the Pacific (Craig-Smith and Fagence, 1992). The island microstates of Oceania are not a homogeneous grouping. Variations in population size, resource bases and relative isolation will clearly affect the ability of such nations to develop robust tourism industries (Milne, 1992a). Unlike their Australian and New Zealand counterparts, the Pacific island microstates have a very small domestic tourism market (Fagence, 1997).

Overviews of ecotourism in Oceania have been made by Hay (1992), Hall (1994), Carter and Davie (1996), Fagence (1997) and Weaver (1998a, b). Hall (1994, p. 137) asserts that 'in few regions around the world has interest in ecotourism been as pronounced as it has been in Australia, New Zealand and the countries of the South Pacific'. He adds that 'the natural environment of the south-west Pacific is a major tourist drawcard'. Ecotourism in Oceania comprises an eclectic assortment of levels of understanding, government commitment, maturity of destination and approach to business. At one end of the spectrum is Australia with its well-established ecotourism industry, demonstrable government support, well-established infrastructure, national strategy and industry association. At the other end lies a number of Pacific island nations which have few of the above elements but have outstanding natural attributes and enthusiastic communities. Boundless opportunities exist for ecotourism development in the Pacific, as recognized by the Cook Islands (McSweeney, 1992) and Samoa (Tourism Resource Consultants, 1991, as cited by McSweeney, 1992). One great advantage of the South Pacific region is that it has considerable natural resources and is still at an early enough stage of tourism development for alternative options to still remain available (Weaver, 1998b). Somewhere in between, and in reality probably closer to Australia than the Pacific island countries, is New Zealand, which markets its 'clean and green image'.

Common to all is their relative isolation from the major population centres and tourist trails of the northern hemisphere. Of course this is the very reason that they are now viewed as opportunities for ecotourism development given their unspoilt environments, diversity of cultures, relative security and friendly people. The countries form one of the world's rapidly growing tourist regions. This growth is from a small base and hence the growth is relative, rather than absolute, in number. However, with increasing ease of accessibility, cost, organizational aspects and favourable exchange rates, many of the countries of the region are emerging as 'new world' destinations for the travellers of Europe and North America. Hall and McArthur (1996, p. 131) state that 'since the early 1980s the number of visits to national parks and reserves has grown dramatically in Australia and New Zealand, and will continue to grow as tourist authorities increasingly market nature-based tourism or ecotourism activities'.

However, the region is not without its challenges. Problems facing the Pacific island countries include their relative isolation, rising population growth, limited resources, lack of infrastructure and financial resources, a limited pool of skilled labour, and the reliance on the export of raw materials (Sofield, 1994). Tourism is attractive to these countries, because it offers a solution to these problems and in recent years tourism has been touted as an emerging industry for the Pacific island countries. Yet, perhaps because most of the Pacific islands are remote, difficult to access and expensive to visit (Craig-Smith, 1994), just over a decade ago the regional performance in tourism attraction was described as 'lacklustre' (Economist Intelligence Unit, 1989). Thus the challenge for ecotourism in the Pacific region is to turn its problems into opportunities.

Sofield (1994) provides a litany of environmentally disastrous tourism developments in the South Pacific. They include mangrove reclamation for resort development in Fiji, coastal construction in Vanuatu, and the clearing of vegetation in Tonga. Barrington (1996) describes a similar situation in the Cook Islands which she argues is a paradise under siege from both tourism and general development. Many have argued that these situations threaten tourism development in the Pacific islands and that ecotourism offers a viable solution to this problem. Craig-Smith (1994) argues that the development of tourism in the small islands of the Pacific must be built around a niche market with an environmentally sustainable product. Under the European Union-funded Pacific Regional Tourism Development Programme, the Tourism Council of the South Pacific (TCSP) has been actively promoting ecotourism and it has been championed for a number of countries in the region. Choy (1998, p. 382) argues that 'the lesson from the experience of the Pacific islands over the last two decades is that the combination of sun, sand and sea in an exotic environment is not sufficient for continued success'.

The use of ecotourism as a solution to tourism problems in the Pacific islands is, however, also problematic. Just under a decade ago when describing an ecotourism conference in the Pacific, Valentine (1993, p. 108) noted that there were less ecologists present than either developers, architects, bankers, administrators or bureaucrats. He argued that there was an urgent need to put ecology back into ecotourism.

There is little obvious homogeneity in Oceanian ecotourism. The countries are vastly different geographically, are physically separated by huge tracts of ocean and comprise differing cultures. The one thing that binds them all together is their position in the Pacific Ocean and their relatively unspoilt natures. A brief review is now made of ecotourism in Australia, New Zealand and the Pacific islands (Fig. 9.1).

Australia

Recent developments in Australian ecotourism have placed it at the forefront of global ecotourism initiatives (McArthur and Weir, 1998). The country has made remarkable accomplishments in ecotourism over the past few years (Department of Industry, Science and Tourism, 1998). These include implementation of a range of ecotourism strategies, setting up national and local ecotourism associations, the publication of an annual industry guide, hosting international ecotourism

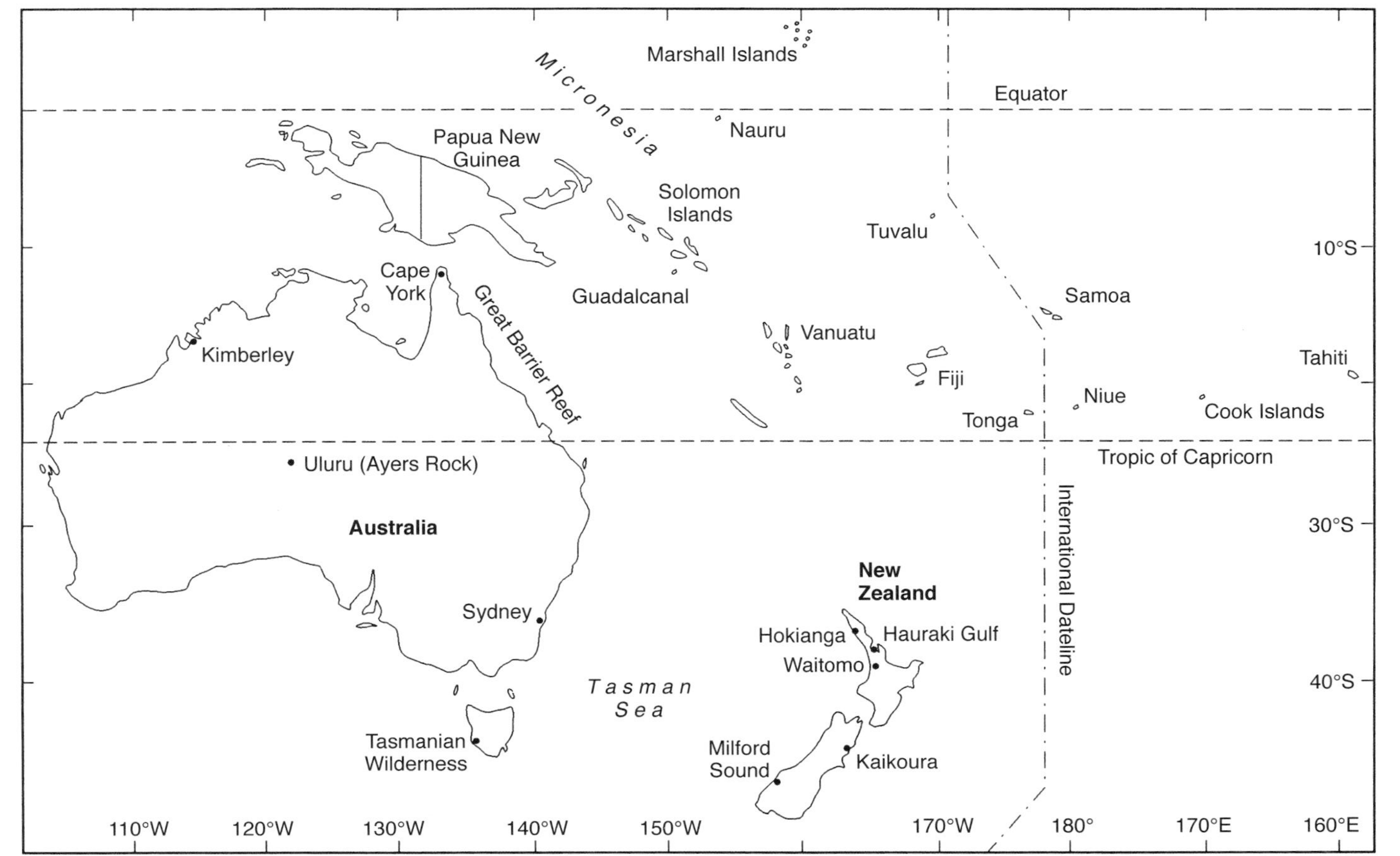

Fig. 9.1. Location map of Oceania (based on outline in Hall, 1994).

conferences, establishing an international research centre, establishing best practice ecotourism techniques, setting up ecotourism education and training courses, and developing the national ecotourism accreditation scheme (EAA and ATOA, 1997).

Ecotourism in Australia is being developed by 600 operators and it is estimated to generate a turnover of AU$250 million per annum. The industry employs approximately 4500 full time equivalent staff, which represents 1% of the total tourism industry employment in the country (Cotterill, 1996). Natural tourist icons include the Great Barrier Reef (1.5 million visitors per year (GBRMPA, 1998)), Uluru (Ayers Rock; Fig. 9.2) (350,000 visitors per year) and the Tasmanian wilderness.

The considerable progress in ecotourism over the past decade is mainly attributed to a clear and demonstrated lead by the Federal Government through its National Ecotourism Program executed between 1994 and 1996. The programme fostered the development of national ecotourism through innovative projects that aimed to increase Australia's competitiveness as an ecotourism destination, enhance visitor appreciation of natural and cultural values and contribute to the long-term conservation and management of ecotourism

Fig. 9.2. Climbers on Uluru (Ayers Rock) central Australia (photo by Jane James).

resources. A National Ecotourism Strategy was launched in March 1994 which provided an overall framework to guide the integrated planning, development and management of ecotourism in Australia on a sensitive and sustainable basis (Department of Tourism, 1994).

Australia's ecotourism development has benefited greatly from the formation in 1991 of the Ecotourism Association of Australia (EAA). In particular, the EAA pioneered, in conjunction with the Australian Tourism Operators Association (ATOA), the National Ecotourism Accreditation Program (NEAP) which was launched in November 1996 (Commonwealth of Australia, 1996a) (see Chapter 29). The programme distinguishes bona fide ecotourism products on the basis of eight principles including best practice environmental management, education, contribution to local communities, sensitivity to different cultures, consistency of product delivery and ethical marketing. The entities eligible for accreditation are nature-based tour companies, natural attractions relating to the regional environment, and accommodation providers in natural areas. Another recent initiative is the development of a National Nature and Ecotour Guide Certification Program. The key components for this are interpretation, education, ecological sustainability, minimal impact techniques, operations and awareness.

Trends

There is strong market demand for ecotourism in Australia (Hatch, 1998). Its rich and diverse natural heritage ensures Australia's capacity to attract international ecotourists and gives Australia a comparative advantage in the highly competitive tourism industry (Department of Industry, Science and Tourism, 1998). Ecotourists are generally from higher education groups that are comfortable with technology. Thus, many of their ecotour enquiries and bookings are transacted via the Internet and email. With publicity generated by the 2000 Summer Olympics held in Sydney, Australia is facing a major tourist influx that will have an impact on prospects for sustainable ecotourism.

The principles of ecotourism have widespread application and they can act as an exemplar for other forms of environmentally responsible tourism through the promotion of best practice in planning, design, management and operations (Department of Tourism, 1995; Tourism Council Australia, 1996). Lessons learned in the ecotourism industry can be applied to all types of tourism, for example, through the greening of hotels and resorts (TCA and CRC Tourism, 1998).

Another major trend in Australian ecotourism is the rapidly expanding knowledge base gained from research and education. A number of Australian universities offer ecotourism educational courses and carry out ecotourism research (see Chapter 40). The knowledge base is also improving and the EAA and the Federal Government's Office of National Tourism are both very active in the dissemination of ecotourism information nationally.

'Partnerships' was one of the buzzwords of the 1990s, especially as it related to government accomplishing its objectives. Partnerships require a shared vision, shared risks and shared benefits in order to be successful (Walters, 1992). The 1998 EAA conference concluded with specific recommendations to pursue improved partnerships with the indigenous and cultural tourism sectors, the conservation sector and the heritage sector (McArthur and Wight, 1999). This vision, if pursued, will enable Australia to maintain its position at the forefront of ecotourism development. Finally, Australian ecotourism is becoming more professional in orientation and whereas in the past many traditional ecotourism operators were involved in it for 'environmental' or 'lifestyle' reasons there is now a focus on increased professionalism and business orientation (McKercher, 1998).

Issues

There are a number of issues emerging in the development and growth of ecotourism in Australia. They include ecotourism's environmental (ecological, cultural and economic) sustainability (Burton, 1998) and the role of indigenous people. Since the natural environment underpins ecotourism, it is essential that it is protected and conserved. While there have been many successful Australian ecotourism operations which have been built on sustainable principles (Commonwealth of Australia, 1996b), some entrepreneurs operating in natural environments under the banner of 'ecotourism' are simply not looking after the environment and are merely pursuing 'easy' profits (Fries, 1998).

The other key issue facing Australian ecotourism is the place of indigenous people and culture (Bisset *et al.*, 1998). A key characteristic of the Australian definition of ecotourism is that it is based on cultural as well as natural values. In this regard the National Ecotourism Strategy states that 'cultural involvement requires consultation and negotiation with local communities (particularly indigenous communities) and organisations responsible for the management of cultural heritage values' (Department of Tourism, 1994, p. 18). The reality is, however, that there is minimal involvement in ecotourism by indigenous people in Australia despite the strong demand for it by international visitors. Yet to leave people out of the equation in tourist visitation to natural areas is merely to replicate the Western view that wilderness areas must be areas where humans do not live. This is errant nonsense and badly needs to be addressed in Australia. Such issues are particularly important for areas in northern Australia such as Cape York and the Kimberley Region, where indigenous people compose a large portion of the population. Other aspects of indigenous ecotourism include the impacts it has on communities and the intellectual property right issues associated with interpretation.

Australia's ecotourism industry is at the leading edge of ecotourism worldwide with its national ecotourism programme and strategy, rapidly growing industry association, the national ecotourism accreditation and guide certification schemes and multitude of training and education courses. Other innovations include built design principles, water and waste minimization practices, marketing strategies such as using the Internet, and the increasing interest in partnerships.

New Zealand

Tourism is well established in New Zealand and it makes a major contribution to the country's gross domestic product, foreign exchange earnings and employment. Traditionally the focus of tours to New Zealand have been on the Maori culture and/or adventure tourism. However, the recent overseas promotion of the country by its tourism board is one of a fresh, green and unspoilt environment and the number of visitors interested in New Zealand's natural features has increased accordingly. Key attractions include the glow worm caves at Waitomo (250,000 visitors per year), the fiords of Milford Sound (200,000) and whale-watching at Kaikoura. Over 55% of the country's 1 million international visitors per year, visit one or more of its national parks (New Zealand Tourism Department, n.d.) (Fig. 9.3).

Much of the tourism industry in New Zealand is now based on the natural environment, and in the mid-1990s concern was expressed about the relationship between tourism and the environment (NZTB and DoC, 1993; Hall, 1994). Notwithstanding ecotourism's potential to support environmental conservation it was also argued that the policies of directing tourists into natural environments was creating disturbance and causing adverse impacts (Warren and Taylor, 1994). At this time it was suggested that visitor pressure on some icon attractions such as the Waitomo Caves and Milford Sound could not be sustained even in the medium term without major attention being given to reducing adverse visitor effects. Ward and Beanland

Fig. 9.3. The Mackay Falls on the Milford Track, southern New Zealand (photo by Ross Dowling).

(1996) suggested that as there was still much to learn about the New Zealand environment it was difficult to anticipate visitor impacts on the environment before they occurred.

In 1997 the government investigated this situation and identified three principal adverse environmental effects associated with tourism (Parliamentary Commissioner for the Environment, 1997). These were the loss of quality of the natural environment, a reduction of amenity values from incremental development and the rising costs to the local communities of establishing infrastructure required for tourism development. The report's principal recommendation to the Minister for Tourism was for the government to implement a national sustainable tourism development strategy. It was envisioned that such a strategy would increase environmental awareness within the tourism industry, maintain high environmental quality, provide appropriately for the indigenous people and the land, and promote industry–government partnerships. This recent development in sustainable tourism is significant for the development of ecotourism as it brought about considerable changes to the traditional types of tourism. Guides are now expected to be knowledgeable about both the natural and the cultural environments as well as proficient in specific adventure activity skills (Ryan, 1997).

While ecotourism has been firmly defined and characterized across the Tasman Sea in Australia, in New Zealand a clear definition of the term still remains elusive. B. Ryan (1998, p. 1) states that ecotourism as a concept remains without a widely agreed definition and he argues the need for both descriptive and prescriptive definitions of the term. Interestingly he discusses ecotourism from a number of stances including its status as a product, a property and a social construction. Following this discussion Ryan paraphrases Budowski's (1976) definition as 'Ecotourism is a symbiotic relationship between conservation and tourism'. His prescriptive definition includes the three 'essential' criteria of it being nature-based, environmentally sustainable and contributing to conservation.

Writing more recently, Simmons (1999) suggests that this operational definition is a positive step toward shifting the focus of the definition from its current product orientation to a process orientation. He argues that there needs to be a more robust definition of ecotourism which acknowledges the environmental resource base on which it fundamentally depends. To achieve this goal he adds that a first step is to establish a set of indicators of environmental change induced by tourism.

Pearson (1998) suggested that there are four key issues facing ecotourism in New Zealand. These are the nature of the industry as well as a lack of clarity, management and coordination. To redress this situation Pearson calls for the industry to embrace a definition, develop a national strategy and establish an accreditation system. A major part of this strategy should be the integration of the natural and cultural environments as primary visitor attractions incorporating the sound industry base built around the Maori culture.

A number of studies have been carried out on the possibility of developing ecotourism ventures with and for Maori people. For example, a survey of the Hokianga District of Northland found that the area has considerable potential for marae-based ecotourism ventures (Mitchell *et al.*, 1998). The relatively unspoilt natural environment combined with the willingness of the local community to develop ecotourism proved to be positive factors. In addition the researchers noted that if the negative aspects of infrastructural and economic limitations were overcome then ecotourism ventures could be developed to form part of the livelihoods of these communities.

However, as with the situation in Australia, for New Zealand ecotourism to truly embrace Maori culture, let alone develop Maori ecotourism, it will need to understand the Maori view of the environment instead of imposing our own interpretation on it. Manuka Henare, a Hokianga Maori, defined Maori tourism as 'essentially a spiritual experience. People come to a place to learn more about themselves. They come not just to see but to be introduced to a world unseen too. The spiritual character is the competitive advantage Maori tourism can claim' (Urlich Cloher, 1998, p. 2).

Thus in New Zealand, ecotourism is at a relatively embryonic stage. Without an agreed definition, formalized national strategy or accreditation scheme, its ecotourism development faces some challenges. This is particularly so because the demand for New Zealand nature-based tourist activities is high, and consequently the industry is demand-led. This is a situation that could have significant impacts on the resources if they are not managed appropriately. C. Ryan (1998, p. 304) asks whether ecotourism in New Zealand is condemned to a duplication of small companies increasingly dependent upon overseas markets and the marketing strategies of intermediaries in the channel of distribution over which they have little or no control. He suggests that one answer is the creation of coalitions of operators offering products in partnership with each other.

South Pacific

Tourism in the South Pacific has been advanced by the TCSP and the South Pacific Regional Environmental Program (SPREP). However, there is a great diversity of environments and level of infrastructure in the nations of the region, so commonalities of ecotourism policy, planning, development and experience are few. For example, the government of Niue has marketed ecotourism based on its fringing reef and caves and chasms (Milne, 1992b). The Wotho Atoll in the Marshall Islands, sometimes described as the world's most beautiful atoll, has been suggested as a prime site for high-value ecotourism development due to its white sandy beaches, fringing coconut palms, and crystal clear waters rich in assemblages of coral and fish (Valentine, 1993). In Tuvalu the country's tourism strategy is low-key and is directed at international visitors who are attracted by its small size, remote nature, extensive marine environment and friendly, relaxed people (TCSP, 1992a).

Tourism development in Samoa is based firmly on its natural and cultural environment (TCSP, 1992b, p. 105). The country's tourism master plan acknowledges the need for strong environmental guidelines for tourism developments while enriching the environmental awareness and experience of the general interest tourist. In addition the country has invoked an ecotourism strategy in part to utilize its protected areas in order to create a representative framework of natural reserves and icon sites. This survey revealed a widespread awareness of ecotourism and a number of ecotourism products already in operation (Fagence, 1997). On six islands in Micronesia (Saipan, Rota, Guam, Palau, Pohnpei, Kosrae) tourism and conservation is being trialled by the US Department of Agriculture's Forest Service Tropical Forestry Program (Wylie, 1994).

In a general review of environmental protection and tourism development, the Solomon Islands were described as having a spectacular array of scenic beauty areas (marine and terrestrial) and interesting and diverse flora and fauna (TCSP, 1990, p. 18). An ecotourism development project on Makira Island has been established by a number of local villages to show how conservation and development can be integrated (Gould, 1995). Ecotours are being run by the locals in a bid to protect their old-growth forests and generate funds to improve health and education services. Another successful project is the Guadalcanal Rainforest Trail which 'offers an opportunity for ecologically sustainable tourism development, controlled and managed by indigenous Solomon Islanders' (Sofield, 1992, p. 99).

Fiji

Almost a decade ago the case was made for the development of nature-based tourism in Fiji in order to adapt to changing markets from short-haul to long-haul tourists (King, 1992) and as an alternative means of protecting the indigenous forest resources (Weaver, 1992). In a survey of the country in the mid-1990s Ayala (1995, p. 39) argued that Fiji is a natural ecotourism destination with its more than 300 islands of diverse geological origin encompassing a variety of species and habitats of outstanding scientific and heritage value, landscapes of exceptional scenic beauty, as well as unique marine features. She noted that its key assets include a relatively undisturbed environment, strong heritage and cultural traditions and existing tourism infrastructure. She concluded that the islands have a unique opportunity to integrate their natural, cultural and heritage assets to deliver a high-quality 'ecoproduct' with distinctive ecological dimensions. These include the sacred red prawns of Vatulele and turtle calling at Nacamaki. However, Ayala also suggested that, to capitalize on ecotourism, Fiji would need to improve its marketing through the potential of joint marketing, improve its training and service delivery, instigate better coordination of its inter-island links, and establish integrated tourism circuits for international visitors.

Today Fiji relies heavily on tourism

based on its natural environment. It provides a major source of jobs, especially for indigenous Fijians living in traditional villages, and it is the principal source of foreign exchange income. As a consequence the government has begun to show concern about the impact of tourism on Fiji's natural environment and its society (Haywood and Walsh, 1996). In 1997 an ecotourism plan was prepared for the country which specified five principles on which ecotourism should be developed (Harrison, 1997). These are environmental conservation, social cooperation, complementarity, centralized information, and infrastructure development. A key point involves the acceptance that ecotourism, with its strong link to village-based community tourism, can only complement, but not replace other forms of tourism. Policies that have been proposed to help implement the plan include environmental sustainability, public awareness campaigns, the embrace of cultural tourism, and the promotion of local economic benefits. Another key suggestion is the establishment of a technically oriented government Ecotourism Committee, to act as a focus for ecotourism development.

The integration of ecotourism and village-based community tourism is illustrated through the development of several grass roots ecotourism operations which have been established (Young, 1992). These include the Bouma ecotourism venture on Tavenui Island (Lees, 1992) and the Abaca Ecotourism Co-operative Society which was established at Koronayitu in 1993 (Gilbert, 1997). This area is the last remaining large area of unlogged tropical montane forest in western Viti Levu, Fiji's largest island. With pressure on the area for logging and mining the local village communities established the Abaca Cultural and Recreation Park. It consists of an ecolodge built from the remains of an old logging village, educational tours, and walk trails to scenic and historic sites. Earnings in the first year were equal to the whole community's total income before the business commenced operation. As a consequence of the success of this initiative a number of other villages in the region are now setting up their own cooperative ecotourism societies.

By way of contrast, Jean-Michel Cousteau, son of the famed French marine ecologist, Jacques Cousteau, has established an upmarket ecolodge on the island of Vanua Levu. Through operating the resort his aim is to show that it is possible to be environmentally friendly and financially successful at the same time (McCabe, 1998). This was something that Valentine (1993, p. 115) had argued 5 years before when he proposed that 'there is a need for research on successful small-scale village-based ecotourism and the design of practical working examples'. The resort contains only 20 bures (traditional Pacific island building primarily used for accommodation) which have been built with local timber and thatch and are oriented in such a way as to take advantage of the trade winds, thus eliminating the need for air conditioning. In addition their elevated nature allows the air to circulate underneath, also keeping them cool as well as preventing rot and deterring bugs. Mosquito control has been advanced by establishing endemic plants to attract birds and by creating a series of ponds to lure mosquitoes away from the resort where fish, toads and shrimp can then eat their larvae. The environmentally friendly resort includes a high-tech video link which allows non-diving guests to communicate with divers, and view the surrounding undersea environment on a giant screen, through the telecommunications Uplink video system (Murphy, 1997).

Discussion

This survey of ecotourism in Oceania raises a number of trends and issues. Across the region there is great diversity of natural attributes, potential ecotourism resources, infrastructure development, involvement of the indigenous people, and levels of understanding. Common to all countries, however, is the relative isolation of the region, the political interest in ecotourism as a perceived generator of economic benefits and desire by local communities to become involved in it.

Hall (1994) suggests that the term ecotourism is commonly used in the Pacific to refer to either 'green' or 'nature-based' tourism, that is, a distinct market segment, or any form of tourism development which is regarded as environmentally friendly. He argues that these two approaches to ecotourism pose distinct management, policy and development problems. Writing from Fiji, Harrison (1997, p. 75) argues that

> in recent years 'ecotourism' has become something of a buzz word in the tourism industry. To put the matter crudely, but not unfairly, promoters of tourism have tended to label any nature-oriented tourism product an example of 'ecotourism' while academics have so busied themselves in trying to define it that they have produced dozens of definitions and little else.

He goes on to suggest that if sustainable tourism development is to occur, trade-offs are inevitable and often nature will be the loser. He adds that ecotourism can not solve all the problems of mass tourism and may in fact generate problems of its own.

Harrison (1997) suggests that it should not be considered a stepping stone to large-scale tourism, though it often proves to be so. He concludes that ecotourism is an ideal, but one worth working towards, because the goals of ecotourism are to foster environmental conservation and cultural understanding.

As Ayala (1995, p. 40) has warned, 'it is becoming increasingly obvious that the small-volume definition of ecotourism is incorrect; the small-scale restriction on accommodation size is not viable; and control of ecotourists' direct contact with destination landscapes and cultures is needed urgently'. She suggests that resorts established in protected areas could play a major part in the advancement of ecotourism business in Pacific countries by providing tourists with unique insights into these places through education, interpretation and experience. An immediate planning concern for many of the South Pacific nations is the need to establish a network of protected areas in order to secure its natural attributes. Once this has been completed then decisions can be made on the potential to convert these attributes to assets, whereby ecotourism can be harnessed to generate the funds necessary to conserve and protect natural areas.

The inclusion of host communities in ecotourism is a contentious issue as many definitions appear to be predicated on the concept that ecotourism occurs only in natural areas which are devoid of people. However, more enlightened approaches to ecotourism include host communities in their characteristics. For example, Sofield (1991, p. 56) argues that 'while tourism has been drawn into current debate about "ecologically sustainable development", the topic of "sustainable ethnic tourism" has largely escaped attention'. He then illustrates this through the Pentecost Land Dive of Vanuatu (also 'naghol' – an annual ceremony in which specially selected initiates leap head first to the ground from a platform approximately 70 ft high, with vines tied to their ankles so that their foreheads just touch the earth). Helu-Thaman (1992) argues that the human elements of ecotourism are better expounded by 'ecocultural' tourism which is more culturally sensitive and integrated and more sustainable in the long term.

Other issues facing the development of ecotourism in the region include the need for increased knowledge and research, the reduction of economic leakage, the need for an ethical approach, the problems of natural and man-made disturbances and the need for planning. Fagence (1997) concludes his survey of ecotourism in the Pacific by stating that many island ecotourism strategies have been based on unrealistic resource appraisal and incomplete knowledge of potential source markets. Niue's heavy dependence on imported goods means that much of the initial expenditure leaks away from the country before it can generate income, employment and government revenue (Milne, 1992b). Dee Pigneguy, who operates a marine-based ecotourism enterprise in the Hauraki Gulf Maritime Park suggests that ecotourism operators should develop conservation ethics. She states that 'ecotourism is a high growth industry and it must be recognised that there will always be a point where economic benefits are outweighed by ecological costs' (Coventry, 1993, p. 56).

The threat of natural disasters such as cyclones, floods, landslips, avalanches and fires will always be present and pose a significant problem for the development and growth of tourism in natural areas (Craig-Smith and Fagence, 1994). Fiji experienced slumps in tourism in 1983 and 1985 because of cyclones (Haywood and Walsh, 1996). Tourism declined sharply in the aftermath of two successful coups in 1987 (Miller and Auyong, 1991). Cyclone Ofa in February 1990, one of the most destructive in the past two centuries, destroyed Niue's hotel, dive operations and coastal walking trails, and severely damaged its guest houses. Cyclones have also created challenges for the ecotourism industry in Samoa (McSweeney, 1992). In Papua New Guinea tourism has not realized its enormous potential due to its unstable law and order situation, the impact of mining and internal conflicts. In the Solomon Islands, tourism development was halted for a number of years over land disputes between foreign investors and traditional owners (Sofield, 1994).

The need for adequate planning for national ecotourism development is common throughout the region. In an early study of the impact of tourism on the Commonwealth of the Northern Mariana Islands, the need for adequate planning was emphasized (University of Hawaii, 1987). Suggestions included the establishment of proper land use controls such as the introduction of zoning; prescribed setbacks, height restrictions and maximum site coverage ratios for buildings; and the protection of natural and cultural sites.

A basic axiom when discussing ecotourism potential is that not all natural sites lend themselves to ecotourism development. Lees (1992, p. 61) states that 'Ecotourism is frequently billed as the panacea in helping conservation pay its way. The reality in the Pacific is that very few areas of conservation value also have the potential to support economic local tourism enterprises'. She goes on to describe the case of the Ubai Gubi region in the highlands of Papua New Guinea. Here an international environment group keen to facilitate the area's protection suggested an upmarket ecolodge for the viewing of the indigenous birds of paradise. The lodge was built but 'the project was a failure. Tourists have not arrived. The lodge remains but the enterprise has folded, yielding no benefits to the landowners or sound protection for the forest' (Lees, 1992, p. 62). As a further example, of the 24 areas suggested for conservation in the Solomon Islands, only three have been recommended for ecotourism development (TCSP, 1987).

Conclusions

It has been argued that the traditional export of Western concepts of tourism planning and development have limited application in the countries of Oceania. Rather than create 'enclave tourism', as in the case of a number of Pacific island nations, there is merit in creating integrated small-scale developments (Craig-Smith, 1994, p. 25). Further, it has been suggested by Hall (1994, p. 154) that the South Pacific region is 'facing a form of ecological imperialism in the region in which a new set of European cultural values is being impressed on indigenous cultures through ecotourism development'. Fagence (1997) argues that commitments to low volume, low technology, special interest tourism need to be based on a clear awareness of both the demand and supply. There is considerable evidence that the implications of a commitment to strategies of ecotourism are not well understood in the Pacific region.

Carter and Davie (1996, p. 71) state that 'like Australia, ecotourism in the Pacific is in a developing phase'. However, the authors suggest that in contrast with

Australia, tourism development in the Pacific is viewed as embracing culture, benefiting local communities and fostering environmental protection and management. They conclude that Australian ecotourism will result in a markedly different product focused on natural icons and largely ignoring the cultural elements of the landscape.

Thus ecotourism in Oceania is undergoing increased demand from travellers, emerging interest by local communities and the involvement of industry and governments. Based on its unspoilt natural terrestrial and marine environments, friendly people, and increased knowledge and infrastructure, the region will continue to foster ecotourism development at an unprecedented rate in the new millennium.

References

Ayala, H. (1995) From quality product to eco-product: will Fiji set a precedent? *Tourism Management* 16(1), 39–47.

Barrington, J. (1996) Paradise lost – Cook Islands under siege. *Forest & Bird* 279, 26–31.

Bisset, C., Perry, L. and Zeppel, H. (1998) Land and spirit: Aboriginal tourism in New South Wales. In: McArthur, S. and Weir, B. (eds) *Australia's Ecotourism Industry: a Snapshot in 1998.* Ecotourism Association of Australia, Brisbane, pp. 6–8.

Budowski, G. (1976) Tourism and environmental conservation: conflict, coexistence or symbiosis. *Environmental Conservation* 3(1), 27–31.

Burton, F. (1998) Can ecotourism objectives be achieved? *Annals of Tourism Research* 25(3), 755–758.

Carter, R.W. and Davie, J.D. (1996) (Eco)tourism in the Asia Pacific region. In: Richins, H., Richardson, J. and Crabtree, A. (eds) *Ecotourism and Nature-Based Tourism: Taking the Next Steps. Proceedings of the Ecotourism Association of Australia National Conference, Alice Springs, Northern Territory, 18–23 November 1995.* Ecotourism Association of Australia, Brisbane, pp. 67–72.

Commonwealth of Australia (1996a) *National Ecotourism Accreditation Program.* The Department of Industry, Science and Tourism, Canberra.

Commonwealth of Australia (1996b) *Projecting Success: Visitor Management Projects for Sustainable Tourism Growth.* The Department of Industry, Science and Tourism, Canberra.

Choy, D.J.L. (1998) Changing trends in Asia – Pacific tourism. *Tourism Management* 19(4), 381–382.

Cotterill, D. (1996) Developing a sustainable ecotourism business. In: Richins, H., Richardson, J. and Crabtree, A. (eds) *Ecotourism and Nature Based Tourism: Taking the Next Steps. Proceedings of the Ecotourism Association of Australia National Conference, 18–23 November 1995, Alice Springs, Northern Territory.* Ecotourism Association of Australia, Brisbane, pp. 135–140.

Coventry, N. (1993) New Zealand eco-tourism under international spotlight. *Accountants Journal* March, 55–56.

Craig-Smith, S.J. (1994) Pacific Island tourism – a new direction or a palimpsest of western tradition? *Australian Journal of Hospitality Management* 1(1), 23–26.

Craig-Smith, S.J. and Fagence, M. (1992) *Sustainable Tourism Development in Pacific Island Countries.* United Nations, New York.

Craig-Smith, S.J. and Fagence, M. (1994) A critique of tourism planning in the Pacific. In: Cooper, C. and Lockwood, A. (eds) *Progress in Tourism, Recreation and Hospitality Management*, Vol. 6. John Wiley & Sons, Chichester, UK, pp. 92–110.

Department of Industry, Science and Tourism (1998) Ecotourism. *Tourism Facts No. 15, June 1998.* Office of National Tourism, Canberra.

Department of Tourism (1994) *National Ecotourism Strategy.* AGPS, Canberra.

Department of Tourism (1995) *Best Practice Ecotourism: a Guide to Energy and Waste Minimisation.* Commonwealth Department of Tourism, Canberra.

EAA and ATOA (1997) *National Ecotourism Accreditation Program.* Ecotourism Association of Australia and Australian Tourism Operators Network, Brisbane.

Economist Intelligence Unit (1989) The Pacific Islands. *International Tourism Report* 4, 70–99.

Fagence, M. (1997) Ecotourism and Pacific Island Countries: the first generation of strategies. *The Journal of Tourism Studies* 8(2), 26–38.

Fries, P. (1998) Purists take on the profiteers in a battle for the great outdoors. In: *Profits in the Wilderness: The Ecotourism Industry's Identity Crisis.* The Weekend Australian Financial Review 11–12 July.

GBRMPA (1998) *State of the Great Barrier Reef World Heritage Area report, 1998.* Great Barrier Reef Marine Park Authority, Brisbane.

Gilbert, J. (1997) *Ecotourism Means Business.* GP Publications, Wellington, New Zealand.

Gould, J. (1995) Protecting Pacific forests. *Habitat Australia* 23(5), 14–15.

Hall, C.M. (1994) Ecotourism in Australia, New Zealand and the South Pacific: appropriate tourism or a new form of ecological imperialism? In: Cater, C. and Lowman, G. (eds) *Ecotourism: a Sustainable Option.* John Wiley & Sons, Chichester, UK, pp. 137–157.

Hall, C.M. and McArthur, S. (1996) Natural places: an introduction. In: Hall, C.M. and McArthur, S. (eds) *Heritage Management in Australia and New Zealand: the Human Dimension*, 2nd edn. Oxford University Press, Melbourne, pp. 128–134.

Harrison, D. (1997) Ecotourism in the South Pacific: the case of Fiji. In: *BIOSFERA World Ecotour '97 Abstracts Volume.* BIOSFERA, Rio de Janeiro, Brazil, p. 75.

Hatch, D. (1998) Understanding the Australian nature based tourism market. In: McArthur, S. and Weir, B. (eds) *Australia's Ecotourism Industry: a Snapshot in 1998.* Ecotourism Association of Australia, Brisbane, pp. 1–5.

Hay, J. (ed.) (1992) *Ecotourism Business in the Pacific: Promoting a Sustainable Experience.* Conference Proceedings, Environmental Science Occasional Publication No. 8. Environmental Science, University of Auckland, New Zealand and the East-West Center, University of Hawaii, USA.

Haywood, K.M. and Walsh, L.J. (1996) Strategic tourism planning in Fiji: an oxymoron or providing for coherence in decision making? In: Harrison, L.C. and Husbands, W. (eds) *Practising Responsible Tourism: International Case Studies in Tourism Planning, Policy and Development.* John Wiley & Sons, New York, pp. 103–125.

Helu-Thaman, K. (1992) Ecocultural tourism: a personal view for maintaining cultural integrity in ecotourism development. In: Hay, J. (ed.) *Ecotourism Business in the Pacific: Promoting a Sustainable Experience.* Conference Proceedings, Environmental Science Occasional Publication No. 8. Environmental Science, University of Auckland, New Zealand and the East-West Center, University of Hawaii, USA, pp. 24–29.

King, B. (1992) Cultural tourism and its potential for Fiji. *Journal of Pacific Studies* 16, 74–89.

Lees, H. (1992) Ecotourism – restraining the big promise. In: Hay, J. (ed.) *Ecotourism Business in the Pacific: Promoting a Sustainable Experience.* Conference Proceedings, Environmental Science Occasional Publication No. 8. Environmental Science, University of Auckland, New Zealand and the East-West Center, University of Hawaii, USA, pp. 61–64.

McArthur, S. and Weir, B. (eds) (1998) *Australia's Ecotourism Industry: a Snapshot in 1998.* Ecotourism Association of Australia, Brisbane.

McArthur, S. and Wight, P. (1999) Strategic issues facing ecotourism into the new millennium: perspectives from conference delegates. In: McArthur, S. and Weir, B. (eds) *Developing Ecotourism into the Millennium.* Proceedings of the Ecotourism Association of Australia 1998 National Conference, pp. 14–20. www.btr.gov/conf-proc/ecotourism

McCabe, C. (1998) True-blue green. *The Weekend Australian* 14–15 February.

McKercher, B. (1998) *The Business of Nature Based Tourism.* Hospitality Press, Melbourne.

McSweeney, G. (1992) Observing nature in action in the Pacific. In: Hay, J. (ed.) *Ecotourism Business in the Pacific: Promoting a Sustainable Experience.* Conference Proceedings, Environmental Science Occasional Publication No. 8. Environmental Science, University of Auckland, New Zealand and the East-West Center, University of Hawaii, USA, pp. 1–5.

Miller, M.L. and Auyong, J. (1991) Coastal zone tourism: a potent force affecting environment and society. *Marine Policy* March, 75–99.

Milne, S. (1992a) Tourism and economic development in South Pacific island microstates. *Annals of Tourism Research* 19(1), 191–212.

Milne, S. (1992b) Tourism development in Niue. *Annals of Tourism Research* 19(3), 565–569.

Mitchell, N., Park, G. and George, F. (1998) Evaluation of sites for ecotourism potential in the Hokianga District of Northland. In: Kandampully, J. (ed.) Proceedings of the Third Biennial New

Zealand Tourism and Hospitality Research Conference, *Advances in Research*, Part 1. Lincoln University, Canterbury, New Zealand.

Murphy, R.C. (1997) Environmental responsibility at the Jean-Michel Cousteau Fiji Islands Resort. In: *BIOSFERA World Ecotour '97 Abstracts Volume.* BIOSFERA, Rio de Janeiro, Brazil, pp. 166–168.

NZTB and DoC (1993) *New Zealand Conservation Estate and International Visitors.* NZ Tourism Board and The Department of Conservation, Wellington.

NZ Tourism Department (n.d.) *New Zealand Tourism and the Environment.* NZ Tourism Department, Wellington.

Parliamentary Commissioner for the Environment (1997) *Management of the Environmental Effects Associated with the Tourism Sector, Summary.* Office of the Parliamentary Commissioner for the Environment, Wellington.

Pearson, S. (1998) An ecotourism strategy for New Zealand. In: Kandampully, J. (ed.) Proceedings of the Third Biennial New Zealand Tourism and Hospitality Research Conference, *Advances in Research*, Part 1. Lincoln University, Canterbury, New Zealand.

Ryan, B. (1998) Ecotourism: what's in a name? In: Kandampully, J. (ed.) Proceedings of the Third Biennial New Zealand Tourism and Hospitality Research Conference, *Advances in Research*, Part 1. Lincoln University, Canterbury, New Zealand.

Ryan, C. (1997) Rural tourism in New Zealand: rafting in the Rangitikei at River Valley Ventures. In: Page, S.J. and Getz, D. (eds) *The Business of Rural Tourism: International Perspectives.* International Thomson Business Press, London, pp. 162–187.

Ryan, C. (1998) Dophins, canoes and marae. In: Laws, E., Faulkner, B. and Moscardo, G. (eds) *Embracing and Managing Change in Tourism.* Routledge, London, pp. 285–306.

Simmons, D.G. (1999) Eco-tourism: product or process? Paper presented to the *Manaaki Whenua (Cherishing the Land) Conference*, Landcare Crown Research Institute, Te Papa, Wellington, 21 April 1999.

Sofield, T. (1991) Sustainable ethnic tourism in the South Pacific: some principles. *The Journal of Tourism Studies* 2(1), 56–72.

Sofield, T. (1992) The Guadalcanal Track ecotourism project in the Solomon Islands. In: Hay, J. (ed.) *Ecotourism Business in the Pacific: Promoting a Sustainable Experience.* Conference Proceedings, Environmental Science Occasional Publication No. 8. Environmental Science, University of Auckland, New Zealand and the East-West Center, University of Hawaii, USA, pp. 89–100.

Sofield, T. (1994) Tourism in the South Pacific. *Annals of Tourism Research* 21(1), 207–215.

Tourism Council Australia (1996) *Tourism Switched On: Sustainable Energy Technologies for the Australian Tourism Industry.* Tourism Council Australia, Canberra.

TCA and CRC Tourism (1998) *Being Green Keeps You Out of the Red – an Easy Guide to Environmental Action for Accommodation Providers and Tourist Attractions.* Tourism Council Australia, Canberra.

TCSP (1987) *Identification of Nature Sites and Nature Subjects of Special Interest – Solomon Islands.* Tourism Council of the South Pacific, Suva, Fiji.

TCSP (1990) *Solomon Islands Tourism Development Plan, 1991–2000.* Tourism Council of the South Pacific, Suva, Fiji.

TCSP (1992a) *Tuvalu Tourism Development Plan.* Tourism Council of the South Pacific, Suva, Fiji.

TCSP (1992b) *Western Samoa Tourism Development Plan, 1992–2000.* Tourism Council of the South Pacific, Suva, Fiji.

University of Hawaii (1987) *The Impact of Tourism on the Commonwealth of the Northern Mariana Islands.* School of Travel Industry Management, University of Hawaii at Manoa, Hawaii, USA.

Urlich Cloher, D. (1998) The sustainability of indigenous tourism. In: Kesby, J.A., Stanley, J.M., McLean, R.F. and Olive, L.J. (eds) *Geodiversity: Readings in Australian Geography at the Close of the 20th Century.* Special Publication Series No. 6. School of Geography and Oceanography, University College, University of New South Wales, Australian Defence Force Academy, Canberra, pp. 491–496.

Valentine, P. (1993) Ecotourism and nature conservation: a definition with some recent developments in Micronesia. *Tourism Management* 14, 107–115.

Walters, W. (1992) Partnerships in parks and preservation. *Trends* 29(2), 2.

Ward, J.C. and Beanland, R.A. (1996) *Biophysical Impacts of Tourism*, Information Paper No. 56. Center for Resource Management, Lincoln University, Canterbury, New Zealand.

Warren, J.A.N. and Taylor, N.C. (1994) *Developing Eco-tourism in New Zealand.* New Zealand Institute for Social Development and Research, Wellington.

Weaver, D.B. (1998a) *Ecotourism in the Less Developed World.* CAB International, Wallingford, UK.

Weaver, D.B. (1998b) Strategies for the development of deliberate ecotourism in the South Pacific. *Pacific Tourism Review* 2, 53–66.

Weaver, S. (1992) Nature tourism as a means of protecting indigenous forest resources in Fiji. *Journal of Pacific Studies* 16, 63–73.

Wylie, J. (1994) *Journey Through a Sea of Islands: a Review of Forest Tourism In Micronesia.* USDA Forest Service, Honolulu, Hawaii.

Young, M. (1992) Ecotourism – profitable conservation. In: Hay, J. (ed.) *Ecotourism Business in the Pacific: Promoting a Sustainable Experience.* Conference Proceedings, Environmental Science Occasional Publication No. 8. Environmental Science, University of Auckland, New Zealand and the East-West Center, University of Hawaii, USA, pp. 55–60.

Chapter 10
Europe

S. Blangy[1] and S. Vautier[2]
[1]*Department of Tourism and Sustainable Development, SECA (Société d'Eco-Amènagement), Parc Scientifique Agropolis, Montpelier, France;* [2]*Wittelsbacherallee, Frankfurt am Main, Germany*

Introduction

Political background

Europe is a vast continent, centred on one main political organization, the European Union (EU). As of 2000, 15 countries were members: Austria, Belgium, Denmark, France, Finland, Germany, Greece, Ireland, Italy, Luxembourg, The Netherlands, Portugal, Spain, Sweden and the UK. Switzerland and Norway chose not to belong to the EU.

The rest of Europe does not benefit from such a tight economic and political network. Several Eastern and Central European countries are applying to join the EU. Some of them hope to join in the near future, including Hungary, Estonia, Poland, Slovenia and the Czech Republic. Others are working on accession. The European Commission (EC) supports several economic projects in Eastern and Central European countries under the TACIS and PHARE programmes. PHARE covers all the former Eastern European countries and TACIS deals with the Newly Independent States (NIS) from the former Soviet Union bloc. Two other organizations unify European countries, the Council of Europe and the Organization for Economic Cooperation and Development. However, they are considered more as discussion forums than political organizations. The Council of Europe was created in 1949 to seal the reconciliation between nations after the Second World War. Located in Strasbourg, France, it now has 41 member states, including the Russian Federation. The oldest and largest political organization in Europe, it became a structure for receiving the new democracies of Central and Eastern Europe after 1990 (Council of Europe, 1997).

In examining patterns of ecotourism in Europe, this chapter adopts the World Tourism Organization (WTO) regional divisions: Central/East Europe, Northern Europe, Southern Europe, Western Europe and the East Mediterranean. However, general issues will also be addressed for protected areas as well as rural, coastal and mountain areas.

Unity of nature and culture

More than any other continent, Europe is a mosaic of relatively small countries. It comprises an important diversity of ecosystems and landscapes, varying from dry land on the Mediterranean coast to humid and cold wetlands on the Atlantic coast, alpine mountains, forests and rivers, and boreal taiga. The culture of the European countries is equally diverse, embracing different languages, traditions and gastronomy. The peculiarity of the European landscape is that it is mainly a product of human activity. The high density of inhabitants on the European continent and the development of human activities almost everywhere have led to the development of specialized landscapes, such as terraces and open land, which have been important in the preservation or increase in biodiversity. This variety makes Europe a very successful but specialized destination for ecotourism. In most European nature destinations, ecotourism activities link nature interpretation with local traditions, architecture and culture.

Tourism success

Tourism flows throughout Europe have increased constantly over the past decade because of the development of the leisure society, reduction in working hours, increased mobility and incomes, and growing numbers of retired people. Europe leads the world in terms of tourism flows (arrivals and receipts), spending and employees. In 1996, Europe accounted for 59% of international tourist arrivals and 51% of international tourism receipts (WTO, 1998). Within Europe, France ranks first in terms of international tourism arrivals (62.4 million in 1996) and second in terms of international tourism receipts (Italy is first with US$30 billion in 1996).

Europeans tend to travel within Europe, the major trend being travel to the south, and especially around the Mediterranean coast. This is likely to continue in the future. However, because of the simultaneous development of increased leisure time and growing competition in the tourism industry, demand will become more critical and specialized. Nature and cultural tourism will increase their share of the market, especially for short holidays and as complementary activities to beach holidays. Competition means that demand will also be increasingly critical on quality, including environmental aspects.

Tourism in Europe is viewed as a very important part of the economy and a tool for land management. It is therefore not just a concern of private businesses, but also of many other stakeholders including the European Council and EC, and the various state and local governments and authorities. Since the Rio Earth Summit in 1992, the priority at every level is to promote sustainable tourism development in Europe. This includes ecotourism, although it is more common in Europe to speak about rural, nature or sustainable tourism (see Chapter 29). Many of these concepts involve nature interpretation together with enjoying the environment and local culture.

Ecotourism and Sustainable Tourism: European Definitions and Markets

Definitions

The term ecotourism is not as widely used in Europe as elsewhere in the world. The term sustainable tourism is preferred, and is applied by the EU as a concept, approach and form of organization. This concept is illustrated through the integrated quality management (IQM) approach described below (see Box 10.1). The WTO definition of *sustainable tourism* has been widely adopted by the European stakeholders, i.e. 'A form of tourism which meets the needs of present tourists and host regions while protecting and enhancing opportunities for the future'. It can be used in the context of all kinds of tourism, including urban as well as rural. Although it implies catering for markets which respect, and are interested in, host environments and culture, and encouraging products which are authentic, low impact, etc., the definition is not explicit in this regard. In contrast, the term *ecotourism* is not taken just as another word for sustainable tourism, but as something quite specific. Rather than describing an approach or philosophy for tourism, it is used more often to describe a type of tourism activity, essentially wildlife/nature-based tourism that is sensitive to environmental and social conditions. It is perhaps more widely used by European outbound tour operators than by destinations. This reflects the relatively more limited amount of wildlife and wilderness tourism in Europe compared with other continents as a proportion of all tourism.

Markets

European statistics do not provide clear data on the ecotourism market, because of important confusion between nature tourism, rural tourism, adventure tourism and nature interpretation activities. This is intimately related to the characteristics of the European natural environment. Very few areas in Europe remain free of human settlements to the extent that they can be considered as wilderness; most of the continent consists of humanized landscape. However, European citizens are demanding more environmentally friendly products. Surveys conducted by the EU have shown that 82% consider that environmental protection is an immediate and urgent problem. Nine out of ten declare that they are quite, or very, anxious about various forms of pollution that are threatening their countries; 67% were ready to pay more for environmentally friendly products (Commission Européenne, 1995). Holidays are no exception to these findings. Demand is evolving dramatically, and changing rapidly towards greater sustainability in all sectors, activities and regions, and on both the demand and the supply side.

An important part of rural tourism consists of visitors seeking peace and quiet in a quality natural environment and wanting participative holidays together with occasional interpretative activities. However, over the last 10 years Europe has witnessed the growth of a very specific market composed of 'nature aficionados'. Northern Europeans are the most important consumers of holidays (German tourists had the highest overall tourism expenditure in the world in 1996 with US$50.8 billion (WTO, 1998)). They also have the highest travel intensity, followed by Switzerland and Denmark (Rein and Scharpf, 1997). These northern countries lead the market for sustainable tourism, and most environmental management schemes and ecotourism tour operators are found here. In Germany for instance, an association for alternative tourism (*Forum Anders Reisen*) is a federation of more than 40 small and medium tour operators and travel agencies fulfilling ecologically sound tourism criteria. A German web site (www.eco-tour.org) provides all definitions, political declarations, label schemes, criteria and projects on sustainable tourism or ecotourism,

Box 10.1. Integrated quality management as a process to help define sustainable tourism.

IQM

Integrated quality management is a management process, not a form of tourism, which has been considered by the EU–DG Enterprise in the context of tourism destinations. The definition adopted by them is:

> The management of a tourist destination in a way which should simultaneously take into account, and have a favourable impact on, the activities of tourism professionals, tourists, the local population and the environment (that is the natural, cultural and man-made assets of the destination). It must have the requirements of tourists as one of its major considerations.
>
> (European Commission, 1999)

It is true that 'sustainability' should be an objective of IQM in a destination, but the emphasis is on *processes* to check and deliver beneficial impacts of tourism in terms of visitor satisfaction, business performance, and social and community impacts, in an integrated way.

together with links to package offers and tour operators. *ITB Berlin*, the main commercial trade show in Europe, is delivering workshops and a forum for discussions on sustainable tourism. Participants and organizers have been paying more attention to sustainable tourism over the recent years, through these seminars on the subject during the fair.

Specialized and small travel fairs are developing around Europe on green, nature and sustainable tourism, for example the *Reise Pavillon* in Hanover, Germany. In the UK, ecotourism has become a significant part of the tourism economy, with more than 50 tour operators and travel agencies operating in this market (Guicherd, 1994). Numerous associations for the protection of the environment, such as the National Trust and Royal Society for the Protection of Birds also offer ecotourism trips. In the UK, birdwatching has proved to be very successful. Volunteering is also included in the ecotourism market. This is mainly developed in Germany, the UK and The Netherlands. The British Trust for Conservation Volunteers now operates 'Natural Break' holidays in most national parks. The organization runs 400 separate projects, each with an average of 12 people. Mixed packages include conservation work, recreation (rambling) and interpretation activities (evening lectures).

Ecotourism in the Tourist Destinations

Protected areas in Europe

Europe has developed a wide and complex network of between 10,000 and 20,000 protected areas. Since 1982, an impressive extra 10 million ha – an area larger than Hungary – has acquired protected area status. The IUCN categories of protected area most used in Europe are II, IV and V (see Chapter 18). These protected areas have many different labels: wilderness areas, nature reserves, marine reserves, nature parks, regional parks, national parks and protected landscapes. They suffer from many pressures, including tourism. In Central and Eastern Europe, the recent political and economic changes could also be the main threats to valuable ecosystems but, in the meantime, there is the opportunity to establish a well-managed protected areas network (IUCN, 1994).

According to the IUCN (1994), the 1990s offered an unprecedented opportunity for protected areas because:

- human populations are relatively stable and affluent;
- there are declining pressures on land in many areas because of agricultural surpluses and reduced military activity;
- there is a high level of public support for conservation; and
- there is a climate of international cooperation.

Parallel to this opportunity, and to encourage its member states to go further in nature conservation, the EU established the *Natura 2000* network. This is based on two European directives: *Birds* (1979) and *Habitats* (1992). Each member state is responsible for the choice of sites and management tools, in accordance with the European requirements. In September 1999, the 15 member states had designated 2492 sites occupying a total area of 169,823 km^2. *Natura 2000* will be a leading programme for environmental protection in Europe. It will promote sustainable management in these areas, promoting activities, including special tourism activities, which are compatible with the protection of the sites.

Tourism in European protected areas

European protected areas are becoming increasingly popular tourism destinations. No international survey of visitors in parks is available. However, some national data give an idea of the importance of this phenomenon. The UK recorded 103 million visitor/days in its 11 national parks during 1987. Figures being monitored by the Tatra National Park in Slovakia show that visitor numbers doubled over the last 20 years.

The approach towards tourism in protected areas varies strongly with the philosophy of nature conservation. Two main approaches can be identified:

- Strict natural reserves or national parks, where the main issue is to control tourism impact and allow specific activities via specific zoning systems, facilities and direction.
- Inhabited protected areas, where the main issue is still conservation. However, the difference is that they are seeking to allow sustainable development, where appropriate, in parallel with this, as they are also concerned with socio-economic issues. Often the issue in these areas is to maintain a balance or mutual support between tourism and specific traditional economic activities that contribute to the quality of the landscape and biodiversity.

Techniques for managing tourism in protected areas have improved considerably over the last 20 years. These include efficient planning, such as zoning; the management of tourism flows, equipment, facilities, interpretation trails and centres; GIS techniques; and the development of proactive tourism products. The publication of the report *Loving them to death?* from the Federation of Nature and National Parks of Europe (Europarc Federation, 1993) was a significant lever for a positive partnership-based approach to tourism management in protected areas. This report recommended local community involvement; a strategic approach; the assessment of carrying capacity; survey and analysis of visitors and the impact of tourism activities; voluntary promotion of products, including educational tourism, targeted to the potential ecotourism market; and cooperation with the private sector. This was supported by the IUCN, which recommended the creation of a European Charter for Sustainable Tourism in Protected Areas, and the establishment of a sustainable tourism service to help implement and monitor the Charter in its action plan for European Protected Areas, *Parks for Life* (IUCN, 1994). The European Charter was officially presented in April 1999 (Europarc Federation, 1999) and piloted in ten European parks. Recommendations for ecotourism development are made on the basis of the overall needs of the area (environmental, economic, social, and the needs of local people and tourists). They are the results of partnerships between private operators and local people. The charter commits signatories to design strategies to: improve the tourism product; raise visitor awareness; train parks' and private employees; preserve and enhance the quality of life for local people; conserve and enhance local heritage, including the natural environment; contribute to social and economic development; and manage tourism flows. The aim of this scheme is to distinguish areas and enterprises for their excellence in the field of sustainable development. Simultaneously, the World Wide Fund for Nature UK launched its *Pan Park* scheme. This aims to create a quality brand, which represents an expanding network of well-managed protected areas, 'must-see' sites for tourists and wildlife lovers.

These programmes are the result of a new philosophy from the European parks, choosing to encourage a certain kind of development, compatible with nature conservation, rather than to restrict all development. It is also the consequence of a new approach to nature conservation in Europe, aimed at better involvement and consideration of the interests of the residents of the buffer zones. This new voluntary approach from the parks has led to the development of various nature products and labelling schemes. The French Natural Regional Parks, for example, have developed specific trademarks for environmentally friendly weekly holidays, a chain of environmentally friendly hotels, *Hôtel Nature*, and *Gîtes Panda*. These gîtes (self-catering accommodation) provide visitors with information about the local fauna and flora, direct access by foot to nature sites, and provide material for observing fauna and flora (binoculars, maps, books, etc.). The gîte owners commit themselves to raising visitor awareness, helping them to better understand the environment. They are encouraged to accompany their guests when they visit the park.

In the Abruzzo National Park, in Italy, the tourism strategy is based on the observation of wild fauna such as wolves, bears and lynx. Its eco-development strategy started in the early 1970s with the creation of a museum of the wolf in a small village. Today, the return on investment reaches 60%. The park directly employs 100 people, but the number of jobs created indirectly is easily ten times that amount. Its guiding principle is to preserve a balance between revenue linked to tourism and that of other sectors. After a very sensitive zoning of the park and a global strategy for the area, the park developed a network of museums of local fauna, and areas to orient visitors to nearby villages, thus increasing the impact on the local economy. Villages are surrounded by short interpretation trails, giving an introduction to nature in the area. For the aficionados, the park offers increasingly successful guided walks and volunteering programmes. Many parks are developing holiday packages, tourism associations and supporting private initiatives for ecotourism products. At present, approaches are not coordinated at the European level. However, schemes such as the European Charter and Pan Park have this aim.

Rural tourism: the raw material for ecotourism

Changes in the European way of life have led to new forms of tourism, including short-break holidays, which favour rural tourism development. At the same time a decrease in agriculture and forestry, together with a rural exodus, have encouraged many rural areas to view tourism as an alternative boost to their economy, creating jobs. Landscape characteristics mean that European rural tourism cannot be compared with the American concept of ecotourism. Rural tourism often includes all forms of tourism taking place in natural areas with a low density of population and where agriculture, forestry and traditional activities are intimately related to tourism. Some definitions refer to the aim of the destination, for example: 'a wish to give visitors personalised contact, a taste of the physical and human environment of the countryside and opportunities to participate in activities, traditions and lifestyles of local people' (LEADER, 1997). The delivery of this tourism experience involves many different players: natural resources, cultural traditions, transport services, a whole range of tourism enterprises as well as public authorities such as parks and local authorities.

With respect to rural tourism destinations, the following products can be identified:

- traditional, popular destinations near sizeable urban areas receiving a high proportion of daily visitors;
- traditional, popular destinations with a significant quantity of visitor accommodation and infrastructure;
- rural areas where a major part of the product is characterized by small historic towns and villages and a rich historic, architectural, cultural or industrial heritage interspersed in the countryside;
- rich agricultural areas where farming provides much of the visitor appeal;
- areas close to the sea, wishing to develop rural tourism in inland locations away from the coast;
- mountain or forest locations;
- protected areas seeking to manage tourism as well as the environment and local economy in an integrated way (IUCN Category V protected areas);
- remote areas with appeal based on wildlife and wilderness (often situated in mountain areas, national parks or natural reserves).

In the past decade, rural destinations have made enormous efforts to increase the quality of standards in services and accommodation, as well as to diversify and specialize their product. Many destinations have started to develop specific products for selected targets of clientele such as families, short-holiday takers, educational or school groups, senior citizens, people with special interests such as walking, cycling, local heritage interpretation, and people with disabilities (European Commission, 1999).

Coastal tourism: a lesson for more environmentally friendly tourism

European coastal areas have always been very attractive tourist destinations. According to the WTO, the Mediterranean coast receives 35% of international tourist arrivals, with 90% concentrated in the coastal areas of France, Spain and Italy (Rein and Scharpf, 1997). Many environmental problems have arisen from this tourism pressure, including over-construction, site coverage, over-consumption of water, and stress on the environment and landscape. These negative impacts have led to numerous reactions from political organizations. The French government, for instance, reacted by passing coastal legislation. In mid-1975, the Conservatoire de l'Espace du Littoral was founded with the aim of buying endangered coastal areas and dedicating them to nature conservation. Today, considerable effort is focused on restoration of the coastal environment through international programmes. At a multilateral level, the Mediterranean Action Plan was founded in the early 1970s under the auspices of the United Nations Environment Programme (UNEP) and today results in active cooperation between countries. The blue flag campaign is a European award scheme encouraging local authorities to maintain clean and safe beaches for the local population and tourists (UNEP *et al.*, 1996). In recent years, it has become a criterion in selecting their destination for many European citizens.

In 1996, the EC set up a demonstration programme to remedy the alarming environmental condition of European Coastal zones. It launched Integrated Coastal Zone Management (ICZM), a proactive and adaptive process of coastal resource management. Between 1997 and 1999, 35 projects in Europe have piloted ICZM. In December 1999, the EC also published a study on the integrated quality management of coastal tourist destinations, including 15 case studies (Commission Européenne, 1999a). In most of these, environmental management was paralleled by the development of nature activities, such as natural and cultural interpretation trails or visits to neighbouring natural areas.

Mountain areas

Mountain areas represent a very important European landscape. The Alpine districts of seven countries, with an area of 191,287 km^2 and 13 million inhabitants, are very attractive mountain tourism destinations. However, mountain tourism has also developed in other areas such as the Scottish Highlands, the Scandinavian fjords, Spanish and French Pyrenees, the mountains of Macedonia and Bulgaria, etc. However, it is the Alps, with now over 100 million visitors a year, which have suffered most from environmental pressures. Alpine tourism began in the mid-18th century, but skiing really escalated in the 1970s and problems ensued, such as the construction of ski lifts, increased urbanization and traffic and landscape destruction. Sociocultural impacts include the proliferation of holiday homes (used only a few weeks each year) and the abandonment of traditional activities which played a key role in mountain landscape management (Manesse, 1994). Consequently, sustainable development, including ecotourism, is being promoted to remedy this situation. International cooperation for the protection of the Alps exists through the International Commission of the Protection of the Alps (CIPRA), an NGO created in 1952. CIPRA aims to reduce environmental impacts on the Alps and supports sustainable development projects. In 1991, the governments of the Alpine countries signed a convention to protect the Alps, *The Alpine Convention*, which includes a tourism protocol. This intends to promote harmonization of policies and programmes between countries and different economic sectors to ensure better environmental protection and land management. Euromontana, a Continental European association targeting economic and social issues for the mountain areas, also promotes international cooperation and exchanges, including research on sustainable tourism in mountain areas.

In the past decade, because of a decline in the skiing industry, the private sector has also launched many environmental protection initiatives. These take the form of 'sustainable tourism resorts' and development of nature interpretation activities. Several popular skiing resorts in Austria, Switzerland and France have recently developed sustainable tourism strategies and action plans, based on quality, environmental management of accommodation, landscaping of ski lifts and interpretation of natural resources. Chamonix and Avoriaz, in France, for instance, are working on a marketing policy based on the sustainability and preservation of mountain ecosystems.

Sustainable Tourism in the Regions of Europe

In 1996, Europe received 349 million international tourist arrivals. The regional shares are: Western Europe 34%, Southern Europe 28%, Central and Eastern Europe 23%, Northern Europe 11% and Eastern Mediterranean 3% (WTO, 1998).

Western Europe

Western Europe has a long tradition of tourism and offers a wide range of destinations from coastal, rural and mountain areas to cities. High population density means that most rural areas of Western Europe are inhabited. It is not surprising, therefore, that this area is the most suited for the development of tourism linking nature interpretation with experience of local lifestyle and cultural heritage. The tourism product is quite well developed and organized, with quality trademarks, national strategies, eco-labelling schemes, etc.

The European Commission plays a key role in rural tourism development through LEADER funds, in a cohesive strategy aimed at restoring a balance in the level of development between different regions of Europe. In 1995, a report from the LEADER European Observatory, *Marketing Rural Tourism*, identified 71 projects, out of a total of 217, where tourism was a major objective in rural development (LEADER, 1997). In 1995, the EC, in partnership with 17 national authorities, established an award for tourism and environment projects. Many countries of Western Europe support the development of high-quality nature tourism or ecotourism.

Austria

In Austria, the self-regulated Holiday Villages in Austria Association was established in 1991 to encourage high environmental standards compatible with growing demands from tourism. There are strict membership criteria concerning village character, minimum ecological standards and minimum social and touristic standards coupled with load thresholds. The Tyrolean Environmental Seal of Quality is an eco-labelling scheme that monitors tourist establishments on an annual basis according to their performance on a set of obligatory criteria (Williams and Shaw, 1996).

Portugal

Portugal recently passed a law setting criteria for all kinds of activities, services and accommodation in parks. The official text defines the type of services, but also the necessary commitment of owners towards environmental management and raised visitor awareness. Increased tourism pressure on protected areas also led to the creation of a National Program for Nature Tourism, a protocol of cooperation between the Minister for Tourism and the Minister of Environment. Several levels of demand for tourist use in protected areas have been established, from areas in great demand to those less popular. This programme has two strategic aims: first, to promote the creation of an integrated product that fits the conservation objectives of each protected area; second, to add potential to the tourist activity (which promotes local development and respects local economic and social aspirations). Several measures and tools are planned towards that end. Regulations already published include, for nature tourism, a nature tourism kitemark, tourism guide, code of conduct, training plan and designation of pilot projects.

Southern Europe

The Mediterranean is the most attractive destination in Europe, but the most adversely affected by tourism, suffering from its popularity. The region is well known for its beautiful weather, its attractive landscapes, its warm-hearted people and their characteristic, relaxed lifestyle (Europarc Federation, 1993). As 'the cradle of western civilisation', the region has an exceptionally rich cultural heritage. The scale of tourism in the region is enormous. This is the world's leading area, accounting for 35% of the international tourist trade. The economies of many Mediterranean countries are highly dependent on tourism. The Mediterranean Action Plan predicts even greater increases; that between 380 and 760 million people will visit the Mediterranean in 2025.

Further tourism development will bring tremendous environmental pressures. Countries such as Italy, Greece and France have already developed very popular holiday resorts all along the Mediterranean coastal zone, but these have suffered from over-visitation in the past, and are experiencing a recent stagnation in tourism flows. Environmental pressure and damage have resulted from badly designed resorts and serious marine pollution. Today, many Mediterranean holiday resorts are falling out of fashion, leading to a corresponding fall in tourist arrivals. As a result, resorts and coastal municipalities are looking for alternatives. Within the IQM programme conducted by the EC, several pilot resorts have been studied. Small islands (the Spanish Canary and Balearic Islands, and the Greek islands) have developed sustainable tourism plans and actions and adopted guidelines though INSULA, a network of sustainable tourism islands. Popular destinations have hosted seminars and conferences on sustainable tourism, for example Lanzarote and Calvia (EcoNETT, 1999). Several European programmes and networks, such as MEDPO and MEDWET, are trying to act as catalysts for sustainable tourism strategies and promotional material.

Central and Eastern Europe

The Eastern European countries are catching up in terms of sustainable tourism and learning fast from the mistakes and successes of Western countries. After the fall of the Berlin wall, CEE countries became attractive and fashionable destinations because of their mystery, high biodiversity, rich wildlife and cultural heritage. Charismatic species such as bison, wolves and bears that have disappeared in Western European countries can still be seen in Eastern countries, for example in most of the natural reserves and national parks of Poland (Hall, 2000).

However, lack of development, environmental pressures and consequent degradation, volatile political situations, and desire for fast profits have all hindered the process. Flows of visitors from Western countries stagnated in some areas, but are now recovering following a trend towards greater sustainability. In some places that were chosen as pilot areas to develop wildlife tourism and boost hunting tourism, tour operators had to withdraw because of lack of professionalism and miscalculation of the tour package prices (Blangy, 1996). This was the case in Berezinski Reserve in Belarus, where Western tour operators stopped visiting the reserve in 1997 after a 100% price rise and a reduction in observation opportunities attributable to increased hunting. Competition can be high between Eastern

European countries with common borders and similar attractions such as wildlife, wilderness, attractive landscapes and cultural tourism. Tour operators will not hesitate to transfer their operations from one country to another, even if they run their business with conservation and sustainable development in mind.

Before the changes, tourism in the countries of the former USSR was centralized and controlled from Moscow through the multi-faceted organization called Intourist (Europarc Federation, 1993). Tourists were mainly domestic, with school groups coming from all over the Soviet Union. Foreign tourism was limited to specific places, mostly cities. Nature tourism was in its infancy. Today, however, many Western European countries have been funding the exchange of information and expertise. The UK Know-How Fund, for example, is assisting protected areas to consider how to develop sustainable tourism. Both the EC and the Council of Europe are funding pilot projects and encouraging cross-border cooperation to develop rural sustainable tourism, conservation and wildlife-watching projects. Environmental protection has been an integral part of the EC TACIS and PHARE programmes to assist economic reform. The use of such funds is being reoriented to assist the development of sustainable tourism related to protected areas. Three cross-border projects sound interesting in terms of sustainable tourism: Karelia (Finland, Russia), Niemen Region (Poland, Belarus) and Carpathia (Romania and Ukraine).

Eastern and Central Europeans are keen to learn, and many want to participate in Western European seminars on conservation and sustainable tourism. Several delegations, such as from Slovenia, Slovakia and Poland, have been invited to visit Western countries and learn from their experience. Field trips are organized around the sustainability of tourism projects in mountain and coastal resorts. The Ukraine and Slovenia assigned much importance to sustainable tourism and environmental requirements. Protected areas, however, need to be reinforced and enlarged, given their incipient nature. An aim is benchmarking to Western norms to ensure adequate product quality and authenticity prior to marketing.

Northern Europe

The UK

In 1989, the English Tourist Board and the Countryside Commission published their principles for tourism in the countryside. Shortly afterwards (in 1991), the government set up a taskforce on tourism and the environment. The principles of rural sustainable tourism that emerged from this were tested by various pilot projects, described in a report *Sustainable Rural Tourism, Opportunities for Local Action* (Countryside Commission, 1996). In the UK, enjoyment of nature and rural landscapes is an important part of the culture. National agencies, local authorities and individual projects have placed considerable emphasis on planning for sustainable tourism and on visitor management. Many initiatives for developing ecotourism take place in the UK national parks. The Peak District National Park is the most visited park in Europe, and has a very dynamic structure for sustainable development. Initiatives include integrated visitor management plans; mechanisms to raise money from visitors for environmental projects; and development of marketing and tourism associations. The Peak Park has a strategy to promote public transport.

Scandinavia

The essential tourism assets of Denmark are on the North Sea coast, which accounts for more than one-half of tourist nights (Ellul and Council of Europe, 1996). This coastal tourism is linked with a wide range of outdoor activities including golf, cycling and fishing. Although coastal tourism has not been as damaging to the North Sea as it has been to the Mediterranean, high concentrations in time and space generate problems such as demand for land, pollution, use of resources and conflicts with local people. To cope with these problems, the government introduced planning policy at national, regional and local levels based on high local participation. In 1994, new legislation was introduced, prohibiting any new development within 3 km of the coast. Denmark has, since 1917, had specific legislation allowing private landowners compensation to protect landscapes and natural areas of national interest (Ellul and Council of Europe, 1996). At the same time, the government promotes free access to nature. Each year new land is purchased and made available to the public, thus also ensuring nature conservation. This strategy allows improved tourism flow management and zoning. It is accompanied by an information strategy aimed at raising public awareness of the value of nature and landscapes, and promoting better understanding and responsible behaviour from visitors.

Tourism in Sweden is mainly based on nature, mountains, culture, and leisure facilities such as zoological parks. As in other northern countries, free access to nature is a Swedish right. Sweden has one of the highest percentages of population (77%) going on holidays in Europe (Ellul and Council of Europe, 1996); Swedes have 6 weeks holiday per year. This amount of leisure time, together with the development of short breaks, has led to many conflicts between private landowners and tourists, as well as environmental impacts such as path erosion and disturbance of wildlife. As nature tourism is a very important cultural characteristic, the national government developed a positive strategy based on development of the tourism product and public information. The Swedish environmental agency is working in close partnership with the tourism industry, making codes of behaviour available for specific destinations and activities.

Finland has experienced a spectacular rise in the European tourism market in the past decade, promoting Lapland, snow, skiing, northern lights, reindeers and Father Christmas. However, this is based on mass tourism and is an inappropriate use of the Sami culture, although Finland has apparently developed a sustainable tourism strategy at the national level (Parviainen and Pöysti, 1995). In Swedish and Norwegian Lapland, several Sami communities are developing community-based tourism linked to the reindeer economy and Sami traditional activities.

East Mediterranean

The East Mediterranean is a relatively new tourist destination, mainly oriented on coastal tourism and receiving Western European citizens (mostly Germans and British).

The desire of governments to boost their economy through tourism has led to the development of tourist resorts that are rarely accompanied by adequate infrastructure such as sewerage or waste water treatment. The destinations remain highly dependent on coastal tourism and foreign investment. However, some countries have recently started to implement better tourism development control and incentives for product diversification. In 1990, the Cypriot government announced a moratorium on coastal tourism development. Planning policy, as well as a national tourism plan, permits new construction only under certain conditions, including compulsory environmental impact assessment. The new strategy aims at promoting the development of new forms of tourism, especially agro-tourism, mountain and nature tourism.

In Turkey, the development of tourism remains an economic priority. However, with the support of the World Bank, the government initiated a project to protect the south-west coast. Priorities are for infrastructure development for waste treatment to stop marine pollution. Simultaneously, tourism is being promoted to more remote areas of the country to encourage a better spread of tourism benefits. These initiatives should soon lead to the development of new nature and cultural products, including ecotourism, in these destinations.

The Main European Organizations Involved in Sustainable Tourism

As discussed in Chapter 29, supra-national organizations are becoming an increasingly important mechanism for implementing tourism policy in Europe. In this regard, both the EC and the Council of Europe are implicitly attaching a high priority to ecotourism by promoting environmentally friendly tourism and sustainable tourism practices.

The European Commission

The EC has taken different measures to promote sustainable tourism especially ecotourism in sensitive areas. For example, tourism and the environment has been one of the main themes of the *Community Action Plan to Assist Tourism* since 1990 (Tzoanos, 1992). One of the Plan's six criteria for selecting measures for Community support was contribution to conserving natural environmental quality and cultural heritage, along with respecting the way of life of local populations. The Action Plan was adopted by the 12 member states in 1993. The argument for intervention was that the sheer size and diversity of the tourism sector necessitates a close collaboration between the Commission, member states, and different sectors of the industry.

The *Community Action Plan to Assist Tourism* includes support for:

- initiatives aimed at making tourists and operators more aware of the interdependence of tourism and the environment;
- innovative pilot projects aimed at maintaining a balance between tourism and the protection of natural environments, in particular coastal zones, upland areas, national parks and nature reserves; and
- initiatives aimed at developing different forms of sustainable tourism.

Within this first Plan, several projects were funded in 1993 and 1995 under the supervision of the General Directorate DG XXIII (Tourism Unit), renamed DG Enterprise in 1999. These 17 pilot projects have informed the 15 member states and some have led to further national projects:

- ECOTOE Biotope protection and ecotourism. Coastal ecotourism case studies.
- The ECOMOST project, EC models of sustainable tourism.
- GRECOTEL, a tourism and environment network of hotels in Greece.
- A common agenda for sustainable golf development and management.
- A handbook of good practice for sustainable tourism in walled towns.
- Green suitcase, the Ökologisher Tourismus in Europa (ÖTE) seal of quality.

The European prize for tourism and the environment in 1995 was awarded to five exemplary destinations, out of 60 applicants. Germany now organizes a similar event each year to reward private operators and initiatives aimed at promoting sustainable tourism. The ReisePavillon trade show was one of projects awarded in 1999.

Various important sustainable tourism networks initiated and supported at the European level within the *Community Action Plan to Assist Tourism* are still running in 2000:

- EcoNett, hosted by the World Tourism and Travel Council (WTTC), is a web site recognized as a focal point for environmental information, good practice, new techniques and technologies.
- Ecotrans, a network of experts working in the field of sustainable tourism.

The Commission has also produced and widely disseminated several booklets which present the most helpful findings and experiences gained from the pilot projects (Alpenforschungsinstitut, 1995). A second programme, Xylophénia, submitted in 1997 was more ambitious, but not adopted by the member states. However, several other pioneering works and research were conducted at the European level:

- The Integrated Quality Management of coastal, urban and rural tourist destinations (see above).

- The visitor payback process. Support for various innovative, transferable, research and pilot projects, such as into the 'visitor payback' process of raising money from visitors to support conservation (The Tourism Company, 1998).

From the environmental perspective as well, sustainable tourism has become a major issue. In the *Fifth Community Programme for Environment and Sustainable Development* (Commission Européenne, 1993) tourism became one of the five priority areas. The specific priorities for tourism are:

- integration of environmental considerations into tourism policy at the most appropriate level, and in land-use planning;
- a framework for the protection of sensitive areas;
- information for environmentally friendly behaviour of tourists;
- management of tourist flows to respect the carrying capacity of tourist sites.

The DG Environment (previously DGXI) Nature Conservation Unit, in charge of EC environmental policy, is playing a major role in this field. DG Environment runs a fund, LIFE (Nature and Environment) supporting conservation-based pilot projects around Europe and coordinating a new network of protected areas, Natura 2000. LIFE Environment has supported several conservation projects aimed at developing specific models of sustainable tourism in natural environments. Among them two major initiatives have been essential for protected areas in Europe:

- the European Charter for sustainable tourism in protected areas;
- the guidelines for sustainable tourism in protected areas and Natura 2000 sites.

In 2000, LIFE Environment will focus on pilot projects aiming at implementing general principles of sustainable tourism in protected areas and conservation areas. Guidelines will be submitted to the different member states for approval, and to the Convention on Biological Diversity. The message which the Commission has been trying to put across, and which the industry is now beginning to understand, is that environmental integrity makes good business sense, and is a necessary response to consumer demands and the market, rather than a strictly altruistic gesture.

The Council of Europe

In 1995, a pan-European Biological and Landscape Diversity Strategy was adopted by the Council of Europe, and within the strategy implementation a committee of experts was created which deals with tourism and the environment. The group's work resulted in the elaboration of several recommendations relating to general policies for tourism development and the environment, and the development of environmental management training for professionals in the tourism sector. Two recommendations concerning tourism development in protected areas (Council of Europe, 1995) and in coastal zones (Council of Europe, 1997) are specifically relevant to this study. These recommendations have been adopted by the Committee of Ministers of the Council of Europe and are being widely disseminated.

In addition, the Council of Europe has developed a specific programme focused on promoting sustainable tourism in two ways:

- Intergovernmental cooperation and technical assistance to pilot projects on sustainable tourism development located in critical regions of Albania, Slovakia, Belarus, Romania, Ukraine and Latvia.
- Conferences and workshops which were held in several Council of Europe member states, i.e. Hungary, Poland, Slovenia, Cyprus, Romania, Bulgaria and Latvia. An international conference on Tourism, Environment and Employment was organized for the end of 2000 in Berlin (Germany).

The Council of Europe has also drawn up two important documents which will have an impact on coastal tourism development:

a Model Law on sustainable management of coastal areas and a European Code of Conduct for coastal zones.

The different measures taken by the EC and the Council of Europe have played a major role in the evolution of European tourism, influencing the national policies of different member states as well as the private sector. Some of the states lead in terms of national strategy, regulations, labelling, incentives and funding for sustainable tourism. None, however, has reached the level of sophistication of the Australian Ecotourism Strategy (see Chapters 9 and 29). However, the different national initiatives combine together to form an interesting blend of experimental tools and policies.

Other European organizations involved in sustainable tourism

In Europe several different organizations (NGOs and not-for-profit) have been instrumental in promoting sustainable tourism and helping support the industry to develop policies and adopt principles. Some of them have an international mandate. However, as they have European headquarters they significantly influence national policies. This is the case for WTO in Madrid and UNESCO and UNEP IE in Paris. Others are specific to Europe.

The main organizations acting for sustainable tourism are the following:

- *Europarc*: The pan-European protected areas organization, whose aim is to improve conservation and quality and effectiveness of protected areas, developed the *European Charter for Sustainable Tourism in Protected Areas* discussed earlier.
- *WWF Pan Parks* aims to provide a nature conservation-based response to the growing market for nature-oriented tourism. This is to be achieved by creating a quality brand which stands for: (i) an expanding network of well-managed protected areas; (ii) areas which are widely known by Europeans as natural capitals, which they know and are proud of; (iii) 'must see' sites for tourists and wildlife lovers; (iv) wider public and political support for protected areas; and (v) new income for parks and new jobs for rural residents. Concrete actions include management organization and logo establishment; draft Pan Parks Principles and Criteria; collaboration with protected areas as pilot areas; Pan Parks workshops; and the information newsletter *Pan Courier*. Pan Parks aims to strengthen and diversify financial support for protected areas from both public and private sectors, in particular through logo attribution (logo holders will pay a proportion of their revenue to the protected area).
- *WTTC with Eco-Nett*, as described above.
- *UNESCO, UNEP IE*. Based in Paris, UNEP IE has produced several publications widely used in Europe as references such as Ecolabel and Environmental Guidelines for the Tourism Industry (UNEP/IIPT, 1995).
- *IUCN 'Parks for life'* coordination unit is based in Slovenia. IUCN is actively participating in international cooperation in sustainable development. It contributed to the development of the *European Charter for Sustainable Tourism in Protected Areas* in partnership with Europarc Federation and promotes the development of sustainable tourism.
- *Tourism Concern* is a UK-based charity with a global membership network, started in 1989. Tourism Concern aims to promote awareness of tourism impacts on people and their environment. It produced ten principles for sustainable tourism to achieve the aims of the Rio Earth Summit and influence policies and programmes adopted by the travel and tourism industry worldwide (WWF, 1992).
- *WTO* has its headquarters in Madrid, Spain and strongly influences European policies.

In Europe, membership networks similar to The Ecotourism Society and the Ecotourism Association of Australia do not exist. Networks that do exist are spread out and do not have the same audience. For instance, ECOTRANS, a European network for sustainable development, consists of 30 experts and is supporting Eco-tip, a European information service on the internet which provides information on eco-label and on good practices in sustainable tourism (www.eco-tip.org).

The Private Sector

Several initiatives are worth mentioning in the private sector. For hotels (independent, resorts and hotel chains) and self-catering accommodation, the International Hotels Environment Initiative (IHEI) (UNEP *et al.*, 1995), the Youth Hostels Association (IYHA, 1994), and Farm Holidays network (European Centre for Eco-Agro Tourism or ECEAT) lead in terms of environmental management. The European Federation of Camping Sites Organisations awards an environmental prize, the David Bellamy Award, and has developed an environmental charter.

The European Federation of Youth Hostels Association has developed an environmental charter and manages several training and pilot programmes for youth hostel managers. Mirrow 21 is the most advanced example of sustainable youth hostels combining specific design, alternative energy, environmental management and education activities.

The Gîtes de France (self-catering accommodation) have developed Gîtes Panda, environmentally sound properties, together with information on fauna and flora and observation opportunities in natural parks. The label is given by WWF following the visit of an expert. Tour operators are also getting the point. The European Tour Operators Association developed environmental guidelines in 1992. Tour operators have also taken many individual initiatives such as the Ethics Charter of a French tour operator, Atalante. This Charter has been adopted by tourism-related private enterprises such as *Lonely Planet*, Swissair, Aigle (sports retailer), and *Trek Magazine*. Some tour operators also run charity programmes to support conservation work where they operate (Allibert, France).

Conclusions

Sustainable tourism is a growing trend at different levels in Europe, under which many implicitly ecotourism-related initiatives are subsumed. The tourism industry is expected to increase its involvement and evolve towards greater sustainability. Many pilot projects and good practices have been identified all over Europe and are being funded and supported by the various organizations mentioned above. Eco-tip has selected the Top 100 best sustainable tourism practices in its web site. These best practices cover the following fields:

- guidelines, charters, recommendations, codes of ethics for developing sustainable tourism;
- planning, sustainable tourism strategies;
- charters;
- certification and accreditation scheme;
- visitor payback;
- product development;
- visitor management;
- soft mobility and transport options.

These trial projects provide the input for further policies, incentives and tools. Further national policies and pilot projects are to be expected in the near future from EU member states and, in particular, from the Eastern European countries.

At the EC, the Tourism and Environment Directorates both aim towards greater integration of sustainable tourism practices in community policies that affect tourism. Synergy between the different departments concerned with quality (employment, enterprise and agriculture) is also being aimed for. Different seminars and workshops point towards a strong need for more formal networks of experts, site managers, ground operators, and actors in the field to exchange information and develop policies. A sustainable tourism network and European association is needed as well as further research and publications in this field.

References

Alpenforschungsins (1995) *Tourism and the Environment.* European Commission, DGXIII, Brussels.

Anon (1995) *Tourism in National Parks: a Guide to Good Practice.* Rural Development Commission, Wales tourist board, English Tourist Board, Countryside Council for Wales and Countryside Commission.

Commission Européenne (1995) Les Européens et l'Environnement, Quelques grands résultats du sondage effectué dans le cadre de l'Eurobarométre 43.1bis, Bruxelles.

Commission Européenne (1999a) Les enseignements du programme de demonstration de la Commission Européenne sur l'aménagement intégré des zones côtières (AIZC), Luxembourg.

Council of Europe (1993) Sustainable tourism. *Naturopa* 84, 48.

Council of Europe, Committee of Ministers (1995) *Recommendation No. R (95) 10 of the Committee of Ministers to Member States on a Sustainable Tourist Development Policy in Protected Areas.* Council of Europe, Strasbourg.

Council of Europe (1997) Questions and answers. *Tourism and Environment.* Vol. 3.

Countryside Commission (1996) *Sustainable Rural Tourism: Opportunities for Local Action.* Countryside Commission, London.

EcoNETT (1999) *Calvia Declaration on Tourism and Sustainable Development in the Mediterranean. The call for, and Regional Authorities.* EcoNett, Brussels.

Ellul, A. and Council of Europe (1996) Tourisme et environnement dans les pays européens, collection Sauvegarde de la nature, 83.

Europarc Federation (1993) *The European Charter for Sustainable Tourism in Protected Areas. Guide to Implementation of the Charter by Protected Areas and Evaluation Process.* The European Commission, The French Federation of natural parks, Paris.

European Commission, Enterprise DG, Tourism Unit (1999) *Towards Quality Rural Tourism, Integrated Quality Management (IQM) of Rural Tourist Destinations.* European Commission, Brussels.

Guicherd, B. (1994) *Les Perspectives de Développement du Tourisme Naturaliste dans les Parcs Nationaux et Régionaux Français.* Ministère de l'Environnement, Région, Rhone Alpes, pp. 21–22.

Hall, R.D. (2000) Evaluating the tourism-environment relationship: central and east European experiences. *Environment and Planning* 27.

International Youth Hostel Association (IYHA) (1994) *Implementing the IYHF Environmental Charter.* IYHF, Brussels.

IUCN (1994) *Parks for Life: Action for Protected Areas in Europe.* IUCN, Gland, Switzerland.

LEADER European Observatory (1997) *Marketing Quality Rural Tourism: the Experience of LEADER* 1, 1997.

Manesse, J. (1994) Les montagnes d'Europe, Présentation, Menaces et solutions. Parcs transfrontaliers. In: *Tourisme de montagne et rôle des parcs naturels régionaux.* Conseil de l'Europe, Strasbourg, pp. 13–65.

Parviainen, J. and Pöysti, E. (1995) *Towards Sustainable Tourism in Finland: the Results of an Eco-audit Experiment in Ten Tourist Enterprises and Suggestions for Further Measures.* Finnish Tourist Board, Helsinki.

Rein, H. and Scharpf, H. (1997) *Biodiversity and Tourism, Conflicts on the World's Seacoasts and Strategies for their Solution.* German Federal Agency for Nature Conservation, Springer, Berlin.

The Tourism Company (1998) Visitor payback. Encouraging tourists to give money voluntarily to conserve the places they visit. Ledbury/London.

Tzoanos, G. (1992) Tourism and the environment: the role of the European Community. *Industry and Environment* 15, 3–4.

UNEP IE/IIPT (1995) *Environmental Codes of Conduct for Tourism.* UNEP, Paris.

UNEP (1995) *Environmental Action Pack for Hotels. Practical Steps to Benefit your Business and the Environment.* The International Hotel Association, The International Hotels Environment Initiative, UNEP, UNEP IE, Paris.

UNEP, WTO and FEEE (1996) *Awards for Improving the Coastal Environment: the Example of the Blue Flag.* United Nations Publications.

Williams, A.M. and Shaw, G. (1996) *Tourism, Leisure, Nature Protection and Agri-tourism.* Tourism Research Group, University of Exeter.

World Wide Fund for Nature UK (WWF) (1992) *Beyond the Green Horizon. A Discussion Paper on Principles for Sustainable Tourism.* WWF, Godalming, UK.

WTO (1998) *Yearbook of Tourism Statistics.* WTO, Madrid.

Chapter 11
Latin America and the Caribbean

D.B. Weaver[1] and R. Schlüter[2]
[1]Department of Health, Fitness and Recreation Resources, George Mason University, Manassas, Virginia, USA; [2]Centro de Investigaciones y Estudios Turisticos, Avenida del Libertador, Beunos Aires, Argentina

Introduction

Among the world's 'macro-regions', Latin America and the Caribbean stands out for the high profile of its ecotourism sector, even though most progress in this respect has been achieved only since the early 1980s. The purpose of this chapter is to survey the status of ecotourism in Latin America and the Caribbean as of 2000. In order to contextualize this theme, the environmental characteristics of the region will also be considered where appropriate, as will the general tourism industry. For discussion purposes, the study area is divided into three regions: the Caribbean and Mexico, Central America, and South America. This division adheres to geographical convention except for the inclusion of Mexico with the Caribbean, which is based upon commonalities of proximity to the Anglo-American market and the dominance of 3S (sea, sand, sun) tourism. However, as will be seen, the dissimilar geographical characteristics of these regions give rise to distinctive patterns of ecotourism activity.

The Caribbean and Mexico

At first perusal, the Caribbean, and to a lesser extent Mexico, may not appear to be a likely venue for ecotourism activity, given the presence of a 3S-based mass tourism industry that dominates most of the states and dependencies in the region (see Fig. 11.1). The insular Caribbean, for example, accounts for just 0.7% of the world's population, but 2.4% of all international stayovers (Weaver and Oppermann, 2000). In addition, the Caribbean is the most important area in the world for cruise ship tourism (Wood, 2000). The dependence of the region on tourism is evidenced by the observation that this sector accounts for at least 10% of GNP in 19 of 24 insular Caribbean states or dependencies (including Bermuda and the Bahamas). A hyper-dependent relationship, moreover, is apparent in the 11 or 12 entities that rely on tourism to generate at least 30% of GNP. The situation in Mexico is very different in relative terms, since tourism accounts for only 2% of the diverse national economy. However, in absolute terms, Mexico is the world's seventh largest destination for international stayovers, receiving over 21 million in 1996 (Weaver and Oppermann, 2000).

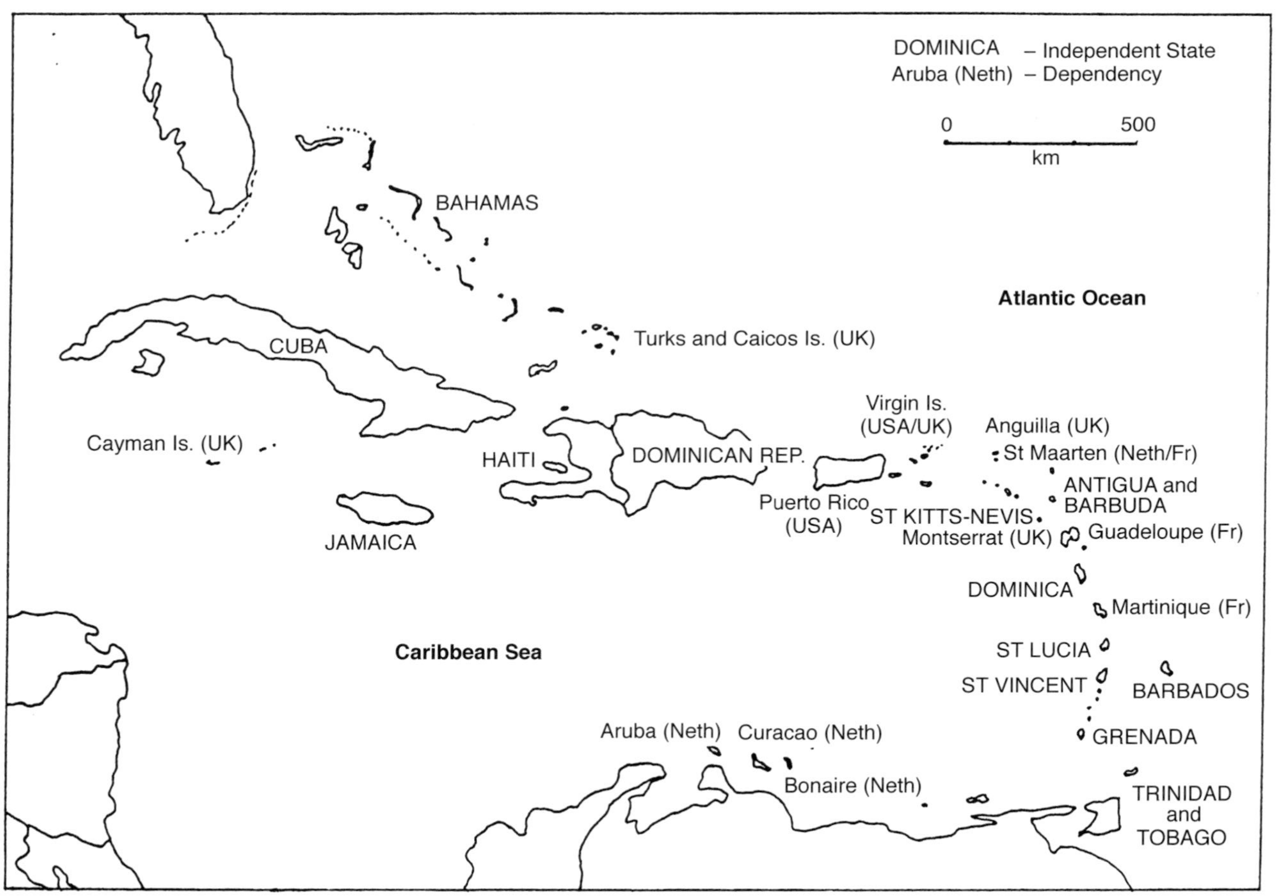

Fig. 11.1. The Caribbean.

Despite this situation, the actual and potential status of ecotourism in the Caribbean and Mexico is considerable. This is in part due to the inherent suitability of island settings for ecotourism-related activity (see Chapter 15). In addition, mass tourism tends to be confined to just a small portion of the littoral in even the most tourism-intensive countries, leaving the remaining terrestrial areas free to accommodate activities associated with alternative tourism, including ecotourism. The following discussion will therefore emphasize the actual and potential status of ecotourism within this 90–95% of the Caribbean land area that is not oriented toward mass tourism.

Specialized ecotourism destinations

Not all Caribbean islands have a tourism sector that adheres to the stereotype of the beach resort. The most notable of these at the national level is Dominica, which has been consciously marketing itself as a nature-based destination since the 1970s. Prior to this time, the government had aspired to develop in the 3S mode, but these aspirations were thwarted by the island's lack of white-sand beaches, mountainous terrain that hindered the development of the necessary infrastructure, political uncertainties, and high levels of rainfall. Accordingly, these perceived disadvantages were re-marketed as assets, and Dominica was promoted as the 'nature island of the Caribbean'. Subsequent promotional references to 365 waterfalls, though probably apocryphal, offer an interesting counterpoint to Antigua's emphasis on its own purported 365 beaches. But more tangibly, the island has encouraged various policies, such as the establishment of small, locally controlled nature lodges in the interior, that attest to its status as a comprehensive ecotourism destination (Weaver, 2001). Arguably, the only other country-level destinations that can make a legitimate claim to this status are St Vincent and the Grenadines, and Montserrat prior to its devastation by a volcanic eruption in 1994 (Weaver, 1995, 1998).

The second type of specialized ecotourism destination in the Caribbean involves peripheral islands of archipelagic states. Examples include Tobago (relative to Trinidad), Nevis (relative to St Kitts), St John (relative to St Thomas), Barbuda (relative to Antigua), and Little Cayman Island and Cayman Brac (relative to Grand Cayman Island). None of these entities, however, is as comprehensively and deliberately ecotourism-oriented as Dominica. At a broader scale, the tourism policy of the Bahamas advocates large-scale 3S tourism for Grand Bahama and New Providence islands, but small-scale, nature-based tourism for the remaining 'Family Islands'.

Other ecotourism venues

On islands oriented toward other types of tourism, numerous opportunities are still available to pursue ecotourism. Three types of venue are especially important. First, many islands contain mountainous areas, usually in the interior, that retain extensive forest cover and sometimes harbour endemic flora and fauna. Among the larger Caribbean islands, such areas include the Sierra Muestra of south-eastern Cuba, the Cordillera Central of the Dominican Republic, Trinidad's Northern Range, the El Yunque region of Puerto Rico and the Blue Mountains of Jamaica. On a smaller scale, mountainous interiors are characteristic of Saint Lucia, St Kitts, Grenada and Martinique. Large areas of the Mexican interior are similarly endowed, giving rise to ecotourism icons such as the Monarch Butterfly Reserves, which attracted approximately 100,000 tourists in 1989 (Hawkins and Khan, 1998).

Second, most Caribbean islands, and portions of the Mexican coastline, are fringed by coral reef formations that offer opportunities for scuba-diving and snorkelling. If such activities in the main are conceded to constitute a form of ecotourism, then the latter activity assumes great importance as an adjunct to the mass tourism of Grand Cayman Island, Barbados, Cancún and several other destinations. In addition, peripheral islands such as Bonaire and Saba (Dixon *et al.*, 1993) are even more solidly positioned as specialized marine ecotourism destinations. Third, and least prevalent of the three, are littoral wetlands. Prominent examples that already accommodate nature-oriented tourist activity include the Caroni wetlands near Port-of-Spain in Trinidad, the Peninsula de Zapata in Cuba, the mangroves of Bonaire and the mouth of the Black River in Jamaica.

Strengths and opportunities

The insular nature of the Caribbean suggests formidable potential in the area of marine and littoral ecotourism, while the potential of mountainous interiors remains relatively untapped. A second major strength is proximity to the North American market, which rivals Western Europe as the world's major ecotourist source region. More controversial is whether the existence of a dominant 3S-based mass tourism industry should be perceived as a strength or liability. The argument for the former is that mutually beneficial relationships, in theory, can be formed between ecotourism and mass tourism. As discussed in Chapter 5, these involve the provision of mainly soft ecotourism diversions for the mass tourist market, which thereby supplies a critical mass of clientele, and an incentive to develop and enhance such nature-based products. In return, the overall holiday package is enhanced through diversification, and a 'greener' product is encouraged throughout the tourism system. To a large extent, the ecotourism product of the Caribbean and Mexico is already structured in such a way, with diversionary, soft ecotourism being dominant except in specialized destinations such as Dominica. Ecotourism-type accommodations have even been established in such unlikely destinations as Bermuda, as described in Chapter 33.

Cuba represents a promising opportunity for the regional ecotourism sector as a whole, but a possible threat to established ecotourism destinations in the Caribbean and Mexico because of its high potential. This potential is based on its post-1959 legacy of small-scale 'socialist' tourism, its extensive natural assets, advanced protected area systems (which cover 17% of the island's land area) and strong political support for the ecotourism sector. The need to cope with the US embargo has also contributed in a bizarre way to ecotourism by fostering the use of low energy, 'soft' technologies. Finally, prototypes such as the Moka Ecolodge have already been established (Honey, 1999).

Weaknesses and threats

Widespread environmental degradation, both marine and terrestrial, is a significant weakness and threat to the regional ecotourism product. Extreme levels of deterioration are evident in Haiti, which retains less than 1% of its original forest cover, but the situation is also serious in the Dominican Republic (25% retention), Cuba (29%) and Jamaica (35%), among the larger countries. In Mexico, 77% of original 'frontier' forests are considered threatened (World Resources Institute, 1998). With respect to offshore resources, the coral reefs of the Caribbean rank among the most endangered in the world due to mass tourism, industrialization, shipping, fishing and sedimentation. The rapidly growing tourism industry is itself regarded as a major contributor to the area's environmental problems, despite its recent embrace of the rhetoric of sustainability (Holder, 1996). It is for this reason that the link between mass tourism and ecotourism can also be perceived as a threat. One example is the Costa Maya project in Mexico's Quintana Roo Province, which purports to be an 'ecotourism' development, yet includes 18-hole golf courses and at least one full-service marina (Ceballos-Lascuráin, 1996). In addition, there is a risk that ecotourism will be used as a marketing ploy by unscrupulous operators who are not genuinely committed to the principles of sustainability. A related weakness is the low proportion of land occupied by protected area systems in most islands, although Cuba and the Dominican Republic (31.5% protected) constitute significant exceptions, at least on paper.

Because of their geographical location, the Caribbean and Mexico are highly susceptible to natural disasters such as hurricanes, volcanic eruptions and earthquakes. Small islands are especially vulnerable, given that a single seismic or climatic event can devastate an entire state or dependency. The destruction of Montserrat by a volcanic eruption in 1994 is the best illustration so far of this devastation scenario. Hurricanes may wreak a greater ubiquity of destruction, but the effects are temporary. Finally, ecotourism suffers from a lack of institutional articulation. Among countries and dependencies in the region, any common understanding of ecotourism is absent, with some islands proffering definitions that can only be described as bizarre. Martinique, for example, includes golfing as a component of ecotourism (Weaver, 2001).

Central America

In terms of ecotourism activities, Central America is notable for the differences among its constituent countries (see Fig. 11.2). On one hand, Costa Rica has achieved, rightly or wrongly, iconic status as an ecotourism exemplar. In contrast, El Salvador and Nicaragua demonstrate little evidence of any such activity. Between these two extremes, Belize and Panama are moving toward the Costa Rica model, while Honduras and Guatemala are incipient. This section will begin with an outline of Costa Rica, then progress to Belize as an emerging competitor to Costa Rica. The remaining countries of Central America will then be discussed as members of the incipient group.

Costa Rica

Costa Rica is arguably the best-known ecotourism destination in the world at a national level (Schluter, 1998; Honey, 1999). This status, as with most high-profile destinations (see Chapter 18), is closely associated with Costa Rica's well-developed national protected area system, which covers approximately one-quarter of the country (about one half of which is strictly protected). Included in this system are well-known public entities such as Poas, Irazu, Carara and Manual Antonio, and private reserves such as Monteverde, Rara Avis and La Selva. These protected areas capitalize on an impressive level of biodiversity that results from Costa Rica's mainly tropical climate, its variability of altitude, and its location astride the North and South American biological provinces. The growing popularity of these protected areas is reflected in the finding that one-half of all international visitors in 1991 spent at least some time in such an area, compared with 20% in 1983 (Epler Wood, 1993). By the mid-1990s, two-thirds of all arrivals had visited at least one protected area. Recent reconfigurations of this system have fostered a greater potential for community-led ecotourism development (see Chapter 18). Other innovations include restoration-oriented units such as Guanacaste National Park, in the exhausted pasturelands of the north-east. In respect to factors external to tourism, the country has experienced a prolonged period of political stability, and was thrust into a positive spotlight in 1987 when then-president Arias was awarded the Nobel Peace Prize.

Yet, there are grounds for contesting Costa Rica's reputation as an ecotourism exemplar. Beyond its protected area system, environmental degradation (and forest clearance in particular) continued unabated throughout the latter half of the 20th century, to a point where the non-deforested portion of the country's land area is essentially co-extensive with that system. The protected areas themselves have been chronically underfunded, relying to a large extent on foreign donations and volunteer

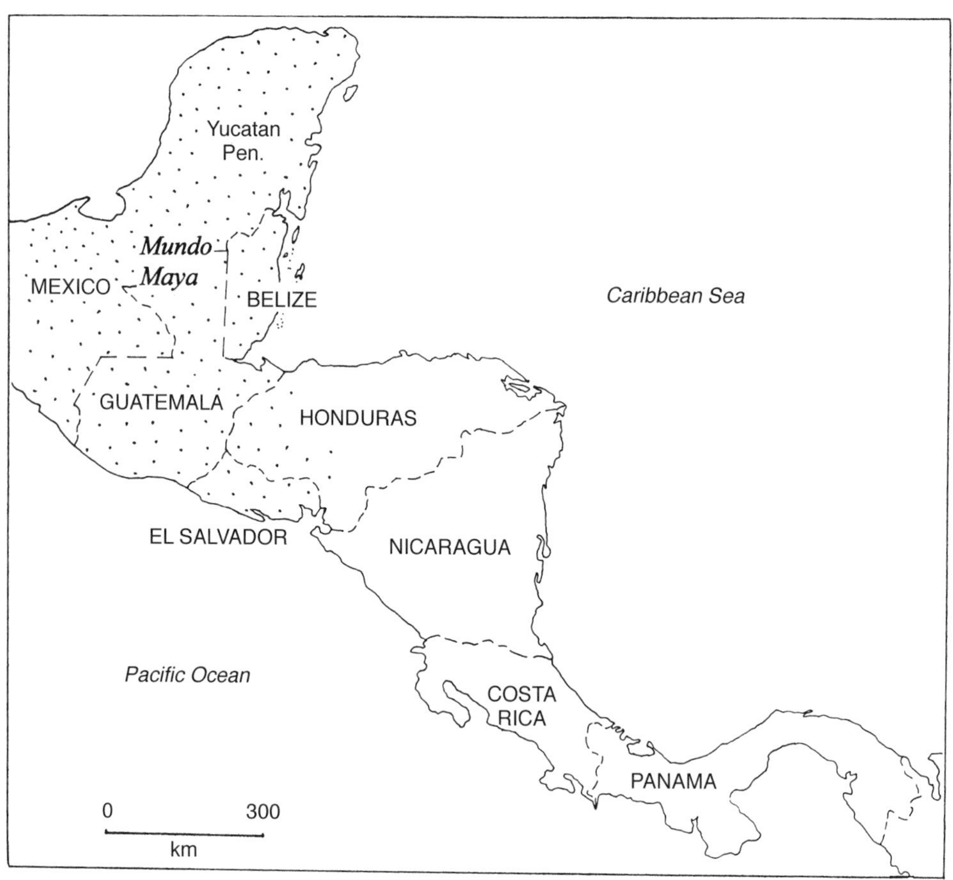

Fig. 11.2. Central America.

activity for their maintenance and management. With respect to the tourism sector, ecolodge-type facilities, despite their high profile, actually account for only a miniscule proportion of all accommodations. Most ecotourism activity in reality involves diversionary, daytime visits by soft ecotourists to a small number of protected areas that are readily accessible to San Juan or the beach resorts of the Pacific coast (Weaver, 1999).

Visitor motivations reflect this multipurpose profile. According to a 1995 survey, 44% of US visitors cited 'sea and sun' as a major purpose for their visit, compared with 42% for 'natural history'. The comparable figures for the European market (excluding Germany) were 45% and 50% (TTI, 1996). As with the Caribbean, this pattern of association can be interpreted as either a threat or an opportunity. In either case, it is clear that the government of Costa Rica has always pursued mass tourism at least as avidly as ecotourism. It is toward the development of the former that most government policy and incentives are oriented, while the impetus for ecotourism-related development has for the most part come from foreign and domestic non-government organizations (NGOs), individuals, and local community associations (Honey, 1999).

Belize

Costa Rica's status as the primary ecotourism destination of Central America is being challenged increasingly by Belize, which was already recognized in the early 1990s for its fledgling ecotourism sector (Boo, 1990). The development of ecotourism in Belize has been assisted by at least eight factors:

- the second longest barrier reef in the world, extending 115 km;
- a largely extant natural environment, with forest cover accounting for about 85% of the country's land area;
- a low population density of about 10 persons per km;
- a well-established protected area system that in theory strictly protects almost 21% of the country;
- extensive complementary cultural assets;
- political and social stability;
- proximity to the North American market and to major 3S destinations such as Cancún; and
- the status of English as the country's official language.

In terms of venue, ecotourism in Belize is associated with high profile community-based protected area initiatives such as the Community Baboon Sanctuary and Rio Bravo (see Chapter 19), and public sector protected areas such as the Hol Chan Marine Reserve and Cockscomb Basin Wildlife Sanctuary (Lindberg *et al.*, 1996). Although the integrity of Belize as an ecotourism destination is assisted by the absence of widespread environmental destruction, the country does resemble Costa Rica in terms of the parallel growth of mass tourism in coastal regions such as Ambergris Cay. As in Costa Rica, it is questionable whether much of this development, and the concomitant growth of visitor arrivals, is complementary to the philosophy and practice of ecotourism (Cater, 1992; Lindberg *et al.*, 1996; Weaver, 1998).

Incipient Central American ecotourism destinations

Of the remaining Central American countries, Panama shows the greatest ecotourism potential, with a high level of strict land protection (about 19%) and a significant inventory of relatively undisturbed forests and other natural environments. Situational factors are also important. Panama is already a well-known destination because of the presence of the Panama Canal, Panama City's status as a regional financial centre, and proximity to Costa Rica, with whom bilateral protected area initiatives are being pursued. Ecotourism opportunities are already being offered by NGOs such as the Smithsonian Tropical Research Institute, and by protected areas within or adjacent to the old Canal Zone buffer zone. Furthermore, the government appears interested in including a high profile ecotourism and conservation component in its planning as an adjunct to the development of a rapidly expanding conventional tourism industry (Ayala, 1997, 1998).

Honduras, Guatemala and Nicaragua all have considerable ecotourism potential that has not yet been realized to any great extent due to a combination of environmental, image, infrastructural and political problems. The situation in Honduras is illustrative. A limited amount of ecotourism already occurs in protected areas such as La Tigra, Pico Bonita and La Muralla National Parks. However, the further evolution of this incipient sector has been hindered by the relative obscurity of these attractions, the devastation caused by Hurricane Mitch in 1998, and the presence of stronger regional competitors. In particular, the Moskitia region in the country's north-east, along with adjacent parts of Nicaragua, is well positioned to become a major ecotourism destination. As for Nicaragua, this country is estimated to contain four times as much forested land as Costa Rica, and, like Panama, is pursuing bilateral protected area initiatives with that country. Guatemala's ecotourism sector will potentially be assisted by the participation of this country in the regional Mundo Maya programme (see below). El Salvador is the only Central American country that appears to have little ecotourism potential, given the extensive deforestation of the country.

Mundo Maya

The Mundo Maya initiative is worthy of mention because of its ecotourism connections and multilateral character, but also because of the controversy that it has generated. The concept was formalized in the early 1990s by Guatemala, Belize, Honduras, El Salvador and the neighbouring states of Mexico to develop a regional tourism product around the theme of the Mayan culture. Related projects that purportedly emphasize ecotourism include the formation of the Toledo Ecotourism Association in Belize and the Rio Bec Ecotourism Corridor in southern Mexico (Mundo Maya Organization, 1996). Yet, despite the high profile status of ecotourism, the Mundo Maya Organization which manages the initiative has been criticized for giving preference to domestic or foreign-controlled mass tourism projects (e.g. Costa Maya), while ignoring or neglecting small operations in its attempts to maximize revenue earnings (Thomlinson and Getz, 1996). Even if this problem is addressed, it seems clear that the Mundo Maya will follow the model that dominates ecotourism worldwide, that is, by providing soft ecotourism opportunities as an adjunct to mass tourism.

South America

South America dwarfs the Caribbean and Central America in terms of land area and available natural assets (see Fig. 11.3). Yet, the development of ecotourism to date has been seriously impeded by South America's distance from major tourist-generating regions, and by the lack of a strong domestic ecotourist market, especially in the northern half of the continent. Regarding the former factor, South America accounts for 5.9% of the world's populations, but only 2.5% of all international stayovers, most of whom are of intra-regional origin (Weaver and Oppermann, 2000). In contrast to the previous material on Central America, the following sections consider South American ecotourism from the perspective of four dominant physical environments that straddle international boundaries in the region: the Amazon basin, the Andes, the Pantanal and Patagonia. Other physical regions, such as the Gran Chaco of Paraguay and Argentina, Brazil's Atlantic rainforest (Fig. 11.4), the Venezuelan savannah, the temperate rainforests of southern Chile and the deserts of the Pacific coast, are not included, although these areas do accommodate a small amount of ecotourism activity. At a country level, Guyana is perhaps the only South American country so far that appears intent on developing a tourism industry dominated by ecotourism. Brazil has also formulated an ecotourism master plan, though national tourism development overall is following a more conventional path (see Chapter 29).

The Amazon basin

The Amazon basin is by far the dominant physical region of South America in terms of size (7–8 million km^2), country presence (Brazil, French Guiana, Suriname, Guyana, Venezuela, Colombia, Ecuador, Peru and Bolivia) and biodiversity (e.g > 2000 species of fish and 2500 orchid varieties). The last characteristic, and the fact that most of the flora and fauna in this region is still extant, suggests that the Amazon basin should be an ecotourism powerhouse. Yet, this is not the case. Brazil, as cited above, has made progress toward the institutionalization of ecotourism in the Amazon and elsewhere. Actual development, however, has been curtailed by poor accessibility, and by the presence of established but potentially incompatible industries such as logging and mining.

Fig. 11.3. South America.

Tourism, ironically, is mostly confined to regional urban gateways such as Manaus and Belem. Accordingly, most ecotourism is found within a relatively short distance of these strategic urban locations. Because most parks in this region are 'paper parks' that lack facilities or effective management practices (SUDAM/OEA, 1995), ecotourism tends moreover to occur within privately owned areas of secondary forest cover

Fig. 11.4. Train ride through the Mata Âtlantica (tropical rainforest), southern Brazil.

(Wallace and Pierce, 1996). This conforms to a wider global trend wherein privately owned protected areas are becoming increasingly important as ecotourism venues (see Chapter 19). The implication is that such experiences may not impart a true appreciation for primary Amazonian ecosystems, and are less likely to reflect the ideals associated with ecotourism. Concerns are thus raised as to whether these 'ecotourism'-labelled products deserve the name.

This said, government and the private sector are both pursuing ecotourism as a development option for Brazil's Amazonian region, and are well aware of the area's potential in this regard. According to a study undertaken by the Superintendency for the Development of the Amazon (SUDAM), and the Organisation of American States (OAS), 2204 distinct attractions were identified in the region, of which 1142 (64.4%) belong to the nature/ecological category. Other categories that are potentially complementary with ecotourism include history/culture (16.1%), folklore (14.4%), scientific research and technical activity (2.3%) and programmed events (1.6%) (SUDAM/OEA, 1997). There is evidence that specialized travel agencies were already capitalizing on the presence of Amazonia's nature-based attraction in the 1960s, although more significant growth did not occur until the late 1980s. An ethical code of conduct for the practice of ecotourism was established in 1996 (Pires, 1999), and research has been conducted into the demographic and behavioural profile of this market. According to this research (Ruschmann, 1997), most ecotourists are between 26 and 55 years of age, and three-quarters have attained higher education qualifications.

Brazil is not the only Amazonian country that is engaged in ecotourism. Several prominent ecolodges have been established in the Peruvian Amazon, and especially in the Madre de Dios region. For example, Manu Lodge, in the Manu World Heritage Site, was hosting 500 visitors per year by 1990 (Roe *et al.*, 1997). In the same year, about 3000 visitors were accommodated by the Cuzco-Amazonico Lodge. However, the viability of these operations has been threatened by encroachment from colonists, and by tenure problems that resulted in ecolodge land being turned over to local indigenous reserves. Of particular concern to the ecotourism sector has been the apparent favouritism shown by the government to the colonists for political reasons, even though that same government is aware of ecotourism's potential to differentiate and diversify the Peruvian tourism product (Yu *et al.*, 1997). Similar dynamics apply to other peripheral Amazonian countries, such as Ecuador (the Galapagos Islands are covered in other chapters), Venezuela, Colombia and Bolivia. Specific and generic information on rainforest tourism that is relevant to the Amazon basin is provided in Chapter 12.

The Andes

The Andes are the second major physical feature of South America and, like the Amazon, this alpine region is shared among a large number of South American countries (i.e. Venezuela, Colombia, Ecuador, Peru, Bolivia, Chile, Argentina) and remains in a largely natural state. However, no country dominates the Andes to the same extent as Brazil dominates the Amazon basin. Another major difference is the adherence to a strong ecotourism/protected area relationship in the Andes. This is most evident in the border region between Chile and Argentina, where well-established protected area systems have combined with good infrastructure, political stability, and relatively strong domestic ecotourist markets to foster significant ecotourism activity. Notable as an ecotourism gateway is the city of San Carlos de Bariloche, at the foot of the Andes mountain range, which emerged in the 1930s following the creation of Argentina's first national park (Schlüter, 1999). In the remaining Andean countries, the development of ecotourism in alpine regions has been hindered by inadequate protected area funding (i.e. the more typical 'paper park' pattern that exists in the Amazon basin and within the less-developed world in general), as well as political and social instability. The well-publicized activities of the Sendero Luminoso in Peru are illustrative, as is the ongoing conflict between the government and leftist insurgents in Colombia. Further discussion of ecotourism in the South American Andes is provided in Chapter 13.

The Pantanal

The Pantanal is a vast, flooded plain of approximately 140,000 km^2 situated in the south-west of the states of Mato Grosso and Mato Grosso du Sul, and extending into Bolivia and Paraguay. Situated between the Amazonian region and the Central High Plain of Brazil, the area exhibits characteristics of both ecosystems. Hence, it is regarded as one of the world's great concentrations of fauna as well as one of its most important natural nurseries. The attractiveness of the Pantanal is enhanced by the presence of an annual rainy season, which gives rise to completely different conditions from those experienced during the dry season (SUDAM/OEA, 1997).

Although the region received sporadic ecotourist visitations in the 1960s, it was not until the 1990s that concerns were voiced over an increasing visitor influx. On one hand, ecotourism was welcome by farmers and ranchers as an opportunity to diversify their revenue sources. However, given the considerable fragility of the Pantanal wetland ecosystem, conservation groups are concerned about the negative impacts that may result from excessive visitation levels. According to a study undertaken by SUDAM/OEA (1994), the rapid and disorganized growth of tourism has been associated with the inadequate treatment of residues and drinking water, animals killed by vehicles, and noise pollution caused by road traffic and motor boats. Adverse impacts have also been noted with respect to local customs and traditions. With the objective of minimizing the negative impacts of tourism, a series of programmes which aim to preserve the environment and train the local population have been implemented. The objective is to transform the Pantanal into an important destination for international ecotourism which facilitates sustainable development for the entire region.

Patagonia

Unlike Amazonia, tourism in the southern Argentine region of Patagonia was always closely linked to the area's protected area system, part of which was established to safeguard the fauna of the Atlantic Coast. Species such as whales, penguins, elephant seals and sea lions suffered from vigorous commercial exploitation in the early 20th century (Figs 11.5 and 11.6). Federal and provincial restrictions implemented in the 1960s did little to prevent the problem of poaching. In 1964, a government agency was established with the dual objectives of introducing more effective protection through an improved system of faunal reserves, and sustainably managing the intake of tourists that was anticipated to be attracted by these reserves (Schlüter, 1999).

By the end of the 1990s, this protected area system extended from the 41st latitude to the extreme southern tip of mainland Argentina. The rapid replenishment of many endangered species, and the fact that this has occurred despite the rapid growth of tourism, is evidence that both objectives have been achieved. Tourism is presently one of the dominant economic activities on the coast of Patagonia (Fig. 11.7). However, because potentially conflicting fishing and oil exploration activity is also carried out in the region, tourism stakeholders are working with these co-existent sectors to manage the coast in such a way as to accommodate all users in an environmentally appropriate manner.

Over-fishing and oil spills are two ways in which these other activities have negatively affected the local fauna, and thus ecotourism. One consequence of all these activities, including ecotourism, has been the alteration of whale behaviour in such a way that fewer boats can make the trip for sightings, and strict guidelines regarding viewing distance thresholds have had to be implemented. The soft nature of ecotourism on the Patagonian coast is indicated by the fact that visitors tend on average to spend between 1 and 1.5 h in a reserve. Ninety per cent of these visitors consider that they are able to learn something new and interesting about nature and conservation during this time (Tagliorette and Lozano, 1996).

Fig. 11.5. Marine elephants on the Patagonian Atlantic coast, Argentina.

Fig. 11.6. Penguins at the Cape Dos Bahías Natural Reserve, Patagonia, Argentina.

Fig. 11.7. Lighthouse Punta Delgada Ecotourism Complex. Valdes peninsula, Patagonia, Argentina.

Conclusion

For discussion and generalization purposes, Latin America and the Caribbean can be divided into three distinct regions based on physical geography and the status of the ecotourism sector. Figure 11.8 summarizes the patterns that are presented in this chapter in terms of the general status of ecotourism, and the strengths, opportunities, weaknesses and threats that are associated with this sector in each of the three regions. Before commenting on these general patterns, it is appropriate to reiterate that ecotourism activity is unequally distributed in all three regions, each of which possesses nodes of well-developed ecotourism, as well as extensive areas where this activity is virtually non-existent. As evident in Chapters 6–10, this pattern is consistent throughout the world. Structurally, the other pattern that is consistent with global trends is the emphasis on soft ecotourism, and its concentration in a small number of protected areas that are accessible to international gateways, major transportation routes, and developed resort areas.

In Latin America and the Caribbean, the general pattern involves moderate activity in the 3S-dominated Caribbean and coastal Mexico, followed by a higher level of engagement in Central America. It is ironic that at least two specialized ecotourism

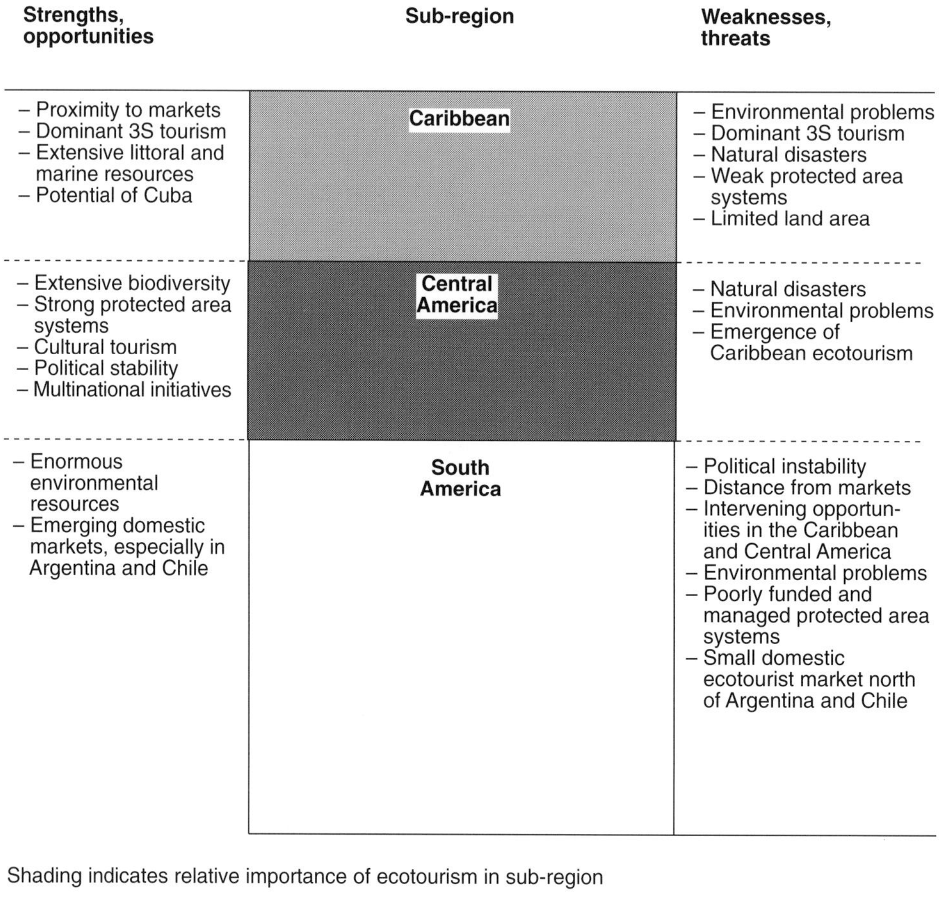

Strengths, opportunities	Sub-region	Weaknesses, threats
– Proximity to markets – Dominant 3S tourism – Extensive littoral and marine resources – Potential of Cuba	Caribbean	– Environmental problems – Dominant 3S tourism – Natural disasters – Weak protected area systems – Limited land area
– Extensive biodiversity – Strong protected area systems – Cultural tourism – Political stability – Multinational initiatives	Central America	– Natural disasters – Environmental problems – Emergence of Caribbean ecotourism
– Enormous environmental resources – Emerging domestic markets, especially in Argentina and Chile	South America	– Political instability – Distance from markets – Intervening opportunities in the Caribbean and Central America – Environmental problems – Poorly funded and managed protected area systems – Small domestic ecotourist market north of Argentina and Chile

Shading indicates relative importance of ecotourism in sub-region

Fig. 11.8. Ecotourism patterns in Latin America and the Caribbean.

countries have emerged in the mass tourism-dominated Caribbean, but this is explained by the very small size of these particular countries. South America, in general, has a lower level of ecotourism than either of the preceding regions, due to a variety of regional weaknesses and threats that counter its formidable array of environmental assets and its emerging domestic and intra-regional tourist markets. Central America benefits from a high level of biodiversity, a well-articulated protected area system, significant multilateral and bilateral tourism and protected area initiatives (e.g. Mundo Maya), growing political stability, and strong complementary products such as cultural tourism. On the weakness/threat side, Central America is experiencing major environmental problems, is subject to natural disasters, and will face competition in future from the larger Caribbean islands. In the Caribbean, strengths and opportunities include access to the North American market, strong marine and littoral environmental resources, and the potential of Cuba as a major ecotourism destination. Weaknesses and threats include environmental deterioration, episodic natural disasters, poorly developed protected area systems, and limited land areas. The presence of a dominant 3S-based tourism industry can be interpreted as both an opportunity and a threat.

The art of prediction is always fraught with uncertainty. However, it appears in the medium term as if the insular Caribbean and Mexico will rival Central America as an ecotourism-providing region, especially as links between mass tourism and soft ecotourism in that region are expanded in the effort to diversify the resort tourism product. South America will probably display two patterns of development. In the more developed southern area occupied by Argentina, Chile, Uruguay and southern Brazil, growing domestic demand and improving infrastructure will generate a pattern similar to North America and Europe. In contrast, the remainder of South America is not likely to progress much beyond incipient ecotourism, given its continuing relative isolation, underdevelopment, and small domestic markets.

References

Ayala, H. (1997) Resort ecotourism: a catalyst for national and regional partnerships. *Cornell Hotel and Restaurant Administration Quarterly* 38(1), 34–45.

Ayala, H. (1998) Panama's ecotourism-plus initiative: the challenge of making history. *Cornell Hotel and Restaurant Administration Quarterly* 39(5), 68–79.

Boo, E. (1990) *Ecotourism: the Potentials and Pitfalls*, Vol. 2. World Wildlife Fund, Washington, DC, USA.

Cater, E. (1992) Profits from paradise. *Geographical Magazine* 64, 16–20.

Ceballos-Lascuráin, H. (1996) Ecotourism or a second Cancun in Quintana Roo? *Ecotourism Society Newsletter* 3.

Dixon, J., Scura, L. and van't Hof, T. (1993) Meeting ecological and economic goals: marine parks in the Caribbean. *Ambio* 22(2–3), 117–125.

Epler Wood, M. (1993) Costa Rican parks threatened by tourism boom. *The Ecotourism Society Newsletter* 3(1), 1–2.

Hawkins, D. and Khan, M. (1998) Ecotourism opportunities for developing countries. In: Theobold, W.F. (ed.) *Global Tourism*, 2nd edn. Butterworth-Heinemann, Oxford, pp. 191–204.

Holder, J.S. (1996) Maintaining competitiveness in a new world order: regional solutions to Caribbean tourism sustainability problems. In: Harrison, L.C. and Husbands, W. (eds) *Practising Responsible Tourism: International Case Studies in Tourism Planning, Policy, and Development.* John Wiley & Sons, Chichester, UK, pp. 145–173.

Honey, M. (1999) *Ecotourism and Sustainable Development: Who Owns Paradise?* Island Press, Washington, DC.

Lindberg, K., Enriquez, J. and Sproule, K. (1996) Ecotourism questioned: case studies from Belize. *Annals of Tourism Research* 23, 543–562.

Mundo Maya Organization (1996) *Mundo Maya: Where Man, Nature and Time are One.* Mundo Maya Organization, Mexico City.
Pires, C. (1999) Management in ecotourism agencies and their insertion in the context of sustainability. *Turismo: Visão e Acão* 1(2), 45–46.
Roe, D., Leader-Williams, N. and Dalal-Clayton, B. (1997) *Take Only Photographs, Leave Only Footprints: the Environmental Impacts of Wildlife Tourism.* HED Wildlife and Development Series No. 10. Environmental Planning Group, London.
Ruschmann, D. (1997) O ecoturismo no Brasil. *Proceedings of the World Ecotour '96.* Rio de Janeiro, Brazil, pp. 1168–1174.
Schlüter, R. (1998) Tourism development: a Latin American perspective. In: Theobold, W.F. (ed.) *Global Tourism*, 2nd edn. Butterworth-Heinemann, Oxford, pp. 216–230.
Schlüter, R. (1999) Sustainable tourism development in South America. The case of Patagonia, Argentina. In: Pearce, D.G. and Butler, R.W. (eds) *Contemporary Issues in Tourism Development.* Routledge, London, pp. 176–191.
SUDAM/OEA (1994) *Diagnóstico do Desenvolvimento do Ecoturismo no Panatanal Brasileiro.* Belem, Brazil.
SUDAM/OEA (1995) *Linhas Básicas Para un Programa de Desenvolvimento do Turismo na Regiäo Amazónica.* Belem, Brazil.
SUDAM/OEA (1997) *Recursos Naturais na Amazonia.* Belem, Brazil.
Tagliorette, A. and Lozano, P. (1996) *Estudio de las Demanda Turística en la Costa Patagónica.* Informe Técnico no. 24, Fundación Patagonia Natural, Puerto Madryn.
Thomlinson, E. and Getz, D. (1996) The question of scale in ecotourism: case study of two small ecotour operators in the Mundo Maya region of Central America. *Journal of Sustainable Tourism* 4, 183–200.
TTI (1996) Costa Rica. *International Tourism Reports* 4, 4–24.
Wallace, G. and Pierce, S. (1996) An evaluation of ecotourism in Amazonas, Brazil. *Annals of Tourism Research* 23, 843–873.
Weaver, D.B. (1995) Alternative tourism in Montserrat. *Tourism Management* 16, 593–604.
Weaver, D.B. (1998) *Ecotourism in the Less Developed World.* CAB International, Wallingford, UK.
Weaver, D.B. (1999) Magnitude of ecotourism in Costa Rica and Kenya. *Annals of Tourism Research* 26, 792–816.
Weaver, D.B. (2001) Mass tourism and alternative tourism in the Caribbean. In: Harrison, D. (ed.) *Tourism in the Less Developed World.* CAB International, Wallingford, UK (in press).
Weaver, D.B. and Oppermann, M. (2000) *Tourism Management.* John Wiley & Sons, Brisbane.
Wood, R.E. (2000) Caribbean cruise tourism: globalization at sea. *Annals of Tourism Research* 27, 345–370.
World Resources Institute (1998) *1998–99 World Resources: a Guide to the Global Environment.* World Resources Institute, New York.
Yu, D., Hendrickson, T. and Castillo, A. (1997) Ecotourism and conservation in Amazonian Peru: short-term and long-term challenges. *Environmental Conservation* 24, 130–138.

Section 3

A Regional Survey by Biome

D.B. Weaver
Department of Health, Fitness and Recreation Resources, George Mason University, Manassas, Virginia, USA

If ecotourism attractions are primarily nature-based, then it is useful to differentiate this sector on the basis of the major biomes that comprise the natural realm. The six chapters in this section examine ecotourism in the context, respectively, of rainforests, alpine areas, polar regions, islands and coasts, deserts as well as grasslands and savannahs, and marine environments. Ecotourism has penetrated all of these ecosystems, but the extent and manner of this penetration, and the management issues that result, depend on the characteristics of each ecosystem and the way they are perceived by relevant stakeholders. The treatment of the ecosystems as separate entities, however, is more a matter of convenience than a reflection of reality, since the divisions between them are often fuzzy, and a single setting, such as a thickly forested mountain slope in Southeast Asia, can belong to more than one category. In addition, ecotourism products often deliberately seek to diversify by establishing synergies among a variety of ecosystems.

Biomes, in the first instance, differ in terms of appearance and species composition. This is critical to ecotourism, since both characteristics help to define the aesthetic appeal of biomes and their potential for providing learning opportunities. Frost, in Chapter 12, describes the impressive and even overwhelming level of biodiversity that is associated with lush, closed-canopy rainforests. High levels of diversity are also often associated, though for different reasons, with alpine areas, savannahs, coral reefs and islands. The opposite end of the biodiversity continuum is occupied by the polar ice caps, which, as described by Stonehouse in Chapter 14, harbour a limited amount and array of terrestrial life, and only on their periphery. Yet, the grandeur and pristine condition of such physical settings ensures a growing level of interest from determined ecotourists. Compared with the above 'poster' settings, grasslands and deserts, as discussed by Weaver in Chapter 16, are 'Cinderella' ecosystems whose charms are often more subtle.

Ecosystems also vary dramatically in terms of their survival in a more or less natural state. On one extreme, the ice caps exist today in much the same form and extent that they have for the past millennium. Deserts may be even more extensive than they once were, though this owes more to the degradation of adjacent ecosystems than to any enlightened attitude toward such areas. In either case, preservation is a matter of default rather than intent; these are settings, in other words, that have long been perceived as having only marginal direct utility for human beings. For ecosystems that are seen as having the potential of such utility, the situation is grim. In Chapter 13, Williams, Singh and Schlüter describe how sensitive alpine environments are being heavily encroached upon by human settlement, resource extraction and, in the more economically developed regions, alpine sporting activity. Similarly, Halpenny indicates in Chapter 15 how coastal and insular ecosystems are threatened by resort tourism and population migration in general. Just offshore of these areas in the tropics, associated threats are being posed to coral reefs, as discussed by Cater and Cater in Chapter 17.

But rainforests, more than any of these areas, have come to symbolize the destruction of the world's ecosystems, whether it is occurring in the Brazilian Amazon or in British Columbia. The amount of intact rainforest has declined by about one-half just in the past century, leading to growing concern about the future ability of rainforests to host ecotourism, and indeed, about the role that ecotourism might increasingly play in arresting this decline. Probably no other ecosystem receives the publicity and research that rainforests receive, including that which is related to ecotourism. In contrast, long- and medium-grass prairies are among the most endangered ecosystems in the world, as Weaver points out, yet have attracted little attention by comparison. This may be due in part to the lack of drama involved in their conversion to farmland, and to the relative ease with which such lands can be rehabilitated.

The inequity in attention is of course partly a function of perception. Most tourists consider rainforests, savannahs and coral reefs, and their associated wildlife, to be far more attractive than grasslands or deserts, a view that is reinforced by incessant publicity. Yet, until recently, all natural ecosystems in Western cultures were regarded in a negative light to the extent that they were spaces not being used for farming, forestry, mining or other 'useful' activities. Despite the continuing degradation of these environments, it is a positive sign that the remaining natural spaces are increasingly being perceived as our most precious assets, rather than wastelands awaiting fitful exploitation.

As stated above, the characteristics and circumstances that attend each of these remaining ecosystems have a bearing on their management as ecotourism settings. In all cases, ecotourism can potentially function as an incentive to preserve and even enhance what is left of the natural component. However, if incorrectly managed, ecotourism can act as one of the forces that contributes to its demise, as many of these chapters show. Diving and viewing pressures, for example, can quickly overwhelm coral reef and cetacean carrying capacities for tourism, while the prevalence of ecotourism hybrids such as trekking in mountain regions can mean the dilution of the sustainability imperative, and overall poor quality of the tourism product. The fact that most rainforests, savannahs and coral reefs are located within less economically developed regions means that sufficient resources are not always available to implement or enforce appropriate management strategies. Moreover, the dichotomy between wealthy, white ecotourists and poor, non-white local communities is often present in such situations, creating challenges for the realization of socio-cultural sustainability. Such, however, is not a concern for uninhabited Antarctica, which is also unique in being subject to an international treaty structure that circumvents the normal hierarchy of municipal, state and national governments.

Many other ecosystems-based management issues can be cited. Frost, for example, describes how visitors to the rainforest usually arrive with preconceived notions as to how these ecosystems operate. Should managers try as much as possible to satisfy visitors by confirming these preconceptions, or should they engage in mass debunking? Rainforests, because of their closed-in character, can also 'hide' large numbers of visitors, whereas even a small number of ecotourists in an open grassland or on an ice cap can give rise to an intrusive and crowded effect. In terms of experiences, Weaver points out that the interaction with nature on the African savannah is primarily a visual experience in which the visitor is almost always guaranteed of seeing at least some interesting wildlife. In contrast, the rainforest experience is more one of feel and sound, and visually more focused on plants than animals. In light of these contrasts, an elevated viewing structure is appropriate for the rainforest, but perhaps not for the grassland. Where ecotourism occurs in coastal regions, Halpenny points out that it is increasingly difficult to dissociate this activity from 3S resort development, wherein visits to sand dunes and mangroves, and excursions to nearby rainforests, are important add-ons to the latter which can foster mutually beneficial relationships. Cater and Cater make the same observation with respect to offshore diving and whale-watching. In sum, the circumstances of each biome will in large part dictate the mode of ecotourism product that is best able to effect sustainable outcomes for that environment.

Chapter 12
Rainforests

Warwick Frost
Department of Management, Monash University, Berwick, Australia

Introduction

For most tourists interested in nature, rainforests are about the most attractive biome on Earth. Tourists often see rainforests as lush, luxuriant, vibrant, immense, mysterious, spiritual and romantic. As most tourists are urban-dwellers from countries without rainforests, a visit to a rainforest is an exotic and rare experience. Rainforests are also associated with other attractive experiences and images. Tropical beaches, islands and resorts are easily associated with rainforests. Rainforests hold a special place for some as the landscapes of the dinosaurs. Today, rainforests are the home of exotic, rare and threatened species, such as the mountain gorillas of Africa and the orang-utans of Sumatra. With increased interest over the last 20 years in preservation of the environment, the fate of the rainforests, especially those of the Amazon basin, has become symbolic of that struggle. For many tourists a visit to a rainforest is an affirmation of their support for the environment.

Rainforests seem easy to understand, certainly at the simple level. Even the most urban-centred tourist can enjoy a short venture into a rainforest (especially if along a well-made path or walkway). In addition to the enjoyment of such pleasant surroundings, such a tourist could easily understand many of the special characteristics and values of the rainforest and how it can be threatened. Massive coverage of rainforest issues, especially through television and educational institutions, has led to most of us feeling we have some expertise in understanding rainforests. In contrast, the special features of other biomes, such as grasslands, are far harder for the ordinary tourist to understand and appreciate.

Increasingly, rainforests are where ecotourism and mass tourism collide. This creates many problems. Should tourism operators and park managers cater for the niche or mass market? Can they satisfy both? If the average tourist already comes armed with a great deal of general knowledge and set expectations about rainforests, how then do we approach the provision of interpretation? Is it acceptable to manipulate the natural environment to better fit preconceptions about rainforests? How do we manage visitors to rainforests to maximize their experience and minimize their impact?

This chapter has two main aims. The first is to provide a general descriptive overview of rainforests and rainforest tourism. That rainforests are seen as so easy to understand is a trap. It is important to fully understand the complexities of their definition, different typology and geographical distribution. Similarly rainforest tourism needs some careful explanation, for it comes in different guises and there are quite marked geographical differences across the world.

The second aim is to provide a discussion of the key issues affecting rainforests and rainforest tourism. These include the difficulties of balancing mass and ecotourism, providing meaningful interpretation and protecting rainforests from excessive visitor impact. Consideration is also given to the increasing trend towards artificial rainforests as tourist attractions.

What are Rainforests?

Defining rainforests generally is simple, defining them exactly is very difficult and has generated much debate. The term rainforest was coined in 1898 by the German botanist Andreas Schimper in his *Plant Geography upon a Physiological Basis* (posthumously translated into English in 1903). He described dense lush tropical forests which he had visited on fieldwork in the Caribbean; South America, Sri Lanka and Indonesia. These forests only occurred in areas of high rainfall. Thus he combined rain and forest for the term *Regenwald*, which was quickly translated into English as *Rainforest* (the alternative *Rain Forest* is mainly used by northern hemisphere writers).

Difficulties arose when other high rainfall forests around the globe were considered. Some were structurally similar to Schimper's tropical rainforest but occurred in subtropical, temperate and even cold temperate climates. Other tall, dense forests occurred in high rainfall areas, but seemed substantially different in structure. In the confusion rainforests were often defined in a negative way (Adam, 1992). Schimper and his fellow botanists were mainly Europeans and North Americans, they already knew about the high rainfall conifer and deciduous forests of their home countries, they were describing different forests, so they defined rainforest in a Eurocentric way as *evergreen broadleaf forest* (i.e. not European). In Australia some botanists defined rainforests as forests in high rainfall areas which were *not eucalypt forests*. Similarly on the west coast of the USA it was decided that the redwood (sequoia) forests were not rainforests.

The problem worsened in the 1960s and beyond as the use of the term expanded beyond scientific circles and into common usage. Ordinary people began to use it generally to describe any high rainfall forest (nowadays a common dictionary definition). Many botanists began to focus on definitions based on density of canopy and other structural differences rather than geography or climate. Some botanists were disturbed by the existence of deciduous rainforest trees, forests that seemed to be rainforests but lacked the diversity which characterized tropical rainforests, and anomalies regarding conifers (araucarian conifers were acceptable, but not others). In addition, there was increasing evidence that species (such as eucalypts) which had been excluded from rainforests had actually evolved from rainforests.

On the other hand, under increased pressure not to log rainforests, foresters and public land managers demanded narrower and narrower 'scientific' definitions of rainforests. This conflict was well illustrated in the instance of Victoria (Australia) where much of the high-rainfall eucalypt forests are intermingled with rainforest species as an understorey and along moister gullies. In 1985 the State Department of Conservation and Environment established a Rainforest Technical Committee consisting of senior botanists. This committee's brief was to finally provide a 'scientific' definition of rainforest. However, when the committee agreed upon a definition which included the mixed eucalypt rainforest forests, the state government deleted these mixed rainforests from the final definition in order to appease forestry interests (Cameron, 1992).

What has developed is a continuum scale of definitions. At one extreme we still have writers who only accept tropical rainforests as rainforest (a common position in many popular works). Moving along the scale there are many who accept a number of different types of rainforest, but draw the line if so-called non-rainforest species are present. Next come a group (seemingly growing) who accept various levels of mixing. Finally we have the other extreme that all high-rainfall forests are rainforests (a position increasingly adopted by some tourist operators trying to attract as many tourists as possible).

Bearing all these difficulties in mind, it is possible to construct a generally acceptable definition of rainforest. A rainforest is a high density grouping of tall trees and other vegetation, in which the tall trees form a dense canopy which significantly reduces light levels at the forest floor. Many of the trees in the rainforest will be evergreens with relatively soft leaves and these trees will have the ability to reproduce under the undisturbed canopy. However, the rainforest may also include numbers of conifers and hard-leaved evergreens. As these species do not usually have the ability to reproduce under the canopy they may be relics from previous disturbances or be found mainly around the edges of the rainforest.

Types of Rainforest

There are a number of different ways of classifying rainforests, the following is probably the most common. Though it uses a nomenclature which suggests typology based on climate, the differences are really much more of a structural nature.

Tropical rainforest

This is the original rainforest as described by Schimper, the most common type and the stereotype of rainforest firmly lodged in the mind of most tourists. Its chief characteristics are:

- A very wide diversity of tree and plant species. Even though tropical rainforests only cover 7% of the Earth's landmass, they provide about 50% of the world's plant species (Whitmore, 1990).
- A very dense canopy, often multi-layered, sometimes with a fairly open understorey at ground level.
- Most trees have very large leaves and often massive buttress roots.
- Large numbers of thick, woody vines, palms and epiphytes.

Tropical rainforest primarily requires warm temperatures (a minimum of 18°C) and secondarily high rainfall (a minimum of 100 mm each month). Half the world's tropical rainforests are in the Americas, particularly Central America, the Caribbean and the northern third of South America. Tropical rainforests are also found in West Africa, Madagascar, western India, Sri Lanka, southern China, Southeast Asia, Papua New Guinea, the far northeastern coast of Australia and many Pacific and Indian Ocean islands (Whitmore, 1990).

Subtropical rainforest

Subtropical rainforest occurs adjacent to tropical rainforest in areas that are slightly cooler due to difference in altitude or latitude. Subtropical rainforest looks very similar to tropical rainforest, but is somewhat less luxuriant and diverse. It is dominated by only a few tree species and it is less layered. Buttresses, figs, palms, large epiphytes and woody vines may be less frequent and there may be more ferns.

An interesting and confusing variation occurs with rainforest on less fertile and acidic soil. Even though it may be adjacent to subtropical rainforest, this depauperate type is called *warm temperate rainforest.* It is typically dominated by one or two species, trees are shorter, leaves are smaller and it has far less tropical rainforest characteristics (for example buttresses, woody vines).

Dry or monsoonal rainforest

Usually contiguous to tropical and subtropical rainforests, these are rainforests markedly affected by a pronounced dry season. They are characterized by species typically found in the other rainforests, but which have adapted to the more seasonal conditions. This adaptation might include dwarfing or a limited growing season. Such rainforests are typically more open with far less luxuriant foliage. They may be dominated by more drought-tolerant rainforest species, such as araucarian conifers. In some instances small patches of dry rainforest may be found along watercourses and in gorges in regions that are normally considered quite arid.

The subtropical, warm temperate and dry or monsoonal rainforests extend significantly outwards from the tropical rainforest cores of Central–South America, West Africa and Southeast Asia. For example, rainforest is found in northern Iran. However, rainforests are usually not regarded as extending into Europe, the USA or Canada.

Cool temperate rainforest

Cool temperate rainforests do not really match the stereotypes of rainforests. They lack the diversity and luxuriance of tropical rainforests. They are usually cold and wet and therefore unattractive to some tourists. They look far more like European forests. Yet, in the last few decades they have come to occupy a special niche for rainforest lovers. In a way that tropical rainforests are not, cool temperate rainforests are seen as *real wild places*, a sort of last frontier. Being cold and wet they are usually not in close proximity to intensive cultivation or large densities of humans. Furthermore, they are viewed as having strong links to the Earth's prehistoric past, they are perhaps seen as *living fossils*. This link is best seen in how modern cool temperate rainforests were used as the background for the 1999 BBC TV series *Walking with Dinosaurs*.

Cool temperate rainforests are geographically distant from the other rainforests. They are mainly found in the southern half of the southern hemisphere, namely Chile, Tasmania and New Zealand. However, very small patches can be found further north, but only at high altitudes (though depending on one's definitional stance, the west coast forests of USA and Canada could be included in this category).

The chief characteristics of cool temperate rainforests are:

- one dominant tree variety; usually *Nothofagus* (southern beech);
- very small leaf size; sometimes deciduous;
- buttresses, palms, figs, large epiphytes and woody vines completely absent;
- abundant ferns, mosses and lichens.

Some writers, mainly from the northern hemisphere, refer to cool temperate rainforests as *montane rainforests* and to tropical and subtropical rainforests as *lowland rainforests*.

Evolution of the Rainforests

Schimper in 1898 only *discovered* rainforests in a Eurocentric sense, by naming them. The far longer history of rainforests has really only been pieced together more recently. Their origins are in the great southern supercontinent Gondwanaland which existed between 160 and 100 million years ago and consisted of modern day South America, southern Africa, India, Australia and Antarctica. In a world much warmer and wetter than today, Gondwanaland developed as a rainforest continent. The Gondwanan dinosaurs grazed in an extensive and lush environment of conifers, ferns, palms and cycads. A striking example of how much more extensive the rainforest coverage was then is the recent uncovering of hundreds of rainforest fossil leaves from the desert at Lake Eyre in South Australia (White, 1994).

About 125 to 100 million years ago flowering plants began to develop, probably as opportunists filling newly created ecological niches as sea levels varied (White, 1994). At around the same time, Gondwanaland began to break apart, some parts drifting off to collide with the northern supercontinent Laurasia and other parts remaining separate. Today's rainforests are either found on former parts of Gondwanaland or in regions of close proximity. The break up of Gondwanaland was accompanied by (and probably caused) global cooling and drying, which was particularly manifested in the development of polar ice caps and irregular Ice Ages. This caused rainforests to evolve their cool temperate form and prevented their spread into Europe and temperate North America. Indeed increased understanding of our botanical history reverses traditional Eurocentric views. The supposedly ancient forests of Europe are really post Ice Age youngsters.

Changing Attitudes to Rainforests

For tens of thousands of years the rainforests of the Americas, Africa, Asia and Australia have been the home of indigenous people. In the last 500 years the expansion of Europeans over the globe has led to nearly all rainforests coming under some sort of colonial administration. In many cases the European colonies were primarily extractive with indigenous people coerced into a colonial labour force. In some instances indigenous people were replaced or supplemented with labour drawn from other locations (such as African slaves or Chinese indentured labourers). In Australia and New Zealand the Europeans created settler societies. In the 20th century European colonialism declined dramatically, with most countries gaining independence. However, in place of colonial powers, domination has passed to a handful of key economic powers. All these different interests have led to a wide range of attitudes towards rainforests.

For many indigenous people the rainforests were their entire world. Certain parts of the rainforest, particularly certain groups of trees, were regarded as sacred and taboo and certain rainforest animals had religious and totemic significance (Flannery, 1998; Boomgaard, 1999). Rainforests provided nearly all their economic wants. Food came from hunting and gathering, often combined with simple slash and burn agriculture. Tropical rainforest was particularly diverse in the range of resources it provided, a diversity which required indigenous peoples to develop and pass on from generation to generation a massive range of local botanical and zoological knowledge (Flannery, 1998). Indeed, so abundant were the resources of rainforests that they may have stifled agricultural development in many regions.

The close relationship between indigenous people and rainforests leads to a common misunderstanding. Because the rainforests were not exotic and therefore (it is sometimes argued) *special* to these peoples, it has been quite easy for Eurocentric commentators to conclude that indigenous people had nothing more than a utilitarian relationship with the rainforests.

Initially, for Europeans, the rainforest was a mask they had to remove. It hid precious minerals and sometimes hid the forces of resistance. Most importantly it hid agricultural opportunities. Removing the rainforest allowed the rain and (sometimes imagined) fertile soil to produce high-value export commodities. Rainforests were cleared for rubber, coffee, tea, sugar and dairying (Aiken and Leigh, 1995; Dean, 1995; Grove, 1995; Frost, 1997). Where overcropping caused declining soil fertility, land was abandoned and more virgin rainforest cleared and planted (Dean 1995; Grove 1995; Frost 1997). Until late in the 20th century, timber-cutting fell far behind mining and agriculture, for with the exception of some very high-value woods, it was not economic to export timber. As a result cleared rainforest was usually burnt as a waste product.

Over time a variety of conservation sentiments developed. Declining stream flow and soil fertility quickly became a major problem on some West Indian and Indian Ocean islands (Grove, 1995). In some colonies Europeans formed conservation societies in order to protect particular animals for their exclusive game hunting (Boomgaard, 1999). However, in other cases there was very little interest in conservation (Dean, 1995).

It was in Australia, the one area where rainforests were successfully converted into family farms by European settlers, that the strongest and most widespread regard for rainforests developed. While clearing large areas, many farming communities took great care to preserve small patches (especially waterfalls and gullies) as parks. Rainforest beauty spots were not only valued by locals; between the 1870s and the 1930s rainforests were seen as especially attractive by urban-dwellers and became a major component of a successful nature-based tourism sector in Australia (O'Reilly, 1945; Ritchie, 1989; Frost, 2000).

After the Second World War the clearing of tropical rainforests quickened due to the massive economic expansion which characterized this period, especially in the Asia-Pacific Region. The buoyant economies of the USA and East Asia increasingly demanded timber, food and minerals. These could be gained cheaply by clearing rainforests in poorer countries which were missing out on industrialization but were still keen to grab a piece of the global action. The development and utilization of modern machinery allowed clearing to occur far more quickly and cheaply than before (Collins, 1990; Whitmore, 1990; Aiken and Leigh, 1995).

The scale of modern clearance is difficult to quantify. Government statistical authorities rarely collect forest clearance data. Estimates may be done on different criteria and for different time periods and comparisons between countries may be very difficult (for example see Salim and Ullsten, 1999). If we are just counting area cleared, the greatest modern clearance has occurred in Brazil, followed by Indonesia and Nigeria. On the other hand if we consider area cleared as a percentage of total rainforest, the highest rates of clearance are in Ivory Coast, Nigeria, Costa Rica and El Salvador (Whitmore, 1990). However we view the statistics, it is clear that in the 1980s and 1990s rainforest clearing in many countries has reached a rate which cannot be sustained if these rainforests are to survive as significant biomes.

The scale and intensity of recent clearance has directly led to a tremendous growth of interest and appreciation in rainforests. Less than 20 years ago some commentators bemoaned that few tourists understood rainforests (Valentine, 1991). However, by 1990 it was confidently proclaimed that 'rainforests have crossed a threshold of perception' (Whitmore, 1990). Writers on heritage argue that many things only come to be seen as heritage when they are under threat (Davison, 1991). That is exactly what occurred with rainforests. Publicity about clearing stimulated anger *and* fascination. Rainforests became a cause célèbre of the 1980s and 1990s. Film, television and popular music reinforced images of rainforest as something worth saving. And as interest in the conservation of rainforests grew, so too did interest in visiting and experiencing rainforests.

Rainforest Tourism Today

Rainforest tourism is very difficult to quantify and package neatly. There are major problems in defining rainforest tourists and counting them. Do we define by interests, activities or attractions visited? Do we count numbers or revenue yielded? Is it right (as is commonly done) to value international tourists as far more important than domestic tourists? How do we deal with comparisons between countries? As we are a long way off quantifying rainforest tourism and there has been very little research specifically on rainforest tourism, the approach taken here is descriptive and somewhat speculative.

The most significant development in rainforest tourism in recent decades has been the growth of high-value package tours. These have been particularly noticeable in Latin America, most notably Costa Rica, but also Guatemala, Honduras, Belize, southern Mexico and Brazil (Thomlinson and Getz, 1996; Wallace and

Pierce, 1996; Lumsdon and Swift, 1998; Weaver, 1998; Honey, 1999; Minca and Linda, 2000). Tours also occur to a lesser extent in most of the rainforested areas of the world, although political instability and warfare have severely limited their development in some countries (Shackley, 1995; Weaver, 1998).

The market for this type of rainforest tourism is typically relatively well-to-do tourists from well-developed countries, especially from the USA and northern Europe. Such tourists fit the classic ecotourism mould, they are generally well-educated, keen to incorporate learning experiences into their holiday and concerned about conservation. Whether or not the tours and experiences they engage in are truly ecotourism is the subject of a lively ongoing debate (Thomlinson and Getz, 1996; Wallace and Pierce, 1996; Lumsdon and Swift, 1998). Nonetheless, many of these tourists would either see themselves as *real* ecotourists or as far more *serious* than the usual sun and sand crowd.

The cost of such tours averages US$100–200 per day per person for the land component only (Shackley, 1995; Thomlinson and Getz, 1996). When airfares are added a rainforest holiday is an expensive proposition. This tends to limit the market to older, high income, experienced, highly motivated travellers. It also tends to preclude domestic tourists (though Costa Rica and Australia do have strong domestic visitor rates). However, it is important to distinguish between the current and potential markets. In recent years there has been the growth of a younger backpacker market. Their tendency is to be more independent, accept cheaper accommodation, meals and transport and splurge on short expensive rainforest tours and experiences (in the same way as normally frugal backpackers still tend to be big spenders on diving experiences on the Great Barrier Reef).

Much of the high-value rainforest tourism is through traditional style group tours. These are typically 7–14 days, all inclusive of food, accommodation and attractions, often cover a large area and a number of countries and are usually built around a strong theme. In addition there has been much development of accommodation properties as self-contained destinations. These often have distinctive themes or styles, for example they may be promoted as eco-lodges or safari camps or as boutique or specialized (Moscardo *et al.*, 1996; Wallace and Pierce, 1996). Some even present themselves as scientific research centres (Weaver, 1998; Honey, 1999). As well as food and lodging they tend to offer exclusive access to local rainforest, guided tours, animal feeding and interpretative talks. Many now have strong links to indigenous groups, utilizing them as guides and interpreters and in some cases these facilities may be owned and operated by local communities (Wallace and Pierce, 1996; Weaver, 1998).

Rainforests are not the sole attraction for these tourists. They may be interested in a range of attractions which are geographically linked to the rainforest, or it may even be that the rainforest is just the background for a far stronger interest. Tours of Central America are packaged and promoted around a number of features, including Mayan ruins, beaches, adventure activities and indigenous culture as well as the rainforests (Thomlinson and Getz, 1996; Lumsdon and Swift, 1998; Weaver, 1998). Most tropical beach resorts have some linkages to rainforests, either having adjoining stands or offering longer tours to nearby forests. While their customers are primarily interested in the beach and resort activities, the rainforest offers variety and exotic glamour. Animal and birdwatching tourism is a particularly significant high-value niche market which often utilizes rainforest environments (Shackley, 1995).

A different form of rainforest tourism has tended to develop in Australia, particularly in Queensland. It is chiefly distinguished by its markets. In Queensland they are chiefly domestic tourists, international visitors from Asia and backpackers. Prices are lower and packages are built around accommodation destinations rather than long tours. These accommodation destinations (and there are hundreds) skilfully use rainforest plantings and views as their setting. Nonetheless, the rainforest is typically a background for the tourists' chief interests in beaches, water activities and adventure experiences. Ecotourism ventures do exist and have grown in recent years, but they are only a small segment. Another significant difference in Australia is that perhaps over half of rainforest visitors are from nearby areas and many tourists are taken to rainforests by friends or relatives whom they are visiting (Parsonson *et al.*, 1989; Valentine 1991).

Rainforest Interpretation

Interpretation aims at providing tourists with explanations about the places they visit and is a very important component of ecotourism (see Chapter 35). Interpretation has two components: the message (or theme) and the method (or mode). Unfortunately, tourism managers often concentrate more on the method than the message and rainforest tourism is no different in this respect.

Determining what are the key messages for tourists in rainforests is difficult for four reasons. First, most tourists come to rainforests already loaded down with preconceptions. Do tourism managers shape the experience to fit and satisfy these preconceptions or do they risk challenging them? Second, as noted earlier in this chapter, there is considerable unresolved debate about certain aspects of rainforests. How should they be handled? Should the interpretation be kept simple or can it include multiple conflicting explanations? Third, at what level should the interpretation be aimed? Tourists range from children to highly educated adults, from the mass tourist seeking a pleasant experience to the dark green ecotourist. How do tourism managers strike a balance? Fourth, different tourism managers will have different messages depending on their own circumstances and beliefs. For example a government agency charged with managing rainforests may give its highest priority to promoting what a good job it is doing in conserving a particular patch. However, down the road, an independent tour operator may explain to their tourists that the same government agency is encouraging logging and agricultural clearance.

It has been suggested that there are certain messages which should be included in interpretation at all rainforests. These are the following (adapted from Frost, 1999).

1. What makes a rainforest and the debate over what is and what is not a rainforest.
2. The different types of rainforest and, in particular, the type that this rainforest belongs to.
3. How indigenous people interacted with this rainforest.
4. How European colonization or settlement affected this rainforest.
5. The major threats today.
6. Plant and tree varieties.
7. Animals, birds and insects.
8. Special growing conditions associated with this rainforest (such as the nutrient cycle or the presence of buttressed roots).
9. The fragility *and* resilience of rainforests in general and of this particular rainforest.
10. Any revegetation or scientific research projects in progress.

Each of these messages varies in terms of complexity, controversy and vested interests. When the interpretative materials at six popular rainforest parks in Australia were surveyed, some strong patterns emerged (Frost, 1999). None of the parks provided any information about current threats to rainforests in Australia. However, some referred to rainforest clearing in Brazil! The likely explanation for these omissions is that the government agencies responsible for these parks were usually parts of larger agencies responsible for timber-cutting in the area. In addition, none of the six attempted to provide a 'scientific' definition of rainforest. Again this may have been due to the broader logging interests of the park managers or it may have been seen as just too complex to try to explain to tourists. Curiously, three of the six had good information on indigenous use of the rainforest and three had nothing on this topic.

In contrast, all six parks provided extensive excellent quality interpretative material regarding the special growing conditions to be found in that rainforest. Five of the six provided information about the animals, birds and insects and labelled the major tree varieties. Such emphases can be explained in two ways. First, this interpretation focused on information which was incontestable and uncontroversial: the trees had buttresses, they were of a particular species, there were epiphytes, nutrients were returned to the soil by rapidly rotting leaves, etc. Second, this interpretation related to the internal dynamics of the rainforest; it did not go beyond the rainforest and consider how it interacted with the rest of the world.

The quality of interpretation is highly dependent on the level of knowledge of its creator, the writer of text for signs or a tour guide. The level of knowledge has become particularly important for rainforest tourism in Latin America. On the high-value tours which characterize this region the tourists are typically well educated, knowledge-hungry and have high expectations. They expect to interact with local (perhaps indigenous) guides. However, they can often be dissatisfied by expecting Western standards from non-Western guides. Examples of problem areas arising from cultural clashes include guides with low levels of scientific or technical knowledge and guides seemingly indifferent to Western ideals of preserving nature (Wallace and Pierce, 1996).

Research into the quality of the messages conveyed to tourists through interpretation is a fairly new area. However, for rainforest tourism it is becoming a vital ingredient in the long-term sustainability of individual operators and regions. Rainforest visitors, especially those we characterize as ecotourists, come to the rainforests to *enhance* and *increase* their existing knowledge. To satisfy such tourists, tourism managers need to be aware of this and prepared to meet these needs.

Elevated Viewing Structures

Rainforest tourists can choose from a wide range of modes of experiencing and understanding the rainforests. Some of these, though used in other biomes, work very well in rainforests. These include small guided walks, trails with signage and night-time spotlight walks. However, there is one particular mode which has become almost exclusively associated with rainforests, this is the *elevated viewing structure*.

The stated logic behind elevated structures is that as the canopy is the most interesting feature of a rainforest, most tourists will wish to go there. As well as being the distinguishing characteristic, the canopy is where tourists can see epiphytes, fruit, flowers and wildlife close up. Such experiences cannot be had at ground level. A second attraction of the elevated structure (though rarely openly stated) is that it gives the tourist a thrill. For that reason many of them are suspension structures which move, swing and shake. They are essentially soft adventure tourism.

Elevated viewing structures are a recent phenomenon (though at least one dates to the 1930s). Their numbers and range have rapidly expanded in the last 10 years in response to the increase in tourist interest in rainforests. A number of types can be distinguished. The first are publicly built structures, usually located in national parks. Typically these are designed to cater for large numbers of mass tourists (for only high attendance can justify their cost). They are easily accessed by good roads and may have visitor centres and catering attached. They provide a short, concentrated experience, generally no more than 1 h (Frost, 1999).

A second type are privately constructed. They are nearly always associated with accommodation. These structures provide an *exclusive* experience for the paying guests.

A third category consists of (usually privately owned and operated) mass viewing structures. Their cost is met by admission fees. The best examples of these are the recently built cable cars running through rainforests in Costa Rica and Queensland (Chapman, 1996; Honey, 1999).

Elevated viewing structures may be seen as examples of *hardening* (the use of tough materials to protect the environment from tourist traffic), *sacrificial sites* (overdevelopment of one site in order to protect other sites) and *concentration* (providing a focal point for tourists to visit, either for the purposes of collecting revenue, managing visitors or providing services). Such structures have also been criticized as possibly being a poor substitute for good quality interpretation. It may be that having invested in the capital works, tourism managers are either unwilling or feel it unnecessary to spend further on tour guides or signage (Frost, 1999). Such structures have also been criticized as providing a sanitized and limited experience for tourists (Evans, 2000).

The Artificial Rainforest

Another recent tourist development is the artificial rainforest. As rainforests are highly attractive, but expensive to visit, some developers have taken the approach of bringing the rainforest to the city rather than vice versa. It is interesting to note that one key tourism textbook only refers to rainforest attractions in this sense, citing the case of the indoor Lied Jungle in Nebraska, USA, which attracts 1.3 million visitors annually (Goeldner *et al.*, 2000). Other examples include the massive tropical rainforest glasshouse opened in 1988 in the Botanic Gardens in Adelaide, Australia and the indoor forest (including living trees 200 feet high) opened in 2000 at the Museum of Victoria, Melbourne, Australia.

Such developments are highly dependent on technology, either to create realistic artificial copies or to keep real specimens alive. Their massive cost requires very large numbers of visitors paying small entry fees. Generally they provide an hour or two of interest and are directly competing with a wide range of similarly priced accessible attractions (including cinemas and other museums). Whether or not they are financially viable in the long term (especially after the novelty has worn off and costly revamping is required) remains to be seen. What is also uncertain is whether or not these urban alternatives affect demand for real rainforest tourism. (See Chapter 20 for a more detailed discussion of ecotourism in modified environments.)

Environmental Impact

The great increase in rainforest tourism has tended to affect the environment in two ways. The first is degradation through increased traffic. The world's surviving rainforests are typically in remote, sparsely populated areas. What remain today are rainforests which lacked either accessibility or fertility and so were not utilized for farming or logging. Increased tourism requires the building of roads and other services in relatively unspoilt areas. Unless carefully managed, tourism may lead to increased erosion, soil compaction, weeds, diseases and pollution. Conversely, taking positive steps to minimize negative impacts may become an attractive selling point for the environmentally conscious tourist. For example, much is made of how the pylons of rainforest cable cars in Queensland and Costa Rica were brought in by helicopters, so negating the need for clearing for permanent access roads (Chapman, 1996; Honey, 1999).

The second negative effect arises from the strong preconceptions of how rainforests should look which tourists bring with them. It may be highly tempting to reshape rainforests in a standardized format in order to satisfy those expectations. As such at a resort on Lindeman Island off the Queensland coast, exotic rainforest species were introduced, 'to assist in reinforcing the tropical island image desired for the resort' (Harris and Walshaw, 1995). At Cable Beach Resort in Broome, Western Australia, the buildings and landscaping were done in a Balinese style and the developers were initially quite ignorant that indigenous dry rainforest existed on their property.

A Place for Tourism?

In the last decade there has been a great deal of discussion about the accelerated destruction of rainforest and measures for preserving and protecting what remains. Unfortunately a great deal of this literature ignores tourism, either as an influence or as a force for preservation (for examples see Collins, 1990; Whitmore, 1990; and most significantly Salim and Ullsten, 1999). Such an omission is indeed worrying. Tourism already exists as an activity in rainforests and is growing. If efforts to preserve rainforests are to be successful, planners, managers and governments need to take account of tourism and its potential, both as a force which could damage rainforests and as a force for promoting interest in and understanding of rainforests. In turn tourism operators and managers need to be actively involved in preservation and education.

References

Adam, P. (1992) *Australian Rainforests.* Oxford University Press, New York.

Aiken, S.R. and Leigh, C.H. (1995) *Vanishing Rainforests: the Ecological Transition in Malaysia.* Clarendon, Oxford.

Boomgaard, P. (1999) Oriental nature, its friends and its enemies: conservation of nature in late-colonial Indonesia, 1889–1949. *Environment and History* 5(3), 257–292.

Cameron, D. (1992) A portrait of Victoria's rainforests: distribution, diversity and definition. In: Gell, P. and Mercer, D. (eds) *Victoria's Rainforests: Perspectives in Definition, Classification and Management.* Monash Publications in Geography, Melbourne.

Chapman, K. (1996) Skyrail: rainforest cableway. In: Charters, T., Gabriel, M. and Prasser, S. (eds) *National Parks: Private Sector's Role.* University of Southern Queensland Press, Toowoomba, pp. 134–139.

Collins, M. (ed.) (1990) *The Last Rainforests.* International Union for Conservation of Nature and Mitchell Beazley, London.

Davison, G. (1991) The meanings of heritage. In: Davison, G. and McConville, C. (eds) *A Heritage Handbook.* Allen & Unwin, Sydney, pp. 1–13.

Dean, W. (1995) *With Broadax and Firebrand: the Destruction of the Brazilian Atlantic Forests.* University of California Press, Berkeley.

Evans, S. (2000) Ecotourism in tropical rainforests: an environmental management option for threatened resources? In: Font, X. and Tribe, J. (eds) *Forest Tourism and Recreation: Case Studies in Environmental Management.* CAB International, Wallingford, UK, pp. 127–142.

Flannery, T. (1998) *Throwim Way Leg.* Text, Melbourne.

Frost, W. (1997) Farmers, government and the environment: the settlement of Australia's 'Wet Frontier', 1870–1920. *Australian Economic History Review* 37(1), 19–38.

Frost, W. (1999) Straight lines in nature: rainforest tourism and forest viewing structures. In: *Tourism Policy and Planning: Proceedings of the International Geographical Union Sustainable Tourism Study Group Conference*, Centre for Tourism, University of Otago, Dunedin, pp. 163–173.

Frost, W. (2000) Nature-based tourism in the 1920s and 1930s. In: *Peak Practice: Papers of the 2000 Council of Australian Tourism and Hospitality Educators Conference.* La Trobe University, Melbourne.

Goeldner, C., Ritchie, B. and McIntosh, R. (2000) *Tourism: Principles, Practices, Philosophies*, 8th edn. John Wiley & Sons, New York.

Grove, R.H. (1995) *Green Imperialism: Colonial Expansion, Tropical Island Edens and the Origins of Environmentalism, 1600–1860.* Cambridge University Press, Cambridge.

Harris, R. and Walshaw, A. (1995) Club Mediterranee Lindeman Island. In: Harris, R. and Leiper, N. (eds) *Sustainable Tourism: an Australian Perspective.* Butterworth-Heinnemann, Sydney, pp. 89–99.

Honey, M. (1999) *Ecotourism and Sustainable Development: Who Owns Paradise?* Island Press, Washington, DC.

Lumsdon, L.A. and Swift, J.S. (1998) Ecotourism at a crossroads: the case of Costa Rica. *Journal of Sustainable Tourism* 6(2), 155–172.

Minca, C. and Linda, M. (2000) Ecotourism on the edge: the case of Corcovado National Park, Costa Rica. In: Font, X. and Tribe, J. (eds) *Forest Tourism and Recreation: Case Studies in Environmental Management.* CAB International, Wallingford, UK, pp. 105–126.

Moscardo, G., Morrison, A.M. and Pearce, P.L. (1996) Specialist accommodation and ecologically-sustainable tourism. *Journal of Sustainable Tourism* 4(1), 29–52.

O'Reilly, B. (*c.*1945) *Green Mountains.* Smith & Paterson, Brisbane.

Parsonson, R., Wearing, S., Anderson, K., Robertson, B. and Veal, T. (1989) *New England-Dorrigo Rainforest Tourism Study.* Centre for Tourism and Leisure Studies, Kuring-Gai College, Sydney.

Ritchie, R. (1989) *Seeing the Rainforests in 19th-century Australia.* Rainforest Publishing, Sydney.

Salim, E. and Ullsten, O. (1999) *Our Forests our Future: Report of the World Commission on Forests and Sustainable Development.* Cambridge University Press, Cambridge.

Shackley, M. (1995) The future of Gorilla tourism in Rwanda. *Journal of Sustainable Tourism* 3(2), 61–72.

Thomlinson, E. and Getz, D. (1996) The question of scale in ecotourism: case study of two small eco-tour operators in the Mundo Maya region of Central America. *Journal of Sustainable Tourism,* 4(4), 183–200.

Valentine, P.S. (1991) Rainforest recreation: a review of experience in tropical Australia. In: Werren, G. and Kershaw, P. (eds) *The Rainforest Legacy: Australian National Rainforest Study,* Vol. 3. Australian Heritage Commission, Canberra, pp. 235–239.

Wallace, G.N. and Pierce, S.M. (1996) An evaluation of ecotourism in Amazonas, Brazil. *Annals of Tourism Research* 23(4), 843–873.

Weaver, D.B. (1998) *Ecotourism in the Less Developed World.* CAB International, Wallingford, UK.

White, M. (1994) *After the Greening: the Browning of Australia.* Kangaroo, Sydney.

Whitmore, T.C. (1990) *An Introduction to Tropical Rain Forests.* Clarendon, Oxford.

Chapter 13

Mountain Ecotourism: Creating a Sustainable Future

P.W. Williams[1], T.V. Singh[2] and R. Schlüter[3]

[1]*School for Resource and Environmental Management, Simon Fraser University, Burnaby, British Columbia, Canada;* [2]*Centre for Tourism Research and Development, Indira Nagar, Lucknow, India;* [3]*Centro de Investigaciones y Estudios Turisticos, Avenida del Libertador, Buenos Aires, Argentina*

Introduction

Tourism has become a primary source of revenue for many mountain areas, providing a rare opportunity for mountain people to participate directly in the global economy. Indeed it is estimated that as much as 15–20% of the tourist industry, or US$70–90 billion per year, is accounted for by mountain tourism activities (Mountain Agenda, 1999). In contrast to the generally small contribution of alpine environments to national economies, the value of mountain tourism to many regions is very significant (Zimmerman, 1995; Price *et al.*, 1997; Ritchie, 1998). Ecotourism represents an emerging and promising option for many mountainous locations that are seeking alternatives to the more traditional forms of tourism development that have characterized alpine tourism in the past.

Given the growing demand for such forms of development, this chapter describes the fundamental attractiveness of alpine areas for ecotourism, the inherent natural and cultural sensitivities associated with developing such areas for ecotourism, and the fundamental strategies which should be considered when developing and managing ecotourism on a sustained basis. It places particular emphasis on describing these challenges from an industry development perspective. It selects this specific focus in order to underline the importance of supporting small and medium-sized enterprises which tend to be the backbone of the fledgling ecotourism sector (Thomlinson and Getz, 1996). Throughout the discussion, examples from major alpine regions are used to highlight innovative approaches to mountain ecotourism development and management.

Alpine Areas as Ecotourism Destinations

The lure of tourists to mountain regions is based for the most part on natural features that are attractive for travellers. These include clean, cool air and water; varied topography with unique blends of biotic and abiotic resources; and ever-changing scenic mountain viewscapes (Price *et al.*, 1997). In many ways, mountains are focal points of global biodiversity, often retaining a greater number of species than adjacent lowlands due to extreme variations in altitude and the presence of micro-environments (Mountain Agenda, 1999). All of these natural attributes are conducive to activities and outcomes, including physical health, wellness, contemplation and meditation, that motivate mountain travellers (Singh and Kaur, 1985; COFREMCA, 1993). However, it is recognized that many of these pursuits also have significant impacts on the natural resources that they use. The management of mountain areas must strive for a careful balance between the protection of these natural resources, the needs of local people, and the desires of tourists (McConnell, 1991; Schlüter, 1993; Gill and Williams, 1994).

Mountains are also home to many diverse traditions and unique cultural landscapes that are linked to indigenous mountain people. Indeed, many mountain communities have rich cultural heritages that are visually expressed in the form of costume, buildings, and a wide range of lifestyle practices (UNEP, 1994). Such features are attractive to travellers seeking opportunities to experience and learn about the cultural and heritage dimensions of more remote and exotic destinations. Local communities and enterprises can stabilize their economies by recognizing the interest tourists have in understanding mountain cultures. For instance, the citizen-founded Hand Made in America organization, which operates in the Appalachian Mountains of the eastern USA, has successfully capitalized on the tourism potential associated with cultural heritage. Through initiatives such as its Craft Heritage Trails programme, it has stimulated enterprise development for craftspeople while concentrating the flow of tourists in areas that do not compromise the area's cultural integrity (Yates-McGill, 1999). Cultural heritage is a key element of the attractiveness of many mountain regions for tourism. It can be a valuable attraction for tourists, but it should not be sacrificed for short-term benefits (Muller and Thiem, 1995) (see Chapter 25).

Geographic Focus of Mountain Ecotourism

The spread of mountain ecotourism development in response to market demand and regional development priorities is concentrated primarily in western North America, Europe, the Himalayas and western South America, each of which has varying levels of development and activity focus. Although not focused upon in these discussions, similar issues, albeit on a smaller scale, are associated with other, relatively minor mountainous areas of the world. These include the Atlas Mountains of North Africa, the Drakensberg range of South Africa, the Australian and New Zealand Alps, the Urals, the Appalachian Mountains, the Caucasus, and the high mountains of New Guinea.

North America

Mountain ecotourism in a North American context is focused primarily in the western cordillera region of the continent. Extending from Alaska to Mexico, this Nearctic area is the largest single mountainous region in the world (Thorsell and Harrison, 1992). Within this zone ecotourism development tends to be concentrated in specific geographic locations especially noted for their high-quality wilderness and protected area attributes. These sites, in particular, are associated with areas of Alaska (e.g. Denali, Katmai, Alexander Archipelago), Yukon (Dawson Range and Pelly Mountains), Northwest Territories (Mackenzie Mountains), British Columbia (Columbia Mountains, Pacific Ranges, Vancouver Island Ranges), the Pacific Northwest and mountain region of the USA (Cascade Range, Rocky Mountains, Olympic Mountains), and California (the Sierra Nevada).

Europe

In Europe, the focus of mountain ecotourism development is centred in the

Western Palaearctic zone. While the main alpine feature is the Alps, the lesser ranges in the region (e.g. the Pyrenees, Cantabrians, Taurus, Apennines, Balkans, Carpathians, Jotunheim and the highlands of Scandinavia) provide the primary foci for the continent's emerging mountain ecotourism industry. The focus of ecotourism activity in these regions reflects the small-scale adventure and culturally oriented character of initiatives found in North America. Unlike the other major regions, European mountain regions are located in close proximity to large population centres, creating added pressure in terms of other tourism activities (and especially winter sports such as skiing) and other modes of resource use.

Asia

In Asia, mountain ecotourism is primarily centred in the South/Central Palaearctic zone which crosses seven countries. Containing the greatest concentration of high mountains in the world (Thorsell and Harrison, 1992), the area comprises the ranges of the Himalaya, Hindu Kush, Karakoram and Pamir. Of these, the Himalayas in particular have developed a nature-based tourism industry that resembles and links with ecotourism in several ways. The Himalayan ranges commence at Naga Parbat (8126 m) in the north-west, and pass through Pakistan and India's Himachal Pradesh state, ending at Namche Barwa (7828 m) in the east. The area embraces the watershed of the Indus and Brahmaputra rivers.

Most of the ecologically based tourist activities in this region are confined to national parks and protected areas that are rich in bio-cultural diversity. Nature tourism of a more 'hard core' variety is found in the higher ranges of the Karakoram and Hindu Kush. In contrast, softer forms, based on ecological resources, are emerging in Ladakh (Jammu-Kashmir), Kulu and Kangara (Himachal), Har ki doon (Uttarakashi), Valley of Flowers (Chamoli), and Kunda (Almora), where indigenous populations, particularly Bhotias, Johars and Marchchyas, have become engaged in such options. This trend has received a further boost from national and regional policy goals favouring ecotourism. The Sherpas of Nepal, particularly in the Kumbu, Langtang, Rolwaling and Annapurna regions, present good examples of ecotourism. Semblances of ecotourism are also evident in Himalayan pilgrimage tourism, with good examples found in the Garhwal Himalayas, and particularly around the geopious pilgrim resorts of Badrinath, Kedarnath, Gangotri and Yamunotri.

By and large, the main ecotourism activities in this mountain region are trekking, rambling, enjoying the scenery offered by mountain panoramas, birding, angling, viewing wildlife, forest-recreation, boating in the lake waters, and photography. In some parts, eco-trekking and conservation holidays have been initiated where hosts and guest have opportunities for interaction. Some of the more appreciable forms of ecotourism are emerging in Bhutan and the Sikkim portion of the Himalayas. In other areas, such as Kashmir and along the Indo-Tibetan border, geopolitical instability hinders the development of ecotourism and other forms of tourism.

South America

In South America, apart from the upland massifs in Brazil and Venezuela, the major mountainous region used for ecotourism activity is the Andean Cordillera. The Andes mountain range rises abruptly from the Caribbean Sea and extends the whole length of South America to Tierra del Fuego. It runs parallel to the west coast of the continent and contains a considerable number of national parks: from the Henri Pittier National Park in Venezuela to the Tierra del Fuego National Park where the mountain range descends into the Antarctic sea. In addition, the Andes mountain range north of Argentina consists of a number of plains that are generally known as the 'Altiplano', or the High Plain, located predominantly in Bolivia and Peru. It is a discontinuous series of plains and basins of variable dimensions and heights, separated by mountain chains and deep canyons. Here is found Lake Titicaca, the highest navigable lake in the world, which attracts ecotourists from different parts of the world.

As in much of the world (see Chapter 18), there is a marked tendency in Andean countries to consider ecotourism as a form of tourism that is practised primarily in national parks. In Venezuela, for example, the national parks are defined officially as tourist products. The first Andean national park was gazetted in 1934 in Argentina (Nahuwl Huapi National Park), with the Henri Pittier National Park in Venezuela created in 1937. In Colombia, the Purace National Park in the south of the country consists of volcanic landscapes and thermal waters that allow for the observation of diverse wildlife, immersion in thermal waters, and long walks. However, as Leitch points out (1993, p. 203), social instability, political problems and the violence associated with drugs have inhibited Colombia in its development of tourism and, as a result, the Purace National Park has not been able to realize its potential and benefit from a healthy flow of ecotourists.

This is not the case with Argentina and Chile, where the Andean-Patagonian national parks constitute one of the continent's principal ecotourism assets (Schlüter, 1993) for both hard and soft ecotourist markets. The Andean-Patagonian Mountain Range in Argentina is the home to seven national parks, or 80% of the country's national park system. The Glacier National Park, in the extreme continental south of Argentina, started to gain in popularity in the 1990s. Surveys of park visitors here show that almost all go to see the Perito Moreno glacier. As is the case with the majority of the Andean National Parks, most of the visitors to the Glacier National Park are Argentinean, although foreign tourists compose more than 40% (Schlüter, 1997), with Europeans the dominant international market. The flow of tourists to Chilean Patagonia intensified from the beginning of the 1990s when the 'Carretera Austral', the Southern Motorway extending from the South of Puerto Mont (Rivas-Ortega and Martinez-Coronado, 1989) to the south of the city of Coihaique, was created (Figs 13.1–13.4).

Fig. 13.1. Glacier at Lake Argentino, Glacier National Park, Patagonia, Argentina.

Fig. 13.2. Los Alerces National Park, Patagonia, Argentina.

Fig. 13.3. Lake Puelo National Park, Patagonia, Argentina.

Alpine Ecotourism Activities

For the most part, alpine ecotourism development tends to be small scale (average of 12–15 per travel party) and primarily focused on low-impact adventure activities. Ecotourism is usually differentiated from adventure tourism by the significant element of risk that prevails in the latter (see Chapter 5). However, it can be argued that alpine ecotourism commonly hybridizes with adventure tourism to the extent that virtually any direct contact with nature in an alpine setting implies a certain

Fig. 13.4. Hanging glacier near Puyuhuapi in southern Chile.

degree of risk in itself. This is the case in Peru, for example, where the most important natural and cultural sites are found at high altitudes where low oxygen levels pose a physical risk for people of advanced age and for those who suffer from precarious health.

Operators in such areas place emphasis on providing opportunities for nature observation, interpretation and photography, while travelling through zones of pristine wilderness and solitude on foot, horseback, or in canoes. Normally, facility development associated with these operations is limited to a few base camp lodges which serve as supply and staging points for outward bound visitors. To varying but substantial degrees, these operations accentuate the use of environmentally sensitive hard and soft technologies to ensure low impacts from their tourism-related activities.

The core activities that are intricately linked to mountain ecotourism are primarily nature and culture-based appreciative pursuits. The nature-based activities are centred on various types of fauna viewing (e.g. bird, bear, mountain goat, elk, moose, etc.) and landscape appreciation (e.g. flora photography, caving and glacier explorations, and astronomical observation). Cultural pursuits are mainly focused on guided visits to traditional aboriginal areas, resource-based wellness sites, and/or various trekking adventures along pioneer travel routes. These core pursuits are often complemented by a range of more clearly adventure-oriented water (e.g. kayaking, rafting, canoeing), land (e.g. mountaineering, trail riding, hut-to-hut skiing, snowshoeing, dog-sledding, and snowmobiling), and air (e.g. aerial wildlife/nature viewing, heli-hiking, paragliding, hot air ballooning) experiences. A good example of product diversity occurs within the Peruvian Andes. There, the Huascaran National Park and the Inca path to Macchu Picchu are the country's two most important sites for both ecotourism and adventure tourism.

Notwithstanding the small scale and environmentally sensitive character of most of these operations, demand for ecotourism and other forms of 'environmentally friendly' recreation opportunities are creating rapidly growing stresses in many mountain regions. These stresses are being expressed in the form of negative effects on wildlife, vegetation, soils, air and water quality. They are derived primarily from the construction, use and maintenance of transportation corridors, accommodation, trails and other recreation facilities designed to cater to the needs of visitors.

While the effects of such physical developments can be significant in their own right, they tend to be magnified when introduced into the fragile ecosystems of most mountain areas. A combination of extreme and rapidly changing weather conditions, unstable and steep slopes, rapid and peaked water runoff, limited flora coverage, slow-growing vegetation and limited habitat all serve to make mountain areas especially susceptible to destabilization processes created by tourism and other forms of human intervention. Furthermore, these interventions have only been exacerbated by new technologies that have increased the ability of tourists to penetrate more remote, rugged and high mountain regions.

This is true not only in the alpine regions of industrialized countries, but also on the highest and most remote mountain ranges of the less-developed world. The advent of larger and more frequent groups of tourists, each with specialized equipment needs for lodging, personal hygiene, cooked food and warmth has led to significant environmental damage in the Himalaya, the Andes and elsewhere (Banskota and Sahrma, 1995). In response, management agencies are beginning to create the institutional frameworks and management strategies needed to allow the industry to develop in a more systematic and environmentally friendly fashion (Mallari and Enote, 1996; BCMELP, 1998; Gerst, 1999). Such initiatives are best articulated at present in the European Alps, through multilateral organizations such as CIPRA and Euromontan (see Chapter 10).

Mountain Ecotourism Stakeholders

Careful use of environments, maintenance of biodiversity, and safeguarding the needs and heritage of local people must be balanced carefully against the growing demand for ever-expanding forms of mountain tourism development. Overall, the tourism industry has great responsibility in this regard, and forms of ecotourism suited to mountain environments offer growing opportunities for some regions. Yet this type of development brings many challenges for a variety of stakeholders. These stakeholders and their constituents include:

- federal, state and community governments responsible for representing the public's interest in tourism, regional development, natural protected areas designations, natural resource allocations, cultural protection, environmental management, and economic development (Howe *et al.*, 1997; Kezi, 1998);
- non-governmental organizations (NGOs), e.g. community groups and indigenous populations concerned with resource allocations and management strategies associated with tourism, economic development, recreation and conservation decisions (May, 1991, Walle, 1993; Price *et al.*, 1997); and
- individual tourism businesses, private and public tour operators and tour wholesalers concerned with the day to day challenges of developing and managing sustainable ecotourism ventures that attract a profitable flow of customers (Sirakaya, 1995; Herremans and Welsh, 1999).

The remainder of this chapter focuses on the challenges that stakeholders concerned with developing and managing mountain ecotourism must address.

Mountain Ecotourism Development and Management Challenges

There are many challenges associated with the development and management of ecotourism in mountain regions, especially in the less-developed world. While well intended, the delivery of programmes in many protected areas is less than optimal. Leitch (1993, p. 161) affirms that 'the system is better developed conceptually than materially'. Only a few protected areas have a suitable infrastructure that includes tourist centres, specialized personnel, a system of paths and descriptive brochures, etc. Some of the more critical issues that tend to be common to most alpine areas include product development and enhancement, packaging and resource protection. These issues are discussed in the following sections.

Product development

It is difficult to establish the overall spatial distribution of ecotourism products in mountain regions primarily due to the fledgling character of this industry and its tendency to hybridize with adventure and cultural tourism. Indeed, the full extent of such operations is continuing to unfold. Generally, the distribution of ecotourism developments is primarily focused on less-developed and remote mountainous areas with outstanding natural resources. Defined by their own cultural rhythms, the destinations of mountain ecotourists tend to have unique attributes which are often associated with formally or informally designated protected spaces. Some of these places are nationally or internationally significant. For example, tourists can view a wide variety of birds of prey and over 40 species of reptiles and amphibians in the Dadia Forest Reserve on the Rhodope Mountains of north-eastern Greece. The development of tourism in this area has helped to transform Dadia from an isolated and impoverished village into a lively community-based ecotourism centre (Valaoras, 1999).

Similarly, some mountain tour operators have developed ecotourism hikes into the Rocky Mountains of northern British Columbia and Yukon. Here, their customers gain a stronger first-hand appreciation of the interconnectedness between people and the unique wildlife (e.g. grizzly bears, eagles and mountain sheep) that inhabit the Tatshensheni and Alsek River valleys (http://mtsobek.com). This situation is mirrored in Asia and South America where visitors, for example, can enjoy the rare sight of high altitude Himalayan pheasants (monal) in protected areas of the upper Beas basin in Himachal Pradesh and Govind Wildlife sanctuary in Uttarkashi (Garhwal Himalaya). They can also witness the spectacle of Himalayan musk-deer in the Kedarnath sanctuary.

Despite the superb locations and product development of the best mountain ecotourism ventures, many of them are less impressive with regard to the latter. Measured against the dimensions of quality (hospitality, reliability, accessibility, variety, etc.) associated with other competitive tourism destinations and businesses, mountain ecotourism developments vary considerably in their quality. To remedy these inconsistencies, a range of alternatives exists, including raising the number of new products, promoting seasonal diversification and upgrading product quality.

Increasing the number of new products

While the demand for mountain ecotourism adventures is growing, appropriate and sustainable infrastructure to support this market expansion is lacking. Few well-developed products and facilities exist to cater to the needs of domestic and foreign travellers. Moreover, in many regions, the focus of most nature-based mountain tourism is still fixed on hunting and fishing activities which may not be compatible with the interests of ecotourism travellers. More products are needed which complement the principles of ecotourism, and are of a quality that will attract domestic and/or international tourists. For instance, some tour operators are introducing small group, low-impact expeditions (1–10 people) into the Coastal and Rocky Mountains of western Canada. Adhering to an environmentally friendly code of conduct referred to as 'Traveler's Commandments', these expeditions provide tourists with personalized opportunities to learn about and respect the ecosystems they are passing through. Furthermore, relatively unusual transportation methods (e.g. dogsledding, goat trekking and hut-to-hut cross-country skiing) are offered to provide a competitive edge in attracting customers (http://www.gorp.com/outeredge). A portion of each trip's fees is also used to support local research and conservation projects.

In Asia examples of innovative mountain ecotourism development include the introduction of electric ropeways in the Huang Shan scenic area of China's famous heritage site. These facilities have allowed visitors to experience the scenery of the local landscape without disturbing the fragile habitat. In the Garhwal Himalaya, the upgrading of eco-pilgrimages along the sacred Alaknanda and Bhagirathi rivers has improved the overall quality of the ecotourism experience.

Despite these examples of good practice, more must be done to create a better 'fit' between ecotourism products offered by mountain tourism operators and the markets that they are trying to attract. Mountain ecotourists seek products that provide unique opportunities to experience outstanding sites. At the same time, these travellers also base their destination decisions on convenience, quality and value for the money spent. Unfortunately, mountain ecotourism suppliers do not always meet these criteria.

Seasonal diversification

Many mountain ecotourism businesses are highly seasonal activities. Often they work at close to overflow capacity during peak summer seasons when alpine climatic conditions make access and activity convenient. In many locations this 'high' season only lasts for a period of 3–4 months. These operations then become almost dormant during traditional 'shoulder' and winter seasons. This leads to low annual facility utilization rates (often less than 40%), which in turn translates into limited financial sustainability for many operations.

Given the willingness of many ecotourism markets to travel at non-peak traffic periods, opportunities exist to develop ecotourism adventures during lower utilized times. For instance, many traditional mountain ecotourism activities could be extended into the shoulder seasons and augmented by activities surrounding the observation of annual life cycle events such as animal migrations, vegetation change, as well as localized spring and autumn cultural festivities. For example, ecotourists in Switzerland can hike in the Engadine Valley and observe the changing vegetation in nearby mountain meadows during the shoulder seasons. Similarly, during the autumn season, environmentally friendly travellers visit the mountain streams of North America's coastal mountains to view the annual migrations of Pacific salmon.

Local communities have capitalized on life cycle rituals by establishing festivals designed to celebrate these events while concomitantly extending their tourism seasons. Ladakh in Kashmir Himalaya offers opportunities for cultural tourism, and especially for visits to Buddhist festivals in the July–August period, when rains block access to other parts of the Himalayas. Soft adventure options are provided to visitors of the Himalayas during the winter as another way of addressing the problem of seasonality. By diversifying the range of experiences provided, ecotourism operations can create stronger capacity utilization and revenue generation levels needed to stabilize many businesses.

Upgrading product quality

Ecotourists tend to be relatively knowledgeable and 'footloose' in their choice of mountain destinations. Not bound to continually return to any one place, they more typically seek high quality products and places to experience on their trips. Product quality in this context tends to be associated with the following criteria:

- range of activities available;
- professionalism and knowledge of those providing guiding/interpretation services;
- hygiene, cleanliness and comfort of accommodation available;
- diversity, healthfulness and indigenous attributes of foods provided; and
- the extent to which these elements are brought together in the form of seamless packages.

Unfortunately, as stated earlier, the quality of mountain ecotourism products is uneven with respect to these criteria. While there are some outstanding examples, many others do not meet the standards expected by today's sophisticated tourists.

Steps are needed to publicize the best examples and to advise operators on how to improve their products (Economic Planning Group of Canada, 1999; Williams and Budke, 1999). For instance, the Canadian Aboriginal Tourism Association and British Columbia-based Native Indian Education Centre have developed training resources designed to assist indigenous groups in developing culturally appropriate tourism businesses. Many of these materials have been used to support the development of aboriginal ecotourism initiatives in the mountain regions of western Canada (White *et al.*, 1998). In Asia, the upgrading of Kulu Dashera in the Indian Himalaya, by the government of Himachal Pradesh, has enabled ecotourists to enjoy a unique experience when tribal gods and goddesses descend from over 360 Himalayan valleys for a colourful pageant (Singh, 1989). The event attracts a large number of domestic and foreign tourists, though continued growth could threaten the quality of the product if not appropriately managed.

Packaging

Well-developed trip packages combine a core 'experience' with components such as transportation, accommodation, meals, guides and/or interpreters. Many excellent ecotourism packages exist for travel to mountain destinations. For example, mountain ecotourism operators offer multi-day ecotourism packages to such exotic destinations as the mountain volcanoes of Ecuador, the glaciers of Argentina and Chile, and the Inca heritage sites of Peru (http://www.wildland.com). Typically, with these packages, a portion of the trip's cost is invested in projects supporting local environmental or cultural resource management initiatives (e.g. tree planting programmes, conservation trusts, rainforest purchases, etc.). Similar well-developed packages exist for ecotourism adventures into Asia, including the practice of ecotrekking in Nepal's Sagarmatha National Park.

For the most part, however, mountain ecotourism packages do not meet the standards of the best. Indeed, throughout most mountain tourism regions, the packages available are quite uneven in terms of quality, the value they represent and the standards to which they aspire. More significantly, many operators only package their products in a minimal way, if at all. Some who advertise as 'outfitters' do little more than run rental operations. Others offer only basic activities, devoid of the transportation, accommodation, meals and other activities that are necessary for success. A number of other problems related to packaging include inadequate programming, a limited number of activity options, and a lack of cultural interpretation programming. To help remedy this situation, some ecotourism operations are developed in a collaborative effort with the host country's conservation organizations (e.g. http://www.treadlightly.com).

Establishing effective connections with the marketplace is another area where local mountain ecotourism operators often fall short. In some cases, difficulties arise because some products being offered are poorly matched to the needs of the marketplace. Others are priced too low to allow the margin necessary to encourage the involvement of tour operators in the marketing of available products and services. In still other cases, due to a combination of limited resources, geographic inaccessibility and a lack of knowledge, mountain ecotourism operators are simply unable to establish links with suitable travel-trade partners. Overall, there is a need to improve the quality of existing packages and to increase their availability in the marketplace.

Resource protection

To remain viable, mountain ecotourism must be environmentally sustainable. This is an especially challenging task in mountain regions where climatic, topographic, vegetation and wildlife factors create fragile environments for most forms of land use and economic activity. Ecotourism developments must be fully aware of the realities when planning for the long-term protection and sustainability of these alpine ecosystems.

Indeed, situations already exist where various forms of ecotourism development are being confronted with the challenges of balancing development with conservation and managed use (Price *et al.*, 1997). In such areas, ecotourism operations are coming into conflict not only with other forms of alpine tourism (e.g. mountain biking, hunting, fishing and adventure tourism), but also with other resource sectors (e.g. forestry and mining) (Williams *et al.*, 1998). In many cases this has led to significant increases in the stresses placed on natural and cultural environments. However, in some instances, ecotourism has been used in coordination with other economic sectors to help stabilize local economies and secure more integrated land uses. For instance, in the southern Mexican State of Oaxaca, Zapotec communities have initiated a community-managed ecotourism strategy. In their approach, revenues from ecotourism are combined with income from a neighbouring community forestry project to provide social security for the families working in the enterprise. Ecotourism has also proved to be profitable enough to pay for a land survey, a first step towards resolving long-standing problems associated with land tenure in the area (Suarez Bonilla, 1999). Similarly, in Peru, NGOs are playing an important role in support of governmental agencies in protected areas management. Through their efforts they have attracted support from international conservation organizations and involve private and government organizations in efforts to manage protected areas more effectively. Emphasis has shifted from strict preservation to sustainable use including ecotourism in appropriate areas (Suarez de Freitas, 1998).

More internally focused concerns have also been raised about the field practices of many mountain ecotourism operators (waste and garbage disposal), the intensity of their resource use and the impact of their operations on wildlife and plant life (Herremans and Welsh, 1999). Lessons concerning best environmental practices and codes of conduct in other mountain environments would serve to alleviate this concern (Hawkes and Williams, 1993; Wight, 1999). For instance, in Asia, the Annapurna Conservation Area Project in Nepal presents a successful story in achieving sustainable ecotourism through community consultation and empowerment. The key elements in its success were multiple land-use methods of resource management which combined environmental protection and community development (Gurung and Coursey, 1994).

Conclusions

This chapter has focused on the four main alpine regions of the world to illustrate trends and issues that pertain to the emerging area of mountain-based ecotourism. As in most other parts of the world, mountain ecotourism is closely associated with protected area systems. However, to a greater extent than many other physical environments, mountain ecotourism may be conceived as a form of tourism that hybridizes readily with allied products such as cultural and adventure tourism. This contributes to a product base of uneven quality, which is problematic not only from a business standpoint, but also from the perspective of a physical environment that is particularly vulnerable due to extremes of seasonality, slope, temperature, etc.

For those concerned with the sustainable development of mountain regions, there are many challenges and opportunities in balancing the local conditions of individual mountain regions and their communities with the demands of tourism. In this regard, mountain ecotourism development provides a useful option for helping to diversify mountain economies while sustaining the resiliency of local communities and their citizens. However, there are many constraints and barriers that must be addressed before this form of development can be considered a truly effective contributor to alpine regions. This chapter has outlined many of the more fundamental and common challenges confronting tourism operators interested in orchestrating the delivery of mountain ecotourism experiences.

To strengthen the viability and longer-term sustainability of mountain ecotourism, a variety of strategic activities must occur. While generally applicable to most forms of small-scale tourism development, these proposed strategies are especially pertinent in a mountain ecotourism context. They are as follows:

1. Enhancing product development by:
- encouraging the exchange of experiences and know-how regarding the development of 'market ready' mountain ecotourism products;
- developing consistent standards and certification for 'market ready' mountain products.

2. Strengthening resource protection initiatives by:
- developing local inventories of unique and/or significant natural and cultural resources to be protected;
- encouraging operators and local communities to conserve and protect local natural and cultural resources through developing community-based codes of conduct;
- establishing community-based techniques for addressing priorities for resource use, resolving conflicts among users and improving the management of natural and cultural resources.

3. Encouraging human capacity building by:
- identifying mountain ecotourism business training needs and priorities;
- developing relevant training programmes, resources and 'how to' manuals;
- coordinating and providing access to relevant training opportunities.

Every mountain region includes a great diversity of stakeholders with specific interests in the local economy and the resources on which it depends. Some of these interest groups are interested in tourism while others are not. From a mountain ecotourism perspective, the long-term sustainability of many operations depends on the support of those groups that make decisions affecting their activities. As a corollary, these stakeholders cannot make informed decisions about ecotourism developments without understanding the diversity of multiple and changing demands confronting these operators. To address this situation, ongoing dialogue and collaboration between industry and community groups should occur. This chapter has presented some of the key industry priorities that should be central to that dialogue.

References

Banskota, K. and Sharma, B. (1995) *Mountain Tourism in Nepal: an Overview.* Discussion Paper MEI 95/7, ICIMOD, Kathmandu.

BCMELP (1998) Commercial Recreation on Crown Land Policy. British Columbia Ministry of Environment, Lands and Parks, Victoria, British Columbia.

COFREMCA (1993) *Pour un Repositionnement de l'Offre Tourisme-Loisirs des Alpes Francaises.* COFREMCA, Paris.

Economic Planning Group of Canada (1999) *On the Path To Success: Lessons from Canadian Adventure Travel and Ecotourism Operators.* Canadian Tourism Commission, Ottawa.

Gerst, A. (1999) Yukon Wilderness Licensing Act: crafting legislation through consultation. In: Williams, P.W. and Budke, I. (eds) *On Route To Sustainability: Best Practices In Canadian Tourism.* Canadian Tourism Commission, Ottawa and the Centre For Tourism Policy and Research at Simon Fraser University.

Gill, A. and Williams, P.W. (1994) Managing growth in mountain tourism communities. *Tourism Management* 15, 212–220.

Gurung, C.P. and Coursey, D. (1994) The Annapurna Conservation Area Project: a pioneering example of sustainable tourism. In: Cater, E. and Lowman, G. (eds) *Ecotourism: a Sustainable Option?* John Wiley & Sons, New York, pp. 177–194.

Hawkes, S. and Williams, P.W. (eds) (1993) *The Greening of Tourism: from Principles to Practice – a Casebook of Best Environmental Practice in Tourism.* Centre for Tourism Policy and Research, Burnaby; Simon Fraser University, in cooperation with Industry, Science and Technology Canada – Tourism Canada and Globe '92.

Herremans, I.M. and Welsh, C. (1999) Developing and implementing a company's ecotourism mission statement. *Journal of Sustainable Tourism* 7(1), 48–76.

Howe, J., McMahon, E. and Propst, L. (1997) *Balancing Nature and Commerce in Gateway Communities.* Island Press, Washington, DC.

Kezi, J. (1998) Formulation of an ecotourism policy framework for Manitoba. Unpublished practicum report, Masters of Natural Resources Management, Winnipeg.

Leitch, W. (1993) *Parques Nacionales de Sudamérica. Guía Para el Visitante.* Emecé Editores, Buenos Aires.

Mallori, A.A. and Enote, J.E. (1996) Maintaining control: culture and tourism in the Pueblo of Zuni, New Mexico. In: Price, M.F. (ed.) *People and Tourism in Fragile Environments.* John Wiley & Sons, Chichester, pp. 9–32.

May, V. (1991) Tourism, environment, and development: values, sustainability, and stewardship. *Tourism Management* 12(2), 112–118.

McConnell, R.M. (1991) Solving environmental problems caused by adventure travel in developing countries: the Everest Expedition. *Mountain Research and Development* 11(4), 359–366.

Mountain Agenda (1999) *Mountains of the World: Tourism and Sustainable Mountain Development.* Centre For Development and Environment, Institute of Geography, University of Berne, Switzerland.

Muller, H. and Thiem, M. (1995) Cultural identity. *The Tourist Review* 4, 14–19.

Price, M.F., Moss, L.A.G. and Williams, P.W. (1997) Tourism and amenity migration. In: Messerli, B. and Ives, J.D. (eds) *Mountains of the World: a Global Priority.* Parthenon, London, pp. 249–280.

Ritchie, J.R.B. (1998) Managing the human presence in ecologically sensitive tourism destinations: insights from the Banff-Bow Valley Study. *Journal of Sustainable Tourism* 6(4), 293–313.

Rivas-Ortega, H. and Martínez-Coronado, J. (1989) *La Carretera Austral: Un Espacio Incorporado al Desarrollo Turístico.* Universidad Austral de Chile, Valdivia.

Schlüter, R.G. (1993) Ecotourism: a blessing or a curse? The case of San Carlos de Bariloche, Argentina. *World's Eye-View* 7(4), 8–11.

Schlüter, R.G. (1997) *Areas Protegidas y Turismo en Argentina.* Ciet, Buenos Aires.

Singh, T.V. (1989) *The Kulu Valley: Impact of Tourism Development in the Mountain Areas.* Himalayan Book, New Delhi and Centre for Tourism Research and Development, Lucknow.

Singh, T.V. and Kaur, J. (1985) In search of holistic tourism in the Himalaya. In: Singh, T.V. and Kaur, J. (eds) *Integrated Mountain Development.* Himalayan Books, New Delhi, pp. 365–401.

Sirakaya, E. (1995) Voluntary compliance of ecotour operations with ecotourism guidelines. Unpublished PhD dissertation, Clemson University, South Carolina.

Suarez Bonilla, A. (1999) Ecotourism: a basis for commitment to the land and opportunities for young people. In: Mountain Agenda (ed.) *Mountains of the World: Tourism and Sustainable Mountain Development.* Centre For Development and Environment, Institute of Geography, University of Berne, Switzerland, p. 12.

Suarez de Freitas, G. (1998) Lessons from twelve years of NGO-government cooperation in protected areas management in Peru. *Environments* 26(1), 58.

Thomlinson, E. and Getz, D. (1996) The question of scale in ecotourism: case study of two small ecotour operators in the Mundo Maya region of Central America. *Journal of Sustainable Tourism* 4(4), 183–200.

Thorsell, J. and Harrison, J. (1992) National Parks and nature reserves in mountain environments and development. *GeoJournal* 27(1), 113–126.

UNEP (1994) *Managing Tourism in Natural World Heritage Sites.* Report of the International Workshop Held in Dakar, Senegal, November 1993. UNEP, Industry and Environment; UNESCO, World Heritage Centre.

Valaoras, G. (1999) Women's co-operatives, rural renewal and conservation. In: Mountain Agenda (ed.) *Mountains of the World: Tourism and Sustainable Mountain Development.* Centre For Development and Environment, Institute of Geography, University of Berne, Switzerland, p. 16.

Walle, A.H. (1993) Tourism and traditional people: forging equitable strategies. *Journal of Travel Research* 31(3), 14–19.

White, S., Williams, P.W. and Hood, T. (1998) Gearing up for aboriginal tourism delivery: the case of FirstHost. *Journal of Hospitality and Tourism Education* 10(1), 6–12.

Wight, P. (1999) *Catalogue of Exemplary Practices in Adventure Travel and Ecotourism.* Canadian Tourism Commission, Ottawa.

Williams, P.W. (1999) Challenges and lessons on the route to sustainability. In: Williams, P. and Budke, I. (eds) *On Route To Sustainability: Best Practices in Canadian Tourism.* Canadian Tourism Commission and Centre For Tourism Policy and Research, Simon Fraser University, Ottawa, pp. 1–4.

Williams, P.W. and Budke, I. (eds) (1999) *On Route To Sustainability: Best Practices In Canadian Tourism.* Canadian Tourism Commission and the Centre For Tourism Policy and Research at Simon Fraser University, Ottawa.

Williams, P.W., Penrose, R.W. and Hawkes, S. (1998) Tourism industry perspectives on the Cariboo-Chilcotin CORE process: shared decision making? *Environments: a Journal of Interdisciplinary Studies* 25(2 & 3), 48–63.

Yates-McGill, K. (1999) Valuing culture in a tourist economy. In: Mountain Agenda (ed.) *Mountains of the World: Tourism and Sustainable Mountain Development.* Centre For Development and Environment, Institute of Geography, University of Berne, Switzerland, pp. 10–11.

Zimmerman, F.M. (1995) The alpine region: restructuring opportunities and constraints in a fragile environment. In: Montanari, A. and Williams, A.M. (eds) *European Tourism: Regions, Spaces, and Restructuring.* John Wiley & Sons, Chichester, pp. 19–40.

Chapter 14

Polar Environments

B. Stonehouse
Scott-Polar Research Institute, University of Cambridge, Cambridge, UK

Introduction

Formerly renowned for their remoteness and cold, the two polar regions were for centuries avoided by all but the hardiest indigenous people, explorers and exploiters. Now both are becoming popular tourist venues. Though fundamentally similar, they differ in interesting ways. Central to the Arctic region lies the deep Arctic Ocean basin, ringed by the northern shores of Alaska, Canada, Greenland, Europe and Asia. A visitor to the North Geographic Pole stands on slowly shifting pack ice close to sea level. Central to the Antarctic region lies the high plateau of continental Antarctica. A visitor to the South Geographic Pole stands on a plateau 2800 m above the sea, on ice that is itself almost 2800 m thick. Tourists may currently visit either pole, travelling north by atomic icebreaker or south in especially arranged flights from South America. However, most polar tourism occurs in peripheral areas of the regions that are more readily accessible and perhaps more interesting.

Both regions are cold and intensely seasonal, with long winter nights and long summer days. Both have dry, anticyclonic climates, frequently invaded by warm, moist air masses that bring snow, sleet or rain to their peripheries. The Arctic supports indigenous people who traditionally live on its natural resources, hunting and gathering, fishing and herding reindeer. Arctic fringe lands have been hunted commercially for furs, and mined for metallic ores and hydrocarbons. Arctic seas have been hunted for walrus ivory, seal skins and oil, whales and fish. The Antarctic region has neither indigenous humans, land mammals nor readily exploitable minerals, but its coasts have been hunted for seals, its seas for whales and fish. Both regions have proved resistant to civilization. Modern humans, children of the tropics, still find polar regions something of an adventure, and until the early 20th century both remained free of industrial artefacts and pollutants. Tourism represents the most recent attempt by an industry to exploit these regions. What is polar tourism? What resources does it exploit, and what are the chances of those resources being exploited sustainably?

Defining Polar Tourism

As Hall and Johnston (1995a) point out in the opening chapter of their keynote symposium volume (1995b), the concept of 'polar tourism' is subject to a conflicting range of definitions, leading to confusion when numerical comparisons are attempted. The two polar regions are defined in different ways for different purposes, and there is no agreed use of the term 'tourism' that generates comparable sets of statistics, either within or between the regions.

While Pearce's (1992) general-purpose definition of tourism as 'journeys and temporary stays of people travelling primarily for leisure or recreational purposes' serves superficially for both the Arctic and the Antarctic, in neither is it rigorously applied in segregating visitors with purely recreational motives. For the Antarctic, where since the Second World War government-sponsored research at scientific stations has been the paramount activity, Hall (1992) defines tourism as 'all existing human activities other than those directly involved in scientific research and the normal operation of government bases'. This satisfactorily includes both commercial tourism operations and recreational activities of government personnel, and includes as 'tourists' the small but growing numbers of sky divers, parties ski-marching to and from the Pole, and others whose activities are neither scientific nor government-sponsored.

The Arctic presents a more complex picture of indigenous populations mingling with government servants, business visitors, and travellers with a variety of motivations. Incomers may be counted and assigned to statistical categories when they cross a political border, but residents who travel for pleasure or recreation within their own state or province generally escape the statistics. Hall and Johnston's (1995a) definition of polar tourism as 'all travel for pleasure or adventure within polar regions, exclusive of travel for primarily governmental commercial, subsistence military or scientific purposes' addresses some but not all of these problems.

Boundaries, Geography and Ecology

The polar circles (66°33′ north and south), though boundaries of little ecological or political significance, enclose equal areas and allow useful geographic comparisons. The Arctic Circle surrounds deep and shallow oceans, lowland plains, forests, tundra, farmland, and towns and cities that support a human population currently numbering over 2 million. The Antarctic Circle surrounds an almost entirely glaciated desert continent ringed by coastal seas. The land is without trees, shrubs or continuous ground cover, and populated only by a few hundred transient scientists and support staff.

Latitude for latitude, the southern hemisphere is uniformly colder than the northern, due mainly to the influence of the high continent (by far the world's highest), of which the ice cap rises almost to 4000 m. Winter and summer alike, surface air temperatures at the South Pole are 25–30°C colder than at the North Pole. Year-round the lowest Antarctic temperatures occur not at the pole itself but at the highest point of the polar plateau. The inland Russian station at Vostok each year records the world's lowest surface temperatures. Lowest Arctic summer temperatures occur high on the Greenland icecap, lowest winter temperatures in central Siberia, both of which are remote from the North Pole and indeed from the polar basin.

In lower latitudes the southern hemisphere is more stable in temperature throughout the year, reflecting its higher proportion of ocean. Southern summers are slightly cooler, and winters slightly warmer, than in equivalent latitudes of the northern hemisphere. In the far north the zonal pattern is disturbed by warm ocean currents that carry heat into high latitudes, indeed well into the polar basin, providing a range of radically different climates. The mean temperature of the coldest winter month in Bergen, Norway, strongly influenced by the North Atlantic Drift, is 29°C higher than at Okhotsk, Russian Federation, in the same latitude, a difference greater than the winter difference between

Bergen and Acapulco, Mexico (Stonehouse, 1989).

Biologists define polar regions not by the polar circles, but by boundaries that more clearly relate to flora and fauna. The Arctic is defined by the tree-line, separating boreal forest from forest tundra, the Antarctic by the Polar Front or Antarctic Convergence, a line in the ocean where cold waters of polar origin disappear below warmer subtropical waters, separating distinctive polar and sub-polar suites of planktonic plants and animals. As one is a land boundary and the other oceanic, it is difficult to effect direct comparisons between them. An alternative is the 10°C isotherm for the warmest summer month, which effectively separates areas of warm and cool summers, on land or at sea. Defined by either of these boundaries, the Arctic region covers almost twice the area contained within the polar circles; the Antarctic region is almost twice as large again (Fig. 14.1).

Manned stations operate in or near the coldest areas of both polar regions. More tolerable living conditions occur mainly in coastal and low-altitude areas. While travel to either end of the Earth is possible all the year round, nearly all scheduled tours are restricted to three or four summer months when weather is most congenial, and wildlife – a major attraction in both regions – is most accessible.

Exploitable Resources

Polar regions contain many natural resources, both renewable and non-renewable, that occur elsewhere in the world. Despite difficulties arising from distance and inhospitable climates, many of these have already been exploited. Both polar regions have been subjected to whaling, sealing and fishing, generally to the serious detriment of stocks. North America and Siberia are important industrial sources of metal ores, petroleum and gas.

Tourism exploits both natural and human-based resources. Natural resources of particular interest to tour operators include wilderness, scenic beauty and wildlife. Human-based resources include native human communities, scientific stations, and artefacts associated with the history of exploration or exploitation, from explorers' graves to abandoned mines, military installations and whaling stations.

Wilderness and scenery

Wilderness – loosely defined as landscape from which human artefacts and influences are absent – is highly valued by tourists jaded by civilization. While both regions include expanses of featureless marine or land ice, of little aesthetic appeal, Svalbard in the north and Antarctic Peninsula in the south especially are valued for their spectacular ice-capped mountains, tumbling glaciers, fjords, rocky headlands and islets. Volcanic action is rare in either region, but where it occurs, for example in central Iceland and Kamchatka in the north, Deception Island (South Shetland Islands) and Ross Island (McMurdo Sound) in the south, it emphasizes wilderness and diminishes further the role of humans. At one point on a well-trodden route in sub-Arctic Iceland, visitors may have the unusual experience of straddling one of the world's major geotectonic features, the Mid-Atlantic Rift.

Terrestrial wildlife

Though neither polar area shows the variety and density of plants and animals to be seen in most non-desert temperate or tropical regions, Arctic visitors welcome the colour and variety of summer and autumn vegetation, and both areas are rich in summer wildlife. Arctic vegetation ranges from forest-tundra, characteristic of the tree line, to tundra and polar desert, three biomes clearly derived by selective spreading from contiguous temperate lands, and representing stages of impoverishment due to increasing cold and aridity. Tundra, dominated by knee-high small shrubs, grasses and other flowering plants, with an

(a)

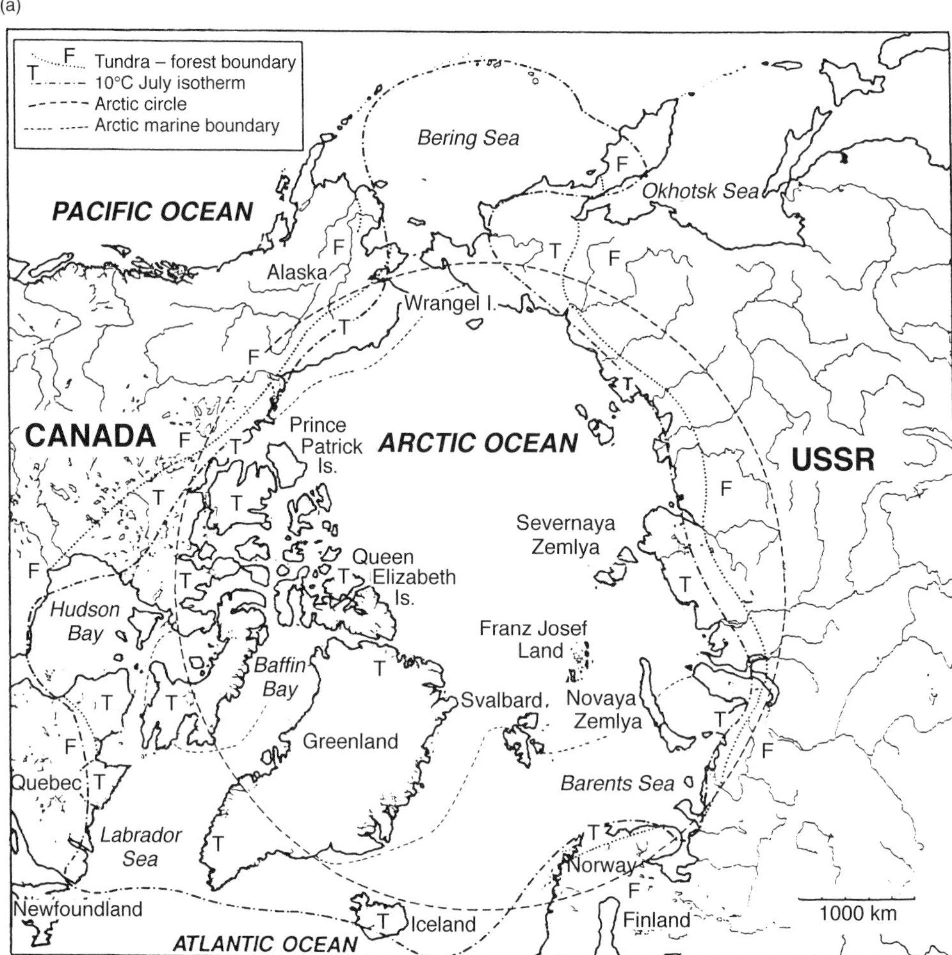

Fig. 14.1. Arctic (a) (*above*) and Antarctic (b) (*opposite*) regions, showing boundaries: the polar circles, 10°C summer isotherms, northern treeline, Antarctic convergence.

understorey of mosses, liverworts and lichens, is protected in winter by blanketing snow. Polar desert, in the far north, is impoverished by the loss of upstanding plants due to cold, aridity, lack of winter snow, and strong winds, which only the hardiest, low-lying species can survive (Chernov, 1985).

As the winter snow melts, Arctic tundra becomes predominantly green, then more colourful with an outburst of spring flowers. Poor drainage, due to underlying permafrost (permanent ice in the subsoil), results in ponds, streams and bogs with emergent vegetation. Spring warming produces flushes of plant growth, soon followed by late-summer crops of berries and seeds that support browsing and grazing birds and mammals. Abundant insects including butterflies, moths, bees, beetles, and several species of mosquitoes and biting flies, provide food for insectivorous birds. Prominent among grazing birds are migrant ducks, geese, swans and waders. Voles, mice, lemmings, hares and musk oxen are year-round tundra residents. Summer migrants include reindeer, caribou and moose that winter in the forests, and polar bears which move into coastal lands from the sea ice (Sage, 1986).

(b)

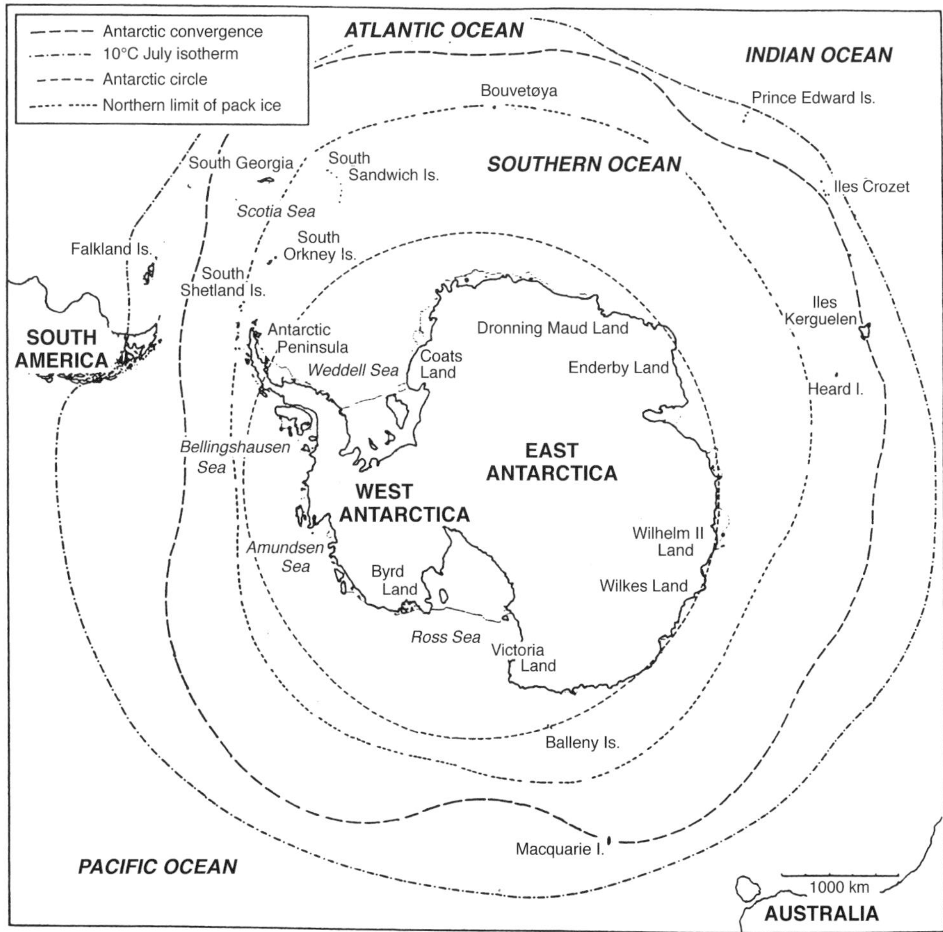

Neither forest-tundra nor tundra occur in the Antarctic region. There is virtually no land to support them in the appropriate latitudinal zone. Such peripheral islands as South Georgia and Iles Kerguelen support treeless floras of coastal tussock grass and upland fell-field, restricted in species due mainly to their isolation. The high Antarctic in summer remains icy (over 95% of the coastline is ice cliff) or, in ice-free areas, predominantly brown, covered by bare rock and scree, with thin polar desert vegetation. Not surprisingly, the Antarctic region has very few indigenous land birds, no indigenous land mammals, and no summer influx of terrestrial migrants.

One reason for tourists to visit the Arctic is to see the colourful spring array of tundra flowers, including gentians, dwarf lupins and saxifrages. A relatively poor area of Arctic tundra may support 50–60 species of flowering plants, and polar desert at least a dozen species. By contrast the whole Antarctic continent supports only two species of flowering plants, both small and insignificant. South Georgia and other relatively mild islands of the Southern Ocean, still within the 10°C summer isotherm, have richer floras, though of curiously restricted composition, with few flowering plants, and none of them colourful or striking.

Marine and coastal wildlife

The Arctic Ocean basin supports little life. Even in summer, primary productivity of its sun-starved waters is low, and seabirds, seals and whales are scarce. Much richer are the marginal areas around it, where sea ice melts in spring and winds and currents stir nutrients up from the depths. Polynyas – areas of sea kept free of sea ice in winter and spring by wind or currents – are especially rich. Surface waters support planktonic (floating) plants and invertebrates, notably copepods and euphausiid shrimps that become the main food of fish, inshore seals and seabirds. Sub-arctic seas off Iceland and south-west Greenland, toward Svalbard, around North Cape, off Alaska, and east of Siberia and Kamchatka, have long been known as good hunting grounds for whales and seals. Their waters include some of the world's richest fishing and shrimping grounds, today supporting huge commercial fisheries. Seabirds, notably auks, guillemots, kittiwakes and other cliff-nesting species, return in their thousands every spring for their share of marine abundance.

The Southern Ocean is generally more dynamic, constantly exchanging waters with other oceans, continuously stirred by winds and currents. The Weddell and Ross Seas, and some other near-continental waters have almost permanent coverings of pack ice, and are biologically poor, though again polynyas (see above) provide anomalous patches of local enrichment. Other seas, with seasonal pack ice, are richer in summer. Away from land between 35°S and 55°S, where the ocean is relatively ice-free, surface waters are driven constantly eastward by westerly winds. Where their flow is diverted or interrupted, for example east of the tip of Antarctic Peninsula and downstream from the islands of the Scotia Arc, nutrients well up to the surface from deeper layers, creating enormous eddies and patches of rich water where plankton proliferates. Squid, fish, birds, seals and whales are attracted to these areas. The birds breed on nearby cliffs and strands, the seals breed on local beaches and sea ice. The Southern Ocean supports some of the world's largest flocks of seabirds, mostly petrels and penguins (Fig. 14.2). Both the 19th century sealing industry and 20th century whaling drew their wealth from the same source.

Fig. 14.2. Emperor penguins with chicks under the ice cliffs of the Weddell Sea, Antarctica. Tourists visit this colony by icebreaker and helicopter (photograph: B. Stonehouse).

Human resources

The Arctic is home to a variety of human populations, predominantly coastal or riverine, mostly nomadic hunters and gatherers dependent on the meagre natural resource base. Few of their modern representatives adhere strictly to traditional ways of life, but villages and camps of Alaskan Indian trappers, Canadian Inuit fishermen and Scandinavian Saami reindeer herders are of considerable interest to tourists. Participation in hunting, herding, kayaking and other traditional activities (including shooting (under licence) of bears and other game), are the focus of many organized tours throughout the Arctic.

Of equal interest are historic artefacts, include the Viking settlements of Greenland and eastern North America, the huts, anchorages, graves and other evidence of early maritime explorers, the camps, trails and settlements of 19th and early 20th century prospectors and miners, and villages and townships that serviced the early 20th century herring industry. More recent remnants of military, scientific and industrial installations, regarded by some as rubbish to be cleared, by others as potentially valuable industrial archaeology, are gradually acquiring historic interest.

Antarctica's lack of indigenous population and short history of human occupation restrict its human artefacts to a few 19th century sealing camps, late 19th and early 20th century explorers' huts, and the moorings, debris, hulks and derelict stations of early 20th century whaling. Of more general interest are the two to three dozen scientific stations currently in operation. A few of these welcome tourist visits and provide additional interest in the form of nature trails and information. Most protect their research programmes by strictly rationing visits, and diverting visitors away from areas where research is in progress.

Sovereignty and Responsibility

Politically, Arctic lands and neighbouring seas are claimed by seven states: Canada, Denmark (Greenland), Finland, Norway, Russia, Sweden and the USA. Each owns, exercises sovereignty over, and takes responsibility for lands and adjacent oceans within its sector. There are few disputes over sovereignty, and none currently affecting tourism or its regulation. Through the mid-20th century, strategic implications of the Cold War imposed severe strains on relationships between the Arctic nations. Most of these, however, have now disappeared, allowing for cooperation and consultation, notably in areas of social, industrial and scientific development.

Each Arctic state exercises direct control over polar tourism within its bounds, generally favouring and encouraging the development of the industry, deploying some responsibilities to local legislation and communities, and employing the kinds of administrative instruments that it uses in its non-polar areas. Remoteness of polar tourist areas from seats of government may place them at a competitive disadvantage: non-polar areas are likely to be more accessible and more popular, to yield higher revenues and be better served and protected by rangers and wardens.

Continental Antarctica too is claimed by seven states: Argentina, Australia, Chile, France, New Zealand, Norway and the UK. Claims of Argentina, Chile and the UK overlap geographically: ownership of these and other sectors is disputed by other nations, for example the USA and Russia, with interests in Antarctica. Governance is effected through the Antarctic Treaty System, based on a treaty to which 12 nations agreed in 1961, and in which 26 nations currently exercise ruling interests. The Treaty System operates within latitude 60°S, but extends for certain purposes to the Antarctic Convergence, the ecological boundary of Antarctica. Some peripheral islands within the Convergence remain under national ownership and governance. Iles Crozet and Kerguelen are administered by France, Macquarie and Heard islands by

Australia, and South Georgia and the South Sandwich Islands are governed by the UK, but claimed also by Argentina.

The prime concern of the Antarctic Treaty was to preserve the continent for peace and science, objectives in which it has so far succeeded admirably. The Treaty contained no references to tourism, but subsequent instruments (collectively called the Antarctic Treaty System), notably the Protocol on Environmental Protection to the Antarctic Treaty (HMSO, 1992), address a broad spectrum of environmental issues, including those raised by tourism. Those most directly involved in administering the Antarctic Treaty System regard all problems arising from tourism as soluble within the terms of the System. Others can only hope that a Treaty System with no direct capacity for wilderness management or law enforcement may eventually succeed in its self-appointed task of managing Antarctica, under the onslaught of a lively, expanding and diversifying tourist industry (Stonehouse, 1994a).

Hall and Johnston (1995a) assert that, at both ends of the world, tourism is currently being used to support territorial claims. They cite as examples the creation of national parks and other protected areas along the Northwest Passage by the government of Canada (procedures that are currently mirrored by Siberian authorities along the Northeast Passage), and the special interest shown by Australia, Argentina and Chile in developing Antarctic tourism. Other ways in which these political issues affect management of polar tourism are discussed more fully below.

The Industry at Work

Tourism has a relatively short history in either polar region. In the Arctic, summer cruises to Svalbard, Alaska and the Canadian north became popular during the mid-to-late 19th century, for trophy hunting and wilderness appreciation. Modern tourism has only recently developed to a point where it begins to overtake more traditional uses of natural resources: hunting, agriculture and mining. Tours are operated both by indigenous groups, usually small and locally based, and by major operators based far from the Arctic. Collectively they offer a wide and diverse menu of recreational experience, from backpacking, skiing, dog-sledging and canoeing to luxury cruising.

In contrast, Antarctic tourism began only in the late 1950s. In the absence of indigenous people and settlements, its operators are based entirely outside the area, and its development has so far been more limited in scope and variety. Currently over 97% of tourists that land in Antarctica are shipborne, using their ships as mobile hotels and leaving them only for brief visits ashore. The rest are airborne, landing for spells of camping, climbing or skiing from field camps. Recent developments include overflights by aircraft from Chile and Australia, and cruises by liners capable of carrying over 1000 passengers, neither of which release tourists ashore.

Forms of tourism

In its simplest forms, Arctic tourism presents natural resources raw and unadorned. Outfitting and guiding small parties of backpackers, for example, are relatively simple, low-cost operations, requiring little capital investment and yielding concomitantly low returns. More complex forms of tourism enhance the resources through capital investment, for example in roads and airstrips that provide access, coaches and cruise liners that bring in more clients, and hotels that provide sophisticated accommodation and amenities of civilization on the wilderness fringe. Enhancements add value to the operations and provide higher returns, but usually result in greater impacts on the amenities exploited.

Antarctica's remoteness and difficulties of access prohibit such simple forms of tourism as independent backpacking. There is no infrastructure of roads and hotels: visitors must first be brought in by cruise ship, yacht or aircraft, all expensive forms of logistic support, involving substantial capital investment and the possibility (as yet fortunately unrealized) of substantial negative impacts.

Shipborne tourism, common to both regions, includes both *scenic cruising* by ocean-going liners, which to increasing extents include polar regions in their worldwide itineraries, but do not normally land passengers, and *adventure cruising* by smaller ships, which operate coastally and make a point of frequent passenger landings. Scenic cruising brings cruise liners with passenger capacity of 1000 and more to the fjords and waterways of Alaska, eastern Canada and Svalbard, and has recently (2000) brought the first liner of this size to the Antarctic Peninsula. Built for speed and comfort, few of these ships are ice-strengthened, and their scope for polar travel is limited to a relatively few safe routes.

Adventure cruising was devised by the travel entrepreneur Lars-Eric Lindblad to lead clients away from the well-trodden tourist paths. Particularly appropriate where there is no infrastructure of hotels or other amenities ashore, it is the main form of tourism practised in Antarctica, and is becoming increasingly popular in the Arctic: the same ships and cruise operators work three to four summer months alternately at either end of the world. Adventure cruising involves small, ice-strengthened ships carrying 50–150 passengers (rarely up to 500), cruising for 8–15 days and landing passengers once, twice or thrice daily, usually by inflatable boats (Stonehouse, 1994b). At each landing the passengers spend 1–3 h ashore, then re-embark and move on to the next venue. Landings may alternate with sightseeing by boat and, in certain ships, with helicopter flights. Between cruises the ship returns to a gateway port to re-fuel, restock and take on new passengers.

Both scenic cruising and adventure cruising offer opportunities for influencing passenger activities and behaviour. Lindblad ensured, through lectures and other on-board presentations, that passengers on his cruises were well-briefed on environmental protection. This ethic particularly suited the prosperous, conservation-minded passengers who made up his clientele, and still characterizes adventure cruising in both polar regions.

Fig. 14.3. Ship-borne tourists enjoy calm seas and summer sunshine along the fjord coast of the Antarctic Peninsula (photograph: B. Stonehouse).

Numbers of tourists

From neither polar region is it easy for researchers to wring the statistics and other essential information needed to judge the impacts of tourism. There is no single library with a comprehensive archive of Arctic, Antarctic or polar tourism, and no single authority responsible for recording basic statistics of the industry (Stonehouse, 1994c).

Compared with similar-sized regions elsewhere in the world, neither region attracts large numbers of tourists. Total numbers visiting the Arctic are impossible to determine. Governments responsible for northern regions that include sectors of the Arctic seldom distinguish 'arctic' from 'non-arctic' areas of their regions. While figures for Svalbard are reasonably accurate, those for northern Siberia are hard to come by. Johnston's (1995) tabulated figures for Greenland, Iceland, Svalbard, northern Scandinavia, Alaska, Yukon and Northwest Territories, for a range of years between 1988 and 1994, indicate an annual total of about 400,000, but emphasize the difficulties of achieving even estimates. In comparison with visitor figures for more popular venues, and in relation to the enormous area involved, this is not a large total: individual national parks in Europe and the USA receive more.

Reliable figures showing trends in numbers are even harder to ascertain. Most Arctic governments within the past 10 years have sought to promote tourism within their sectors, and increasing numbers of ships and aircraft each year carry passengers to Svalbard, Iceland, Greenland, northern Canada and Siberia. Slow but steady increases seem likely in all sectors. For the Antarctic, statistics are more reliable and readily available. Commercially organized cruises to Antarctica began in 1958, and tours ships have visited Antarctica every year since 1966 (Stonehouse, 1994b). The industry has grown slowly and irregularly. Numbers of shipborne passengers rose to a peak of over 3500 in 1974/75, declined to fewer than 1000 in 1980, and rose again by the mid-1980s to over 2000. Figures for the past 12 years, published annually by the International Association of Antarctica Tours Operators (IAATO), indicate a slow, erratic increase from 5000 to 9000 throughout the late 1980s and 1990s, and a current value (1999/2000) exceeding 12,000. Numbers of airborne passengers landing in Antarctica remain small in comparison, possibly amounting to 200–300 per year.

Fragility and Sustainability

Developments in tourism, as in other fields, are deemed sustainable if they use resources in ways that meet the needs of the current generation without compromising those of future generations. The natural assets on which polar tourism so heavily depend, notably scenery, wilderness and wildlife, are particularly vulnerable to human activities. Because of their relative emptiness, polar regions are especially vulnerable to changes, whether purposeful or unintentional, brought about by development. Scars remain far longer than in temperate or tropical latitudes.

Polar ecosystems have been characterized as 'fragile', i.e. more susceptible to human disturbance, and less capable of self-regeneration and recovery, than temperate or tropical systems. This concept, of key importance in the management of polar tourism, contains elements both of truth and of ignorance. Changes occur all the time in polar ecosystems, whether or not humans are present. When the cause of some particularly striking change is unknown, the currently fashionable but anthropocentric view prevails that humans are to blame, and must be restrained. Ignorance can thus be used to restrict tourist activities of all kinds, from visiting penguin colonies to walking on vegetation.

Some ardent conservationists go so far as to argue that the 'environmental fragility' of polar regions demands nothing less than their complete, unconditional protection from all forms of human development. Others no less concerned with protection perceive dangers in thus setting parts of the world off-limits to humanity. What we cannot visit, we are unlikely either to cherish or seek to understand. Areas excluded from human interest will effectively be excluded from environmental responsibility and protection, and perhaps vulnerable to clandestine and irresponsible development.

As our knowledge of polar ecosystems increases, the concept of fragility is giving way to an alternative view, that polar ecosystems, existing in some of Earth's most testing environmental conditions, are tough enough to withstand major perturbations, including many of those brought about by tourists. They have considerable natural capacity for regeneration, which may sometimes be enhanced or manipulated to speed recovery. Invoking 'environmental fragility' to protect polar environments should be recognized as an admission of ignorance, for which the remedy is applied research.

An example from polar soils

Polar soils and vegetation in particular are regarded by many as 'fragile': for a review of relevant literature and arguments see Stonehouse (1999, 2001). Walker *et al.* (1987) and Strandberg (1997), have created a terminology useful in considering disturbance and damage to soil ecosystems, which can equally be applied to a range of other ecosystems that, within the context of tourism, are subject to human interference and damage.

- *Disturbance* denotes a change in conditions which displaces an ecosystem beyond its normal limits of variation, or interferes with the normal functioning of a biological system.
- *Resistance* denotes the ability of a system to withstand displacement from a given state.
- *Resilience* measures the degree, manner and pace of recovery of an ecosystem to its original state after disturbance. *High resilience* describes a system that recovers rapidly after disturbance, *low resilience*, one that recovers slowly or not at all.
- *Recovery* is the process by which an ecosystem achieves biological and physical stability following disturbance. *Recovery potential* is a measure of the capacity of an ecosystem to recover. *Complete recovery* denotes recovery to an original state. *Functional recovery* denotes recovery to stability at a level different from the original.

In these terms, 'fragile' ecosystems would be those that demonstrate low resistance to disturbance, low resilience, and low rates of recovery, perhaps only to a limited state of functional recovery.

Mechanical disturbance of soils, by walking, excavating or moving loads across them, interferes with normal soil processes, but results in damage only if severe enough to overcome the soil's inherent resistance and resilience, preventing either complete or functional recovery. Intentional disturbance, for example by laying roads or erecting buildings, is usually sanctioned on grounds of expediency. It is, however, usually accompanied by incidental disturbance, for example campsites, rubbish dumps and vehicular tracks across the tundra, which is often responsible for more of the total damage.

Disturbance of either kind often leads to subsequent damage by natural causes. Rutted roads may become drainage channels, bulldozed surfaces extended by windstripping. Walker *et al.* (1987) present an index of severity of disturbance, essentially the area of final disturbance, divided by the area originally disturbed by human activity. Their highest reported ratio, up to 2.6, resulted from mechanical stripping of soil and vegetation. Trampling by humans, often cited as a cause of severe environmental damage, generally produced ratios of one or less, indicating no spread and an element of natural recovery.

Polar soils and vegetation have evolved in the presence of considerable non-human or natural disturbances, including climatic instability, landslides, trampling and scraping by animals, and pollution by salt water, much of it far more disruptive than anthropogenic damage. Clearly, they have natural means of recovery, which must be of interest to those who seek to repair man-made changes. Walker *et al.* (1987), Crawford (1997) and Iskandar *et al.* (1997) have reviewed possibilities of remediation by enhancing the soil's natural capacity for self-regeneration. Self-restoration occurs, though usually at rates that are too slow to be generally acceptable. Artificial enhancement of natural processes, to speed up regeneration, is usually an option in practical site management.

In managing areas where polar soils are likely to be disturbed by humans, the first consideration must be conservation: wise use that, so far as possible, avoids damage altogether. Where damage occurs, it is necessary to identify the source, if possible to halt or abate it, alternatively to decide on a level of damage that can be tolerated. This may require radical re-thinking of management policies or objectives. Then it becomes necessary to monitor the rate of damage and, if the level of toleration is exceeded, to stop further disturbance and decide on a level of rehabilitation. Should the site be restored completely to its pristine condition, or partially to a point that facilitates functional recovery? How quickly is restoration required, and what intermediate stages are regarded as evidence of progress?

Much the same principles apply to other aspects of polar environments where human activities effect disturbance. Sound management, based on understanding of natural processes, cannot fail to provide more effective environmental protection than invocations of 'fragility'.

A model for development

Butler's (1991) discussion of sustainable development in relation to sensitive environments is directly relevant to polar regions, and McKercher (1993) provides a simple model of development that is equally applicable. The concept of sustainable development concedes that part of a natural resource base may be expended in order to create a man-made asset of at least equivalent value. In McKercher's model, should a natural asset of value V be modified by development to create a man-made asset of value M, leaving a residue of value R, then the development may be deemed sustainable if M + R equals or exceeds V.

An example involving Arctic tourism could be a wilderness area of particular beauty, perhaps in Svalbard or northern Alaska, which is becoming increasingly popular among backpackers and campers. In the absence of a designated campsite, visitors camp all over the area, spoiling the amenity for each other. Designating a campsite, however tactfully, forfeits some of the wilderness value and natural beauty, but enables campers to use the area without further loss of amenity. Applying McKercher's model, V represents the value of the original area to mankind, M that of the campsite, and R the remaining beauty. If M + R equal V, the development is sustainable.

Should the demarcated site prove poor or ill-considered, then M + R falls short of V, denoting a decline in value. Should no action be taken to control the camping, then the original value V would decline as amenity deteriorated. Should the authority prohibit camping, fewer people would visit the site and V – the value to mankind – would decline accordingly. If both camping and visiting were prohibited, V would be completely devalued. Should provision of the campsite allow more campers to use the area without loss of amenity, then V would increase accordingly.

This simple model of sustainable development assigns no units to M, R and V, and is subject to abuse by conservationists and developers alike. However, it contains an important underlying principle. Conservationists may fear that virtually any expenditure of natural assets in the name of sustainable development, without an ecological imperative, must sooner or later degenerate to non-sustainable development. Developers, including tour operators who provide facilities for people to travel and enjoy natural assets, are equally fearful that some aspects of their work – wilderness and adventure tourism especially – will be curtailed if the conservationist view alone is considered. So McKercher's model, with all its limitations, is worth keeping in mind when considering polar tourism or any other form of development that aspires to sustainability.

The model may be applied not only to areas of tundra and forest, but equally to human values and traditions. It is especially relevant in Arctic areas where tourism brings income to poor communities. In Antarctica, tourists welcome a lack of infrastructure ashore: wilderness is expected, and may be enjoyed without enhancement. In the Arctic, it would be an unwise operator who landed passengers in a simple fishing village without ensuring some infrastructure – coaches, a coffee shop, toilets, information or other amenities – to soften impacts and bring compensatory benefits to the hosts.

Managing Polar Tourism

Though polar regions at either end of the world tend to be dominated by governments, they tend also to be neglected politically and starved of funds. Voters are few, and revenues for management are scarce. Earlier exploiters, whether whalers, sealers or mining consortia, worked virtually unhampered by regulation, with devastating results. Some recent forms of development, for example provision of pipelines for transporting hydrocarbons in Arctic North America (though less so in Siberia), have been subject to stricter controls. Can tourism, the most recent and rapidly developing industrial exploitation of polar regions, be managed effectively?

In the Arctic, every area of land and inshore water of interest to tour operators comes directly under the control of a sovereign government that has both the responsibility and the capacity to regulate tourism within its boundaries. Drawing upon experience gained in warmer latitudes, each government has set up national parks and reserves, providing appropriate legislation and systems of governance. Several have published short-term and long-term strategic plans, including provision for tourism, based on established principles of wilderness management. All have provided some level of ranging and policing to ensure that regulations are respected. However, revenues from tourist visits are small: even with subsidies, there is seldom provision for more than token field staff, to cover enormous areas of park or reserve and deal with the many problems that tourists – particularly naive, irresponsible or ill-managed tourists – can bring.

In Antarctica the claimant states have effectively surrendered sovereign rights and responsibilities to the Treaty System, including rights to set up and manage national parks and reserves in which tourism might effectively be controlled. Governments that are signatory to the Treaty can control the activities of their own nationals in Antarctica under domestic legislation drawn up in accord with Treaty deliberations. Neither individually nor collectively can they directly manage areas that are being visited annually by thousands of tourists, areas that, in the Arctic or elsewhere, might be judged to be at risk. In a few instances this has been over-ridden by common sense. New Zealand does not permit tourists to enter the historic huts of British explorers Scott and Shackleton, dating from the early 20th century and lying within its sector, except in the presence of a government-appointed warden.

Development of self-regulation

Absence of legislation during the early days of Antarctic tourism produced an interesting form of self-regulation. When ship-borne tourism began, almost simultaneously with the Antarctic Treaty, principles of management were developed by the tour operators themselves, notably by Lars-Eric Lindblad (see 'Forms of tourism' above). As the industry grew, cruise operators (again inspired by Lindblad) formed a coalition, the International Association of

Antarctica Tour Operators (IAATO), which issued guidelines and codes of conduct both for its members and for their clients ashore. Later the Treaty issued recommendations for visitors and similar guidelines of its own. However, it was the industry, not the regulating body, that first rose to the occasion.

While accepting the overall authority of the Treaty System and working consistently within it, the industry in practice continues to exercise both self-regulation and a keen sense of environmental responsibility. The more forward-looking companies go beyond the IAATO guidelines in producing their own manuals of good practice, conforming to Treaty regulations and where possible anticipating future legislation.

Self-regulation and sovereignty

In theory at least, Arctic tourist areas under sovereignty should be better protected than Antarctic areas under a well-intentioned but clumsy international regime. Self-regulation by an industry must be preferable to anarchy, but regulation by a sovereign power must surely be better than either. Is not Svalbard, for example, better protected from impacts of ship-borne tourism than the Antarctic Peninsula? In practice it may not be. Svalbard is well endowed with parks for recreation and reserves for environmental protection, which land-based tour operators, on the whole, respect. However, not all of its ship-borne tour operators follow the Lindblad pattern of indoctrinating their clients, or subscribe to IAATO-style guidelines and codes of conduct. During recent visits to landing sites in Svalbard I have met tourists who have been put ashore from cruise ships without benefit of education or briefing, seen litter on beaches, and witnessed disturbance of bird nesting colonies, all rare occurrences in Antarctica.

The government of Svalbard has powers to police its beaches, and provides both rangers and law enforcement officers throughout the tourist season. Yet numbers of ships, cruises and tourists are increasing rapidly: already the archipelago receives almost three times as many tourists annually as the Antarctic Peninsula. A small corps of enforcement officers, even one equipped with zodiacs (inflatable boats) and a helicopter, cannot be everywhere at once.

In this respect at least, Antarctica seems better protected under guidelines drawn up by a responsible industry and endorsed by a government that lacks powers of enforcement, than Svalbard under sovereign legislation with inadequate powers of enforcement. A recent initiative by World Wide Fund Arctic (Humphreys *et al.*, 1998) has sought to remedy this situation, drawing up guidelines and codes of conduct for Arctic tour operators similar to those that have proved successful in Antarctica.

Conclusion: a Principle for Management

This example illustrates a general problem in development of polar tourism. Polar regions are ripe for development, but thinly populated. In well-regulated communities and in wilderness, entrepreneurs of tourism consistently force the pace on regulating authorities, far outstripping the capacities of government servants to modify legislation, or field officers to provide effective policing.

While responsible self-regulation in the tourist industry is a blessing, it cannot be relied on entirely to protect polar regions. Not all tour operators follow codes of conduct, even their own. Self-regulation in the best companies does not compel good behaviour among the less responsible, and cowboys, though generally held at bay by high costs, may move in at any time. Should we even consider leaving regulation largely to the industry, when for every Lindblad there are a dozen opportunists?

Management of polar tourism cannot be adversarial: developers and regulators need to find roles that are complementary, and divide responsibilities as even-handedly as possible between them. The essential role for tour operators, whether ship-borne or land-based, is to draw up sensible guidelines and codes of conduct for themselves, that reflect the best and most effective practices within their industries. It is best that the entrepreneurs themselves, or very well-informed and experienced third parties, draw up the codes, for they alone have practical knowledge of what is needed. It is in their interests to produce sound rules, and to abide by them. If they do not, they may be sure that governments will draw up alternative rules, designed to suit their own convenience of administration, with little insight into working practices.

The role for regulating bodies is equally clear. They need to scrutinize and critically endorse the operators' codes of conduct, ensuring that they are consistent with legislation, and to provide inspection to ensure that the codes are fully and properly observed. Fears that this policy gives too free a hand to the operators, and withholds too much responsibility from the regulators, are unfounded. A code of conduct must necessarily conform to existing law, and the regulating body has the right to confirm that it does. Inspection, rather than policing, then becomes the rational follow-up.

Is there a role for enforcement? While few would dispute the need, policing these extensive and thinly populated regions is barely practicable, and less to be relied on than responsible self-regulation. In the Arctic police and rangers are few and far between: under existing legislation they have no equivalent in the Antarctic. Effective enforcement is a professional business, which cannot be left to amateurs. Collecting evidence of violations requires knowledge and skills equal at least to those of good defence lawyers.

As we have witnessed elsewhere in the world, tourism is a substantial force with strong powers for good or evil. Responsible as they are for vulnerable environments and communities, polar authorities and tour operators alike need to take it very seriously, and harness its powers as efficiently as possible for good. The lessons they learn from managing this new and burgeoning industry may well be applicable in other fields of development, from which neither end of the Earth is likely to be immune in the very near future.

References

Butler, R.W. (1991) Tourism, environment and sustainable development. *Environmental Conservation* 18(3), 201–209.

Chernov, Yu. I. (1985) *The Living Tundra.* Cambridge University Press, Cambridge.

Crawford, R.M.M. (ed.) (1997) Disturbance and recovery in Arctic lands. *Proceedings of the NATO Advanced Research Workshop on Disturbance and Recovery of Arctic Terrestrial Ecosystems, Rovaniemi, Finland, 24–30 September 1995.* NATO Advanced Science Institutes, Series 2, Vol. 25. Kluwer Academic Publishers, Dordrecht.

Hall, C.M. (1992) Tourism in Antarctica: activities, impacts and management. *Journal of Travel Research* 30(4), 2–9.

Hall, C.M. and Johnston, M. (1995a) Introduction: pole to pole: tourism impacts and the search for a management regime in polar regions. In: Hall, C.M. and Johnston, M. (eds) *Polar Tourism: Tourism in the Arctic and Antarctic Regions.* John Wiley & Sons, Chichester, pp. 1–26.

Hall, C.M. and Johnston, M. (eds) (1995b) *Polar Tourism: Tourism in the Arctic and Antarctic Regions.* John Wiley & Sons, Chichester.

HMSO (1992) Protocol on Environmental Protection to the Antarctic Treaty, with Final Act of the Eleventh Antarctic Treaty Special Consultative Meeting. Cmd 1960, Miscellaneous No. 6. HMSO, London.

Humphreys, B.H., Pedersen, Å.Ø., Prokosch, P.P. and Stonehouse, B. (eds) (1998) Linking tourism and conservation in the Arctic. *Proceedings from Workshops 20–22 January 1996 and 7–10 March 1997, Longyearbyen, Svalbard.* Norsk Polarinstitutt Meddelelser No. 159, pp. 49–58.

Iskandar, I.K., Wright, E.A., Radke, J.K., Sharratt, B.S., Groenevelt, P.H. and Hinzman, L.D. (eds) (1997) *Proceedings of the International Symposium on Physics, Chemistry and Ecology of Seasonally Frozen Soils, Fairbanks, Alaska, 10–12 June 1997.* CRREL Special Report, pp. 97–100.

Johnston, M. (1995) Patterns and issues in Arctic and Sub-Arctic tourism. In: Hall, C.M. and Johnston, M. (eds) *Polar Tourism: Tourism in the Arctic and Antarctic Regions.* John Wiley & Sons, Chichester, pp. 27–42.

McKercher, B. (1993) The unrecognized threat to tourism: can tourism survive 'sustainability'. *Tourism Management* 14(2), 131–136.

Pearce, D.G. (1992) *Tourist Development*, 2nd edn. Longman Scientific and Technical, Harlow, UK.

Sage, B. (1986) *The Arctic and its Wildlife.* Croom Helm, Beckenham, UK.

Stonehouse, B. (1989) *Polar Ecology.* Blackie, Glasgow.

Stonehouse, B. (1994a) Tourism and protected areas. In: Lewis Smith, R.I., Walton, D.W.H. and Dingwall, P.R. (eds) *Improving the Antarctic Protected Areas.* IUCN, Cambridge, pp. 76–83.

Stonehouse, B. (1994b) Ecotourism in Antarctica. In: Cater, E. and Lowman, G. (eds) *Ecotourism: a Sustainable Option.* John Wiley & Sons, Chichester, pp. 195–212.

Stonehouse, B. (1994c) Polar tourism: do library resources meet researchers' information needs? In: Walton, D.W.H., Mills, W. and Phillips, C.M. (eds) *Bipolar Information Initiatives: Proceedings of the 15th Polar Libraries Colloquy.* Scott Polar Research Institute, Cambridge, pp. 25–28.

Stonehouse, B. (1999) Biological processes in cold soils. *Polar Record* 35(192), 5–10.

Stonehouse, B. (2001) Remediation and restoration of frozen ground: a terminology. *Polar Record* 37, (in press).

Strandberg, B. (1997) Vegetation recovery following anthropogenic disturbance in Greenland. In: Crawford, R.R.M. (ed.) *Disturbance and Recovery in Arctic Lands. Proceedings of the NATO Advanced Research Workshop on Disturbance and Recovery of Arctic Terrestrial Ecosystems, Rovaniemi, Finland, 24–30 September 1995.* NATO Advanced Science Institutes, Series 2, Vol. 25. Kluwer Academic Publishers, Dordrecht, pp. 381–390.

Walker, D.A., Cate, D., Brown, J. and Racine, C. (1987) *Disturbance and Recovery of Arctic Alaskan Tundra Terrain.* CRREL Report 87–11, Academic Publishers, Dordrecht.

Chapter 15
Islands and Coasts

E.A. Halpenny
Nature Tourism Solutions, Almonte, Ontario, Canada

Introduction

The presence of ecotourism in coastal areas and islands has grown rapidly during the last 30 years, in concert with the progression of the field and the explosive growth of tourism. This chapter explores the phenomenon of ecotourism within these two similar and yet unique environments. First, coastal areas and islands will be defined, and the differences and similarities that characterize ecotourism activity in these two environments will be outlined. An overview of how these two environments differ from other environments where ecotourism takes place will also be undertaken. This will be followed by an examination of the spatial distribution of ecotourism within coastal and island settings, as well as an analysis of recent trends in the growth and magnitude of ecotourism in these environments. Finally special attention will be paid to development issues arising from ecotourism's appearance and expansion in coastal and island destinations. This will include an exploration of the role of planning, financing, marketing, product development, players involved, and the challenge of maintaining or improving cultural, social and economic integrity.

Characteristics of Coastal and Island Ecotourism

Coastal and island environments share many of the same challenges and opportunities. This understanding is reflected in Article 25 of the Programme of Action from the United Nations Conference on the Sustainable Development of Small Island Developing States: 'Sustainable development in small island developing States depends largely on coastal and marine resources, because their small land area means that those States are effectively *coastal entities*' (1994). These shared characteristics play a leading role in determining the success of ecotourism in such settings.

Shared characteristics

Islands and coastal areas share a maritime tradition, and the most dominant force affecting them is the sea, be it an ocean, coastal estuary, or large inland lake. Historically the dominant human activities in these settings have been extractive activities such as fishing and forestry and the trade and shipping that followed. In the latter part of the 20th century, coasts and islands became preferred venues for sun, sand and sea tourism. This more venerable form of tourism can often come into conflict with ecotourism, one of the most recent activities that has arisen as an alternative livelihood for island and coastal communities.

Both islands and coastal areas can be labelled as critical, vulnerable environments. They both face similar challenges. Coasts, situated on an island or mainland receive all the runoff and waste from inland sources such as manufacturing, extractive and agricultural industries. Rivers flowing out to sea carry sediments and pollutants that can have a profound impact on the health of coastal ecosystems. Coastal elements such as coral reefs and sea grass beds are under increasing pressure from these land-based sources of pollution. Sea-based threats, for instance illegal dumping of waste from cruise ships and accidental oil spills, play an equally salient role. Direct negative impacts by marine recreationists, for example scuba-divers breaking off coral, have been well documented but play a relatively minor role when compared with the threat posed by activities such as dynamite fishing or improperly treated sewage outflows from tourism centres and cities.

Coasts and islands are also impacted upon by maritime sources. Storms are major builders and destroyers of coasts. This natural process in recent years seems to be accelerating, as exhibited by larger, more frequent and prolonged storm systems. Many believe that this may be a trend resulting from global warming, a phenomenon that puts coasts and islands under increased threat and perhaps eventual annihilation as polar ice caps continue to melt, contributing to rising sea levels. Popular ecotourism destinations are under threat from this phenomenon. For example, Shah (1995) states that 70% of the Seychelles land area may be lost due to climate change.

Overuse and contamination of natural resources on islands and mainland coasts are other common concerns. Water is a valuable and vulnerable commodity for both islands and coasts. Coastal areas have become a popular destination not only to visit but also to settle and work. A high portion of the world's population lives within 100 km of the world's coastlines. Depletion of surface and underground sources is an especially serious and prevalent concern, particularly in areas of low rainfall. Depletion of island aquifers can lead to salinization of the groundwater. Freshwater aquifers can also be contaminated by human waste from underground waste-management systems. Contamination by waste products not only ruins potable water, but can result in nutrient loading problems for coastal features such as coral reefs through underground migration of wastes to the ocean. Some tourism destinations are taking steps to address these issues. One example is in Akulmal, Mexico, which is the gateway to the southern coast of the Yucatan Peninsula. Here, above-ground, hydroponic, human waste treatment facilities called 'created wetlands' are being used by residents and hotels as local geology makes underground waste-management facilities such as septic systems a danger to coastal marine systems.

Differences between islands and coasts

In general, islands face a greater threat to their natural resources than mainland coasts, as island resources are generally more limited in number and size. In the Seychelles, a popular ecotourism and mass tourism destination, water is collected from rainwater catchments and wells. The groundwater aquifer on many of the coral islands is in the form of a lens-like body where fresh water which has percolated through the island rock floats on the more dense saline sea water beneath. Excessive use of the freshwater can lead to sea water intrusion. Within the archipelago the island of Coetivy has experienced this disaster (Shah, 1995).

Islands exhibit further dissimilarities, or perhaps more accurately, special characteristics which make them different from mainland coasts. This in turn of course affects their ability to host ecotourism sustainably. As indicated above, islands suffer from limited resources, and the balance of resource use can be much more critical than that found in mainland coastal environments due to isolation. Islands are also characterized by limited space. They also feature fewer ecosystems and species diversity than mainlands. However, a high degree of endemism, characteristic of many islands, has developed due to isolation (Hall, 1993). Species endemism or uniqueness, where animals and plants can be found nowhere else in the world, is the main attraction at ecotourism destinations such as the Galapagos and Madagascar. For example in the Galapagos 95% of the reptiles, 50% of the birds, 42% of the land plants, 70–80% of the insects and 17% of the fish live nowhere else in the world (Honey, 1999).

Spatial limitations on islands often encourage the development of ecotourism rather than mass tourism, as ecotourism's small scale is more suited to this context. An example of this is the size of the airport landing strip in Dominica. The short landing strip, in part due to lack of flat terrain, has contributed to the slow growth of tourism on the island and has encouraged the development of niche market tourism rather than the more mainstream sun, sand and sea tourism. Limited space also plays a big role in the disposal of wastes on islands. Solid waste disposal is an especially vexing problem as landfill sites are often not possible due to financial or space limitations. Recycling, due to scales of economy, is also not economically feasible in most cases.

The limitations of diverse habitats or ecosystems would seem to be an inhibitor to ecotourism success, but certain islands are exceptions to this generalization. For example, the islands that make up New Zealand are famed for the diversity of their nature tourism opportunities and for the high potential for quality ecotourism experiences, all within a small area. Continental coastal zones on the other hand generally have greater opportunity to provide a more diverse ecotouring experience, in part due to transportation links with the inland. The rise of Belize as an ecotourism destination in the last 10 years exhibits this strength, combining excellent eco-diving opportunities on its coasts with outstanding natural and cultural itineraries in its interior.

Other differences between mainland coasts and islands can be linked to socio-economic considerations such as the prevalence of commercial and industrial activity on mainland coasts. Hence, a greater pool of skilled workers is available on the mainland. Because of their size and lack of industrial development it can be theorized that it is easier to plan and implement integrated community and economic development programmes on islands. This ease of planning, in theory, and how this impacts the development of ecotourism on islands, will be explored later in this chapter.

Finally, due to their isolation, islands are particularly vulnerable to air travel routing and multinational airline policy changes. The development of airports, nearby air transportation hubs, and company policies all influence the success of tourism in island destinations. This is especially true for small island developing states, whose visitors generally come from distant destinations. For example the Pacific Islands receive an estimated 95% of their tourist arrivals by air (R. Smithies, 1998, Lanzarote, personal correspondence).

Spatial Distribution, Magnitude and Growth

In the 1960s and 1970s ecotourism was very much limited in distribution, and represented only a small percentage of the international travel market. There were exceptions such as the Galapagos to which organized ecotourism began in the late 1960s (Honey, 1999) and East Africa, although many forms of safari tourism such as hunting did not qualify as ecotourism. In the developed countries ecotourism was a popular pastime domestically among small, specialized groups of animal watchers or park visitors. Today, estimating the size of the ecotourism market remains a challenge, and understanding the size and distribution of ecotourism in coasts and islands must be largely based on anecdotal reports and mainstream tourism data. Table 15.1 reflects a summary of recent growth and distribution of ecotourism in coastal and island areas, based on selected regional summaries (Milne, 1992; Hall, 1993; Valentine, 1993; Weaver, 1993; Ayala, 1995; Fagence, 1997; Royle, 1997; Wilkinson, 1997; B.N. Devi, 1998, unpublished; Lindberg *et al.*, 1998; Weaver, 1998a, b; Honey, 1999) and the author's own observations.

In general, ecotourism experiences and facilities need to be separate from mainstream mass tourism. Destinations in the Caribbean such as Dominica and the US Virgin Islands (USVI) are struggling to balance visitation by cruise passengers and the 'stayover' tourist market. Ecotourists

Table 15.1. The growth of ecotourism and mainstream tourism in selected island and coastal destinations 1995–1999.

Region	Selected island and coast ecotourism destinations	Ecotourism growth (1995–1999)	Mainstream tourism growth (1995–1999)
Sub-arctic and Antarctica	Greenland, Ellesmere, Campbell and Auckland Islands, Macquarie, Heard Island, Iceland	Moderate increase	None
Pacific	Fiji, Solomon Islands, Papua New Guinea, Hawaii, Midway, Micronesia, Palau, Samoa, Vancouver Island, Alaska, Russia's eastern coast	Increasing	Increasing
Southeast Asia	Bali, Lombok, Sulawesi, Komodo, Sulu Sea, Cebu	Increasing	Moderate increase
Australia/New Zealand	Great Barrier Reef, Western Australia	Increasing	Moderate increase
Indian Ocean and Africa	Seychelles, Mauritius, Madagascar, Zanzibar, Ghana, Red Sea	Moderate increase	Moderate increase
Caribbean	Dominica, St Lucia, Nevis, Trinidad, Florida Keys, Texas Coast, Southern Quintana Roo coast, Mexico; Bay Islands, Honduras; Meso-American Reef, Belize; Kuna Yala, Panama; Bahamas, Guyana, Suriname	Moderate increase	Decline
Mid and South Atlantic Ocean	Falkland Islands, St Helena, Atlantic coast of northern Brazil	Stable	None
North Atlantic	Sweden, Newfoundland, Canada; New England, USA; Chesapeake and Eastern Shore, USA; Canadian maritime (east) provinces, Ireland, Portugal	Increase	Increase

are included in this latter category. Trunk Bay in St John, USVI, is a classic example of this challenge. Cruise ship passengers typically go to the park for a half-day visit, lying on the beach and exploring the nearby reef with its underwater trail. Billed as one of the most beautiful beaches in the Caribbean, the park is also popular with stayover tourists and locals. The US Park Service has invested a significant amount of money to harden the site at the park to accommodate the estimated 10,000 monthly visitors (M. Morrison, St John USVI, 1998, personal communication). Clashes in desires will invariably arise as these two visitor types mix, since their differing expectations (e.g. tolerance of crowding and expectation of quality interpretation) are not easily combined. The pristine and uncrowded conditions that an ecotourist craves have long since evaporated from the site. Ecotourism destinations that face similar challenges of accommodating cruise ship excursions to delicate sites include Belize, St Lucia, Alaska and Dominica. Regional planning must address this. In areas such as Dominica and Samoa (Weaver, 1998b), ecotourism is being used as an alternative development strategy to mass tourism. In other island-states, both types of tourism are being pursued simultaneously. For example, mass tourism will continue in the main destination islands of Nassau and Grand Bahamas in the Bahamas, and Veti Levu, Fiji. However, ecotourism is being encouraged as a strategy on some of the naturally and culturally rich outer islands of these archipelagos, as seen on Exuma, Inagua and Andros in the Bahamas and Taveuni and Kadavu in Fiji (Wilkinson, 1997; Harrison, *c.*1998; Weaver, 1998b).

Due to their isolation, islands are more likely to provide the remote experience that 'hard' ecotourists are looking for. The recent development of 'island trails' in the USA, Canada and the Pacific is a phenomenon that can be linked to this search for exclusive, pristine, peaceful destinations (Wylie and Rice, 1998). In Canada and the USA these trails are organized and maintained mostly by non-profit volunteer groups. Trails ranging from Florida to Alaska are used by recreationists as well as nature tourists, many of whom are ecotourists on self-guided expeditions or accompanied by local tour operators. The Cascadia water trail in Washington state, USA, earned fame as a British Airways for Tomorrow award recipient in 1999. In the Asia-Pacific region water trails are more often developed by ecotour entrepreneurs. However, the issues of management and overuse are often neglected, sometimes to the point of resource destruction, as illustrated by the case of Sea Canoe's pioneering of Thailand's Phang Nga Bay sea caves (Anon, 1999b; Rome, 1999). The issues associated with development of coastal and island ecotourism destinations and products will be explored in the following section.

Ecotourism Development Issues

Transport

Few islands are able to finance their own national airlines or build airports large enough to receive large, long-haul passenger planes. The capacity of island airports is based on factors such as length of runway, refuelling facilities, lighting and navigational aids (R. Smithies, 1998, unpublished). The type of airport, and thus aircraft, plays a major role in determining the type of tourism market that can be developed for a particular island. Some islands choose to focus on high-end tourism, such as St Vincent and the Grenadines, which have seen extensive tourism development without building intercontinental airport structures. Others choose nature or 'discovery tourism', such as Gomera in the Canary Islands (F. Vellas, 1998, unpublished) and the Galapagos (both of which are accessible mainly by boat) and Dominica, which relies on a small airport. Airport upgrades have been completed in recent years in the following island ecotourism destinations: Barbuda, Mauritius, Micronesia (Pohnpei), St Kitts and Nevis, Lucia, Seychelles, Palau

(Koror), and Trinidad and Tobago. However many air routes to these island-states suffer from low traffic volumes, especially in countries where tourism activities are not well developed. This can be an impediment to the provision of viable airport and air navigational services, as the revenues derived from user fees are not sufficient to cover local government costs of running the airports (International Civil Aviation Organization, 1999).

Small island governments want to see profitable airports as well as airlines. National airlines were once envisioned as an important alternative to international carriers; a buttress against dependency. At one time the subsidy for Air Nauru from the South Pacific was said to be one-quarter of the national budget. States can no longer support airlines in this way. Many national airlines have been privatized, including Air Jamaica and BWIA in the Caribbean (R. Smithies, 1998, unpublished). This privatization can lead to benefits such as an increased ability to attract capital and the avoidance of politically motivated government requirements to perform uneconomical services (R. Smithies, 1998, unpublished). However, it also exposes the airline to the whims of the global economy and forces policy makers in local government and at the airline to make tough choices regarding tourism development for the destination. Vellas explores the choice of deregulation further in his 1998 report to the Congress on Sustainable Tourism for Small Island States. He also examines the role of cooperation between airlines, and the choice between charter and regular air transport.

With the exception of atolls, and small islands close to the mainland coasts that rely on regular ferry services, land transport is essential for ensuring the development of ecotourism in coastal and island settings. Causeways, bridges and roads can have a profound effect on coastal environments, yet they are necessary to transport ecotourists to the destination. Therefore, adequate impact assessment and planning must accompany their construction.

A form of transport that is increasingly used for nature and cultural tourism is expedition cruise boats. This form of touring has grown rapidly in the last decade, but has had a long history in areas such as the Galapagos. The ability of small sail and cruise boats to slip in and out of anchorages in remote, wilderness areas gives ecotourists access to pristine, unique environments. Many of these companies combine this experience with education of tourists, including extensive interpretation programmes and calls to environmental action. However, some ecotourism proponents still debate their ecotourism status, as less revenue is generated by boat tourism for coastal and island destinations than terrestrial forms of tourism. In Ecuador, the boat company Lindblad Special Expeditions is trying to address this by initiating a donation programme for their clients, in addition to the US$100 park fee that they pay to enter the Galapagos National Park. In its first 2 years the Galapagos Conservation Fund collected US$400,000 for the Galapagos islands and its inhabitants. This year the company plans to invite other tour boat companies in the region to participate in the fund programme (O'Brien, 1999; T. O'Brien, Kota Kinabalu, 1999, personal communication).

Finance

The development of infrastructure such as lodges or roads, or skills such as nature guiding expertise or hospitality training, can all be linked to the availability of capital. Yet, this is one of the major hurdles to developing ecotourism, since such alternative forms of tourism have often been shunned by private and public investors as being too risky or elusive. In general there have been two main sources of funding for ecotourism development in island and coastal areas: (i) private entrepreneurs, using their own or family money; and (ii) aid programmes run by multilateral or bilateral donors. The former was and is used to finance small businesses such as ecolodges or in-bound tour companies. The latter provides capital to mainstream tourism products such as airports or is tied to biodiversity conservation or community development programmes in less developed countries. Ecotourism was and still is thought to be a more benign use of natural resources, and a potential source of income which can support further sustainable development efforts in coastal and island environments. In the 1980s and 1990s buy-in by the donor community was mixed. Some development assistance agencies chose to ignore ecotourism entirely while others flirted with the concept, generally in rainforest settings. With the increased urgency attached to preserving coral reefs and fishery stocks in the mid to late 1990s, donor agencies also finally injected more aid and investment packages into island states and coastal areas in developing countries. Today nearly all international development agencies, such as The World Bank Group, United Nations Development Program, USAID, and UK Department for International Development (DFID) provide aid or investment packages to developing countries or non-government organizations (NGOs) and large- to medium-scale nature tourism entrepreneurs who are working in these environments. The main objective of this financing remains biodiversity conservation and community development (Halpenny, 1999).

A successful example of bilateral aid to an island state for ecotourism development can be found in the Solomon Island's Marovo Lagoon. The New Zealand Official Development Assistance Programme enabled a strategy developed by the communities from the Marovo Lagoon region to develop small-scale industries, including ecotourism, as an alternative to logging. Since 1993/94 when proposals were first drafted and 1996 when the Marovo Lagoon Ecotourism Association was formed, nine rustic lodges have been built with a combined capacity of 84 guests along with 20 bungalows providing sweeping views of the lagoon and its outer islets. Visitors are given the opportunity to participate in a wide range of activities including bush walks, snorkelling and birdwatching (B.N. Devi, 1998, unpublished). Despite the positive beginning the programme still faces many challenges. Simple logistical arrangements and communications continue to be problematic, and upgrading hospitality and business training is an ongoing process (B.N. Devi, Madrid, 1998, personal communication). Ensuring the success of this project will be a 10–15 year investment of time and funding.

The players

Those involved in coastal and islands ecotourism are not essentially different from the stakeholders in other ecotourism settings. They include governments, donors, industry, NGOs, communities, schools and academics. In the past, as with other settings, individuals and companies rather than government strategies have often powered ecotourism in coastal and island areas. Destinations have largely been developed through the efforts of a handful of tour companies or hotels interested in the particular region. As cited earlier, there have been exceptions to this, such as Dominica and Samoa with their 'deliberate' ecotourism (Weaver, 1991, 1993, 1998a, b) and Fiji and the Bahamas' use of ecotourism as a deliberate or complementary development strategy (Wilkinson, 1997; Harrison, 1998; Weaver, 1998b). However, many researchers still warn that island and coastal government officials often lack a clear understanding of tourism, let alone ecotourism (Ayala, 1995; Fagence, 1997). Some regional governments and their agencies have been grappling with the questions of ecotourism for years. The Great Barrier Reef Marine Park Authority is one of the more famous examples. The efforts of the states of western and southern Australia to manage their coastal resources by balancing ecotourism with other uses is another example (Commonwealth Coastal Action Program, 1997; Nature Based Tourism Advisory Committee, 1997; Wachenfeld *et al.*, 1998; Western Australia Tourism Commission, 1999).

As mentioned earlier, development organizations have a profound effect on ecotourism development in island and coastal settings in developing countries. Aid programmes pay for human resource training, protected areas purchase, infrastructure development and marketing. The challenge remains to ensure that the development efforts are appropriate. A current example of the debate surrounding appropriate development can be found in the Pacific where it was recently announced that European Union aid for marketing tourism products in the Pacific would be eliminated. The money has instead been slated for '"human resource development", private sector assistance and environment preservation projects'. The Tourism Council of the South Pacific warned that this will have devastating consequences for the islands 'given the isolation and smaller size of Pacific nations' (Pacific Islands Report, 1999). Ecotourism will no doubt benefit from the emphasis of funding on human resource development because of its comparatively higher training needs than mainstream tourism. Local communities and small-scale entrepreneurs who characterize the ecotourism field often have less market experience, fewer hospitality and business skills, and the ecotourism product often requires high information content. The same can be said for ecotourism benefiting from contributions to the private sector and environment preservation projects. However, this change in funding should not come at the expense of eliminating all marketing efforts for a region, as the announcement for donors originally implied, especially for a long-haul destination such as the Pacific.

Industry plays a major role in developing an island or coast as an ecotourism destination, and in managing the destination as well as the experiences that visitors receive. To maintain the ecological integrity of a destination, including both socio-cultural and natural aspects, tour operators, lodge owners, outfitters, and other ecotourism-related businesses must take steps to reduce their footprint on local environments, and increase positive impacts. In the Galapagos, Lindblad Special Expeditions agreed to adhere to a set of sustainable harvest guidelines set by the Marine Stewardship Council. The company will buy only Council-certified local fish products, and runs a guest awareness campaign on fish harvesting practices. Sufficient time will be given to training and raising awareness among local fishers to ensure they are not penalized by this new standard (J. Novy, Washington DC, 1998, personal communication; T. O'Brien, Kota Kinabalu, 1999, personal communication).

NGOs have also played an important role in the development of ecotourism on islands and coasts. International NGOs such as The Nature Conservancy and Conservation International supply expertise and act as brokers for private investors or donor organizations (see Chapter 30). International NGOs also lead travel programmes to island and coastal ecotourism destinations. Membership programmes are an important source of revenue for organizations such as World Wide Fund for Nature, city zoos and university alumni associations. These often utilize exotic, rich and diverse locations such as those provided by coastal and island ecotourism destinations to satisfy the educational and experiential motivations which characterize NGO members (Young, 1992).

Local NGOs such as the Marovo Lagoon Ecotourism Association in the Solomon Islands (J. Devi, 1998, unpublished), the Association of Galapagos Tour Operators in Ecuador (Honey, 1999) and the Alaska Wilderness and Recreation Tourism Association in the USA (AWRTA, 1999) all act as voices for responsible tourism, conservation of biodiversity and community development. Local organization around community and conservation issues is essential for an island or coastal destination if ecotourism is to be monitored and developed sustainably. The National Ecotourism Accreditation Program administered by the Australian Ecotourism Association is an excellent example of such an outcome. The existence of such organizations also plays an important part in giving ecotourism businesses a voice within the much larger national tourism sector.

Communities are another essential component of the equation that results in successful ecotourism in island and coastal destinations. Benefits to local communities are an essential component of ecotourism. Achieving these benefits is often a great challenge. However, mechanisms such as participatory planning, socio-cultural impact assessments, enterprise development, hospitality skills training, and micro loan programmes can be used to achieve ecotourism-inspired benefits for local communities over a long period of time (Sproule, 1996; Ashley and Roe, 1998; Epler Wood, 1998; A. Poon, 1998, unpublished; Anon, 1999a). Some communities choose to isolate tourism from most of the destination's population (e.g. Maldives) or take control of all tourism development within their territories (e.g. Gwaii Hanas, Canada or Arnhem Land, Australia). The San Blas islands in Panama provide a good example of community control of ecotourism development. For more than 25 years the Kuna Indians of the Kuna Yala territory have banned foreign ownership and development of ecotourism or mainstream tourism products (Smith, 1998). The only current form of foreign-owned tourism businesses have been the transient cruise ships that visit weekly for cultural and marine excursions. This policy has led to a fairly stunted tourism industry, but one that has preserved the gem-like quality of the region with its unsullied coral reefs and homespun cultural museum.

Finally, schools and researchers play an important role in the development of ecotourism on islands and coastal areas. The role of academics and researchers is clear, with suggestions for their future actions detailed later in this book (see Chapters 40 and 41). Schools on islands and coastal areas are still in need of improvement. This is especially true of training institutions in island regions such as the Pacific and Caribbean, where hospitality and business training are essential but often undercapitalized components of ecotourism development. In an example of the private sector taking the initiative to find creative solutions to this problem, Pacific-based tour operator Sea Kayak Tonga made a commitment to their Tongan partners by obtaining a NZ$15,000 grant from the Pacific Islands Industry Development Scheme to provide further business skills training in New Zealand (D. Spence, Vancouver, 1999, personal communication). An additional effort must also be made at the primary school level to raise environmental awareness among the next generation and foster a deeper understanding of the tourism sector in general, as often this sector is the dominant economic activity in the region (Halpenny, 2001). The ecotourism industry is an essential partner in this process, as are local governments.

Products: experiential and physical

The products that are associated with coastal and island ecotourism are varied. They range from physical features, such as ecolodges, canopy walkways and underwater trails to the experiential, such as emotions elicited or knowledge gained from the visit. Box 15.1 and following paragraphs discuss these products briefly. Further information can also be found in Chapter 17.

Overall these activities are growing at a rapid, if unknown pace along with the larger phenomena of nature and cultural tourism. Worldwide, there are more than 15 million certified scuba-divers (DEMA, Anaheim, 1998, personal communication).

Box 15.1. Coastal and island ecotourism activities and products.

Sea kayaking and canoeing
Scuba-diving and snorkelling
Wildlife watching (including whales, birds and dolphins)
Trails: underwater, coastal hiking and boating
Marine museums and interpretation centres
Catch and release fishing[a]
Expedition cruising

[a] Ecotourism status is debated.

However, in the USA it has been estimated that 90% of these divers cease being active divers (defined as at least one dive per year) after 3 years, and growth in the sport in the USA has therefore levelled off. Elsewhere, especially in Southeast Asia, the number of divers is still increasing rapidly. Sea kayaking has also witnessed tremendous growth (see Table 15.2) as documented in the North American-based *Specialty Travel Index*.

Animal watching is also at an all-time high. An example documented at the global scale is the estimated US$504 million generated through souvenir, hotel and tour sales associated with whale-watching in 1995 (Hoyt, 1995). At a more local level, Butler and Hvenegaard (1988) found that birdwatching in Point Pelee National Park, Canada, generated CDN$3.8 million during the peak season month of May, 1987. Interestingly, a majority of birders, when asked about expenditures, were willing to pay twice as much for their experience at the important migratory stopover, resulting in a net economic value of CDN$7.9 million.

The essential factor to hold in mind in developing ecotourism products in island and coastal destinations is to take advantage of the setting, capitalizing on the strengths of the coastal zone such as the presence of endemic flora and fauna. Diverse activities such as scuba-diving and sea kayaking can be combined with interpretative shore hikes or coastal forest birdwatching.

Islands and coasts also present challenges to the developers and users of tourism infrastructure. Hurricanes and tsunamis are examples of the more extreme challenges that ecotourism infrastructure faces. Some developers (such as Stanley Selengut in the US Virgin Islands) have chosen to build semi-permanent structures such as the Concordia Eco-tents and Maho Bay tents which may lose screens and mesh sidings during a hurricane but can be restored long before conventional hotels reopen. This subject is explored in greater depth in *Marine Ecotourism: Guidelines and Best Practice Case Studies* (Halpenny, 2001).

Islands in a region must also seek to develop complementary products, developing ecotourism products and experiences that complement those of neighbouring islands (Weaver, 1991; Ayala, 1995). For example, Nevis's hiking opportunities complement St Kitt's more urban, historical experience. By creating a network of different island opportunities, visitors stay longer, leaving more money in the region.

Table 15.2. Sea kayaking activity (Wylie and Rice, 1998).

Year	Tour companies	Trip venues (includes US states and international countries)
1991	25	35
1996	61	112

Planning

Integrated planning is an essential component in ensuring the success of ecotourism in an island or coastal setting (Jackson, 1986; Inskeep, 1987; Wong, 1993; Gunn, 1994; Boyd and Butler, 1996; Commonwealth Coastal Action Program, 1997; Manidis Roberts Consultants, 1997). Planning must not only occur to ensure that negative impacts on local environments are mitigated, but also to ensure positive impacts such as community development and conservation of natural and cultural elements. Planning also increases the probability that ecotourism can coexist with or act as a substitute for other land and water uses such as shipping, forestry, mari- and aqua-culture, as well as mainstream tourism developments such as golf courses and water recreation parks. Ecotourism businesses and destinations must plan and coordinate with other ecotourism providers to ensure that a consistent, quality product is available. For example, in the Galapagos a fixed number of tourism boats is allowed to operate at any one time. This ensures that the numbers of tourists arriving at any particular island in the Galapagos chain is limited, resulting in the best possible experience for the visitor, and minimizing visitor impacts (Honey, 1999).

As mentioned earlier some islands and coastal destinations such as Dominica and Samoa have actively planned the island's tourism development to be dominated by ecotourism (Weaver, 1993, 1998b). Others, such as Fiji and the Bahamas treat ecotourism more like a complement to the main island's mass tourism product (Weaver, 1993, 1997, 1999; Harrison, 1998). Regardless of which path is pursued, all destinations must employ various planning tools to ensure successful, sustainable implementation of ecotourism. Some of these planning tools include zoning, limits of acceptable change (LAC) (Stankey *et al.*, 1985), the tourism optimization management model (TOMM) (Manidis Roberts Consultants, 1997) and the ecotourism opportunity spectrum (Boyd and Butler, 1996). Two classic examples of pioneer efforts to balance tourism with other coastal uses through the application of zoning can be found in Pulau Seribu, Indonesia (Soegiarto *et al.*, 1984) and the Great Barrier Reef Marine Park, Australia (Wachenfeld *et al.*, 1998). Both sites today reflect the need to go beyond zoning and utilize additional methods such as enforcement, park fees, community enterprise development, mooring buoy programmes, environmental incentive programmes, and GIS-assisted view scapes assessments. Recent experimental applications of LAC in the Lesser Antilles island of Saba (McCool, Vermont, 1999, personal correspondence), and TOMM on Kangaroo Island, Australia (Manidis Roberts Consultants, 1997) and Gwaii Hanas, Canada (Wight, Vermont, 1999, personal correspondence) have provided valuable learning experiences on the challenges of balancing community needs and wants with industry realities and government abilities/jurisdiction.

Another essential ingredient for ensuring that ecotourism remains a sustainable activity in a region is the development of codes of conduct for tourists, and codes of practice for industry. Sometimes government will be involved in the development of these guidelines, as will community members. In the formulation of codes of practice, it is essential that the ecotourism industry is an integral player in the creation of relevant guidelines. Gjerdalen's (1999) examination of adherence to a code of practice developed by and for whale-watching companies who operate orca (killer whale) tours within Johnstone Strait, British Columbia, Canada, provides a good case study. In this example, buy-in by all companies was essential to ensure self-enforcement by the industry.

Social, cultural and economic criteria and impacts

Cultural impacts resulting from tourism have been well documented (Smith, 1989; McLaren, 1998). The danger that ecotourism poses is perhaps even greater due to its intimate nature as ecotourists travel to a destination to learn about and interact with other cultures. The impact of this interaction can result in the theft of cultural property such as songs or plant knowledge, loss of cultural integrity and homogeneity, and erosion of cultural values with the introduction of the tourists' values. Leaders of the Maldives feel that these threats warrant the isolation of tourism activities to a select number of islands, thereby maintaining cultural purity throughout most of the Maldives archipelago. Ecotourism can also support cultural integrity through the actions of ecotourists who demonstrate an interest in local culture, their valuing of local culture in turn instilling a sense of pride among community members. Ecotourism also produces an additional reason for maintaining craft traditions because the ecotourist wishes to learn how to make the local crafts, or wishes to make an actual purchase. For example ecotourists visiting the Marovo Islands may learn from community members how to fish using traditional methods (B.N. Devi, 1998, unpublished).

Social impacts are just as difficult to decipher. The increased wealth generated by tourism can unbalance traditional social structures. In some societies this could be interpreted as a positive development as many women begin to have greater opportunities for financial freedom. In the Galapagos income generated by tourism has created a magnet for Ecuadoreans from the mainland. In 1997 tourism in the Galapagos was directly or indirectly providing income for an estimated 80% of the people living on the Galapagos Islands and generating 60% of all tourism revenues earned by the Ecuadorean government (Honey, 1999). In recent years a major influx of non-Galapagos community members has taken up residence on the islands to capitalize on the ecotourism jackpot. Many of these mainlanders do not share similar cultural values and knowledge with native Galapaguenos such as respecting harvesting limits of land and marine resources (Honey, 1999) and this has resulted in some social unrest.

Often ecotourism has been looked to as a tool for achieving political strength and justifying or reinforcing land or territory rights. Many indigenous groups have been working, especially in Australia and Canada, to achieve self-government and reclaim tribal territory. Ecotourism is being explored as one of the new development strategies that will make self-government and autonomy a reality (Anon, 1999a). Other coastal and island communities, whose traditional economies have collapsed, e.g. fisheries, are also looking to ecotourism as a community development and autonomy building tool (Newfoundland: Woodrow, 1999; Western Ireland: White, 1999; Norway: Hallenstvedt, 1999; Texas: Richardson, 1999). The danger in such an approach is over-reliance on ecotourism; in all destinations ecotourism must only be one of several economic and social tools for achieving sustainable development.

Economic impacts of ecotourism have been largely positive, but not large enough in many instances. Aside from support for local communities, income from ecotourism can also be used to support protected areas. The Galapagos National Park is self-sufficient, deriving much of its revenue from tourist visitation fees and donations (Vieta, 1999) whereas in the Komodos National Park in Indonesia, Goodwin *et al.* (1997) found that the park covered only 7.17% of its 1994/95 management costs through entrance fees. Further analysis, utilizing the idea that tourism should at least pay for its related expenses such as visitor management and infrastructure, found that in 1994/95, 109.05% of tourism-related expenditure by park managers was covered by tourism revenues. In Saba and Bonaire (Dixon, 1993; Dixon and van't Hof, 1997; Vieta, 1999) and Belize (J. Gibson, Belize City, 1999, personal communication) visitor fees have been used to pay for staff salaries, moorings for dive and fishing boats, maintenance of all park facilities and equipment, interpretative materials, local volunteer groups, educational talks and law enforcement activities. Vieta (1999) stresses that a significant portion of ecotourism revenue should be '"earmarked" for reinvestment into ecotourism products and its natural, cultural and social support systems rather than being sent to a national treasury for redistribution'. Visitor fees are not only levied by protected areas. South of New Zealand, the Campbell and Auckland islands charge a visitor fee for 'management purposes' (Hall, 1993). Other coastal and island destinations collect hidden visitor fees in the form of bed, airport and sales tax (e.g. Belize's PACT: B. Spergel, 1996, unpublished).

The challenge that all tourism destinations face is to retain as much revenue in the country or region as possible. While leakage of tourism-generated revenue is a major problem for regions such as the Caribbean, it is also thought to be less prevalent with ecotourism as many of the products used to deliver ecotourism such as building materials and foodstuffs are derived from local sources (Weaver, 1991). This debate is illustrated in Epler's (n.d.) research which found that although the total value of tourism in the Galapagos islands equalled US$32.6 million in both

1990 and 1991, of which approximately 85% was paid to vessels and airlines (mostly owned by mainland Ecuadoreans and foreigners), a mere 3% went to on-land hotels and park entrance fees. Other statistics collected by Epler show that 92% of the tourist dollar was spent on floating hotels (most not owned by Galapaguenos) and only 8% on day boats and land-based hotels (mostly locally owned) (as cited in Honey, 1999). Goodwin *et al.* (1997) found that in Komodo National Park, at least 50% of all visitor expenditure leaked out of the local economy because of importation of goods from outside the region (e.g. bottled water). In addition, the transport system is in large part owned by the government or run by external operators. However, in one town leakage from revenue generating activities was found to be negligible, since revenue was based on the provision of labour and primary produce. Much of the revenue remained in the village.

Marketing

Misuse of the term ecotourism in marketing products that are unsustainable, benefiting neither communities nor the natural environment, has been one of the greatest challenges ecotourism advocates face. Another problem has been the preoccupation of government and business personnel with marketing, rather than the development of quality products (Ayala, 1995).

Islands and coastal destinations must learn to jointly market their ecotourism products with nearby competitors. This is especially important for the tourism-intensive island nations in the Pacific and Caribbean (Fagence, 1997; Weaver, 1998a, b). Within nations there have been efforts to piggyback ecotourism on the mainstream tourism marketing campaign, but with mixed results. Certainly the most inexpensive form of advertising, the Internet, must be better understood and utilized by developing countries (Hokstam, 1999). Several recent shortfalls in donor and government funding for marketing has created panic in both the Caribbean (Hokstam, 1999) and Pacific (Pacific Islands Report, 1999). Hokstam (1999) states that because regional experts have observed a decline in Caribbean tourist arrivals, which experienced a 20% drop in 1999 compared with 1998, industry leaders such as the Caribbean Hotel Association are proposing multilateral marketing efforts. This decline may not be as sharp for the niche ecotourism sector, as much of the Caribbean's tourism woes may be tied to a declining mass tourism market as sun, sand and sea destinations lose appeal.

Conclusion

Coasts and islands as settings for ecotourism are proving to be an exciting option for both tourists and managers. Their many unique characteristics such as rich cultural traditions, high occurrence of endemic species, remoteness and slower, holiday-paced atmosphere make coasts and islands ideal ecotourism destinations. The unique environmental characteristics of islands and coasts, such as their vulnerability to externally produced pollution and limited natural resources, create greater challenges for coastal managers and businesses charged with making ecotourism a success. Ecotourism's potential as a tool for community and economic development and its capacity for generating revenue for conservation and environmental awareness among visitors and hosts creates a positive future for ecotourism in island and coastal regions. As in other ecotourism destinations the challenge that government, industry, communities and NGOs now face in islands and coasts is the implementation of ecotourism in a sustainable manner, maintaining both socio-cultural and environmental integrity. This has been initiated successfully in select locations with innovative, collaborative and adaptive coastal management tools, planning, infrastructure, marketing and funding schemes. However, much more work needs to be accomplished if ecotourism is to achieve its promise of sustainable development in coastal and island settings.

References

Anon (1999a) Ecotourism, sustainable development, and cultural survival: protecting indigenous culture and land through ecotourism. *Cultural Survival Quarterly* Summer 1999, 23(2), 25–59.

Anon (1999b) Reaction: sea canoe controversy. *Action Asia* April/May 1999, 2(8), 9–11.

Ashley, C. and Roe, D. (1998) *Enhancing Community Involvement in Wildlife Tourism: Issues and Challenges.* Wildlife and Development Series No. 11, International Institute for Environment and Development, London.

AWRTA (1999) *Guidelines Newsletter.* Alaska Wilderness and Recreation Tourism Association. http://www.alaska.net/~awrta/meminfo/newsletter/index.html

Ayala, H. (1995) From Quality Product to Eco-product: Will Fiji set a Precedent? *Tourism Management* 16(1), 39–47.

Boyd, S.W. and Butler, R.W. (1996) Managing ecotourism: an opportunity spectrum approach. *Tourism Management* 17(8), 557–566.

Butler, J.R. and Hvenegaard, G.T. (1988) The economic values of bird watching associated with Point Pelee National Park, Canada, and their contribution to adjacent communities. *Second Symposium on Social Science in Resource Management. University of Illinois, Urbana, 6–9 June 1988.*

Commonwealth Coastal Action Program (1997) *Coastal Tourism: a Manual for Sustainable Development.* Prepared for Tourism Council Australia, Australian Local Government Association, Royal Australian Planning Institute, in collaboration with Portfolio Marine Group, Environment Australia Office of National Tourism. Based on a draft prepared by Southern Cross University. Department of the Environment, Sport and Territories, Canberra.

Dixon, J.A. (1993) Meeting ecological and economic goals: marine parks in the Caribbean. *AMBIO* 22(2–3), 117–125.

Dixon, J.A. and van't Hof, T. (1997) Conservation pays big dividends in Caribbean. *Forum for Applied Research and Public Policy* Spring 1997, 12(1), 43–48.

Epler Wood, M. (1998) *Meeting the Global Challenge of Community Participation in Ecotourism: Case Studies and Lessons from Ecuador.* American Verde Working Papers Number 2. The Nature Conservancy, Arlington, Virginia.

Fagence, M. (1997) Ecotourism and Pacific island countries: the first generation of strategies. *Journal of Tourism Studies* 8(2), 26–38.

Gjerdalen, G. (1999) The Johnstone Strait Code of Conduct for Whale Watching: Factors Encouraging Compliance. In: Williams, P.W. and Budke, I. (eds) *On Route to Sustainability: Best Practice in Canadian Tourism.* Prepared by Industry Competitiveness, The Canadian Tourism Commission and The Center for Tourism Policy and Research, Simon Fraser University, Ottawa, pp. 15–19.

Goodwin, H.J., Kent, I.J., Parker, K.T. and Walpole, M.J. (1997) Tourism, Conservation and Sustainable Development: Vol. 1, Comparative Report. Unpublished report for Department for International Development, UK.

Gunn, C.A. (1994) *Tourism Planning: Basics, Concepts and Cases*, 3rd edn. Taylor & Francis, Washington, DC.

Hall, C.M. (1993) Ecotourism in the Australian and New Zealand Sub-Antarctic Islands. *Tourism Recreation Research* 18(2), 13–21.

Hallenstvedt, A. (1999) Marine recreation and integrated coastal management. Presented at the 1999 International Symposium on Coastal and Marine Tourism, 26–29 April 1999, Vancouver, British Columbia.

Halpenny, E. (1999) The state and critical issues relating to international ecotourism development policy. Presented at the 1999 Ecotourism Association of Australia National Conference, Kingfisher Bay, Queensland, 13–17 October 1999.

Halpenny, E. (2001) *Marine Ecotourism: Guidelines and Best Practice Case Studies.* The Ecotourism Society, North Bennington, Vermont.

Harrison, D. (ed.) (*c.*1998) *Ecotourism and Village-based Tourism: a Policy and Strategy for Fiji.* Based on work prepared by Sawailau, Samisoni and Malani, Manoa. Ministry of Tourism and Transportation, Fiji.

Hokstam, M.A. (1999) Caribbean-tourism: concern as tourism figures dip. From Sustainable Tourism Listserve http://www.egroups.com/list/sustainable-tourism/ By Inter Press Service listed at http:www.oneworld.org/ips2/june99/23_10_097.html

Honey, M. (1999) *Ecotourism and Sustainable Development: Who Owns Paradise?* Island Press, Washington, DC.

Hoyt, E. (1995) *The Worldwide Value and Extent of Whale Watching: 1995.* Whale and Dolphin Conservation Society, Bath, UK. Presented as IWC/47/WW2 to International Whaling Commission (IWC) 1995 AGM, Dublin, Ireland, pp. 1–34.

Inskeep, E. (1987) Environmental planning for tourism. *Annals of Tourism Research* 14, 118–155.

International Civic Aviation Organization (1999) Sustaining Development of Air Transportation in Small Island States. An addendum (E/CN.17/1999/6/Add.15) to Progress in the Implementation of the Programme of Action for the Sustainable Development of Small Island Developing States. Report of the Secretary-General for Commission on Sustainable Development Seventh Session 19–30 April 1999. United Nations Economic and Social Council, New York.

Jackson, I. (1986) Carrying capacity for tourism in small tropical Caribbean islands. *Industry and Environment* January/February/March, 7–10.

Lindberg, K., Furze, B., Staff, M. and Black, R. (1998) *Ecotourism in the Asia-Pacific Region: Issues and Outlook.* Forestry Policy and Planning Division, Rome Regional Office for Asia and the Pacific, Bangkok and US Department of Agriculture, Forest Service, with The Ecotourism Society, N. Bennington, Vermont.

Manidis Roberts Consultants (1997) *Developing a Tourism Optimization Management Model (TOMM): a Model to Monitor and Manage Tourism on Kangaroo Island,* Final Report. South Australian Tourism Commission, Adelaide.

McLaren, D. (1998) *Rethinking Tourism and Ecotravel: the Paving of Paradise and What You Can Do to Stop It!* Kumarian Press, West Hartford, Connecticut.

Milne, S. (1992) Tourism and development in South Pacific microstates. *Annals of Tourism Research* 19, 191–212.

Nature Based Tourism Advisory Committee (1997) *Nature Based Tourism Strategy for Western Australia.* Western Australia Tourism Commission, Perth, 32 pp.

O'Brien, T. (1999) Ship-based marine ecotourism: itineraries, interpretation, and potential. Presented at World Ecotourism Conference and Field Seminars: The Right Approach, Kota Kinabalu, Sabah, Malaysia, 17–23 October 1999.

Pacific Islands Report (1999) European Union Tourism Decision Threatens Pacific Economies. From Sustainable Tourism Listserve http://www.egroups.com/list/sustainable-tourism/ By Pacific Islands Report listed at http:pidp.ewc.hawaii.edu/PIReport/1999/July/07-22-04.html

Richardson, S.L. (1999) Balancing community and regional interests in tourism development: lessons from the Texas coast. Presented at the 1999 International Symposium on Coastal and Marine Tourism, 26–29 April 1999, Vancouver, British Columbia.

Rome, M. (1999) Shooting to kill. *Action Asia* Feb/March 1(8), 19–25.

Royle, S.A. (1997) Tourism to the South Atlantic Islands. In: Lockhart, D.G. and Drakakis-Smith, D. (eds) *Island Tourism: Trends and Prospects.* Pinter, London, pp. 323–344.

Shah, Nirmal Jivan (1995) Managing coastal areas in the Seychelles. *Nature & Resources* 31(5), 16–33.

Smith, V. (ed.) (1988) *Hosts and Guests: the Anthropology of Tourism.* University of Philadelphia Press, Philadelphia.

Soegiarto, A., Soewito and Salm, R.V. (1984) Development of marine protected areas in Indonesia. In: McNeely, J.A. and Miller, K.R. (eds) *National Parks, Conservation, and Development – the Role of Protected Areas in Sustaining Society. Proceedings of the World Congress on National Parks, Bali, Indonesia, 11–12 October 1982.* Smithsonian Institution Press, Washington, DC.

Sproule, K. (1996) Community-based ecotourism development: identifying partners in the process. In: Malek-Zadeh, E. (ed.) *The Ecotourism Equation: Measuring the Impacts.* Bulletin Series, Yale School of Forestry and Environmental Studies, Number 99. Yale University, New Haven, Connecticut, pp. 233–250.

Stankey, G.H., Cole, D.N., Lucas, R.C., Petersen, M.E. and Frissell, S.S. (1985) *The Limits of Acceptable Change (LAC) System for Wilderness Planning.* General Technical Report INT-176. US Department of Agriculture, Forest Service, Intermountain Forest and Range Experiment Station, Ogden, Utah.

Valentine, P.S. (1993) Ecotourism and nature conservation: a definition with some recent developments in Micronesia. *Tourism Management* 14(2), 107–115.

Vieta, F. (1999) Ecotourism and its Role in Socioeconomic Development and Environmental

Protection. Background Paper No. 16 (DESA/DSD/1999/16) prepared by the United Nations Department of Economic and Social Affairs and Division of Sustainable Development for the Commission on Sustainable Development Seventh Session, 19–30 April 1999, New York.

Wachenfeld, D.R., Oliver, J.K. and Morrissey, J.I. (1998) *State of the Great Barrier Reef World Heritage Area.* Great Barrier Reef Marine Park Authority, Townsville, Queensland.

Weaver, D.B. (1991) Alternative to mass tourism in Dominica. *Annals of Tourism Research* 18, 414–432.

Weaver, D.B. (1993) Ecotourism in the small island Caribbean. *GeoJournal* 31(4), 457–465.

Weaver, D.B. (1995) Alternative tourism in Montserrat. *Tourism Management* 16(8), 593–604.

Weaver, D.B. (1998a) Strategies for the development of deliberate ecotourism in the South Pacific. *Pacific Tourism Review* 2, 53–66.

Weaver, D.B. (1998b) *Ecotourism in the Less Developed World.* CAB International, Wallingford, UK.

Western Australia Tourism Commission (1999) *Environmental and Planning Guidelines for Tourism Development on the North West Cape.* Department of Environmental Protection and Ministry of Planning, Western Australia, Perth, 33 pp.

White, F. (1999) New perspectives on the role of tourism in marginal coastal communities in the West of Ireland. Presented at the 1999 International Symposium on Coastal and Marine Tourism, 26–29 April, 1999, Vancouver, British Columbia.

Wilkinson, P.F. (1997) *Tourism Policy and Planning: Case Studies from the Commonwealth Caribbean.* Cognizant Communications Corporation, New York.

Wong, M. (ed.) (1993) *Tourism vs. Environment: the Case for Coastal Areas.* Kluwer, Amsterdam.

Woodrow, M. (1999) Fisher communities in Newfoundland: generating interpretation programs for tourism. As presented at the Tenth Society for Human Ecology Conference, 31 May 1999, Montreal.

Wylie, J. and Rice, H. (1998) Sea kayaks as vehicles for sustainable development or coastal and marine tourism. In: Miller, M.L. and Auyong, J. (eds) *Proceedings of the 1996 World Congress on Coastal and Marine Tourism, 19–22 June 1996, Honolulu.* Washington Sea Grant Program and the School of Marine Affairs, University of Washington and Oregon Sea Grant College Program, Oregon State University, Seattle, Washington, pp. 131–138.

Young, M. (1992) Should international conservation organizations like WWF be involved with travel clubs? In: Weiler, B. (ed.) *Ecotourism Incorporating the Global Classroom.* Bureau of Tourism Research, Canberra, pp. 221–223.

Chapter 16

Deserts, Grasslands and Savannahs

D.B. Weaver

Department of Health, Fitness and Recreation Resources, George Mason University, Manassas, Virginia, USA

Introduction

The purpose of this chapter is to examine the status of ecotourism within desert, grassland and savannah-type ecosystems. Other non-woodland environments, such as tundra, alpine areas and polar deserts, are considered elsewhere in this section (see Chapter 13 and 14). While the boundaries between these ecosystems tend to be transitional rather than sharp, each is treated as a discrete category for discussion purposes. Figure 16.1 depicts the generalized spatial distribution of these ecosystems as they are considered in this chapter. Various parameters of precipitation and rates of evapotranspiration have been proposed in the literature to differentiate these biomes. However, since it is the resultant vegetation cover that is more relevant to ecotourism, this particular criterion will be used to make the distinction.

Each section of this chapter begins with a short description of the relevant biome, including its spatial distribution and condition (i.e. extent to which it has been modified by human activity). This leads into a discussion of the hypothetical strengths and weaknesses of each ecosystem as a venue for ecotourism, and then into a consideration of its actual utilization as such. This chapter does not take into account ecotourism attractions that have no direct relationship to the biome in which they occur. For example, the Undara lava tubes of northern Queensland (Australia) are popular geological attractions that just happen to be located within a savannah-type ecosystem (Sofield and Getz, 1997). In contrast, some geomorphological features form only under conditions of extreme aridity, and thus should be discussed under the category of 'desert'.

Deserts

'Deserts' are often associated in the public image with a non-vegetated landscape dominated by sand dunes. However, this stereotype, properly referred to as an *erg*, is just one type of desert, occupying about 30% of the Sahara and Arabian deserts, but only about 1% of North America's arid lands (Huber *et al.*, 1988). More characteristically, deserts consist of rocky, sandy or stony lands hosting a discontinuous cover of short grass, cactus and/or shrubs. Trees are found only in oases, along permanent rivers, or in other areas where a reliable water supply is locally available. A useful distinction can be made between 'hot' deserts such as the Sahara, Arabian, Mojave, Atacama, Great Indian, Kalahari,

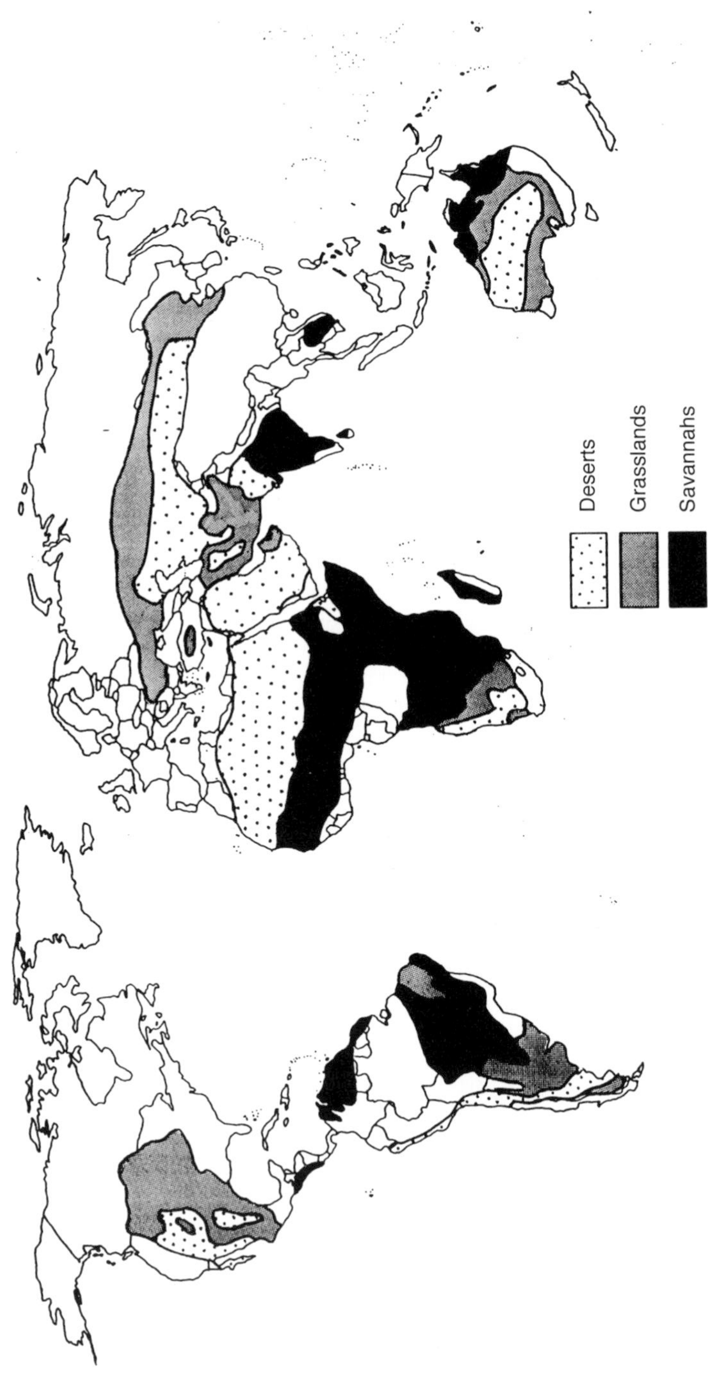

Fig. 16.1. Generalized distribution of deserts, grasslands and savannahs.

Namib and the deserts of Australia (Great Sandy, Gibson, Simpson), and 'temperate' deserts such as the Gobi, northern Patagonia, and much of the Great Basin of the western USA.

As a potential venue for ecotourism, deserts have the advantage of being largely unaltered by direct human intervention. This is not due to some litany of enlightened attitude toward arid lands, but rather to the more mundane reality that deserts, by their nature, are largely unsuited for agriculture or permanent human settlement. Having said this, there are significant areas of desert, such as in the southwestern USA, that have been altered by irrigation, mining, military uses and urbanization. This contraction, however, may be offset by the desertification-induced encroachment of arid ecosystems on adjacent grasslands and savannahs. A second inherent advantage of deserts might be termed the 'visibility factor', wherein wildlife is more likely to be sighted in a sparsely vegetated landscape due to the availability of a broad viewing range and the lack of cover for larger animals.

On the negative side, deserts are commonly perceived, at least in the Western mind, as a lifeless and dangerous environment beset by extremely high temperatures and extremely low precipitation; it is useful to point out that the word 'desert' derives from the Latin *desertum*, or something that has been abandoned. Furthermore, while endowed with a generous visibility factor, the number of larger animals that can be seen in most deserts is restricted by low inherent carrying capacities. The fauna that *are* present tend again to be stereotyped, especially by images of unpleasant and dangerous reptiles and arachnids such as rattlesnakes, scorpions and tarantula spiders.

Desert ecotourism

The above evaluation may lead one to suspect that ecotourism is rare or absent altogether within arid ecosystems. This, however, is not the case, although desert ecotourism does admittedly suffer by comparison with more accessible and more vegetated ecosystems. In terms of attractions, a cursory examination of current desert ecotourism activity shows a distinct pattern of association with seven factors:

1. Exceptional geological features associated with arid climatic conditions; these include the Grand Canyon (Arizona, USA), the ancient sand dunes of the Skeleton Coast (Namib desert, Namibia) and Uluru (Ayer's Rock) in central Australia.
2. Wildflower and other episodic floral displays; examples include the desert regions of Western Australia, Cape Province (South Africa) and other areas where heavy rainfalls induce an ephemeral blossoming of desert flora.
3. Ancient, large or unusual vegetation; examples include the 2000-year-old *Welwitschia* plants of the Namib desert, and the giant saguaro cacti of the southwestern USA.
4. Caravans or other desert trekking; e.g. the Tuareg camel trek offered during the early 1990s in Algeria's Sahara desert by a private adventure travel company (Daniel, 1993).
5. Indigenous inhabitants; given that traditional indigenous cultures are often inextricably linked to their surroundings, they may constitute an 'ecotourism' attraction in their own right, or at least in terms of their interaction with their surroundings. This of course is a debatable point. Examples include the above-mentioned Tuareg trek, and activity affiliated with Australia's arid land Aborigines (as for example at Uluru), and the Bushmen of the Kalahari desert (Hitchcock, 1997).
6. Oases; there are a growing number of ecotourism sites that are affiliated with luxury resorts situated in an oasis environment. One of the best instances is the Al-Maha resort in the United Arab Emirates, which includes a 16 km^2 nature reserve stocked with reintroduced Arabian oryx and sand gazelle.
7. Protected areas; desert ecotourism is to a very large extent associated with formally protected areas (see below).

The first three of these factors reveal a direct ecotourism focus, with the first being entirely predictable (i.e. the geological attraction is guaranteed to be there), and the second being largely unpredictable (the appearance and quality of the wildflower display depends upon the occurrence and type of precipitation). The next three factors are more indirect given that they incorporate ecotourism as a supplement to adventure, cultural and luxury resort tourism, respectively.

The final factor, association with protected areas, is probably the most important and encompassing of all. As in other ecosystems, most desert ecotourism occurs within accessible public or private protected areas. These provide suitably impressive natural attractions retained in a more or less natural state, appropriate services and facilities (e.g. interpretive centres, tracks, roads, infrastructure) and sometimes a high public profile associated with national park, world heritage, or similar status (see Chapter 18). In a desert context, this phenomenon is best illustrated in south-western USA, where federal protected areas such as Grand Canyon, Death Valley and Joshua Tree each attract more than 1 million visitors each year (see Table 16.1). Not all visitors to these parks merit classification as ecotourists, but an analysis of available services and activities suggests that 'soft' ecotourists probably constitute a strong majority. A typical pattern of activity involves the use of private vehicles to participate in 'scenic drives' that include periodic stops and, optionally, short interpretive walks and talks, at points of interest including interpretive centres. Hence, the pattern of visitor concentration that is so apparent in the more popular protected areas worldwide is evident in desert parks.

A distinct variation of the desert protected area is the relatively small site in which selected desert flora are planted and displayed for educational and scientific purposes in a way that emulates their natural surrounds. This is illustrated by the Boyce Thompson Arboretum State Park east of Phoenix, Arizona, which covers only 120 ha, but hosted over 95,000 visitors during the 1997/98 fiscal year (personal communication, L. Soukup, Boyce Thompson Arboretum). A similar Australian example is Alice Springs Desert Park, which opened in 1997 and expected 100,000 visitors in 1999. (It should not be assumed, however, that all of these visitors are ecotourists.) At the core of the 13 km^2 site is a 50 ha core area that is meticulously designed to display and interpret a variety of Australian desert plant communities. The facility also engages in the captive breeding of the resident reptiles, birds and mammals (Brown, 1999).

The USA and Australia (where Uluru National Park hosts approximately 300,000 visitors per year) are the two countries where desert ecotourism is best represented and longest established. Hence, they may be described as the 'top tier' of desert eco-

Table 16.1. Visits to major federal protected areas in desert ecosystems of the USA, 1998 (National Park Service, 1999).

State	Protected area	1998 visitation
Arizona	Grand Canyon National Park	4,239,682
Arizona	Organ Pipe Cactus National Monument	182,126
Arizona	Petrified Forest National Park	816,506
Arizona	Saguero National Monument	716,160
California	Death Valley National Monument	1,177,746
California	Joshua Tree National Monument	1,410,312
California	Mojave National Preserve	374,378
New Mexico	White Sands National Monument	592,957

tourism. This status is explained by the presence of high-profile desert protected areas (such as those mentioned above) that are accessible to affluent domestic populations with a high proclivity to engage in soft ecotourism activities. In addition, inbound tourists also compose a significant component of protected area visitors. By comparison, ecotourism in the deserts of most other countries is incipient. Among the best developed of these 'lower tier' desert ecotourism destinations is Namibia, which accommodates some 35,000–40,000 visitors each year in the Namib-Naukluft and Skeleton Coast Nature Reserves. Contributing to this relative maturity is proximity to the white South African market (which is similar in affluence and culture to the USA and Australia), a well-developed national road and air network, and an extensive and well-serviced protected area network in its desert areas. In addition, the deserts of Namibia are characterized by unique and interesting features such as the ancient and highly unusual dunes of the Namib desert, quivertree (*Aloe dichotoma*) 'forests', the aforementioned *Welwitschia mirabilis* plants, and robust populations of larger and easily observed desert mammals (Weaver and Elliott, 1996).

More typical of the lower tier is the status of ecotourism in the great belt of desert that extends almost continuously from Mauritania in western Africa through the Middle East and the former Soviet central Asian republics to north-central China and Mongolia. In the first instance, relatively few protected areas have been established within this desert belt. Countries dominated by desert tend to have among the lowest portion of land set aside for such purposes (e.g. as of 1997, United Arab Emirates and Iraq had 0%; Libya, 0.1%; Egypt, 0.8%; Kuwait, 1.4%; Mauritania, 1.7%; Uzbekistan, 2.1%; Jordan, 3.4%; World Resources Institute, 1998). Secondly, those protected areas that do exist contain few if any facilities to accommodate soft ecotourists. Thirdly, the transport network necessary to access these parks is rudimentary. Fourthly, the proclivity of domestic tourists in these countries to engage in ecotourism is minimal, as is the inflow of Western non-business tourists who would constitute the most likely market for such a product.

Not surprisingly, desert ecotourism in most of the lower tier destinations (exceptions such as Al-Maha notwithstanding) is an informal 'hard' variety that intersects with adventure tourism (see Chapter 5) and involves such disparate elements as scientific expeditions, exclusive adventure packages, and individual 'exploration' by four-wheel drive or other means. As an organized activity, there does not appear to be any history of ecotourism at all in Uzbekistan or any of the other former Soviet Central Asian republics that contain significant amounts of desert (Sievers, 1998). A very small amount of ecotourism appears to be occurring in the Egyptian desert, where local oases are used as bases of operation for private companies offering day-long or multi-day desert excursions. As part of its tourism diversification campaign of the early 1980s, consideration was given to the development of the desert as an ecotourism destination. However, this was rejected in favour of the coastal 3S-type tourism after it was determined that little market existed either domestically or internationally for such a product (Cockerell, 1996). In the Tunisian desert, an international non-governmental organization is incorporating ecotourism into a project to develop the archaeological and historical resources of the Douiret town site into a major tourist attraction (Ouessar and Belhedi, 1998).

Threats to desert ecotourism

In the top tier destinations of the USA and Australia, the primary internal threat to desert ecotourism derives from the concentration of high visitor levels within certain portions of some public protected areas, which can lead to the deterioration of both the natural resource base and the visitor experience. A major external threat is the indiscriminant use of all-terrain vehicles, which contributes to the degeneration of non-protected desert habitat and is directly disruptive to the ecotourism experience. For lower tier destinations, the threats are entirely different. One significant internal threat within the few protected areas that *have* been established is the lack of proper management owing to resource scarcities and the low priority given to desert areas. Externally, unregulated 'consumptive' activities such as big game hunting are increasing in unprotected desert areas (and sometimes within the protected areas themselves) because of their income-generating potential. This is especially evident in the former Soviet central Asian republics. While not inherently unsustainable, problems can occur when hunting depletes local wildlife populations or interrupts those attempting to pursue ecotourism. An additional threat is warfare, the devastating environmental effects of which were demonstrated in the wake of the Gulf War in both Kuwait and Iraq.

Grasslands

As the name implies, grasslands are ecosystems dominated by a continuous cover of grasses and other non-woody plants. Depending on the amount of moisture available, grasslands can range from tallgrass to medium grass and shortgrass subtypes. Major grassland ecosystems include the North American Great Plains or Prairies, the Eurasian Steppe, the Sahel of Africa, the Pampas of Argentina, the South African Veld and the grasslands of Australia. Unlike deserts, grassland ecosystems have been extensively altered for agricultural purposes, given the suitability of underlying soils for the cultivation of grains and other crops, or for supporting grazing animals in marginal grassland areas. The situation is especially critical in the case of tallgrass prairie, which harbours the most productive soil structures. The American state of Illinois typifies the demise of tallgrass ecosystems. Once covering 5 million ha, these grasslands have been reduced to 800 ha distributed among a few scattered patches that do not allow for the operation of a viable representative ecosystem (Robertson, 1999). Similarly, most of the world's tropical lowland grasslands have been destroyed because of their suitability for farming or grazing (Neldner *et al.*, 1997). The status of medium grass prairie is less dire, although it has been estimated that 82% of Canada's native mixed grass prairie have been converted to agriculture (Weaver, 1997). The situation is better for the shortgrass prairie, where the introduction of grazing by domestic animals does not often entail the removal of native plants, and does not automatically preclude ongoing grazing by native wildlife.

Whereas deserts are perceived by many as a dangerous environment, grasslands are handicapped by the widespread perception that they are nondescript and uninteresting. With a few exceptions, grasslands do not support the charismatic wildlife that is associated with savannahs, rainforests or oceans. Furthermore, most visitors cannot readily appreciate the subtle differences that often differentiate one grassland plant from another. It is the prevalence of such attitudes, of course, that has facilitated the conversion of natural grassland to agriculture and pasture over the past two centuries.

Grassland ecotourism

It might be assumed from the above description that the prospects and actual status of ecotourism in this ecosystem are highest in the shortgrass areas, and lowest in the tallgrass areas. This pattern holds true to a large extent, but must be qualified. No doubt, the potential of ecotourism in the tallgrass has been seriously curtailed by the almost complete annihilation of this biome. However, this same rarity has stimulated an array of preservation and restoration initiatives. Concurrently, the public in areas such as the US Midwest is being educated to appreciate the fragility and value of the remaining tallgrass areas, a trend which is assisted by dramatic images of native plants growing 3 m in height or higher. As a result, the 'remnant gems' of tallgrass prairie have become the objects of considerable ecotourism interest, including education and scientific research. The Living Prairie Museum in the Canadian province of Manitoba is an example of this phenomenon. Because this 12 ha remnant of unbroken tallgrass prairie is located within the city limits of Winnipeg, the Museum has become a major urban ecotourist attraction. Another high profile initiative is underway in the Midewin National Tallgrass Prairie in Illinois, where 6000 ha are gradually being restored to a natural tallgrass state on land that is being made surplus by the military. Unlike the Winnipeg site, the Midewin project is an example of restorative ecotourism taking place within the context of a modified landscape (see Chapter 20). In 1998, 36 escorted public tours with 850 participants were conducted at the site, and it is expected that ecotourism will eventually emerge as the major activity in this protected area (Illinois Nature Preserves Commission, 1999). Europe's tallgrass areas are just as fragmentary, although Hungary's large Hortobagy National Park does preserve a significant portion of the *puszta* ecosystem, and provides an array of ecotourism opportunities to a growing number of domestic and international visitors.

As mentioned above, the impact of human settlement on shortgrass prairie has been less traumatic, given its adaptability to grazing by domestic animals. In contrast to the tallgrass, a number of large protected areas have been established in the shortgrass areas of North America, albeit more recently than in woodland or desert areas. Again, it is in North America that this trend is most apparent. Shortgrass-dominated national parks in the USA such as Theodore Roosevelt in North Dakota and Badlands in South Dakota are extremely popular despite their relative remoteness, recording 448,226 and 1,021,049 visitors in 1998, respectively. An incipient creation is Grasslands National Park in the Canadian province of Saskatchewan, which accommodated less than 4000 visitors (essentially all ecotourists) during 1997/98. Far more spatially extensive are the US national grasslands, which are similar to the US national forests in their multi-use orientation. Established in 1960 as a vehicle for conserving and restoring degraded shortgrass prairie, the 1.5 million ha of national grassland include both public and private property, and allow for grazing, mining and hunting, as well as ecotourism. From the latter perspective, these areas are notable for their thriving and easily observable ungulate populations, which include antelope, mule deer and elk (USDA Forest Service, 1999).

As with deserts, the prevalence of well-developed, grasslands-based protected area systems, and associated ecotourism activity, diminishes in the less developed countries, including the former Soviet Union. One exception is southern Africa, where ecotourism is relatively well developed in the public and private protected grassland areas of South Africa, Botswana and Namibia, and especially in associated wetland areas such as the Etosha Pan (Namibia) and the Okavango Swamp (Botswana). Among other less developed regions, the potential of Mongolia is notable. This country contains some of the world's most extensive remaining natural grasslands, which support among other wildlife over 2 million Mongolian gazelle, one of the largest herds of migrating mammals left in the world. Also of interest in terms of incipient and potential ecotourism is the last wild population of Przewalski's horse, located in the 89,000 ha Khustain Nuruu Nature Reserve. To exploit this potential, the Institute of Mongolian Biodiversity and Ecological Studies has been established with the support of the Mongolian parliament to encourage ecotourism in this and other Mongolian protected areas.

Threats to grasslands ecotourism

Because most grassland areas are highly suited for cropping and/or grazing, protected areas' managers will find it difficult to avoid strategies that emphasize a multiple use approach to the management of relevant protected areas. However, concurrent activities such as grazing, mineral extraction and hunting could prove to be incompatible with ecotourism under some circumstances. Related threats include the encroachment of cropping on lands usually used for grazing, and the encroachment of deserts (desertification) in marginal areas as a result of inappropriate farming or grazing practices. The less developed countries in particular are likely to face these problems. For tallgrass and some medium grass prairie environments, a major threat in all parts of the world will be posed by the 'island effect', wherein small remnants of habitat are too small to foster a viable grassland ecotourism sector, and are particularly vulnerable to invasion by exotic flora and fauna. More broadly, a smaller proportion of the temperate grassland (i.e. 0.98%) is protected than any other biome, reflecting the extent to which such areas have been extensively modified (Green and Paine, 1997).

Savannahs

'Savannah' is a generic term that describes tropical or semi-tropical grasslands that are interrupted by a discontinuous cover of trees. Where precipitation is relatively low, savannahs tend to be shortgrass, and trees few and far between. At the other extreme, tallgrass savannahs with extensive tree cover occur in humid regions adjacent to the true forests. Major areas of savannah include the *llanos* of Venezuela, the *campos* of Brazil, much of India and parts of northern Australia. However, by far the most extensive and best known savannahs occur in sub-Saharan Africa. With its large and readily observable populations of attractive large mammals (e.g. lion, leopard, elephant, giraffe, cheetah, rhinoceros), these savannahs offer an ideal potential ecotourism venue. Other savannahs are impoverished by comparison. The ability to attract ecotourists in all savannahs, however, is influenced by the pronounced seasonality of precipitation, which reaches its extreme in monsoonal India. As an ecosystem, the African savannahs have been more modified by human activity than those of Venezuela or Australia, but less than the widespread modification that has occurred in India.

Savannah ecotourism

The African savannahs, without doubt, are the most developed in terms of existing ecotourism activity. However, the distribution within this region is uneven. Savannahs to the north of the Equator are not currently being utilized for ecotourism to any great extent. Nigeria's Yankari Game Reserve, with about 30,000 visitors a year in the late 1980s, is considered to be one of the major ecotourism sites in the western African savannahs (Olokesusi, 1990). Among the factors that account for this paucity are the lack of accessible and serviced protected areas with a high profile, overall tourism sectors that are poorly developed, and a relative paucity of high demand wildlife compared with eastern and southern Africa (Weaver, 1998).

But the pattern of uneven distribution is apparent even in the latter regions. Tourism in general is almost non-existent in Angola and Mozambique due to their 30-year legacies of liberation struggle and civil war. Most activity, in contrast, is concentrated in a 'safari corridor' that extends from the northern part of South Africa to southern Kenya (Fig. 16.2). Within this corridor, the greatest amounts of ecotourism activity occur in those southern and northern extremities, that is, South Africa, Botswana and Zimbabwe in the south, and Kenya/northern Tanzania in the north. Smaller amounts of activity are found in southern Tanzania, Zambia and Malawi.

It is within the northern and southern core areas that one finds the savannah-based protected area 'crown jewels' where safari-based ecotourism is well represented. The core southern parks with significant visitation levels include Kruger, Hwange and Chobe National Parks, while

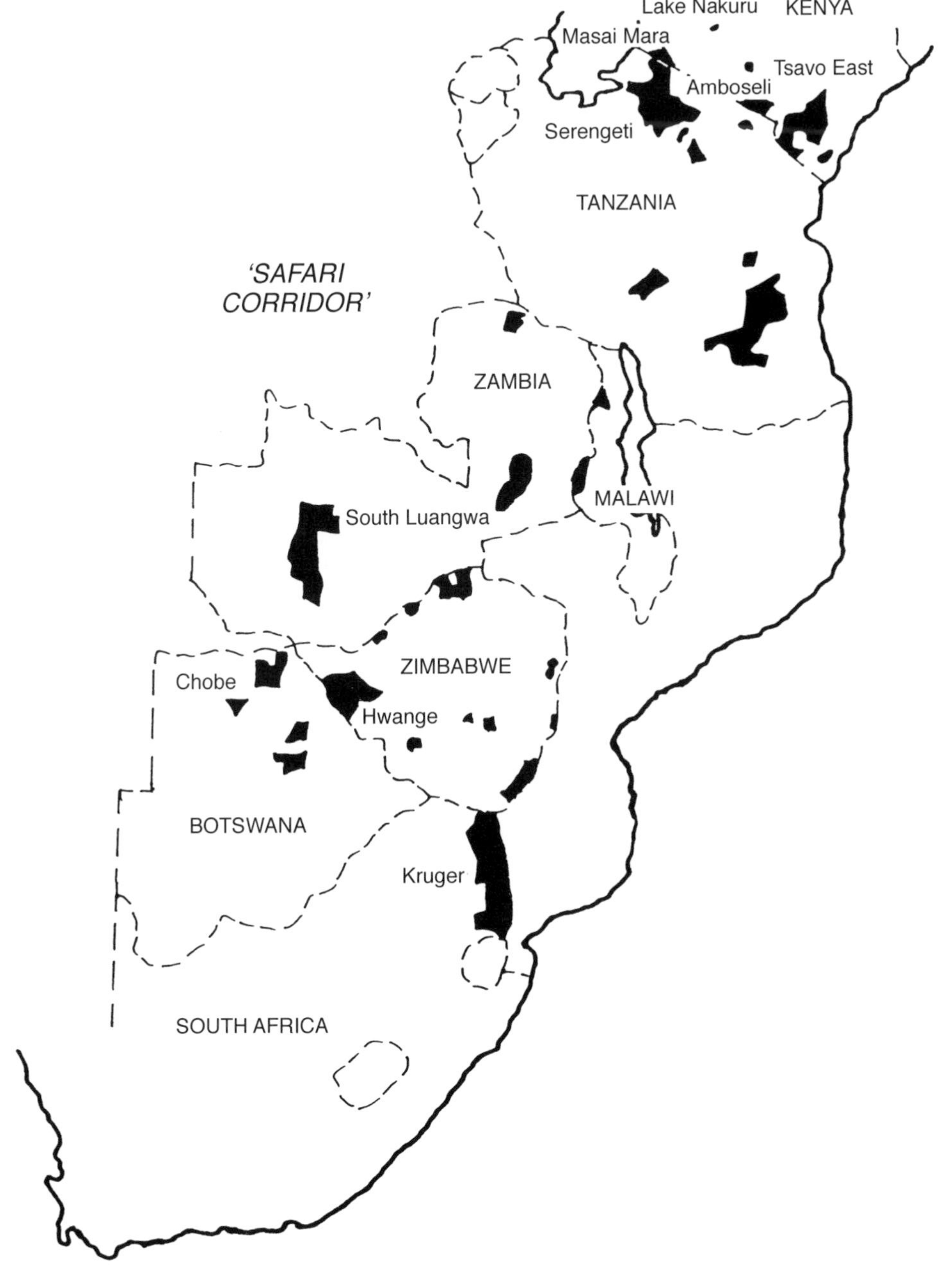

Fig. 16.2. The 'safari corridor' of the East African savannah.

the north is represented by Amboseli, Masai Mara, Tsavo East and West, and Serengeti (Table 16.2). Several factors account for the popularity of these parks as ecotourism destinations.

- Large concentrations of 'charismatic' wildlife such as lion, cheetah, elephant, hippopotamus, zebra and giraffe.
- High visibility factor due to the nature of the savannah landscape, the density

of faunal populations, and the tendency to concentrate predictably at certain locations, such as waterholes; furthermore, many animals have become habituated to being viewed from safari vehicles at a close distance (Roe *et al.*, 1997).
- Strong public profile in major ecotourism markets, due to the savannah image disseminated by magazines such as *National Geographic* and films such as *Born Free*.
- Relatively good protection of the resource base and the implementation of appropriate management practices to cope with tourist arrivals.
- Accessibility of these parks to major urban centres such as Nairobi and Johannesberg, and to good accommodations and other facilities and services.
- Relatively high level of political and social stability in countries offering major savannah-based tourism opportunities.

In short, these parks offer the stereotypical safari experience that visitors expect, and do so with a relatively high level of services.

The importance of these parks to their respective countries is illustrated by the finding that 80% of inbound visitors to Kenya and Zimbabwe are attracted primarily by the wildlife-viewing opportunities provided mainly in these high-profile locations (Risk and Policy Analysts Ltd, 1996). Evidence further suggests that most ecotourism activity in the northern and southern core is of the soft variety. This is apparent from a pattern of overnight accommodation that emphasizes urban or 3S resorts as well as luxury safari lodges closer to the actual parks (Weaver, 1998). Also, most of the wildlife viewing takes place in comfortable safari vehicles that travel along a relatively small number of roadways within the parks. Of over 950,000 visitors to South Africa's Kruger National Park in 1997, only 4654 participated in wilderness trail walks (Ferreira and Harmse, 1999). The high profile public protected areas cited above are supplemented by a growing number of private game reserves that offer some ecotourism opportunities, but concentrate primarily on trophy hunting (which is forbidden in the national parks).

Outside the two cores (i.e. the middle of the corridor), there is significantly less ecotourism activity. For example, Malawi's five national parks cumulatively hosted only 17,136 visitors in 1989 (Burton, 1995), while Zambia's most popular park, South Luangwa, hosted only just over 15,000 tourists in 1995 (Table 16.2). Aside from South Africa, Kenya and Zimbabwe, Tanzania has some of the best prospects for becoming a major ecotourism destination. This is due to the widespread view that Tanzania's wildlife resources are more

Table 16.2. Visitation to selected protected areas in the savannah corridor of Africa.

Protected area	Country	Year	Visitors
Masai Mara	Kenya	1995	133,000[a]
Tsavo East	Kenya	1995	229,000[a]
Amboseli	Kenya	1995	115,000[a]
Lake Nakuru	Kenya	1995	167,000[a]
Kruger	South Africa	1997	954,398[b]
Serengeti	Tanzania	1998	20,000[c]
South Luangwa	Zambia	1995	> 15,000[d]
Hwange	Zimbabwe	1993	98,000[e]

[a] Kenya, 1996.
[b] Ferreira and Harmse, 1999.
[c] Honey, 1999.
[d] Roe *et al.*, 1997.
[e] Potts *et al.*, 1996.

extensive and more pristine than those of Kenya. Furthermore, Tanzania not only attracts a growing number of European and North American ecotourists from neighbouring Kenya, but is enhancing its status as a destination in its own right through improvements in infrastructure and services (Anon, 1998).

The incidence of ecotourism within savannah regions outside Africa is more limited. A notable exception, however, is Australia's 20,000 km^2 Kakadu National Park, where a savannah and wetland-dominated landscape attracted about 230,000 visitors annually during the early 1990s, the vast majority of whom could be described as ecotourists (Australia, 1997). The high levels of management and services available at Kakadu attest to Australia's status as the only developed country with a significant area occupied by savannahs.

Threats to savannah ecotourism

Persistently high levels of population growth in countries dominated by savannah is a long-term threat to the viability of the ecosystem, and hence to ecotourism itself. These pressures are especially acute in sub-Saharan Africa, and may translate not only into the conversion of unprotected savannahs into cropland and pasture, but also to encroachments (both legal and illegal) on protected areas. Accordingly, it is essential that the revenue derived from savannah-based ecotourism (and other forms of tourism) is substantial enough to constitute a viable alternative to more destructive land uses, such as cropping. The high visitor numbers recorded in many of the higher profile parks suggests that such levels are being attained at some sites. However, this in turn has resulted in various tourism-related internal threats to the natural environment. As with virtually any high profile protected area, visitation in the African savannah park system tends to be concentrated within just a few individual parks, and just in certain locations within those parks. In locations that have not been appropriately site-hardened or otherwise managed, this has led to negative environmental impacts and a concomitant deterioration of the ecotourism experience. Well-chronicled problems in places such as Kruger and Amboseli include landscape degradation caused by off-road vehicular travel, the congregation of wildlife at garbage tips, traffic congestion, and the 'mobbing' of cheetahs, lions and other predators by excessive and invasive numbers of safari vehicles (Weaver, 1998; Ferreira and Harmse, 1999). In some cases, the stresses arise due to improper management practices that allow wildlife populations to exceed normal carrying capacities. For example, Zimbabwe's Hwange National Park cannot cope with its growing population of 30,000 elephants, which are wreaking havoc with the ecosystem in many areas (Potts *et al.*, 1996). Fortunately, savannah flora and fauna appear to have considerable resilience in the face of such problems; research in Masai Mara and Amboseli showed that off-trail damage by vehicles was quickly reversed once the practice was stopped, especially in the wet season (Onyeanusi, 1986).

Discussion

While this chapter attempts to identify general patterns that apply to desert, grassland and savannah ecotourism, one must never lose sight of the fundamental diversity that characterizes each ecosystem. Hence, an impact or strategy that applies to one particular desert or savannah area might not be applicable to another. Nevertheless, there are various patterns and management issues that should be assessed as being more likely to pertain to these ecosystems. It is worth reiterating that deserts, grasslands and savannahs all repeat the global pattern whereby ecotourism, in its 'soft' manifestation, tends to concentrate in a limited space within just a few high profile and more accessible protected areas. All three ecosystems also possess a high visibility factor that increases the likelihood of successful wildlife viewing, yet can also negatively affect visitor satisfaction by increasing the probability of viewing other tourists as well. The opportunity for both soft and hard ecotourism beyond these protected spaces is influenced by the condition of each ecosystem. Deserts tend to be relatively unaltered, whereas grasslands (and tallgrass in particular) have experienced the greatest amount of degradation and conversion; savannahs as a whole are in an intermediate position. Another influence is the extent to which a market exists for ecotourism in the protected areas and beyond. As the domestic market for ecotourism in less developed countries is generally incipient, it is apparent that domestic ecotourists are significant as a market only in North America (deserts, grasslands), Australia (deserts, grasslands, savannahs), the western grasslands of Europe (e.g. Hungary) and, to a more limited extent, in South Africa and Namibia.

With respect to management issues, the question of site-hardening and other strategies to deal with visitor concentrations within protected areas is as important to deserts, grasslands and savannahs as it is to all other ecosystems. A related issue is access to water, given that deserts and grasslands are areas of inherent moisture deficit, and savannahs are usually subject to seasonal deficiencies. Similarly, managers of ecotourism in grasslands and savannahs must cope with the reality of fire as a normal part of ecosystem dynamics. Pertinent questions include the extent to which these can and should be initiated and controlled by managers (as opposed to being left to the devices of nature), and to what extent these will interfere with or enhance the ecotourism experience in both the short and long term. The migratory behaviour of many desert, grassland and savannah larger mammals is also an important issue, since ranges often extend beyond protected area boundaries. This can lead to a re-assessment of wildlife as pests in those unprotected areas, or as objects for consumptive tourism (i.e. hunting). A sort of informal stratification appears to be emerging in the safari belt where ecotourism is promoted in the higher-order protected areas, while hunting is given priority in lower-order and private protected areas, as well as in communal lands. Zimbabwe's well-known and controversial CAMPFIRE programme illustrates the continuing importance that is attached to big game trophy tourism in the region (Butler, 1995). The issue is not whether hunting is sustainable or not (this depending on how it is regulated), but instead whether a peaceful coexistence between the two sectors can be achieved, or whether ecotourism interests will 'concede the field' to consumptive tourism outside just a few of the crown jewels.

References

Anon (1998) *International Tourism Reports* 1, 67–82.

Brown, B. (1999) Desert Park defies belief. *The Australian* 7 July, p. 6.

Burton, R. (1995) *Travel Geography*, 2nd edn. Pitman, London.

Butler, V. (1995) Is this the way to save Africa's wildlife? *International Wildlife* 25(2), 34–39.

Central Bureau of Statistics (1996) *Economic Survey 1996*. Central Bureau of Statistics, Nairobi.

Cockerell, N. (1996) Egypt. *International Tourism Reports* 2, 5–20.

Commonwealth of Australia (1997) *A Plan of Management in Respect of Kakadu National Park*. Commonwealth of Australia, Canberra.

Daniel, J. (1993) *The Buzzworm Magazine Guide to Ecotravel*. Buzzworm Books, Boulder, Colorado.

Ferreira, S. and Harmse, A. (1999) The social carrying capacity of Kruger National Park, South Africa: policy and practice. *Tourism Geographies* 1, 325–342.

Green, M.J. and Paine, J. (1997) State of the world's protected areas at the end of the twentieth century. Paper presented at the IUCN World Commission on Protected Areas Symposium, Albany, Australia, 24–29 November.

Hitchcock, R. (1997) Cultural, economic, and environmental impacts of tourism among Kalahari

Bushmen. In: Chambers, E. (ed.) *Tourism and Culture: an Applied Perspective*. State University of New York Press, Albany, pp. 93–128.

Honey, M. (1999) *Ecotourism and Sustainable Development: Who Owns Paradise?* Island Press, Washington, DC.

Huber, T., Larkin, R. and Peters, G. (1988) *Dictionary of Concepts in Physical Geography*. Greenwood Press, New York.

Illinois Nature Preserves Commission (1999) Midewin National Tallgrass Prairie. http://www.fs.fed.us/mntp/

National Park Service (1999) Recreation visits by State for 1998. http://www.aqd.nps.gov/stats/bystate

Neldner, V., Fensham, R., Clarkson, J. and Stanton, J. (1997) The natural grasslands of Cape York Peninsula Australia. Description, distribution and conservation status. *Biological Conservation* 81, 121–136.

Olokesusi, F. (1990) Assessment of the Yankari Game Reserve, Nigeria. *Tourism Management* 11, 153–162.

Onyeanusi, A. (1986) Measurements of impact of off-road driving on grasslands in Masai Mara National Reserve, Kenya: a simulation approach. *Environmental Conservation* 13, 325–329.

Ouessar, M. and Belhedi, H. (1998) The role of NGOs in promoting drylands ecotourism: an ongoing project in southern Tunisia. *Aridlands Newsletter* 43 (Spring/summer). http://ag.arizona.edu/OALS/ALN/aln43/tunisia

Potts, F., Goodwin, H. and Walpole, M. (1996) People, wildlife and tourism in and around Hwange National Park, Zimbabwe. In: Price, M. (ed.) *People and Tourism in Fragile Environments*. Wiley, Chichester, UK, pp. 199–219.

Risk and Policy Analysts Ltd (1996) *The Conservation and Development Benefits of the Wildlife Trade*. Department of Environment, London.

Robertson, K. (1999) The Tallgrass Prairie in Illinois. http://www.inhs.uiuc.edu/~kenr/tallgrass.html

Roe, D., Leader-Williams, N. and Dalal-Clayton, B. (1997) *Take Only Photographs, Leave Only Footprints: The Environmental Impacts of Wildlife Tourism*. HED Wildlife and Development Series No. 10. International Institute for Environment and Development, London.

Sievers, E. (1998) The potential for ecotourism in Uzbekistan. *Aridlands Newsletter* 43 (Spring/summer). http://ag.arizona.edu/OALS/ALN/aln43/uzbekistan

Sofield, T. and Getz, D. (1997) Rural tourism in Australia: the Undara Experience. In: Page, S. and Getz, D. (eds) *The Business of Rural Tourism: International Perspectives*. ITP Press, London, pp. 143–161.

USDA Forest Service (1999) Dakota Prairie Grassland. http://www.fs.fed.us/r1/dakotaprairie/index.html

Weaver, D. (1997) A regional framework for planning ecotourism in Saskatchewan. *Canadian Geographer* 41, 281–293.

Weaver, D. (1998) *Ecotourism in the Less Developed World*. CAB International, Wallingford, UK.

Weaver, D. and Elliott, K. (1996) Spatial patterns and problems in contemporary Namibian tourism. *Geographical Journal* 162, 205–217.

World Resources Institute (1998) *1998–99 World Resources: a Guide to the Global Environment*. World Resources Institute, New York.

Chapter 17

Marine Environments

C. Cater[1] and E. Cater[2]

[1]*School of Geographical Sciences, University of Bristol, Bristol, UK;*
[2]*Department of Geography, University of Reading, Reading, UK*

Introduction

Unity in diversity

The IUCN's definition of the marine environment as 'Any area of intertidal or subtidal terrain, together with its overlying water and associated flora, fauna, historical and cultural features' (IUCN, 1991) reflects the enormous complexity and diversity present in this realm. This is hardly surprising since the area of sea and seabed is over two-and-a-half times as great as the total area of land masses of the world (IUCN, 1991). Marine ecosystems vary from coral reefs (the most species diverse of all marine habitats, approaching tropical rainforests in their species richness) to coastal mangrove wetlands; species range from sperm whales to sea horses; and marine tourism embraces a multiplicity of activities from whale-watching to scuba-diving. The purpose of this chapter is to highlight the distinctive features of marine ecotourism, while setting it into an overall context. It examines various forms of marine ecotourism for their key attributes of sustainability across various dimensions, and also discusses the role of different types of marine protected areas (MPAs). While over-generalization is counterproductive it is evident that there are certain recurring themes from across the globe.

The context of marine tourism

If outer space is the 'final frontier', it may be argued that the marine environment is the penultimate, with advanced technology enabling an increasing number of marine tourists to literally reach new depths. There is, however, an inherent danger of merely regarding marine ecotourism as one of the latest developments in the never-ending search for new touristic experiences, thus isolating it from other forms of economic activity. The context in which it is set as a process and as a principle is all important, because that context has a vital role to play in prospects for sustainable outcomes.

To borrow the terminology of the Swedish economist, Gunnar Myrdal, there are marked spread (or positive) and backwash (or negative) effects between the various sectors, levels and interests (Fig. 17.1). Indeed the relationships are not entirely unrelated to his overall thesis as many of these interdependencies are bound up with centre–periphery relationships; in this case

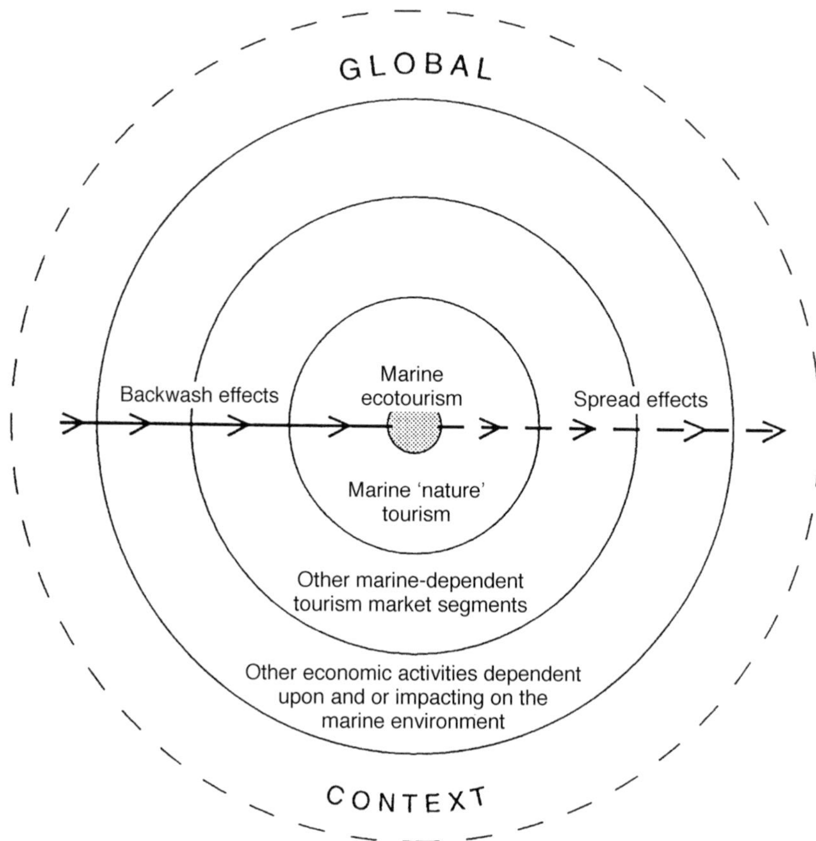

Fig. 17.1. Spread and backwash effects in marine ecotourism.

a result of patterns of dominance implicit in the international organization of the tourism industry. The spread effects from marine ecotourism include raising environmental awareness and disseminating an understanding of the coincidence of good environmental practice with advantages to business. The backwash effects hinge around the fact that other, often competing, activities, are frequently prejudicial to the success, if not the existence, of marine ecotourism. It is also vital to consider the overall, global context. The implications of global climatic change for coral bleaching and consequent destruction of a significant marine ecotourism resource is a case in point.

There are various levels to consider. First, it is imperative that marine ecotourism is viewed in the context of marine tourism as a whole. Any one marine location is likely to host a variety of frequently incompatible recreational pursuits. There is, for example, conflict between scuba-diving and high speed watercraft. Even marine nature-based tourism may compromise genuine marine ecotourism. Conscientious operators, such as SeaCanoe, may find their efforts constantly thwarted by the unsustainable activities of other 'nature' tour operators whose businesses may be ecologically based, but far from ecologically sound. SeaCanoe began its kayaking operations in the tidal sea caves of Phang Nga

Bay, Thailand, in 1989. The company has won a number of awards for its low environmental impact/high local benefits cave kayaking experiences in Southeast Asia. In Thailand, however, the very success of SeaCanoe in an unregulated scenario inevitably spawned unscrupulous imitators. Cave visitation has been taken to four figures a day, with dozens of kayaks waiting in line to beat the tide. Illegal extortion of tourist entry fees to the caves also occurs. Inevitably the caves have become degraded by these high volume, environmentally unaware entries (Gray, 1998a, b).

The second contextual level is that of marine ecotourism with respect to other tourism market segments which are dependent on, and consequently impact on, the marine environment. The development and operations of coastal resorts, for example, have manifest implications for the success or otherwise of marine ecotourism, as indeed does the burgeoning growth of the cruise industry. As Milne (1998, p. 47) suggests:

> in attempting to achieve more appropriate forms of tourism, it is also essential that we steer away from creating a dichotomy between 'alternative' and 'mass' tourism. Such a division serves little real purpose and diverts our attention away from the interlinked nature of all types of tourism development.

Third, with regard to the overall picture of sustainability, it is vital to consider the interactions that occur with all other forms of economic activity. As Butler (1998, p. 34) asserts, 'tourism is part of the global system and cannot be tackled in isolation, spatially, economically or temporally'. It is vital that a move is made beyond a tourism-centric view, as it is 'inappropriate to discuss sustainable tourism any more than one might discuss any other single activity ... we cannot hope to achieve sustainability in one sector alone, when each is linked to and dependent upon the others' (Butler, 1998, p. 28). The interplay with forest management, for example, is clear when it is considered that destructive logging practices, which result in extensive runoff from the land and consequent siltation of coastal waters, have serious repercussions for marine life and, in turn, for marine ecotourism.

There is also the vexed question of ethnocentricity when the contextual aspect of marine ecotourism is closely examined; not only do the needs of visiting tourists and host populations frequently diverge, but also foreign and domestic tourists often have markedly different, incompatible agenda. At Moalboal and Pescador Island, Philippines, for example, local divers practise spearfishing, which obviously compromises both the safety and sensitivity of those who have come to appreciate the underwater life. It is vital, therefore, that all these areas of difference and potential discord are identified at an early stage in the planning for more sustainable outcomes.

The distinctive features of marine tourism

There are several distinctive features of marine tourism which have a bearing on prospects for sustainability. First is the open nature of the marine environment, which brings with it considerable problems of management. The high degree of connectivity in the seas facilitates the transmission of substances and effects (IUCN, 1991). Sea currents carry sediments, nutrients, pollutants and organisms through and beyond a specific location. Consequently, actions taken in one locality, by whatever form of activity, tourism or otherwise, marine or terrestrial, may affect another hundreds of kilometres distant and often nations apart. This openness also means that, despite designations of marine protected areas, or indeed of territorial waters (while territorial limits are defined as 12 km out, the United Nations (UN) Convention on the Law of the Sea gives coastal states jurisdiction over all resources, including living resources, in an exclusive economic zone that can extend up to 200 nautical miles from their coasts (UN, 2000)) there are few physical barriers to accessibility. This renders coastal waters, particularly in more remote locations, notoriously difficult to police. The assumption that oceans and seas are common property resources further compounds the problem. They are considered to be owned in common by everybody, not subject to individual or private ownership, and can therefore normally be used without payment. This is the 'tragedy of the commons'. Under such circumstances there is a positive incentive for individual users to exploit the resource to the maximum, even if destruction of marine resources is the inevitable result.

Second, marine tourism takes place in an environment in which humans do not live, and consequently in which they are dependent on equipment to survive (Orams, 1999). While this dependence may engender a sense of humility and respect for the unfamiliar, it may, equally, result in serious physical damage from careless handling or inappropriate use of technical support and facilities.

Third, increasing interest in the marine environment has meant that the growth rate of marine tourism exceeds that of most of the rest of the tourism industry. Whale-watching, for example, has displayed average annual growth rates of around 10% during the 1990s (compared with an average annual 4.3% increase in world tourist arrivals worldwide). Dive tourism to Zanzibar more than doubled between 1990 and 1995, with a concomitant increase of dive operators from one to 11 over the period (Cater, 1995).

A key elements approach

Weaver (Chapter 5) sets ecotourism into the context of other tourism types, describing the problems involved in defining ecotourism on the basis of discrete categories when the reality is one of a continuum with considerable overlap and blurring of boundaries. In particular, there is the question of scale of operation. While purist, ecocentric analysts would confine the definition of marine ecotourism to small-scale, low volume, visitation, it has become increasingly recognized that a spectrum of participation and involvement can be discerned (see Chapter 2). This spectrum ranges from hard-core, specialist groups undertaking scientific observation and research, to more casual marine-based activities such as snorkelling, glass-bottom boat observation or whale-watching. It is not only irrational to deny the designation of ecotourism to large-scale marine tourism if it adheres to all the requirements of sustainable tourism, but also, as Weaver suggests, it neither makes economic sense nor acknowledges the potent lobbying force constituted by increased participation. It is not necessarily a case of 'big is bad; small is beautiful'. Nor is it a foregone conclusion that hard marine ecotourism is any more sustainable than soft marine ecotourism. It makes more sense, therefore, to examine each case on its own merits rather than to attempt a heavy-handed allocation to broad categories of any crude, all-embracing definition. The key elements approach of Bottrill and Pearce (1995), therefore, has much to commend it as a means of identifying and classifying ecotourism ventures in the field. While the specifics of these key elements are likely to be open to debate, such assessment allows a flexible approach which can take into account scale and local circumstance. Of course the questions of what constitutes a reasonable standard and whether all, most or just some of the elements are essential qualifications for ecotourism, are debatable issues (Bottrill and Pearce, 1995). There is also the question of 'value added'. Ought there to be some way of measuring progress towards more sustainable outcomes?

The specific examples of various marine ecotourism activities described in this chapter are examined for their key attributes of sustainability across the dimensions of environmental responsibility; socio-cultural integrity; and economic viability (with the specific requirement to benefit local livelihoods). An essential part of this examination is also to draw attention to particular constraints and problems that occur.

Scuba-diving

Snorkellers and divers as ecotourists?

Snorkelling and scuba-diving are, without doubt, the most popular forms of underwater visitation. Shackley (1998) suggests that snorkelling is a far less intrusive activity than scuba-diving, but this is a dubious claim for several reasons. The number of snorkellers at any one location at any point in time is likely to be higher than scuba divers; little or no training is required and snorkellers are rarely given guidelines about responsible behaviour; snorkellers' activities are not normally monitored by either the presence of a co-diver (buddy) or divemaster; and snorkellers are far more likely to come into physical contact with reefs, either through treading water or resting.

Many might argue that the size of the scuba-diving industry, with estimates of about 14 million divers worldwide (Viders, 1997, quoted in Shackley, 1998), means that it cannot be considered as a true form of ecotourism. Indeed, as Weaver shows in Chapter 5, diving is an example of what may be considered 'mass eco-tourism' since many of the participants are actually on a 3S (sea, sand, sun) format holiday. However, if ecotourism is interpreted as a set of principles rather than being confined to small-scale activities, scuba-diving may be considered one of the original ecotourism practices. Schuster (1992) contends that ecotourism is neither a new word nor concept in diving, as 'from the beginning dive travel has been a form of ecotourism since diving involves observing nature'. This contention, however, assumes responsible behaviour, which is by no means automatic.

A growing pastime

There has been significant growth in the number of qualified divers over the last 30 years. The world's largest diving organization, the Professional Association of Diving Instructors (PADI), has issued in excess of 8 million certifications since 1967 (Table 17.1), with 779,967 new certifications (494,750 were the basic Open Water Diver qualification) in 1998 (PADI, 1999). PADI trains 55% of the world's divers, while other major global diving organizations include BSAC, NAUI and SSI. In the UK alone PADI registered 56,000 new basic

Table 17.1. Growth in PADI dive training certifications worldwide (PADI, 1999).

Year	Certifications per year	Cumulative certifications	Average annual % increase
1967	3,226	3,226	
			79.3
1972	51,842	135,904	
			7.1
1977	69,771	443,482	
			15.3
1982	141,429	998,070	
			17.4
1987	315,468	2,203,079	
			13.8
1992	529,463	4,376,821	
			7.3
1997	753,157	7,728,695	
1998	779,967	8,508,662	

qualifications in 1996–1998, an annual average of just over 18,500. However, the total number of newly qualified UK divers each year is considerably in excess of this figure as the UK is firstly BSAC (the British Sub Aqua Club) territory, and many UK divers also increasingly choose to qualify while on holiday abroad. The Recreational Scuba Training Council of Europe (RSTC) estimates that there are 3.2 million active European divers, and that an estimated 825,000 of these travel to their diving destinations abroad while on holiday each year (RSTC, 1997). According to RSTC the expenditure from this travel alone may amount to $US3 billion.

Statistics from PADI show that 80% of newly qualified Open Water divers have a college education, but this is not to say that they may be more ecologically aware. Instead it is more likely to illustrate the fact that diving is an expensive hobby, as suggested by Orams (1999). He contends that marine activities are patronized, relative to land-based activities, by upper socio-economic groups because of the significant cost of such pursuits. A typical open water training course might cost up to US$200, and a day of diving about US$50. Both thus add considerably to the cost of a holiday.

The attractions of scuba-diving

What is behind this significant growth in the numbers of qualified divers around the globe? The reasons are many. On one hand, there are factors that have enabled access to diving to a much broader population. The cumulative effects of global tourism have brought dive operations to an increasing number of destinations, so the correct facilities are more likely to be on hand, particularly the 'safety net' of a nearby compression chamber. Schuster (1992) claims that eco-diving 'could create a new wave of interest for potential divers and keep certified divers involved in the sport'. On the other hand, there are a number of push factors that have led to the increasing popularity of dive tourism. It has been argued that tourists increasingly want more than sights alone, they wish to 'participate with their own skins' (Moeran, 1983). Diving is the only marine activity that offers complete immersion within the environment itself, as the alternatives might only be visual, as in submarines or glass-hull boats, or only partial, as in snorkelling.

In addition, scuba-diving is a prime illustration of a concept that has become increasingly used in tourism literature (Ryan, 1997; Orams, 1999), that of flow, originally suggested by Csikszentimihalyi (1975), to explain the appeal of certain leisure activities. The nature of the flow experience demands the setting of a challenge, the meeting of this challenge and the completion of the task that leads to satisfaction with the experience. During the activity there may be intense concentration on the task at hand, often resulting in the loss of any sense of time, and feelings of competency and satisfaction, both during and after the experience. Diving clearly provides a challenge in so far as training is required to undertake such an activity. Also a typical dive offers a host of challenges in a specific time frame, the completion of which lead to feelings of satisfaction. These include donning all the equipment which also requires checking; following correct procedures for ascent and descent; monitoring of gauges to ensure adherence to correct profiles and times; maintaining an awareness of location underwater relative to the boat; and finding, observing, and possibly photographing, wildlife of interest beneath the water. All this assumes that these challenges can be met by the diver, so there is an important element of matching skills and tasks. If they are out of balance the diver may find a dive either boring or, alternatively, too stressful, with the inherent danger that he/she might not cope. In the latter situation the presence of external factors becomes increasingly important, such as the presence of more-experienced buddies, and the provision of advanced facilities mentioned above.

It is suggested that males find the opportunity for risk and adventure more attractive than females (Orams, 1999). This could be borne out by statistics which show that of PADI divers in 1992, 65% were male (PADI, 1994). However, these figures should be treated with caution, as it is undeniable that they are likely to even out over time: diving holds appeal irrespective of gender. Personal observation confirms that, particularly in the initial training stages, diving is as popular with women as with men.

Diver behaviour

The fact that scuba training is necessary to practise the sport means that, unlike other marine-based activities, it is much easier to educate the participant in terms of sensitive environmental behaviour. One of the most important elements of diver training is that of buoyancy control, meaning that the diver can rest at any point in the dive without either rising or sinking in depth. While ensuring that divers are able to control themselves underwater, it is also stressed, during the training, that this will minimize contact between divers and any sensitive marine life. In addition, all of the large dive agencies have environmental education programmes which are integrated into the dive training programme, for example PADI's Project AWARE (Aquatic World Awareness, Responsibility and Education). Sometimes divers may provide an early warning of ecological crisis, as they are in a unique position to observe the environment at close hand, and are encouraged to report anything unusual to local environmental protection agencies.

Diving procedures mean that divers always dive in pairs as 'buddies'. This is obviously for safety, but it also influences diver behaviour. The advantages when observing a marine environment are clear, as two pairs of eyes working together will find a greater number of interesting things than one pair. In addition, the continual monitoring of another person reduces the likelihood of damage to marine environments, as buddies may be able to warn each other of unintentional harm that an action might cause to that environment. Examples might include hitting the reef with a fin or oxygen tank, as it is difficult for an individual to appreciate how much further both these extend outside the body space. Personal observation also highlights both buddy-to-buddy disapproval and the individual guilt that such an incident provokes within the diving fraternity. In addition, most divers now demand a sensitive environmental operation from the dive companies themselves. With forces from above and below, most successful companies will have sound environmental policies, such as the establishment of shared permanent moorings off a reef by different dive operators. While positive for the environment this also makes business sense as operators can advertise their eco-credentials, often to significant effect.

Management of divers

Despite increased ecological awareness present within the diving community, the careful management of divers is extremely important. Although aware of the fragility of the underwater environment, an underwater holiday maker still has human curiosity and the desire to make the 30–45 min experience as worthwhile as possible. This leads to instances such as that reported at Sharm el Sheik, Egypt, in March 1999, when an estimated 30 divers were chasing one turtle. Clearly the potential stress caused by such an incident must be avoided wherever possible, but the relative invisibility of such an occurrence to all but the participants makes it difficult to police.

The most obvious method for doing so is to limit the numbers of divers at any one site, though this requires the establishment of thresholds. Dixon *et al.* (1993) show how diver thresholds have been set for the marine reserve of Bonaire in the Netherlands Antilles. Results from interviews with divers together with data on coral cover and species diversity suggest that the threshold stress level for any one dive site at Bonaire is between 4000 and 6000 dives per year. Multiplying by the number of individual sites gives an upper limit of maximum theoretical capacity within the park. This would still, however, accommodate unacceptably high visitation levels at the more popular sites, so this upper limit is then halved to give a more realistic threshold. However, each location will have different capacity levels, meaning that these calculations need to be tailored to the individual case, as shown by Hawkins and Roberts (1992). In addition Bonaire is fortunate in that it has been a reserve since the early 1980s, and consequently there is a historical record of the condition of the reefs. In many of the emerging ecotourism destinations of the less developed countries, there is little scientific record of the marine environment, and marine parks are often being set up well after the diving operations have been in place.

Some of the difficulties of this 'catching-up' are highlighted by Shackley (1998) in her discussion of the world famous Stingray City in the Cayman Islands. At present there are no controls over the high visitation levels to this site where divers may hand-feed stingrays, as the area is outside present marine reserves. There is little data, beyond observation at the site, on how the feeding may have influenced the natural behaviour of the stingrays.

One of the more extreme measures taken in setting thresholds for a dive site is that taken at Pulau Sipadan, off the eastern coast of Sabah, Malaysia. In early 1998 the Malaysian Ministry of the Environment and Tourism introduced restrictions to the numbers of visitors, many of whom were divers, allowed to the island. Effective limits were set at a quarter of the previous peak daily number (Cochrane, 1998). Restrictions have been enforced, ostensibly to reduce the impacts that divers were having on this tiny island's population of turtles and a dwindling supply of fresh groundwater. More cynical commentators, however, suggest that this radical action may be more related to a territorial dispute over the island between Malaysia and Indonesia. Irrespective of the exact reason, the plan should have important implications for the local marine environment, although the island now has a further degree of exclusivity, with diver operations raising their prices to over US$1000 for 5 days' diving.

Some marine parks pay for their management through the use of fees, although this is still a relatively untapped source of potential revenue. Evidence suggests that divers are willing to pay extra levies to ensure the continued preservation of the reef ecosystems that they enjoy. The work carried out by Dixon *et al.* (1993) showed that 92% of respondents agreed that the US$10 user fee in Bonaire was reasonable. A willingness-to-pay survey conducted in Zanzibar yielded comparable findings, with 82% of divers prepared to pay US$10 for visitation to an individual marine site (Cater, 1995).

It is important to note that the large majority of dive schools are operated and staffed by Western dive instructors. Frequently this is not a question of ability, but of cost and the difficulty of getting the correct training. A dive operator in Zanzibar, lamenting that he would like to train local staff, stated: 'PADI do not produce a training manual in Swahili' (C. Golfetto, Zanzibar, 1995, personal communication). However, while not overt, in an activity such as diving where personal risks may be higher, trust is likely to be an issue. Western tourists are likely to feel safer with a Western instructor. Although this picture is changing, it is important when considering the local socio-economic impacts of a dive operation in relation to other ecotourism ventures.

There are, undeniably, still far too many cases of degradation of marine environments attributable to over-visitation and insensitive behaviour in dive tourism. A further problem is the fact that the vast majority of diving occurs within only 0.025% of the marine environment, i.e. around coral reefs. However, it is suggested that scuba-diving may be at the forefront of changing attitudes and a more responsible ethos.

Whale-watching

While whale-watching as a commercial activity began in 1955 along the southern Californian coast, there were still only around a dozen countries conducting commercial whale-watching activities by the early 1980s. This form of marine observation accelerated during the 1990s, so that by 1994, 5.4 million tourists went whale-watching worldwide, generating over US$500 million in revenue. It is significant to note that this growth parallels the global decline in commercial whale hunting. It has been estimated that currently 295 communities in 65 countries host whale-watching (Orams, 1999). High profile destinations include Tofino, British Columbia; Hervey's Bay, Queensland; Kaikoura, New Zealand; and Puerto Piramide in Argentinian Patagonia. The last two locations have registered a 15–20-fold increase in visitation over the past 10 years (Table 17.2). Whale-watching has undoubtedly brought an economic turnaround for small coastal settlements, such as Puerto Piramide's 90 residents (Orri, 1995) and the 3000-strong town of Kaikoura. However, it is undeniable that such a rate of growth has brought with it considerable problems of management, and there are reasons for concern in many areas. Duffus and Dearden (1993) describe the scientific uncertainty and institutional inertia surrounding killer whale viewing on the north-east coast of Vancouver Island which mean that even this high-profile marine mammal is inadequately protected.

Table 17.2. The recent growth in whale-watching in Argentina and New Zealand (Orri, 1995; Vinas, 1999, personal communication; Whale Watch, 1999).

	Total whale-watching visitors	
Year	Puerto Piramide	Kaikoura[a]
1987	5,214	
1988	10,519	
1989	12,336	3,500[b]
1990	16,524	NA
1991	17,446	[c]
1992	29,121	25,000
1993	33,772	NA
1994	44,829	NA
1995	NA	NA
1996	NA	40,000
1997	72,000	50,000
1998	79,481	60,000

[a] Approximate figures only.
[b] Kaikoura Tours.
[c] Whale Watch established.

In the case of Kaikoura the situation is being closely monitored. The town was badly affected by recession during the 1970s, and post-1984 restructuring resulted in the loss of 170 jobs in the town (McAloon *et al.*, 1998). Commercial whale-watching began as a result of a partnership between an American researcher and a local fisherman; they established Naturewatch in 1988. The venture offered a range of whale-watching products from 2 h trips to 3–10 day packages. In 1989 local Maori began trading as Kaikoura Tours. While the two operators worked well together, Naturewatch sold out to Kaikoura Tours in 1991, and the award winning Whale Watch was born, which to this day holds the monopoly of sea-borne whale viewing in the area (Horn *et al.*, 1998). The operation has evolved from an initial small-scale operation to large scale, carrying 60,000 passengers in 1998. This scale of operation has brought undoubted economic benefits for Kaikoura. A recent survey found that a quarter of respondents worked either full or part time in tourism, and that 80.6% of respondents felt the 'community as a whole' benefits from tourism (Horn *et al.*, 1998). Furthermore, through a range of tourist developments in Kaikoura, including Whale Watch, local Maori moved from a position of relative powerlessness, and low socio-economic status, to become a major employer and economic force in the community (Horn *et al.*, 1998). It has been estimated that 70% of Maori in Kaikoura have been involved in tourism (Simmons and Fairweather, 1998).

The level of visitation, however, inevitably raises the question of environmental change, but whale-watching at

Kaikoura is regulated and closely monitored by the New Zealand Department of Conservation (DoC). They use the precautionary principle of not issuing any further whale-watching permits at Kaikoura, and Whale Watch are also not allowed to increase the number of trips that they operate per day. Four other operations, however, offer scenic flights to view whales and dolphins along the Kaikoura coast. A strong regulatory framework is in place as all marine mammals around New Zealand are fully protected under the Marine Mammals Protection Act 1978, amended in 1990 to introduce regulations specifically for the control and management of marine mammal watching. These regulations were reviewed in 1992 when the Royal New Zealand Navy provided technical advice on the impact of noise on whales and dolphins. As a result a minimum set of conditions were established. Boats are required to approach a whale from a direction parallel to, and slightly to the rear of, the whale. No more than three (including airborne) vessels are allowed within 300 m of a whale at any one time and sea vessels are required to travel at a 'no wake' speed inside this distance. A minimum approach distance of 50 m has also been set and vessels are required to keep out of the path of any whale (Baxter and Donoghue, 1995).

However, whale-watching at Kaikoura is not wholly without problems. Residents recognize the negative impacts that tourism brings, the most commonly cited being pressure on existing infrastructure including water, sewage disposal and car parking space. The monopolistic nature of Whale Watch operations has been criticized as unfair. Maori use their position as Maori to defend their monopoly, which unfortunately adds a political and racial, focus to this strategy, whereby any criticism of this position is construed as racist (Horn *et al.*, 1998). The extent to which tourism can remain under local control as it grows has also been brought into question. Some observers feel that outside investment is inevitable, but this, in turn, implies outside control. It is essential to maintain local ownership and management of key facilities, and to retain local control in decision making (Horn *et al.*, 1998; Simmons and Fairweather, 1998). In terms of impact on the whales themselves, the cumulative impacts of this burgeoning activity have, perhaps, yet to be realized. DoC recognize that many questions remain unanswered about the long-term effects of marine mammal watching. Driven solely by conservational objectives, and not required to balance commercial development against the protection of marine mammals, the department is likely to continue to err on the side of caution. It is not difficult to perceive a state of economic vulnerability on behalf of the resident population.

Sea Kayaking

Within the past decade there has been a surge of interest in sea kayaking. Wylie and Rice (1998) document how the number of companies listed in the *Specialty Travel Index* jumped from 25 (with 35 different venues) in 1991 to 61 (with 112 venues) in 1996. While North and Central America dominate the scene, with 72 different destinations listed in 1996, there is an increasing number of operations in Europe, Oceania and Asia.

Sea kayaking is potentially the most environmentally benign of all marine tourism as, providing waste is taken back, it is non-polluting, 'a canoe across water leaves no trace' (SeaCanoe, 1999). It also is less intrusive to wildlife: birds and animals tend to be curious rather than frightened (N. Johnson, North Uist, 1999, personal communication). As the infrastructural demands are low it also offers the considerable potential for increased local input and consequent benefits.

SeaCanoe is a pioneering example of a sustainable combination of adventure tourism and ecotourism (Fennell, 1999). It has now extended its operations from southern Thailand to northern Vietnam, the Philippines, Lao PDR and the South Pacific. Local people are selected to staff, and eventually own, local operations (SeaCanoe Thailand is now majority owned by local people and employs over 50 staff). It is estimated that 90% of SeaCanoe's budgets stay in the host communities. Their human resources programme provides full benefits to all employees, including training and education (SeaCanoe, 1999).

Underwater Observation

Technological change has facilitated relatively passive means of viewing the diversity of marine life below the surface. The Milford Sound Underwater Observatory in Harrison Cove in Milford Sound, New Zealand (Fig. 17.2) was opened in December 1995. The north side of Milford Sound, where the observatory is located, was gazetted as a Marine Reserve, with World Heritage status, in 1993. The observatory consists of a cylindrical, 450 t, viewing chamber that is completely submerged beneath a main reception area. Comprehensive environmental impact assessments were conducted between 1987 and 1995 before permission from the various authorities was granted to the facility. The whole ethos behind the observatory is one of educating the visitor about the complex ecology of the fjord environment, making the underwater experience accessible to all, not just divers. An interpretation centre in the reception area is complemented by clear species keys above each viewing window, and visitors receive a talk from a marine scientist. As the observatory is in a Marine Reserve it complies with the strict environmental regulations laid down in that designation. In the first 3 years of its operation the observatory received between 41,000 and 55,500 visitors per year (Hamilton, 1999, Milford Sound, personal communication). Owned by a group of South Island business people, and managed by Milford Sound Red Boats, the observatory is accessible only by boat.

Underwater viewing of marine life is also possible from glass-bottomed boats or from larger vessels with specially constructed

Fig. 17.2. The Milford Sound Underwater Observatory, Harrison Cove.

underwater viewing galleries. The Kyle of Lochalsh based *Seaprobe Atlantis*, the UK's only such craft, began operations in July 1998 (Fig. 17.3). It accommodates as many as 24 passengers at a time on a variety of excursions ranging from short, 35 min trips, to see seals and kelp forests, to extended tours at certain times of the year to view dolphins. Two thousand passengers were carried in the few summer months of operation in 1998, but before the 1999 season, the craft was chartered for a special exercise in community education by the Loch Maddy Marine Special Area of Conservation (SAC), North Uist, Scotland (M. Smith, Kyle of Lochalsh, 1999, personal communication). The management scheme for the SAC is being developed by the local community and government agencies, and special legislation gives locals the opportunity to influence how the status can benefit them in terms of opportunities to develop business ventures such as ecotourism. As part of this programme of involvement, 281 local residents were taken on half-hour trips in March 1999 to view the underwater ecology of this sea loch (A. Rodger, North Uist, 1999, personal communication). However, the degree of local interest is still disappointingly low, illustrated by a lack of participation of local schools in the Kyle area (N. Smith, Kyle of Lochalsh, 1999, personal communication).

Underwater observation from semi-submersibles, such as Le Nessee in Mauritius, or from tourist submarines, is also rapidly growing. In 1985, the first tourist submarine was lauched, in Grand Cayman. By 1997 there were 45 underwater vehicles in operation worldwide, carrying over 2 million passengers and taking US$150 million in revenue. As the average price of a dive is between US$65 and 85, underwater sorties are accessible to an increasing number of tourists (Newbery, 1997). With over 15 submarines, Atlantis Submarines is the world's biggest tourist submarine operator, employing over 500 people and catering to almost a million tourists each year in Aruba, Grand Cayman Island, the British and US Virgin Islands, the Bahamas, St Thomas, Cancun, Guam, Barbados and Hawaii (Orams, 1999). There are justifiable claims of environmental integrity as the tourist submarines are entirely non-polluting, with battery-powered electric thrusters that emit no effluent. They also operate at

Fig. 17.3. Kyle of Lochalsh-based *Seaprobe Atlantis*.

low speeds with high manoeuvrability so they never come into contact with coral reefs or marine life. Indeed, many are approved to operate in marine parks and reserves. They also arguably promote environmental stewardship; observing and appreciating marine life in its natural setting will motivate an increasing number of people to protect the marine environment (Newbery, 1997). However, the practice of underwater feeding to attract fish ('chumming') by scuba divers swimming alongside the Atlantis tourist submarines, undoubtedly affects the marine ecology. Also the very considerable capital costs of entry (a minimum of US$4.5 million for a tourist sub), coupled with stringent maintenance and safety requirements, put this form of entrepreneurship way beyond the realms of truly local involvement.

Marine Protected Areas

A symbiotic relationship

It is implicit, from the presentation of the results of their British Columbian test cases, that Bottrill and Pearce (1995) follow writers such as Kutay (cited in Ziffer, 1989) in confining ecotourism to legally protected areas. Indeed, such areas do constitute a popular if not exclusive venue for ecotourism in most parts of the world (see Chapters 18 and 19). However, it is felt that this is too restrictive a criterion for marine ecotourism, particularly as less than 1% of the world's marine area is currently within established protected areas (IUCN, 1991). Marine protected areas (MPAs) are a relatively new concept; most sites were established within the past two decades (World Resources Institute, 1997). However, it is undeniable that a symbiotic relationship can more readily occur in such areas, with environmental protection resulting both *from* and *in* enhanced local livelihoods, sustained visitor attraction, continued profits for the industry and revenue for conservation (Cater, 1997).

Colwell (1998) examines the role of small-scale MPAs as an essential complement to the ambitious proposal by the IUCN, the World Bank and others to create a worldwide network of primarily large-scale MPAs to ultimately protect 10% of all marine and coastal areas. He suggests that small-scale MPAs may be particularly appropriate in coral reef areas, where nearby reefs can be managed not only by local communities and non-government organizations (NGOs) but also by tourism entrepreneurs who have a vested interest in promoting abundant marine life, such as dive resorts.

Community-based MPAs

The small island of Balicasag in the Philippines was the target of a community-based marine resource management project in the mid-1980s which assisted the island community to establish a municipal marine park through the local government. The community endorsed a sanctuary area of 8 ha, and the entire coral reef was included within a marine reserve stretching to 0.5 km offshore. Guidelines were adopted by the community to manage the sanctuary and reserve areas, while the Philippine Tourism Authority initiated its first 'backyard tourism' pilot project. This includes a small-scale beach hotel for scuba-divers. Villagers are employed in the resort and involved in running it; the profits are directed at the maintenance of the marine park and divers are charged extra to dive in the sanctuary area of the park. Overall there has been a significant net contribution of marine tourism in terms of environmental quality, raising community awareness and increasing local incomes, although the distributional effects are not wholly equitable (White and Dobias, 1990).

The role of NGOs in MPAs

NGOs can perform a very important facilitative role in not only raising local capacity to manage and benefit from MPAs, but also in their initial designation. The non-profit organization Coral Cay Conservation (CCC) recruits paying volunteers to survey tropical reefs in several locations across the globe. The data and information collected on reef ecosystems not only enhances local knowledge and understanding of the fragility of such systems but also furnishes an all-important base-line upon which to base future decision making. It facilitates the identification of zones of particular vulnerability and therefore points towards those areas where tourism and other forms of economic activity in the future will do least damage. The data furnished by Coral Cay were instrumental in the designation of the Belize Barrier Reef as a World Heritage Site in 1996, and in its subsequent management. The conferral of that status has had an undeniable impact on enhancing the image of Belize as an ecotourism destination. In the Philippines, 3 years after CCC joined forces with the Philippine Reef and Rainforest Conservation Foundation to survey the coral reefs of Danjugan Island, the island has become a world-class marine reserve.

The organization responds to requests for collaboration and assistance from the host country rather than imposing itself on a destination. This ensures that the local populations are active participants rather than passive recipients. CCC's aim of building up local capacity so that the local population can eventually run their own projects has already been fulfilled in Belize, where CCC ceased direct involvement in December 1998 (Raines and Ridley, 1999). Since 1993, up to 50 Belizeans a year have benefited from the joint Belize Fisheries Department/CCC Scholarship programme to provide crucial skills in scuba-diving, marine life identification and survey techniques. CCC also provided the funds to build the Marine Research Centre, which volunteers helped to construct, on Calabash Cay (Fig. 17.4). This was handed over to the University College of Belize at the end of 1998 as a fully functioning and self-financing research centre.

Entrepreneurial MPAs

Many MPAs lack sufficient funding and management, and therefore do not provide any real protection. Colwell (1998) suggests that in certain instances small-scale, commercially supported, entrepreneurial MPAs

Fig. 17.4. The University College of Belize Marine Research Centre, Calabash Cay.

may provide the best form of protection and that such support may come from dive resorts, or similar commercial entities. Such entrepreneurs can, in certain circumstances, act as the primary stewards of coral reef resources as managers of small-scale MPAs, using tourism to achieve long-term economic and environmental sustainability. Among the essential features of truly successful entrepreneurial MPAs are the inclusion of local stakeholders together with the provision of necessary training and consultation to increase local capacity.

One such example of an entrepreneurial MPA is The Chumbe Island Coral Park Project (CHICOP), Zanzibar, Tanzania. Reidmiller (1999) describes the long, uphill struggle against bureaucratic and legislative constraints from the inception of the project in 1991 through to the arrival of the first marine ecotourists on Chumbe in 1997. While CHICOP is not yet economically viable, its achievements are considerable. After commissioning ecological baseline surveys on the flora and fauna to establish the conservation value of Chumbe island and its fringing reef, the reef sanctuary was gazetted as a protected area in 1994. It became the first functioning marine park in Tanzania. The seven visitors' bungalows and the Visitors' Centre were all constructed according to state-of-the-art eco-architecture (rainwater catchment, greywater recycling, compost toilets and solar power generation). Former fishermen from adjacent villages have been employed and trained as park rangers by volunteer marine biologists and educationists (Reidmiller, 1999). The educational component of CHICOP is also important. Capacity building and the raising of local awareness has occurred via the training of the rangers and their ongoing interaction with other local fishers. The project has also helped to raise conservation awareness and understanding of the legal and institutional requirements among government officials both of departments involved in initial negotiations and of the three departments (Fisheries, Forestry and Environment) who continue to be represented on the project's Advisory Committee. Free excursions are offered to local schoolchildren during the off-season and a Visitors' Centre provides information and guidelines for both day and overnight visitors.

There have, however, been a number of problems. Substantial bureaucratic delays tripled project implementation from 2 to 7 years. The innovative eco-architecture, coupled with considerable logistical problems, extended building operations from an initially envisaged 1 year to 4. These delays caused initial cost estimates to quadruple (Reidmiller, 1999). While the overall investment may be lower than would have been the case with a donor-funded project through the government machinery, cost recovery is an undoubted problem. The project is placed in the invidious position of having to attempt to market itself as an up-market location. As such, it is confronted with what is suggested to be 'unfair competition' from unmanaged nature destinations, where no management costs occur, or from donor-funded projects which effectively subsidize the tourists and tour operators, with little or no management costs being passed on (Reidmiller, 1999).

Conclusions

It is evident, from the foregoing discussion, that marine ecotourism embraces an enormous diversity of activities. These activities are not only the expression of, but also impinge upon, the interests of a range of stakeholders at a variety of scales and from markedly different circumstances. There are, however, certain recurring themes from across the globe. One is, obviously, how sustainable operations are continually frustrated in the absence of regulatory and legislative frameworks. The contrast between whale-watching operations at Kaikoura, New Zealand, and Puerto Piramide, Argentina, illustrates this point. Another is the pitfall of over-generalization. While it is evident that not all marine tourists are ecotourists, it is equally true that they cannot be distinguished as such on the basis of income, socio-economic class or education. A survey conducted of fur seal visitation on the Kaikoura peninsula, for example, found that visitors' behaviour could not be differentiated according to socio-economic characteristics, and that most failed to read the on-site interpretation signs and thus failed to keep 5 m from the seals (Barton *et al.*, 1998). A further recurrent theme is that of the problems faced by small-scale, locally based operations in marketing themselves. As Colwell (1998) suggests, while it is essential to guard against surrendering too much control to commercial entities, it makes sense to utilize the management potential, and indeed access to markets, of commercial partners. As with any partnership, however, the choice of the correct partner to create a working relationship that recognizes the interests of all stakeholders is crucial to success. Colwell describes the non-profit conservation organization CORAL (the Coral Reef Alliance) in the US who work with dive resorts, scientists, educators, governments, conservationists and experts in MPA and community-based management to promote small-scale coral reef MPAs in developing countries. CORAL develops educational material, identifies opportunities, determines the need for technical and material assistance and provides necessary training and consultation. It also provides microloans of US$3,000–15,000 to support entrepreneurial MPAs and networks so that experiences and lessons are shared.

As described earlier, it is vitally important to examine marine ecotourism in context. Its prospects for sustainability, together with the contribution that it can make towards sustainable development in general are determined by a myriad of economic, socio-cultural, political, ecological, institutional and technical forces. These forces are endogenous and exogenous as well as dynamic. Thus there are very important two-way relationships operating at various scale levels, with the positive spread effects, as described earlier, of marine ecotourism ideally diffusing through the hierarchy. A prime example of this is the recognition that whale-watching has given considerable impetus, and economic rationale, to whale-watching as opposed to whale hunting as a commercial activity. A similar scenario is emerging with regard to an incentive for coral reef conservation, as nations such as the Philippines and Indonesia recognize the loss of significant ecotourism resources through destructive fishing practices such as blasting and cyanide poisoning. It is undeniable that negative backwash effects, of not only the activities of unsustainable marine or coastal tourism but also those of other economic activities, marine based or land based, which impinge on the marine environment, prejudice truly sustainable outcomes (Fig. 17.1). The message for a holistic overview of the future of the oceans and seas of the planet is therefore clear. The widely publicized International Year of the Reef in 1997, together with the declaration of 1998 as the International Year of the Ocean by UNESCO have recently played an important role in raising awareness of how much the human population depends on a healthy marine environment for sustainable livelihoods. The quality of these livelihoods is also increasingly dependent on leisure opportunities that allow us to appreciate, understand, and thus help to safeguard, the remarkable diversity of marine environments.

References

Barton, K., Booth, K., Ward, J., Simmons, D. and Fairweather, J.R. (1998) Visitor and New Zealand fur seal interactions along the Kaikoura Coast. *Tourism Research and Education Centre Report* No. 9. University of Lincoln, Lincoln, New Zealand.

Baxter, A.S. and Donoghue, M. (1995) *Management of Cetacean Watching in New Zealand* (online). Department of Conservation, Auckland. http://www.physics.helsinki.fi/whale/newzeala/manage/html

Bottrill, C.G. and Pearce, D.G. (1995) Ecotourism: towards a key elements approach to operationalising the concept. *Journal of Sustainable Tourism* 3(1), 45–54.

Butler, R. (1998) Sustainable tourism – looking backwards in order to progress? In: Hall, C.M. and Lew, A.A. (eds) *Sustainable Tourism: a Geographical Perspective.* Longman, Harlow, UK.

Cater, C.I. (1995) Dive tourism in Zanzibar: new depths. BSc dissertation, University of Bristol, Bristol, UK.

Cater, E. (1997) Ecotourism or ecocide? *People and the Planet* 6(4), 9–11.

Cochrane, C. (1998) Sipadan's last chance? *Action Asia* 7(1), 17–19.

Colwell (1998) Entrepreneurial marine protected areas: small-scale, commercially supported coral reef protected areas. In: Hatziolos, M.E., Hooten, A.J. and Fodor, M. (eds) *Coral Reefs: Challenges and Opportunities for Sustainable Management.* World Bank, Washington, DC.

Csikszentimihalyi, M. (1975) *Beyond Boredom and Anxiety.* Jossey-Bass, San Fransisco.

Dixon, J.A., Scura, L.F. and van't Hof, T. (1993) Meeting ecological and economic goals: marine parks in the Caribbean. *Ambio* 22(2–3), 117–125.

Duffus, D.A. and Dearden, P. (1993) Recreational use, valuation, and management of killer whales on Canada's Pacific coast. *Environmental Conservation* 20(2), 149–156.

Fennell, D. (1999) *Ecotourism: an Introduction.* Routledge, London.

Gray, J. (1998a) SeaCanoe Thailand – lessons and observations. In: Miller, M.L. and Auyong, J. (eds) *Proceedings of the 1996 World Congress on Coastal and Marine Tourism.* University of Washington, Seattle, and Oregon Sea Grant Program, Oregon State University, pp. 139–144.

Gray, J. (1998b) Update on SeaCanoe Wars. *Trinet* (online) 2 December 1998. http://www.caveman@seacanoe.com

Hawkins, J.P. and Roberts, C.M. (1992) Effects of recreational SCUBA diving on fore-reef slope communities of coral reefs. *Biological Conservation* 62, 171–178.

Horn, C., Simmons, D.G. and Fairweather, J.R. (1998) Evolution and change in Kaikoura: responses to tourism development. *Tourism Research and Education Centre Report* No. 6. University of Lincoln, Lincoln, New Zealand.

IUCN (1991) *Guidelines for Establishing Marine Protected Areas.* IUCN, Gland, Switzerland.

McAloon, J., Simmons, D.G. and Fairweather, J.R. (1998) Kaikoura: historical background. *Tourism Research and Education Centre Report* No. 1. University of Lincoln, Lincoln, New Zealand.

Milne, S.S. (1998) Tourism and sustainable development: the global-local nexus. In: Hall, C.M. and Lew, A.A. (eds) *Sustainable Tourism: a Geographical Perspective.* Longman, Harlow, UK.

Moeran, B. (1983) The language of Japanese tourism. *Annals of Tourism Research* 10, 93–100.

Newbery, B. (1997) In league with Captain Nemo. *Geographical Magazine* 69(2), 35–41.

Orams, M. (1999) *Marine Tourism; Developments, Impacts and Management.* Routledge, London.

Orri, D. (1995) *A Case Study in Patagonian Whale Watching* (online). Hervey Bay, Encounters with Whales Conference. http://www.physics.helsinki.fi/whale/argentina/orri/cas.html

PADI (1994) *PADI Certification Statistics.* PADI, Santa Ana, California.

PADI (1999) *PADI Certification Statistics.* PADI, Santa Ana, California.

Raines, P. and Ridley, J. (1999) Belize: twelve years of teamwork. *Coral Cay Conservation Newsletter, 1999.*

Reidmiller, S. (1999) The Chumbe Island Coral Park Project. http://www.chumbe.island@raha.com

RSTC Europe (1997) *Facts and Figures.* RSTC, Switzerland.

Ryan, C. (1997) *The Tourist Experience.* Cassell, London.

Schuster, B.K. (1992) What puts the eco in ecotourism? *The Undersea Journal* (1), 45–46.

SeaCanoe (1999) *'Eco' Development* (online). Thailand: SeaCanoe. http://www.seacanoe.com/seamore2.html

Shackley, M. (1998) 'Stingray City' – managing the impact of underwater tourism in the Cayman Islands. *Journal of Sustainable Tourism* 6(4), 328–338.

Simmons, D.G. and Fairweather, J.R. (1998) Towards a tourism plan for Kaikoura. *Tourism Research and Education Centre Report* No. 10. University of Lincoln, Lincoln, New Zealand.

UN (2000) *Oceans and Law of the Sea* (online). United Nations, New York. http://www.un.org/Depts/los/los_mrl.htm

Whale Watch (1999) *Company Profile.* Whale Watch, Kaikoura, New Zealand.

White, A.T. and Dobias, R.J. (1990) Community marine tourism in the Philippines and Thailand: a boon or bane to conservation? In: Miller, M.L. and Auyong, J. (eds) *Proceedings of the 1990 Congress on Coastal and Marine Tourism.* National Coastal Resources Research and Development Institute, Newport, Oregon, USA.

World Resources Institute (1997) *World Resources 1996–1997.* Oxford University Press, Oxford.
Wylie, J. and Rice, H. (1998) Sea kayaks as vehicles for sustainable development of coastal and marine tourism. In: Miller, M.L. and Auyong, J. (eds) *Proceedings of the 1996 World Congress on Coastal and Marine Tourism.* University of Washington and Oregon Sea Grant Program, Oregon State University, Seattle, pp. 131–138.
Ziffer, K. (1989) *Ecotourism: The Uneasy Alliance,* Working Paper No. 1. Conservation International, Washington, DC.

Section 4

Ecotourism Venues

D.B. Weaver
Department of Health, Fitness and Recreation Resources, George Mason University, Manassas, Virginia, USA

In addition to the type of biome that accommodates ecotourism, it is also necessary to consider the generic venues that play host to this activity. As pointed out by Lawton in Chapter 18, one particular kind of setting, the 'protected area', has attained a virtual monopoly with respect to the provision of ecotourism opportunities, at least if the literature is any indication. This, perhaps, is not too surprising, considering that many of the over 30,000 protected areas currently in existence facilitate the three basic criteria of ecotourism by preserving a usually outstanding component of the natural environment from activities that are deemed harmful to this environment. Moreover, learning and appreciation opportunities are usually included in the mandate of protected areas, which often offer various services and facilities to expedite the activity of visitors.

However, Lawton's description of ecotourism as an activity that is highly concentrated within and among these protected areas is something that must be considered in depth. If only a few areas within a few protected areas are accommodating most of the visitors, this could have both positive and negative implications for managers. On one hand, high concentrations suggest the possibility that existing site carrying capacities may be breached. However, as Lawton recognizes, these same concentrations offer economies of scale that justify sophisticated site-hardening measures as well as the comprehensive services and facilities that are desired by soft ecotourists. At a system-wide level, these same soft ecotourists tend to congregate in protected areas that are accessible to coastal resort areas and international gateways. This skewed pattern allows limited resources to be focused on just a few parks, and creates opportunities for synergy between ecotourism and resort or business tourism. Yet, it also means that local communities throughout most of the country cannot capitalize on the economic opportunities afforded by an appreciable influx of ecotourists.

The situation described by Lawton is likely to intensify further, given the rapid rate at which unprotected natural environments are disappearing. In turn, this will lead to even greater pressures on cash-starved public protected area authorities. One increasingly popular response is the establishment of private protected areas, as discussed by Langholz and Brandon in Chapter 19. One general advantage of private protected areas is their provision of environmental protection in a way that does not require public subsidy. However, this same non-governmental role can mean that the profit motive takes priority over sustainability, and that protected status can give way to some other less benign land use, should the controlling party deem this to be warranted. In other words, the public interest may not be perceived as a paramount consideration in a privately controlled protected area. Also, private reserve managers may not have the skills or funds to cope with such necessary tasks as policing, providing services, etc. Yet, perhaps to a greater extent than in public areas, the profit motive (or ethical motives in the case of non-government organization run parks) may induce managers of private protected areas to provide quality ecotourism experiences that will ensure the continued safeguarding of the natural environments that they harbour.

An ideal arrangement may involve the establishment of private protected areas as a buffer zone surrounding a publicly controlled core area, as long as this is implemented in a way that avoids competition and encourages synergy. Such buffer zones can experience quite a bit of change and modification without causing serious impacts on the core, and are often the most appropriate location for the establishment of overnight accommodations and other tourism-related facilities and services. More generally, ecotourism planners need to pay greater attention to the possibility of accommodating ecotourism within spaces, whether controlled by the private or the public sector, that have already experienced modification to the extent that they cannot be classified as natural areas. This is the basis, for example, of Lawton's recommendation that ecotourism should be more vigorously promoted in lower order protected areas under the IUCN system. In Chapter 20, Lawton and Weaver carry this logic further still by suggesting that highly modified environments can provide high-quality opportunities for observing certain types of wildlife. Examples include the use of farm fields and landfill sites by migrating waterfowl, the colonization of artificial reefs by marine life, and even the establishment of peregrine falcon populations in the central business districts of large metropolitan areas. The authors of this chapter argue that such extensions not only relieve tourism pressures on vulnerable natural spaces, but do so in a way that does not threaten to undermine the already-modified setting, and may even instil the desire to further enhance the capacity of such areas to accommodate wildlife.

At the other extreme, there is considerable debate about the appropriate role of ecotourism in wilderness settings. Hammitt and Symmonds' examination of this topic in Chapter 21 recognizes that wilderness is a subjective and largely Eurocentric concept, but usually entails the absence of significant levels of activity and modification by non-indigenous groups. As such, wilderness can encompass vast tracts of space (as with Antarctica), or can exist as pockets in the midst of heavily settled landscapes. From an ecotourism perspective, the challenge is not just to ensure sustainable outcomes, but to accommodate levels of activity that do not detract from the very qualities that define the area as wilderness in the first place. Hammitt and Symmonds illustrate some of the attendant problems with examples from protected areas in Africa and the UK.

The issue of using wilderness for ecotourism, however, is complicated by the presence of indigenous groups around the world which have been vigorously asserting their traditional rights to a substantial portion of the world's remaining natural environments areas, furthermore, that they would not consider to constitute 'wilderness'. As this process continues, indigenous people are emerging as one of the major stakeholder groups in the ecotourism sector, as is already evident in areas such as New Zealand, Australia and much of Canada. Widely held ecotourism ideals of environmental sustainability, as discussed by Hinch in Chapter 22, may not be complementary with the reality of indigenous culture and its special relationship with the land, leading to disappointment, if not resentment, among some visitors. But even if expectations cohere with the reality, growing numbers of satisfied ecotourists may inadvertently lead to negative socio-cultural impacts within these communities. Despite the threat, indigenous people often support this mode of development because it is perceived to be more benign than the alternatives for raising much needed revenue, and because it may assist in attempts to establish political control over their territories. To be successful, Hinch refers to the centrality of community empowerment in implementing ecotourism on indigenous lands, which includes having control over the land itself. As in all other venues, the hoped-for scenario is mutual benefit, wherein ecotourism provides the incentive to enhance the environmental and socio-cultural sustainability of the destination, which in turn make for quality ecotourism products.

Chapter 18

Public Protected Areas

L.J. Lawton
School of Business, Bond University, Gold Coast, Queensland, Australia

Introduction

A 'protected area' is defined by the World Conservation Union (IUCN, 1994) as 'an area of land and/or sea especially dedicated to the protection and maintenance of biological diversity, and of natural and associated cultural resources, and managed through legal or other effective means'. With their emphasis on preserving the natural environment, protected areas have obvious appeal to the ecotourism sector, which is based primarily on natural attractions. Such areas, in practice, constitute by far the most important venues for ecotourism activities, a status that is reflected in the ecotourism literature (e.g. Ceballos-Lascuráin, 1996; Weaver, 1998; Butler and Boyd, 2000) and within this encyclopedia. The goal of this chapter is to outline the actual and potential relationships that exist between ecotourism and protected areas, focusing only on those that are publicly controlled. The first section in this chapter introduces the topic by outlining the growth, distribution and major categories of public protected areas. Subsequently, the compatibility between these areas and ecotourism is considered. This is followed by a more in-depth discussion of the relationship, which is based on the desire to maximize tourism-derived revenues while minimizing the environmental impacts. The final section examines a number of issues that are particularly relevant to the evolving relationship between ecotourism and public protected areas. Examples from all parts of the world are used to illustrate the content of this chapter.

Growth, Distribution and Types of Public Protected Areas

Although publicly protected areas have been in existence for at least 3000 years, it is only within the past century that they have accounted for a significant proportion of the world's landscape. Yellowstone, established in 1870, is widely regarded as the world's first national park, and since then, the overall number of national parks and other protected areas has proliferated. By 1993, the World Conservation Union estimated a global population of 8619 public protected areas, covering 792 million ha or 5.9% of the world's land area (IUCN, 1994). This takes into account only those entities of at least 1000 ha. Just 4 years later, the number had increased by 20% to 10,401, and the area to 840 million ha, or 6.4% of the world land surface. If entities of less than 1000 ha are added, then the number and land area increase to 30,350 and 1.32 billion ha, respectively (8.8% of the world's land area) (Green and Paine, 1997). The ubiquity of protected areas is reflected in the fact that only 11 countries (including Yemen, United Arab Emirates and Syria) reported no such areas as of 1997, while 38 had designated 10% or more of their territory in this way. For Denmark, the Dominican Republic, Ecuador and Venezuela, the proportion is in excess of 30% (World Resources Institute, 1998).

The status of marine protected areas (MPAs) is incipient by comparison, with relatively few established to date, and little data available on characteristics and visitation levels. One recent estimate suggests the existence of approximately 1300 MPAs, a figure that is inadequate to achieve even basic conservation objectives (Boersma and Parrish, 1999). The situation in Canada is a case in point. Of 29 recognized marine ecoregions, only three were represented by MPAs as of 1999 (Parks Canada, 1999a). Several factors account for this neglect, including the limited state of knowledge on marine ecosystems, lingering perceptions that marine resources are limitless and thus do not require protection, and the fact that most marine resources do not stay within imposed administrative boundaries. Functional boundaries, therefore, are difficult to establish, demarcate, and police. In addition, only a small portion of marine space lies within the clear jurisdiction of states and dependencies, that is, geopolitical entities which are in a position to establish national MPA systems.

The above statistics, in any event, are misleading for both land and marine environments, in so far as protected areas do not offer a single, homogeneous level of 'protection', nor focus only on protection as a management objective. According to Green and Paine (1997), there are 1388 different types of 'protected areas' in the world. Because most jurisdictions tend to use their own idiosyncratic systems of classification, the World Conservation Union has devised a standard international classification system of protected areas, consisting of six categories (see Table 18.1). Such a scheme, for example, appropriately distinguishes the highly modified national parks of the UK (Category V) from the semi-wilderness national parks of the USA and Canada (Category II). It should be noted, with some qualification, that higher levels of allowable human intervention are generally associated with higher category numbers.

Compatibility of Protected Areas with Ecotourism

Within the IUCN classification scheme, it is possible to assess the hypothetical compatibility between various protected area types and ecotourism. Figure 18.1 provides such a generalized assessment, taking into account both hard and soft ecotourism (see Chapter 2) as well as other types of tourism activity. Conventional tourism activities become more compatible in the higher-numbered categories, in line with the above statement regarding allowable levels of human activity. The status of ecotourism is more complex, with its soft and hard manifestations displaying very different trends. Soft ecotourism is incompatible with Category I areas, but highly compatible with Categories II and III. For the remaining categories, the compatibility is reduced, but still high, given the nature of this type of ecotourism. Hard ecotourism has a qualified place in Category I and, like its soft counterpart, displays high compatibility in Categories II and III. However, this declines greatly in the remaining categories as a consequence of the high degree of modification that these categories allow. In general, low compatibility in the first three categories is attributable to prohibitions on certain types of activities, while low compatibility in the next three categories owes to the unsuitability of the landscape. That is, there is nothing to legally prevent hard ecotourism from occurring in a Category VI protected area, but little incentive to use such spaces for that purpose. The linkages between ecotourism and the IUCN categories will now be considered in greater detail.

Table 18.1. IUCN (World Conservation Union) protected area categories.

Category	Designation, number and area (1997)	Description
Ia	Strict Nature Reserve 4389: 97.9 million ha	Area of land and/or sea possessing some outstanding or representative ecosystems, geological or physiological features and/or species, available primarily for scientific research and/or environmental monitoring
Ib	Wilderness Area 809: 94.0 million ha	Large area of unmodified or slightly modified land, and/or sea, retaining its natural character and influence, without permanent or significant habitation, which is protected and managed so as to preserve its natural condition
II	National Park 3384: 400 million ha	Natural area of land and/or sea, designated to (a) protect the ecological integrity of one or more ecosystems for present and future generations, (b) exclude exploitation or occupation inimical to the purposes of designation of the area and (c) provide a foundation for spiritual, scientific, educational, recreational and visitor opportunities, all of which must be environmentally and culturally compatible
III	Natural Monument 2122: 19.3 million ha	Area containing one, or more, specific natural or natural/cultural feature which is of outstanding or unique value because of its inherent rarity, representative or aesthetic qualities or cultural significance
IV	Habitat/Species Management Area 11,171: 246 million ha	Area of land and/or sea subject to active intervention for management purposes so as to ensure the maintenance of habitats and/or to meet the requirements of specific species
V	Protected Landscape/ Seascape 5578: 106 million ha	Area of land, with coast and sea as appropriate, where the interaction of people and nature over time has produced an area of distinct character with significant aesthetic, ecological and/or cultural value, and often with high biological diversity. Safeguarding the integrity of this traditional interaction is vital to the protection, maintenance and evolution of such an area
VI	Managed Resource Protected Area 2897: 360 million ha	Area containing predominantly unmodified natural systems, managed to ensure long-term protection and maintenance of biological diversity, while providing at the same time a sustainable flow of natural products and services to meet community needs

Categories I–III

Category I protected areas, such as strict biological reserves, with their strict prohibitions on human activity, accommodate at best a small amount of 'hard' ecotourism. These are likely to entail scientific and/or educational activities. In contrast, the national parks of Category II and, to a lesser extent, Category III protected areas, are highly compatible with ecotourism, and dominate the empirical literature as high profile ecotourism venues. There are several factors that have contributed to the close relationship between Category II and III protected areas and ecotourism, as follows.

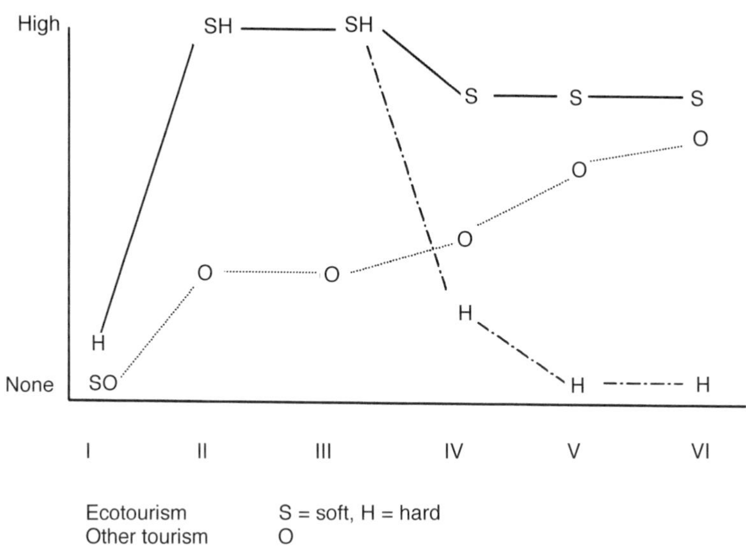

Fig. 18.1. Compatibility of tourism and ecotourism with protected area categories.

High level of protection, but allowance for tourism

While not as proscriptive as Category I, the management criteria for Category II and III protected areas are still strict, and designed to ensure the maintenance of the area's ecological integrity. Thus, with their largely unspoiled natural environments, they provide an extremely high quality venue for ecotourism-related pursuits, and one that tolerates and even encourages such activity. Moreover, by minimizing or prohibiting altogether the presence of potentially incompatible resource users, and by establishing a range of other environmental regulations, there is greater assurance that a basic criterion of ecotourism, sustainability, is achieved (see Chapter 1). Another basic criterion of ecotourism, i.e. educational and/or appreciative interface with the attraction, is also given a high priority. Category II protected areas, in particular, are better positioned to accommodate the dual objectives of tourism and environmental protection because of their size, which averages 118,000 ha, and because of zoning provisions that allocate certain activities to certain areas.

Special qualities

Designation as a national park or national monument usually occurs because an area contains some outstanding natural attribute or attributes in addition to the presence of relatively unspoiled surroundings. These include:

- outstanding natural scenery (e.g. Yosemite and Banff);
- exceptional representation of a particular biome (e.g. the savannahs of Kruger and Serengeti);
- rare or unusual flora and/or fauna (e.g. the mature sequoia and redwood groves of several California national parks, the panda bear populations of certain protected areas in China's Sichuan province, the monarch butterfly overwintering reserves in Mexico);
- rare and/or unusual geological features (e.g. Undara and Uluru in Australia, or the gorges and caves of Malaysia's Gunung Mulu National Park).

The presence of such exceptional features make national parks or national monuments even more attractive to ecotourists.

Market awareness

With their outstanding qualities, Category II and III protected areas are often well known as tourist attractions, and sometimes even achieve iconic status within their destination countries or regions. Thus, Banff and Jasper are virtually synonymous with the Canadian Rockies, Kruger with South Africa, Mount Fuji with Japan, and Uluru and the Great Barrier Reef with Australia. This iconic status is often applied more generically to entire park systems. Thus, most nature-oriented tourists visit countries such as Costa Rica and Kenya because of exemplary protected area systems, rather than any one particular entity (Weaver, 1998). The public profile of certain protected areas has been further enhanced by the introduction of prestigious international designations, the best known of which is the World Heritage Site. As of 1999, 145 Natural Heritage Sites (i.e. World Heritage Sites emphasizing a natural attraction) had been established by UNESCO (WHIN, 1999).

Tourism facilities

Category II and III protected areas incorporate tourism as an integral part of their management role, as indicated earlier. Given this, and the fact that managers often encourage tourism because of its revenue-earning potential, these entities often provide services and facilities, such as interpretation centres, hiking trails, camping facilities, etc., that attract large numbers of 'soft' ecotourists. However, the assumption that all park visitors are ecotourists must be avoided. In the absence of hard data, Fig. 18.2 proposes that the majority of visitors to Category II and III protected areas can be classified as 'soft' ecotourists, as in Chapter 2. Hard ecotourists, in contrast, constitute a much smaller portion of visitors. The remaining non-ecotourist component consists of two main segments, namely other types of tourists, and local visitors who do not meet the technical thresholds that denote a 'domestic tourist' even though they may consume ecotourism products. In the less developed countries, a higher proportion of 'other tourists' can be assumed, since domestic visitors have less proclivity to engage in ecotourism in comparison to international visitors.

The actual number of park visitors, and hence ecotourists, is indeed formidable. While no statistics are available for the world as a whole, the figures from individual countries illustrate the importance of Category II and III protected areas as visitor attractions. For example, the US National Park Service (NPS, 1999) forecast 64.5 million visits in 2000 just to the nature-oriented

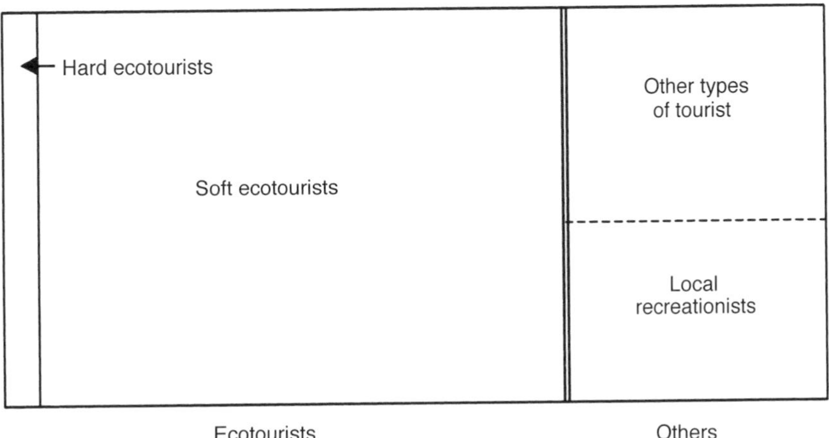

Fig. 18.2. Visitor segments in category II and III protected areas.

component of its national park system. The much smaller Canadian national park system recorded over 15 million visitors during the 1998/99 fiscal year (Parks Canada, 1999b), while Australian national parks accommodated an estimated 17 million visits in 1991 (Carter, 1996).

Categories IV–VI

The remaining IUCN categories are not readily associated with ecotourism, due mainly to the tolerance of significantly large amounts of potentially incompatible human activity, the presence of landscapes that are significantly modified by those activities, and lower market awareness. However, there are many exceptions, given the heterogeneity of these categories. Examples of protected areas in these categories that do accommodate significant amounts of ecotourism include the national parks of England and Wales (Category VI), the Ngorongoro Conservation Area (Category VI) in Tanzania and Inverpolly National Nature Reserve (Category IV) in Scotland. Beyond such examples, the status of ecotourism is more one of potential rather than practice, as discussed later in this chapter. But even where ecotourism is known to exist in these categories, visitation levels and their impacts are seldom monitored. In this sense, these IUCN categories are similar to MPAs.

Relationship between Ecotourism and Public Protected Areas

The historical relationship between tourism and higher-order protected areas is characterized by an ambivalence that stems from doubts about the actual compatibility between tourism and environmental preservation. Many public park systems, such as those in Costa Rica, were founded strictly on a non-profit environmental mandate, yet are becoming increasingly reliant on tourism-based revenues. The US national park system, for example, generated US$3 billion in tourism revenues during 1991 (Norris, 1992). The irony here, and thus the essential basis of the ambivalence, is that the continued integrity of the parks is becoming dependent upon higher levels of an activity that potentially threatens this integrity. Increasingly, the revenues from tourism are large enough to constitute an incentive for maintaining the parks in the face of growing pressure from competing resource users such as farmers, loggers, ranchers and miners. Yet, intensive levels of tourism activity can result in impacts that are equally detrimental, as demonstrated by the overuse of Amboseli National Park in Kenya, and Manuel Antonio National Park in Costa Rica (Weaver, 1998).

Short of prohibiting tourism altogether and assuming reliance on alternative sources of funding, the logical solution to this dilemma would be to ensure that tourism within the protected area is carried out in an environmentally and socially sustainable manner. The ideal relationship is then symbiotic, where parks provide a high-quality venue and freedom from incompatible competitors to tourism, while sustainable tourism (mostly in the form of ecotourism) provides the revenues to maintain this quality, and the exposure to the public that contributes to continued popular support. This, of course, is easier to achieve in theory than in practice, but it is vital that the goal of sustainability is pursued. In order to discuss appropriate protected area management strategies, it is helpful to imagine a hypothetical environmental carrying-capacity threshold for any given protected area that, if exceeded by visitation levels, indicates an unsustainable situation. The specific strategies that are subsequently adopted to keep tourism below this threshold depend on whether the latter is perceived to be stable or flexible.

Stable carrying-capacity thresholds

In certain circumstances, it is appropriate to assume that environmental carrying-capacity thresholds are stable. This is a useful assumption, for example, if the carrying capacities are unknown, or if no changes are made to the area or infrastructure of the park. In this case, an appropriate strategy is to establish and enforce conservative visitor quotas on an annual, seasonal, monthly, weekly, daily or hourly basis, as warranted, and irrespective of market demand (Fig. 18.3a). Although this can reduce potential revenues in situations where the demand is high, such policies are justified on the basis that public protected areas are not normally mandated to generate profits, but rather to achieve some broader public 'good', such as environmental preservation. Hence, quota systems are much more prevalent in public protected areas than in privately controlled sectors of the tourism industry.

It has become increasingly popular in recent years to modify the quota principle through the introduction of escalating user fees (assuming that these increased revenues are used for the parks and not just to swell general government coffers). Thus, instead of prohibiting further entry once the quota is reached, visitation levels are controlled by a de facto quota based on the willingness of the market to pay for the visitation privilege. Assuming that demand is robust, this has the advantage of deriving a larger revenue intake from the same or even a lesser number of users (Lindberg, 1991). A danger, however, is the perception that the site, presumably a public good, is available in practice only to the wealthy elite. The government of Costa Rica addressed this problem in 1988 by maintaining a nominal national park entry fee

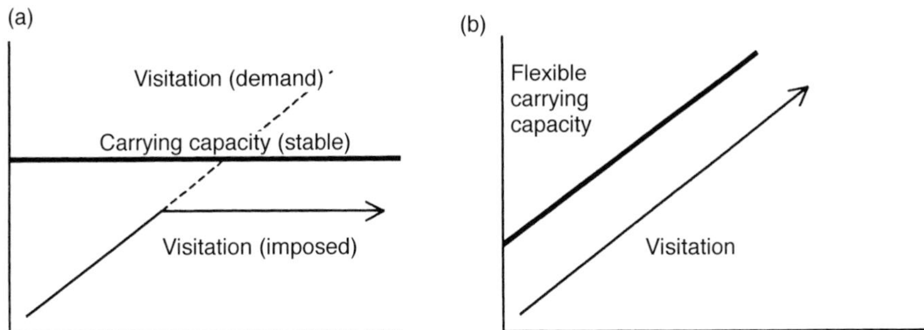

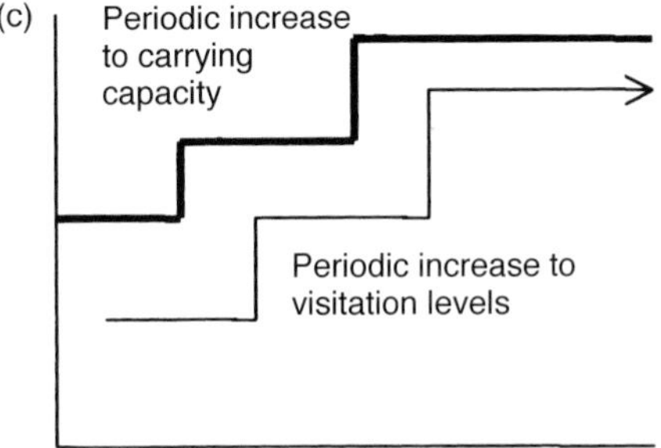

Fig. 18.3. Stable and flexible carrying-capacity thresholds.

for Costa Rican nationals, while raising the fee for non-nationals from about US$1 to US$15 (Weaver, 1998). According to Dixon and Sherman (1990), such increases, where the base level is so low, provides almost no deterrent to foreign tourists, who have invested several thousands of dollars in their travel experience. The actual elasticity of these escalated entry fees can be quite impressive in situations where demand is high, as with mountain gorilla ecotourism in Rwanda (Lindberg, 1991; Shackley, 1995), wildlife viewing in the Galapagos National Park (Weaver, 2000) and high peak climbing in Nepal (Robinson, 1994). In the case of Rwanda's Parc National des Volcans, visitation remained stable at about 6000 tourists per year between 1980 and 1988 despite the escalation in entry fees from US$14 to US$170 per person. However, it must also be borne in mind that the availability of new competing opportunities (e.g. the provision of mountain gorilla viewing opportunities in Uganda and Congo) can suddenly reduce demand and, hence, user willingness to pay such high fees.

Flexible carrying-capacity thresholds

There is no inherent reason for assuming that environmental or social carrying-capacity thresholds are inflexible, and many scenarios can be cited in which an adjustment in such thresholds is warranted. Once these adjustments are made, then visitation levels can be increased accordingly (Fig. 18.3b). The opposite situation, where the thresholds are increased *in response to* increased visitation, is a reactive approach that should be avoided. Two major strategies for effecting higher thresholds are site-hardening measures and consumer education.

Site-hardening measures

'Site-hardening' simply implies the establishment of facilities, services, etc. so that a location is capable of accommodating a larger number of users without detriment to its environmental integrity. Walking trails provide a good illustration. As a dirt track, a particular trail might have a carrying capacity of 100 users per day, but if lined with cobblestones, that same trail may safely accommodate 1000 users per day. At least two associated considerations, however, must be taken into account. First, the site-specific consequences of the site-hardening measures need to be assessed; extensive vegetation clearance and the use of chemical-laced paving materials, for example, would not normally be deemed acceptable. Second, while the trail itself may be capable of accommodating much higher levels of use as a result of the site-hardening, these levels must adhere to the carrying capacities of the surroundings, so that resident nesting birds and other flora and fauna are not distressed. Returning to the first point, a conscious decision may be made to 'sacrifice' a 2-ha site for a modern waste treatment facility, so that the effects of contaminated effluents are minimized in the remainder of the park. Other types of facilities that are amenable to site-hardening include viewing platforms or towers, parking lots, campgrounds, interpretation facilities and fixed accommodations (e.g. ecolodges). While these kinds of facilities may be seen as intrusive by 'hard' ecotourists, they will serve to enhance the experience and appreciation of the dominant 'soft' ecotourist segment.

Consumer education

Even if no site-hardening measures are implemented, larger numbers of visitors can be accommodated within a protected area if the behaviour of those visitors can be modified to minimize their negative environmental and social impacts. For example, park users may be required to be entirely silent in certain locations, to travel in small groups only, and/or to stay on certain footpaths, and not walk within a certain distance of certain sites. Well-trained guides are usually required to ensure enforcement. Such restrictions may result in a certain amount of visitor discontent, but as with escalated entry fees, these regulations are warranted in a publicly controlled site where the 'common good' of environmental protection takes priority over the maintenance of client satisfaction. Moreover, the approach can be rationalized from a commercial perspective by the extent to which the quality and 'authenticity' of the attraction is maintained, and by indications that these kinds of restrictions are tolerated and even welcomed by an increasingly 'green' tourist market (Poon, 1993).

In practice, consumer education (i.e. modification of the market) is often combined with site-hardening (i.e. modification of the product) in order to best effect a sustainable tourism sector within the protected area. Similarly, the stable and flexible threshold options are not mutually exclusive, and can also operate conjunctionally. As depicted in Fig. 18.3c, a 'stairway' effect occurs when carrying capacity thresholds are periodically increased in response to the implementation of appropriate managerial tactics. This pattern is evident in many public protected areas, and the example of the Galapagos Islands is illustrative.

The Galapagos Islands National Park

The archipelago, most of which is protected under several layers of overlapping protected area status, is renowned for its extremely high level of endemic flora and fauna. The area reflects the great paradox of ecotourism, which is that the rarest attractions are simultaneously the most in demand and the most vulnerable to the visitation levels that stem from that demand. Park managers have attempted to compromise between the preservation of the archipelago's ecological integrity and the need to obtain operating revenue by practising a strategy of incremental access similar to Fig. 18.3c (Weaver, 2000). A visitor ceiling of 12,000 was established in 1973, but this was raised to 25,000 in 1981 and to 50,000 in the early 1990s, as new areas were designated to accommodate visitors. These Intensive and Extensive Visitor Zones, as appropriate, are site-hardened to withstand visitor pressure. At the same time, quotas are enforced on the number of visitors allowed in any particular zone (i.e. 90 at one time in an Intensive Zone, and 12 in an Extensive Zone). Concurrently, the behaviour of visitors is subject to a host of constraints. For example, no tourist can go anywhere ashore without being accompanied by a licensed guide, must remain within a given distance from that guide, and cannot wander beyond a rigorously defined network of narrow footpaths.

On paper, the regulations that govern tourism in the Galapagos National Park are among the most stringent in the world, and the authorities are generally praised for their management of the Park. Yet, many problems are still apparent. To some extent, the upward adjustment of visitor ceilings is as much the outcome of political as ecological considerations, and even these have been frequently breached. For example, annual visitation levels in the late 1990s have exceeded the 50,000 limit by anywhere from 5000 to 10,000 visitors. At a site level, the following problems have been observed.

- Nesting birds such as boobies suffer distress even with the strict regulations.
- Some trails have been erroneously constructed on highly erosive sandy soils rather than nearby lava beds that are largely impervious to erosion.
- No matter how high their environmental standards, tourism ships have been identified as a significant agent for the inadvertent dispersal of insects.
- Although visitor quotas are based on a given inventory of visitor zones, many of the latter are closed at any given time, increasing the pressure on the remaining sites.
- Underqualified local residents are sometimes hired as guides for political reasons.
- Potentially incompatible activities such as sport fishing have been introduced.
- Park authorities have been unable to prevent illegal incursions into the park by local residents and others.

Relevant Issues

The Galapagos case study is instructive in demonstrating the gap between sustainable theory and practice, a situation that persists to a greater or lesser extent in virtually every protected area, and hinders the attainment of the ideal synergy between these entities and the tourism sector. With reference to this broader goal, this chapter will now focus on a number of issues that will significantly influence the evolving relationship between ecotourism and public protected areas.

Protected area configurations

The spatial configuration of protected areas can exercise a significant effect over the variety and quality of the ecotourism experiences that they offer, just as it can affect their ability to safeguard the area's ecological integrity. The 'island effect', for example, where small and isolated protected areas are scattered throughout a region, is widely regarded as a hindrance to the maintenance of biodiversity as well as sustainable and high-quality ecotourism products (Fig. 18.4a). In recent years, considerable attention has been focused on the re-configuration of individual protected areas and protected area systems to optimize their ability to meet specified environmental objectives. Admittedly, tourism is usually a secondary consideration, though the implications for the latter can be significant.

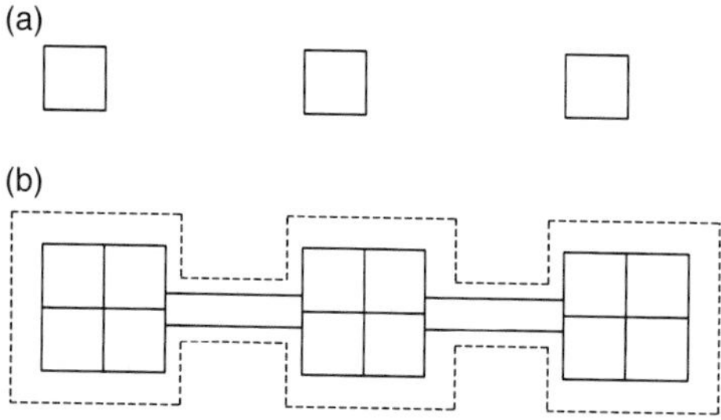

Fig. 18.4. Contrasting spatial configurations for protected areas: (a) island effect; (b) clusters, corridors and buffer zones.

Relevant strategies include the interrelated concepts of clusters, corridors (MacClintock *et al.*, 1977) and buffer zones (Wells and Brandon, 1993). It is widely believed that the effectiveness of individual protected areas is amplified when they are located adjacent to other protected areas; potentially, the combination of contiguous protected areas can then function as a single ecological entity. The concept is being operationalized in Costa Rica through the establishment of 'Conservation Areas', which are groupings of protected areas that are divided into one or more 'core zones' controlled by the state, and surrounding zones consisting of privately or community owned protected areas (IUCN, 1992). The latter areas provide a buffer for more sensitive core zones. For the ecotourism sector, such clusters offer a greater range of attractions and activities, and the buffer zones are suitable for providing accommodations and other services. This model illustrates how the public protected areas can maintain a synergistic relationship with their privately owned counterparts (see Chapter 19). At an international level, the Biosphere Reserve programme of UNESCO, which involved 350 Reserves as of early 2000, operates on similar principles (UNESCO, 2000).

In some situations, protected areas or protected area clusters have been connected through the establishment of corridors. One type of corridor is evident in the eastern part of Victoria, Australia, where the Alpine National Park is configured as four major areas joined together by narrow tracts of land that are included in the park jurisdiction. The result is a continuous 200 km arc of public protected space. Such corridors will facilitate long-distance walking opportunities, given that walking trails on private property remain there only by the continued consent of the landowner.

In contrast to the island effect of Fig. 18.4a, the spatial configuration generated by the combination of clusters, corridors and buffer zones presents, in principle, a much more effective venue for achieving the dual directives of environmental preservation and the provision of tourism opportunities (Fig. 18.4b). Traditionally, international boundaries have been considered an impediment to the formation of such entities, but this is being redressed by the increasingly popular concept of the transboundary protected area (World Conservation Monitoring Centre, 1999), which is really only a practicality when they involve state-owned rather than privately owned lands. As of 1997, 136 transboundary protected area complexes had been identified (Green and Paine, 1997). Costa Rica is again a pioneer in this regard, having established La Amistad International Park on the border with Panama, which is also in the process of setting up an adjacent protected area.

Optimal utilization of protected area systems

Ecotourism is not only associated with Category II and III protected areas, but also with only a few high-profile units within most of the national park systems that account for those categories. For example, just four parks in Costa Rica (Poas, Manuel Antonio, Irazu and Santa Rosa) account for two-thirds of all visitors. Similarly, the top five parks in Kenya (Tsavo East, Animal Orphanage, Lake Nakuru, Masai Mara and Amboseli) account for 62% of all park visitors in that country (Weaver, 1999). This 'drought–deluge' pattern can lead to a situation where the carrying capacity thresholds of some parks are being breached, while other parks are not even close to fulfilling their tourism potential. Hence, investigations into the feasibility and appropriateness of dispersing tourists more widely are warranted in most countries.

This still leaves the question of visitor concentration within the Category II and III units. Beyond these relatively high profile designations, most jurisdictions contain an array of more obscure protected areas that could, to a greater or lesser extent, accommodate ecotourism-type activities. Thus, if pressure is being brought to bear on higher profile spaces, it is logical to explore such options. Lawton (1993, 1995) has identified 27 protected area designations that have IUCN status in the Canadian province of Saskatchewan. As depicted in Table 18.2, some of the categories have low potential for ecotourism because of user restrictions, the nature of the resource being protected or, most commonly, the presence of potentially incompatible resource users. The areas that have medium to high compatibility with ecotourism comprise 57% of all public protected areas, or about 3.7% of the province. Of that amount, about one-half are accounted for by national and provincial parks that accommodate, and are managed to accommodate, various types of ecotourism activity. Yet, there are only 23 such entities, compared with 390 other compatible protected areas, indicating that the latter are much smaller. The point here is not to go into detail over specific strategies for integrating these areas into the ecotourism sector, but rather to emphasize their potential as ecotourism venues in the face of increasing market demand and growing congestion in the more popular protected areas. An additional factor is the potential of such areas to contribute to the regional dispersion of ecotourism-related revenues.

Integration with mass tourism

However much the ecotourism clientele is dispersed throughout a protected area system, and however much those systems are expanded, one may anticipate that visitation pressure on public protected areas will continue to increase as global tourism maintains its relentless rate of growth. One common response to this pressure is to regard the tourist fundamentally as an enemy or a necessary evil, and the introduction of mass tourism as the worst fate that can befall the protected area. Such attitudes among park managers are still surprisingly common, and owe not only to a biocentric tendency in training and philosophy, but also to a compelling body of evidence that documents the negative impacts of rapid tourism growth within protected

Table 18.2. Protected areas in Saskatchewan 1997 and their ecotourism compatibility (Lawton, 1993, 1995; N. Cherney, Saskatchewan Environment and Resource Management, personal communication).

Category	IUCN designation	No.	Area	% of province	Ecotourism compatibility
International					
RAMSAR sites	IV	2	12,700	0.02	high
Federal					
Migratory Bird Sanctuaries	IV	15	56,851	0.09	high
National Historic Parks	V	4	1,700	< 0.01	low[b]
National Historic Sites	V	7	177	< 0.01	low[b]
National Parks	II	2	429,487	0.66	high
National Wildlife Areas	IV	34	21,254	0.03	high
PFRA Pastures	VI	64	708,091	1.09	medium[d]
Federal/Provincial					
Canadian Heritage River	V	1	0	0	high
Provincial					
Ecological Reserves	I	2	815	< 0.01	low[a]
Game Preserves	IV	24	108,859	0.17	high
Historic Sites	V	8	22	< 0.01	low[b]
Protected Areas	III, IV, VI	21	4,996	< 0.01	medium[a,b]
Provincial Community Pastures	VI	56	249,375	0.38	medium[d]
Provincial Heritage Property	III, V	3	134	< 0.01	low[b]
Provincial Parks		(33)	(1,130,544)	(1.73)	
Historic	V	9	271	< 0.01	low[b]
Natural Environment	II	11	677,263	1.04	high
Recreation	V	10	14,704	0.02	med. high[e]
Wilderness	I	3	438,307	0.67	low[a]
Recreation Sites	V	150	42,147	0.06	med. high[e]
Wildlife Development Fund Land	IV	NA	54,277	0.08	medium[d]
Wildlife Habitat Protection Land		NA	(1,365,334)	(2.10)	low[d]
Act	VI	NA	1,311,477	2.01	
Regulation	VI	NA	50,856	0.08	
Policy	VI	NA	3,001	< 0.01	
Wildlife Refuges	IV	24	2,560	< 0.01	high
Municipal					
Municipal Heritage Property	III, V	16	505	< 0.01	low[c]
Regional Parks	V	101	8,043	0.01	med. high[e]
Urban Parks	V	5	7,878	0.01	low[c]
Total			4,205,749	6.45	

[a] Restricted to hard ecotourists.
[b] Protects historical resources.
[c] Urban location.
[d] Open to potentially incompatible resource use (e.g. hunting, grazing).
[e] Open to range of recreational activities.

areas. Park managers are often heard to say that they do not wish to become another Yosemite, Amboseli, Banff or Yellowstone.

Yet, it may be argued that such problems are more a matter of mismanagement than any inherent flaw with mass tourism. As argued elsewhere in this encyclopedia (see Chapter 5), there is no essential contradiction between ecotourism and mass tourism, soft ecotourism itself being in effect a variant of mass tourism. Managers should consider the revenue-generating opportunities that arise from increased visitor flows, and concentrate their efforts on managing these flows in a sustainable way. Quotas, site-hardening measures and consumer education have already been outlined as appropriate management strategies under such circumstances. Zoning is another relevant strategy.

Combining these approaches, it is possible to accommodate very large numbers of visitors in a sustainable and satisfying way within a very small area (say for example 1%) of a protected area. Typically, the vast majority of visitors in most of the more accessible protected areas already happily confine themselves to the 1 or 2% of the park that is zoned for intensive visitor use. It is, after all, in the nature of the soft ecotourist to prefer a high level of services and comfort, and to enjoy mediated contact with the natural environment such as provided by well-designed interpretive centres and trails. These visitors rarely extend their visit into the 99% of the park that is maintained in a more or less undisturbed state. In essence, the message here is that many sustainably managed protected areas are *already* mass tourism venues or will become so, and that this scenario should be regarded as an opportunity rather than a threat that should be resisted at all costs.

Privatization

The notion of an unequivocally public protected area is being challenged by an increased tendency toward privatization and private sector involvement (Charters *et al.*, 1996). Only the most extreme free market supporters advocate a complete privatization of protected areas, but a surprisingly large number of stakeholders accept the principle that at least some elements of the operation are best left to the private sector. This assumes that the public sector does not have the resources or expertise to efficiently and sustainably manage all aspects of a protected area operation, while concurrently providing a high level of visitor satisfaction. This is especially true for less developed countries. The argument for private sector involvement also considers the substantial contributions already being made toward wildlife conservation, including the growing tendency toward the establishment of private protected areas (see Chapter 19). Concessions, accommodations, tour operations and services such as waste collection are examples of areas where the private sector is already being conceded a larger role. Areas of potential expansion include sponsorships, interpretation, policing and other facets of resource management.

While the concept of a partnership between the public and private sectors is supported in the literature (e.g. Carter, 1996), there are concerns that the profit motive will take precedence over environmental sustainability in those aspects of the protected area controlled by the private sector. Figgis (1996), for example, expresses concern over the level of degradation in Australia's Kosciusko National Park and the Victorian Alps associated with the ski industry. The rapidly growing town site in Banff National Park in the Canadian Rockies is another oft-cited illustration of private sector involvement run amok.

Relationships with local communities

It is often contended that the long-term survival of protected areas and tourism depends on the maintenance of community goodwill (Nelson *et al.*, 1993). This is due to the ability of disaffected residents to sabotage both sectors through hostile and destructive acts such as poaching, both within and outside park lands. The latter is significant given the dependency of wildlife on non-protected lands in some regions. In Kenya, for example, 60–75% of wildlife are found in these areas at any given time (Honey, 1999). Yet, there are numerous examples where ecotourism in public protected areas has contributed to the alienation of adjacent local communities through resentment caused by resource use restrictions, and/or the insufficient generation of alternative revenue through involvement in ecotourism. The extent to which ecotourism can foster a viable economic base for these local communities is often exaggerated. For example, less than 1100 of 87,000 working-age residents in the vicinity of Nepal's Royal Chitwan National Park, one of Asia's most visited protected areas, are employed directly in park-related ecotourism. In all, only 6% of working-age residents earned at least some income directly or indirectly from this source (Bookbinder *et al.*, 1998). Well-publicized problems have also occurred in the Galapagos National Park (Weaver, 2000) and in the Maasai communities adjacent to certain protected areas in Kenya and Tanzania (Honey, 1999).

Ecotourism, therefore, has not proven itself to be the godsend for local communities that it is claimed to be by some supporters. This is a matter of great concern, given that the size and expectations of local communities, and hence the demand for resources, is continuing to increase throughout most of the world. An earlier section in this chapter emphasized the potential importance of buffer zones to ecotourism, yet these could become zones of hostility to the sector, and to the public protected areas in general, if the issue of community involvement is not adequately addressed. For national governments at least, the issue is in part finding a compromise between the 'good of the nation' and the welfare of local residents, which do not always coincide. A special case, especially within more developed countries such as Canada, Australia and New Zealand, is the involvement and interests of indigenous communities (see Chapter 22).

Conclusions

The area occupied by public protected areas is continuing to expand at an impressive rate, at the same time as the total area occupied by relatively undisturbed lands continues to contract. Hence, it is not difficult to envisage a convergence in the next 20 or 30 years wherein the world's natural landscapes will essentially be confined to protected areas of one type or another. One implication is that these spaces will even further consolidate their status as the dominant venue for ecotourism-related activities. The relationship between protected areas and tourism, accordingly, is likely to become even more complicated and ambivalent, as managers are compelled to maintain the ecological integrity of their parks while accommodating an ever-increasing level of visitor demand. Yet, these two objectives, both related to the attainment of the 'public good', are not mutually exclusive. In theory, ecotourism can reinforce the environmental mandate of the parks by providing sufficient revenues and public support to fend off incursions from competing resource users. The challenge is to manage ecotourism so that it *remains* ecotourism: educational, nature-based tourism that is environmentally and socially sustainable. Relevant management strategies include the implementation of escalating user fees, sustainable site-hardening measures and consumer education. Greater synergies between ecotourism and protected areas may also be achieved in other ways, including spatial reconfigurations that better facilitate the two objectives. In addition, protected area systems need to be better utilized. Currently, ecotourism activity in many countries is largely confined to just a few, high-profile Category II and III entities, despite the presence of extensive Category IV, V and VI spaces that have great potential to accommodate the softer varieties of ecotourism. Concurrent issues that will have to be engaged include privatization and the retention of positive relationships with local communities, including indigenous groups.

References

Boersma, P. and Parrish, J. (1999) Limiting abuse: marine protected areas, a limited solution. *Ecological Economics* 31, 287–304.

Bookbinder, M., Dinerstein, E., Rijal, A., Cauley, H. and Rajouria, A. (1998) Ecotourism's support of biodiversity conservation. *Conservation Biology* 12, 1399–1404.

Butler, R.W. and Boyd, S. (2000) *Tourism and National Parks: Issues and Implications.* John Wiley & Sons, Chichester, UK.

Carter, B. (1996) Private sector involvement in recreation and nature conservation in Australia. In: Charters, T., Gabriel, M. and Prasser, S. (eds) *National Parks: Private Sector's Role.* USQ Press, Toowoomba, Australia, pp. 21–36.

Ceballos-Lascuráin, H. (1996) *Tourism, Ecotourism, and Protected Areas.* World Conservation Union, Gland, Switzerland.

Charters, T., Gabriel, M. and Prasser, S. (eds) (1996) *National Parks: Private Sector's Role.* USQ Press, Toowoomba, Australia.

Dixon, J. and Sherman, P. (1990) *Economics of Protected Areas: a New Look at Benefits and Costs.* Island Press, Washington, DC.

Figgis, P. (1996) A conservation perspective. In: Charters, T., Gabriel, M. and Prasser, S. (eds) *National Parks: Private Sector's Role.* USQ Press, Toowoomba, Australia, pp. 54–59.

Green, M.J. and Paine, J. (1997) State of the world's protected areas at the end of the twentieth century. Paper presented at the IUCN World Commission on Protected Areas Symposium, Albany, Australia, 24–29 November.

Honey, M. (1999) *Ecotourism and Sustainable Development: Who Owns Paradise?* Island Press, Washington, DC.

IUCN (1992) *Protected Areas of the World: a Review of National Systems: Neoarctic and Neotropical,* Vol. IV. World Conservation Union, Gland, Switzerland.

IUCN (1994) *1993 United Nations List of National Parks and Protected Areas.* World Conservation Union, Gland, Switzerland.

Lawton, L.J. (1993) *Protected Areas in Saskatchewan: a Statistical Report.* Technical Report 93-2. Saskatchewan Environment and Resource Management, Regina.

Lawton, L.J. (1995) *A Status Report of Protected Areas in Saskatchewan.* Policy and Public Involvement Technical Report 95-1. Saskatchewan Environment and Resource Management, Regina.

Lindberg, K. (1991) *Policies for Maximizing Nature Tourism's Ecological and Economic Benefits.* World Resources Institute, Washington, DC.

MacClintock, L., Whitcomb, R. and Whitcomb, B. (1977) Evidence for the value of corridors and minimization of isolation in preservation of biotic diversity. *American Birds* 31, 6–16.

Nelson, J.G., Butler, R.W. and Wall, G. (eds) (1993) *Tourism and Sustainable Development: Monitoring, Planning, Managing.* University of Waterloo, Waterloo, Canada.

Norris, R. (1992) Can ecotourism save natural areas? *National Parks* 1–2(66), 30–34.

NPS (1999) National Park Service forecast of recreation visits 1999 and 2000. http://www.aqd.nps.gov/stats/forecast9920.pdf

Parks Canada (1999a) National Marine Conservation Areas Program. http://parkscanada.pch.gc.ca/nmca/nmca/program.htm

Parks Canada (1999b) Parks Canada attendance 1994/95 to 1998/99. http://parkscanada.pch.gc.ca/library/DownloadDocuments/DocumentsArchive/Attendance_e.pdf

Poon, A. (1993) *Tourism, Technology and Competitive Strategies.* CAB International, Wallingford, UK.

Robinson, D.W. (1994) Strategies for alternative tourism: the case of tourism in Sagarmatha (Everest) National Park, Nepal. In: Seaton, A.V. (ed.) *Tourism: the State of the Art.* John Wiley & Sons, Chichester, UK, pp. 691–702.

Shackley, M. (1995) The future of gorilla tourism in Rwanda. *Journal of Sustainable Tourism* 3, 61–72.

UNESCO (2000) CI-UNESCO Biosphere Reserves Partnership – Biosphere Reserves. http://www.conservation.org/science/cptc/capbuild/unesco/reserves.htm

Weaver, D.B. (1998) *Ecotourism in the Less Developed World.* CAB International, Wallingford, UK.

Weaver, D.B. (1999) Magnitude of ecotourism in Costa Rica and Kenya. *Annals of Tourism Research* 26, 792–816.

Weaver, D.B. (2000) Tourism and national parks in ecologically vulnerable areas. In: Butler, R.W. and Boyd, S. (eds) *Tourism and National Parks: Issues and Implications*. Wiley, Chichester, UK, pp. 107–124.

Wells, M. and Brandon, K. (1993) The principles and practice of buffer zones and local participation in biodiversity conservation. *Ambio* 22, 157–162.

WHIN (World Heritage Information Network) (1999) *World Heritage Newsletter*. http://www.unesco.org/whc/nwhc/pages/sites/main.htm

World Conservation Monitoring Centre (1999) Transboundary protected areas. http://www.wcmc.org.uk/protected_areas/transboundary/index.shtml

World Resources Institute (1998) *World Resources: a Guide to the Global Environment*. Oxford University Press, New York.

Chapter 19

Privately Owned Protected Areas

J. Langholz[1] and K. Brandon[2]

[1]*Department of Natural Resources, Cornell University, Ithaca, New York, USA;*
[2]*Hyattsville, Maryland, USA*

Introduction

Privately owned protected areas continue to proliferate throughout much of the world. Despite this expansion, there has been little coverage of them in the literature. Research into their ecotourism aspects has been even rarer. This chapter begins to fill that information gap by: (i) describing the current state of knowledge regarding ecotourism and private nature reserves worldwide; and (ii) highlighting key issues and problems relating to private reserves and ecotourism. Specific examples of private reserves appear throughout the chapter. Our intent is to shed much needed light on this little understood but increasingly important conservation trend. Given limited public resources available for biodiversity conservation, and growing interest in both ecotourism and private sector conservation initiatives, it is imperative that we begin a systematic examination of this emerging partnership.

Background

Private nature reserves have been quietly proliferating throughout the world, yet descriptions and analysis of them have generally been indirect (e.g. Adams, 1962; Zube and Busch, 1990; Sayer, 1991; IUCN, 1992; Schelhas and Greenberg, 1993; Barborak, 1995; Murray, 1995; Borrini-Feyerabend, 1996; Uphoff and Langholz, 1998). A few authors have conducted case studies highlighting various aspects of specific reserves (e.g. Horwich and Lyon, 1987; Horwich, 1990; Glick and Orejuela, 1991; Wearing, 1993; Echeverria *et al.*, 1995; Alyward *et al.*, 1996). Additional studies have verified the private sector's increasingly large role in biodiversity conservation (Bennett, 1995; Edwards, 1995; Merrifield, 1996). Although the 63 reserves surveyed in one study were protecting more than 1 million ha (Alderman, 1994), we still lack hard data on the number of private reserves worldwide and the amount of land they are protecting. Even the World Conservation Union, IUCN, has launched projects to learn more about this relatively new conservation tool (Beltran, 1998, personal communication). In effect, what Dixon and Sherman (1990) described as a small but important development in protected area management a decade ago has evolved into a notable new direction in conservation.

While there is a standardized nomenclature for protected areas (IUCN, 1994) (see Chapter 18) there is no widely accepted typology of private protected areas worldwide. For the purposes of this chapter, we consider a private nature reserve to consist of lands that are: (i) not owned by a governmental entity at any level; (ii) larger than 20 ha; and (iii) intentionally maintained in a mostly natural condition. Langholz and Lassoie (unpublished) have proposed categorization of private reserve types based on their objectives, ownership patterns, and the considerable overlap among them. Types of reserves include: formal parks, biological stations, hybrid reserves, farmer-owned forest patches, personal retreat reserves, non-government organization (NGO) reserves, hunting reserves, corporate reserves and 'ecotourism reserves', where nature conservation is combined with tourism. Depending on definitions, indigenous and community-owned reserves can also be in the category of 'private' provided they meet the criteria given above.

Reasons for the rapid increase in private parks remain relatively unstudied. Langholz (1999a) has suggested three closely related factors:

- government failure to adequately safeguard biodiversity;
- rising societal interest in biodiversity;
- ecotourism expansion.

The first factor, government failure, stems from governments' unwillingness or inability to meet society's demand for nature recreation and conservation. Despite the creation of thousands of protected natural areas, biodiversity continues to disappear at alarming rates (World Resources Institute *et al.*, 1998). Government failure to protect biodiversity has occurred both outside protected areas and in the poor quality of protection of lands designated as protected. A large percentage of these latter areas are un- or under-protected 'paper parks' (e.g. Machlis and Tichnell, 1985; Amend and Amend, 1992; van Schaik *et al.*, 1997; Brandon *et al.*, 1998).

The second factor contributing to private nature reserve proliferation, rising societal interest in biodiversity conservation, stems from biodiversity's emergence on to the world stage over the last two decades, culminating in the 1992 Biodiversity Convention. Supplementing this legal mandate has been extensive documentation of biodiversity's value to humanity. Values include ecological, genetic, social, economic, scientific, educational, cultural, recreational, aesthetic and others (e.g. Wilson, 1992; Heywood and Watson, 1995; Kellert, 1996). Closely related to the rising awareness of biodiversity loss has been an awareness of scarcity of many species. In many parts of the world, there have been long-standing traditions of 'private' reserves or parks. These private areas have often been hunting grounds for titled nobility and elites. As recognition has increased that species numbers are dwindling, many private and undeveloped landholdings have increasingly been left as wilderness areas to meet the owners' designated conservation objectives. For example, areas that an owner had purchased with the intent to log may be left forested once the owner becomes aware of the impact of clearcutting on nearby lands. In other cases, communities have turned natural areas into community-managed reserves, often for combinations of reasons, such as keeping outsiders out, ensuring long-term access to wildland resources, and generating income through projects viewed to be compatible with local social and environmental objectives. Ecotourism is often viewed in this final category.

The final interrelated factor behind the proliferation of private parks has been the ecotourism explosion, as described in other chapters. The relationships between ecotourism and private reserves are often mixed. In some cases, nature conservation occurs primarily as a vehicle for promoting profit-motivated tourism. In other cases the opposite is true, and tourism occurs principally as a means to support nature conservation. Occupying the middle ground between these extremes are reserves where the goal is to combine profit with conservation. Lastly there have been numerous sites worldwide where owners have viewed ecotourism as an economically competitive land use compared with conversion to other uses. In some cases, the revenue from tourism alone has been sufficient to stop conversion of lands and resources to other uses. In other cases, the potential for income generation through multiple land and resources use has been a major factor in the establishment of private protected areas (Brandon, 1996). For example, Africa has a long-standing tradition of using private lands for tourism as a supplement to agriculture, ranching and other activities. There are also examples of such mixed-use ecotourism development in Latin America. Hato Pinero is a 170,000 ha Venezuelan cattle ranch that includes an 80,000 ha section used for ecotourism and wildlife protection. The reserve has a great diversity of large and readily observed birds and mammals. In the dry season from December to April, when pools are drying, birds and caiman concentrate at the remaining water sources, offering easy and spectacular wildlife viewing opportunities. The success of Hato Pinero has led to the opening of a new ranch called Chinea Arriba, just 4 ha south-west of Caracas. The ranch covers 1000 ha and is situated on the Guarico and Orituco Rivers. Costs in January 1993 for a combination of trail riding and birding were US$120 per person per day (Brandon, 1996). Ecotourism has been shown to be a profitable land use.

Prevalence of Ecotourism at Private Nature Reserves

The level and type of ecotourism at private nature reserves varies substantially from site to site. As noted in Langholz (1999), ecotourism reserves range in size and purpose from large resorts with a reserve as an added attraction for guests, to small family-run lodges that depend exclusively on tourism revenues. The prevalence of ecotourism at private reserves is highly correlated with the type of reserve and its objectives. For example, a personal retreat protected for the owner's interest in biodiversity conservation may have few, if any, visitors. Reserves managed by NGOs have highly variable levels of tourism depending on both site characteristics and the NGO's management and monetary objectives. There are no worldwide data on the prevalence of ecotourism at private nature reserves by category. However, studies which have been done indicate that ecotourism is an extremely common activity at many privately owned protected areas. Alderman (1994) showed that in 1989 tourism accounted for 40% of operating income among 63 private reserves in Latin America and sub-Saharan Africa. By 1993, a follow-up study of the same private reserves (Langholz, 1996a) showed that this dependency on tourism had increased to 67% of their operating income. Nearly half of all respondents (n = 15) said they depended on tourism for 90% or more of their revenues, and slightly over one-third (n = 12) were completely dependent on tourism (Langholz, 1996b).

In a study focusing exclusively on private reserves in Costa Rica, Langholz (1999a) found that ecotourism was an extremely common activity, with more than half of all reserves engaging in it on some level. Sixty per cent of 68 reserve owners studied hosted overnight visitors either 'sometimes' or 'often', and 46% hosted daytime visitors. Yet equally interesting was the finding that 40% of reserves hosted tourists 'rarely' or 'never'. This surprisingly high percentage refutes a common perception that all private reserves are involved in the ecotourism industry. It is a signal to policy makers that private reserves can thrive even in the absence of a well-established tourism industry.

Emergence of NGO and Community-managed Reserves

As described earlier, there is an array of categories of private reserves. While 'private' is often thought of as equivalent to ownership by a limited number of people or a corporation, it can also include lands owned by communities or indigenous groups, provided that these groups are non-governmental. Throughout Latin America, there are numerous examples of private reserves being established and maintained by NGOs or communities as a means to protect biodiversity. This trend is of sufficient importance that it merits separate mention here. Three different kinds of private reserves, which all include ecotourism as a key component, are described. The first is a large reserve managed by an NGO primarily for biodiversity conservation. The second is a reserve created by contiguous small holders joining sections of their lands to be managed in a common manner. The third category, an outgrowth of the 'community' emphasis of joint smallholder plots, is of areas owned, leased, or managed through usufruct rights by a community as reserves, with the objective of tourism. The tourism represented in this spectrum included a variety of types of tourists, from ecotourists, to those principally interested in indigenous culture, or archaeology, ethnobotany, and a few adventure tourists. What each of these three models of private reserves has in common is that all tend to be supported by national or international conservation organizations. Examples of each category are given below.

One of the best known cases of a large reserve managed by an NGO primarily for biodiversity conservation is the Rio Bravo Conservation and Management Area in Belize. An NGO called 'Programme for Belize' (PfB) legally owns and manages this large (92,614 ha) reserve, equivalent to 4% of Belize's terrestrial area, in perpetuity for the public good. PfB is a Belizean non-profit organization established in 1988 to promote the conservation and wise use of Belize's natural resources. Ecotourism development at the reserve has been viewed as one of the primary mechanisms for supporting the reserve's management costs. By the end of fiscal year 1995, total revenues earned from tourism-related activities in Rio Bravo covered 45% of the operating expenses of PfB (Wallace and Naughton-Treves, 1998). Another well-known example of an NGO managed reserve is the Monteverde Cloud Forest Reserve, established in Costa Rica in 1973 and operated by the Tropical Science Center in San Jose. Ecotourism to the reserve has been so high that the reserve has been on the cutting edge of using pricing policies to regulate visitation (Church and Brandon, 1995; Honey, 1999).

The Community Baboon Sanctuary, which is located in a rural community 53 km north-west of Belize City, provides an example of contiguous smallholders joining sections of their land for common management. This reserve was established in 1985 to protect the black howler monkey, *Alouatta pigra*, through the efforts of an expatriate and 12 landowners. Local landowners were asked to follow a land-use plan which would maintain a skeletal forest from which howlers and other species could use the regenerating cut forests, while helping landowners reduce riverbank erosion and reduce cultivation fallow time. The sanctuary at the last report included over 120 landowners. Ecotourism has been a significant component of the programme, with over 6000 tourists visiting in 1990. Yet the most significant programme impact has been to entice over 30 communities to undertake 'community-based ecotourism and resource management programs' (Horwich and Lyon, 1987). These range from 'entirely private lands to private and public land mosaics to entirely public lands' (Horwich and Lyon, 1987).

A survey conducted by The Nature Conservancy in the Peten in Guatemala and Belize identified ten community-based tourism projects (J. Beavers, 1995, unpublished). What all of these groups had in common was a desire to obtain economic benefits from tourism through some site-based activity. Half of the groups were renting the land that they used as the equivalent of a 'reserve'. The majority of these areas were located in or near a protected area, including archaeological sites. The majority of these groups desired to capture economic benefits that they felt were bypassing them, and created reserves that drew in tourism (Norris *et al.*, 1998). There are comparable examples, although more limited in number, of indigenous communities in South Africa designating portions of communally owned lands for tourism and conservation (Honey, 1999).

Non-tourism Activities at Private Reserves

As suggested above, ecotourism is not the only activity occurring in many privately owned protected areas. It is simply one land use among a broad portfolio of options. There are virtually no broad-based studies of non-tourism activities at private reserves, with the exceptions of the studies that follow. Alderman (1994) found that ranching and agriculture provided for 17% of reserves' income. Non-government reserves such as those related to CAMPFIRE in Zimbabwe, often rely on hunting as a revenue source (Metcalfe, 1993). Langholz (1999) showed that 84% of private reserve owners in Costa Rica said they used reserves for their own 'personal enjoyment' above all else, either 'sometimes' or 'often'. Other common activities, in descending order of frequency, were: research projects (44%), harvesting logs for construction or artesanal products (25%), collecting firewood for home use (24%), harvesting decorative plants (24%), harvesting medicinal plants (22%), grazing cattle or horses (19%), and harvesting wild food plants for nursery or home use (19%).

Owners engaged in several additional activities, but to a much lesser extent. These additional activities included mining for rocks or sand, and harvesting wild food plants or logs for sale. Note that except for 'personal enjoyment', all of the activities either produce revenue directly or serve to avoid household expenditures. Likewise, many reserve owners in Costa Rica are now receiving cash payments, technical assistance, property tax breaks, and other government incentives in exchange for carbon sequestration and other environmental services provided by their reserves (Langholz *et al.*, 2000a). The clear message from these data is that private reserve owners rely on a broad array of revenue-producing activities that can supplement or replace ecotourism revenues.

Profitability among Private Nature Reserves

Sustainability requires not just the maintenance of ecological integrity, but also attention to social and economic factors. Regarding economics, Alderman (1994) showed 38% of reserves to have been profitable during the year preceding the data collection (1988). Furthermore, her study group optimistically predicted a 26% rise in profitability within the following 5 years. Their optimism proved justified, but not to the extent expected. Langholz (1996a) documented that by 1993 profitability had risen by 21% among private reserves. Given ecotourism's ongoing expansion, it is unlikely that this trend toward higher profitability will reverse. Wells (1997) correctly notes, however, that the profitability of such operations is difficult to estimate, partly because these private businesses are under no obligation to disclose financial information, and partly because no serious effort has been made to study them from a financial or economic perspective. For a more detailed discussion of economic considerations at privately owned parks, readers are referred to Langholz *et al.* (2000b).

Many of the newer private reserves worldwide have been established to specifically cater to the increasing ecotourism market. One such ecotourism reserve in Costa Rica had cleared over US$1.4 million in profit over its first 6 years in operation. The amount of profit would be even higher had the owners not decided to pay off, ahead of schedule, the US$1.2 million borrowed to purchase the land. While the sensitive nature of such financial information precludes us from providing specific details on the reserve (e.g. name, location, owners), we can discuss several factors contributing to this reserve's high profitability. Important among them is the fact that the reserve provides a high-value wilderness experience at a high price. The guests who occupy the reserve's 14 bungalows enjoy excellent luxury service and lodging within a jungle setting and pay more than US$160 per night. By catering to an upscale clientele, the owners assure financial viability while keeping to a minimum the number of tourists who use the reserve. Catering to the affluent also minimizes competition with most of Costa Rica's other private nature reserves. Although this particular reserve is exceptional, it reveals the substantial revenue that ecotourism can provide for a private reserve.

Key Issues and Problems

Main problems at private nature reserves

Private reserves are subject to the overall trends and prevailing conditions in the countries in which they are located. It seems apparent that the success of private ecotourism ventures depends on the tourist's perception of general environmental quality in the region. Like public parks, private protected areas face a wide variety of problems. Alderman (1994) found that budget deficiencies, poaching and lack of cooperation from government entities were considered by owners to be the biggest threats to their reserves. The results 4 years later (Langholz, 1996a) were similar, except that budget deficiencies had dropped from first to third place. This shift, combined with the increase in profitability discussed earlier, supports the assertion that private reserves are experiencing improved financial health. Other problems mentioned by landowners included: political unrest in the country, community opposition to loss of access to reserve's resources, squatters, and community opposition to tourism. Ecotourism could provide some benefits by serving as a justification for combining conservation and ecotourism with other management objectives (Brandon, 1996).

Langholz (1999) examined the most pressing problems faced by reserve owners in Costa Rica. Based on poaching's high ranking in previous studies, this particular topic was divided into three separate subgroups: 'poaching of birds', 'poaching of logs', and 'poaching of other plants and animals'. Such segmentation clarified this important issue, with 'poaching of logs' proving to be only a minor problem, ranking near the bottom of the list. 'Poaching of birds' presented more of a problem, ranking fifth overall. The most severe problem was 'poaching of other plants and animals'. According to owners, poachers take mammals mostly. Examples included: white-tailed deer (*Odocoileus virginianus*), peccary (*Tayassu tajacu* and *Tayassu pecari*), agouti (*Dasyprocta punctata*), jaguars (*Felis onca*), pumas (*Felis concolor*), ocelots (*Felis pardalis*) and, especially, a large rodent known as the tepiscuintle (*Agouti paca*).

Problems specifically related to ecotourism were mentioned by private reserves in Costa Rica, but were not common. For example, one reserve owner on the Osa Peninsula had problems with a visitor who was an illegal wildlife trafficker in disguise. The 'guest' used his stay as an opportunity to capture exotic birds for transport to the USA. Similarly, another 'ecotourist' captured colourful winged insects so he could photograph them. Captured insects were dropped into a bowl containing hot water, causing them to open their wings and allowing the photographer to get the shots he wanted. Finally, an owner in Sarapiqui reported that tourists were to blame for excessive trail erosion, a problem that the owner should be able to prevent.

Fluctuations in the ecotourism market

Those reserves dependent on ecotourism are subject to vagaries of the ecotourism market. Although ecotourism continues to expand globally, there is no guarantee that this trend will continue indefinitely. Those reserves located in politically unstable countries are especially vulnerable. Peruvian reserves, for example, suffered financially during the 1980s, when the Shining Path was actively terrorizing many parts of the country. Reserve owners in the Republic of South Africa were uneasy during that country's recent and potentially explosive political transition. Similarly, war-torn Colombia remains unattractive as an ecotourism destination, despite its incredible natural heritage and large number of private nature reserves. Most Colombian reserves, therefore, depend on the domestic tourism market or on other sources of financing. Clearly, political instability remains a troublesome wildcard for private reserves, a factor that lies largely beyond their control yet greatly influences their destiny.

If having too few ecotourists can be a problem, then having too many is also a concern. Rapid growth in tourism can lead to overly ambitious development of lodging, thus causing a rebound effect when the supply of ecotourism destinations outpaces demand. This may have happened in Costa Rica, where steady and dramatic tourism growth in the 1980s and early 1990s finally came to a halt in the mid-1990s. The ensuing slump crippled several private reserves, many of which have yet to recover. Compounding the situation was the kidnapping of a foreign tourist. The kidnapping made headlines in the tourist's home country, causing several thousands of trip cancellations. Competition among private reserves in Costa Rica is likely to increase as more and more ecotourism reserves are established.

These and other fluctuations in the tourism market highlight private reserves' precarious position, and limits to the private-sector approach. In short, biodiversity conservation is such an important and long-term endeavour that it should not be subjected to tourism's short-term and potentially lethal fluctuations. To the extent that they are dependent on ecotourism, private conservation efforts should continue to play a supplementary role to larger governmental efforts to protect natural areas (see Chapter 18). They should be relied upon as a supplement to a nation's conservation strategy, rather than its mainstay.

Conflicts of interest

The single biggest challenge for ecotourism at private nature reserves lies with conflicts of interest. Reserve owners may be tempted to cut corners, pitting profit versus protection. This temptation appears to be rising, as newcomers, who are attracted to private reserves as a financial investment, join older conservation-minded private reserve owners. The financial success of private reserves runs the risk of attracting new entrants who are business people first, and conservationists second, people willing to make conservation trade-offs in the interest of making or saving money. An example of this would be taking wildlife from a wild area and transporting it to the reserve to augment the reserve's wildlife populations. Even worse, there are cases where wildlife have been captured in protected areas and transported to private reserves to establish a small zoo or penned area so tourists can see and photograph the wildlife easily. Protected area and wildlife laws often make such actions illegal. Similarly, laws governing private reserves may include restrictions as well; maintaining captive animals is strictly forbidden among members of Costa Rica's network of private nature reserve owners. However, with few park guards, wildlife agents, or other means to limit wildlife trade and capture, the response of tourists is often either a significant deterrent to, or determinant of, such behaviour, depending on the values of private reserve owners (see Chapter 41).

While maintaining captive animals may be a particularly egregious example of placing economics above ecology, it is by no means the only one. Reserve owners can also be tempted to overbuild facilities. This includes constructing excessive roads, buildings and other infrastructure within the reserve. Quality is also an issue, in terms of the type and amount of construction. Roads, buildings, sewage systems and power facilities all need to be built with an eye toward minimizing their adverse impacts on the reserve. In places where hunting serves as a revenue source, there exists an immense temptation to harvest wildlife at unsustainable rates. Finally, the sheer volume of tourists visiting the reserve needs to be carefully monitored. Larger numbers of visitors may mean greater revenue for the owners, but they can stress a reserve's natural features by eroding trails and frightening wildlife. Owners should use existing tools to determine the human carrying capacity of their reserves.

A related problem is 'piggy-backing' by non-conservationists. As Yu *et al.* (1997) explain, ecotourism reserves can get away with doing little conservation. Owners can rely on an area's conservation reputation without contributing to it. This is especially possible in regions or countries considered attractive as ecotourist destinations. In effect, owners can get away with protecting only the surprisingly small amount of habitat needed to stage a nature walk, relying on a typical tourist's inability to recognize substantially degraded habitat. As Yu *et al.* note, this presents a classic free-rider situation that is not in conservation's best interest.

Similarly, private reserve owners may allocate their lands into officially recognized conservation status for reasons other than biodiversity. In the Republic of South Africa, for example, several wealthy white landowners placed their property into conservation status as power transferred to that country's black majority. They were concerned not so much about protecting biodiversity, but rather about losing their large holdings to government land redistribution schemes (Brinkate, 1996). A similar phenomenon appears to be occurring in Zimbabwe. Even in Costa Rica, some private reserve owners have joined conservation programmes principally for economic reasons. These included improved marketing and publicity resulting from being an officially recognized conservation area (Langholz, 1999). Reserves owned by large corporations, whose principal activities degrade nature, sometimes fall into this category. What better way for a mining, logging, or oil company to draw attention away from the nature it destroys than by creating a token nature reserve, photos of which can adorn the company's newspaper advertisements and annual reports? In such cases, private nature reserves serve more for green-washing than conservation.

Finally, successful private reserve owners may find it very difficult to sell their reserve, including an ecotourism business, to new owners who will carry on their vision. For example, owners of a successful rainforest reserve with an ecolodge on the Osa Peninsula in Costa Rica have combined educational tourism, conservation, sustainability and responsibility to local communities. After 6 years of successful operations, they want to sell their preserve and lodge, but they want to ensure that the delicate balance between rainforest protection and tourism is maintained, and that there is strong cooperation with local communities. Legal protection to carry out this vision is difficult; should much of the land be turned over to the government and merged with an adjacent park, even though the government's resources to manage it are low? Should the owners sell the land along with the lodge in the hope the rainforest will in fact be protected? What price should be charged, the fair market value of increasingly desirable property for hotel or housing construction, or a vastly lower price typically reserved for 'undeveloped' or 'unmanaged' land? As yet, there is virtually no experience on how private reserves managed under one owner's vision can be successfully transferred and managed by others.

Competition with public parks

Privately owned protected areas tend to be located directly adjacent to larger public parks. Alderman (1994), for example, found 46% of the 63 reserves in her study to be bordering public parks. In Costa Rica, an estimated 51% of private reserves are adjacent to public parks (Langholz, 1999). This close proximity has important ecological and economic implications, which are detailed in Langholz (1999) and discussed in Chapter 18. From an ecological perspective, private reserves abutting publicly protected areas help to extend ecosystem functions and stabilize habitats. Being adjacent can also enhance tourism. The national park can attract tourists to the area, many of whom stay at the private park. Likewise, a well-known private park can attract tourists to the area, who then visit the national park as well, creating revenue for the park system. A similar exchange occurs with respect to tourism activities. An adjacent national park adds to the range of natural history options offered by the private park (e.g. a museum or education centre). Similarly, a private park can add to the range of natural history options available in the area (e.g. a canopy walkway).

However, private reserves can also drain benefits from publicly managed parks. For example, a private reserve adjacent to a large national park might be able to lure wildlife from the park, by having saltlicks or water holes. High fees for lodging, upscale tourism, and low management costs lead to high revenue. But this private reserve may in fact draw visitors away from national parks. Other examples include the better potential of private reserves to use outside tour operators and engage in more sophisticated marketing than park management authorities. At the same time, high visitation to private reserves may be another way to demonstrate to governments that wildlife and wilderness areas have good earning potential if properly structured and managed. For example, one study found that a private reserve, Monteverde Cloud Forest Reserve, generates more income from tourism than is generated by all Costa Rican national parks together (Church and Brandon, 1995). A final problem is that private reserves which promote ecotourism may not necessarily meet the criteria of conservation and local benefits that most ecotourists would like to believe are taking place. Honey (1999) provides an excellent review of a variety of ecotourism projects. Her book highlights the difference between 'green lodges and ecotourism' (Reichert in Honey, 1999). Such lodges adjacent to parks, even if classified as private reserves, may in fact compete with or drain resources from public areas.

Thus far, private nature reserves have operated largely in a vacuum. Most governments do not even know how many private reserves exist within their country, or where they are located, let alone attempt to monitor reserves' performance. Private reserves are often protected informally, with no official government recognition or regulation. This makes it especially difficult to ensure that biodiversity conservation is really occurring. It makes it easier for landowners to cut conservation corners, as described earlier. Even if governments did know of private reserves' whereabouts, many developing country governments have difficulty maintaining and monitoring conditions within their public park systems. Governments incapable of overseeing their public parks are even less likely to regulate private ones. As private reserves and ecotourism continue to expand, it will become increasingly important to ensure that conservation is really taking place.

Conclusion

This chapter has explored the little-known relationship between privately owned protected areas and ecotourism. It has described the current state of knowledge regarding ecotourism and private nature reserves worldwide, and has highlighted key issues and problems. We have emphasized background information on private reserves, the prevalence of ecotourism and other activities within them, and their profitability. For key issues, we have focused on reserves' problems, fluctuations in the ecotourism market, conflicts of interest, relationships with public parks, monitoring and evaluation, and social impacts.

As ecotourism and privately owned protected areas continue to expand throughout the tropics, it is increasingly important that we learn more about them. We have barely scratched the surface in terms of understanding their unique role and relationship. Fortunately, a follow-up study to those of Alderman (1994) and Langholz (1996a) will be available soon, which should cast more light on this topic (Mesquita, unpublished). Even with the additional perspective offered by the new study, however, we will still lack even the most basic information about private reserves and ecotourism. The conservation community should initiate a systematic examination of ecotourism and private nature reserves, assessing not just their status and niche relative to public protected areas, but also their key inherent strengths and weaknesses as conservation and development tools. Given ongoing habitat loss in the tropics and recent reductions in public expenditures for protected areas, such an effort would make a major contribution to protecting the world's natural heritage.

References

Adams, A. (ed.) (1962) *First World Conference on Parks.* National Park Service, Washington, DC.

Alderman, C. (1994) The economics and the role of privately-owned lands used for nature tourism, education, and conservation. In: Munasinghe, M. and McNeely, J. (eds) *Protected Area Economics and Policy: Linking Conservation and Sustainable Development.* IUCN and The World Bank, Washington, DC, pp. 273–305.

Alyward, B., Allen, K., Echeverria, J. and Tosi, J. (1996) Sustainable ecotourism in Costa Rica: the Monteverde cloudforest preserve. *Biology and Conservation* 5, 315–343.

Amend, S. and Amend, T. (1992) Human occupation in the national parks of South America: a fundamental problem. *Parks* 3, 4–8.

Barborak, J. (1995) Institutional options for managing protected areas. In: McNeely, J. (ed.) *Expanding Partnerships in Conservation.* Island Press, Washington, DC, pp. 30–38.

Bennett, J. (1995) Private sector initiatives in nature conservation. *Review of Marketing and Agricultural Economics* 63, 426–434.

Borrini-Feyerabend, G. (1996) *Collaborative Management of Protected Areas: Tailoring the Approach to the Context.* World Conservation Union, Gland, Switzerland.

Brandon, K. (1996) *Ecotourism and Conservation: a Review of Key Issues.* Global Environment Division, Biodiversity Series, Paper No. 033. World Bank, Washington, DC.

Brandon, K., Redford, K. and Sanderson, S. (eds) (1998) *Parks in Peril: People, Politics, and Protected Areas.* Island Press, Washington, DC.

Brinkate, T. (1996) People and parks: implications for sustainable development in the Thukela Biosphere Reserve. Presentation at *The Sixth International Symposium on Society and Natural Resource Management.* The Pennsylvania State University, University Park.

Church, P. and Brandon, K. (1995) *Strategic Approaches to Stemming the Loss of Biological Diversity.* Center for Development Information and Evaluation, US Agency for International Development, Washington, DC.

Dixon, J.A. and Sherman, P.B. (1990) *Economics of Protected Areas: a New Look at Benefits and Costs.* Island Press, Washington, DC.

Echeverria, J., Hanrahan, M. and Solorzano, R. (1995) Valuation of non-priced amenities provided by the biological resources within the Monteverde cloudforest reserve. *Ecological Economics* 13, 43–52.

Edwards, V. (1995) *Dealing in Diversity: America's Market for Nature Conservation.* Cambridge University Press, Cambridge.

Glick, D. and Orejuela, J. (1991) La Planada: looking beyond the boundaries. In: West, P. and Brechin, S. (eds) *Resident Peoples and National Parks: Social Dilemmas and Strategies in International Conservation.* University of Arizona Press, Tucson, pp. 150–159.

Heywood, V.H. and Watson, R.T. (eds) (1995) *Global Biodiversity Assessment.* Cambridge University Press, Cambridge.

Honey, M. (1999) *Ecotourism and Sustainable Development: Who Owns Paradise?* Island Press, Washington, DC.

Horwich, R.H. (1990) How to develop a community sanctuary – an experimental approach to the conservation of private lands. *Oryx* 24, 95–102.

Horwich, R.H. and Lyon, J. (1987) Development of the community baboon sanctuary in Belize: an experiment in grassroots conservation. *Primate Conservation* 8, 32–34.

IUCN (1992) *Protected Areas of the World: a Review of National Systems,* vol. 4: *Neoarctic and Neotropical.* IUCN, Gland, Switzerland.

IUCN (1994) *1993 United Nations List of National Parks and Protected Areas.* IUCN, Gland, Switzerland.

Kellert, S.R. (1996) *The Value of Life: Biological Diversity and Human Society.* Island Press, Washington, DC.

Langholz, J. (1996a) Economics, objectives, and success of private nature reserves in sub-Saharan Africa and Latin America. *Conservation Biology* 10, 271–280.

Langholz, J. (1996b) Ecotourism impact at independently-owned nature reserves in Latin America and sub-Saharan Africa. In: Miller, J. and Malek-Zadeh, E. (eds) *The Ecotourism Equation: Measuring the Impacts.* Yale School of Forestry and Environmental Studies Bulletin Series, No. 99, New Haven, Connecticut, pp. 60–71.

Langholz, J. (1999) Conservation cowboys: privately-owned parks and the protection of tropical biodiversity. Dissertation, Cornell University, Ithaca, New York.

Langholz, J., Lassoie, J.P. and Schelhas, J. (2000a) Incentives for Biodiversity Conservation: Lessons from Costa Rica's Private Wildlife Refuge Program. *Conservation Biology* 14(6), 1735–1743.

Langholz, J., Lassoie, J., Lee, D. and Chapman, D. (2000b) Economic considerations of privately owned parks. *Ecological Economics*, 33(2), 173–183.

Machlis, G. and Tichnell, D. (1985) *The State of the World's Parks: an International Assessment for Resource Management, Policy, and Research.* Westview Press, Boulder, Colorado.

Merrifield, J. (1996) A market approach to conserving biodiversity. *Ecological Economics* 16, 217–226.

Metcalfe, S. (1993) The Zimbabwe Communal Areas Management Programme for Indigenous Resources (CAMPFIRE). In: Western, D., Wright, M. and Strum, S. (eds) *Natural Connections: Perspectives in Community-based Conservation.* Island Press, Washington, DC, pp. 161–192.

Murray, W. (1995) Lessons from 35 years of private preserve management in the USA: the preserve system of The Nature Conservancy. In: McNeely, J. (ed.) *Expanding Partnerships in Conservation.* Island Press, Washington, DC, pp. 197–205.

Norris, R., Wilber, J. and Marin, L. (1998) Community-based ecotourism in the Maya Forest: problems and potential. In: Primack, R., Bray, D., Galleti, H. and Ponciano, I. (eds) *Timber, Tourists, and Temples: Conservation and Development in the Maya Forest of Belize, Guatemala, and Mexico.* Island Press, Washington, DC.

Sayer, J. (1991) *Rainforest Buffer Zones: Guidelines for Protected Area Managers.* IUCN's Forest Conservation Program, Gland, Switzerland.

Schelhas, J. and Greenberg, R. (1993) *Forest Patches in the Tropical Landscape & the Conservation of Migratory Birds*, Migratory Bird Conservation Policy Paper No. 1. Smithsonian Migratory Bird Center, National Zoological Park, Washington, DC.

Uphoff, N. and Langholz, J. (1998) Incentives for avoiding the tragedy of the commons. *Environmental Conservation* 25, 251–261.

Van Schaik, C., Terborgh, J. and Dugelby, B. (1997) The silent crisis: the state of rainforest nature reserves. In: Kramer, R., van Schaik, C. and Johnson, J. (eds) *Last stand: Protected Areas and the Defense of Tropical Biodiversity.* Oxford University Press, New York, pp. 64–89.

Wallace, A. and Naughton-Treves, L. (1998) Belize: Rio Bravo Conservation and Management Area. In: Brandon, K., Redford, K. and Sanderson, S. (eds) *Parks in Peril: People, Politics, and Protected Areas.* Island Press, Washington, DC.

Wearing, S. (1993) Ecotourism: the Santa Elena rainforest project. *The Environmentalist* 13, 15–135.

Wells, M. (1997) *Economic Perspectives on Nature Tourism, Conservation, and Development.* Environmental Economics Series, No. 55. The World Bank, Washington, DC.

Wells, M. and Brandon, K. (1992) *People and Parks: Linking Protected Area Management with Local Communities.* International Bank for Reconstruction, Washington, DC.

West, P. and Brechin, S. (eds) (1991) *Resident Peoples and National Parks: Social Dilemmas and Strategies in International Conservation.* University of Arizona Press, Tucson.

Western, D. (1989) Conservation without parks: wildlife in the rural landscape. In: Western, D. and Pearl, M. (eds) *Conservation for the Twenty-first Century.* Oxford University Press, New York, pp. 158–165.

Western, D. (1993) Ecosystem conservation and rural development: the case of Amboseli. In: Western, D., Wright, M. and Strum, S. (eds) *Natural Connections: Perspectives in Community-based Conservation.* Island Press, Washington, DC, pp. 15–52.

Western, D. and Pearl, M. (eds) (1989) *Conservation for the Twenty-first Century.* Oxford University Press, New York.

Wilson, E.O. (1989) Conservation: the next hundred years. In: Western, D. and Pearl, M. (eds) *Conservation for the Twenty-first Century.* Oxford University Press, New York, pp. 3–7.

Wilson, E. (1992) *The Diversity of Life.* Belknap Press of Harvard University Press, Cambridge, Massachusetts.

World Resources Institute, The United Nations Environment Programme, The United Nations Development Programme, and The World Bank (1998) *World Resources 1998–99.* Oxford University Press, New York.

Yu, D.W., Hendrickson, T. and Castillo, A. (1997) Ecotourism and conservation in Amazonian Peru: short-term and long-term challenges. *Environmental Conservation* 24, 130–138.

Zube, E.H. and Busch, M.L. (1990) People-park relations: an international review. *Landscape and Urban Planning* 19, 115–132.

Chapter 20

Modified Spaces

L.J. Lawton[1] and D.B.Weaver[2]
[1]*School of Business, Bond University, Gold Coast, Queensland, Australia;*
[2]*School of Tourism and Hotel Management, Griffith University Gold Coast Campus, Queensland, Australia*

Introduction

The purpose of this chapter is to outline the actual and potential scope for ecotourism within landscapes that have been extensively modified as a result of human intervention. The first section provides the theoretical rationale for considering modified spaces as an appropriate ecotourism venue. Subsequent sections consider the actual and potential status of ecotourism within various categories and subcategories of such space. All of the Earth's landscapes, of course, have been modified to some extent by human activity, but the focus here is on those spaces that have been fundamentally altered by related processes. Categories considered for discussion purposes include agricultural land, urban and peri-urban land, artificial reefs, service corridors and devastated spaces.

The Rationale for Ecotourism in Modified Spaces

Because ecotourism fundamentally relies on the natural environment for its attraction base, it is not surprising that an emphasis is placed on relatively 'natural' or 'unmodified' landscapes as the most appropriate venue for ecotourism-related activities. This linkage is perhaps most apparent in the extent to which this form of tourism is associated with protected areas (see Chapters 18 and 19). If, however, the attraction base of ecotourism is construed as including the natural environment *or some component thereof*, then there is no inherent reason for neglecting, as a potential setting, spaces that have been more modified by various forms of human activity. Many species of wild animal, for example, make extensive use of altered environments, and it is this adaptability factor that is a core element of this chapter's focus on the suitability of modified spaces as viewing sites for such creatures. Ecotourism, then, does not have to be restricted to the context of a wild animal's natural habitat. As illustrated in subsequent sections, modified spaces can actually harbour higher wildlife populations and diversity than natural environments. Furthermore, modified spaces may provide better opportunities and scope for viewing because of their greater accessibility to population concentrations.

Beyond the occurrence and visibility of wildlife within modified spaces, these spaces should also be considered more seriously for ecotourism purposes for other reasons. These include the fact that such lands occupy a large and ever-expanding portion of the Earth's land surface (World Resources Institute, 1998). Being already altered from their natural state, they have greater resilience in terms of their carrying capacity for accommodating ecotourism activities. In addition, the mobilization of such lands for ecotourism will serve to alleviate the pressure that is being exerted by this rapidly growing sector on the ever-shrinking inventory of natural landscapes. Ecotourism may even provide an incentive for the restoration of modified spaces, or at the very least for the management of these lands so that they remain viable as wildlife habitat. Finally, it should be emphasized that most of the world's population has ready access to open modified spaces, so that ecotourism in such locations is less costly and effectively more egalitarian than ecotourism that is confined to natural environments, and wilderness landscapes in particular.

The literature that considers the management and status of wildlife within modified spaces is extensive. However, despite the rationale provided here, the literature that explicitly examines the actual or potential linkages between such spaces and ecotourism is negligible. This chapter is one of the first concerted attempts to 'take stock' of the issue, and is thus primarily exploratory in character.

Ecotourism on Agricultural Lands

Lands used for agriculture account for over half of the world's land surface, and therefore constitute an important potential venue for ecotourism. (See Chapter 27 for further discussion of the link between ecotourism and rural areas.) For discussion purposes below, 'agricultural land' is subdivided into croplands, grazing land and areas of shifting cultivation.

Cropland

'Cropland', which includes temporary and permanent crops (including orchards and other tree crops), temporary meadows, and market and kitchen gardens, occupies about 10% of the Earth's land surface. However, such lands are spatially concentrated, with just four countries (USA, India, Russia, China) accounting for 40% of this total (World Resources Institute, 1998). Although an extreme example of environmental modification, croplands are capable of sustaining significant wildlife populations, especially when they are located in proximity to remnant or extensive natural spaces that provide shelter and adequate breeding habitat. Some types of wildlife, such as white-tailed deer, can even experience significant population increase through the opportunistic exploitation of agricultural landscapes.

Traditionally, the perceptions of crop farmers toward wildlife have ranged from ambivalence to outright hostility. Whether as a result of the trampling of maize crops by elephants in Kenya, or the destruction of newly planted wheat by migratory waterfowl in the North American prairie, some types of wildlife are regarded as pests that inflict major crop damage. It is therefore not surprising that 80% of US farmers report crop damage each year, and that 77% allow hunters access to their land. The ambivalence is indicated by the supplementary fact that 51% of US farmers undertake deliberate management practices to attract or sustain wildlife, including the provision of cover or water, and the practice of leaving crop residue or unharvested crops in their fields. American farmers, cumulatively, invest an estimated 120 million h and US$2.5 billion in expenditure on such measures (Conover, 1998).

Although the tourism/recreation interface between cropland and wildlife has traditionally been dominated by hunting, non-consumptive forms of activity are becoming more important both in absolute and relative terms. Recreational trends in the USA, in fact, show a strong increase in participation among ecotourism-related activities such as birdwatching, while the greatest decline was recorded in the proportion of American adults who engage in hunting (Cordell *et al.*, 1995).

It is premature to say whether these

changes indicate that cropland-based tourism is in a state of transition from a consumptive to a non-consumptive emphasis. However, hunting and ecotourism do appear to be occurring concurrently in some areas, leaving open the potential for conflict among the various types of participants and landowners. Evidence for such concurrence is available from the Canadian province of Saskatchewan, where wildlife viewing and hunting, respectively, are the two most popular client activities reported by vacation farm operators. Of particular interest are the 16 operations where both activities were regarded as 'very important' client opportunities (Weaver and Fennell, 1997). This research further indicated that wildlife viewing occurred on lands that were farmed as well as in adjacent natural areas. Because of seasonal limitations, viewing occurred mainly in the spring and summer, and birds (perching species, migratory waterfowl and birds of prey in particular) were the major observed types. However, despite the importance of wildlife viewing, very few operators had any background or training in ecotourism, while only about 20% had prior qualifications in any aspect of tourism at all (Weaver and Fennell, 1997).

In one of the very few attempts to formally link heavily modified farmland regions with ecotourism, Weaver (1997) has advocated the recognition of Saskatchewan's grain-growing 'agricultural heartland' as a distinct ecotourism 'context zone'. This suggestion not only recognizes the potential of such areas to accommodate a viable ecotourism sector, but acknowledges that such areas will be subject to different management-related issues from those that affect protected areas and other relatively natural settings. These include:

- competition and coexistence with hunters, who will remain a significant user of wildlife resources into the foreseeable future despite the pattern of declining participation;
- the maintenance of remnant natural lands and habitat corridors as a means for supporting viable local wildlife populations;
- a tenure pattern characterized by a large number of privately owned properties, as opposed to most higher-order protected areas, which are wholly controlled by a particular government body;
- the transfer of owner/wildlife problems from hunters to ecotourists; for example, the demand for compensation related to wildlife crop damage will be re-focused toward non-consumptive wildlife-oriented tourists and organizations;
- attaining a coexistence between farm-related activities and ecotourism, which could be negatively affected by machinery noise, pesticide applications, predatory cultivation, etc.; and
- facilitating the involvement of declining agricultural service centres in ecotourism.

Depending on the specific geographic context, circumstances can be identified in some cropland-dominated destinations that could support the possibilities for ecotourism. For example, the English countryside has an extensive network of public footpaths that are already heavily used for non-consumptive recreational and tourism purposes. Similarly, well-developed farm tourism networks in the UK and other parts of western and central Europe already offer a form of accommodation that is complementary to rural ecotourism. In terms of landscape features that support wildlife populations in cropland environments, the hedgerows of England and France are notable, as are the thousands of artificial farm ponds that have been established in the USA beneath the Central and Mississippi flyways (Adams and Dove, 1989). Some plantation landscapes in the Caribbean and in other parts of the less developed world are noteworthy for their ecotourism potential because of the extent to which they offer a forest-like setting for wildlife. The Asa Wright Nature Preserve on the island of Trinidad is a major ecotourism facility that relies primarily on such plantations for its attraction base.

Grazing land

According to the World Resources Institute (1998), 'permanent pasture' is land occupied by natural or cultivated grasses that have been used for at least 5 consecutive years as forage. An estimated 23% of the world's land surface is classified as permanent pasture under these criteria, with Australia, China, the USA, Kazakhstan and Brazil accounting for about 40% of the world total.

When converted from a grassland biome, grazing land differs from cropland in resembling more closely its former natural state, especially if native rather than exotic grasses are maintained for forage purposes. This suggests an enhanced potential to accommodate native wildlife species. A further advantage for ecotourism can be termed the 'visibility' factor. Wildlife is more visible on grazing lands due to the absence or sparseness of woody cover, which also allows the viewer to observe a large amount of space in a single gaze.

The body of literature that links ecotourism-related activities to grazing land is not extensive, though larger than that which relates to cropland. Weaver and Fennell (1997) found that native ungulates such as mule deer were an important wildlife viewing resource on vacation farms located within the range lands of Saskatchewan. Bryan (1991) describes the increasing importance of ecotourism within the guest ranch industry in Wyoming, Montana and Idaho, which has arisen through a combination of factors, including the need for economic diversification, recognition of the importance of sustainable agriculture, and market shifts away from consumptive tourism. As with Saskatchewan's vacation farms, the revenue obtained from ecotourism is not large, but in many cases makes the difference between continued viability and bankruptcy for the overall ranch operation.

The evidence is not confined to North America. In South Africa, approximately 10,000 of 60,000 commercial farmers qualify as 'game ranchers'. According to Pauw and Peel (1993), the animals are used for trophy hunting, sport hunting, live sales and meat sales, as well as non-consumptive tourism. A similar situation, although on a much smaller scale, pertains to the commercial ranchlands of Namibia (Weaver and Elliott, 1996).

On semi-commercial and subsistence-based communal grazing lands in sub-Saharan Africa, the ecotourism/rancher relationship is even more complicated. In such situations, wild animals have long been regarded as unwelcome pests that compete for forage with domesticated animals and introduce disease, especially in lands adjacent to protected areas, where intrusions by wildlife are more frequent. By one estimate, between 65 and 80% of Kenya's wild animals may be found outside of the country's protected area network at any given time (JICA, 1994). The Kenyan government has long been aware of the problem, and has experimented with a variety of policies, including strict anti-poaching measures and compensation payments. Tourism revenue-sharing is a more recent innovation. Since wildlife hunting is currently banned in Kenya, the latter initiative is almost entirely ecotourism-based, and has been implemented with some success in the Maasai-dominated grazing regions (Weaver, 1998).

In Zimbabwe, the wildlife/tourism connection on communal grazing lands is controversial due to the emphasis that the well-known CAMPFIRE programme places on hunting. Over 90% of all associated community revenues are derived from trophy fees, and ecotourism is widely perceived as an option that requires more investment in infrastructure (Weaver, 1998). The Zimbabwe situation, however, is unusual is so far as most grazing-oriented destinations (as with cropland) appear to be gradually moving away from hunting, and toward ecotourism. The tension between hunting and ecotourism that seems to be present in most farming situations is in turn indicative of a broader trend away from farming activity in areas of more marginal production. According to Holmes (1996), the marginal range lands of

Australia are undergoing a radical restructuring due to declining commodity values. He characterizes this restructuring as a shift from a productionist to a post-productionist era in rural resource use. Less productive land, according to this argument, is now surplus in terms of commodity output, but more in demand for its amenity values (i.e. meeting human needs and wants). These amenity values include tourism, recreation, biodiversity preservation and Aboriginal land use. Within the tourism component, moreover, the non-consumptive element is more in keeping with the post-productionist ethos, whereas hunting is more complementary to the productionist mode. Europe has been experiencing a similar transition within its agricultural lands, in a process that has been described as the 'consumption of the countryside'.

If the above thesis is true, then one might expect not just the increased incidence of ecotourism on grazing lands, but the possible conversion of commercial grazing land back to its original state. One of the most audacious and ambitious proposals for range land restoration is the Buffalo Commons scenario that has been put forward for the agriculturally marginal portions of the North American prairie (Gauthier, 1994). This idea suggests that conventional commercial farming should be abandoned altogether across a broad contiguous swath of land large enough to accommodate the reintroduction of the great buffalo herds and associated wildlife. Most versions of the scenario support ecotourism as the economic activity that would be most suited to this re-created pre-European (but eminently post-modern) landscape.

Areas of shifting cultivation

By some estimations (e.g. Getis *et al.*, 1996), shifting cultivation is practised to a greater or lesser extent on as much as one-fifth of the world's land surface, and especially within the tropical rainforest biome. The most compelling argument for ecotourism in these farming regions is the fact that most of the land base is occupied by various stages of forest succession at any given time. As a result, a variety of adjacent niches is available to accommodate a very broad range of wildlife. Furthermore, the practice of shifting cultivation, if carried out in a sustainable fashion, offers an interesting ancillary cultural attraction for ecotourists. Some community-based ecotourism sites in the tropical rainforests of Papua New Guinea and Fiji incorporate areas of shifting cultivation, though the full potential of such areas is clearly not being realized.

Urban and Peri-urban Areas

As in farmland situations, the relationship between wildlife and urban areas has traditionally been perceived as one of incompatibility. This is due in large part to the problems created by proliferating non-domesticated animals such as rats, wild dogs and coyotes, starlings, bats and pigeons. However, in the post Second World War era, the philosophy of urban planning in the more developed countries has increasingly emphasized the association between enhanced quality of life and the provision of semi-natural environments that accommodate desirable species of wildlife (Laurie, 1979a). Accordingly, the field of urban ecology has evolved considerably since the 1970s (Adams and Dove, 1989), and substantial resources have been invested to make the urban environment more conducive to desirable wildlife species. In the USA alone, residents of metropolitan areas spend an estimated 1.6 billion h and US$5.5 billion per year on urban wildlife management (Conover, 1998).

While the notion of 'urban ecotourism' may be considered an oxymoron by some, there is no inherent reason for precluding urbanized areas as legitimate ecotourism venues, as long as the basic criteria for this activity are met (see Chapter 1). Reflecting the resources that have been invested in wildlife management and other facets of urban ecology, the modern city has much to offer in terms of potential ecotourism settings and attractions. Contrary to their popular image as 'concrete jungles', urban areas in the more developed world are actually dominated by green space. For example, only 23% of the Japanese city of Chiba (population 800,000) is occupied by actual buildings and concrete, while 85% of the city of Waterloo, Canada (population 90,000) can be considered 'green' (Dorney, 1986). Only a small portion of this green space bears a resemblance to natural habitat, but this land nonetheless provides valuable wildlife habitat and contributes to the urban oasis effect in farmland or desert settings. To this factor must be added the fact that urban areas usually attract large numbers of tourists associated with VFR (visits with family and friends) and business tourism, as well as various historical and cultural attractions. The potential for exposing these tourists to the semi-natural spaces of the city is considerable.

To date, few efforts have been made to recognize and exploit the formidable ecotourism potential of urban areas. One notable example is the Canadian city of Fredericton in the province of New Brunswick, which is incorporating ecotourism opportunities into its 'Riverfront Master Plan' (personal communication, Joel Richardson, New Brunswick Department of Economic Development, Tourism and Culture). In this case, however, much of the emphasis is placed on relatively natural lands beyond the built-up area that happen to fall within the political boundaries of the municipality. The discussion below, in contrast, emphasizes extensively modified settings within the built-up urban environment *per se*, and similar urban-related settings within the urban–rural fringe.

Parks

Urban parks are one of the more obvious settings that have ecotourism potential, in part because of their accessibility to the public. Where they occupy a large amount of space and are already well-known as tourist attractions, this potential is already being realized, although the term 'ecotourism' might not be explicitly used. Examples in North America include New York's Central Park, High Park in Toronto, Griffith Park in Los Angeles, and Vancouver's Stanley Park. In Europe, the scope is even more extensive in some metropolitan areas due to the presence of large forest parks (e.g. the Tagel and Grunewald in Berlin and the Bois de Vincennes in Paris). The same applies to the surviving urban 'commons' of England, which are now used primarily for non-consumptive recreational purposes (Laurie, 1979b).

Cemeteries

Urban cemeteries have also been recognized as a valuable wildlife habitat. To a similar or even greater extent than parks, cemeteries are a 'guaranteed' green space in which the vegetation is allowed to mature to a degree seldom encountered outside mature natural forests. Unlike parks, however, wildlife viewing in a cemetery setting is a 'monopoly' recreational activity that does not compete with other leisure pursuits or facilities. Mount Auburn Cemetery in Cambridge, Massachusetts has long been famous as a wildlife viewing location for both tourists and local residents (Adams and Dove, 1989).

Golf courses

Golf courses tend to be more of a peri-urban rather than strictly urban land use, and have long been criticized as environmental disaster zones because of their emphasis on exotic vegetation and excessive applications of fertilizers and pesticides. However, in keeping with broader social trends in the direction of environmental sustainability, growing interest is being shown in the development of 'naturalized' golf courses that also serve as high-quality wildlife habitat. Admittedly, this trend is also motivated by the obvious long-term cost savings that are associated with the reduced use of pesticides and fertilizers, and with reductions in the area of manicured terrain.

The issue of golf courses as wildlife habitat is not a trivial one, given that the average course occupies 54 ha, and given that the USA alone now hosts some 25,000 golfing facilities, with one new course on average being opened every day. Legitimate questions can be raised as to whether either wildlife or golfers will be attracted to naturalized courses. Threats to wildlife include the potential danger from golf balls and chemical pesticides that are used on remaining manicured areas. Possible problems for golfers include the higher probability of losing golf balls in rough terrain, and the presence of distracting and/or dangerous wildife (Terman, 1997). In addition, it is not clear how ecotourists would be accommodated on golf courses, given restricted access to the latter and their primary role of providing golfing opportunities. In cooler climates, this possible incompatibility could be resolved by using golf courses primarily for wildlife viewing during the golf off-season.

Terman (1997) suggests that the number of naturalized golf courses is still very small in proportion to the total number of such facilities. Furthermore, there is no clear indication that these or other types of golf course are actually being used for ecotourism to any extent. However, as to the question of whether naturalized golf courses are attractive to both wildlife and golfers (and thus constitute a viable option), the Prairie Dunes Country Club in Kansas provides some positive evidence. Three-quarters of the course (i.e. roughs, out-of-play areas, buffer zone) is being used to re-establish native prairie grasses, and research has shown that the course supports a species-rich bird population, and a higher density of birds than occurs in nearby natural areas. From a golfer perspective, the course rates as one of the most popular in the country.

Sewage lagoons and stormwater control ponds

Sewage treatment facilities in most cities include settling ponds that facilitate the removal of solid wastes and other undesirable materials. Later-stage settling ponds are often very attractive to birds and other types of wildlife, as demonstrated by the Tinicum Marsh near Philadelphia (Adams and Dove, 1989). With 118 recorded bird species, the Hornsby Bend settling lagoon in Austin Texas, on the Central flyway, is a major site for birdwatching, including educational field trips (Bonta, 1997).

Stormwater control ponds are another artificial water-retention facility that may be conducive to ecotourism. According to Tourbier (1994), there is an increased tendency to replace standard dry basins with permanent 'extended detention basins' and 'wet basins' as part of the changing perception of stormwater as a resource rather than a nuisance. When augmented by natural vegetation plantings to discourage erosion and encourage water retention and penetration, and incorporated into urban greenway systems, such sites tend to be highly attractive to wildlife. Adams and Dove (1989), for example, found that artificial stormwater control ponds are actually preferred by some species of duck over natural bodies of water.

Landfill and waste disposal sites

Active urban landfill and waste disposal sites are associated with wildlife pests such as gulls (as in North America) and sacred ibises (as in Australia), and in most cases it is only the 'retired' sites that merit consideration for their ecotourism potential. A possible exception is the town of Churchill, on Canada's Hudson Bay, where the local garbage dump is considered to be a choice site for the viewing of polar bears. However, as this type of feeding activity is highly questionable in terms of its implications for environmental sustainability, it will not be discussed any further as evidence of ecotourism.

The consideration of retired landfill sites as ecotourism venues has been given impetus by improvements made since the 1970s in the technology for safe landfill disposal and closure, which has increased the suitability of these sites as wildlife habitat. One interesting example is the Saugus landfill site in Massachusetts. The plastic liner that caps the landfill means that the area can only be rehabilitated as natural grassland; however, as natural grassland is extremely rare in this part of the USA, the site has attracted a large number and variety of unusual bird species. Trails and observation blinds have been established, and the site is widely known as a quality venue for wildlife viewing (Anon, 1996).

The Leslie Street Spit in Toronto is also indicative of the ecotourism potential of landfill sites. The site is an artificial peninsula that extends 5 km into Lake Ontario as a result of landfill and rubble deposition during the 1960s and 1970s. Like its natural counterpart at Point Pelee, the Spit is very attractive to migratory birds (at least 290 species have been observed) and other types of wildlife. Moreover, the site's management strategy has emphasized its importance as a laboratory for plant succession. Organized bus tours of the Spit began in 1973, and by 1978, 18,000 visitors were recorded, including a high proportion of naturalists. Wildlife viewing activity on the site has been supported by the establishment of observation blinds and by the production of educational brochures (Carley, 1998).

High-rise and other structures

The link between urban high-rise structures and ecotourism is focused on the peregrine falcon, a rare species whose numbers declined dramatically this century due to the influence of DDT and other pesticides. The recovery of this species in the USA from 62 pairs in 1975 to 875 pairs in 1994 is due in large part to the use of urban high-rise buildings as surrogate falcon nesting habitat. This artificial environment actually improves on natural conditions by providing a location relatively safe from great horned owls, the falcon's main predator. In addition, feral pigeons provide an abundant food supply (Line, 1996). From an ecotourism perspective, the falcons are easily observed in their nests and in flight from the interior of the buildings.

Aside from these deliberate programmes of species establishment, wildlife has adopted to urban buildings in other ways. For example, about 50 peregrine falcons now winter along power lines and 'concrete rivers' in metro Los Angeles (Line, 1996). Even more of an unintended consequence is the heat island effect, which has caused insects to fly at a higher than normal altitude. The availability of this food supply means that nighthawks and chimney swifts now nest on rooftops and chimneys, as opposed to their usual nesting sites in grass or tree hollows close to ground level. These species are now more common in urban areas than in the countryside, and constitute another urban ecotourism resource (Dorney, 1986). In Europe, roof-nesting storks serve as a tourist attraction in some urban areas.

Zoos and botanical gardens

These two elements of the urban environment are raised here mainly as a point of discussion. In tourist visitation terms, zoos and botanical gardens are by far the most frequented wildlife-related urban sites. However, because of the captive nature of their wildlife attractions, they are not usually associated with ecotourism, even if the three basic criteria are met. Of interest then would be the determination of the degree of captivity that separates a non-ecotourism wildlife attraction from one which can be classified as ecotourism.

Given the heavily modified and heavily populated character of cities, urban ecotourism will of course differ fundamentally from rural ecotourism with respect to its character and management. An obvious implication is how such activities can be sustainably accommodated in the midst of so many potentially incompatible urban distractions. One possible response is to avoid the fragmentation of appropriate sites, and to concentrate instead on managing contiguous sites as unitary ecotourism venues. For example, it is common in urban environments to find parks, cemeteries, university campuses and golf courses in close proximity to each other, or linked through urban greenway networks. These areas can reinforce each other in terms of critical habitat mass and in screening undesirable urban land uses. Another issue more germane to urban areas is distinguishing ecotourists (as in those who meet the travel threshold criteria associated with tourism) from local residents engaging in the same activities, since the latter are likely to constitute by far the largest user group.

Artificial Reefs

Artificial reefs can be unintentional or intentional. The best example of the former is a shipwreck, while the latter can result from the deliberate submersion of a decommissioned ship or other material suitable for reef formation. Among the prominent American examples, Florida's Dade County has one of the most comprehensive artificial reef programmes, with 89 ships, bridges and barges submerged since 1981. The state of Texas has a 'rigs to reefs' programme to dispose of old or obsolete oil rigs (Fritz, 1994).

While artificial reefs are usually created as a way of improving fish habitat for commercial fishing, it is not clear whether these reefs actually increase fish populations, or merely aggregate existing populations, and hence make them more susceptible to harvest. As a result, there is an increased movement toward the use of artificial reefs for non-consumptive purposes such as ecotourism (Brock, 1994). An example of the artificial reef/ecotourism connection is the Atlantis Waikiki artificial reef off Honolulu, which covers 1.85 ha. This site consists of a sunken oiler, concrete terrace reefs and several surplus aircraft. Atlantis submarine tours began using this site as an attraction in 1989, and current volume now exceeds 200,000 passengers per year, and 20,000 divers, for annual gross revenues of US$8 million. This amount greatly exceeds earnings that would have been obtained had the site been used only for fishing purposes (Brock, 1994).

In the Gulf of Aqaba, off the Israeli port of Eilat, artificial reefs constructed from surplus aircraft and dead coral heads are being established specifically to reduce diving pressure on existing reefs, given that divers like to visit wrecks. In 1996, 16,000 users were reported on just one of these reefs (Wilhelmsson *et al.*, 1998). Low scrap metal prices are giving impetus to similar initiatives in other parts of the world. The HMAS *Swan* attracted 4000 divers in the first 4 months following its submersion off Western Australia in 1997 (O'Brien, 1998). The Artificial Reef Society of British Columbia, founded in 1986, has completed five projects involving the sinking of decommissioned navy ships. Each is estimated to create an average of CDN$1 million per year in revenue generation if properly managed as a wildlife-based viewing attraction (Jones and Welsford, 1997).

Service Corridors

In keeping with the principles of the post-productionist landscape described earlier, rail lines and other infrastructure that are no longer required to facilitate the movement of goods are gradually being closed down or, in an increasing number of cases, converted into recreational 'greenways'. In the USA, this process of functional adaptation has led to the establishment of 1003 'rail-trails' as of late 1998, accounting for 16,635 km of surplus line. Another 1239 trails, accounting for 29,692 km of line, are in the process of being established. These trails were estimated to have accommodated 100 million users in 1998, a significant portion of which were tourists engaged in wildlife viewing (Rails to Trails Conservancy, 1999).

Because of their linear character, rail-trails and other converted service corridors present their own unique set of management challenges as ecotourism resources. These include the possibility that they pass through more than just one municipality and that their narrow buffer zones are inadequate to accommodate wildlife or to minimize the influence of adjacent land uses. It is in respect to the latter issue that rail-trails should be managed as part of the 'farmland ecotourism' possibility discussed earlier in this chapter.

Devastated Landscapes

A final category that should be considered for discussion purposes incorporates landscapes that have been devastated as a result of human intervention. One sub-category consists of landscapes seriously altered by mining activities, such as are found in the interior of Nauru (a phosphate-mining island in the South Pacific), in the strip-mined counties of the American Appalachians, areas occupied by mine tailings and, more ubiquitously, in gravel pit operations. Often too degraded to be used for other economic activities, these devastated landscapes may offer a relatively safe habitat for various types of wildlife, especially if excavation activity results in the presence of artificial ponds. The consequent potential for ecotourism could be even further enhanced by efforts to selectively rehabilitate such landscapes, which leaves open possibilities for the pursuit of 'restoration ecotourism', as demonstrated earlier by the Leslie Street Spit example.

In addition to mining and excavation, devastated landscapes may result from warfare or predatory farming or logging practices. An intriguing example of the latter, which raises many questions about the 'ground-rules' of ecotourism and sustainability, is the 8000 ha Sarigua National Park on the south coast of Panama. This park is popular among ecotourists because of the desert-like terrain and wildlife that are found nowhere else in the country. Far from being a natural ecological anomaly, however, the area is actually an example of severe desertification resulting from deforestation and over-grazing. Yet, this highly modified landscape is valued enough to be accorded status as a high-level protected area (Navarro, 1998).

Areas devastated by warfare have much less obvious potential for ecotourism due to the presence of minefields, unexploded shells and other war-related detritus, and also because of the geopolitical tensions and negative destination image that may persist after the cessation of actual hostilities. The possibilities for ecotourism in such an unlikely environment, however, are illustrated by attempts to have the demilitarized zone (DMZ) between North and South Korea designated as a wildlife reserve. The 1000 km^2 area was almost entirely farmed prior to 1950, devastated by the Korean War, and then allowed to revert to a virtual wilderness that harbours many rare species of wildlife due to its status as a de facto protected area. The conversion of this corridor into an ecotourism-themed 'peace park' is one of several related options being suggested for the DMZ once tensions ease between the two Koreas.

Conclusions

The intent of this chapter has been to illustrate, in an exploratory and anecdotal manner, the potential of modified spaces to provide opportunities for the pursuit of ecotourism. This discussion should not be interpreted as an attempt to justify the continuing conversion of natural and semi-natural landscapes into modified spaces, nor is it intended to imply that these modified spaces can substitute for those natural spaces. Rather, it is emphasized that areas already modified by human intervention can accommodate a significant wildlife presence, and that this in turn creates possibilities for ecotourism-related activity. There are several clear advantages for doing so. First, potential ecotourism revenues provide a financial incentive to maintain and expand wildlife habitat within modified spaces, and thus to further engage the issue of coexistence between humans and nature in urban, peri-urban and farming/grazing environments. Second, ecotourism in modified spaces can serve to relieve the pressures of this rapidly expanding form of tourism on relatively undisturbed environments that are less resilient and declining in area. Third, these venues are readily accessible to most of the world's population, thereby facilitating a more affordable and practical mode of ecotourism activity, while at the same time offering wildlife-viewing experiences for local residents. The next logical stage is to create a more detailed inventory of actual ecotourism opportunities within the various categories of modified space, and to use this as a basis for encouraging the expansion and refinement of such activity and, indeed, its acceptance by the more conventional ecotourism sector. This should include the inter-linking of the various modified ecotourism spaces, and the linkage of these areas with ecotourism in relatively undisturbed areas, and with the non-wildlife based tourism industry.

References

Adams, L.W. and Dove, L.E. (1989) *Wildlife Reserves and Corridors in the Urban Environment: a Guide to Ecological Landscape Planning and Resource Conservation.* National Institute for Urban Wildlife, Columbia, Maryland.

Anon (1996) Turning landfills into wildlife habitat. *American City & Country* 111(10), 16–17.

Bonta, M. (1997) Turning wastewater into wetlands. *Living Bird* 16(4), 22–26.

Brock, R.E. (1994) Beyond fisheries enhancement: artificial reefs and ecotourism. *Bulletin of Marine Science* 55, 1181–1188.

Bryan, B. (1991) Ecotourism on family farms and ranches in the American West. In: Whelan, T. (ed.) *Nature Tourism: Managing for the Environment.* Island Press, Washington, DC, pp. 75–85.

Carley, J. (1998) The Leslie Street Spit: the creation and preservation of a public urban wilderness. http://www.interlog.com/~fos/history.fos.html

Conover, M. (1998) Perceptions of American agricultural producers about wildlife on their farms and ranches. *Wildlife Society Bulletin* 26, 597–604.

Cordell, H., Lewis, B. and McDonald, B. (1995) Long-term outdoor recreation participation trends. In: Thompson, J., Line, D., Gartner, B. and Sames, W. (eds) *Proceedings of the Fourth International Outdoor Recreation and Tourism Trends Symposium and the 1995 National Recreation Resource Planning Conference.* University of Minnesota, St Paul, pp. 35–38.

Dorney, R. (1986) Bringing wildlife back to cities. *Technology Review* 89, 48–54.

Fritz, S. (1994) The oceans: octopus inns. *Popular Science* 244, 30.

Gauthier, D. (1994) The buffalo commons on Canada's plains. *Forum for Applied Research and Public Policy* (Winter), 118–120.

Getis, A., Getis, J. and Fellmann, J. (1996) *Introduction to Geography,* 5th edn. Wm. C. Brown Publishers, Dubuque, Iowa.

Holmes, J. (1996) Diversity and change in Australia's rangeland regions: translating resource values into regional benefits. *Rangeland Journal* 19(1), 3–25.

JICA (1994) *The Study on the National Tourism Master Plan in the Republic of Kenya (Interim Report).* Japan International Cooperation Agency, Nairobi.

Jones, A. and Welsford, R. (1997) Artificial reefs in British Columbia. http://www.artificialreef.bc.ca/research/tjones.html

Laurie, I. (1979a) *Nature in Cities.* John Wiley & Sons, Chichester, UK.

Laurie, I. (1979b) Urban commons. In: Laurie, I. (ed.) *Nature in Cities.* John Wiley & Sons, Chichester, UK, pp. 231–266.

Line, L. (1996) Symbol of hope? *National Wildlife* 34, 36–41.

Navarro, J.C. (1998) *Panama National Parks.* Ediciones Balboa, Panama City.

O'Brien, B. (1998) Wreck creation. http://www.divernet.com./wrecks/creation1298.htm

Pauw, J. and Peel, M. (1993) Game production on private land in South Africa. *Proceedings of the XVII International Grassland Congress,* Vol. III, pp. 2099–2100.

Rails to Trails Conservancy (1999) Facts about rail-trails. http://www.railtrails.org/genfact.html

Terman, M.R. (1997) Natural links: naturalistic golf courses as wildlife habitat. *Landscape and Urban Planning* 38, 183–197.

Tourbier, J. (1994) Open space through stormwater management: helping to structure growth on the urban fringe. *Journal of Soil and Water Conservation* 49(1), 14–21.

Weaver, D. (1997) A regional framework for planning ecotourism in Saskatchewan. *Canadian Geographer* 41, 281–293.

Weaver, D. (1998) *Ecotourism in the Less Developed World.* CAB International, Wallingford, UK.

Weaver, D. and Elliott, K. (1996) Spatial patterns and problems in contemporary Namibian tourism. *Geographical Journal* 162, 205–217.

Weaver, D. and Fennell, D. (1997) The vacation farm sector in Saskatchewan: a profile of operations. *Tourism Management* 18, 357–365.

Wilhelmsson, D., Öhman, M.C., Ståhl, H. and Shlesinger, Y. (1998) Artificial reefs and dive tourism in Eilat, Isreal. *Ambio* 27(8), 764–766.

World Resources Institute (1998) *1998–99 World Resources: a Guide to the Global Environment.* Oxford University Press, New York.

Chapter 21

Wilderness

W.E. Hammitt and M.C. Symmonds

Department of Parks, Recreation and Tourism Management, Clemson University, Clemson, South Carolina, USA

Introduction

When the terms 'wilderness' and 'ecotourism' are combined one inevitably thinks about environmental impacts and the potential, or lack of, for such pristine environments to support any form of wilderness recreation use. Impacts of ecotourism are noted here in the context of wilderness but are also discussed in more detail, and in more general terms, in other chapters (see especially Chapters 23 to 25). The following sections will focus on present wilderness tourism use in the developing and developed world, the potential of wilderness to support ecotourism, and future management and conservation issues.

What is Ecotourism and What is Wilderness?

Ecotourism has been defined by a number of researchers. For example, Ceballos-Lascuráin (1988, p. 5) stated that ecotourism is travel 'to relatively undisturbed or uncontaminated natural areas with the specific object of studying, admiring, and enjoying the scenery of its wild plants and animals, as well as any existing cultural aspects found in these areas'. He also suggested that 'the person who practices ecotourism will eventually acquire a consciousness that will convert him into somebody keenly interested in conservation issues'. More recently, Honey (1999, p. 25) attempted to synthesize definitions to form the following definition:

> Ecotourism is travel to fragile, pristine, and unusually protected areas that strives to be low impact and (usually) small scale. It helps educate the traveller; provides funds for conservation; directly benefits the economic development and political empowerment of local communities; and fosters respect for different cultures and for human rights.

With relation to wilderness use for ecotourism, the above components of ecotourism will be used as a premise for the following discussions.

There is no global definition of 'wilderness'. As with ecotourism, many definitions have been suggested. What constitutes wilderness ultimately depends upon the value placed on an area by people and institutions, and the area itself relative to the surroundings and alternatives. However, the two major aspects of all definitions of wilderness that distinguish it from other environments are degree of 'naturalness' and 'solitude-primitiveness'. Ecotourism and other forms of wildland recreation in wilderness must be dependent on the natural processes and solitude experiences of wilderness areas. Manipulation of ecological processes to restore naturalness, and of social processes to restore solitude are permissible in wilderness, but the forces of nature must dominate those of humans. The purpose of this chapter is not to debate the definition of wilderness. Instead, wilderness will be defined within the broad terms that follow, and its place in ecotourism and its potential to support the ecotourism industry will be discussed.

'Wilderness' derives from Norse and Teutonic languages. In rhetorical terms, 'wild' was derived from 'willed', meaning self-willed or uncontrollable. The word 'deor', from Old English meaning animals not under the control of man, was combined with 'wild' to form 'wilderness'. Thus, 'wild-deor-ness' means 'place of wild beasts' (Nash, 1973). Physically, wilderness refers to places or regions that are uncultivated and uninhabited such as swamps, forested areas, grass plains and savannah, and mountains. Other areas such as the oceans have also been classified as wilderness (Hill, 1994). Socially and psychologically, wilderness is a place that provides 'outstanding opportunities for solitude or a primitive and unconfined type of recreation' (Wilderness Act, 1964). Thus, in terms of ecotourism, wilderness is a place where one can obtain a primitive travel and recreation experience away from society and the built environment.

The term 'wilderness' will not be used interchangeably with 'wilderness area'. A wilderness area, although pertaining to the components of wilderness as described above, is by definition an area of the USA set up under the provisions of the Wilderness Act, 1964. A wilderness area, as defined by the Act, has additional determinants such as minimum recommended acreage and permitted uses. For the purpose of describing ecotourism on a global scale, we will use the term 'wilderness' as defined by the physical, social-psychological components described above and not only by Western criteria.

Some remote, large portions of the Earth's land surface qualify as wilderness. Much, if not most, of its ocean areas also qualify as wilderness. Wilderness exists as very large areas of minimal human intervention (e.g. Antarctica, Arctic, Siberia, Amazon basin, Central Asia, interior Australia) and as much smaller areas that offer some kind of reclusive experience in a relatively undisturbed environment (e.g. urban interface areas in the eastern USA and Europe). A small, but isolated island in the ocean may qualify as wilderness. Thus, size is not necessarily a qualifying limitation of wilderness, unless certain associated ecosystem functions are qualifying limitations.

Neither is the presence or lack of people a qualifier for an area to be wilderness. In traditional cultures, 'wilderness' areas are commonly intimate, lived-in spaces. But, in the Western perception of wilderness there is a dissociation with people. This dual notion of wilderness has led to management conflicts, including the deliberate expulsion of native people from protected areas, only to see ecotourism development occur. Such practices have significant implications concerning the sacrifice of socio-cultural sustainability in order to purportedly bring about environmental sustainability. However, if managed correctly, conservation-based ecotourism can serve to sustain socio-cultural as well as environmental resources of certain protected areas and wilderness. The following sections discuss the use and potential use of world wilderness for ecotourism.

Present Use of Wilderness for Ecotourism

Coccossis (1996) asks: Is there any limit to the future expansion of tourism? As tourism has expanded, the concern over the interaction of tourism and the environment has grown (OECD, 1980). Tourism depends on the local environment, whether it involves natural, social or cultural resources. Thus, the quality of environmental assets is often the key determinant of travellers' choice of destination (Coccossis, 1996). However, the environment is not infinite and there is a limit to the expansion of tourism. Wilderness has a value and therefore a tolerance on the amount of use it can sustain.

The use of wilderness for ecotourism is a highly debated issue. In developed countries, wilderness is generally protected by stringent management policy, usually determined at a national level. Thus, wilderness is often used for tourism in an appropriate and sustainable manner. In the USA, the Wilderness Act 1964 established strict guidelines for the designation and use of wilderness areas. At present, more than 1 million acres have been designated and protected under the Act (Rosenburg, 1994). Wilderness such as the Bob Marshall Wilderness Area, Montana, The Eagle Cap Wilderness Area, Oregon, and the Okefenokee Swamp Wilderness Area, Georgia, are all managed by site-specific guidelines and those guidelines specified by the Act. Similarly, in the UK, the National Parks and Access to the Countryside Act 1949 and later, the Wildlife and Countryside Act 1981 established National Parks and Sites of Special Scientific Interest (SSSI).

All of these areas are nationally protected from change for the benefit of future generations. Thus, the character of wilderness is retained so that solitude and a primitive experience can be sustained, as well as an environment dominated by natural forces. Land is usually acquired by governments or conservation organizations before restrictions and policy are implemented to avoid disputes over who is at 'loss' from the conservation or preservation of an area. Economics are usually the basis for preservation disputes (Rosenburg, 1994). When areas are set aside they deprive someone of potential income, whether from forestry, agriculture, mining or other sources. Thus, governments in developed nations have realized that one of the easiest ways to preserve pristine areas is to purchase and acquire land. Ecotourism can then be controlled in many of the country's most pristine environments.

On a global scale, much wilderness is not protected. In the developing world, areas of Africa, Asia and Latin America are used for tourism with minimal protection. Antarctica is an area with essentially no on-site management or protection, yet is receiving an increasing amount of tourship visitation. The following case studies highlight the use of wilderness for ecotourism and some of the issues encountered in safeguarding the environment for future users and upholding the values of wilderness ecosystems. These values include the potential for adventure and exploration, education, scientific discovery and study and, ultimately, conservation.

Wilderness in the developing world

Case Studies. Africa: a comparison of Cameroon and Kenya

BACKGROUND. The Republic of Cameroon's wilderness is largely unprotected. It has six national parks, five of which are savannah ecosystems. The remaining area, Korup National Park, is a rainforest ecosystem, designated in 1986 (Fig. 21.1). Korup was formulated with the objective of conserving biodiversity. It is home to more than 400 species of trees and is an important habitat for African forest elephants and a diversity of primates (more than a quarter of the world's primate species live in Korup), birds and fish, and also more than 3000 species of rare plants (Topouzis, 1990). Korup National Park is a conservation area with strict sustainable development principles. However, the local community is restricted from using the park for hunting (Gilbert *et al.*, 1994).

In contrast, Kenya has long been promoted as an ecotourism destination. Tourism in Kenya largely grew out of the foundations of big game hunting (Honey, 1999). However, since the late 1970s when a ban was placed on game hunting for commercial profit, the orientation of tourism has adapted to more sustainable principles. In the early 1990s Kenya was hailed as 'the world's foremost ecotourist attraction' by some observers (e.g. Perez, 1991).

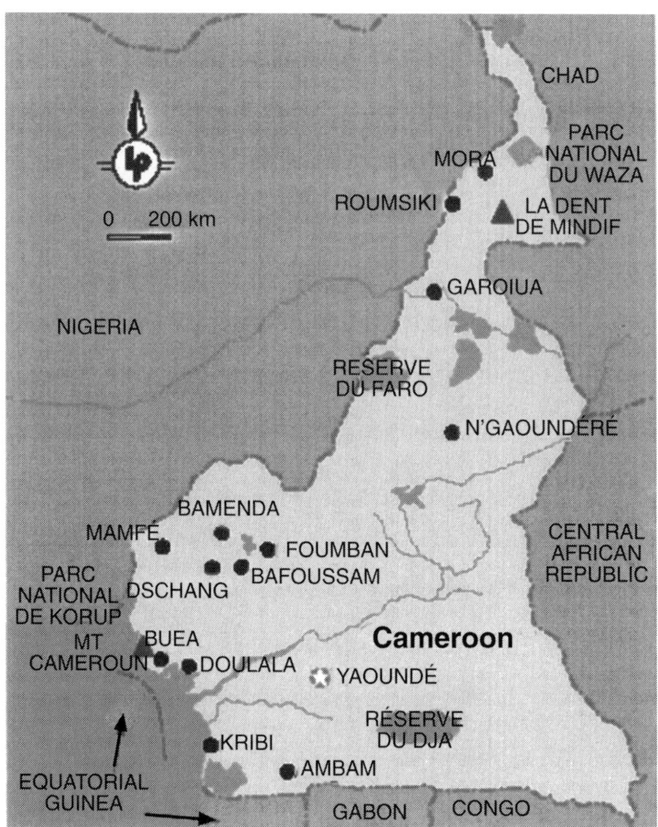

Fig. 21.1. Map of Cameroon with Korup National Park located on the western border (web site: http://www.lonelyplanet.com/dest/afr/graphics/map-cam.gif).

National parks occupy 7% of the country and tourism accounts for 25% of the national income (Gilbert *et al.*, 1994). In the early 1990s, no other African country earned as much revenue from tourism as Kenya (Honey, 1999). The amount of revenue generated from tourism amounts to 40% or US$400 million of total foreign income. However, only US$13 million (3%) is directed to the Kenyan Wildlife Service (Pearce, 1995). Areas of wilderness such as the Masai Mara and Tsavo National Park are marketed for wildlife safaris (Fig. 21.2). Although such tourism is verging on mass tourism in some cases as will be detailed later, these areas remain wilderness. They are untouched and uncultivated for the most part, where large game still inhabit the prairies as they have done for thousands of years. Between the Masai Mara and Serengeti (in adjacent Tanzania), these areas offer the largest concentration of wildlife anywhere in Africa, and the largest land migration of animals anywhere in the world (Honey, 1999). Thus, the opportunity still exists for a wilderness experience although it is debatable whether most tourists achieve or seek this.

Almagor (1985) discussed the upsurge in African vacationing due to the search for 'first hand communion with nature' which is the 'vision quest' of people going to Kenya for safari experiences. These motivations are quite similar to those contained in the definition of wilderness and the experiences it provides. Another factor that has led to the upsurge in ecotourism demand in Kenya is the country's marketing strat-

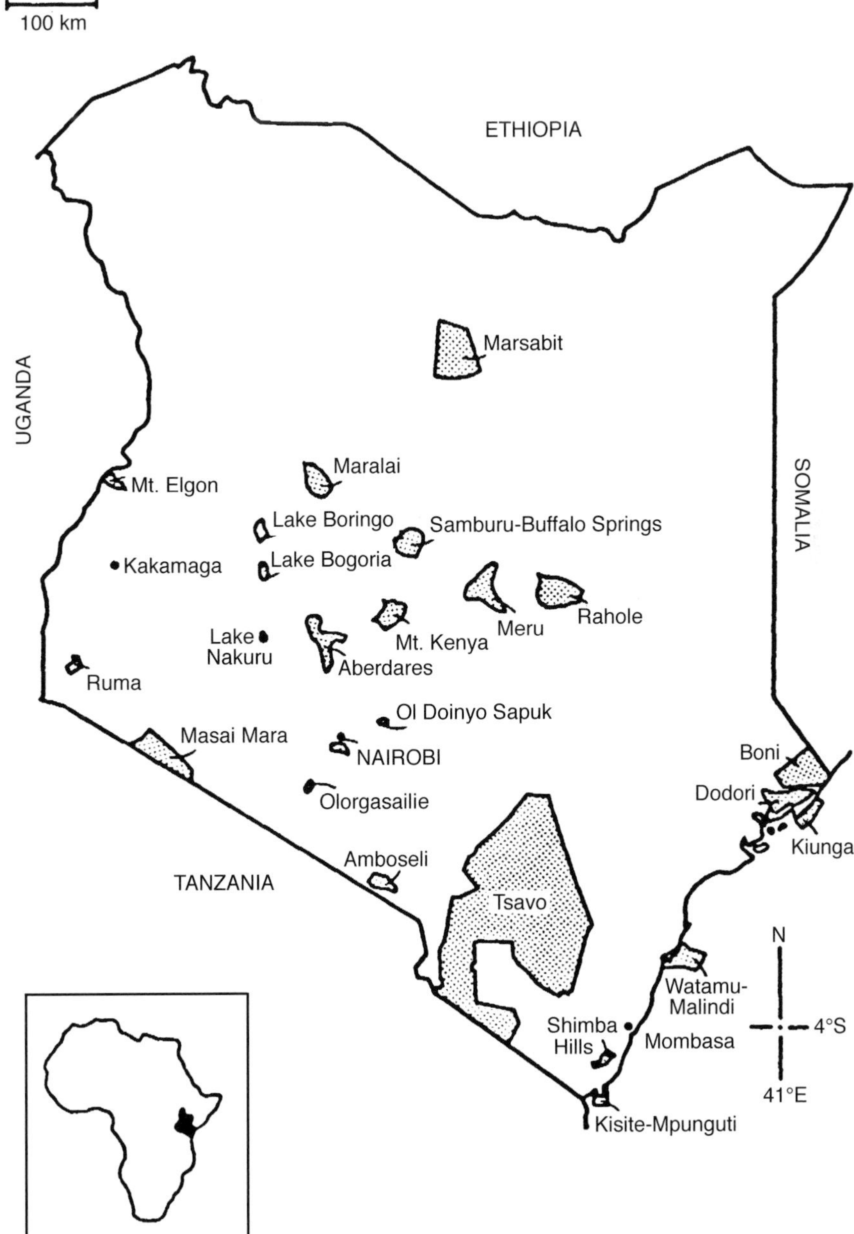

Fig. 21.2. Map of selected Kenya National Parks and Reserves, with Amboseli, Masai Mara and Tsavo National Parks located on the southern border.

egy. Beach and wilderness safaris are combined to offer value for money to the tourist (Rajotte, 1987; Gamble, 1989). These packages are often available at inexpensive charter prices, especially from many areas of Europe. More than 60% of all visitors spend time on Kenya's coast. However, the primary motivations for ecotourism are still defined, with more than 90% of tourists visiting a game park, and 79%

citing 'nature and wildlife' as primary motivations for visiting the country (Kenya Wildlife Service, 1995 cited in Honey, 1999).

The establishment of areas for conservation in Africa is quite different from the developed world. Governments in Africa have generally designated park areas on community land. This has led to land disputes, deprivation of ownership and rights, resettlement, and tension between the local population and the ecotourist. In other areas, managers have faced the paradox in which the revenue generated by ecotourism supports conservation. The processes of irresponsible tourism that can often work to the detriment of the conservation process, are in essence also providing the funds for conservation to take place. Thus, tolerable limits for ecotourism and detrimental impacts have been difficult to define.

ECOTOURISM WILDERNESS USE. In Korup National Park, Cameroon, ecotourism development has not been largely recognized. Visitation was only 500 people in 1989, and was estimated to reach only 1000 by 2000 (Gilbert *et al.*, 1994). Other estimates indicate that in 1990, 300–350 tourists generated revenue of only US$2800. The low visitation is largely due to poor access to the park. The main park entrance is a 1-day drive on unpaved roads from the nearest town, Douala (Alpert, 1996). However, the country expects Korup to gain international recognition as a place to visit nature and experience wilderness, and has therefore planned for a sustainable future.

Ecotourists visiting the Masai Mara, Kenya, pay US$20 to enter the national park. The total cost of visiting the park per European visitor does not often exceed even 1% of the total cost of the vacation. Visitors stay in the park's 12 lodges and campsites and tour the wilderness in minibuses provided by commercial operators. Visitation increased by 260% between 1977 and 1989, reaching 190,000 per year (Gilbert *et al.*, 1994), and further increased to 250,000 in 1992 (Honey, 1999).

Ecotourism has had a number of negative impacts on the Masai Mara. In 1994, a survey of tourist behaviour found that more than 1200 t of local firewood were used by tourists each year (Pearce, 1995). Waste disposal has also been a problem due to permanent tent camps localized in two small areas of the reserve (Honey, 1999). In addition, minibuses observing cheetah hunting prey have been reported to interfere with the wilderness setting to such an extent that the animals often give up their hunt and attack farm animals on local ranches outside the park (Pearce, 1995). Disturbance from this off-track driving is banned but very common (Gakahu, 1992). In 1991, a survey of visitors to the Masai by Wildlife Conservation International (WCI) also found a deterioration of visitor experiences in addition to the physical environment. Tourists stated that they received insufficient information about their safaris, and only 15% of those surveyed were provided a regulations leaflet prior to their visits. Ecotourists also expressed dissatisfaction with guides stating that they were not knowledgeable enough to provide sufficient environmental interpretation (Gakahu, 1992).

Ecotourists visit the Masai Mara to experience the wild game such as cheetah and wildebeest that inhabit the wilderness. However, Pearce (1995) noted that ecotourist visitations have had other measurable impacts on wildlife in addition to those noted above. Conservation of the park, including restricting the indigenous Masai people from subsistence hunting, cattle grazing and farming, has been associated with increases in some wildlife populations. For example, a 400% increase in the wildebeest population was noted in the Greater Serengeti between 1969 and 1995. Although living up to the expectations of the wilderness visitor, the rise in the population is ultimately destroying vegetation within the park.

The wealth of biodiversity in the Masai Mara National Park attracts ecotourists seeking wilderness. They come to view the large game such as rhinoceros and lions that live on the savannah. However, the local people have been displaced by ecotourism to 'conserve' the natural biodiversity. In the Ngorongoro Crater, Masai people were relocated outside the area. As a result, they returned and killed wild animals because they viewed them (and the ecotourism that resulted) as the cause of their own suffering (Pearce, 1995).

Ecotourists seeking a wilderness experience have contributed to mass tourism in some cases. Physical problems such as crowding, loss of vegetation and congestion have diminished the wilderness experience until it is mostly non-existent in places. Animals are often forced to be grouped in areas due to these impacts, making them more accessible to tourists. As a result, some tourist operators are now donating funds to the national park. Abercrombie and Kent and Africa Bound (based in the UK) donate approximately US$8 to an anti-poaching fund for each visitor that they accommodate (Gilbert *et al.*, 1994). However, this amount is still very small, relative to the amount that such tour operators charge the ecotourist.

WILDERNESS MANAGEMENT. Korup National Park is not managed by the Government of Cameroon but by the Korup Forest Project. The Integrated Conservation Development Project is run by the World Wide Fund for Nature with support from the British Overseas Development Administration. The Wildlife Conservation Society from the USA is also involved in the management of the park.

The national park plan states that physical, socio-cultural and economic impacts must be reviewed while working with the community in the construction of accommodation and provision of transport to cater for the future needs of tourists. Strict sustainable development principles are practised. Meetings with the local community are held to discuss issues of conservation, and buffer zones have been created in order to establish an equilibrium between potential ecotourism growth and the needs of the local society.

Infrastructure development is the responsibility of local tourism managers in Korup National Park. In ecotourism areas, accommodations are often designed with a minimum bed occupancy to achieve profitability, thus permitting tourists to focus in certain areas (Gilbert *et al.*, 1994). Poor design and location of accommodation and access roads in the early years of wilderness conservation could be detrimental to Korup, causing tourist congestion. In this sense, management is largely reactive to the problems encountered, as opposed to proactive. Planning must be designed for capacity or tolerable limits, even if this point is decades in the future. Korup National Park has not yet encountered congestion. The park plan for sustainability has a suggested life span of 30 years, allowing for inevitable change to take place. Hopefully, this management approach is not a reflection of an early, anaemic state of the ecotourism industry in the area. Will the park sacrifice these principles if tourist numbers and revenues begin to increase?

Cameroonian law does not permit permanent settlement in Korup National Park. The reason is that conservation seeks to sustain the wilderness for the unimpaired benefit of future generations. Thus, a buffer zone was created around the park in order to support local populations and sustain their way of life. To reduce the environmental impacts of potential ecotourism, alternative natural attractions are being promoted. Korup Link Tours has located areas in the Ndian Division in Cameroon to serve as 'honeypots' to alleviate visitation to Korup National Park.

Many of Kenya's National Parks were established during the 1950s and 1960s and are relatively older than those in Cameroon. Management and conservation progression has therefore been quite different. Zoning has been applied in Kenya in order to conserve wilderness, thus areas have been classified according to the desired type of access and use. In doing so, a framework has been designed to set aside areas for preservation whereas others are set aside for recreation and visitor facilities (Murphy, 1985). Core zones are used to protect wilderness and divert mass tourism away from these areas (Brown, 1982). Central zones are set aside solely for wildlife-viewing tourism (Honey, 1999).

Zoning has also had negative impacts. In some cases, the traditional rights of the indigenous population have been disturbed. Under the objectives of sustainability, locals were restricted from hunting game for subsistence. This action has disregarded the idea of conservation in its widest sense due to the disrespect of the socio-cultural aspect of the sustainability concept.

Gilbert (1993) has criticized the zoning system for conservation. He argues that zoning provides an excuse for concentrated development in areas that should be preserved. Over time, this leads to pockets of development, changing the character of the wilderness as a whole. Ecotourist spending in these pockets often only benefits a zone adjacent to the wilderness. Thus, Gilbert asks: Should populations within the protected or uncompensated zones be compensated? For example, in Tsavo National Park, Kenya, local populations have resettled outside the park and have formed squatter settlements. Before the park was formed they survived by subsistence hunting of elephants until their rights were removed by conservation restrictions. In Lake Nakuru National Park, local people were banned from using pesticides and fertilizers on their land in order to protect the ecological stability within the national park. Zoning also disregards animal movement and thus conservation. Wild game often migrate thousands of miles to access food and water, thus zoning has the potential to disregard natural movements and actually has a negative impact on conservation.

CONCLUSION AND FUTURE CONSIDERATIONS. The two African case studies highlight contrasting uses of wilderness for ecotourism. Kenya has benefited financially but has not adopted a sustainable approach to wilderness ecotourism. The large amount of revenue generated has not been recirculated into conservation. Thus, ecotourists are not paying the price for conservation.

In contrast, Cameroon has adopted a cautious approach to the use of wilderness for ecotourism. Korup National Park is a good example of a sustainable approach to ecotourism in wilderness. Cameroon has adopted an international approach by including non-government organizations (NGOs) from the USA and UK. Thus, conservation has been achieved by adopting an institutionalized framework.

Relative to the other case studies in this chapter, Kenya and Kenyan wilderness areas would probably benefit from the adoption of a more sustainable approach to ecotourism. The 1990s have not been positive for Kenya. Tourism declined due to the Gulf War, and Kenya also faced increased competition as an ecotourist destination from other African countries such as South Africa, Tanzania and Zimbabwe. Competition was augmented due to unrest in Kenya during elections, the bombing of the US embassy in Nairobi, and reports of hostility towards tourists including muggings and corruption from government officials involved in the tourism industry. In 1995 there was a 20% drop in tourist visitation, and by mid-1998 almost 50% of Kenya's coastal hotels closed or were forced to reduce staff numbers. Approximately 50,000 workers (30% of tourism sector employment) lost their jobs due to such closures (Honey, 1999).

At present, wilderness is being used for the profit of the country, and conservation is not living up to its potential. Ecotourism is resulting in intolerable levels of impact on wilderness resources. In fact, many ecotourists were unaware of how little of their entrance fee and expenditure went to conservation. Pearce (1995) states that many tourists said that they would be willing to pay extra admission fees if revenue was used for the conservation of wildlife. The untapped revenue from this willingness to pay was estimated at between US$46 million and 450 million. If all of this revenue went to the conservation of the Masai Mara National Park, the present conservation fund would be increased between 4- and 40-fold. Higher entrance fees might also reduce visitation numbers. A 1991 WCI study (Gakahu, 1992) found that visitors were better educated about conservation following their visit to the Masai Mara and willingness to pay increased due to this education.

Pearce recognizes that there is often a difference between willingness to pay and how much an individual will actually pay. However, the surplus does suggest that ecotourism could provide more revenue for the conservation of wilderness. The primary problem is not revenue but the direction of revenue. Allocation of more ecotourism revenue back into conservation and not central government funds might benefit Kenyan ecotourism. It has been debated whether tourism in Kenya still constitutes 'ecotourism'. Honey (1999) summarized seven principles of ecotourism and how they relate or do not relate to Kenya at present. These principles are: (i) 'travel to natural destinations'; (ii) 'minimizes impact'; (iii) 'builds environmental awareness'; (iv) 'provides direct financial benefits for conservation'; (v) 'provides financial benefits and empowerment for local people'; (vi) 'respects local culture'; and (vii) 'supports human rights and democratic movements'. Following these categorical principles it is evident from the above debate that ecotourism is becoming less defined in areas such as the Masai Mara.

Both countries have adopted zoning approaches to conserve wilderness. However, Kenya appears to have encountered more problems with wilderness conservation. This was primarily due to the enforcement of policy on indigenous people. Cameroon adopted a more democratic approach where local communities were involved in the planning and conservation process. Buffers were created as an alternative for indigenous people, whereas in Kenya alternatives were almost nonexistent, which resulted in social conflict.

Due to the timing of conservation and the methods adopted, Kenya has encountered problems with balancing conservation and ecotourism. Cameroon has adopted a more effective approach, at least for now. Learning from the problems encountered by other African nations, the government has adopted strict policies. Korup National Park was established in 1986, and thus it has benefited from the emergence of the idea of ecotourism and sustainability. However, there are some distinct similarities between the two examples that should be considered when managing wilderness for ecotourism. As with any wilderness, ecosystems in Africa are very fragile and susceptible to change. Probably more so than other world wilderness parks, the Masai Mara attracts global recognition and thus pressure for ecotourism. In addition, host nations often feel the need to use such areas as a source of national income. Indirect profits from ecotourism sometimes outweigh the value of wilderness conservation. Thus, as seen in Kenya, revenue from ecotourism is not re-circulated back into the sustainable management of wilderness. The value of wilderness to the government does not outweigh the monetary value that could be applied on more primary concerns such as industrial development. Thus, who should set the value of wilderness, such as the Masai Mara and Tsavo National Parks? Ecotourism is a threat to the natural biodiversity and experiences that such areas provide because governments often do not have the funds to allocate and purchase land for conservation. There is often less priority on conserving wilderness in developing than in developed nations. African ecosystems are globally unique and thus attract tourists from around the world. They should not only be conserved for the values of host country governments. Instead, planning and management needs to adopt a more institutionalized global approach as seen in Cameroon.

Wilderness in the developed world

Case study: the Gower Peninsula, Wales

BACKGROUND. The Gower Peninsula is located on the south coast of Wales, 32 km west of Swansea (Fig. 21.3). A coastal wilderness, it was the first Area of Outstanding Natural Beauty (AONB) in Great Britain designated in 1956 under the National Parks and Access to the Countryside

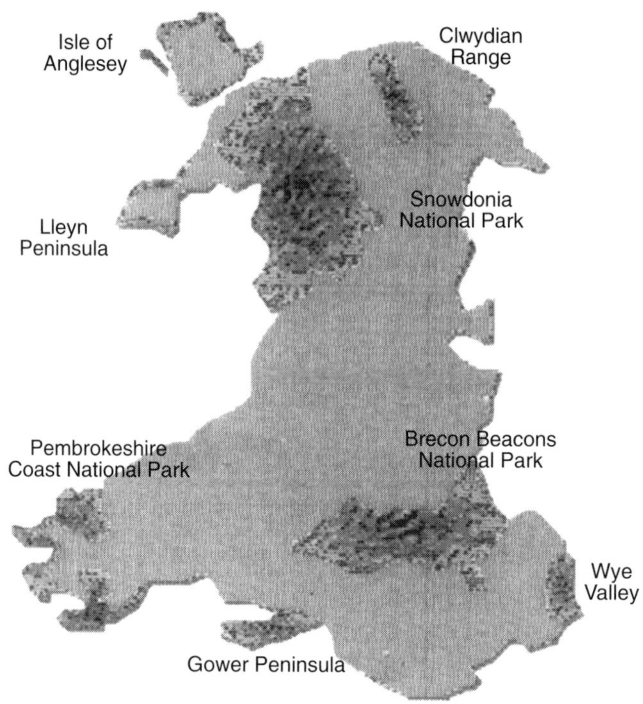

Fig. 21.3. The Gower Peninsula, located on the south coast of Wales, is an example of an urban proximity natural area that offers ecotourism-wilderness opportunities (web site: http://www.data-wales.co.uk/parkmap.gif).

Act 1949. The peninsula is a Category V protected area, as defined by the World Conservation Union (see Chapter 18). Although the peninsula is cultivated in part, other areas including the AONB provide for wilderness experiences and exhibit uninhabited and uncultivated natural settings. The coast is lined with carboniferous limestone cliffs and sand dune ecosystems, home to a diverse population of wildlife both terrestrial and marine (Ballinger, 1996). Because of the area's close proximity to urban areas of South Wales and southwest England, it receives much ecotourist attention. Wilson (1990) estimated that 18 million people live within 4 h of the peninsula. The most popular areas of the peninsula receive between 250,000 and 300,000 visitors per year. The following sections discuss the use and wilderness management of the Gower Peninsula.

ECOTOURISM WILDERNESS USE. Ecotourism on the peninsula is largely determined by its geography and limited infrastructure. A small network of roads limits tourists to a few 'honey-pot' areas where use is concentrated. Other areas such as the north of the peninsula are remote, due to the limited road network and signposting. However, the impacts of ecotourism are evident in some areas. Gower contains several diverse ecosystems ranging from salt marshes to limestone cliffs and sand dunes. These ecosystems attract many visitors. They are also home to a diverse range of flora and fauna, making them fragile and susceptible to even small amounts of ecotourism. Some of these impacts include erosion at sites with most concentrated use. Sand dune ecosystems have been negatively affected, however management has tried to address use types such as off-road cyclists (Gower Countryside Service, 1994). Dune systems not only have conservation, aesthetic and wildlife value, but for centuries the dune systems on the peninsula have provided protection from tide-water flooding for surrounding settlements. Therefore, some of the most valuable areas have been fenced, and use restricted. Marram grass was planted in order to stabilize the system and allow it to support ecotourism. By the early 1980s, 90% of the dune system had been stabilized by management techniques. However, due to this management some of the floral and faunal interest was lost due to the over-fixation of sand (Ballinger, 1996). At present the management of dune ecosystems focuses on three central objectives: the maintenance of stability, the promotion of diversity, and access for educational purposes. Rock climbing has also had an impact on the naturalness of the area. Some recreationists visit the area to use the carboniferous limestone cliffs for recreation. This has led to the defacement of certain areas due to the use of bolts for climbing, sometimes destroying areas of special geologic interest. Climbing has also had an impact on the wildlife population. Many birds use the cliffs for nesting, thus a ban was placed on climbing between March and September and a total ban was enforced on bolt usage. Angling has also had an impact on the area. Fishermen use crabs found on the coastline for bait and often disturb rare species (Davies, 1989). Even the turning of rocks from beach combing has been hypothesized to affect microbial populations and thus the wider food chain.

A greater problem is the increasing number of educational parties that visit the area. Although one of the objectives of the wilderness is to conserve the environment for educational purposes, overuse has been a problem. The Gower Outdoor Network has gone as far as to establish a code of conduct for educational school groups. Water pollution has also been a problem on Gower. However, the pollution of sea water is not a direct result of ecotourism. Sewage is disposed into the sea from outlets further up the coastline from developed areas. Although not a direct result of visitors to the peninsula, this example still provides evidence for the indirect impacts of ecotourism. When an area such as Gower attracts a large number of tourists, the surrounding area must have the capacity to support such an influx. In some cases, this leads to more development on the perimeter of a wilderness. Direct pollution in the form of littering is not a problem on Gower. However, in some areas anglers have frequently discarded fishing equipment such as weights and lines which can cause harm to bird populations and reduce the aesthetic character of the beach. The Countryside Council for Wales (CCW) has attempted to combat this problem by recruiting anglers as wardens to police the area. Despite these impacts, Gower for the most part has been effectively conserved. The natural character of the area has been sustained due to the support of a number of agencies and organizations. In fact, the character of the area is better than it was when it was first designated as an AONB, in 1956.

WILDERNESS MANAGEMENT. Most of the peninsula is covered by the AONB and other environmental conservation designations. The primary purpose of the AONB is the protection and enhancement of natural beauty. The coast also exhibits 55 km of Heritage Coast, 19 SSSIs, 3 National Nature Reserves and 2 Local Nature Reserves. Like much wilderness conservation in Western countries, much of the peninsula land area has been purchased and is governed by NGOs or government. The National Trust, which owns 1295 ha of Gower coastline, and Glamorgan Wildlife Trust have both played a role in the management of Gower. The funding for the purchase of land came from the Enterprise Neptune programme, launched in the 1960s with the objective of buying unspoilt coastline for conservation.

CONCLUSIONS AND FUTURE CONSIDERATIONS. The Gower Peninsula is an example of the sustainable interaction of ecotourism and wilderness. The area has been conserved due to strong support from a variety of political standpoints. It is accepted that pressures from tourism must be effectively incorporated into the wilderness while not losing the original objectives of the AONB and other designated conservation areas. The two development plans for the area, the West Glamorgan Structure Plan (Mullard, 1995) and the Swansea Local Plan (Wilson, 1995), have noted this and aim to provide positive conservation. This effectively involves three administrative zones on Gower: remote, intermediate and intensive. Remote sites are restricted from development of any kind, to retain their wilderness character. Intermediate zones allow for limited recreation and service facilities, and in intensive zones, appropriate visitor facilities are permitted. Site-level zoning policy has also been implemented in order to protect smaller areas not covered by the wider zoning system. Policy is designed to restrict access from mass tourism in order to safeguard the natural environment. These zones were incorporated into the Gower Management Plan (Wilson, 1990). The success of the plan has been dependent upon the coordination of bodies involved with the management of Gower. This has ensured that aspects of sustainability were incorporated into decisions that have affected the Gower Peninsula. It is hoped that special interest and ecotourism-dependent activities that are compatible with sustainable tourism will be promoted in the area. These will develop attractions based on the natural characteristics of the area including out of season birdwatching, and walking on Gower. Public transport has also been promoted. A shuttle bus operation was proposed in order to reduce the indirect impacts of ecotourism. However, external factors have recently increased pressures to protect the Gower Peninsula. A second bridge crossing of the River Severn was completed in 1996, and offshore aggregate dredging is suspected to be the cause of the recent loss of beach areas on Gower. This highlights the need not only for area specific policy to protect areas for ecotourism, but also regional policy considerations.

Issues in Wilderness Ecotourism Management

Wilderness manipulation impacts

The three case studies described in this chapter have highlighted the fact that some impact on wilderness is inevitable when ecotourism is present. Thus, the question arises: How much impact is too much and when does wilderness cease to be wilderness as a result of these impacts? In short, there must be tolerable limits to change. If wilderness is to be used for ecotourism in a sustainable manner then these limits need to be objectively stated to ensure effective wilderness management.

Coccossis (1996) discusses the interdependence of tourism and the environment. Tourism is dependent on both natural wilderness and socio-cultural resources and thus generates multiple impacts. Some wildernesses are, in turn, increasingly dependent on tourism to justify keeping them in a relatively undeveloped state. Evident in all the case studies, ecotourism can have significant indirect impacts in addition to direct impacts on the actual wilderness itself. Sustainable approaches need to address more than just the wilderness resource. Ecotourism not only affects the wilderness resource, but also other resources outside the conserved area, such as land prices for the construction of accommodation and local populations.

Seasonality is also an issue to be addressed. Demand can overload environmental resources at peak visitation times (Pearce, 1989). In addition, certain wilderness ecosystems are more vulnerable to unacceptable change than others. Coccossis (1996) identifies coastal zones and islands as most vulnerable. Coastal tourism is often highly seasonable and intensive. These environments are also fragile and susceptible to change. In addition, the indirect impacts of coastal ecotourism are often highly evident. There is almost always a need for tourist facilities. These often detract from the character of an area as a whole leading to honey-pot sites and infrastructure surrounding the wilderness. For example, there are often fewer access routes to a coastal wilderness than there are to an inland wilderness. Thus, indirect infrastructure development can be detrimental to the character of the entire area and detract from the conserved wilderness core. Island ecosystems are also susceptible to ecotourism impacts due to their closed natural and socio-cultural systems. Thus, they must be subject to sustainable development if wilderness is to be used for ecotourism (see Chapter 15).

In summary, wilderness can be effectively protected from ecotourism impacts through sustainable planning, and through use restrictions where necessary. However, indirect impacts will prevail in adjacent non-wilderness areas that are not as strongly conserved. For example, Ballinger (1996, p. 51) notes that the real problem with the conservation of the Gower Peninsula is 'determining the correct balance between tourist development and conservation, not only at the site level, but also over the peninsula as a whole'.

Ecotourism is not only a site or area specific activity, it is a global phenomenon and sustainable management should seek to conserve wilderness on a global scale. However, the paradox remains that wilderness management is usually heavily influenced by local or national bodies.

Wilderness ecotourism opportunities

Management of wilderness for ecotourism needs to adopt a proactive approach. The management of the world's most pristine environments cannot afford to adopt a reactive approach to change. In some cases, a proactive approach may lead to tourism being restricted from some wilderness. However, if restrictions are the only way to conserve wilderness then they must be enforced. A primary objective of wilderness ecotourism management is to conserve the natural biodiversity of an area while secondarily providing for an acceptable level of tourism or recreation.

Another major opportunity for ecotourism in developing nations is a proactive approach toward wilderness ecotourism conservation. Ecotourism, when based on the biological-cultural conservation of wilderness ecosystems, can help to save wilderness. Conservation-based ecotourism in wilderness can be a legitimate land use and serve as a source of revenue for managing and sustaining wilderness areas. However, the management of wilderness and protected areas for ecotourism is complex, for the intervention of ecotourism, itself, could fundamentally change the nature of the wilderness so that it might become something else altogether. This should not happen.

Finally, wilderness ecotourism opportunities will not be the same, or even similar, for the variety of wildernesses in the world. A 'spectrum of wilderness ecotourism opportunities' must serve as the basis for planning and managing ecotourism in wilderness. The spectrum must be based on the biological, cultural and social sustainability of wilderness, with a spectrum of ecotourism opportunity classes suited for different types of wilderness–ecotourism interactions. Of course, one of the opportunity classes can include 'no ecotourism'. The concept of a wilderness ecotourism opportunity spectrum will be developed more in the next section.

Management frameworks for wilderness ecotourism

Several planning issues need to be addressed in the management of wilderness for ecotourism. Relating back to the definition of wilderness, it is important to note that wilderness is not only a physical place; it is also a psychological place. As well as primarily conserving the environment, it is important to conserve wilderness experiences of users. Thus, the physical management and designation of wilderness must be addressed. In addition, the management of opportunities and experiences must also be addressed.

In recreation resource management a recreation opportunity spectrum (ROS) has been applied to manage experiences (Clark and Stankey, 1979; Driver *et al.*, 1987). Opportunities are conditions and situations demanded by recreationists, or in this case ecotourists. The ROS framework involves specifying recreational goals in terms of classes of recreation opportunity. Like other recreation planning frameworks described later, the ROS involves the specification of goals, specific standards and measurable indicators of those standards to ensure that they are met. It appears that this planning framework could quite feasibly be applied to wilderness ecotourism areas. In fact, an opportunity spectrum approach has been applied in the Gower Peninsula to ensure the conservation of core areas.

Butler and Waldbrook (1991) proposed a tourism opportunity spectrum (TOS). The spectrum, adapted from Clarke and Stankey's original concept, proposes 'a context in which proposed change can be placed', allowing for wider market penetration and greater compatibility of tourism in natural environments. TOS is adapted to cover aspects of tourism and ecotourism including access (difficulty of access, access system, means of access); other uses (from incompatible to compatible); tourism plant (extent, visibility, and complexity of development, level of facilities); social interaction (between hosts and guests); acceptability of visitor impacts (degree of impact, prevalence of impact); and acceptability of regimentation (from minimum to strict). This spectrum offers potential to planners and conservationists involved in ecotourism. It provides a base that is more specific to ecotourism, one that includes tourism developments and types of tourist access that might not be fully accounted for in the ROS.

There is a lack of understanding of the complex relationship between tourism and the environment (Coccossis, 1996). Not only must the acceptability of ecotourism in wilderness be addressed, but the areas as a whole must be incorporated into planning. Ecotourism not only has the potential to have a direct impact on wilderness, but also has the potential to have an indirect impact on the character of the surrounding area. Therefore, tolerable limits of wilderness manipulation and public access must be defined. There are a number of ways to define tolerable limits. However, many have been criticized due to their ultimate reliance on value judgements.

The carrying capacity framework has often been quoted in studies of ecotourism and the environments that support it. However, in wildland recreation the carrying capacity framework has been criticized for its lack of specificity and need to formulate a specific number of people an area can feasibly support. An alternative framework for the management of wilderness is the limits of acceptable change (LAC) (Fig. 21.4). Unlike the carrying capacity framework, LAC does not seek to specify a number of people that an area can ultimately support. Numbers are often not reliable predictors of impacts (Washburne, 1982), thus a different management approach is needed if ecotourism use in wilderness is to be managed sustainably. LAC is a method of condition assessment and identification of specific management goals and measurable objectives. Stankey *et al.* (1985) propose nine steps for wilderness management when using the LAC process (see Fig. 21.4).

LAC is a proactive planning tool, essential for wilderness management. Hendee *et al.* (1990) describe the application of LAC in the Bob Marshall Wilderness Area, Montana, USA. The framework incorporates more scientific judgement into the planning process than the carrying capacity framework does. By inventorying existing conditions (Step 4) and specifying tolerance limits or standards for both physical and social conditions (Step 5), wilderness is more efficiently managed for the protection of environment and conservation of user-ecotourist experiences. The process also involves evaluation and monitoring of management actions (Steps 8 and 9) to ensure that acceptable limits are not surpassed. McCool (1994) used LAC to study the limits of acceptable change for nature dependent tourism development. The application of LAC for tourism is discussed in the context of the economic and output orientation of tourism planning.

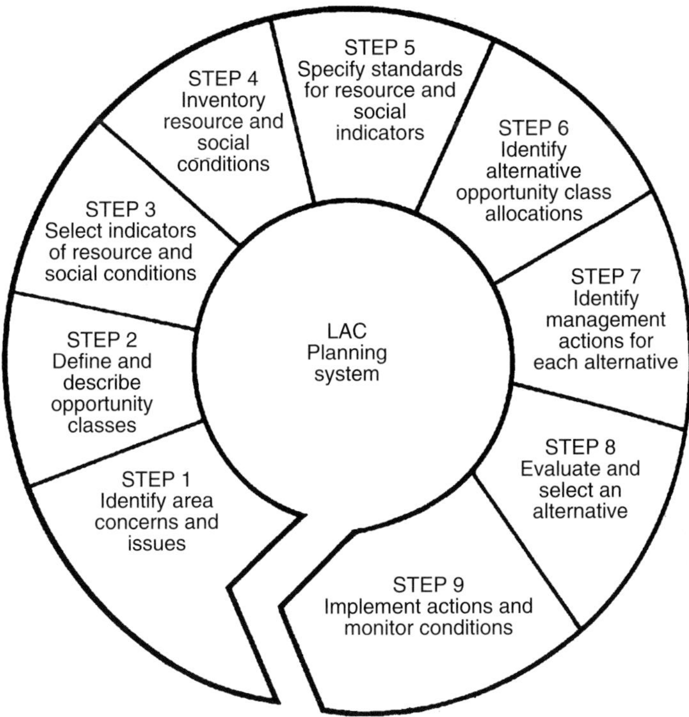

Fig. 2.4. The nine steps within the limits of acceptable change (LAC) planning model for recreation resource management (Stankey *et al.*, 1985).

Conclusion

Wilderness does provide a feasible resource for ecotourism. Some purists would disagree with this statement, but it is evident that wilderness environments around the world are presently being used for ecotourism. However, wilderness is fragile and highly susceptible to change. There remains a need for effective management and conservation frameworks as they relate to wilderness ecotourism. These frameworks must allow for a varied spectrum of wilderness ecotourism opportunities, ranging from the exclusion of any ecotourism to the provision of considerable, but sustainable levels of ecotourism. Many of the world's large, remote wildernesses (e.g. Antarctica) demand minimal human intervention, while other small and accessible wildernesses may tolerate conservation-based ecotourism. In addition to management, there also remains a need for more conservation support in wilderness. Without global cooperation, the world's unprotected and under-protected areas might not withstand the demands of ecotourism. Areas in the developing world need to be cautious about wilderness use for ecotourism. It is not only the responsibility of developing nations to protect these areas. Developed nations need to become more active in the conservation and appropriate use of world wilderness. The root of much wilderness and tourism conflict appears to be financial. With more funds designated for the purchase of lands to be conserved, and with compensation of those who 'lose' from the protection of such areas, the future of wilderness management and ecotourism activity will be brighter. This will in turn make wilderness accessible to future generations of ecotourists who will benefit from the natural diversity and unique experiences that such environments can provide.

References

Almagor, U. (1985) A tourist's vision quest in an African game reserve. *Annals of Tourism Research* 12(1), 31–47.

Alpert, P. (1996) Integrated conservation and development projects: examples from Africa. *BioScience* 46(11), 845–855.

Ballinger, R.C. (1996) Recreation and tourism management in an area of high conservation value: the Gower Peninsula, South Wales. In: Priestly, G.K., Edwards, J.A. and Coccossis, H. (eds) *Sustainable Tourism? European Experiences.* CAB International, Wallingford, UK, pp. 35–53.

Brown, C. (1982) Resource planning for recreation and tourism. In: *Essays in National Resource Planning.* CURS, University of Birmingham, UK.

Butler, R.W. and Waldbrook, L.A. (1991) A new planning tool: the tourism opportunity spectrum. *Journal of Tourism Studies* 2(1), 2–14.

Ceballos-Lascuráin, H. (1988) The future of ecotourism. *Mexico Journal* 17, 13–14.

Clark, R. and Stankey, G. (1979) *The Recreation Opportunity Spectrum: a Framework for Planning, Management and Research.* USDA Forest Service General Technical Report PNW-98. Pacific Northwest Forest and Range Experiment Station, Portland, Oregon.

Coccossis, H. (1996) Tourism and sustainability: perspectives and implications. In: Priestly, G.K., Edwards, J.A. and Coccossis, H. (eds) *Sustainable Tourism: European Experiences.* CAB International, Wallingford, UK.

Davies, A. (1989) Areas of natural beauty: recreation and management. Postgraduate Diploma thesis, University of Wales, Cardiff, UK.

Driver, B.L., Brown, P.J., Stankey, G.H. and Gregoire, T.G. (1987) The ROS planning system: evolution, basic concepts, and research needed. Leisure Sciences 9(3), 201–212.

Gakahu, C.G. (1992) Visitor Dispersal Strategies in Ecotourism Management. Paper presented at *Fourth World Congress of National Parks and Protected Areas, Caracas, Venezuela, February, 1992*, p. 11.

Gamble, W.P. (1989) *Tourism and Development in Africa: Case Studies in the Developing World.* Murray, London.

Gilbert, D.C. (1993) Issues in appropriate rural tourism development for Southern Ireland. *Leisure Studies* 12(2), 137–146.

Gilbert, D.C., Penda, J. and Firel, M. (1994) Issues in sustainability and the national parks of Kenya and Cameroon. In: Cooper, C.P. and Lockwood, A. (eds) *Progress in Tourism, Recreation and Hospitality Management.* John Wiley & Sons, Chichester, UK, pp. 30–45.

Gower Countryside Service (1994) *Annual Report of the Planning Department, Gower Countryside Service.* Swansea City Council, Swansea, UK.

Hendee, J.C., Stankey, G.H. and Lucas, R.C. (1990) *Wilderness Management.* North American Press, Golden, Colorado.

Hill, B.J. (1994) Concepts of wilderness valuation. MSc thesis, Clemson University, Clemson, South Carolina.

Honey, M. (1999) *Ecotourism and Sustainable Development.* Island Press, Washington, DC.

McCool, S.F. (1994) Planning for sustainable nature dependent tourism development: the limits of acceptable change system. *Tourism Recreation Research* 19(2), 51–55.

Mullard, J. (1995) Gower: a case study in integrated coastal management initiatives in the UK. In: Healey, M.G. and Doody, J.P. (eds) *Direction in European Coastal Management.* Samara Publishing Limited, Cardigan, UK, pp. 259–284.

Murphy, P. (1985) *Tourism: a community approach.* Methuen, New York.

Nash, R.F. (1973) *Wilderness and the American Mind.* Yale University Press, New Haven, Connecticut.

Organization for Economic Cooperation and Development (OECD) (1980) *The Impacts of Tourism on the Environment.* OECD, Paris.

Pearce, D. (1989) *Tourist Development.* Longman, Harlow, UK.

Pearce, F. (1995) Selling wildlife short. *New Scientist* 147(1993), 28–31.

Perez, O. (1991) The old man of nature tourism. In: Whelan, T. (ed.) *Nature Tourism: Managing for the Environment.* Island Press, Washington, DC, pp. 23–38.

Rajotte, F. (1987) Safari and beach resort tourism, the costs to Kenya. In: Britton, S.G. and Clarke, W.G. (eds) *Ambiguous Alternatives: Tourism in Small Developing Countries.* University of South Pacific, Suva, Fiji, pp. 78–90.

Rosenburg, K.A. (1994) *Wilderness Preservation: a Reference Handbook.* ABC-CLIO Inc., Santa Barbara, California.

Stankey, G.H., Cole, D.N., Lucas, R.C., Petersen, M.E. and Frissell, S.S. (1985) *The Limits of Acceptable Change (LAC) System for Wilderness Planning.* USDA Forest Service General Technical Report INT-176. Intermountain Forest and Range Experiment Station, Ogden, Utah.

Topouzis, D. (1990) Cameroon's Korup rainforest struggles to survive. *Africa Report* 35(2), 8.

Washburne, R.F. (1982) Wilderness recreational carrying capacity: are numbers necessary? *Journal of Forestry* 80(11), 726–728.

Wilderness Act (1964) Act of September 3, 1964. Public Law 88-577.78 Stat. 890.

Wilson, D.M. (1990) *Gower Management Plan.* Swansea City Council, Swansea, UK.

Wilson, D.M. (1995) *Swansea Local Plan Review, Written Statement Consultation Draft, April 1995.* Swansea City Council, Swansea, UK.

Chapter 22

Indigenous Territories

T. Hinch

Faculty of Physical Education and Recreation, University of Alberta, Edmonton, Alberta, Canada

> Indigenous peoples, peoples who have lived on and from their lands for many generations and who have developed their own culture, history, ways of life, and identities grounded in these places, inhabit vast areas of Asia, Africa, the Americas, and the Pacific. Native Americans, the aboriginal peoples of Australia, the Maasai and the Maoris, and thousands of other indigenous peoples control traditional territories from the Canadian and Alaskan Arctic to the Amazon and Andes, from Siberia to South Africa, and from the high ground of the Himalaya and Tibet to the atolls of the South Pacific and the outback of Australia. In many parts of the world, these homelands of indigenous peoples are the best – and often the last – remaining places of rich wildness and biological diversity.
>
> (Stevens, 1997, p. 1)

Indigenous peoples and ecotourists both value the land. Ecotourists seek out 'relatively undisturbed or uncontaminated natural areas with the specific objectives of studying, admiring, and enjoying the scenery and its wild plants and animals, as well as any existing cultural manifestations (both past and present) found in those areas' (Ceballos-Lascuráin, 1987 in Boo, 1990, p. xiv). This search, in turn, often leads them to indigenous territories: lands that are under the legal control of indigenous peoples as defined by the sovereign state where these lands exist, or more commonly, lands of 'aboriginal title' derived from a history of occupation and use. In an ideal world, ecotourism is an activity in which entrepreneurs, governments and tourists establish sustainable relationships with the environment while improving the welfare of the indigenous people who occupy these territories. Yet even moderate success in this regard is not likely to happen by accident.

The purpose of this chapter is to highlight the unique character of indigenous territories as venues for ecotourism activity (see also Zeppel, 1998). This will be done by addressing the current status of indigenous land claims, the significance of land within indigenous cultures, the motivations of indigenous peoples who have adopted ecotourism within their territories, the prominent issues that are associated with ecotourism on these lands, and finally by presenting basic principles to guide the planning and management of ecotourism in these areas. A major caveat attached to this discussion is that generalizations about indigenous peoples have been used to make this chapter relevant to a broad international audience. Examples from a variety of geographic and cultural backgrounds have been presented but the author's experience is primarily with the indigenous peoples of Canada and New Zealand. While it is true that there are parallels between many indigenous cultures, each one is also unique. Similarly, individuals and groups within a specific indigenous culture or community may share many characteristics but they too, are unique. Readers should consider these differences as they deliberate on the relevance of the points made in this chapter in terms of their own specific situations.

Indigenous Land Claims

It is only relatively recently that the notion of legitimate tenure rights in land for indigenous peoples gained widespread legal recognition. Early claims by indigenous people related to 'aboriginal title' were generally rejected in North America and Australia under the doctrine of *territorium nullius* (empty land), which asserted that 'lands occupied by foraging peoples at the time of settlement by Europeans became the sole property of the "original [European] discoverers" because native people were deemed to be even more primitive than others encountered in European expansion' (Wilmsen, 1989, p. 2). A similar rationale was often presented to justify the displacement of indigenous peoples in other regions of the world.

In those instances where a sovereign state has recognized specific indigenous land titles in law, there will be explicit *de jure* or legal obligations associated with ecotourism operations on these lands. However, by judging these claims on the basis of Western legal concepts and institutions, non-Western societies have been at a definite disadvantage. The majority of indigenous territories remain under dispute. Increasingly, however, the tide has changed. A landmark point in this change occurred in 1957 with the passing of Article 11 of the International Labour Organisation Convention 107 which affirms in international law: 'The right of ownership, collective or individual, of members of the populations concerned over the lands which these populations traditionally occupy' (Colchester, 1994, p. 7). Of particular importance under this article was the establishment of land rights based on the historic occupation of such lands and the recognition of the legitimacy of 'collective' as well as 'individual' land rights (McCullum and McCullum, 1975; Cordell, 1993). Legal rights and settlements associated with these land claims are increasingly being awarded to indigenous peoples throughout the world.

In those instances where legal title is not held but 'aboriginal title' is claimed, there are still *de jure* and ethical obligations associated with ecotourism operations on these lands. The success of ecotourism ventures in such areas is dependent on the support of their unofficial indigenous hosts. Failure to obtain the support of local indigenous people is likely to be self-defeating as they are an integral part of these environments and their consent, advice and hospitality, are therefore critical. In an extreme situation, if indigenous peoples are openly hostile to ecotourists, it is unlikely that these tourists will have satisfying experiences. Even in the absence of any active opposition from indigenous peoples, ecotourism operators have a moral obligation to recognize and protect the interests of indigenous peoples as these people often represent an integral part of the environment of the destination territory.

The Significance of Land within Indigenous Cultures

The Canadian National Aboriginal Tourism Association's video entitled *The Stranger, the Native, and the Land* highlights the relationship between these three elements within a tourism context (Notzke, 1996). Produced and distributed to promote tourism as a development option within indigenous communities, the video describes indigenous tourism as a cultural encounter between 'the Stranger' and 'the Native'. At the nexus of this encounter is 'the Land'. If satisfying ecotourism experiences are to be produced for the strangers to indigenous territories, an appreciation of what the land means to indigenous peoples is required.

Indigenous peoples have an inherent kinship with the land. In a traditional context, indigenous people consider that they belong to the Earth rather than the land belonging to them (Hollinshead, 1992). The concept of the land as the 'mother' on whom survival depends is a recurring one in indigenous cultures. For example, the Gagudju people of Australia express this view in their oral histories through verse such as:

> The rocks remain.
> The Earth remains.
> I die and put my bones in the cave or the Earth.
> Soon, my bones will become the Earth.
> Then will my spirit return to my land, my Mother.
>
> (cited in Martin, 1993, p. xv)

Traditionally, indigenous people have drawn much of their self-identity from the land of which they are a part. To many, the land is the essence of their life. The singular importance of land to the indigenous peoples of northern Canada was also emphasized by Notkze (1994) who cites the settlement of comprehensive land claims in this region of the world and the subsequent designation of the new Inuit-governed territory of Nunavut as a case in point. Control of land from the perspective of indigenous people is different from that of non-indigenous people because the relationship between indigenous people and the land is unique.

There are two aspects to this relationship. On the one hand indigenous peoples' relationship with their territories rests on the importance of resources to the continuing existence of the group. On the other hand, the territory is an area deeply associated with the identity of the people as a whole, which each generation keeps in trust for the future.

> The first element is an indigenous expression of the use or ecological relationship between indigenous peoples and the environment ... while the second aspect covers the cultural reproduction and management of an ecosystem. Whereas for non-indigenous peoples these factors are separated into 'economic' and 'cultural' domains, for indigenous peoples they are aspects of the same phenomenon, where time, space, resource use, management and conservation are all part of the same complex, linking identity to production and reproduction.
>
> (Gray, 1991, p. 22)

The typical depth of the attachment of indigenous people to their territories is difficult to exaggerate. It is much deeper than the connection that non-indigenous people normally have with the land. Two implications of this attachment are particularly important in the context of ecotourism. First, because indigenous people do not see the land as a possession, they are very wary of treating it as a commodity, even in the purportedly benign context of ecotourism. Second, because of their deep attachment to the land, indigenous people see the landscape differently. They attach unique and often complex meanings to place that go beyond its physical properties. Often, these complex meanings include a spiritual dimension that presents fascinating interpretive opportunities for ecotourism. It also may present some limitations as local peoples may choose not to share these insights with visitors, or they may choose to restrict visitors from certain sacred places within this landscape.

Historic land uses such as subsistence food gathering in Australia and hunting and trapping in northern Canada remain important activities in many of these territories (Young, 1995). Yet despite their deep attachment to land, contemporary indigenous people have not restricted themselves to traditional land uses. In a harsh indictment of the impacts of primarily non-native land use in northern Canada, native peoples recognize that their former way of life on the land has been irrevocably altered.

> We are telling you again and again that without the land Indian people have no soul, no life, no identity, no purpose. Control of the land is essential for our cultural and economic survival.
>
> We are people of the land. We love the land but the land is no longer what it was.
>
> We see the land as having been taken into the white man's society, his economy. It is covered in asphalt, surveyed, scarred, tracked in the search for oil and natural gas and minerals that lie under it.
>
> Sadly, we know that we cannot use it much in the old ways [subsistence]. The animals are hiding or dead, the fish are poisoned, the birds fly away. So we must seek a way to live in this new way, but we must not sell our land or allow it to be taken away. It is only for our use, even if the use is a new use. The land is for our children, not for sale. The land is still part of us and we are part of it.
>
> (amalgam of comments from Natives in Northern Canada as quoted in McCullum and McCullum, 1975, pp. 9–10)

Despite their reluctance to treat the land as a commodity, indigenous peoples are increasingly open to non-traditional uses of the land as long as these uses do not compromise their sense of the land's and, therefore, their own integrity. Indigenous peoples see themselves as caretakers of the land but they are continually trying to find ways of protecting their 'mother' while surviving as indigenous cultures in a changing world. Ecotourism is one of the new activities that has been introduced to these traditional lands as a way of seeking this balance. Underlying this search for appropriate land use in indigenous territories is the maintenance and reaffirmation of indigenous control.

The question being asked by indigenous peoples about ecotourism is whether it is a friend or a foe. Will it help them or will it hinder them in their quest for cultural survival? The answer to this question seems to hinge on the concept of control. Despite the increasing levels of development found in many indigenous territories, they still contain 'many of the best – and often the last – remaining places of rich wilderness and biological diversity' (Stevens, 1997, p. 1). There are at least three possible explanations for this. First, historically, indigenous people were often systematically marginalized to peripheral areas that were not originally felt to offer strong development opportunities to colonizing cultures. Second, indigenous peoples have tended to live in ways that have left the lands that they inhabit with much biodiversity. Third, in many cases, the indigenous peoples of these areas have been successful in warding off outside threats to the integrity of these lands. It is, therefore, particularly galling for indigenous people to have conservationists or others seek to limit their control over these lands. Such indignation is reflected in the comments of one Karen facing eviction from the Thung Yai wildlife sanctuary in Thailand:

> When we moved into these forests over two centuries ago, Bangkok was just a small village surrounded by lush vegetation. Over these many years, we Karen have protected our forest lands out of respect for our ancestors and our children. Maybe if we had cut down the forests, destroyed the land, and built a great city like Bangkok, we would not now be faced with possible eviction.
>
> (Thongmak and Hulse, 1993, p. 167)

While the above comments were made in the face of restrictive controls associated with the designation of a 'protected area', they hold significance for ecotourism operators in general. Indigenous peoples need to retain control if ecotourism is to be successful on indigenous lands. Their willingness to entertain ecotourism as a land use in their territories is a reflection of a variety of motivations that they harbour.

Motivations for Adopting Ecotourism on Indigenous Territories

To a large extent the motivations for indigenous peoples to develop ecotourism opportunities within their territories are just as diverse as those of non-indigenous peoples. While economic benefits feature prominently for both groups, there is generally an underlying desire that the net impacts, inclusive of social and environmental as well as economic impacts, are positive. Yet, indigenous peoples also have distinct motivations related to political issues and distinct variations of the economic, environmental and socio-cultural motivations for tourism development in general. The Little Red River Cree First Nation provides a good example of an indigenous group possessing a complex range of motivations in their desire to develop ecotourism in northern Canada (Colton, 2000).

Political motivations

The decision for indigenous peoples to open up their territories to ecotourism is often a political one in that it is based on issues of control. This is particularly true in the context of unsettled land claims whereby indigenous people are trying to demonstrate that their territories are not 'empty lands', that they are occupied and used by indigenous peoples. A similar strategy has been pursued by indigenous peoples when they see the establishment of protected areas within their territory as being in their best interest (Stevens, 1997). A classic example of this political strategy is provided by the Haida Nation and their placement of 'watchmen' at key heritage sites throughout the Haida Gwaii (Queen Charlotte Islands in Canada). These 'watchmen' acted as hosts to ecotourists and their presence eventually contributed to the designation of the territories as a 'National Park Reserve' thereby protecting the land from other types of development and leaving room for the eventual settlement of the land dispute (Guujaaw, 1996). Another example of political motivations underlying tourism development on indigenous territories is provided by the Sa people of the island of South Pentecost in Vanuatu. In this instance the Sa people manipulate tourists and the tourism industry through their control of the *gol* or land dive as a tourist attraction. During the *gol*, men dive head first from a 70-ft tower with vines attached to their feet to break their fall. Spectator access is limited, with the constraints and operational procedures determined by the Sa themselves, thereby reaffirming their culture and independence (de Burlo, 1996).

Economic motivations

The essence of economic motivation lies in the anticipated benefits associated with economic diversification, job creation and increased income, which are particularly valued given the relative poverty found in many indigenous communities. The basis of the economic logic is that income generated through tourism represents a fair exchange of value for value between indigenous and non-indigenous people. In fact, given that their culture is supported by one of the products of ecotourism, indigenous peoples have a competitive advantage in an economic context (Notzke, 1996). By increasing economic independence, a higher degree of self-determination and cultural pride will be generated as the shackles imposed by poverty and social welfare are broken (Hinch and Butler, 1996). Whale Watch Kaikoura in Aotearoa (New Zealand) provides a good example of this motivation in that economic benefits have been one of the fundamental objectives of the company since its inception as a marine wildlife viewing company. Its success has rejuvenated Kaikoura and has ensured that local Maori not only share in this economic rejuvenation but that they play a leading role (Warren and Taylor, 1994). This same motivation for economic benefits has also driven several First Nation groups in North America to open casinos on their lands (Stansfield, 1996). While casinos and ecotourism initiatives represent very different approaches to the use of indigenous lands, they do share the same economic rationale.

Environmental motivations

Environmental motivations are generally a response to real and perceived threats to the land. In the Haida example cited above, one of the main reasons for initiating the 'watchman' programme and encouraging ecotourism was to present an alternative to the forestry industry which was destroying the forests and contributing little to the Haida First Nation. By focusing on non-consumptive activities like ecotourism, indigenous people are able to participate in the global wage economy while conserving their land. Not surprisingly, these environmental motivations are closely related to the political motivations raised earlier. This is illustrated through the comments by an adviser to the Little Red River Cree First Nation in the Caribou Mountains of northern Canada:

> We want to refocus non-Indian use to non-consumptive uses. Tourism development such as ecotourism would allow us to offset these non-sustainable uses of the Caribous. By replacing ... activities such as forestry and oil and gas exploration with sustainable ... ones which are culturally relevant, we can control and strengthen our claim to our traditional land.
>
> (Colton, 2000, pp. 110–111)

Socio-cultural motivations

Finally, socio-cultural motivations exist at two levels: a cross-cultural level and one that is internal to the indigenous group. At a cross-cultural level, it is hoped that ecotourism will foster understanding between guests and hosts. Ultimately, increased understanding is anticipated to result in a more just and equitable relationship between non-indigenous and indigenous peoples (D'Amore, 1988). Within indigenous cultures, important socio-cultural issues include the increasing schisms that have emerged between generations and between the people and the land. Traditionally, the older members within indigenous communities have been looked upon with great respect for their guidance. Given the turbulence of the last few decades, this bond between elders and young people has deteriorated. Part of the reason for this deterioration has been the struggle that young indigenous people experience in trying to fulfil the demands of dominant culture to function independently within a wage economy. When coupled with declining land-based resources and traditional markets for the harvest of these resources, the wisdom and experience of the elders is often lost on the youth. Ecotourism on indigenous lands is seen as an opportunity for young people to integrate contemporary and traditional lifestyles. To do this successfully, these young people need to reconnect with their elders. In reconnecting with their elders they are reconnecting with their culture. Contemporary lifestyles and pressures have also weakened the relationship that indigenous youth have with the land. Given the sense of identity that indigenous people historically derive from the land, these youth have, in effect, lost part of their identity. Ecotourism is seen as an effective strategy for re-establishing this connection. For example, the owners of Tamaki Tours, which offers a culturally based tourism experience on Maori territory in Aotearoa (New Zealand), feel that their staff members have made this reconnection to their Maori heritage. 'There is not one member of staff that hasn't really gone hard out to learn a lot more about themselves, about their culture since working with us ... it doesn't stop here as a performance thing, it's not a superficial "turn-on turn-off" thing' (cited in A.J. McIntosh, T. Hinch and T. Ingram, unpublished data).

Prominent Issues

A broad range of issues will have to be addressed for indigenous peoples to capture the full potential of ecotourism on their territories. At the heart of these are political concerns related to the control of land, power relationships between the various partners throughout the industry, and the relationship between non-indigenous visitors and indigenous hosts. These issues are manifest in the economic, environmental and socio-cultural realms. By exploring the dynamics within each of these three realms, the political issues will emerge.

Economic realm

The starting point in considering economic issues associated with ecotourism on indigenous lands is that a significant portion of the economic benefit should accrue to the indigenous peoples associated with their territory. Not to direct a fair share of the benefits to these people would be unethical under the central principles of ecotourism articulated elsewhere in this book. What is perhaps more difficult to deal with is the distribution of these benefits. The dominant economic framework, which guides the tourism industry, can be classified as laissez-faire or free enterprise. It is based on the concepts of competition and individual initiative. In contrast, indigenous cultures tend to emphasize cooperation over competition and communal initiatives over independent ones. Smith (1989) has highlighted the dangers of marginalization for those individuals of indigenous descent who adopt a Western-based approach in their home communities. Others have argued that indigenous cultures have adapted to the realities of a global economy to the point of accepting competition and independent action (Wuttunee, 1992). Clearly, however, if it is agreed that an indigenous community is an important part of the attraction for ecotourism then the community as a whole must be seen to benefit from it, not just the more entrepreneurial-minded individuals found within the community. Mechanisms need to be put in place to ensure that an equitable distribution of these benefits is achieved.

At a direct level, if indigenous peoples are to enjoy the economic benefits of tourism, they need to be employed within the industry. The basic hospitality skills that are taken for granted by many non-indigenous peoples may not exist in communities that are typically defined by their poverty (Haywood, 1991). One of the challenges in this area is to recognize the difference between hospitality skills that can be taught versus character traits that tend to be ingrained in a particular culture. For example, the following quote from the field notes of a tourism researcher in northern Canada illustrates the potentially deep-seated cultural differences between his Dene host and himself:

> I am conditioned by my society to talk, even though there may be nothing of value to speak of. Andrew seldom talks unless it is related to an actual event of the day or some plans for the next day. I sometimes feel that he forces himself to talk when he would rather just sit and think, and contemplate.
>
> (Colton and Hinch, 1999, pp. 7–8)

In another study, despite the fact that tourists reported an overall high degree of satisfaction with their aboriginal tourism experience, they also voiced frustration with the way that some aboriginal tour companies were operated in the Canadian Arctic (Notzke, 1998). They cited concerns related to tour cancellations and scheduling, inefficiency, disorganization, and poor sales approaches.

Indigenous peoples have also been characterized in terms of their genuine warmth and hospitality, like the Maori of Aotearoa with their *mamaaki*. These types of cultural traits need to be positioned carefully in promotions and in the delivery of ecotourism in indigenous territories. Notzke (1998) extends this issue with her comments that the 'authenticity' of an ecotourism experience represents both a challenge and an asset for tourism on indigenous territories.

Environmental realm

While ecotourists and indigenous peoples both tend to share genuine interests for the sustainability of the environment, their different approaches to achieving this end represent significant issues that must be addressed. One of the most important issues in this area is the consumptive orientation of indigenous people versus the non-consumptive orientation of ecotourists (Hinch, 1998). Given their traditional lifestyles and values, indigenous peoples are very protective of their right to harvest the resources in their territories. Although there are exceptions (Stevens, 1997), indigenous people have traditionally tended to harvest their resources in a sustainable fashion. A central provision of most land claims is that this right to harvest resources should continue. In contrast, most ecotourists explicitly seek out non-consumptive activities while travelling. They are inclined to be wildlife viewers rather than wildlife hunters. The difference may be that ecotourists see themselves as separate from the environment that they visit while their indigenous hosts see themselves as part of the environment. Given these contrasting perspectives, conflict is likely to occur should a group of ecotourists stumble across the harvesting of wildlife while they are visiting an indigenous territory. To avoid this conflict, management strategies can be implemented to separate these activities in time and space (e.g. restricting ecotourism operations to certain times of the year or locations within indigenous territories). However, the basic philosophic differences in world view also need to be considered by both parties.

Potential problems also exist in terms of the apparent contradiction between the espoused environmental values of indigenous people and the litter that is found in many indigenous communities. Again, this discrepancy may be due largely to socio-economic conditions that characterize poverty, but it may also reflect fundamental cultural differences. An example of the former situation exists in the form of the Moken of Rawai Beach in Thailand. A visit to the settlement of these formerly nomadic coastal dwellers exposes tourists to the harsh realities of indigenous peoples' poverty. 'The tourists are often taken aback during such visits by the unexpected squalor and poverty of these people, who are pictured in the touristic periodicals, the promotional literature and on postcards, as free-roaming, primitive boatdwellers' (Cohen, 1996, p. 243). The reality of this problem is also highlighted by the fact that the clean-up of litter was one of the five most needed improvements listed in a survey of visitors to the Northwest Territories in Canada (Acres International Limited, 1990). Yet, do the skeletal remains of butchered game in an Arctic community constitute litter even though Inuit and Dene people have treated these remains in a similar fashion for hundreds of years? More fundamentally, would a sanitized community be authentic in terms of the host culture?

Socio-cultural realm

At the core of the socio-cultural realm is the question of whether ecotourism in indigenous territories will erode the integrity of the host cultures. Ecotourism is a commercial activity and to the extent that indigenous cultures are part of the attraction, these cultures will be commoditized in the process of producing an experience for the tourist. Critics of the commercialization of culture for tourism argue that in the process of commoditization, the hosts' culture will be eroded as it becomes an economic activity devoid of deeper meaning (Greenwood, 1977). King and Stewart (1996) extend this criticism to the change of relationship between indigenous people and the land associated with commoditization:

> For indigenous people, the commodification of nature implies a change in the meaning of their environment from a source of direct sustenance with a use value to a commodity with an exchange value. This change expresses a shift in the relationship between the indigenous people and their environment, from one of working *with* the land to one of working *for* tourists (who observe the land).
>
> (King and Stewart, 1996, p. 296)

Yet Cohen (1988) has argued that this process does not necessarily destroy the meaning of cultural products, although it may change it or add new meaning to old ones. Decisions do, however, have to be made about the nature and extent of change that is acceptable to the indigenous hosts. Control over this process is critical. It is interesting to note that the Ngai Tahu Maori who own and operate Whale Watch Kaikoura, a major ecotourism company in Aotearoa, have exercised this type of control by making a conscious decision not to present their traditional culture as part of their successful ecotourism interpretive programme. This decision was made based on the disappointing experiences of guides who experimented with more culturally based interpretations earlier in the company's development. As one of the guides explained:

> you are not going to 'hand out' your beliefs system for them to 'cheapen' or 'bastardize', and by that I mean that quite often people [tourists] will not accept or even acknowledge your reality in terms of its mythological substance, that myths and legends are part of our history.
>
> (cited in McIntosh *et al.*, unpublished)

The Ngai Tahu are considering more culturally based products in the future but if these products are developed they will be carefully targeted to tourists who are seeking this kind of experience. In general, indigenous people need to consider how their culture is likely to change as a result of ecotourism within their territory. Knowing the potential cultural impacts of various scenarios will enable indigenous people to make decisions that are in line with their goals and aspirations.

A second major socio-culture issue concerns the fact that indigenous cultures are dynamic. The romantic image that non-indigenous people have of natives living by traditional means is no longer accurate and in many cases never was. There are a multitude of factors that are initiating change in these communities, including advances in health, education and communications. Much of this change is consciously pursued by indigenous people who feel that they will be better off as a result. The contrast between tourist expectations of indigenous people living traditional lifestyles and the preferences of many indigenous people for modern lifestyles presents challenges for ecotourism within indigenous territories. These challenges are illustrated in the case of the Sami people in Scandinavia:

> There is an expectation that the Sami hosts shall perform in a traditional way. This is the image known from books and marketing materials. To satisfy their customers the Sami hosts try to fulfil these expectations and perform much more traditionally than they would usually. For example, they wear traditional clothing, use tents, and use reindeer transportation. At the same time they feel that their activities are counter-productive with regards to another important aim: to become a respected and integrated part of the world.
>
> (Viken, 1998, pp. 46–47)

The tension that is described in the context of the Sami is common among most indigenous groups involved in tourism. One of the challenges facing ecotourism operators on indigenous lands is, therefore, how to align the expectations of the ecotourist with the contemporary preferences and lifestyles of the indigenous hosts.

Guiding Principles for Ecotourism on Indigenous Lands

The fundamental principles of ecotourism, if addressed in earnest, will work well in the context of indigenous territories. A genuine respect for indigenous peoples must form the foundation of ecotourism operations on their lands. Part of this respect is recognition of the rights of indigenous people under 'aboriginal title' as well as 'legal title'. This means that significant control must be located with the indigenous hosts whether they are the actual operators of the ecotourism operation or not. Fennell (1999) has identified five key points that need to be considered if productive relationships are to be established between non-indigenous and indigenous peoples in a tourism context. These points have been adapted to articulate basic principles for ecotourism within indigenous territories.

1. Community involvement

If ecotourism activities are occurring within indigenous territories then indigenous people should be involved. Although the level and nature of their involvement may be negotiated between stakeholders, the ultimate decision as to whether to proceed with the ecotourism initiative should remain with the indigenous hosts. Other forms of direct involvement that should be considered include ownership, management and employment.

2. Community benefit

A fair share of the benefits should accrue to the indigenous hosts. These include benefits owed to the community as a whole as well as those earned by individual indigenous entrepreneurs. The collective nature of indigenous communities is distinct from that found in most non-indigenous communities and therefore should be recognized in terms of explicit communal benefits. As a general rule, benefits that indigenous people accrued from the land before the introduction of ecotourism should be maintained or compensation should be negotiated if this stream of benefits is reduced in some way. Access to the territories for subsistence purposes should be continued although management strategies may need to be developed to avoid conflicts between consumptive and non-consumptive users. Ecotourism practices should be designed to reinforce rather than contradict the values that are important to the indigenous hosts, such as respect for elders and respect for the land.

3. Scale

Ecotourism operations should be relatively small scale and the temptation to continually expand these operations to meet demand should be resisted. The marketplace dictates that demand and supply are the key determinants of growth, but in the case of ecotourism on indigenous lands, interventions into this marketplace should be made to limit growth. By doing so, the scale of operations is more likely to remain sustainable given the resources of the indigenous hosts.

4. Land ownership

The legal status of the territories in question should be clarified in order to foster effective partnerships between stakeholders. At a minimum, there needs to be agreement as to whether the lands are under 'aboriginal title' or not. If title is claimed by an indigenous group but not recognized by non-indigenous ecotourism operators, these operators may find themselves in a very awkward position. Operators who do not have the support of the local indigenous peoples will have a difficult time providing their clients with a satisfying tourism experience. Clarification of the degree of indigenous interest in a territory needs to be achieved. If aboriginal title is recognized, or if some other level of association with the territory can be agreed upon, then solid partnerships can be formed. This does not mean that the negotiations will necessarily be easy, but by having the question of land ownership clarified, indigenous people and other stakeholders will be in a position to make decisions and follow through with them.

5. Sensitivity to the needs of area residents and visitors

Ecotourists and indigenous hosts need to possess a well-developed understanding and respect for each other. This is consistent with the rhetoric of ecotourism in which the development of a true understanding of a place, including its people, is a key objective. The standard expectation is that ecotourism operators and, indeed, ecotourists themselves will take the initiative to become educated about their indigenous hosts. Less recognized but equally important is the need for the indigenous hosts to become educated about their visitors. By developing this understanding, indigenous peoples will not only be in a better position to extend their hospitality, they will also be in a better position to make decisions about the directions that they prefer to take in terms of future ecotourism development.

Various mechanisms exist that can assist in the implementation of these principles. At the forefront of these mechanisms is the establishment of protected areas as discussed elsewhere in this book (see Chapters 18 and 19). These areas can be managed in a variety of ways but one of the most common is to develop a joint management agreement that is tailored to the unique needs and resources of the relevant stakeholders (Colchester, 1994). Clearly, however, the involvement of the indigenous hosts must be real rather than cosmetic. King and Stewart (1996) have argued that the goal of such agreements should be ultimate transfer of these responsibilities to the indigenous hosts.

Conclusion

Ecotourists are increasingly seeking out indigenous lands to pursue their travel motivations. Ideally, the outcomes of these interactions between visitors and hosts will be beneficial to all: ecotourists will have satisfying touristic experiences, the natural resources will be sustained, and the indigenous host will enjoy significant net benefits. In reality, there is no guarantee that these outcomes will be achieved. Key topics that were covered in this chapter included the ownership and significance of land to indigenous people, their motivations for involvement and key issues that exist in the economic, environmental and the socio-cultural realms of ecotourism on indigenous lands. Five guiding principles were then presented as strategies for meeting the challenges of ecotourism on indigenous lands. These principles addressed involvement, benefits, scale, land ownership and a recognition of the needs of all stakeholders.

The underlying theme of the chapter was that indigenous people should retain control of their lands and make decisions about the nature of the ecotourism activities that are conducted in these territories. It must, however, be appreciated that the question of who 'should' make these decisions is a different one from who 'can' make these decisions. A variety of structural barriers exist both within a tourism context, and within a societal context, that militate against indigenous control. Despite these barriers, it is in the best interests of sustainability that control migrates towards the indigenous owners of these territories. While this control is a prerequisite of sustainability, it is not necessarily sufficient. If indigenous peoples are to make appropriate decisions about ecotourism in their territories, they must develop a clear understanding of ecotourism as a complex, sophisticated, global industry. These decisions cannot be made in isolation but must be positioned within the broader context of the industry and their community as a whole.

References

Acres International Limited (1990) *Northwest Territories Visitors Survey Summer 1989*. Prepared for the Department of Economic Development and Tourism, GNWT, Yellowknife.

Boo, E. (1990) *Ecotourism: the Potentials and Pitfalls,* Vol. 1. World Wildlife Fund, Washington, DC.

Cohen, E. (1988) Authenticity and commoditization in tourism. *Annals of Tourism Research* 15, 371–385.

Cohen, E. (1996) Hunter-gather tourism in Thailand. In: Butler, R.W. and Hinch, T.D. (eds) *Tourism and Indigenous Peoples*. International Thomson Business Press, London, pp. 227–254.

Colchester, M. (1994) Salvaging nature: indigenous peoples, protected areas and biodiversity conservation. Discussion paper, United Nations Research Institute for Social Development, Geneva, Switzerland.

Colton, J. (2000) Searching for sustainable tourism in the Caribou Mountains. Doctoral dissertation, The University of Alberta, Edmonton, Alberta.

Colton, J. and Hinch, T.D. (1999) Trap-line based tours as indigenous tourism products in Northern Canada. *Pacific Tourism Review* 3(1), 1–10.

Cordell, J. (1993) Boundaries and bloodlines: tenure on indigenous homelands and protected areas. In: Kemf, E. (ed.) *Indigenous Peoples and Protected Areas: the Law of Mother Earth.* Earthscan Publications, London, pp. 61– 68.

D'Amore, L. (1988) Tourism – the world's peace industry, In: D'Amore, L. and Jafari, J. (eds) *Proceedings of the First Global Conference – Tourism a Vital Force for Peace.* Lou D'Amore Associates, Montreal, pp. 7–14.

de Burlo, C. (1996) Cultural resistance and ethnic tourism on South Pentecost, Vanuatu. In: Butler, R.W. and Hinch, T.D. (eds) *Tourism and Indigenous Peoples.* International Thomson Business Press, London, pp. 255–277.

Fennell, D. (1999) *Ecotourism: an Introduction.* Routledge, London.

Gray, A. (1991) Between the spice of life and the melting pot: biodiversity conservation and its impact on indigenous peoples. IWGIA Document No. 70. International Work Group on International Affairs, Copenhagen.

Greenwood, D. (1977) Culture by the pound: an anthropological perspective on tourism as cultural commoditization. In: Smith, V. (ed.) *Hosts and Guests: the Anthropology of Tourism.* University of Pennsylvania Press, Philadelphia, pp. 17–31.

Guujaaw (1996) The Haida Nation approach to tourism. In: *Canadian National Aboriginal Tourism Association (CNATA) Nature-based Tourism Manual, (Appendix B).* http://www.vli.ca/clients/abc/cnata/cnata3.htm

Haywood, K.M. (1991) A strategic approach to developing hospitality and tourism education and training in remote, economically emerging and culturally sensitive regions: the case of Canada's Northwest Territories. In: *New Horizons,* Conference proceedings, The University of Calgary Press, Calgary.

Hinch, T.D. (1998) Ecotourists and indigenous hosts: diverging views on their relationship with nature. *Current Issues in Tourism* 1(1), 120–123.

Hinch, T.D. and Butler, R.W. (1996) Indigenous tourism: a common ground for discussion. In: Butler, R.W. and Hinch, T.D. (eds) *Tourism and Indigenous Peoples.* International Thomson Business Press, London, pp. 3–22.

Hollinshead, K. (1992) 'White' gaze, 'red' people – shadow visions: the disidentification of 'Indians' in cultural tourism. *Leisure Studies* 11, 43–64.

King, D.A. and Stewart, W.P. (1996) Ecotourism and commodification: protecting people and places. *Biodiversity and Conservation* 5, 293–305.

Martin, C. (1993) Introduction. In: Kemf, E. (ed.) *Indigenous Peoples and Protected Areas.* Earthscan Publications, London, pp. xv–xix.

McCullum, H. and McCullum, K. (1975) *This Land is Not For Sale.* Anglican Book Centre, Toronto.

Notzke, C. (1994) *Aboriginal Peoples and Natural Resources in Canada.* Captus University Publications, North York.

Notzke, C. (1996) Partners in conservation: co-management, protected areas and aboriginal ecotourism development in the North, paper presented at *The Canadian Association of Geographers Annual Meeting 1996, 11–16 May, University of Saskatchewan, Saskatoon, Saskatchewan.*

Notzke, C. (1998) Indigenous tourism development in the Arctic. *Annals of Tourism Research* 26(1), 55–76.

Smith, V.L. (1989) *Hosts and Guests: the Anthropology of Tourism,* 2nd edn. University of Pennsylvania Press, Philadelphia.

Stansfield, C. (1996) Reservations and gambling: native Americans and the diffusion of legalized gaming. In: Butler, R.W. and Hinch, T.D. (eds) *Tourism and Indigenous Peoples.* International Thomson Business Press, London, pp. 129–149.

Stevens, S. (1997) Introduction. In: Stevens, S. (ed.) *Conservation Through Cultural Survival: Indigenous Peoples and Protected Areas.* Island Press, Washington, DC, pp. 1–8.

Thongmak, S. and Hulse, D.L. (1993) The winds of change: Karen people in harmony with world heritage. In: Kemf, E. (ed.) *Indigenous Peoples and Protected Areas: the Law of Mother Earth.* Earthscan Publications, London, pp. 162–168.

Viken, A. (1998) Ethnic tourism – which ethnicity? In: Johnson, M.E., Twynam, G.D. and Haider, W. (eds) *Shaping Tomorrow's North: the Role of Tourism and Recreation.* Northern and Regional Studies Series No. 7, Lakehead University, Thunder Bay, Ontario, pp. 37–53.

Warren, J.A.N. and Taylor, C.N. (1994) *Developing Eco-tourism.* NZ Institute for Social Research and Development Ltd, Wellington.

Wilmsen, E.N. (1989) *We Are Here: Politics and Aboriginal Land Tenure.* University of California Press, Berkeley.

Wuttunee, W.A. (1992) *In Business For Ourselves.* McGill-Queen's University Press, Montreal and Kingston.

Young, E. (1995) *Third World in the First.* Routledge, London.

Zeppel, H. (1998) Land and culture: sustainable tourism and indigenous peoples. In: Hall, C.H. and Lew, A. (eds) *Sustainable Tourism: a Geographical Perspective.* Addison Wesley Longman Limited, Harlow, UK.

Section 5

Ecotourism Impacts

P.F.J. Eagles
Department of Recreation and Leisure Studies, University of Waterloo, Waterloo, Ontario, Canada

---

Arguably, all consideration of ecotourism is dependent on the data that are derived from impact measurement. The determination of the size, scale and impact of a phenomenon requires the determination of a measurement goal, a measurement device and a methodology for measurement. In impact assessment, the size of the phenomenon to be measured is typically so large, that a sample must be chosen. Often, it is desirable to choose an indicator to represent the larger phenomenon. An indicator is *that which serves to indicate or give a suggestion of something; an indication of*. The chapters in Section 5 each deal with various aspects of the identification, measurement and management of ecotourism impacts.

The chapters found in Section 5 use many indicators to represent some larger state. One might question whether the indicators chosen are the most appropriate. In addition, it is important to note whether the proper measurement device and methodology was applied. The understanding of data, indicators and impacts must be carefully weighed. It is important not to simply accept impact conclusions based on indicators without caution.

Ercan Sirakaya, Tazim Jamal and Hwan-Suk Choi tackle the substantial problem of the determination of indicators in Chapter 26. They outline the development of the concept of indicators over time. They identify the characteristics that lead to the choice of better indicators. They point out the stakeholder involvement in the development and application of indictors. The role of monitoring and reporting is identified. The authors note that: 'indicators have to be selected so that they are robust, credible, efficient (in time and cost for obtaining the data), and useful to decision makers'. The chapter makes the point that the data from indicators are only inputs to decision making. The importance of the indicators is dependent on the ability of the decision making structure to use the information in an effective and competent manner.

In the understanding of ecotourism, one key factor is its impact on people, communities and environments. The authors of three chapters in Section 5 have identified impacts on the more obvious categories of study: economics, socio-cultural relations and the physical/natural environment. It is critical to recognize that all impact identification and determination is dependent on value judgements. Who makes the judgements is a critical element of the decision process. The process used to involve people and to make the judgements must be identified and must be clearly understood by all who use the outcomes.

Paul Eagles, in a chapter in Section 8, points out that all decisions are dependent on information. The better the information available to the planner and manager, the better the chance for a good decision. Kreg Lindberg, in Chapter 23, summarizes an extensive literature on the economic impact of ecotourism. This chapter identifies economic impact in three categories: jobs, income and profit. Lindberg is careful to identify the methods used to measure the impact and to clarify the extent to which individual studies can be applied elsewhere. Interestingly, the chapter leads the reader to conclude that the economic impact of ecotourism, as important as it is, is frequently underestimated, under-reported and poorly calculated. This suggests that other social factors, possibly environmental protection or community development, are more important and receive more emphasis in the political decision making surrounding the phenomenon. However, it is also clear that until defensible economic impact estimates are done for ecotourism, it will continue to be treated by many in government and in the business community as a niche activity without substantial importance.

In Chapter 24, Ralf Buckley tackles the huge problem of summarizing the extensive literature on environmental impact. Whereas it is relatively clear in economic impact where the positive values lie, typically towards larger impact, it is not nearly so clear where the positive values lie in environmental impact. Is it better to have more or less of a species? How does one know when ecological integrity is intact? How much soil erosion is bad? Professor Buckley makes the important point that the environmental impact of ecotourism must consider the travel to and from the activity destination. So often only the impacts at the visitation site are identified. The chapter points out that the consideration of impacts goes well beyond the measurement of impact. The chapter concludes that often the 'lack of scientific knowledge is less of an impediment than lack of management funds or political support'.

Professor Buckley's chapter provides a broad coverage of the current knowledge of the environmental impact of ecotourism. But the chapter does not identify the environmental impact of the *lack* of ecotourism. The common assumption is that the environmental impact of outdoor recreation or ecotourism should use as its benchmark no human use or no human impact on the environment. This is an invalid assumption, because typically in the absence of outdoor recreation or ecotourism some other economic activity will take place in that environment. If the site is not a national park catering to tourists, it will be supporting a logging industry, a grazing industry or some other resource-based economic activity. Therefore, the environmental impact of ecotourism should be compared with the most likely alternative economic activity, not to some unrealistic utopia without any use.

Stephen Wearing, in Chapter 25, identifies the range of socio-cultural impacts that have been identified for local communities. Wearing concentrates his comments on smaller, rural communities, typically occupied by peoples somewhat marginalized in the large social fabric. Such people are very vulnerable to the social impacts of ecotourism. The biggest issues in socio-cultural impact identification are the assignment of value and the identity of the person who assigns the value. In addition, the political climate that determines the decisions made after value identification is critical to the application of socio-cultural impact identification. Therefore, so much of socio-cultural impact application lies in the field of politics.

Richard Butler looks to rural areas and the bases upon which their involvement in ecotourism is appropriate. He deals in Chapter 27 with the landscape that contains a high degree of agricultural activity. Professor Butler deals insightfully with the identification of value, and the determination of the role of the rural people in the determination of value. The identification of the role of food provision for ecotourism and the resultant economic and social impact is a useful factor that is too often forgotten in ecotourism analysis. The chapter concludes with the important statement that ecotourism in rural areas is:

> just as crucial in terms of environmental conservation and nature appreciation as when it occurs in remote tropical or polar areas, and in terms of fulfilling its role in providing local economic benefits, is infinitely more successful in a rural setting than an unpopulated wilderness one.

The chapters in Section 5 reveal that the principles underlying impact identification, indicator use, data needs, planning form and management functions are not unique to ecotourism. All of these principles are well known and well documented in the relevant fields of management theory, economic theory and planning theory. The identified information and impacts of ecotourism are found in the sociology, leisure studies and environmental studies literature, but the underlying principles are cross-disciplinary and outlined in a fundamental fashion in other fields.

Chapter 23
Economic Impacts

K. Lindberg
School of Tourism and Hotel Management, Griffith University, Gold Coast, Queensland, Australia

Introduction

The jobs generated by ecotourism provide an important reason for interest in, and support for, the phenomenon. These jobs often occur in areas relatively untouched by traditional development efforts and represent tangible economic benefits from natural areas. Several studies have assessed the local employment benefits of ecotourism; not surprisingly, the level of benefits varies widely as a result of differences in the quality of the attraction, access and other factors. In some cases, the number of jobs created will be low, but in rural economies even a few jobs can make a big difference.

Aside from its contribution to development generally, there are at least three reasons why local job creation is important in ecotourism. First, it is equitable in so far as conservation of an area for ecotourism may reduce or eliminate traditional resource use. Second, the ecotourists, as consumers, may support the importance of tourism benefiting local residents (P.F.J. Eagles, J.L. Ballantine and D.A. Fennell, 1992, unpublished). Third, when residents receive benefits, the extractive pressure on natural resources is lessened, and residents are more likely to support tourism and conservation, even to the point of protecting the site against poaching or other encroachment. For example, Lindberg *et al.* (1996) found that ecotourism-related benefits were an important basis for positive resident attitudes toward adjacent natural areas (see also Wunder, 1996, 1998). Conversely, if residents bear the costs without receiving benefits, they may turn against tourism and conservation, and may intentionally or unintentionally damage the site. Whether ecotourism benefits lead to increased support for conservation and, ultimately, to changes in resource use is dependent on a variety of circumstances (Brandon and Wells, 1992; Brandon, 1997).

Although this chapter focuses on ecotourism in particular, it is worthwhile to 'set the stage' by describing the economic impact of tourism in general. Tourism statistics are of variable, and sometimes low, quality. Nonetheless, the methods and quality of the data are improving, and available statistics provide at least a rough idea of tourism's economic impacts. Table 23.1 presents estimates from the World Travel and Tourism Council (WTTC). Tourism's current impact is expected to grow over the next decade, with WTTC estimating that the industry will create over 5.5 million jobs per year during that period. This growth will occur on top of significant recent growth in tourism, with

World Tourism Organization (WTO) estimates of growth in the decade from 1985 to 1994 as follows: Africa 89%, South America 86%, Central America 91%, the Caribbean 71%, East Asia and the Pacific 142%, and South Asia 48% (Mowforth and Munt, 1998, p. 93).

As these figures reflect, in the economic impact arena most attention is paid to the jobs, income and profit that ecotourism generates; these will also be the primary focus of this chapter. Nonetheless, there are important additional economic impacts, both positive and negative, associated with development of tourism in general and ecotourism in particular (economic, environmental and socio-cultural impact groupings can overlap at times, and the present focus is on impacts typically classified into the economic category).

Fiscal impacts (taxes, fees, expenditures)

Tourism not only generates government revenue through business and other general taxes, but also through industry-specific channels, such as payment of occupancy and departure taxes. Conversely, tourism generates fiscal costs in the form of, for example, funding for infrastructure. In an evaluation of tourism in Belize, which is heavily oriented toward ecotourism, Lindberg and Enriquez (1994) note that this revenue covers specific tourism-related costs, such as tourism promotion and maintenance of the airport, but also generates net profits for the government (see also Borden *et al.*, 1996).

Of particular interest in the ecotourism context are fiscal impacts on protected areas. This issue is treated more fully elsewhere (e.g. Lindberg and Enriquez, 1994; Laarman and Gregersen, 1996; Lindberg, 1998; Van Sickle and Eagles, 1998). In brief, ecotourism has substantial potential to financially contribute to the creation and maintenance of protected areas, and this potential has been increasingly realized during the past decade. However, many areas still charge little or no fees, and at such sites ecotourism may cause a net negative fiscal impact due to the costs involved in providing the ecotourism experience.

Reduced access to resources

Tourism utilizes various resources as inputs into the products and services provided to visitors. In the case of ecotourism, one of these products is nature in a partially or totally preserved state. Preservation of nat-

Table 23.1. WTTC economic impact estimates (1999).

Region	GDP			Employment	
	Billions of US$	% of total in region	Annual % growth[a]	Millions of jobs	% of total in region
World	3550	11.7	3.0	192.3	8.2
North Africa	20	6.8	6.0	2.2	7.4
Sub-Saharan Africa	26	11.2	5.2	9.6	7.4
North America	1171	11.8	2.5	21.2	11.9
Latin America	90	5.6	6.1	8.9	6.0
Caribbean	29	20.6	5.5	3.6	15.8
Oceania	68	14.7	3.8	2.1	16.0
Northeast Asia	537	10.0	2.8	57.2	7.1
Southeast Asia	81	10.6	5.5	15.2	7.3
South Asia	27	5.3	9.1	22.3	5.4
Europe	1461	14.0	2.6	47.8	13.2
Middle East	41	7.3	5.2	2.0	6.1

[a] 1999–2010 estimated, adjusted for inflation.

ural areas often involves reduced local access to resources, such as wood or medicinal plants. In so far as tourism is a partial or sole rationale for preserving an area, it also causes reduced access to resources.

Inflation

Many destinations have experienced increased prices for goods, services, and land due to tourism development, and this is a cost borne by residents of the area who purchase these items.

Effects on income distribution

In some cases, tourism development exacerbates existing income inequalities within destination communities, while in others it generates new financial elites.

Revenue sharing

At some ecotourism destinations, residents benefit from revenue-sharing programmes that either provide cash payments or, more commonly, funding for community projects such as wells or schools. For example, Nepal's Wildlife Conservation Act provides for the distribution of 30–50% of protected area fee revenue to surrounding communities (Brandon, 1996).

Whether the above impacts are good or bad will depend on one's perspective. For example, some may desire continuity in local economic (and political) relationships, while others may desire reductions in income inequalities. Persons wishing to sell land would welcome increased land prices, while those who wish to buy land or to retain land they own (and on which they may pay property taxes) would oppose increased prices. Likewise, tourism is said to compete with other sectors, notably agriculture, for land, labour and finance. The desirability of this competition depends on one's perspective; workers earning a higher wage or investors receiving a higher return from tourism may disagree with members of the community who lament the transition away from traditional agricultural activities.

Leakage is often listed as a negative impact, but it is more appropriately viewed as the absence of a positive impact. Rather than causing economic harm, it simply does not provide the benefit of the foregone jobs. Similarly, the instability and, in some cases, undesirability of tourism jobs is often seen as a negative impact, but can alternatively be viewed as the lack of positive impacts (stable, desirable employment). Regardless of how they are classified, these are important considerations in the development of tourism, whether ecotourism or otherwise. Leakage is discussed further below, and Sinclair and Stabler (1997) and Weaver (1998) provide additional treatment of these issues.

The debate over leakage also raises a more general issue, that of the motivation and reference point for evaluating ecotourism, or general tourism. It is true that tourism typically involves high levels of leakage, but that does not necessarily mean it is undesirable as a development strategy. Appropriate questions in this context are: (i) whether leakages can be reduced and, if so, at what expense; and (ii) given current or reduced leakages, combined with other benefits and costs, whether tourism remains more desirable than alternative development options.

Though the diverse impacts of tourism are increasingly being recognized, the traditional impacts of jobs and income (from employment, rather than from revenue-sharing programmes) tend to be the most discussed and researched, and they will be the focus of the remainder of this chapter. The present focus is on concepts and methods for estimating impacts. Tools for enhancing impacts are discussed elsewhere (e.g. Butler, Chapter 27, this volume; Lindberg, 1998). To the extent possible, ecotourism-based examples and applications will be used. However, examples and applications from general tourism or other sectors will be used when necessary to illustrate techniques and principles. In addition, though issues and examples relevant to both developed and developing countries are presented, the discussion is weighted toward the latter.

Expenditure, Linkage and Leakage: a Basic Description of Ecotourism's Money Flows

An understanding of ecotourism's contribution to economic development requires an understanding of the ecotourism 'industry'. Ecotourism is, of course, tremendously variegated; it can encompass everything from paying travel agents thousands of dollars for trips to the furthest reaches of the globe to simply walking to a nearby park. However, to simplify matters, it is useful to think of ecotourism as comprising three components. The first is the outbound operator that sells tours directly to international tourists in the source country. The second is the inbound (ground) operator that actually organizes and leads the trip in the destination country. The third is the attraction that is being visited.

Consider the example of an American tourist wishing to visit Amboseli National Park in Kenya. She might buy a tour from a US outbound operator, which in turn has arranged for an inbound operator to lead the trip in Kenya. The inbound operator will in turn purchase admission to the park, which is managed by the Kenya Wildlife Service. Alternatively, the tourist may choose to arrange the trip directly with an inbound operator, either to save money or because she is already in Kenya. Or, she might forgo using an operator in favour of travelling to the park by herself.

Many observers voice the concern that much of the trip cost, and thus the economic benefit, remains with outbound operators and source-country airlines. To some extent, this is simply due to the nature of the tourism industry; substantial funds are spent on marketing, commissions and transport before tourists even reach the destination. For example, Sorensen (1991) presents a case study of Overseas Adventure Travel (OAT), an outbound operator in Massachusetts, USA. In 1989, OAT sales totalled US$4,525,000 (all figures are rounded), of which US$1,400,000 was for air transport and US$3,027,000 for land tours. The land tours cost US$1,962,000 to supply, with a resulting gross profit from this product of US$1,065,000 (approximately 86% of the total company gross profit). Much of this gross profit remained in the USA through allocation to salaries and related (US$714,000), sales and marketing (US$496,000) and administrative/general (US$264,000). Using preliminary 1990 budget figures, the major sales and marketing budget items were media advertising (6% of sales and marketing budget), catalogue and other sales tools (43%), postage (10%), telephone (6%), and travel agent commission (18%). Though the proportion of total sales revenue actually spent 'in country' at destinations is not estimated, the revenue allocated to land tours represents less than half of total sales.

Similarly, Brown *et al.* (1995) estimate that 40% of foreign visitor expenditure for trips to the Hwange and Mana Pools National Parks in Zimbabwe is lost to the country because of international air travel costs. Noland (1988) in Lindberg (1991) provides a breakdown of trip costs for a Mountain Travel African trek. Of the US$4105 trip price, US$150 (4%) was profit, US$1125 (27%) went to administration and commissions, US$350 (9%) went to the trip leader, US$350 (9%) paid for hotels, and US$2130 (52%) went to field costs, such as the inbound operator and park entrance fees. In this case, more than half of the trip cost was spent in country for field costs and hotels. However, airfare is not included in the price, and inclusion may reduce the in-country proportion to less than half. The catalogue alone for one nature tourism operator cost US$350,000 to produce. When divided by the number of clients who booked tours, the average cost came to US$116.67 per tourist.

In order to understand the issue of leakage, and the associated concept of multipliers, a brief description of economic flows is provided here. Tourism's economic contribution depends not only on how much comes into the region of interest (a country, a state/province/county, or a local community), but also on how much of what comes in stays in the region, thereby producing multiplier effects. The impacts of tourism, or any economic activity, can be grouped into three categories: direct, indirect and induced. *Direct impacts* are those arising from the initial tourism spending, such as money spent at a restaurant. The restaurant buys goods and services (inputs) from other businesses, thereby generating *indirect impacts*. In addition, the restaurant employees spend part of their wages to buy various goods and services, thereby generating *induced impacts*. Of course, if the restaurant purchases the goods and services from outside the region, then the money provides no indirect impact to the region, and *leaks* away. Figure 23.1 is a simplified illustration of some of these impacts and leakages.

A consistent finding of economic impact studies, particularly in developing countries, is the high level of leakage. Much of the initial tourist expenditure leaves the destination country, and especially the destination site itself, to pay for imported goods and services used in the tourism industry. The following examples are estimates of the percentage of tourism spending leaking away from destination country economies (Smith and Jenner, 1992; Brown *et al.*, 1995; Brandon, 1996; Sinclair and Stabler, 1997, p. 141; Lindberg, 1998; Mowforth and Munt, 1998, p. 194; and references cited within these sources):

- 70% for the average Caribbean country (up to 90% in the Bahamas, as low as 37% in Jamaica),
- 70% in Nepal,
- 60% in Thailand,
- 55% for the typical developing country,
- 55% in The Gambia,
- 53% for Zimbabwe,
- 45% in Costa Rica,
- 45% in St Lucia.

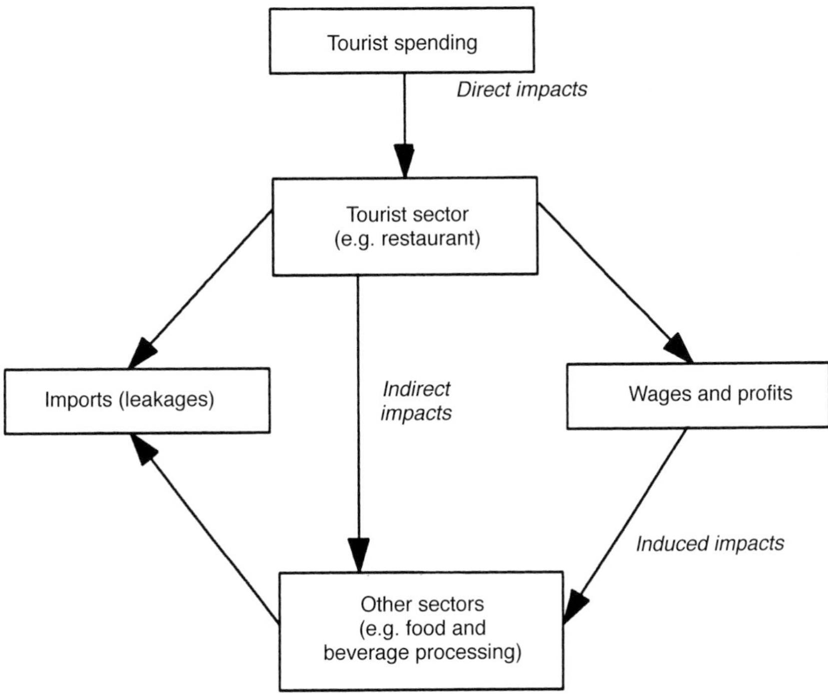

Fig. 23.1. Impacts and leakages.

More than 90% of tourism spending is thought to leak away from communities near most nature tourism sites. For example, Baez and Fernandez (1992) estimate that less than 6% of the income generated by tourism at Tortuguero National Park in Costa Rica accrues to the local communities. Similar figures have been estimated for the Annapurna region of Nepal (Panos, 1997) and lower figures for whale-watching in Baja California, Mexico (Dedina and Young, 1997). In Tangkoko DuaSudara in Indonesia, the benefit distribution is: 47% to the major tour company, 44% to hotels, and only 7% to guides (of which the head reserve guard gets 20%). Guides and food are usually brought from the provincial capital, so few benefits are retained at the village level (Kinnaird and O'Brien, 1996, p. 70). Box 23.1 provides a further example of linkages and leakages at a lodge in Zimbabwe.

The wide variation in leakage estimates across sites is partly a result of differing assumptions, definitions and methods used. However, it also is affected by the size and sophistication of the economy being evaluated (and thus also by its geographic scope), the type of tourists and tourism development, and the policies and efforts of individual tourism businesses. Smaller economies generally will have more leakage because a lower diversity of goods and services is produced in them than in large economies.

The issue of leakage is very complex, and comparisons across sites and types of trips can be misleading. In addition, the ultimate level of local economic benefit depends not only on the level of leakage, but also on the amount of spending. It is conventional wisdom that small-scale tourism development involves less leakage than does large-scale tourism, and there is

Box 23.1. Mana Pools Lodge.

Brown *et al.* (1995) estimate the distribution of revenues for trips involving the Mana Pools Lodge in Zimbabwe. The following figures show how revenues from a typical Harare–Mana Pools–Harare trip costing US$700 are used to purchase various local/national and international inputs (trip cost does not include international airfare to Harare). The leakage column shows the percentage of payment for each item that leaks away from the Zimbabwean economy.

Item	Cost (US$)	Leakage as % of item cost
Retail agent commission	140	72
Staff	82	0
Food/drink	68	2
Administrative overhead	60	0
Advertising and marketing	60	80
Repairs and maintenance	47	24
Energy	42	43
Depreciation	28	43
Communications	5	0
Insurance	5	0
Housekeeping	3	0
Freight and transport	1	50
Printing and stationery	1	2
Travel[a]	1	20
Taxes	27	0
Profit	130	0
Total	700	27

Of the total leakage, slightly more than half, in dollar terms, results from commissions.

[a] Though not specified in the report, this presumably represents staff travel.

empirical evidence from various studies supporting this assertion (Lindberg and Enriquez, 1994, pp. 60–61). However, small-scale tourism typically also involves less visitor expenditure, such that the total economic impact may be less than that for large-scale tourism. This phenomenon is illustrated by the Ecuadorean Amazon Napo region described by Wunder (1998) citing the work of Drumm (1991). The upper Napo region received US$357,000 per year in local income from tourist spending of US$1,340,000 per year. Due to a higher level of leakage, the exclusive and pristine lower Napo region received less local income (US$339,000 per year) despite higher tourist spending (US$3,860,000 per year).

On the other hand, Wunder (1998) presents the case of the Madre de Dios region in the Peruvian Amazon, based on Groom *et al.* (1991). In this case, there is a relatively high local share (25%) in tourism expenditure in the 'backpacker' area of Puerto Maldonado, and a relatively low local share (11%) in the pristine but remote Manu Biosphere Reserve. Nonetheless, Manu generates so much more tourism revenue than does Puerto Maldonado (US$1,700,000 vs. US$172,000, respectively) that it also generates more local income (US$192,695 vs. US$42,910, respectively), despite having higher levels of leakage (lower local share).

Estimating Economic Impacts: Concepts and Methods

This section discusses the most common approaches for estimating economic impacts within the ecotourism context. An issue that arises in each of these approaches is that the definition of ecotourism needs to be operationalized, i.e. to be defined such that a specific person/activity can be classified either as an ecotourist/ecotourism or as a general tourist/tourism, with 'general' being all tourism not defined as ecotourism. This is difficult to do, and most classifications are subject to debate.

The first, and crudest, approach is to adjust impact data for tourism as a whole using the proportion of all tourism that can be considered ecotourism. For example, the WTO forecasts that the East Asia and Pacific region will receive 229 million international arrivals by 2010. They also estimate that nature tourism generates 7% of all international travel expenditure. Assuming faster growth for nature tourism than for general tourism, this proportion may reach 10% by 2010; it is also assumed that the nature tourism proportion of expenditure equals the proportion of arrivals. Using these figures, Lindberg *et al.* (1998) estimated that there would be 22.9 million (229 million multiplied by 10%) international nature/ecotourism arrivals to the region in 2010. Though this example is based on arrivals, similar calculations can be undertaken for expenditure, employment or other variables. Such estimates are clearly rough, given the challenge of deriving a reasonable estimate of the proportion of all tourism that is constituted by ecotourism. However, this approach is often the only available basis for estimates at the national or regional level.

The second approach goes to the opposite spatial extreme and focuses on job creation or other variables at the site level, where a site often involves a natural area and surrounding communities. This approach is typically based on surveys of households and/or tourism businesses. For example, Lindberg and Enriquez (1994) used household surveys to estimate the percentage of households in four Belizean communities that benefited from tourism in various ways. Results are shown in Tables 23.2 and 23.3. A discussion of these results within the context of each community is provided in Lindberg and Enriquez (1994) and Lindberg *et al.* (1996). Although the tables focus on tourism-related jobs in general, it is also possible to focus on specific services or products. For example, Lindberg and Enriquez (1994) also report that tourism-related handicraft sales in Maya Center generated an average of BZ$2336 (US$1168) per household for the year ending March 1993. This revenue is particularly impressive when one considers that Belize GDP

Table 23.2. Tourism's direct local economic impact (percentage of households receiving each benefit, as reported by respondents).

Type of economic benefit from tourism	Community San Pedro (n = 75)[a]	Caye Caulker (n = 31)	Gales Point (n = 34)	Maya Center (n = 12)
Wage-paying job	41	19	21	8
Other job	5	10	0	50
Other income-generating activity	0	0	3	25
One or more of these benefits[b]	44	26	24	67

[a] n, number of households surveyed in each community.
[b] May be less than sum of individual benefits because some households receive multiple benefits.

Table 23.3. Tourism's additional local economic impact (percentage of non-tourism jobs that depend on tourism, as reported by respondents).

Level of dependence	Community San Pedro (n = 75)[a]	Caye Caulker (n = 31)	Gales Point (n = 34)	Maya Center (n = 12)
Totally dependent	22	28	22	20
Partially dependent	48	30	12	30
Total	70	58	34	50

[a] n, number of households surveyed in each community.

per capita was BZ$3124 at the time, and the fact that most of the materials used to construct the crafts were collected locally.

Site-level approaches such as this one typically address the definitional issue by assuming that all visitation in the area, and thus all tourism-related economic activity, is ecotourism. This assumption will be quite reasonable in some cases (such as Maya Center from the above example), yet may be less tenable in others (such as San Pedro). However, the site level is arguably the most interesting one, as much debate focuses on the extent to which ecotourism creates employment in local communities. Nonetheless, the site-level approach frequently suffers from several limitations. First, there is often interest in estimating indirect and induced impacts (multiplier effects), and the data necessary to do so are rarely available at such disaggregated levels. In developed countries, such data may be available down to the state/province or county/shire level. However, in developing countries, they are rarely available at the subnational level. Thus, though one can identify perceived dependence of non-tourism jobs on tourism (as in the Belize example), there are typically little or no data available to verify this dependence.

Second, and related to the first, the site-level survey approach often provides data simply on number of jobs, while the analyst may also be interested in income, profits, taxes paid to government, and other variables. In theory, such information could be gathered through a survey of residents and businesses. However, in practice, respondents may be unwilling to provide it. Most countries have secondary data, of varying quality, for these variables, and the methods used to estimate multiplier effects incorporate such data and provide the relevant estimates. Third, as noted in the case of San Pedro, not all tourism may reasonably be considered ecotourism. Unlike urban destinations, where a given hotel guest might be either a person travelling on business or a person visiting a cultural monument, ecotourism destinations are often geographically remote, with visitors there solely to view natural (and possibly cultural) attractions. None the less, many locations, especially regional centres, cater to both ecotourists and general tourists, such that a simple count of hotel employees may overestimate the impact of ecotourism in particular.

These limitations can be overcome by undertaking more complex analyses. The most common technique within general tourism is input–output (IO) analysis (e.g. WTO, 1985; Fletcher, 1989; Briassoulis, 1991; Wagner, 1997). IO begins with the construction of a transactions table that shows how much each industry, or sector, produces and how much it pays to other sectors to buy the inputs necessary to make its products. The transactions table is then converted into the technical coefficients matrix that shows the same information *per dollar of sales* (reference is made here to dollars, but the technique is the same regardless of currency). To identify indirect and induced effects, additional mathematical manipulation is necessary to create what is known as the Leontief inverse matrix. Readers interested in details of this process can refer to an IO text, such as Miller and Blair (1985); a numerical example for tourism in Belize is provided in Lindberg and Enriquez (1994).

The calculation and use of multipliers is subject to substantial confusion, in part due to the numerous types of multipliers (relatedly, there is confusion in terminology; this chapter uses simplified terminology for ease of reading, including general reference to multipliers rather than differentiation between multipliers and coefficients). One dimension of this multiplier typology is based on the variables for which multipliers are calculated. Promoters of tourism, or other industries, often focus on the sales multiplier because it is inevitably larger than the income multiplier. However, sales *per se* are usually not of interest. Rather, the amount of personal income (payments to households) generated is of interest, so the income multiplier tends to be the most useful from the policy viewpoint. Multipliers for other variables, including employment, can also be calculated. Another dimension is based on what is included (endogenous) within the model. For example, some multipliers exclude (treat as exogenous) wages and/or profits. Such multipliers are conservative because they omit induced impacts.

To determine the total impact of tourism, it is necessary to identify not only the impact of each dollar spent (indicated by the multiplier), but also the number of dollars spent. This can be done either by asking tourists how much they spend or by asking businesses how much they earn from tourists. When the region of interest attracts both ecotourists and general tourists, then only spending by ecotourists should be used to calculate ecotourism's impact. When a given person visits both ecotourism and general tourism attractions, the researcher must determine how to allocate the person's expenditure across these two activities. Johnson and Moore (1993) illustrate one method for allocation and discuss the more general issue of treating expenditures within a with-or-without framework that also recognizes substitution behaviour. Visitor spending is then broken down into the sectors present in the IO model. These amounts are multiplied by the relevant multipliers to derive impact estimates.

Basic IO can be extended in the form of 'social accounting matrices' or SAMs, which provide more detail regarding the distribution of monetary flows, such as the amount of income generated in different income categories within society (Pyatt and Round, 1985; Pyatt, 1988). Due to terminological inconsistency, several IO analyses may be more appropriately viewed as partial or full SAM analyses. Numerous IO studies have been undertaken for general tourism and, to a lesser degree, ecotourism or recreation at natural areas. The following examples illustrate applications and estimates, from various developing and developed countries. Lindberg and Enriquez (1994) used IO to estimate multipliers and economic impacts for tourism in the country of Belize. An estimate of US$100.25 million per year was used for tourism spending (the direct impact). Combining this figure with the IO model led to an estimate of US$211 million in sales each year throughout the Belizean economy due to tourism. More importantly, tourism generated US$41 million each year in payments to households, mostly in the form of wages (both figures are based on inclusion of induced impacts). Wagner (1997) developed a SAM model for the Guaraqueçaba region of Brazil. Based on an estimate of 7500 visitor days per year in the region, tourism was estimated to generate annually US$244,575 in output (sales), US$19,425 in labour payments (income), and 32 jobs (full-time equivalent).

Powell and Chalmers (1995) used visitor surveys and IO analysis to estimate the impact of visitor spending at two national parks in New South Wales, Australia. The study generated an estimate of AU$3.2 million in annual visitor expenditure plus AU$342,000 in annual agency expenditure at Dorrigo National Park (AU$1.00 ≅ US$0.55). Accounting for indirect and induced impacts, it was estimated that Dorrigo, with approximately 160,000 visitors per year, contributes almost AU$4.0 million in regional output, AU$1.5 million in regional household income, and payments to 71 employees. These represent 7–8% of regional totals for each category.

Several evaluations of the impact of natural area visitation have been made in the USA, though many are unpublished. For example, Smyth (1999) estimates that visitors to Glacier National Park generated US$74 million in sales, US$41 million in income and 2531 jobs in 1990. This represented 4% of the region's income and 7% of the region's jobs (cf. Stynes, 1992; Stynes and Rutz, 1995; Moore and Barthlow, 1997). As noted in the introduction, local economic impacts will be highly variable across sites, and the above figures for Glacier are not matched by sites with lower levels of visitation and opportunities for visitor spending. For example, Dawson *et al.* (1993) found that the economic impact of visitation at Great Basin National Park represented only 0.5% of output and 0.7% of employment in their study region.

Before turning to extensions of the basic multiplier concept, and its application through IO analysis, it is worth stressing that IO analysis rests on several assumptions. Although a detailed discussion of those assumptions is beyond the scope of this chapter, Box 23.2 briefly describes some of them to help readers better understand and evaluate the IO method and estimates. Moreover, multiplier analysis has frequently been applied and/or reported in misleading ways. Crompton (1995) provides a good summary of misapplications, though many other critiques have appeared in the literature (e.g. Archer, 1984; Hughes, 1994). Crompton notes that sales multipliers are often provided, when income multipliers are more meaningful (discussed above in the context of the Lindberg and Enriquez, 1994, study in Belize). In addition, employment multipliers may be misrepresented or misunderstood in so far as additional visitor spending may lead to more work for current employees (e.g. a shift from part-time to full-time) rather than hiring of new employees.

Moreover, 'incremental' or 'ratio' multipliers are sometimes used when 'normal' multipliers are appropriate; for example, an income multiplier that includes induced impact should be calculated as (direct + indirect + induced income)/(visitor expenditure), rather than as (direct + indirect + induced income)/(direct income). More generally, although multipliers and impact estimates are often used loosely for illustrative purposes, as in this chapter, any calculations or policy decisions based on them should involve reviewing the methods and assumptions used by the original analyst.

Estimating Economic Impacts: Extensions

Basic IO modelling remains perhaps the most common tool in tourism economic impact analysis, and holds promise also

Box 23.2. IO assumptions.

IO analysis relies on several assumptions concerning the structure of production processes within the economy. These assumptions include the following:

1. All businesses within each sector produce a single, homogeneous product or service, and the input procedures used in the production process are identical. That is, the economy should be disaggregated so that each sector is producing a single good or service. In practice, disaggregation is often performed for the sectors of particular interest, in this case tourism, and aggregation of other sectors is accepted.
2. An increase of production will always lead to purchase of inputs in the ratios shown in the technical coefficients matrix. In technical terms, the production function is linear and homogeneous. This assumption precludes economies of scale; for IO analysis to be accurate, a business will always use the same proportion of inputs regardless of how much it grows. As with the first assumption, this restriction can be overcome in part by using different sectors for businesses of different sizes. In the case of tourism, this could mean creating sectors for small, medium and large hotels.
3. When households are included in the analysis, their spending patterns (consumption functions) must also be linear and homogeneous. Again, this restriction can be overcome in part by disaggregating households into different groups.
4. The structure of the economy will not change. Many IO models are static in nature. They are based on data from a single year yet are often used to estimate impacts in other years. It is possible to construct dynamic IO models, but the data and analysis requirements are substantial.
5. If the analyst is interested in forecasting the effect of future increases in final demand (e.g. visitor expenditure), there must be unemployed resources available to be brought into the sector as inputs. This will often be the case because analysts make such forecasts specifically to identify opportunities for using unemployed resources like labour. However, there will be some cases in which resources are constrained, such as capital or skilled labour. In these cases, the resources will need to be drawn from other sectors or imported, and an unadjusted model will overestimate benefits.

A fuller discussion of IO assumptions can be found in the various general references on IO analysis, as well as the references to IO applications in tourism (e.g. Bulmer-Thomas, 1982; Miller and Blair, 1985; WTO, 1985). The assumptions are often violated in reality. Nonetheless, the fundamental structure of IO theory generally holds true, and economists have come to rely on IO despite the obvious breaches of assumptions.

within ecotourism, especially when most or all of the tourism within the region of interest can be viewed as ecotourism. Nonetheless, there is the opportunity, and sometimes the need, to extend or replace IO modelling with alternative approaches. The first set of 'extensions' utilizes the basic IO model to examine particular issues of interest.

As noted above in the discussion of leakage, there has been debate within ecotourism (and tourism in general) regarding the desirability of various forms of development. For example, is small scale better than large scale, basic better than luxury, and locally owned better than foreign owned? Are 'budget' travellers better than 'luxury' travellers, because the former buy more local products even though they spend less in total? Though these issues are complex, the information provided by multiplier analysis can provide important feedback. For example, it is well recognized that different tourist segments spend different amounts of money while on holiday (McCool and Reilly, 1993; Pearce and Wilson, 1995; Leones *et al.*, 1998). Multiplier models also allow one to evaluate whether, for a given dollar of spending, one segment has more of an impact than another, with the difference due to the pattern of expenditure. For example, Liu (1986) found that Japanese visitors to Hawaii generated higher income multipliers than did visitors from other source markets, a result attributed to proportionally higher spending on retail goods and lower spending on hotels and restaurants (cf. West and Gamage, 1997).

The second utilization of multiplier analysis involves modelling policy effects. Because IO analysis quantitatively models the structure of the economy, it can be used to model changes in that structure, and thus the impact of selected policies and programmes. For example, Lindberg and Enriquez (1994) found that approximately 50% of food and beverage purchases by the hotel and restaurant sectors in Belize were for imported products. Using the IO tables, they estimated that an increase in local purchases from 50% to 75% (a decrease in imports from 50% to 25%) would generate an economy-wide increase in sales of almost US$9 million and in income of US$1.4 million (cf. Telfer and Wall, 1996; Wagner, 1997).

Another policy issue is the economic impact of increases in entrance or other tourism-related fees. For example, the Costa Rican National Chamber of Tourism (CANATUR) estimated that national park fee increases led to reduced visitation and, thereby, a national income loss of US$65 million in the mid-1990s (Inman *et al.*, 1998). As noted by Lindberg and Aylward (1999), the reduced visitation was probably due to a variety of factors, including many unrelated to fee increases. In another example, Krakauer (1998, pp. 26–27) describes how increased fees and limitations on expedition numbers for climbing Mt Everest in Nepal led to a shift from Nepal to Tibet, thereby leaving hundreds of sherpas out of work. However, the shift turned out to be caused by the limitations, rather than the fee. A further increase in the base fee from US$50,000 to US$70,000 per group did not seem to deter groups from Nepal. Despite this example, and many less extreme ones, substantial fees generally will have some effect on visitation levels. Multiplier analysis can be used to estimate how resulting reductions in visitor spending will affect jobs, income and sales within the economy.

The following extensions involve more significant departures from the base IO approach, and generally require significant additional data and/or mathematical development. Therefore, they may be beyond the range of many ecotourism applications. Alternatively, the additional information provided may not outweigh the cost relative to simpler techniques, especially given the low level of inter-industry linkages in local economies. However, they are discussed here briefly given their potential value in the ecotourism context. The first extension involves applying or adapting IO models to a sub-regional level. Although IO analysis can (and, ideally, should) involve substantial primary data collection, cost considerations mean that many IO models depend heavily on secondary (existing) data. These data are usually available only at geographic levels such as counties/shires or larger units, which are broader than the local level typically of interest in ecotourism. The challenge, then, is to develop an IO model at the sub-regional level. Robison (1997) illustrates how this might be done, both with respect to mathematical model development and with respect to using sub-regional data sources to adjust regional (in this case, county)-level IO data.

Simpler alternatives to IO at the sub-regional level include economic base or Keynesian income multiplier methods. Conversely, computable general equilibrium (CGE) models offer a more theoretically appealing, but also more computationally difficult, method for estimating impacts (Adams and Parmenter (1995) and Zhou *et al.* (1997) illustrate recent applications in general tourism). As with IO models, CGE models typically are estimated only for large areas, due to the data and expertise needed to estimate them. However, Taylor and Adelman (1996) illustrate how CGE (and SAM) can be applied at the village level in developing countries (see also Robinson, 1989 for an overview of CGE, as well as IO and SAM models).

Lastly, the basic structure of IO, which shows linkages between different parts of the economy, can be extended to show linkages between the economy and the environment. For example, Johnson and Bennett (1981) incorporated biological oxygen demand, total suspended solids, and carbon dioxide into their IO model of Darlington county, South Carolina, USA (see also Borden *et al.*, 1996 for a simpler evaluation of resource impacts within the tourism sector).

Summary

This chapter has reviewed concepts and methods for estimating ecotourism's economic impacts. Although there has been continuing controversy surrounding impact estimates, it is clear that tourism is a major economic force around the world. Moreover, economic impact, and especially job creation in communities living near natural areas, plays a critical role in the ecotourism context.

The issue of leakage is discussed and illustrated using examples. Though efforts to reduce leakage are worth pursuing, the dominant focus on this issue may distract attention away from a more fundamental issue: given the economic realities within the ecotourism system, modified to the extent possible through efforts to reduce leakage, does ecotourism remain a desirable activity from the perspective of job and income generation? Although ecotourism's benefits may be frequently overstated, it is likely that the answer to this question is, in most cases, 'yes'.

In calculations of tourism's economic impacts, IO has been the 'workhorse' method. It will probably remain important in general tourism, as well as gain importance in ecotourism, given its ability to evaluate linkages and its ease of use relative to other approaches, such as CGE. Nonetheless, the Taylor and Adelman (1996) examples illustrate that there remains potential to apply CGE in the context of rural communities involved in ecotourism. More generally, it is hoped that this chapter has illustrated the complexity of issues in economic impact estimation, and the value of utilizing models such as IO to better understand the issues and, ultimately, to guide ecotourism policy.

Despite the focus of this chapter (and much policy debate) on job creation, there are several other economic impacts that should be considered in ecotourism. In addition, there are important non-economic impacts, including impacts falling into the traditional categories of environmental and social (see Chapters 24 and 25). It is vital to incorporate all of these impacts into decisions involving ecotourism if it is to live up to its ideals.

Acknowledgement

This chapter was written while the author was at the Institute of Transport Economics, Oslo, Norway. The author thanks Edward Waters for comments on the draft manuscript.

References

Adams, P.D. and Parmenter, B.R. (1995) An applied general equilibrium analysis of the economic effects of tourism in a quite small, quite open economy. *Applied Economics* 27, 985–994.

Archer, B.H. (1984) Economic impact: misleading multiplier. *Annals of Tourism Research* 11, 517–518.

Baez, A.L. and Fernandez, L. (1992) Ecotourism as an economic activity: the case of tortuguero in Costa Rica. Paper presented at the First World Congress of Tourism and the Environment, Belize.

Borden, G.W., Fletcher, R.R. and Harris, T.R. (1996) Economic, resource, and fiscal impacts of visitors on Washoe County, Nevada. *Journal of Travel Research* 34, 75–80.

Brandon, K. (1996) *Ecotourism and Conservation: a Review of Key Issues.* World Bank Environment Department Paper No. 033. World Bank, Washington, DC.

Brandon, K. (1997) Policy and practical considerations in land-use strategies for biodiversity conservation. In: Kramer, R., von Schaik, C. and Johnson, J. (eds) *Last Stand: Protected Areas and the Defense of Tropical Biodiversity.* Oxford University Press, Oxford.

Brandon, K. and Wells, M. (1992) Planning for people and parks: design dilemmas. *World Development* 20, 557–570.

Briassoulis, H. (1991) Methodological issues: tourism input-output analysis. *Annals of Tourism Research* 18, 485–495.

Brown, G., Ward, M. and Jansen, D.J. (1995) Economic value of National Parks in Zimbabwe: Hwange and Mana pools. Report prepared for the ZWMLEC project, coordinated by the World Bank.

Bulmer-Thomas, V. (1982) *Input-Output Analysis in Developing Countries: Source, Methods and Applications.* John Wiley & Sons, New York.

Crompton, J.L. (1995) Economic impact analysis of sports facilities and events: eleven sources of misapplication. *Journal of Sport Management* 9, 14–35.

Dawson, S.A., Blahna, D.J. and Keith, J.E. (1993) Expected and actual regional economic impacts of Great Basin National Park. *Journal of Park and Recreation Administration* 11, 45–59.

Dedina, S. and Young, E. (1997) Conservation and development in the Gray Whale Lagoons of Baja California Sur, Mexico. http://www.scilib.ucsd.edu/sio/guide/z-serge.html

Drumm, A. (1991) Integrated impact assessment of nature tourism in Ecuador's Amazon region. Study for FEPROTUR-NATURALEZA, Quito.

Fletcher, J.E. (1989) Input-output analysis and tourism impact studies. *Annals of Tourism Research* 16, 514–529.

Groom, M.J. *et al.* (1991) Tourism as a sustained use of wildlife: a case study of Madre de Dios, southeastern Peru. In: Robinson, J.G. and Redford, K.H. (eds) *Neotropical Wildlife Use and Conservation.* University of Chicago Press, Chicago.

Hughes, H.L. (1994) Tourism multiplier studies: a more judicious approach. *Tourism Management* 15, 403–406.

Inman *et al.* (1998) The case of ecotourism in Costa Rica. In: von Moltke, K. *et al.* (eds) *Global Product Chains: Northern Consumers, Southern Producers, and Sustainability.* United Nations Environment Program (UNEP), Nairobi.

Johnson, M.H. and Bennett, J.T. (1981) Regional environmental and economic impact evaluation: an input-output approach. *Regional Science and Urban Economics* 11, 215–230.

Johnson, R.L. and Moore, E. (1993) Tourism impact estimation. *Annals of Tourism Research* 20, 279–286.

Kinnaird, M.F. and O'Brien, T.G. (1996) Ecotourism in the Tangkoko Duasudara Nature reserve: opening pandora's box. *Oryx* 30, 65–73.

Krakauer, J. (1998) *Into Thin Air.* Anchor, New York.

Laarman, J.G. and Gregersen, H.M. (1996) Pricing policy in nature-based tourism. *Tourism Management* 17(4), 247–254.

Leones, J., Colby, S. and Crandall, K. (1998) Tracking expenditures of the elusive nature tourists in southeastern Arizona. *Journal of Travel Research* 36, 56–64.

Lindberg, K. (1991) *Policies for Maximizing Nature Tourism's Ecological and Economic Benefits.* World Resources Institute, Washington, DC.

Lindberg, K. (1998) Economic aspects of ecotourism. In: Lindberg, K., Epler Wood, M. and Engeldrum, D. (eds) *Ecotourism: a Guide for Planners and Managers*, Vol. 2. Ecotourism Society, North Bennington, Vermont.

Lindberg, K. and Aylward, B. (1999) Price responsiveness in the developing country nature tourism context: review and Costa Rican case study. *Journal of Leisure Research* 31(3), 281–299.

Lindberg, K. and Enriquez, J. (1994) *An Analysis of Ecotourism's Economic Contribution to Conservation and Development in Belize*, Vol. 2. Comprehensive Report. World Wildlife Fund, Washington, DC.

Lindberg, K., Enriquez, J. and Sproule, K. (1996) Ecotourism questioned: case studies from Belize. *Annals of Tourism Research* 23, 543–562.

Lindberg, K., Furze, B., Staff, M. and Black, R. (1998) *Ecotourism in the Asia-Pacific Region: Issues and Outlook.* Ecotourism Society, North Bennington, Vermont.

Liu, J.C. (1986) Relative economic contributions of visitor groups in Hawaii. *Journal of Travel Research* 3, 2–9.

McCool, S.F. and Reilly, M. (1993) Benefit segmentation analysis of state park visitor setting preferences and behavior. *Journal of Park and Recreation Administration* 11, 1–14.

Miller, R.E. and Blair, P.D. (1985) *Input-Output Analysis: Foundations and Extensions.* Prentice Hall, Englewood Cliffs, New Jersey.

Moore, R.L. and Barthlow, K. (1997) *The Economic Impacts and Uses of Long-Distance Trails.* Report prepared for the US Department of the Interior National Parks Service.

Mowforth, M. and Munt, I. (1998) *Tourism and Sustainability: New Tourism in the Third World.* Routledge, London.

Noland, D. (1988) Why this trek cost $3,890. *New York Times* 4 December, p. 14.

Panos (1997) Ecotourism. http://www.oneworld.org/panos/panos_eco2.html

Pearce, D.G. and Wilson, P.M. (1995) Wildlife-viewing tourists in New Zealand. *Journal of Travel Research* 34, 19–26.

Powell, R. and Chalmers, L. (1995) *Regional Economic Impact: Gibraltar Range and Dorrigo National Parks.* New South Wales National Parks and Wildlife Service, Hurstville, New South Wales.

Pyatt, G. (1988) A SAM approach to modeling. *Journal of Policy Modeling* 10, 327–352.

Pyatt, G. and Round, J. (eds) (1985) *Social Accounting Matrices: a Basis for Planning.* The World Bank Press, Washington, DC.

Robinson, S. (1989) Multisectoral models. In: Chenery, H. and Srinivasan, T.N. (eds) *Handbook of Development Economics.* Elsevier, Amsterdam, pp. 886–947.

Robison, M.H. (1997) Community input-output models for rural area analysis with an example from central Idaho. *Annals of Regional Science* 31, 325–351.

Sinclair, M.T. and Stabler, M. (1997) *The Economics of Tourism.* Routledge, London.

Smith, C. and Jenner, P. (1992) The leakage of foreign exchange earnings from tourism. *Travel and Tourism Analyst* 3, 52–66.

Smyth, D. (1999) *Economic Impact of National Park Visitor Spending–Gleaning Visitor Spending Patterns from Secondary Data, Glacier National Park – 1990.* Department of Park, Recreation and Tourism Resources, Michigan State University, East Lansing, Michigan.

Sorenson, R. (1991) Overseas Adventure Travel, Inc. Case Study 9-391-068. Harvard Business School, Cambridge, Massachusetts.

Stynes, D.J. (1992) *Visitor Spending and the Local Economy: Great Smokey Mountains National Park.* Department of Park, Recreation and Tourism Resources, Michigan State University, East Lansing, Michigan.

Stynes, D. and Rutz, E. (1995) *Regional Economic Impacts of Mammoth Cave National Park.* Department of Park, Recreation and Tourism Resources, Michigan State University, East Lansing, Michigan.

Taylor, J.E. and Adelman, I. (1996) *Village Economies: the Design, Estimation, and Use of Villagewide Economic Models.* Cambridge University Press, Cambridge.

Telfer, D.J. and Wall, G. (1996) Linkages between tourism and food production. *Annals of Tourism Research* 23, 635–653.

Van Sickle, K. and Eagles, P.F.J. (1998) Budgets, pricing policies and user fees in Canadian parks' tourism. *Tourism Management* 19, 225–235.

Wagner, J.E. (1997) Estimating the impacts of tourism. *Annals of Tourism Research* 24, 592–608.

Weaver, D.B. (1998) *Ecotourism in the Less Developed World.* CAB International, Wallingford, UK.

West, G.R. and Gamage, A. (1997) Differential multipliers for tourism in Victoria. *Tourism Economics* 3, 57–68.

World Tourism Organization (WTO) (1985) *Tourism's Place in the Input-Output Tables of the National Economy.* WTO, Madrid.

World Travel and Tourism Council (WTTC) (1999) Travel and Tourism's Economic Impact. http://www.wttc.org/

Wunder, S. (1996) *Ecoturismo, Ingresos Locales y Conservación: El Caso de Cuyabeno, Ecuador.* Unión Mundial Para La Naturaleza (UICN/IUCN), Quito.

Wunder, S. (1998) *Forest Conservation Through Ecotourism Incomes? A Case Study from the Ecuadorean Amazon Region*, CIFOR Special Paper. CIFOR, Bogor, Indonesia.

Zhou, D., Yanagida, J.F., Chakravorty, U. and Leung, P. (1997) Estimating economic impacts from tourism. *Annals of Tourism Research* 24, 76–89.

Chapter 24

Environmental Impacts

R. Buckley

International Centre for Ecotourism Research, School of Environmental and Applied Science, Griffith University Gold Coast Campus, Parklands Drive, Southport, Queensland, Australia.

Introduction

The impacts of ecotourism depend on what ecotourism is. Definitions and characteristics have been reviewed extensively in this volume (Chapters 1–5) and elsewhere (Buckley, 1994; Honey, 1999; Fennell, 1999). The critical issue is that ecotourism should involve deliberate steps to minimize impacts, through choice of activity, equipment, location and timing; group size; education and training; and operational environmental management. Under these circumstances, which are regrettably more of an ideal than a practical reality in most cases, the impacts of ecotourism should therefore be those of nature tourism and recreation which incorporates best-practice environmental management; i.e. sustainable nature-based tourism with an environmental education component. Ecotourism has a variety of interrelated impacts on a destination, though for discussion purposes these are often divided into specific categories. Economic and socio-cultural effects are considered in Chapters 23 and 25 respectively, and the intent of this chapter is to focus on environmental impacts.

As with any form of tourism, ecotourism typically involves three components: travel to and from the site; accommodation on site or on tour; and specific recreational activities that may involve local travel by various means. Accommodation may be integrated into the recreational activity, as in an overnight backcountry hiking tour or a stay in a backcountry ecolodge; or it may be quite distinct, as when the tourist stays in a lodge or local accommodation, and takes day tours. There are broader environmental issues relating to the impacts of long-distance air and ground travel to and from an ecotourism destination, and to accommodation in urban hotels before and after an ecotour (Anderek, 1995; Stabler, 1997; McLaren, 1998; Mowforth and Munt, 1998; Hall and Lew, 1999; Honey, 1999). However, these are beyond the scope of this chapter, and are considered only in those situations where they are integrated into the recreational experience.

From an ecological perspective, it is critical to note that the impacts of ecotourism on the natural environment depend on the ecosystem as well as the activity. Different activities, under various management regimes, cause different impacts in different ecosystems; and the ecological significance of these impacts differs greatly among ecosystems. For example, damage to plants by hikers' boots is far more significant on an alpine meadow than in subtropical rainforests, but weed seeds and soil pathogens in mud on hikers' boots are more significant in rainforest than alpine environments. Small alpine lakes or desert waterholes are far more easily polluted by human waste than an ocean or a large turbid river. Human voices can be a major disturbance to fauna in forests and woodland, but not on bare mountain-tops.

The impacts of ecotourism can be classified by many different criteria: by the type of activity, such as hiking or helitouring; by the type of ecosystem, such as forest or feldmark; by ecosystem component, such as wildlife or water quality; or by the scale, duration and significance of impact. In addition, some types of impact are very commonplace, direct and obvious, such as trampling of vegetation. Others are indirect and far less obvious, so it is largely unknown how significant they may be. For example, snowmobiles compress snow, which then provides less insulation, so the soil gets colder, which affects soil arthropods and microfauna. This in turn may affect vegetation and wildlife. Tourists may carry pathogenic microorganisms in their gut fauna, which may be leached into watercourses from human waste, and thence transmitted by native wildlife to other watercourses where they are ingested by other people. Weeds spreading along hiking tracks may compete for insect pollinators with native plants, or may support herbivorous insects that also attack native plants, producing impacts well beyond the immediate vicinity of the track. Repeated disturbance to wildlife, from bears to bighorn sheep, waders to whales, eagles to turtles, may interfere with their ability to feed and breed, causing long-term population decline. Few of these more complex and often inadvertent impacts have been studied in any detail; few have even been recognized and identified. There is now quite an extensive literature on impacts such as trampling, which are easy to quantify experimentally. However, very little is known about impacts such as noise disturbance, soil and water-borne pathogens, and interference with plant and animal population dynamics and genetics, which are likely to have far greater ecological significance.

Of necessity, this chapter presents more detailed data for impacts that are more heavily studied. Future research, however, would be more valuable if it focuses on less obvious impacts.

Major Impacts of Travel, Accommodation and Activities

Travel

Nature-based tourism often involves travel in a variety of motorized vehicles, by land, sea and air. In some cases there is little alternative means of transport available, e.g. in a 1000-km four-wheel drive (4WD) safari across central Australia. As long as minimal-impact practices are followed, this may still be considered as ecotourism. In other cases, there is a readily available non-motorized alternative: hiking instead of an off-road vehicle (ORV), skiing instead of a snowmobile, sea kayaking instead of a jetski or motorboat. We can distinguish between:

1. Cases where motorized transport is used to reach a site for a low-impact activity; e.g. a boat shuttle for a sea kayak trip, or a car shuttle for a river rafting trip;
2. Cases where motorized travel is used to transport physically impaired clients who would be unable to use an unmotorized alternative, as in some boat and coach tours; and
3. Cases where using the motorized vehicle is in itself the recreational activity, as in jetskis, jetboats, snowmobiles and 'bash-the-bush' 4WD tours.

Clearly category 3 is not ecotourism, while categories 1 and 2 may be, depending on the definition adopted and the way the tour is run.

Impacts, however, depend on how and where a vehicle is operated, not why. Roads and formed tracks, if not well designed and constructed, can interrupt surface drainage and cause soil erosion, sometimes on a massive scale. They cause local vegetation clearance and can act as barriers to some animals, particularly smaller vertebrates and non-flying invertebrates. Road verges provide a disturbed habitat often preferentially colonized by weeds. Mud on vehicle tyres may contain weed seeds and fungal spores. Fast-moving vehicles cause roadkill, and engine and tyre noise can disturb animals a considerable distance away. Vehicles driving off-track cut and crush the soil, damage vegetation, crush burrowing animals such as crabs and worms, muddy streams at creek crossings, and so on. Snowmobiles compact the snow, crushing buried plants and the snow and subsnow tunnels made by small mammals, and changing spring snowmelt patterns. Motorized boats and aircraft can cause widespread noise and visual disturbance to wildlife. Even balloons may cause feeding wildlife herds to scatter. Boat engine exhausts cause air and water pollution. Antifouling paints also cause water pollution. Soil and vegetation damage is common at launching and landing sites. Leaks or spills at helicopter and light-aircraft refuelling depots can cause severe soil and water contamination.

Accommodation

Ecotourism accommodation may range from the barely detectable overnight bivouac by the bushwalker or climber to large ecolodges and ecoresorts. The latter are simply hotels that have adopted best principles and practices of environmental design and management. Between these extremes lie a wide range of accommodation types: hiking tents, car tents, tented camps, yurt-like non-fixed accommodations, huts, cabins and lodges. Some have ancillary infrastructure such as access roads, car parks, maintenance plant, generators and sewage treatment systems; others do not. Depending on scale and components, impacts may include:

- crushing or clearance of vegetation;
- soil modifications;
- introduction of weeds and pathogens;
- water pollution from human waste, spent washing and cleaning water, engine fuel and oil residues, and cleaning products;
- air pollution from generator exhausts; noise from machinery, vehicles and voices;
- visual impacts; and
- disturbance to wildlife through all of the above, and through foodscraps and litter, etc.

The most significant issues which apply for all scales and types of accommodation are location and degree of disturbance; access and quarantine against weeds and pathogens; energy sources; disposal of human waste and used washing water; and noise.

Activities

Precisely which types of tourist activity may constitute ecotourism, and under what circumstances, is always debatable. Criteria for inclusion here are: (i) little or no fixed infrastructure; and (ii) motorized vehicles, if any, used for transport only, not as the primary recreational activity. Thus a group of tourists travelling slowly in a 4WD to view and learn about wildlife may be considered ecotourism, if other criteria are satisfied; whereas tourists driving a 4WD off-road for the excitement of the drive, would not. The latter would more appropriately be classified as adventure tourism.

There is a wide range of outdoor activities that may variously be considered as outdoor recreation, outdoor education, outdoor sport, adventure tourism, or ecotourism, depending on how they are carried out. These include; abseiling, birdwatching, boating, bushwalking, canoeing and kayaking, climbing, fishing and hunting, hang-gliding and parapenting, mountain biking, off-road driving and touring, photography, sailing and yachting, skiiing, whale-watching, white-water and blackwater rafting, wildlife viewing, and other activities which take place in protected areas, other public lands and natural environments. Some of these may use vehicles such as Sno-cats, snowmobiles, helicopters, balloons, floatplanes and other light aircraft and motorized boats. Others use horses, mules, burros or llamas. Additional options include mountain bikes, kayaks, rafts, sea kayaks, skis, etc. All of these, even hiking boots, leave some impact. Many of these activities may also involve overnight camping, including cooking, heating and washing, with associated impacts.

Type and Degree of Impacts from Ecotourism Activities

Soils

Tourist vehicles, livestock and hiking both on and off trails can modify soils in a number of ways, for example by removing litter and reducing organic matter and nutrient content; disintegrating soil aggregates; reducing porosity, permeability, penetrability and infiltration; and increasing surface runoff and erosion. Soil compaction can also modify soil temperature profiles. These changes affect soil microbes and invertebrates, plant roots and animal burrows; and these in turn affect aboveground vegetation and animals.

The degree of impact depends on soil type, slope, weather, vegetation cover and other factors. It also depends on the type and scale of activity. Horses' hooves typically exert ground surface pressures of 1000–4000 g per cm^2, for example, compared with 1000–1700 g per cm^2 for 4WD vehicles, 150–400 g per cm^2 for hikers, and 7–10 g per cm^2 for snowmobiles (Liddle, 1998). These values apply for steady pressure on level ground; values up to 10 times greater in each case can occur during acceleration, breaking and sideways skidding. On trails, horses and trail bikes cause more bare ground and more soil erosion than hikers, typically up to 15 times as much; horses also cut deeper trails, especially on steeper slopes (Weaver and Dale, 1978). Soil damage also varies with hikers' footwear and walking technique (Kuss and Morgan, 1986). Soil erosion ranges up to 25,000 cm^3 per m^2 per year on hiking trails (Ketchledge and Leonard, 1970); up to 250,000 cm^3 per m^2 per year for 4WD vehicles on dunes (Eckert *et al.*, 1979); and up to 450,000 cm^3 per m^2 per year for recreational vehicles in Alaska (Rickard and Slaughter, 1973).

In some soils, erosion can continue even once the initial disturbance ceases: e.g. on steep downhill tracks under heavy rainfall; crustose sandy soils in windy areas; and arctic permafrost where insulating vegetation is damaged. In temperate loamy soils, heavy trampling typically reduces porosity by up to one-third (Chappell *et al.*, 1971). Infiltration rates, however, may be reduced by over 50% (Cole, 1982) and in some cases up to 97% (Brown *et al.*, 1977) even in granitic soils. Bulk density may be increased by over 70% (Brown *et al.*, 1977). Waterlogged soils are more susceptible to damage and, in addition, as compaction reduces infiltration, this increases waterlogging and damage further still, especially in fine-grained soils (Bellamy *et al.*, 1971). Soil compaction typically reduces oxygenation, reduces concentrations of nitrifying bacteria by up to 10 times, and increases concentrations of denitrifying bacteria by a similar factor (Duggeli, 1937). In wet arctic-alpine soils, compaction by snowmobiles delays soil warming and hence bacterial activity in spring. Trampling and compaction commonly reduces the number of worms (Cluzeau *et al.*, 1992), and the number of springtails, mites and other small arthropods, sometimes down to a few per cent of their original levels (Yur'eva *et al.*, 1976).

All of the factors outlined above affect the soil's ability to support vegetation, and change the relative abundance of different plant species. In addition, soil trampling, erosion and compaction cause direct damage to plant roots. The effects of recreational trampling on plant roots were recognized over 70 years ago as a result of research in the Californian redwood forests (Meinecke, 1928).

Vegetation

The effects of trampling on vegetation have been studied far more intensively than any other recreational impact, probably because they are easily quantified experimentally. The degree of damage depends on the pressure applied, the number of passes, the time of year, the type of vegetation, and the individual plant species concerned. Even a very low-intensity impact, such as people brushing against plants as they walk by, has been shown to produce physiological and biochemical changes, which may be delayed some time after the initial impact (Hylgaard and Liddle, 1981). The effects of trampling on plant biomass, cover, height, growth form, phenology, physiology, flowering, etc. have all been studied at various levels of detail.

In most vegetation types, even relatively light trampling causes a considerable initial reduction in plant cover, as the more susceptible species are killed. Heavier trampling eventually removes even the most resistant species, but more slowly. Many studies, therefore, have examined the number of passes, whether of wheels or boots, required to reduce plant cover to 50% of its initial value. For hikers, this index ranges from around 12 passes for subtropical eucalypt woodland (Liddle and Thyer, 1986) and 40 for mountain snowbank vegetation (Bell and Bliss, 1973), to around 1500 for pasture grasses (Liddle, 1973; Kendal, 1982). Horses and trail bikes cause significantly more damage; typically, similar impacts are produced by a far smaller (e.g. 5–30 times fewer) number of passes (Cole, 1993, 1995a, b).

The major conclusions from all this work on trampling seem to be the following.

1. We still do not have enough information to predict or model the types and intensities of impacts from different types of trampling in different types of ecosystem in any general sense.
2. The sensitivity of different ecosystems to trampling varies enormously.
3. If trampling is heavy enough in any ecosystem, plant cover will die and local soil erosion, sometimes to considerable depth, will occur.
4. If trampling ceases, soil and vegetation will generally recover at least to some degree, over various timescales, which may be very long.
5. 4WD vehicles, trailbikes, mountain bikes and particularly horses cause vastly greater impacts than hikers.
6. With very few exceptions, the direct impacts of trampling itself do not extend far beyond the actual track, and do not continue to grow if trampling ceases.

The overall conclusion is that the total area of soil and vegetation affected by trampling on tracks is a minuscule proportion of the total area of wilderness. In addition to trampling, direct vegetation damage occurs around campsites, where branches are often broken either to clear space or to collect firewood. Again, the total area is usually small.

Far more significant than direct damage in most cases, however, is secondary vegetation damage (Buckley and Pannell, 1990; Buckley, 1994). This can occur through changes to fire frequency (e.g. from unextinguished cigarette butts and campfires), introduction of plant pathogens such as the jarrah dieback fungus in parts of Australia, and the introduction of weed species. Weed seeds are commonly introduced in mud on tyres and vehicle bodies (Wace, 1977; Lonsdale and Lane, 1992), and to a lesser degree on boots and tent pegs. Weed seeds may also be introduced in gravel used for track and site hardening by land management agencies, and in fodder carried for recreational livestock (Cole, 1993). Weeds and pathogens can spread well beyond the extent of the tracks themselves, and are generally impossible to eradicate once introduced.

Equally significant are secondary impacts on the population of rare or endangered animal species, whether through noise, visual disturbance, barriers to movement, or the introduction of pathogens. Again these occur over a far greater area than the tracks themselves.

Invertebrates

The impacts of ecotourism and recreational activities on invertebrates are relatively unknown. Populations of insects such as the sand scarab beetle are reduced in areas used by off-road vehicles (Luckenbach and Bury, 1983). Populations of ghost crabs on the beaches of Assateague Island in

Virginia, USA, were reduced by 98% after only 100 passes by 4WD vehicles (Woolcott and Woolcott, 1984) and similar effects probably occur on sandy beaches worldwide. Populations of shoreline worms, molluscs and crustaceans are reduced by bait collecting for recreational fishing in many areas (Cryer *et al.*, 1987; Heiligenberg, 1987). Coral reefs are damaged by pollution from coastal resorts, trampling on intertidal flats, and collecting and accidental damage by divers (Kay and Liddle, 1984; Hawkins and Roberts, 1993). Assemblages of terrestrial insect species are modified by the introduction of exotic plants, whether accidental weeds or deliberate plantations. Insect species can also be transported into new habitats on tourist vehicles, as has apparently occurred on tourist boats in the Galapagos Islands (Silberglied, 1978).

Tourism and recreation can also affect the interactions between insect species, and between insects and plants. Introduced plant species may compete for insect pollinators with native plant species, for example. Introduced plant species may also provide a reservoir for insect parasites, increasing their populations and hence their impacts on native plant species. Relatively subtle, indirect and initially inconspicuous impacts such as these may well prove far more significant for the conservation of biological diversity than the more gross, obvious but geographically restricted impacts such as trampling. To date, however, they remain almost entirely unstudied.

Reptiles and amphibia

There seem to be remarkably few, if any, studies on the impacts of ecotourism and other forms of tourism on frogs, toads, newts, salamanders and other amphibia, though in view of their extreme sensitivity to water pollution it seems likely that they would be excellent bio-indicators of water quality impacts. Off-road vehicles in the Mojave Desert have been shown to cause major reductions in lizard populations (Vollmer *et al.*, 1976), and major or complete hearing loss in the fringe-toed lizard (Brattstrom and Bondello, 1983). ORVs in California also killed desert tortoises and destroyed their burrows (Bury and Marlow, 1973). Disturbance by tourists in some areas of the Galapagos Islands led to the collapse of the feeding and mating systems of the Galapagos land iguana (Edington and Edington, 1986). Disturbance by tourists also causes freshwater caiman, alligators and crocodiles to leave their nests, and increases egg predation by coatis in Paraguay, raccoons in the Mississippi, and lizards and hyena in Uganda (Crawshaw and Schaller, 1980; Jacobsen and Kushlan, 1986).

Populations of beach-nesting marine turtles are also affected by tourism. Egg-laying females may be disturbed by wildlife viewers (Jacobson and Lopez, 1994). Hatchlings are disoriented by lights, while vehicles on beaches crush some hatchlings and impede the progress of others, thereby increasing predation rates (Hosier *et al.*, 1981; Witherington, 1997). Water pollution and recreational boats damage seagrass beds in which the marine turtles feed (Williams, 1988), and adults are killed by shark nets in South Africa and Australia (Dudley and Cliff, 1993; Wild, 1994). Habitat destruction and egg collection, however, may be more significant in most areas than any of the above impacts.

Birds

Numerous studies worldwide have shown that a wide range of bird species, in a wide range of environments, may be disturbed by noise or visual sightings of tourists, even at low intensity. Some species are more susceptible than others, and while some may become habituated to disturbance, others do not (e.g. Blakesley and Reese, 1988). Bird species assemblages, populations and behaviour may also be changed in areas used for camping, fishing and recreational boating (e.g. Bell and Austin, 1985; Keller, 1989), and in areas used for hunting and by recreational vehicles and aircraft (e.g. Belanger and Bedard, 1989).

Repeated disturbance by tourists causes substantial reduction in the breeding success of a wide range of shorebirds, often to < 50%. Examples include relatively restricted and endangered species such as brown pelican and Herman's gull (Anderson and Keith, 1980). Ground-nesting bird species are particularly susceptible to damage by ORVs and hikers, particularly if they have dogs (de Roos, 1981; Yalden and Yalden, 1990). Nesting success of large hawks and eagles, which are often major tourist attractions, is also reduced dramatically by tourist disturbance. In many cases, these species cease nesting completely in areas frequented by tourists, even if this forces them from their preferred habitat into less favourable areas. Examples include bald eagles and ospreys in the USA (Bangs *et al.*, 1982; Buehler *et al.*, 1991), imperial eagles in Spain (Gonzalez *et al.*, 1992), golden eagles and peregrines in Europe and the USA (Bocker and Ray, 1971; Watson, 1976) and various species in The Netherlands (Saris, 1976).

Individual eagles may fly 1 km or more before alighting, once disturbed by tourists (e.g. Grubb and King, 1991). Flight distances for disturbed waterbirds typically range from 100 to 800 m (Hume, 1972; Batten, 1977; Klein, 1993; Burger and Gochfeld, 1998; Fitzpatrick and Bouchez, 1998), with greatest distances for disturbance by power boats, and least for disturbance by walkers. Individual nesting birds with eggs or chicks may remain on their nests even with much closer approaches, but under stress. Human disturbance can also alter birdsong patterns, critical to territorial and breeding behaviour (Gurtzwiller *et al.*, 1997; Hill *et al.*, 1997).

Mammals

In many parts of the world, the impacts of tourism on mammals include hunting as well as wildlife watching and inadvertent disturbance. Typically, individual species are far more wary and easily disturbed in areas where they are hunted than in areas where they are not; and in areas where hunting occurs, they are more easily disturbed by behaviour typically associated with hunters. This applies, for example, to bears in Canada (McLellan and Shackleton, 1988), caribou in the Arctic (Calef, 1976), and dall sheep in the Rocky Mountains (MacArthur *et al.*, 1982).

Even in areas without hunting, tourism can cause significant disturbance to large mammals. In the Sierra Nevada, for example, bears abandon their winter dens in areas used heavily by skiers, even if the dens contain cubs (Goodrich and Berger, 1994). In Scandinavia, deer were so disturbed by orienteering events that some died (Sennstam and Stalfelt, 1976; Jeppesen, 1987). In the USA, various studies have shown that elk, mule deer and white-tailed deer are disturbed by hikers, skiers and, particularly, snowmobiles (Eckstein *et al.*, 1979; Freddy *et al.*, 1986). Helicopter overflights in the Grand Canyon reduced feeding of bighorn sheep by 45%; and in Alaska, dall sheep run in panic from helicopters, sometimes over 800 m (Price and Lent, 1972). Even where animals do not run, they may still suffer stress from approaching hikers. A classic study by MacArthur *et al.* (1982) on dall sheep in Alberta, Canada, used cardiac telemetry to show that their heart rate increased by up to 20 beats per minute when hikers approached. The increase in heart rate was triggered when hikers approached to within 50 m, if they approached from a road; but at 150 m, if they approached from the side away from a road, or with a dog. In areas where the survival of overwintering individuals depends critically on their energetic balance, any increase in metabolism or unproductive activity, and any decrease in feeding time or the quality of feed available, may lead to the death of part of the overwintering population.

The impacts of tourism on smaller mammals, not subject to hunting, have been relatively little studied. Populations of marmots on Vancouver Island, Canada, are threatened by a combination of forestry and recreational activities (Dearden and

Hall, 1983); and marmot behaviour in the Swiss Alps is strongly affected by hikers, particularly those hiking with dogs or off-trail (Mainini *et al.*, 1993). Marmot populations also suffer where snow compaction reduces insulation of their overwintering burrows (Schmid, 1970). Similarly, over-snow vehicles crush the winter runways of the northern bog lemming (Layser and Burke, 1973). In the campgrounds of Yosemite National Park, generalist-feeding deer mice have increased in numbers whereas specialist-feeding mountain mice decreased (Garton *et al.*, 1977). Chimpanzees in Uganda, and rainforest wildlife in Sumatra are disturbed by hikers, but can become habituated (Griffiths *et al.*, 1993; Johns, 1996).

Aquatic biota

Water-based recreation can cause a wide variety of impacts (Arthington *et al.*, 1989). Propeller-driven boats affect aquatic plants through propeller damage, wash, increased turbidity, and exhaust and petroleum residues from outboard motors. This is particularly evident at launch, landing and turning areas. Larger recreational boats may cause impacts from antifouling paints, which may contain herbicides (Pearce, 1995), and from discharge of sewage. Water pollution by nutrients and microorganisms also occurs from discharge of sewage and human waste from boat toilets and waterside accommodation and campsites. Even relatively small numbers of recreational swimmers may increase the concentrations of bacteria in small, pristine streams (Warnken, 1996), and backcountry hikers are also implicated in the distribution of certain waterborne pathogenic bacteria and protozoa (Buckley *et al.*, 1998; W. Warnken and R.C. Buckley, unpublished). Recreational fishing causes impacts through bankside trampling, damage to fish eggs by wading in streams (Roberts and White, 1992), the introduction of exotic fish species specifically for recreational angling, and introduction of diseases to native fish populations (Langdon, 1989).

Marine mammals

Whales, dolphins and other marine mammals such as manatee and dugong now support a large-scale tourism industry worldwide. This industry has assisted in the conservation of these species by alerting people to the depredations of commercial whaling, and marine mammal mortality in by-catch from commercial fishing operations. In areas with an intensive marine-mammal tourism industry, however, populations may now be affected by the impacts of tourism. Various species of whales are disturbed by boats in Glacier Bay, Alaska (Watkins and Goebel, 1984), the Canadian Arctic (Breton, 1996), and Australia's Hervey Bay (Stevens and Chaloupka, 1992). Whales are also disturbed by aerial viewing. This includes helicopter viewing of sperm whales and grey whales, and light-aircraft viewing of bowhead whales, which typically crash dive if aircraft fly below 300–450 m (Richardson *et al.*, 1985). In Florida, USA, where manatees support a major recreational boating industry, over 10% of the total population were killed by propeller cuts and boat impacts in 1989; this was more than the population replacement rate (Shackley, 1992). Wild dolphins, some habituated to human interactions, also form significant tourism attractions in many parts of the world, with associated effects on dolphin behaviour and perhaps populations (e.g. Orams, 1997).

Environmental Management by Tour Operators

One of the core defining criteria for ecotourism is best-practice environmental management. A nature or adventure tour is not an ecotour unless it ranks with industry leaders in its efforts to minimize negative impacts on the natural and cultural environment, whether through planning and design, equipment and activities, training and education of guides and clients, or a combination of these approaches. Of course, this is not a straightforward criterion. For example, if commercial horsepackers stay on designated trails, use only weed-free fodder, travel only in small groups, control noise, and use minimal-impact camping practices, can they be considered as ecotourism operations? Or are they disqualified by the mere fact that horses have so much greater impact than hikers, and by the fact that the tour operators could have taken their clients on foot?

Similar considerations could apply, for example, to the use of motorized watercraft where yachts or sea kayaks could provide an alternative. Of course, in areas like Hervey Bay, Australia, 200 whale-watchers in a single vessel may well provide significantly less disturbance to the whales than 200 individual sea kayaks, even assuming that sea kayaks could travel far enough offshore in the time available. In areas such as Glacier Bay and Prince William Sound, Alaska, however, cruise boats and light aircraft often use drifting sea kayakers as a cue to locate whales and bears; and motorized craft cause much greater disturbance than sea kayaks, often causing the wildlife to flee.

In attempts to reduce the impacts of ecotourism, various associations and organizations have produced a range of environmental guidelines, minimal-impact training materials, best-practice handbooks, etc. These range from the highly specific, e.g. for watching particular wildlife species at particular sites, to the very general, such as introductory manuals produced for the Australian tourism accommodation and tour sectors (Talacko, 1998; Basche, 1999). Between these extremes lies a wide range of guidelines produced by research organizations (e.g. Buckley, 1999a), conservation groups (e.g. Roe *et al.*, 1997), National Parks Services (e.g. Australian Alpine Parks, 1993–1997), ecotourism associations, not-for-profit outdoor education organizations (e.g. National Outdoor Leadership School, 1994) and individual tour companies (e.g. Willis's Walkabout, 1994).

There are also popular texts, videos and interactive computer demonstrations with titles such as *Soft Paths* (Hampton and Cole, 1988) and *How to Shit in the Woods* (Meyer, 1994; Clevermedia, 1999). The most recent of these provide quite detailed instructions for specific activities in specific ecosystems, such as hiking in the Australian Alps, or horse-riding in the Pacific Northwest of the USA. In the USA in particular, minimal-impact guidelines produced by the non-profit Leave-No-Trace Inc. (LNT) have been adopted, endorsed and widely distributed by land management agencies. Basic LNT materials are intended to improve the environmental awareness of all visitors to parks and wilderness areas. More advanced LNT materials (Buckley, 1996) endeavour to teach backcountry travellers and ecotour clients not only these specific techniques for particular activities in particular environments, but also how to minimize impacts in new environments.

The effectiveness of such codes of practice depends on how widely they are read and how closely they are followed. At best, however, they can reduce the impacts as outlined above from the upper to the lower end of the ranges quoted. Ecotourism, and particularly large-scale commercial ecotourism, almost invariably still has impacts and environmental costs that must be weighed against its potential conservation benefits.

Public Land and Visitor Management

Ecotourism is an industry, and it operates in the real world of practical politics and past land-use patterns. In particular, ecotours need land or water on which to operate, and somebody owns that land. The impacts of ecotourism on the natural environment are determined not only by the activity itself and the environmental management practices of the ecotour operator, but also by the land, water and visitor management practices of the public agencies and private landholders where the ecotour is operating (see especially Chapters 31 and 32).

The degree to which different landholders and land management agencies regulate, monitor and manage tour operators varies enormously between countries, between different types of land tenure in the same country, and between different areas under similar tenure. At one extreme lies Antarctica, where there is an international treaty governing land use and impacts, but no monitoring or enforcement, so the only environmental management controls are those established by the operators themselves under IAATO, the International Association of Antarctic Tour Operators (see Chapter 14). At the other extreme are private landholders who operate tours on their own land with complete control over what they do and where they go. Most ecotours, however, operate on public lands where they are subject to some form of concession agreement or licensing arrangement. Typically, these impose conditions that are intended to limit negative impacts on the natural environment. The actual aggregate impacts of ecotourism in a given area, therefore, depend strongly on the resource and visitor management strategies adopted by the land management agency. In addition, it places impacts of ecotourism in context, relative to potential impacts from other possible land uses or from uncontrolled tourism.

Land management agencies have two main approaches to controlling impacts (Manning, 1979; Pigram and Jenkins, 1999; Buckley, 2000a, c). Funds permitting, they can harden the natural environment against impacts, typically through construction of infrastructure such as tracks or toilets. This localizes tourists' impacts, but at the cost of impacts from the infrastructure itself. It is also expensive, though costs can sometimes be defrayed from entrance, permit and licence fees. Alternatively, they can control visitors so as to limit the area, timing and type of impacts. There are three broad approaches. The most common are direct prescriptive regulations such as bans or quotas on access, pets, fires, firearms or motorized vehicles in particular areas at particular times. Alternatively, similar results may be obtained by charging differential fees for various activities at different times and places, from park entry fees to seasonal fishing or camping permits. Or finally, land managers can control visitors indirectly through permit conditions or partnerships with commercial tour operators.

A variety of different land and visitor management systems and protocols have been put forward over recent decades, each incorporating a slightly different set of tools and indicators (Buckley, 1998, 2000a). Most of these incorporate, either explicitly or implicitly, the concepts of recreational opportunity spectra and limits of acceptable change (or LAC) (Stankey *et al.*, 1985). The former implies that different visitor activities are either encouraged or discouraged, e.g. through facilities or prohibitions, in different areas and/or at different times. The latter implies that the land management agency identifies specific measurable parameters to act as indicators of environmental quality and the impacts of tourism, and defines thresholds or limits within which the primary conservation goals of the protected area are met.

The idea is that the indicators are to be monitored routinely, and if they transgress the limits of acceptable change, the management agency will deploy one or more of the various tools and techniques at its disposal to reduce impacts. In practice, however, this may be far from straightforward. It is often difficult to identify indicators where visitor impacts can be distinguished from natural ecological fluctuations; where there is adequate warning before ecological changes become irreversible (W. Warnken and R.C. Buckley, unpublished); and where effective remedial actions can be prescribed if LAC are exceeded. These difficulties are by no means confined to the tourism sector (Buckley, 1993).

Irrespective of technical issues in the use of LACs and an associated set of monitoring and management tools, 'M&M toolkit' (Buckley, 1998, 1999b), there are political issues as to who should define the related parameters. In cases where LACs have been employed in practice, they sometimes seem to have been set quite arbitrarily, with quite inadequate knowledge of baseline variation and of the stress–response relation between the impacts of tourism and the value of the indicator parameters (Warnken and Buckley, 2000).

Eagles (personal communication) has identified various possible constituencies who might reasonably claim some interest in setting LACs for tourism in protected areas, but notes that in practice, all these groups are part of larger political processes. They include:

- parks staff, because of expertise and on-site experience;
- independent experts, because of broad technical knowledge;
- local communities, because of local concerns;
- park visitors, since they are the most direct users;
- potential visitors, since they have equal rights to actual visitors;
- entire provinces, nations or the global population, any of whom may visit, or at least value the area for its option and existence benefits.

As mentioned above, economic issues and impacts in ecotourism are beyond the scope of this chapter. However, it is worth reiterating that the precise design of fees and charges for individual visitors, non-profit groups and commercial ecotours are important not only to raise funds for managing impacts on the natural environment, but also as tools in themselves to manage visitor numbers, activities and hence indirectly, impacts.

For many protected areas and other public land and waters, the proportion of visitors on commercial ecotours is increasing relative to those on private recreation. This provides opportunities for land managers to use private tour operators as another means to control environmental impacts. For example, land managers may grant permits only to operators who have particular equipment, qualifications or accreditation, or who undertake specified training programmes. They may specify permit conditions that set quotas, control activities or require specific interpretive programmes, and use tour guides as surrogate rangers to ensure compliance. They may enlist ecotour operators, guides and clients to assist in routine or one-off monitoring and management exercises. Or they may lease operating rights for particular areas or facilities, such as campgrounds, heritage buildings, or equipment rentals or guiding facilities, to private concessionaries, under appropriate conditions. Many other forms of partnership are also possible (Buckley, 2000b, c), though all involve risks of environmental impact if conditions are not followed, as well as risks of legal liability if commercial clients are injured or even dissatisfied.

No matter how well ecotourism is managed, it will still produce negative impacts on the natural environment. With continuing growth in the number of visitors, especially commercial tourists, to national parks and other protected areas, endangered species and ecosystems that were once believed safely protected may now be threatened again. In these areas the environmental impacts, monitoring and management of tourism are critical for conservation. Ecotourism is preferable to uncontrolled tourism, but still of concern for conservation.

Outside protected areas, however, ecotourism has the potential to make a major positive contribution to conservation of natural environments, by displacing other land uses with much greater local impacts on the natural environment, such as forestry, farming, fishing, mining or hydroelectric power generation (Eagles and Martens, 1997; Buckley, 2000c). Similarly, in countries without effective management and enforcement in protected areas, ecotourism may provide a local incentive to displace destructive land-use practices which, though illegal, are still widespread. These may include clearing for agriculture, timber cutting for firewood or construction, grazing of domestic livestock, burning off, poaching of wildlife and collecting of endangered plants. Many of these practices occur illegally in the national parks of developed nations such as the USA, Canada and Australia as well as the developing nations of Africa, Asia and South America.

Conclusions

All forms of tourism produce negative impacts on the natural environment. Ecotourism, if it is more than a marketing label, has lower per capita impacts than other forms of tourism, but these impacts tend to be concentrated in areas of highest conservation value, especially in protected areas. Impacts can be reduced by the environmental management practices of ecotour operators, environmental education of clients by ecotour guides, and land and visitor management practices by landholders and land management agencies. The precise impacts of different ecotourist activities, with different equipment, in different ecosystems, at different seasons are not well known. The effectiveness of different management tools in reducing these impacts is even less well known. In many cases, however, lack of scientific knowledge is less of an impediment than lack of management funds or political support. The ecotourism industry has a responsibility to minimize its impacts in protected areas. Indeed, the degree to which it does so is one of the main litmus tests of ecotourism. The ecotourism industry also has a role in changing land and water use patterns outside protected areas, from higher-impact uses to lower-impact ecotourism.

Acknowledgements

I thank Caroline Kelly, Katie Lawrance, Karen Sullivan, Wiebke Warnken, Jan Warnken, Clyde Wild, and Tatia Zubrinich for assistance in compiling the database of impact studies which we have maintained since 1991. I also acknowledge my debt to Michael Liddle's book *Recreation Ecology*: the summary of different impact types broadly follows the layout of his book, updated as relevant.

References

Anderek, K.L. (1995) Environmental consequences of tourism: a review of recent research. In: McCool, S.F. and Watson, A.E. (eds) *Linking Tourism, the Environment, and Sustainability*. US Forest Service Intermountain Research Station. General Technical Report INT-GTR-323. Ogden, Utah.

Anderson, D.W. and Keith, S.O. (1980) The human influence on seabird nesting success: conservation implications. *Biological Conservation* 18, 65–80.

Arthington, A.H., Miller, G.J. and Outridge, P.M. (1989) Water quality, phosphorus budgets and management of dune lakes used for recreation in Queensland (Australia). *Water Science Technology* 21, 111–118.

Australian Alps National Parks (1993) *Car Camping Code*, (1993) *Snow Camping Code*, (1994) *Horse Riding Code*, (1997) *Bushwalking Code*, (undated) *River Users Code*. AANP, Canberra.

Bangs, E.E., Spraker, T.H., Berns, V.D. and Baily, T.N. (1982) Effects of increased human populations on wildlife resources of the Kenai Peninsula, Alaska, USA. In: Sabal, V. (ed.) *Transactions of the North American Wildlife and Natural Resources Conference*, Vol. 47. Wildlife Management Institute, Washington, DC, pp. 605–616.

Basche, C. (1999) *Being Green is Your Business*. CRC Tourism, Gold Coast and Tourism Council Australia, Sydney.

Batten, L.A. (1977) Sailing on reservoirs and its effects on water birds. *Biological Conservation* 11, 49–58.

Belanger, L. and Bedard, J. (1989) Responses of staging greater snow geese to human disturbance. *Journal of Wildlife Management* 53, 713–719.

Bell, D.V. and Austin, L.W. (1985) The game fishing season and its effects on overwintering wildfowl. *Biological Conservation* 33, 65–80.

Bell, K.L. and Bliss, L.C. (1973) Alpine disturbance studies: Olympic National Park, USA. *Biological Conservation* 5, 25–32.

Bellamy, D., Radforth, J. and Radforth, N.W. (1971) Terrain, traffic and tundra. *Nature* 231, 429–432.

Blakesley, J.A. and Reese, K.P. (1988) Avian use of campground and non-campground sites in riparian zones. *Journal of Wildlife Management* 52, 399–402.

Bocker, E.L. and Ray, T.D. (1971) Golden eagle population studies in the south west. *Condor* 73, 463–467.

Brattstrom, S.P. and Bondello, M.C. (1983) Effects of off-road vehicle noise on desert vertebrates. In: Webb, R.H. and Whilshire, H.G. (eds) *Environmental Effects of Off-Road Vehicles*. Springer, New York, pp. 167–206.

Breton, M. (1996) *Guide to Watching Whales in Canada*. Department of Fisheries and Oceans, Ottawa.

Brown, J.M. Jr, Kalisz, S.P. and Wright, W.R. (1977) Effects of recreational use on forested sites. *Environmental Geology* 1, 425–431.

Buckley, R.C. (1993) Biodiversity and EIA: the Mt Todd case study. *Australian Environmental Law News* 3/93, 46–47.

Buckley, R.C. (1994) Ecotourism: a framework. *Annals of Tourism Research* 21, 661–669.

Buckley, R.C. (1996) Principles for best-practice leave-no-trace training. *Masters Network* 11, 3, 15.

Buckley, R.C. (1998) Tourism in wilderness: M&M Toolkit. In: Watson, A.E., Aplet, G.H. and Hendee, J.C. (eds) *Personal, Societal, and Ecological Values of Wilderness: Sixth World Wilderness Congress Proceedings on Research, Management, and Allocation*, Vol. 1. RMRS-P-4. USFS, Rocky Mountain Research Station, pp. 115–116.

Buckley, R.C. (1999a) *Green Guide to White Water: Best-Practice Environmental Management for Whitewater Raft and Kayak Tours*. CRC Tourism, Gold Coast, Australia.

Buckley, R.C. (1999b) Tools and indicators for managing tourism in parks. *Annals of Tourism Research* 26, 207–210.

Buckley, R.C. (2000a) Tourism in the most fragile environments. *Tourism and Recreation Research* 25, 31–40.

Buckley, R.C. (2000b) *Voluntary Contributions by Tourism to Conservation*. CRC Tourism, Gold Coast Australia.

Buckley, R.C. (2000c) Tourism in wilderness: dancing with the messy monster. In: McCool, S. (ed.) *Wilderness Science*. University of Montana, Missoula.

Buckley, R.C. and Pannell, J. (1990) Environmental impacts of tourism and recreation in national parks and conservation reserves. *Journal of Tourism Studies* 1, 24–32.

Buckley, R.C., Clough, E. and Warnken, W. (1998) *Plesiomonas shigelloides* in Australia. *Ambio* 27, 253.

Buehler, D.A., Mersmann, T.J., Fraser, J.D. and Seegar, J.K.D. (1991) Effects of human activity on bald eagle distribution on the Northern Chesapeake Bay. *Journal of Wildlife Management* 55, 282–290.

Burger, J. and Gochfeld, M. (1998) Effects of ecotourists on bird behaviour at Lozanhatchee National Wildlife Refuge, Florida. *Environmental Conservation* 25, 13–21.

Bury, R.B. and Marlow, R.W. (1973) The desert tortoise: will it survive? *Environment Journal* June, 9–12.

Calef, G.W. (1976) Numbers beyond counting, miles beyond measure. *Audubon* 78, 42–61.

Chappell, H.G., Ainsworth, J.F., Cameron, R.A.D. and Redfern, M. (1971) The effect of trampling on a chalk grassland ecosystem. *Journal of Applied Ecology* 8, 869–882.

Clevermedia (1999) How to crap in the woods. www.flasharcade.com/crapinthewoods.html

Cluzeau, D., Binet, F. and Vertes, F. (1992) Effects of intensive cattle trampling on soil–plant–earthworms system in two grassland types. *Soil Biology and Biochemistry* 24, 1661–1665.

Cole, D.N. (1982) *Wilderness Campsite Impacts: Effects of Amount of Use*. Research Paper INT-3-3 USDA Forest Service, Intermountain Forest and Range Experiment Station, USFS, Ogden, Utah.

Cole, D.N. (1993) *Trampling Effects on Mountain Vegetation in Washington, Colorado, New Hampshire and North Carolina*. Research Paper; INT-464. USDA Forest Services Intermountain Forest and Range Experimental Station, Ogden, Utah.

Cole, D.N. (1995a) Experimental trampling of vegetation. Relationship between trampling intensity and vegetation response. *Journal of Applied Ecology* 32, 203–214.

Cole, D.N. (1995b) Experimental trampling of vegetation. Predictors of resistance and resilience. *Journal of Applied Ecology* 32, 215–224.

Crawshaw, P.G.J. and Schaller, G.B. (1980) Nesting of Paraguayan caimen *Caimen yacare* in Brazil. Papers *Auulsos Zoologica* (Sao Paulo) 33, 283–292.

Cryer, M., Whittle, G.N. and Williams, R. (1987) The impact of bait collection by anglers on marine intertidal invertebrates. *Biological Conservation* 42, 83–93.

Dearden, P. and Hall, C. (1983) Non-consumptive recreation pressures and the case of Vancouver Island marmot (*Marmot vancouverensis*). *Environmental Conservation* 10, 63–66.

Dudley, S.F.J. and Cliff, G. (1993) Some effects of shark nets in the Natal nearshore environment. *Environmental Biology Fisheries* 36, 243–255.

Duggeli, J. (1937) Wie wirkt das oftere Betreten des Walbodens auf einzelene physikalische und biologishe Eigenschaften. *Schweizerische Zeitschrift fur Furstwesen* 88, 151–165.

Eagles, P.F.J. and Martens, J. (1997) Wilderness tourism and forestry: the possible dream in Algonquin Provincial Park. *Journal of Applied Recreation Research* 22, 77–79.

Eckert, R.E. Jr, Wood, M.K., Blackburn, W.H. and Petersen, F.F. (1979) Impacts of off-road vehicles on infiltration and sediment production of two desert soils. *Journal of Rangeland Management* 32, 394–397.

Eckstein, R.G., O'Brien, T.F., Rongstad, O.J. and Bollinger, J.G. (1979) Snowmobile effects on movements of white-tailed deer: a case study. *Environmental Conservation* 6, 45–51.

Edington, J.M. and Edington, M.A. (1986) *Ecology and Environmental Planning*. Chapman & Hall, London.

Fennell, D.A. (1999) *Ecotourism: an Introduction*. Routledge, London.

Fitzpatrick, S. and Bouchez, B. (1998) Effects of recreational disturbance on the foraging behaviour of waders on a Rocky Beach. *Bird Study* 45, 157–171.

Freddy, D.J., Bronaugh, W.M. and Fowler, M.C. (1986) Responses of mule deer to disturbance by persons afoot and snowmobiles. *Wildlife Society Bulletin* 14, 63–68.

Garton, E.O., Bowen, C.W. and Foin, T.C. (1977) The impact of visitors on small mammal communities of Yosemite National Park. In: Foin, T.C. (ed.) *Visitor Impacts on National Parks: the Yosemite Ecological Impact Study*, Vol. 10. Institute of Ecology, University of California, Davis, pp. 44–50.

Gonzalez, L.M., Bustamante, J. and Hiraldo, F. (1992) Nesting habitat selection by the Spanish imperial eagle *Aquila adalberti*. *Biological Conservation* 59, 45–50.

Goodrich, J.M. and Berger, J. (1994) Winter recreation and hibernating black bears *Ursus americanus*. *Biological Conservation* 67, 105–110.

Griffiths, M., Schaik, C.P. and Van-Schaik, C.P. (1993) The impact of human traffic on the abundance and activity periods of Sumatran rainforest wildlife. *Conservation Biology* 7, 623–626.

Grubb, T.G. and King, R.M. (1991) Assessing human disturbance of breeding bald eagles with classification tree models. *Journal of Wildlife Management* 55, 500–511.

Gurtzwiller, K.J., Krose, E.A., Anderson, S.H. and Wilkins, C.A. (1997) Does human intrusion alter the seasonal timing of avian song during breeding periods? *Auk* 114, 55–67.

Hall, M. and Lew, A. (eds) (1999) *Sustainable Tourism: a Geographical Perspective*. Longman, London.

Hampton, B. and Cole, D. (1988) *Soft Paths*. Stackpole, Harrisburg, Pennsylvania.

Hawkins, J.P. and Roberts, C.M. (1993) Effects of recreational scuba-diving on coral reefs: trampling on reef flat communities. *Journal of Applied Ecology* 30, 25–30.

Heiligenberg, van den T. (1987) Effects of mechanical and manual harvesting of Lugworms *Arenicola marina* L. on the benthic fauna of tidal flats in the Dutch Wadden Sea. *Biological Conservation* 39, 165–177.

Hill, D., Hockin, D., Price, D., Tucker, G., Morris, R. and Treweek, J. (1997) Bird disturbance: improving the quality and utility of disturbance research. *Journal of Applied Ecology* 34, 275–288.

Honey, M. (1999) *Ecotourism and Sustainable Development: Who Owns Paradise?* Island Press, Washington, DC.

Hosier, P.E., Kockhar, M. and Thayer, V. (1981) Off-road vehicle and pedestrian effects on the sea approach of hatchling loggerhead turtles. *Environmental Conservation* 8, 158–161.

Hume, R.A. (1972) Reactions of goldeneyes to boating. *British Birds* 69, 178–179.

Hylgaard, T. and Liddle, M.J. (1981) The effect of human trampling on a sand dune ecosystem dominated by *Empetrum nigrum*. *Journal of Applied Ecology* 18, 559–569.

Jacobsen, T. and Kushlan, J.A. (1986) Alligators in natural areas: choosing conservation policies consistent with local objectives. *Biological Conservation* 36, 667–679.

Jacobson, S.K. and Lopez, A.F. (1994) Biological impacts of ecotourism: tourists and nesting turtles in Tortuguero National Park, Costa Rica. *Wildlife Society Bulletin* 22, 414–419.

Jeppesen, J.L. (1987) Impact of human disturbance on home-range movements and activity of red deer *Cervus elaphusis* in a Danish environment. *Danish Review of Game Biology* 13, 1–38.

Johns, B.G. (1996) Responses of chimpanzees to habituation and tourism in the Kibale Forest, Uganda. *Biological Conservation* 78, 257–262.

Kay, A.M. and Liddle, M.J. (1984) *Tourist Impact on Reef Corals.* Great Barrier Reef Marine Park Authority, Townsville, Queensland.

Keller, V. (1989) Variations in the response of great crested grebes *Podiceps cristatus* to human disturbance – a sign of adaptation? *Biological Conservation* 49, 31–45.

Kendall, P.C. (1982) The effect of season and/or duration of trampling in an open forest of South East Queensland. MSc thesis, Griffith University, Brisbane.

Ketchledge, E.H. and Leonard, R.C. (1970) The impact of man on the Adirondack high country. *Conservationist* 25, 14–18.

Klein, M.L. (1993) Waterbird behavioural responses to human disturbance. *Wildlife Society Bulletin* 21, 31–39.

Kuss, F.R. and Morgan, J.M. (1986) A first alternative for estimating the physical carrying capacities of natural areas for recreation. *Environmental Management* 10, 225–262.

Langdon, J.S. (1989) *Prevention and Control of Fish Disease in the Murray-Darling Basin.* Proceedings of the Workshop on Native Fish management. Murray-Darling Commission, Canberra.

Layser, E.F. and Burke, T.E. (1973) Northern bog lemming and its unique habitat in nortn eastern Washington. *Murrelet* 54, 7–8.

Liddle, M. (1973) The effect of trampling and vehicles on natural vegetation. PhD thesis, University College North Wales.

Liddle, M. (1998) *Recreation Ecology.* Chapman & Hall, London.

Liddle, M.J. and Thyer, N. (1986) Trampling and fire in a subtropical dry sclerophyll forest. *Environmental Conservation* 13, 33–39.

Lonsdale, W.M. and Lane, A.M. (1992) Vehicles as vectors of weed seeds in Kakadu National Park, in plant invasions: the media of environmental weed in Australia. *Kowari* 2, 167–169.

Luckenbach, R.A. and Bury, R.B. (1983) Effects of off-road vehicles on the biota of the Algodones Dunes, Imperial Country, California. *Journal of Applied Ecology* 20, 265–286.

MacArthur, R.A., Geist, V. and Johnston, R.H. (1982) Cardiac and behavioural responses of mountain sheep to human disturbance. *Journal of Wildlife Management* 46, 351–358.

McLaren, D. (1998) *Rethinking Tourism and Travel.* Kumarian Press, Connecticut.

McLellan, B.N. and Shackleton, D.M. (1988) Grizzly bears and resource extraction industries: effects of roads on behaviour, habitat use and climography. *Journal of Applied Ecology* 25, 451–460.

Mainini, B., Neuhaus, P. and Ingold, P. (1993) Behaviour of marmots (*Marmota marmota*) under the influence of different hiking activities. *Biological Conservation* 64, 161–164.

Manning, R. (1979) Strategies for managing recreational use of national parks. *Parks* 4(1), 13–15.

Meinecke, P. (1928) *The Effect of Excessive Tourist Travel on the California Redwood Parks.* California Department of Natural Resources, Division of Parks, Sacramento.

Meyer, K. (1994) *How to Shit in the Woods*, 2nd edn. Ten Speed Press, Berkeley, California.

Mowforth, M. and Munt, I. (1998) *Tourism and Sustainability: New Tourism in the Third World.* Routledge, London.

National Outdoor Leadership School (1994) *Leave-No-Trace Outdoor Skills and Ethics.* NOLS, Lander, Wyoming.

Orams, M.B. (1997) Historical account of human–dolphin interactions and recent developments in wild dolphin-based tourism in Australasia. *Tourism Management* 18, 317–326.

Pearce, F. (1995) Alternative antifouling widespread in Europe. *New Scientist* 15 Jan 1995, 7.

Pigram, J. and Jenkins, J. (1999) *Outdoor Recreation Management.* Routledge, London.

Price, R. and Lent, P.C. (1972) Effects of human disturbance on Dall sheep. *Alaska Co-operative Wildlife Research Unit Quarterly Report* 23, 23–28.

Richardson, W.J., Fraker, M.A., Wursig, B. and Wells, R.S. (1985) Behaviour of bowhead whales (*Balaena mysticetus*) summering in the Beaufort Sea: reactions to industrial activities. *Biological Conservation* 32, 195–230.

Rickard, W.E. and Slaughter, C.W. (1973) Thaw and erosion on vehicular trails in perma frost landscapes. *Journal of Soil and Water Conservation* 28, 263–266.

Roberts, B.C. and White, R.G. (1992) Effects of angler wading on survival of trout eggs and pre-emergent fry. *North American Journal of Fish Management* 12, 450–459.

Roe, D., Leader-Williams, N. and Dalal-Clayton, B. (1997) *Take Only Photographs, Leave Only Footprints.* IIED Wildlife Development Service, 10. International Institute for Environment and Development, London.

Roos, G. Th. de (1981) *The Impact of Tourism upon some Breeding Wader Species on the Isle of Vlieland in The Netherlands Wadden Sea.* Mededelingen Landbouwhogeschool, Wageningen, pp. 81–114.

Saris, F.J.A. (1976) *Breeding Populations of Birds and Recreation in the Duivelesbert.* Free University Institute of Environmental Studies Working Paper 66. Vrieje Universitat, Brussels.

Schmid, W.D. (1970) Modification of the Sunivian microclimate by snow mobiles. In: Haugen, A.O. (ed.) *Proceedings Symposium on Snow and Ice in Relation to Wildlife.* Iowa Coop. Wildl. Res. Unit, Iowa State University, pp. 251–257.

Sennstam, B. and Stalfelt, F. (1976) Rapport angaende 1975 ars femdanger-sorienterings inverkan pa klouviltet. *Rapport* 12, 35.

Shackley, M. (1992) Manatees and tourism in southern Florida: opportunity or threat? *Journal of Environmental Management* 34, 257–265.

Silberglied, R. (1978) Inter-island transport of insects aboard ships in the Galapagos Islands. *Biological Conservation* 13, 273–278.

Stabler, M.J. (ed.) (1997) *Tourism and Sustainability: Principles to Practice.* CAB International, Wallingford, UK.

Stankey, G.H., Cole, D.N., Lucas, R.C., Petersen, M.E. and Frissell, S.S. (1985) *The Limits of Acceptable Change (LAC) System for Wilderness Planning.* USDAFS, Ogden, Utah.

Stevens, T. and Chaloupka, M. (1992) *Whale Watching in Hervey Bay Marine Park and the Whitsundays: Implications for Management. Abstracts, 5 Annual Conference Australian Wildlife Management Soc*iety, Queensland University of Technology, Brisbane.

Talacko, J. (1998) *Being Green Keeps You Out of the Red.* CRC Tourism and Tourism Council Australia, Sydney.

Vollmer, A.T., Maza, B.G. and Medica, P.A. (1976) The impact of off-road vehicles on a desert ecosystem. *Environmental Management* 1, 15–129.

Wace, N. (1977) Assessment of dispersal of plant species – the car-borne flora in Canberra. *Proceedings of the Ecological Society of Australia* 10, 167–186.

Warnken, J. and Buckley, R.C. (2000) Monitoring diffuse impacts: Australian tourism developments. *Environmental Management* 25, 453–461.

Warnken, W. (1996) *Threshold detection of ecotourism impact: microbiological and chemical indicators of recreational effects on water quality in a subtropical rainforest conservation reserve.* PhD thesis, Griffith University, Gold Coast, Australia.

Watkins, W.A. and Goebel, C.A. (1984) Sonar observations explain behaviours noted during boat manoeuvres for radio tagging of humpback whales (*Megaptera novaea ngliae*) in the Glacier Bay area. *Cetology* 48, 1–8.

Watson, A. (1976) Human impact on animal populations in the Cairngorms. *Landscape Research News* 1, 14–15.

Weaver, T. and Dale, D. (1978) Trampling effects of hikers, motorcycles and horses in meadows and forests. *Journal of Applied Ecology* 15, 451–457.

Wild, C. (1994) Shark control an environmental problem. *Griffith Gazette* 9, 3.

Williams, S.L. (1988) *Thalassis testudinum* productivity and grazing by green turtles in a highly disturbed seagrass bed. *Marine Biology* (Berlin) 98, 447–456.

Willis's Walkabouts (1994) Personal hygiene. In: *Guide to Bushwalking in North and Central Australia.* Willis's Walkabouts, Millner, Northern Territories, p. 23.

Witherington, B.E. (1997) The problem of photopollution for sea turtles and other nocturnal animals. In: Clemmons, J.R. and Buchholz, R. (eds) *Behavioural Approaches to Conservation in the Wild.* Cambridge University Press, Cambridge, pp. 303–325.

Woolcott, T.G. and Woolcott, D.L. (1984) Impact of off-road vehicles on microinvertebrates of a mid-Atlantic beach. *Biological Conservation* 29, 217–240.

Yalden, P.E. and Yalden, D.W. (1990) Recreational disturbance of breeding golden plovers *Pluvialis apricarius. Biological Conservation* 51, 243–262.

Yur'eva, N.D., Matveva, V.G. and Trapido, I.L. (1976) Influence of recreation on soil invertebrate groups in birch woods round Moscow. *Lesovedenie* 2, 27–34.

Chapter 25

Exploring Socio-cultural Impacts on Local Communities

S. Wearing

School of Leisure and Tourism Studies, University of Technology, Sydney, Lindfield, New South Wales, Australia

Introduction

This chapter investigates the socio-cultural impacts of ecotourism. It particularly focuses on those communities that are living in marginal or environmentally threatened areas and take an economic interest in the preservation of these areas while also attempting to provide an ecotourism experience for travellers. Some claim that ecotourism is mass tourism in its early pre-tourism development stage. However, if the criteria used to describe the various components of ecotourism are applied to each particular tourism situation (see Chapter 1), it becomes clearer if the type of tourist activity being undertaken conforms to what Wallace (1992, p. 7) describes as 'real' ecotourism. Essential to this is a two-way interactive process between host and guest, and therefore 'the culture of the host society is as much at risk from various forms of tourism as physical environments' (Sofield, 1991, p. 56). From definitions of ecotourism proposed in earlier chapters comes the aim of sustaining the well-being of local communities. Ecotourism can therefore be viewed as a development strategy leading to sustainable development and centring on the conjunction of natural resource qualities, host community and the visitor that all benefit from tourism activity.

Host Communities

A 'host community' refers here to a group of people who share a common identity, such as geographical location, class and/or ethnic background. They may also share a special interest, such as a concern about the destruction of native flora and fauna. The host community provides support services for tourism and may have involvement in its management, but some tourism theories postulate that the sustainability of the community in a peripheral area tends to decline as tourism intensifies. According to Murphy (1985), the long-term success of the tourism industry depends upon the acceptance and support of the host community. Therefore, to ensure that ecotourism is able to be maintained, it is essential to ensure the sustainability of both the natural and cultural environments of the destination.

Tourism destinations usually involve a series of separate elements such as landscapes, wildlife, specific activities, etc. The people who best know and understand how these elements function, are the people in the host community who are exposed to them on a regular basis. However, the community is rarely asked by private operators about their vision for the area. Nor have they been traditionally part of the planning process. Decisions relating to the likely impacts on the area are often made by planners who do not understand the intricacies or functions of the host community and local tourism resources. Consequently, the tourism industry that evolves does not suit community needs or use the resources to their best advantage, thereby creating unnecessary social pressure on the host community. Clearly, a process is needed whereby *direct knowledge, experience and understanding* from the community forms the basis for the management of socio-cultural impacts so that these communities can engage in ongoing development and enhancement through ecotourism. One avenue that allows this to occur is through socio-cultural planning appraisals, wherein the community itself has direct involvement and thus influences the process and outcomes.

There are a number of reasons why host communities may consider an ecotourism approach to tourism development. The main principles or elements of ecotourism are designed to maximize the social benefits of tourism while minimizing the socio-cultural impacts. Ecotourism can, in an ideal circumstance, provide the following benefits to the socio-cultural environment:

- increase demand for accommodation houses and food and beverage outlets, and therefore improve viability for new and established hotels, motels, guest houses, farm stays, etc.;
- provide additional revenue to local retail businesses and other services (e.g. medical, banking, car hire, cottage industries, souvenir shops, tourist attractions);
- increase the market for local products (e.g. locally grown produce, artefacts, value-added goods), thereby sustaining traditional customs and practices;
- use local labour and expertise (e.g. ecotour guides, retail sales assistants, restaurant table waiting staff);
- provide a source of funding for the protection and maintenance of natural attractions and symbols of cultural heritage;
- provide funding and/or volunteers for field work associated with wildlife research and archaeological studies; and
- create a heightened community awareness of the value of local/indigenous culture and the natural environment.

These benefits suggest that ecotourism is about attracting visitors for the 'right' reasons and not just promoting tourism for the sake of the 'tourist dollar' at the expense of a community's natural and cultural attributes. In essence, the overall objective of an ecotourism-based approach should be a process that a supportive community wants and controls. It follows that the result would be an environment that is more receptive to tourists. The information gained from socio-cultural consultation can also be used by planners to guide decision making in matters such as landscape enhancement, and by the community to investigate projects they would like to undertake or even operate themselves. Yet, while a planning process at the community level may appear simple in theory, it is complicated by many factors, including conflicting interests among stakeholders and lack of prioritization of resource allocation to areas that people feel need it most. If communities can be involved in the planning process from the beginning, this can reduce the future likelihood of conflict and misinformation.

Communities and Socio-cultural Impact

There are many challenges that host communities experience in pursuing ecotourism, and as with all forms of development there are both costs and benefits. Notably, Boo (1990) and Valentine (1987) have voiced their concern with the leakage of income that occurs when resources have to be imported. Boo (1990), however, found that ecotourists in particular, when compared with other market segments, are more likely to appreciate local tradition, customs and cuisine. Therefore, it is considered important that the community is involved in setting up ecotourist projects, as these types of projects benefit from taking into account factors that relate to each community's lifestyle.

Box 25.1 illustrates the issues that need to be considered when examining the impacts resulting from tourism. The majority of literature relating to socio-cultural impacts is found in the general tourism literature, but these need to be understood in order to assess and manage socio-cultural impacts of ecotourism projects. Basically, the impacts are similar, yet the emphasis is different. Tourism can lead to an increase in the cost of living in the local community, as is expected for tourism destination areas (see McNeely and Thorsell, 1989). This exemplifies the way in which resources can gradually be rendered inaccessible to the local people. Inflationary pressures lead to increases in the costs of consumer goods and real estate, making it difficult for some local people to remain in the area (Cater, 1987). This also illustrates how economic impacts (see also Chapter 23) can have a profound bearing on subsequent socio-cultural impacts. The difference in consumption patterns of visitors and nationals may have the outcome of further widening the gap between the more and less developed regions (Cater, 1987; Cater and Lowman, 1994). Further, local community members could possibly view the areas as being developed exclusively for foreign interests (McNeely and Thorsell, 1989). One solution to this problem is the establishment of differential entry fees, one for tourists and another for nationals. This would be applicable in the case of publicly owned sites but may be problematic on private sites. This policy has already been implemented in some host communities, with entrance fees to local entertainment facilities or protected areas. The larger problem is whether entrance fees are retained by the attractions themselves. Boo (1990), for example, has shown how revenues generated by parks in Costa Rica are diverted to other sources such as hospitals.

Box 25.1. Socio-cultural impacts of tourism (Figuerola, cited in Pearce, 1989).

1. Impact on population structure
   - Size of population
   - Age/sex composition
   - Modification of family size
   - Rural–urban transformation of population
2. Transformation of types of occupation
   - Impact on/of language and qualification levels
   - Impact on occupation distribution by sector
   - Demand for female labour
   - Increase in seasonality of employment
3. Transformation of values
   - Political
   - Social
   - Religious
   - Moral
4. Influence on traditional way of life
   - Art, music and folklore
   - Habits and customs
   - Daily living
5. Modification of consumption patterns
   - Qualitative alterations
   - Quantitative alterations

Socio-cultural impacts are those 'influences that come to bear upon the host society as result of tourist contact' (Prasad, 1987, p. 10). These impacts can both benefit and impose costs on the community. In the case of ecotourism, benefits may stem from contact that ecotourists have with the host culture. The longer duration of stay may promote a deeper understanding between the individual ecotourist and individual community members. In turn, this may increase the tourist's understanding of the host community. Travis (1982) suggests that this contact can also lead to an improved reputation and visibility of the host community in the society of the ecotourist. Given that advanced industrial societies need to explore less resource-exploitative modes of lifestyle, this contact can be beneficial to those societies. The global benefits that may accrue will appear less tangible but no less beneficial and far-reaching. As Cater (1987) suggests, one possible scenario would be the creation of a wider understanding between nationalities. The influence that a host culture can have on the guest culture is illustrated by the following anecdote, related by an ecotourist:

> I think the biggest impact was what I was talking about earlier, and the way that the experience affected me. I think my tolerance level and my acceptance levels, aren't as black and white any more. It doesn't necessarily have to be about environmental things, just generally. You may not agree with the way someone lives or someone else's value system, but you've got no right to sit there and judge it, or try and change it unless you really see some benefit in re-educating people to another way of thinking. It doesn't matter that they have a totally different system to the way we do things, just enjoy the differences (Amy – ecotourist).
>
> (Wearing, 1998, p. 210)

According to Cater (1987), the dynamic state of culture means that change is inevitable. However, tourism, with its direct contacts between host and guest, accelerates these changes, especially in the host culture. Butler (1990) suggests that as ecotourism penetrates deeper into destinations, long-term development results, thus contributing to impacts on the host culture. In some cases, long-term personal contacts can be overly intrusive, and thus harmful in their own right despite the best of intentions among tourists. Yet, without some shift in the agricultural economic base (or other single economic-based rural communities), younger people may be forced to leave the area to find work, which itself initiates its own set of impacts.

Creating Supportive Communities

In principle, ecotourism ensures a supportive host community by investing control within the community itself. This might be achieved by ensuring the highest level of local participation ('Initiating Action' as identified by Paul, 1987). Community attitudes can then be monitored to guarantee that the opinions of residents are considered (Bjorklund and Philbrick in Mathieson and Wall, 1982). Monitoring can occur through a range of formal and informal means, including informal conversations, group discussions and questionnaires. Regular meetings can also be held with community interest groups, as this allows members of the community to voice their opinions concerning the consequences of ecotourism, while also allowing for the discussion of strategies that will help to manage these impacts. A community management plan could be developed which, ideally, would lead to a more integrated form of management.

To ensure that impacts can be managed, the community must be involved in the complete tourism development process, from the planning stage to the implementation of tourism projects, through avenues of consultation. Consultation, in this context, involves 'a process which aims to reconcile economic development with the broader interests of local people and the potential impact of development on their natural, social and cultural environment' (WWF, 1992, p. 25). Methods such as participatory rural appraisal (PRA) offer an example of how this might be achieved. According to the World Bank Participation Sourcebook (1998), PRA is a label given to a growing family of participatory approaches and methods that emphasize local knowledge and enable local people to make their own appraisal, analysis and plans. It uses group animation and exercises to facilitate information sharing, analysis, and action among stakeholders. The Institute of Development Studies (IDS, 1996) broadens this definition when it additionally describes PRA as something that enables communities *to monitor and evaluate* the results of these initiatives. Mukherjee (1993) further defines PRA as a means of collecting different kinds of data, identifying and mobilizing intended groups and evoking their participation and opening ways in which intended groups can participate in decision making, project design, execution and monitoring. It involves a set of principles, a process of communication and a menu of methods for seeking villagers' participation in putting forward their points of view about any issue, including ecotourism. In enabling them do their own analysis with a view to make use of such learning, PRA initiates and sustains a participatory process.

Communities can achieve a range of benefits from ecotourism in their region, including employment and infrastructure developments. This may engender a positive predisposition within the host community, and particularly among the businesses that see a possible increase in revenue. It is an aim of ecotourism that cultural attractions not only benefit the visitor but also the host population. The interactive dimension of ecotourism, that is, relations between the ecotourist, the environment, and the host community, will be influenced by both the ecotourist and community representatives. This interaction may not always be extensive or congenial, but it does offer opportunities for the examination of the influences of the interaction on the socio-cultural values of the community. Too often cultural attractions through commodification become overtly commercialized in nature, satisfying the visitors' needs but losing all meaning and significance for the local and/or indigenous population. This conservation of cultural integrity continues the idea of sustaining the well-being of the local people, as highlighted in the definition of ecotourism. To illustrate, a group of Aborigines in central Australia saw 'involvement in tourism as a possible means of re-educating and re-establishing a pride, and sometimes even a knowledge, of traditional skills and values amongst their younger generations' (Burchett, 1992, p. 6). Thus the development of cultural attractions can be seen to benefit the local people here as well as the tourist.

Another aim of ecotourism is to ensure that profits from such programmes flow back into local communities (O'Neill, 1991). For example, in the case of indigenous populations generally, they will encourage ecotourism and will be more likely to participate in conservation programmes if they can benefit from such activities and are included in the management process. While it is often important for ecotourism that traditional values are maintained, Wallace (1992) suggests indigenous people must not be asked to maintain their traditional practices simply for the sake of tourists. As cultures undergo a constant process of change and it is this process of genuine culture change and exchange that is a component of the experience of ecotourism, it is important that tourism does not inhibit this change. In keeping with the idea of sustaining the well-being of the local people, Wallace suggests that ecotourism is a 'useful tool for locally directed rural development and wildland protection' (1992, p. 6). With this change can come modifications in attitudes of the indigenous people to the use of protected areas, such as the discontinuation of slash and burn agriculture.

Local communities should be involved in the entire tourism development process through avenues of consultation, but participation by local communities in tourism must not be limited simply to employment opportunities. Two examples of where participation levels are significantly high are the villages of South Pentecost in Vanuatu, and Aboriginal tourism in the Northern Territory, Australia. In South Pentecost, the Pentecost Land Dive is a traditional ceremony of the villages in this area that occurs annually in April/May. In response to increasing negative cultural impacts as a result of tourism, the local chiefs of the villages established 'The South Pentecost Tourism Council' to manage the occasion with its 'primary responsibility to safeguard the cultural integrity of the event' (Sofield, 1991, p. 59). This involves maintaining customs with tourist visits, preventing filming of the event and limiting numbers of tourists attending the performance. This not only provides the tourist with an authentic cultural experience but also maintains the cultural significance of the ritual to the villagers themselves and allows them some degree of control over the activity of tourism.

Similarly, the influx of tourists to the Northern Territory, Australia, encouraged indigenous groups to participate in tourism in an attempt to control visitation on to their lands (Burchett, 1992). An example of this is the Umorrduk area in North Western Arnhem Land. The aboriginal tribe residing in this area allows a tour operator of aboriginal origin to conduct safari tours to selected areas allocated by the local aboriginal people. Control procedures implemented by the aborigines, such as entry permits and the prohibition of photography at some sacred sites, ensures that the numbers of tourists are limited and the cultural integrity of the aboriginal people is maintained. This type of operation 'reinforces the privileged nature of the opportunity to enter this area and of the capacity of the traditional owners to maintain absolute control on who enters, where they travel and what information they get in regards to traditional matters' (Burchett, 1992, p. 12). In this way, both the visitors and the hosts benefit from the tourism experience while at the same time avoiding negative cultural impacts on the local and/or indigenous population. Participation of local communities in the activity of tourism, therefore, is an essential element to sustaining the well-being of the local people.

As stated earlier, one of the economic benefits of tourism associated with local communities is the increase in employment opportunities and income generation for the host region. However, 'flourishing employment, living standards and consumption levels for some, added to the unequal distribution of benefits to a portion of the population, can contribute to social tensions and hostility' (WWF, 1992, p. 19). Increasing recruitment of local staff at all levels of the industry would benefit the host population, but more importantly, less foreign ownership and more locally owned operations or vested interests in local operations would see greater economic benefits accruing to local communities. However, this is provided that the latter are not restricted to the existing local elite.

Returning to the case of the Australian Aborigines, tourism was perceived as being able to 'offer some employment in remote parts of the Territory where alternative economic opportunities were few' (Burchett, 1992, p. 6). Guided walks, demonstrations of tracking skills and food processing techniques and other aspects of aboriginal life are carried out at Uluru or Ayers Rock. Performances of traditional dance are undertaken by three different groups who maintain control of their dance routines (Burchett, 1992). Specialized, small group tours are undertaken by the Tiwi community of Melville Island who see 'the development of an isolated, comfortable safari camp as being an ideal way for them to combine their needs for employment, cash flow and cultural underpinning' (Burchett, 1992, p. 7). The produce from traditional hunting and fishing activities undertaken by tourists is returned to the local community with only sufficient amounts left at the safari camp for tasting by the tourists (Burchett, 1992). Tourists experience the traditional and authentic activity and can taste the 'catch' but these vital resources, necessary food stocks for the aborigines, are not depleted just for the sake of the tourist.

Conversely, the Gagudju people of Kakadu National Park prefer not to actively participate in tourism activities but have instead invested in tourism infrastructure (Burchett, 1992). They own two major hotels in Kakadu National Park and have an investment in a third property. The Gagudju 'maintain the control and the direction of their own involvement in the tourist industry' (Burchett, 1992, p. 14) and generate income for the benefit of themselves. The Aboriginal people are receiving direct economic benefits and employment from tourism, while presumably conserving their perception of what their cultural identity should be, allowing them to challenge any form of cultural hegemony that may result from tourism. By developing an appreciation of local communities and their customs and traditions 'a process of mutual respect and understanding between societies can be greatly enhanced' (Burchett, 1992, p. 10) and the achievement of successful interaction between hosts and guests will only benefit and sustain the well-being of local communities.

Ecotourism insists on the understanding and appreciation of both nature and different and/or indigenous cultures, and also their relationship with each other. Through this interactive process, the visitor and the host population both benefit experientially from ecotourism. Local communities can benefit from ecotourism economically if they play a greater participatory role in the tourism process. The greater the control over tourism in their region, the more culturally sustainable they will become.

Exploring Employment Further

The primary employment opportunities through ecotourism appear to be in jobs such as hotel servicing, craft making, shop ownership, tour operations, government agency staff, and park rangers. However, in many circumstances there are limited employment benefits from tourism development because infrastructure such as accommodation establishments are already established and staffed. Also, in sectors such as tour operators, most businesses are only small, one or two person companies. This is further compounded by the fact that ecotourism depends on a lack of infrastructure as its attraction to tourists is partly due to the 'pristine' environment.

Despite this, there are still long-term employment opportunities resulting from ecotourism that are potentially open to locals. These are evident in the description of ecotourism as 'labour intensive' as opposed to other types of tourism which are capital intensive. For example, the training of small guiding operations offers employment opportunities for many rural communities with knowledge of the area. Still, in light of the lack of capital intensity of ecotourism, it is still not a viable economic possibility for many local populations to enter the market. Joy and Motzney (1992) suggest that locals should buy and manage small accommodation establishments, yet this may not be an economic possibility for many people.

The Central American country of Belize is attempting to counter this problem by developing policies that provide feasible financial avenues and 'competitive advantages' for locals to invest in small to medium scale private tourism enterprises, such as food and beverage outlets, accommodation establishments and sporting operations. Government enforcement of the policies is through the restriction of trade licences, concessions or duty exemptions, and vetting procedures (Maguire, 1991).

Currently, a general lack of host community skills and resources has meant that many ecotourism ventures are often owned and operated by expatriates (Weiler and Hall, 1992). Clark and Banford (1991) claim that it is unfeasible to expect the local population to automatically assume employment positions within ecotourism, and state that 'the hard truth is that a local farmer, fisherman or plantation worker cannot always be changed overnight into a tourist guide or hotel manager' (1991, p. 9). Wearing (1998) believes the planning, staff and management of accommodation and parks by expatriates in developing countries may have dire effects on the local population and culture. This can lead to a 'homogenization' of cultures and the overlooking of local and traditional methods of managing the natural resources, in addition to host-community hostility and anger toward tourism.

It is generally accepted that training and education of locals is needed before they can gain meaningful employment from ecotourism (see Wearing, 1993; Wearing and Larson, 1996). Weiler and Johnson (1991) claim that skills need to be developed (in particular language, environmental and natural history skills), while Clark and Banford (1991, p. 9) pose a longer-term solution. They propose a 'sensible partnership' between the tourism industry and the local population, which would ensure that local people benefit from tourism (even before they gain employment) despite expatriate domination. Furthermore, they would develop an education system for the children of local people so that they may later participate fully in tourism operations (Clark and Banford, 1991).

Clark and Banford, however, may be overlooking more ideal forms of employment for local people. The skills for running private business enterprises may not be available within the local community, but local expertise and knowledge can be a powerful tool for tourist guides and park wardens in protected areas. Bunting (1991, p. 3) supports this by claiming:

> proper management of protected areas requires employment of park rangers and guards, as well as workers to maintain park buildings, roads and trails. Ecotourism in protected areas creates demand for guide services ... providing employment for ... local people familiar with the flora and fauna of the area.

According to Ceballos-Lascuráin (1992, p. 5) local people possess the 'practical and ancestral knowledge of the natural features' of the area. They have the incentive to become dedicated to ecotourism in positions such as park rangers, since 'their subsistence would depend in a major degree on the sustained preservation of the natural qualities of their environment'.

Bates (1991) believes that with or without ecotourism, Papua New Guinea is the most rapidly Westernizing nation on Earth and, as a consequence, there are growing social problems, unemployment and a rapidly diminishing culture. The Ambua Lodge in the highlands of Papua New Guinea is an ecotourism establishment providing employment opportunities to local people and therefore stopping the urban drift towards the crime-ridden major cities. As such, it provides the incentive to preserve not only the natural environment, but the unique features of their culture as well. The construction and operation of Ambua Lodge provides a diverse range of long- and short-term employment positions for the local community such as construction workers, art and crafts makers, performers, waiters, cooks, guides, gardeners, room cleaners, laundry operators, maintenance personnel and vegetable growers (Bates, 1991).

Generally, in the employment sector, the widening opportunity for women often reduces their dependence on family and may affect family relationships. Also, the employment opportunities are very limited, and the jobs have relatively low pay (Cater, 1987). There appear to be insufficient training programmes that provide the necessary language, natural history and environmental skills for the local community. A large number of the operators therefore come from outside the host community, increasing economic leakage (Valentine, 1991), even though there is an augmentation in economic opportunities for communities.

Access and Research

As a result of tourism, local people in many areas have lost access to land and resources they had previously enjoyed. Ecotourism can lead to a change in resource ownership and management that, while being beneficial to the tourism industry, is detrimental to the local people. Subsequent local resentment toward national parks and designated conservation and protected areas can arise when the area is viewed to be of principal benefit to tourists rather than locals (Ceballos-Lascuráin, 1992). Extreme cases can lead to the destruction of natural areas or flora and fauna. For example, poaching in African and Costa Rican protected areas demonstrates that spiteful destruction may occur if local communities believe that they are not receiving the benefits from their lands that are singled out for protection (West and Brechin, 1991; Western and Wright, 1994).

This scenario has been countered in Costa Rica by the establishment of a joint UNESCO-MAB and Costa Rica National Park project that intentionally prefers residents to foreign tourists. Other attempts have also been made to incorporate local communities into protected area management with an emphasis on Costa Rican recreational and educational needs, and the employment and training of local community members in the areas of park maintenance, interpretation, management and habitat restoration (see Wearing and Larson, 1996). These sorts of programmes have the long-term benefit of the gradual transfer of control of tourism from expatriate or developed country influences. The ideal for planning ecotourism development is cited by Clark and Banford (1991, p. 7): 'There is no reason why countries or communities should not decide what type of tourism they are willing to accept and set limits to the amount of change they are prepared to put up with. This applies to ecotourism'.

Goal setting, policy development and management for socio-cultural impacts are needed at the national, regional and local level. Unfortunately, in practice, the planning system is often set up in a way that gives local people little or no opportunities for input (West and Brechin, 1991; Western and Wright, 1994). These events occur and are often designed and implemented in a political context in which local and/or indigenous people have minimal voice. Countries such as Bhutan and Nepal, however, have developed a system (through resource management plans) that specifically benefits local people by giving them increased power and a greater role in decision making. Increasing access to information for local and/or indigenous people provides them with more scope for involvement in planning and decision making. In Panama and Costa Rica, meetings between various groups of indigenous peoples have been arranged in which problems and difficulties are discussed, and potential solutions, such as PRA, are proposed.

Small, rural communities have certain social, cultural and historical relationships that bind them in the form of identifiable units. The 'nature' of these social relationships is, therefore, important to identify and understand in helping to define a community and the socio-cultural values it holds. Tourism has a profound influence on the socio-cultural environment through its potential to contribute to the change of lifestyles and behaviour of local communities. It can thus be examined within a dialogue that allows for struggles and reformulations, as host cultures or subcultures challenge the usual dominant cultural forms of the tourist. For example, MacCannell (1992) refers to the interaction between 'moderns' (that is, tourists) and 'ex-primitives' or host peoples, as a cannibalistic endeavour. The invading tourists whose dominant white, Western culture empowers them, are able to consume, devalue and ultimately eliminate the host culture.

Attempts to research the socio-cultural values of the community and the impacts of tourism on these host communities usually focus on community attitudes toward tourism. Questionnaires are often used to gauge community opinion of the perceived positive and negative impacts of tourism (e.g. Liu and Var, 1986; King *et al.*, 1993; Getz, 1994; Haralambopoulos and Pizam, 1996). However, these surveys frequently do not take into account non-tourism factors, such as the modernization process in general, that could be influencing these attitudes and impacts. Research needs to employ techniques, such as those embodied in PRA, that can provide a rapid survey of community values and attitudes, and then link these with ecotourism. In general, a more systematic range of techniques and methodologies, as found within the social sciences, is necessary to identify and analyse local attitudes (see Furze *et al.*, 1996). Particular attention should be paid to local attitudes toward the natural environment, since this is the core attraction base for ecotourism.

In general, research needs to focus on the many impacts of tourism to provide insight into how ecotourism can adopt a different and more benign approach. Among the impacts that require further investigation are the following.

- A high degree of leakage resulting from foreign ownership of infrastructure (major hotels for example: see Mathieson and Wall, 1982).
- Tourists bringing their own social values and behaviours which can distort social habits and customs, especially if they are adopted by locals through the

demonstration effect (see Mason, 1990), an example of which is the number of evening entertainment venues which begin to proliferate.
- Domestic economy impacts, such as the price of vegetables and other staple food items rising while luxury items, such as cheese and chocolate, are increasingly becoming a staple food which places higher expectations and pressure on the local residents in terms of purchasing.
- Community members perceiving that the area has been developed for foreigners only and thus a feeling of resentment towards them develops (McNeely and Thorsell, 1989, p. 31). Discontent can occur among communities who find it difficult to coexist with tourists because the tourists are on holidays while the community members must continue with their work (Mason, 1990).

The general literature on socio-cultural impacts (Pearce, 1989; van Doorn, 1989; Craik, 1991, 1995; Pearce, 1994, 1995; Sharpley, 1994; Faulkner and Tideswell, 1997) provides a negative view. Ecotourism for the most part is aimed at avoiding some of these impacts, through the use of local hotels, participating in local cultural events, eating traditional dishes and in some cases working with local community members (Wearing, 1998). Yet as with most forms of tourism, ecotourism is only one form that exists in the host area and the scale of impact needs to be addressed in comparison with other market segments.

With regard to the demonstration effect, Murphy (1985) claims that the members of the community who are most susceptible are the youth, who may feel dissatisfied with local opportunities and are prepared to imitate the lifestyle of visiting tourists as a way of seeking alternatives to their current lifestyles. Still, ecotourists are generally aged between 20 and 40, or over 55 (Ballantine and Eagles, 1994; Wearing and Neil, 1999). They expect a quieter lifestyle with longer-term interchange which allows for a more substantial context for the exchange of lifestyle and behaviour, and can therefore lessen what has been identified as a primary impact.

Related to demonstration effect is acculturation. Acculturation is the process whereby people borrow from each other's cultural heritage (Lea, 1988). In the case of tourism it is common for the Western ideas and practices introduced by tourism to be accepted by the host culture (for example, see Nash, 1996). In the case of ecotourism this problem is directly addressed in that the ecotourist's stay is of a markedly longer duration than that of conventional tourism. Thus the potentiality in everyday interaction exists for a more equal transfer of cultural values, though this in itself raises a range of issues relating to such transfers.

There is a need to understand ecotourism and its socio-cultural impacts within the complexities of the exchanges that occur between host and guest. Hence, questions should be asked that will push theory beyond its current limitations. By examining the tourist destination as a site and space in which dynamic social interactions occur, rather than as a static and bounded place, research can link tourist experiences to possibilities for extending the self beyond the limitations of culturally specific discourses. This expands and modifies the concept of the universalizing gaze, suggesting that the ecotourist in this specific context is interacting in a creative way with the tourist space so that the self is involved. An asymmetrical perspective focusing on the one-way action of the tourist upon the host culture gives way to a more symmetrical approach which recognizes the two-way interaction between tourist and host, and includes the subjective experiences of both.

Awareness and Education

Education also plays a powerful role in increasing local involvement and contribution in ecotourism. Wearing (1998) claims that initiatives such as Costa Rica's university and high-school ecotourism programmes will eventually lead to greater local involvement in protected areas and, eventually, the tourism industry. This is achieved when local people are involved in researching, studying, discussing and devising strategies to gain control over the development decision making process. Concern and consideration for the local culture can be incorporated into the planning and marketing of ecotourism destinations and products. The local community can help to ensure that tourists treat them with respect by developing and imposing social guidelines. According to Blangy and Epler Wood (1992), social guidelines could incorporate desirable and acceptable behaviour in the following areas:

1. Local customs and traditions
2. Permission for photographs
3. Dress
4. Language
5. Invasion of privacy
6. Response to begging
7. Use and abuse of technological gadgetry
8. Bartering and bargaining
9. Indigenous rights
10. Local officials
11. Off-limits areas

Blangy and Epler Wood (1992) recommend that government agencies, tourism boards, the tourism industry and local inhabitants could all play a role in the education of tourists about cultural issues through the implementation of social guidelines. They suggest government should be responsible for developing guidelines, but recommend significant input from the local community. The local community can be incorporated into the development of these guidelines by using government funding (if available) to get assistance with the preparation and editing of brochures for distribution. Alternatively, the local community could collaborate with 'international and local non-governmental organisations' (Blangy and Epler Wood, 1992, p. 4) and become involved with environmental education projects. Another source of potential assistance is the tourist boards. Blangy and Epler Wood (1992, p. 4) suggest boards should allocate funds for all stages of the education process, through the 'generation, printing and distribution of local guidelines'. In addition to the distribution of brochures and printed matter at tourist centres, tour guides could play an important role by briefing tourists on what is acceptable and unacceptable in the region being visited.

Several examples exist where local people have taken measures to ensure they benefit from ecotourism both personally and as a community. In many small communities living in remote areas such as Easter Island, beds within local houses are open to tourists and provide the major source of accommodation. The additional income gained is often spent increasing the quality of life within the local community. In Papua New Guinea's highlands, villagers have a source of income from the accommodation huts that they have built on their property (Bates, 1991). By collaborating with local tour operators, these huts provide accommodation to groups of tourists. However, a major problem with increased local involvement in ecotourism is the dependency that may result. As tourism is often seen as the most economically lucrative form of employment, other avenues of resource utilization may be discarded. This in turn leads to the replacement of more traditional industries such as hunting, fishing and forestry with ecotourism (Maguire, 1991, p. 6).

Despite the often good intentions of tourists and some tour operators, it is apparent that 'ecotourism can damage the natural assets on which it rests. The outcome depends on how it is managed' (Lindberg, 1991, p. ix). Thus the implications for management are enormous. For example, park managers must find a way to 'capitalise on its potential without jeopardising the special features of natural areas' (Boo, 1990, p. xiv). This would enable a proactive interaction with the tourism industry where they can ensure that capacity is set and managed by their own standards, rather than by a continual unmanaged growth of tourism numbers and operators.

Any form of ecotourism depends on the use of the host community's available tourist facilities and infrastructure such as accommodation, roads, transport, car parking and ancillary services such as food and other retail outlets. Ecotourism also depends on the host community's goodwill and positive attitude to tourism. Local communities (in many cases via elected representatives) make decisions about the provision of such resources for tourist use. Decisions about the allocation of scarce resources ultimately favour one or more groups of residents, or instead, aid the local tourist industry in its ability to attract tourists to the area. This dynamic relationship needs to be understood, since local and/or indigenous communities usually comprise groups with different and potentially conflicting interests (see Fig. 25.1). That is, not all groups want the same things and so they influence the socio-cultural environment created through their attitudes.

To expand, ecotourism requires a healthy business environment that provides:

- financial security;
- a trained and responsible workforce; and
- attractions of sufficient quality to ensure a steady flow of visitors who stay longer and visit more often and are environmentally and culturally aware.

Those interested in the natural environment and cultural heritage issues seek:

- protection of the environment through prevention, improvement, correction of damage and restoration; and
- to motivate people to be more aware and subsequently 'care for' rather than 'use up' resources.

Community members seek a healthy place in which to live with:

- adequate food and clean water;
- health care;
- rewarding work for equitable pay;
- education and recreation;
- respect for cultural traditions; and
- opportunities to make decisions about the future.

Socio-cultural impact is one concern that all these groups have in common. Others may include:

- issues of access, such as when, where and how tourists visit and move from place to place;
- host and guest issues, such as cultural impacts or common use of infrastructure;
- land use issues, such as hunting/wildlife habitat, agriculture/recreation and preservation/development, etc.

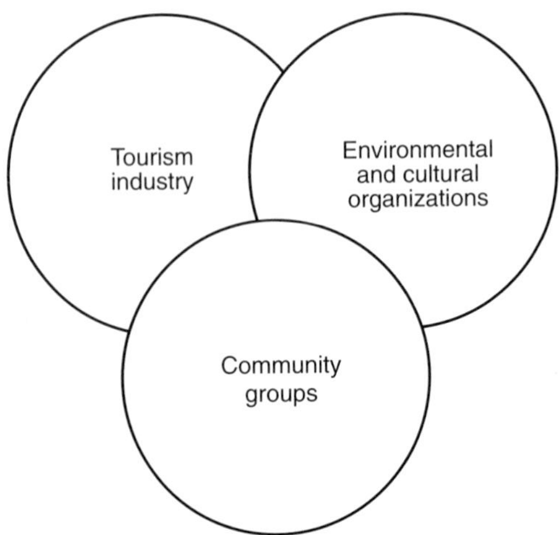

Fig. 25.1. Stakeholders in the local community that might influence the socio-cultural environment.

Conclusion

In order to create a socio-cultural environment that is conducive to the ideals of ecotourism, the environment needs to be managed according to accepted standards where the needs and preferences of all groups are respected. If approaches are adopted that stem from the philosophical ideals inherent to ecotourism (see Wearing and Neil, 1999) all local interest groups in perfect circumstances should be provided with the opportunity to 'have their say' early in any policy, management or planning process. This community input will make it possible to benefit from tourism without local residents, business people, park managers and environmental organizations feeling that their needs have been ignored which creates a dynamic that produces negative attitudes to ecotourism from the outset.

At the risk of stating the obvious, most experiences sought by ecotourists, and indeed many experiences sought by mainstream tourists, depend on the natural environment being in its unaffected (by humans) state. Many ecotourists are sensitive to decreases in water quality and air quality, loss of vegetation, loss of wildlife, soil erosion and a change in the character and visual appeal of an area due to development. The degradation of the natural environment will severely reduce visitor demand in the long term as the natural attributes on which ecotourism depends will be perceived as less attractive, less legitimate and less able to provide satisfying ecologically based experiences. The most important feature in the preservation of a community's nature-based visitor attractions requires that the community has a positive attitude to tourism, as this will affect the visitor's experience to a much larger degree.

Thus, ecotourism is identified as not only concerned with experiencing natural environments but also with understanding the cultural heritage of indigenous and/or local communities. 'Carefully designed tourism programs can make protected natural areas a focus for fostering local knowledge, skills and lifestyle to perpetuate traditional values among indigenous people and to educate outsiders about their culture' (Kutay, 1990, p. 38). Because local people often inhabit or depend on the environments in the most sought after areas for visitation by ecotourists, any attempts to establish ecotourism must incorporate the inhabitants living on or near the natural resource, as they will, in a sense, become an integral part of that resource.

Local communities are significantly vulnerable to the deleterious impacts of tourism development – particularly indigenous cultures – as they directly experience its socio-cultural effects. The subsequent impact of tourism's dynamic growth on communities has in some cases precipitated strong protests by community groups, which, being sensitive to the impacts of tourism, have actively opposed large-scale tourism developments for their locality. Other community groups have been more accepting of a gradual growth in tourism to their region over many years, only to become aware of the negative impacts at a later date when these impacts cannot easily be ignored.

Knowledge of socio-cultural impacts tells us that the interdependence of tourism and the social/physical environment is fundamental to the future of each. Seeking a way to accommodate the needs of all parties, without control being external to those who experience its effects most directly, is therefore essential. Features of the natural and cultural environment and supportive host communities are the foundations of a successful ecotourism industry. Neglect of conservation and quality of life issues threatens the very basis of local populations and a viable and sustainable tourism industry.

References

Ballantine, J. and Eagles, P. (1994) Defining Canadian ecotourists. *Journal of Sustainable Tourism* 2, 210–214.

Bates, B. (1991) Ecotourism – a case study of the lodges in Papua New Guinea. In: *PATA '91 Bali, Indonesia: Conference Record 40th Annual Conference.* Pacific Asia Travel Association, San Francisco, pp. 228–234.

Blangy, S. and Epler Wood, M. (1992) *Developing and Implementing Ecotourism Guidelines for Wildlands and Neighbouring Communities.* The Ecotourism Society, North Bennington, Vermont.

Boo, E. (1990) *Ecotourism: the Potentials and Pitfalls.* World Wildlife Fund, Washington, DC.

Bunting, B. (1991) Nepal's Annapurna Conservation Area. In: *PATA '91 Bali, Indonesia: Conference Record 40th Annual Conference.* Pacific Asia Travel Association, San Francisco, pp. 209–218.

Burchett, C. (1992) A new direction in travel: Aboriginal tourism in Australia's Northern Territory. Paper presented at the Northern Territory Tourist Commission Environmental Conference, April 1992.

Butler, R.W. (1990) Alternative tourism: pious hope or Trojan horse? *Journal of Travel Research* 28, 40–45.

Cater, E. (1987) Tourism in the least developed countries. *Annals of Tourism Research* 14, 202–206.

Cater, E. and Lowman, G. (1994) *Ecotourism: a Sustainable Option?* John Wiley & Sons, New York.

Ceballos-Lascuráin, H. (1992) Tourism, ecotourism and protected areas: national parks and protected areas. Paper presented to the IVth World congress on national parks and protected areas, 10–12 February, Caracas, Venezuela, International Union for Conservation of Nature and Natural Resources.

Clark, L. and Banford, D. (1991) Ecotourism potentials and pitfalls. In: *The Promotion of Sustainable Tourism Development in Pacific Island Countries, Seminar Proceedings, Regional Seminar on the Promotion of Sustainable Tourism Development in Pacific Island Countries, 18–22 November,* Suva, Fiji, pp. 56–63.

Craik, J. (1991) *Resorting to Tourism.* Allen & Unwin, Sydney.

Craik, J. (1995) Are there cultural limits to tourism? *Journal of Sustainable Tourism* 3, 87–98.

Doxey, G.V. (1975) A causation theory of visitor-resident irritants: methodology and research inferences. In: *The Impact of Tourism, Sixth Annual Conference Proceedings of The Travel Research Association.* The Travel Research Association, San Diego, pp. 195–198.

Faulkner, B. and Tideswell, C. (1997) A framework for monitoring community impacts of tourism. *Journal of Sustainable Tourism* 5, 3–28.

Furze, B., De Lacy, T. and Birkhead, J. (1996) *Culture, Conservation, and Biodiversity – the Social Dimension of Linking Local Level Development.* John Wiley & Sons, Sydney.

Getz, D. (1994) Residents' attitudes towards tourism: a longitudinal study in Spey Valley, Scotland. *Tourism Management* 15, 247–258.

Haralambopoulos, N. and Pizam, A. (1996) Perceived impacts of tourism: the case of Samos. *Annals of Tourism Research* 16, 503–506.

Institute of Development Studies (IDS) (1996) *The Power of Participation.* IDS policy briefing, Issue 7, August. Sussex.

Joy, A. and Motzney, B. (1992) Ecotourism and ecotourists: preliminary thoughts on the new leisure traveller. In: *Seminar proceedings of the AMA Winter Educator's Conference.* American Marketing Association, Chicago.

King, B., Pizam, A. and Milman, A. (1993) Social impacts of tourism: host perceptions. *Annals of Tourism Research* 20, 650–665.

Kutay, K. (1990) Ecotourism: travel's new wave. *Vis a Vis* July, 34–80.

Lea, J.P. (1988) *Tourism and Development in the Third World.* Routledge, London.

Lindberg, K. (1991) *Policies for Maximising Nature Tourism's Ecological and Economical Benefits.* International Conservation Financing Project Working Paper. World Resources Institute, Washington, DC.

Liu, J. and Var, T. (1986) Resident attitudes towards tourism impacts in Hawaii. *Annals of Tourism Research* 13, 193–214.

MacCannell, D. (1992) *Empty Meeting Grounds: the Tourist Chapters.* Routledge, London.

McNeely, J.A. and Thorsell, J. (1989) *Jungles, Mountains and Islands: How Tourism can Help Conserve the Natural Heritage.* World Leisure and Recreation Association, Geneva.

Maguire, P.A. (1991) Ecotourism development policy in Belize. In: Veal, A.J., Jonson, P. and Cushman, G. (eds) *Leisure and Tourism: Social and Environmental Change, Papers from the World Leisure and Recreation Association Congress, Sydney, Australia, 16–19 July.* Centre for Leisure and Tourism Studies, University of Technology, Sydney, pp. 624–631.

Mason, P. (1990) *Tourism: Environment and Development Perspectives.* World Wide Fund For Nature, Surrey.

Mathieson, A. and Wall, G. (1982) *Tourism: the Economic, Physical and Social Impacts.* Longman, Harlow, UK.

Mukherjee, N. (1993) *Participatory Rural Appraisal: Methodology and Applications.* Concept Publishing, New Delhi, India.

Murphy, P. (1985) *Tourism: a Community Approach.* Methuen, London.

Nash, D. (1996) *Anthropology of Tourism.* Pergamon, Oxford.

O'Neill, M. (1991) Naturally attractive. *Pacific Monthly* September, 25.

Paul, S. (1987) *Community Participation in Development Projects: the World Bank Experience*, World Bank Discussion Paper 6. The World Bank, Washington, DC.

Pearce, D. (1989) *Tourist Development.* Longman, Hong Kong.

Pearce, P. (1994) Tourist-resident impacts: examples, explanations and emerging solutions. In: Theobald, W.F. (ed.) *Global Tourism: the Next Decade.* Butterworth-Heinemann, Oxford, pp. 103–123.

Pearce, P.L. (1995) From culture shock and culture arrogance to culture exchange: ideas towards sustainable socio-cultural tourism. *Journal of Sustainable Tourism* 3, 143–153.

Prasad, P. (1987) The impact of tourism on small developing countries – an introductory view from Fiji and the Pacific. In: Britton, S. and Clarke, W.C. (eds) *Ambiguous Alternative – Tourism in Small Developing Countries.* University Press of the South Pacific, Suva, Fiji, pp. 9–15.

Sharpley, R. (1994) *Tourism, Tourists and Society.* ELM Publications, Huntingdon.

Sofield, T.H.B. (1991) Sustainable ethnic tourism in the South Pacific: some principles. *Journal of Tourism Studies* 2, 56–72.

Travis, A.S. (1982) Managing the environmental and cultural impacts of tourism and leisure development. *Tourism Management* 3, 256–263.

Valentine, P. (1987) Towards an alternative view of tourism. *Magnus Taurus*, 16–18.

Valentine, P. (1991) Nature-based tourism: a review of prospects and problems. In: Miller, M.L. and Auyong, J. (eds) *Proceedings of the 1990 Congress on Coastal and Marine Tourism.* National Coastal Resources Research and Development Institute, Newport, Oregon, USA, pp. 475–485.

van Doorn, J.W.M. (1989) A critical assessment of socio-cultural impact studies of tourism in the third world. In: Singh, T., Theuns, H. and Go, F. (eds) *Towards Appropriate Tourism: the Case of Developing Countries.* Peter Lang, Frankfurt, pp. 71–91.

Wallace, G. (1992) Real ecotourism: assisting protected area managers and getting benefits to local people. Paper presented to the IVth World Congress on National Parks and Protected Areas, February 10–12, Caracas, Venezuela, International Union for Conservation of Nature and Natural Resources.

Wearing, S.L. (1993) Ecotourism: the Santa Elena rainforest project. *The Environmentalist* 13(2), 125–135.

Wearing, S.L. (1998) The nature of change through ecotourism: the place of self, identity and communities as interacting elements of alternative tourism experiences. PhD thesis, School of Environmental and Information Sciences, Charles Sturt University, Albury, Australia.

Wearing, S.L. and Larson, L. (1996) Assessing and managing the socio-cultural impacts of ecotourism: revisiting the Santa Elena rainforest project. *The Environmentalist* 16(2), 117–133.

Wearing, S.L. and Neil, J. (1999) *Ecotourism: Impacts, Potentials and Possibilities.* Butterworth-Heinemann, Oxford.

Weiler, B. and Hall, C. (eds) (1992) *Special Interest Tourism.* Belhaven Press, London.

Weiler, B. and Johnson, T. (1991) Nature based tour operators – are they environmentally friendly or are they faking it? In: Stanton, J. (ed.) *Proceedings of the National Tourism Research Conference: The Benefits and Costs of Tourism, Marie Resort, Nelson Bay, 3–4 October.* University of Newcastle, Newcastle, Australia, pp. 115–126.

West, P.C. and Brechin, S.R. (eds) (1991) *Resident Peoples and National Parks: Social Dilemmas and Strategies in International Conservation.* University of Arizona Press, Tucson.

Western, D. and Wright, M.R. (eds) (1994) *Natural Connections: Perspectives in Community-based Conservation*. Island Press, Washington, DC.

World Bank Group (1998) *World Bank Participation Sourcebook: Participatory Rural Appraisal, Collaborative Decisionmaking – Community-based Method*. World Bank Group, New York.

World Wide Fund For Nature (WWF) (1992) *Beyond the Green Horizon: a Discussion Paper on Principles for Sustainable Tourism*. WWF (UK), Surrey.

Chapter 26

Developing Indicators for Destination Sustainability

E. Sirakaya, T.B. Jamal and H.-S. Choi

Department of Recreation, Parks and Tourism Sciences, Texas A&M University, College Station, Texas, USA

Tourism should be developed in a way so that it benefits the local communities, strengthens the local economy, employs local workforce and wherever ecologically sustainable, uses local materials, local agricultural products and traditional skills. Mechanisms, including policies and legislation should be introduced to ensure the flow of benefits to local communities. Tourism activities should respect the ecological characteristics and capacity of the local environment in which they take place. All efforts should be made to respect traditional lifestyles and cultures.

(The Berlin Declaration, 1997)

Introduction

Rapid change in transportation technology along with rising incomes and leisure time in Western nations has enabled a democratization of travel, the growth of mass tourism and new tourism forms such as ecotourism (Mowforth and Munt, 1998). The globalization of capitalism, finance, labour, technology, transportation and communication in the 20th century has enabled tourism to become the world's largest industry. The World Tourism Organization (WTO) estimates that international tourism arrivals in 2000 were in the vicinity of 637 million, up from 443 million in 1990 (US Department of Commerce, 1998). Yet, until recently, tourism and ecotourism have been mostly evaluated in terms of their potential for economic growth in many countries and communities around the world. Their contribution to regional and national economies through income generation, hard currency earnings, new jobs and infrastructure has been celebrated as the saviour of many local, regional and national economies. Not surprisingly, many destinations have been caught off-guard in dealing with the adverse impacts of tourism on natural, social and cultural resources[1]. Meanwhile, ecotourism has been gaining prominence

[1]Negative impacts associated with tourism and mass tourism include congestion, pollution, inflation, acculturation and loss of cultural identity, high income-leakage, demonstration effect, and degradation of physical and natural environments (Butler, 1990; Gunn, 1990; Farrell and Runyan, 1991; Cater, 1993; Valentine, 1993; Wight, 1993; Allcock and Bruner, 1995; Sirakaya, 1997). For additional social impacts such as conflict between host and guest, commodification of culture, seasonal low-paying jobs, and loss of social cohesion see (De Kadt, 1979, p. 65; Britton, 1982; Mathieson and Wall, 1982; Cater, 1985; Murphy, 1985; Holder, 1988; Greenwood, 1989; Smith, 1989; Butler, 1990; Mowforth and Munt, 1998).

since the 1980s as a more benign form of tourism that is considered to have characteristics of a universal remedy for ameliorating the developmental ills associated with traditional mass tourism.

This chapter discusses the development of indicators for monitoring tourism impacts, an essential requirement of strategic planning and management of ecotourism destinations, given the essential ecotourism criterion of sustainability. To this end, both the *content* of ecotourism indicators and the *process* of developing and using indicators in strategic planning are addressed. A brief overview of environmental considerations and the growth of tourism is provided first below, followed by a discussion of sustainability based indicators, including a brief historical overview of indicators, benefits, definitions and types of indicators. Next, we identify several sets of sustainability indicators for ecotourism based on their policy relevance, analytical soundness and measurability. Following this, we discuss the process of indicator development, since this is a critical aspect of ensuring successful monitoring of impacts, and conclude with a call for developing a set of robust, integrated and holistic indicators which can be adapted by ecotourism destination managers, planners and tour operators to their particular context.

An Indicator-based Approach to Sustainability

Environmental concerns can be traced back through a lineage of well-known writings such as Thoreau's *Walden* (originally published in 1854) and Leopold's *A Sand County Almanac* (1949), as well as forward to the markers of the modern environmental movement, such as Rachel Carson's *Silent Spring* (1962) and the 1970 Earth Day event (Nash, 1967, 1968). The Club of Rome's report *The Limits to Growth* (Meadows *et al.*, 1972), and a series of other well-known writings have also served to highlight the environmental issues facing our planet and the need for action towards sustaining the Earth for future generations. The Brundtland Commission's report *Our Common Future* (WCED, 1987) and a host of related works offer 'sustainable development' advice to commercial practice and governments. Interestingly, the Brundtland Commission's report makes little reference to the role of tourism in the battle for ecological sustainability, despite its ability to create significant positive and negative impacts on global economies, resources and the environment.

In addition to positive economic and socio-cultural impacts, tourism is also a major force of social and cultural change, as well as a major contributor to environmental degradation and habitat fragmentation. Despite the criticisms levelled at 'sustainable development' (see Wheeller, 1993, 1997; Peterson, 1997), we find the concepts and principles of sustainable (tourism) development useful and thus employ them sensitively in this chapter. As defined by the WTO (1995) (in Agenda 21 for the Travel and Tourism Industry, p. 30):

> Sustainable tourism development meets the needs of present tourists and host regions while protecting and enhancing opportunity for the future. It is envisaged as leading to management of all resources in such a way that economic, social, and aesthetic needs can be fulfilled while maintaining cultural integrity, essential ecological processes, biological diversity, and life support systems.

Ecotourism has strong correlates with sustainable tourism, so we draw upon indicators based on sustainable tourism (development) as well as ecotourism in the tables presented below. The Ecotourism Society (Lindberg and Hawkins, 1993) defines ecotourism as 'purposeful travel to natural areas to understand the culture and natural history of the environment, taking care not to alter the integrity of the ecosystem while producing economic opportunities that make the conservation of natural resources beneficial to local people'. Definitional issues, and evidence for an emerging consensus on the meaning of ecotourism and the inclusion of a sustainability component, are pursued further in Chapters 1 and 2 of this volume. Yet, with a few exceptions (e.g. Herremans and Cameron, 1999), tour operators and marketers seem to rush into 'eco-selling' the regions and tourism resources for short-term economic gains without considering adverse long-term impacts, both ecological and socio-cultural (see Wight, 1993; Dann, 1996)[2]. Ecotourism can be the thin edge of a wedge which opens the door to mass tourism, unless policies and measures are put in place early on to manage the potential impacts of introducing an ecological area to tourism (see Bookbinder *et al.*, 1998; Mowforth and Munt, 1998). The degradation of tourism-related attractions in turn threatens the economic viability of the tourism industry in general, and points to the critical need for tracking changes in the socio-economic, cultural, economic and political systems as they relate to tourism. As yet, there are no agreed-upon general measurement and monitoring systems in place to guide tourism decision makers in creating policies and strategies to minimize further degradation and destruction of natural, social and human resources around the globe. To evaluate the past, guide the action of the present, and plan for the future, we need to know what to monitor, what data to collect and what to measure. In other words, to track changes in social, natural, cultural, economic and political arenas of ecotourism destinations, we need several sets of sustainability-centred ecotourism indicators based on their policy relevance, analytical soundness and measurability.

The quest for sustainable ecotourism indicators requires that we have a reference point with respect to the multitude of interpretations of sustainable tourism and ecotourism (Orams, 1995; Diamantis, 1997). Accordingly, for the purposes of this chapter we define ecotourism as the type of primarily nature-based tourism that is ecologically, culturally, politically, as well as economically responsible, and specifically promotes the stewardship of natural and cultural resources. Gunn (1994, p. 87) recommends the following guidelines for a well-designed planning approach to sustainable tourism:

- developing tourism goals and objectives linked to the broader comprehensive plan for a region and/or community;
- formulating a set of indicators reflecting the objectives of tourism development;
- implementing management strategies designed to direct tourism toward the achievement of the stated objectives;
- monitoring the performance of tourism with respect to these indicators;
- evaluating the effectiveness of selected management strategies in influencing the performance of tourism with respect to these indicators;
- developing strategic policies for tourism management based upon the monitored effectiveness of these techniques.

Monitoring and evaluation is the final step in the planning process. This step is an integral part of the planning process for evaluating the performance and impacts of the development or project during the implementation phase, and developing mitigation measures and new indicators as required (Inskeep, 1991). In the late 1960s, indicators were introduced as a major product of the 'Social Indicators

[2]Negative socio-cultural, environmental and economic impacts of ecotourism are now being reported in many ecologically sensitive destinations of the world stretching from the Caribbean to Antarctica. Increased land prices and inflation, high leakage of tourism related revenues, cultural degradation, introduction of exotic species, damage to cultural heritage sites, destruction of coral reefs brought about by tourist use (e.g. in the Caribbean), disturbance of breeding habits of birds in Antarctica by tourist activity and actions, and pollution through waste and sewage disposal in popular eco-destinations are just a few alarming signals of increasing ecological destruction (for specific examples see Erize, 1987; Holder, 1988; Wilkinson, 1989; Cater, 1993; Healy, 1994; Place, 1995; Hall and McArthur, 1998).

Movement' (see Carley, 1981). Since then, the use of indicators has steadily increased and is now more commonly employed by managers and researchers who monitor social and biophysical changes in social and natural settings (Wallace, 1993).

What is an indicator and how is it used? We can answer these questions by using simple examples from our daily lives. In everyday life, we encounter many types of indicators that reflect a person's health such as the colour of the face, body temperature, feeling, and so on. Although we recognize these kinds of symptoms and make inferences based upon them (e.g. 'I have a cold coming on'), our decision falls into the category of the subjective judgement. This implies that it may result in a bad decision, perhaps causing us to use a wrong over-the-counter type of medicine and ending up with no improvement in health condition and concomitant waste of financial resources. Thus, in order to get an objective evaluation, most people consult an expert, a physician. He or she uses a set of indicators based on tests, scientific knowledge and experience with similar symptoms. Hence, indicators in various forms have been with us for a long time, whether subjective or objective, including indicators for social life, the natural environment, economics, climate, political domains, etc. Since tourism is a major social phenomenon of post-modern society, social indicators are discussed in some detail in the historical overview provided below.

Historical Overview of Indicators

In the early 20th century, William Ogburn, a well-known sociologist at the University of Chicago, was experimenting on measurements in social exchange theory. The objective of his study was to create statistical measures that could be used to detect trends and change in the society. In 1929, President Hoover appointed Ogburn as the Director of the Research Committee on Social Trends and his role was to issue an annual report of 'Recent Social Trends in the US' (Rossi and Gilmartin, 1980). This comprehensive statistical report covered virtually every aspect of American life such as childhood, youth, medical care of children, school, recreation, religious education and so on. However, despite Ogburn's tremendous effort and the findings related to the measurement of social change, these issues/social trends did not receive much attention until the mid-1960s, the only exception being the use of macroeconomic data to evaluate national economic conditions (Rossi and Gilmartin, 1980). In 1962, President Kennedy commissioned the Social Advisory Committee to establish a systematic collection of 'basic behavioral data' for the USA (EPA, 1973).

The mid-1960s saw growing dissatisfaction among researchers and government decision makers with the amount of available social information and its quality. This gave rise to the 'Social-Indicators Movement'. According to Carley (1981), this phenomenon can be explained as a reaction against one-dimensional measures of economic performances as indicators of quality of life and social change. Bruce Russett's book, entitled *World Handbook of Political and Social Indicators* (1964, see Miles, 1985) had a strong impact on the field of political science (but was less influential in practice). In 1966, the term 'social indicators' came into common use. For instance, the National Aeronautics and Space Administration financed a collection of essays by Bertram Gross, Albert Biderman, Robert Rosenthal and Robert Weiss, entitled *Social Indicators* (Miles, 1985). This research provoked considerable interest and expanded the original scope of the social impact of the space programme to predict the impact of technology and to evaluate the costs and benefits of government programmes. This volume of papers edited by Bauer (see Miles, 1985) contributed to the creation of some critical guidelines for the development of indicators, established principles for evaluation of specific social programmes, developed a system of social accounts and raised various issues pertaining to social policy development (Miles, 1985). Another contribution to the social indicators movement was a collection of articles in two volumes, *The Annals of the American Academy of Political and Social Science* (1967, see Land, 1975) in which authors such as Etzioni and Lehman criticized uni-dimensional measurements of social phenomena, asserting that using single measures of indicators could be dangerous and result in distortion of the indicator system (Land, 1975). They thus called for the creation of a multi-dimensional system for measuring social change.

The term 'social indicators' also embraced diverse attempts to establish specific indicators of socio-economic well-being that ranged from specific measurement of childcare quality to expanded measurement of the quality of life. In 1972, Wilcox, Brooks, Beal and Klonglan published an annotated bibliography, which listed over a thousand entries of social indicators (Carley, 1981). Adding to this complexity in indicator development is the lack of consensus on the definition of social indicators. Several definitions are provided below to illustrate the multi-faceted properties of this concept. The US Department of Health, Education and Welfare, in *Toward a Social Report*, defined a social indicator as follows:

> A statistic of direct normative interest, which facilitates concise, comprehensive and balanced judgments about the condition of major aspects of a society. It is in all cases a direct measure of welfare and is subject to the interpretation that, if it changes in the 'right' direction, while other things remain equal, things have gotten better or people are 'better off'.
>
> (US HEW, 1969, p. 97)

Generally, 'an indicator is meant to indicate something beyond the property it expresses *prima facie*, otherwise the term forfeits its conceptual relevance. An index, on the other hand, indicates the thing itself' (Mukherjee, 1975). Furthermore, indicators can be aggregated into composite indicators, as described by the Organization for Economic Cooperation and Development (OECD) (1997):

> An indicator is an empirical interpretation of reality and not reality itself. Indicators are commonly used to present a quantitative account of a complex situation or process. They can also be used to point out or identify something, which is not immediately visible, audible or perceived in a precise situation. Indicators usually translate data and statistics and can be aggregated and attributed weighted values in order to produce composite measure known as indices. Finally, three major functions of indicators are simplification, quantification, and communication.
>
> (OECD, 1997, p. 14)

As the use of indicators has grown, its dimensions have been broadened to include broad technical indicators (i.e. indirect/direct, descriptive/analytical and subjective/objective) and discipline-based indicators (e.g. economic indicators, social indicators, tourism indicators or psychological indicators). According to Rossi and Gilmartin (1980), direct indicators refer to a measure of the variable itself. For example, if the interest of concern is the economic health in a certain country, the change in the rate of per capita income or gross national product for a certain period could be the basis for a direct indicator. On the other hand, indirect indicators refer to 'a [proxy] measure of some other concerns that is assumed (based on experience or theory) to be closely related to the variable of interest' (Rossi and Gilmartin, 1980). Using the previous example, war in a neighbouring country would be an indirect variable that would affect the indicator measuring the economic health for the country, if that country was trading with its neighbour.

Subjective indicators reflect the comments that people make about their emotions, attitudes, attributes and personal evaluation, while objective indicators refer to 'counts of behaviors and conditions associated with a given situation' (Rossi and Gilmartin, 1980). For instance, if a researcher were measuring resident attitudes in a community toward potential ecotourism development, this would be a subjective indicator. Subjective indicators are usually developed based on information obtained from interviews or survey research. In contrast, counts of the number of political protests against ecotourism development fall into an objective indicator category, in which can be included measures for behaviour, environmental conditions and physiological attributes, among others. Many social science disciplines such as sociology, economics, ecology, tourism, psychology and political science have created their own indicators and though different understanding of value and ideas abound, some commonalities may be identified. For instance, some major tasks of economic indicators are to measure national economic health and quality of life, predict economic fluctuations, and to control economic cycles. Psychological indicators usually focus on subjective matters and the range of their measurement is focused on individual well-being. Environmental indicators represent specific interests in natural and human environments such as related to biodiversity, habitat fragmentation, ecosystem management, pollution, etc. The major concerns of political indicators revolve around issues of effectiveness, efficiency, performance and party evaluation. Finally, sociological indicators cover a variety of social elements from individual behaviour to institutional organizations (Liu-Chieh, 1976).

In short, indicators have been used for a wide range of purposes to evaluate changing conditions of a society. According to Rossi and Gilmartin (1980), an indicator:

- assists to describe a state of a society such as the quality of life in various social strata (quality of life of poor, quality of life of elderly, etc.);
- is used to conduct analytical studies of social changes including identifying sources of influence on an indicator's trend and suggesting the causal sequence and importance of relevant variables used to estimate future values of selected indicators (i.e. forecasts or projections);
- is used to evaluate the effectiveness of social programmes;
- can help policy makers and managers to set goals and establish priorities;
- can be used to develop social accounting systems.

The growing concern about environmental and cultural sustainability has led to an increased need in tourism studies to develop indicators for monitoring the sustainability of tourism-related resources. This is discussed below, commencing with a brief discussion on sustainable development.

Special Type of Indicators: Sustainability Indicators

Growing worldwide concern over the preservation of environmental and sociocultural resources is a driving force for sustainable development at the global, regional and local levels. Based on the concepts advanced in the Brundtland Report, *Our Common Future*, the Globe 1990 conference on sustainable tourism held in Vancouver, Canada in March 1990 produced the following definition of sustainable tourism development (Wheeler, 1995): 'Sustainable tourism development is envisaged as leading to management of all resources in such a way that we can fulfill economic, social and aesthetic needs while maintaining cultural integrity, essential ecological processes, biological diversity and life support systems.' Among the policy implications forwarded by the conference was the recommendation that sustainable tourism development should be given policy definition and direction for each country, region and locality where it occurs. Managing the actual process of tourism planning and development is, however, a complex task due to various tourism and destination-specific factors. As discussed by Jamal and Getz (1996), a tourism destination has multiple stakeholders, and no individual stakeholder is able to fully control development and planning. Furthermore, all the main stakeholder groups are not necessarily located within the destination. Transportation companies such as airlines are extremely important to a destination's success, but may be headquartered in a large city or tourist-originating countries some distance from the destination. Fragmentation of control is exacerbated by the element of public and social good contained in tourism development and marketing, which enhances public sector involvement in tourism. In addition, changing consumer demographics and preferences, rapid technological advances, globalization of labour and trade, population mobilization and migration as well as increasing pressure on the natural environment all contribute to making tourism a global phenomenon and a major agent of social and cultural change. The domain of destination planning and management is a complex and challenging one due to the interdependencies and spatio-temporal dynamics outlined above.

Increasing awareness of the negative impacts of tourism (Smith, 1977; O'Grady, 1990) and the call for growth management strategies (Gill and Williams, 1994) within the carrying capacity of the natural and socio-cultural environment (Getz, 1983; Gunn, 1988) has led to a much greater focus on developing indicators for monitoring the sustainability of the natural and socio-cultural environment. However, sustainable development has more dimensions than were reported in the World Commission on Environment and Development report. Sustainable development contains ecological, social, economic, institutional, cultural and psychological dimensions at all levels – international, national, regional and community – within various fields such as agriculture, tourism, political sciences, economics and ecology (Bossell, 1999). How can these complex dimensions of sustainability and the global tourism system be incorporated within a sustainability framework for measuring, monitoring and managing the impacts of tourism, both positive and negative? In the rest of this chapter, we concentrate on two aspects of this issue: the content and the process of indicator development.

Nieto (1996) gave a philosophical answer to the above question. He argued that sustainable development should be treated differently from traditional approaches to development because traditional approaches to development emphasize growth and not progress. Growth is a quantitative measure of human development and a source of many intentional or unintentional socio-economic and environmental problems, whereas 'progress' is a qualitative concept indicating an improved state of being. In other words, economic and technological development should not degrade and destroy the very resource upon which the development is based (Gunn, 1994). Moreover, the welfare of future generations becomes the centre of an ethical debate. From an operational perspective, sustainable progress can be measured by a certain set of pre- and post-development indicators, with threshold levels set to provide warnings of when limits are being reached in the availability of various resources at the destination (see the *Local Agenda 21 Planning Guide* for an excellent sustainable development planning framework for communities and destination areas (International Development Research Centre, 1996).

There is a plethora of studies that have dealt with developing various types of indicators (Liverman *et al.*, 1988; Kuik and Verbruggen, 1991; OECD, 1994). These efforts are criticized because of their heavy focus on environmental (ecological) aspects and almost total ignorance of the relationships that exist between communities and ecosystems (Azar *et al.*, 1996) (see Chapters 22 and 25). While not exhaustive in scope, Table 26.1 illustrates the diversity of works on sustainability indicators in regions and areas other than tourism.

Considering the above arguments and the existing literature on sustainability, a new way of thinking, a new paradigm, is needed to alleviate some of the problems associated with traditional approaches to sustainability. First of all, sustainability indicators need to be included in monitoring and managing all forms of tourism development. These indicators should involve environmental, technological, social,

Table 26.1. The focus of some indicator areas for sustainability (Azar *et al.*, 1996).

Reference	Indicated area	Societal activities[a]	Environmental pressure[b]	State of the environment[c]
Adriaanse (1993)	The Netherlands	X	X	
Alfsen and Sabo (1993)	Norway			X
Ayres (1995)	Mainly USA	X	X	
Ten Brink (1991)	Specific ecosystem			X
Brown *et al.* (1994)	The world	X	xx	xx
Carlson (1994)	Sweden	X	X	
ECE (1985)	ECE member nations	xx	X	X
Environment Canada (1991)	Canada		X	X
Gilbert and Feenstra (1994)	Specific ecosystem		xx	X
Holdren (1990)	The world		X	xx
Holmberg and Karlsson (1992)	Not specific	X	X	
Miljominsteriet (1991)	Denmark	xx	X	X
Nilsson and Bergstrom (1995)	Municipality and company	X	X	X
OECD (1994)	OECD countries	xx	X	xx
Opschoor and Reijnders (1991)	Not specific		X	X
SNV (1994)	Sweden			X
Haes *et al.* (1991)	Specific ecosystem			X
Vos *et al.* (1985)	The Netherlands		X	X
Azar *et al.* (1996)	The world	X	X	

[a] Social activity indicator measures activities occurring within a society (i.e. the use of extracted minerals, the production of toxic chemicals and recycling of material).
[b] Environmental pressure indicator measures human activities that will have a direct impact on the state of environment (i.e. emission rates of toxic substances).
[c] Environmental state or quality indicators indicate the state of environment (i.e. the concentration of heavy metals in soils and pH levels in lakes) (adopted from Azar *et al.*, 1996).
Note: The symbol 'X' indicates the key focus area and 'xx' indicates a minor role in the area.

economic, political and psychological aspects and must be planned and implemented at all levels: local, regional, national and international. Second, indicators of sustainability need to be used in such a way as to provide an understanding not only of individual impacts, but also of the *cumulative* effects of various impacts. This requires an *integrated* approach to examining, monitoring and managing the impacts of tourism (development), and hence an integrated approach to developing sustainability indicators.

Indicators of sustainability for tourism differ from traditional development indicators because they take into consideration the web of complex interrelationships and interdependencies of resources and stakeholders in the tourism system. For example, variables such as gross domestic product (GDP), unemployment, number of companies and number of jobs created are traditionally used as economic indicators to measure wealth of the country or community. However, a single indicator like GDP does not have the ability to capture the vital aspects of *sustainable progress.* It may capture all flow of goods/services and some social costs produced in a domestic economy. However, it cannot measure how much and how fast natural resources are being consumed to produce that GDP, since it considers natural resources as free goods whose costs are externalized rather than internalized. Some of the positive externalities (or spillover effects) produced by society, like vaccinations and/or environmental beautification (see Daly and Cobb, 1989, pp. 52–54; National Grid for Learning, 2000) are used by many of us without any compensation but the society and individuals are not worse off because of the availability of such services. However, a chemical plant discharging its pollutants into a river where recreational fishing occurs, can produce disbenefits that will leave the recreational fisherman downstream worse off. Moreover, society is left having to deal with the clean-up and other costs, unless the plant is held directly responsible for the full costs of its actions. Unfortunately, current accounting systems do not take these interrelated aspects of resource use/abuse into consideration. Thus, indicators like GDP as currently defined cannot be objective measures of national wealth and well-being within a sustainability-based framework, although they might help us to see certain problem areas (see Daly and Cobb, 1989, for full-cost accounting discussions). In tourism, plenty of examples can be thought about in a similar manner, where the sustainability of tourism-related resources requires full-cost accounting, and where all costs are internalized rather than externalized. For example, environmental enhancement through community beautification or restoration programmes are usually not paid for by many users of the resultant benefits, even though society overall may be better off because of these programmes. However, if a tourism provider discharges untreated sewage into the natural environment this will cause disbenefits to the community unless the costs are fully accounted for and assumed by the stakeholder(s) involved.

A number of criteria for developing sustainability indicators can be identified from the vast amount of research and writing in this area (Liverman *et al.*, 1988; Kuik and Verbruggen, 1991; Jamieson, 1997; Hart, 1998; Bossell, 1999). Some key guidelines are listed here.

- Indicators must be created to guide policies and decision at all levels of society and cover the entire spectrum of socio-economic, cultural, natural and political environments at the local, regional, national and international levels.
- The number of indicators must be manageable and be implemented with ease at the destination and community level in a timely fashion.
- Community participation must be maximized in order to reflect the visions and values of a community-based destination, and a long-term welfare-view of the destination is required in order to facilitate long-term sustainability.
- Indicators must have a high degree of reliability, predictive capacity and integrative ability.
- The process of developing indicators cannot be haphazard in that it requires a systematic approach to developing indicators that are robust, measurable, affordable and able to provide an integrated view of specific and overall conditions pertaining to the sustainability of the destination and its natural and cultural resources.

Table 26.2 shows that sustainability indicators are different from traditional indicators. The above discussion suggests that more detailed indicators are needed that take into consideration interdependencies in the system, intangible and cumulative effects, and full-cost accounting of impacts. For example, diversity and vitality of the local job base, number and variability in size of employers, and the diversity of skill levels required for jobs are important to monitor since economic sustainability entails addressing factors such as resilience of the job market, and the ability of the job market to be flexible in times of economic change.

As stated above and in previous chapters, ecotourism is meant to be non-consumptive of nature, as well as an educational, low impact and responsible form of tourism (a new tourism form as Mowforth and Munt, 1998, discussed). It aims to bring travellers to relatively undisturbed natural areas that may hold ecological, cultural and historical significance, for the purpose of understanding and appreciating the natural and socio-cultural history and setting of the host destination. It is a form of tourism that is expected to result in: (i) minimal negative impacts on the host environment; (ii) increased contribution to

Table 26.2. Traditional vs. sustainable indicators (Hart, 1998).

Economic indicators			Environmental indicators			Social indicators		
Traditional indicators	Sustainable indicators	Emphasis of sustainable indicators	Traditional indicators	Sustainable indicators	Emphasis of sustainable indicators	Traditional indicators	Sustainable indicators	Emphasis of sustainable indicators
Median income Per capita income relative to USA	Number of hours of paid employment at the average wage required to support basic needs	What wage can buy Need to define basic needs in terms of sustainable consumption	Ambient levels of pollution in air and water, generally measured in parts per million of specific pollutants	Biodiversity Number of individuals of key species, such as salmon in a stream or birds in a given area	Ability of the ecosystem to process and assimilate pollutants	Number of registered voters	Number of voters who vote in elections Number of voters who attend town meetings	Participation in democratic process Ability to participate in democratic process
Unemployment rate Number of companies Number of jobs	Diversity and vitality of the local job base Number of variability in size of employers Number of variability of industry type Variability of skill levels required for job	Resilience of the job market Ability of the job market to be flexible in times of economic change	Tons of solid waste produced	Amount of material recycled per person as a ratio of total solid waste generated	Cyclical use of resource			
Size of the economy by GNP and GDP	Wages paid in the local economy that are spent in the local economy Dollar	Greatest possible local financial independence	Per capita energy use	Ratio of renewable energy used to non-renewable energy used Total amount of energy used from all sources	Use of renewable energy Conservation			

environmental protection and conservation of resources; (iii) creation of necessary funds to promote sustained protection of ecological and socio-cultural resources; (iv) enhancement of interaction and understanding between visitors and locals; and (v) contribution to the economic (monetary profits and job opportunities) and social well-being of the local people. Thus, ecotourism is fundamentally based on sustainable use, management and conservation/preservation of natural and cultural resources for present and future generations. Ecotourism incorporates the coexistence of the natural environment and people (tourists and local inhabitants), and encourages the active involvement of tourists and the local population in preservation efforts (Sirakaya *et al.*, 1999). The current literature in this area suggests that a fair amount of progress has been made in several areas of ecotourism research, but not in the area of monitoring. However, in order to develop a set of general ('universal') indicators of ecotourism for destinations to apply (with modifications) to their own context, clear goals, objectives and principles must first be delineated (Boo, 1992).

The goal of ecotourism refers neither to saving disappearing ecosystems nor revitalizing every community from poverty (Whelan, 1991). The goal of ecotourism should be to improve the quality of life for both host and guest, provide quality experience for the visitors, and protect the natural and human environment including cultural, social and political dimensions (McIntyre, 1993). This overall goal (which is in keeping with sustainability principles) is reflected in several guidelines for ecotourism development.

1. Protection of natural and cultural resources.
2. Generation of economic benefits to local communities.
3. Environmental education (including socio-cultural and ecological dimensions).
4. Provision of a high quality tourism experience.
5. Including local community participation and stakeholder collaboration in destination planning and tourism management.

The ecotourism literature cites a number of similar or related guidelines (see Boo, 1992; Nelson, 1993; Scace, 1993; Wight, 1993; Wallace and Pierce, 1996), from which we have attempted to synthesize key ecotourism principles.

- Ethical responsibilities and codes of conduct need to be undertaken by ecotourism stakeholders, including governments, tourists, ecotourism destinations, tour operators and other tourism businesses (Fennell, 1999; Herremans and Cameron, 1999; Jamal and Everett, unpublished) (see also Chapter 41).
- Participation in ecotourism planning and development requires multi-stakeholder involvement at all levels of planning and policy making, bringing together governments, non-government organizations, residents, industry and professionals in a partnership process.
- Ecotourism and destination managers/planners need to provide educational information to residents, visitors, tour operators and other stakeholders about the planning and conservation of ecotourism resources. This requires offering some reflexive discussion that raises stakeholders' awareness of their preservation and/or conservation values.
- Tour operators and other ecotourism ventures involving the interpretation of 'nature' in an ecotourism destination should provide quality information and dialogue on the value of the area's natural and cultural environment. This requires sharing the various interpretations and stories of the natural and cultural environment that may be present, rather than controlling what interpretations are to be provided to the visitor as 'authentic' representations of a natural experience (see Chapter 35).
- Ecotourism should provide a sustained, long-term economic linkage between destination communities and industries (in order to be beneficial to communities), and also minimize negative

impacts to the natural environment and to the social and cultural well-being of communities affected by ecotourism (see Chapters 23–25).

In the recent past, there have been some sporadic efforts to develop indicator sets concerning ecotourism (e.g. Marsh, 1993; Nelson, 1993; Payne, 1993). Table 26.3 presents a sample of some commonly cited indicators, based upon Nelson's eight indicators emphasizing economic perspectives (1993) and Payne's (1993) indicators used to measure impacts in parks, wilderness and natural resources from a tourism management perspective. Marsh (1993) forwards a useful sustainability index to assess ecological, economic, institutional and social sustainability. His framework is shown in Table 26.4 in order to point out the importance of taking institutional factors into consideration in the overall sustainability framework. Institutional indicators could include indicators for: (i) degree and quality of non-government organization involvement; (ii) percentage of ecotourism profits (e.g. from tour operators) funding non-profit organizations and other sustainable projects in the ecotourism destination; and (iii) green labelling and classification of ethically responsible ecotourism operators, guides, destinations, based on criteria such as application of energy efficient building materials in construction, recycling and reducing waste and pollution, designing lodges and tours that minimize pressure on natural habitats, etc.

The WTO has developed a set of indicators to assist local managers and decision makers to work effectively with ecotourism. Manning (1999, p. 180) said that the objectives of the WTO were to:

> identify a small core of indicators set which is likely useful in almost any situation; to supplement these with additional indicators known to be useful in particular ecosystem or types of destinations; and to additionally require a scanning process for risks not covered by the aforementioned indicator sets, which produces further indicators critical to the management of the particular site/destination.

As Table 26.5 reveals, the WTO's indicator set is an excellent start to further research and development of sustainability indicators, i.e. indicators for monitoring the sustainability of natural and socio-cultural resources in ecotourism development. This broad set of sustainability indicators may need to be tailored to a particular destination's ecological and cultural context, in order to produce effective monitoring methods (Manning, 1999).

Another useful framework to adapt might be the pressure–state–response framework developed by the World Resources Institute, for environmental issues. Pressure indicators are indicators of pressure or stress from human activities that cause environmental change. State indicators are indicators of changes or trends in the physical or biological state of the natural world. Response indicators reflect the policy and other measures adopted in response to environmental problems (Hammond *et al.*, 1995). Table 26.6 (from Jamieson, 1997) demonstrates the use of this framework.

The Process of Indicator Development and Use

Integral to the development and use of indicators for monitoring ecotourism impacts is a process for stakeholder participation in the development and application of sustainability related indicators. Like that of tourism, the context of ecotourism development is a highly political one, played out on a global stage. It involves a number of stakeholders (e.g. national park managers, ecotourism operators and marketers, destination area residents, tourists, etc.), many of whom are interdependent and not all of whom may see eye to eye on the value or use of a natural site (see Jamal, 2000). Furthermore, since these stakeholders are generally not all located within the destination, the player who has the most legitimate say in decision making on issues pertaining to value, heritage and development becomes a contested issue.

Table 26.3. Indicators of tourism management (modified from Nelson, 1993; Payne, 1993).

Themes	Key issues	Indicator
The tourism industry	Providers of opportunity Policy coordination	• Changes in number and characteristics of employees, revenue, capital expenditures, operating budgets • Change in male/female ratio for available jobs in tourism • Change in resident to non-residents jobs in tourism • Change in formal to informal types of employment in tourism • The extent of conflict, competition and cooperation at all levels • Change in ownership of tourism operation • Change in ownership among groups including government, private non-resident groups, franchises, residents • Change in ownership of land • Change in ratio of non-resident to resident ownership of land: change in ratio of private land ownership to community land ownership • Degree of control maintained by community where tourism is in conflict with or threatens
Tourism opportunities	Opportunities/use of opportunities	• The percentage changes over 5-year period in the components of tourism opportunities. Opportunities can be classified by activity, setting and experiences • Local access to training and other forms of support for tourism • Percentage change in use by activity, setting and experience on all levels and by public, private and non-private providers
Sustainability	Ecological sustainability	• Changes in biodiversity, naturalness at all levels
	Cultural sustainability	• Degree to which initiative builds on cultural heritage of community and is culturally appropriate, not conflicting with community vision or plan
	Social sustainability	• Changes in social relations and organization – qualitative changes in attributes of social sustainability including variables of the nature subsistence activities, family and/or kinship structures and decision making structures for allocating resources
	Economic sustainability	• Changes in the structure of employment opportunities • Income distribution at national, provincial and regional levels • Regional and community balance of trade impacts of tourism initiatives, strategies or plans • Backward and forward linkage between tourism activity and other formal and informal activity in the community

Table 26.4. Sustainability index (Marsh, 1993).

Issues	Indicator
Ecological	Species demographics
	Water quantity, quality and use
	Air quality
	Recycling practices
	Efficiency of resource use
	Scenery degradation
	Others
Economic	
Community	Income from tourism, and who receives it
	Costs of tourism, and who pays them
	Investment in tourism by community
	Others
Tourism industry	Profits and losses
	Business initiation
	Business bankruptcy
	Others
Social	
Community	Jobs, quantity and quality
	Migration in and out of community
	Complaints about tourism
	Others
Tourists	Number of visitors, and trends
	Proportion of repeat visitors
	Length of stay
	Tourist satisfaction and complaints
Institutional	Laws and regulations regarding tourism
	Infractions and court cases
	Recognition of tourism in official plans
	Existence of tourism plans
	Tourist and interpretive information
	Government and private tourism organizations
	NGO response to tourism
	Existence of codes of ethics for tourists and industry

Eagles (1999) points out that natural resource decision making has to take into account the values and knowledge of the public, and that natural heritage destinations such as national parks need a decision-making system that allows stakeholders (especially affected publics) to influence the direction of ecotourism development in their destination domain.

An effective monitoring programme will therefore require an organizational structure and process for ensuring that impacts are monitored, evaluated and acted upon. What set of indicators should a community use to ensure that it keeps on top of developments and changes that affect the residents' quality of life and well-being? Some impacts can take a long time to emerge, and realization may come too late to save the resource by the time severe degradation becomes visible. Choosing what should be monitored in tourism planning and development needs to be guided by the community's tourism vision, goals, objectives and action plans, as well as the principles of sustainable (eco)tourism development (see Inskeep, 1991; Jamal and Getz, 1996; Ross and Wall, 1999). Once it is agreed that monitoring is needed, monitoring strategies need to be developed and implemented.

Table 26.7 offers one framework for designing a process for indicator development and use in the monitoring process. Essential to the success of a monitoring process is the establishment of a monitoring body that is ongoing, and the development of a centralized database in which indicator data can be stored and updated so that the destination data bank contains historical and up-to-date information which stakeholders can draw upon to make development, planning and management decisions.

As discussed in Jamieson (1997), constant monitoring and adapting of the strategic plan is crucial to ensure that community goals are met, while ensuring the sustainable development of vital resources. This is facilitated by establishing an organizational body (structure and process) for developing and implementing the monitoring strategy, such that it works effectively through changing leaderships in local elections, i.e. its survival is not dependent on who gets elected into local government. An ongoing threshold and monitoring body to monitor the tourism-related impacts on the community (both positive and negative) is one possible structure for such a body, as has been implemented in Canmore, Alberta (see Jamal and Getz, 1999). Jamieson's (1997) sustainable tourism workbook offers several considerations with respect to the indicator and monitoring body:

- Selection of threshold and monitoring body/committee participants: how much council involvement to include? How much direct community involvement in the structure and process?
- Involvement of scientific and planning experts: Who? When? How?
- Funding for the threshold and monitoring committee and for the monitoring activities: consider sources such as council or the regional government; obtain portion of accommodation, hotel or other tourism taxes or expenditures; raise funds through local/regional campaigns, etc.
- Duties and remuneration of threshold and monitoring committee: volunteer or paid? Permanent or temporary members? Length of term of duty?

The importance of an *ongoing* and community-based organizational structure for developing and implementing the indicator and monitoring process cannot be emphasized enough. Not only does this enable a strategic ecotourism plan to remain dynamic and responsive to short-term issues, it also enables the destination community to direct development through local and regional political and economic shifts over the long term. The ecological soundness of the planning decision and the quality of indicator development is dependent upon the stakeholders involved having good knowledge and information about the issues being addressed.

Conclusions

The use of indicators to monitor impacts can assist destination managers and planners to achieve ecotourism goals, and can signal social trends and changes in resources related to ecological health and quality of life in the host destinations. In addition, a set of effective indicators enables a more holistic approach to the planning and management of tourism destinations (Manning, 1999). Indicators are not a panacea for poor development and planning; their effectiveness is dependent on the quality of the indicators themselves and the effectiveness of their use. Good indicators provide destination decision makers with information that enables them to identify, evaluate and make timely decisions on critical changes being caused by tourism to the natural environment, communities and other resources in the destination (see Buckley, 1999). Indicators have to be selected so that they are robust, credible, efficient (in time and cost for obtaining the data), and useful to decision makers. It may be best to select a key number of indicators that provides the most relevant information, since a large number of impacts and indicators could potentially be

Table 26.5. Sustainable tourism indicator at various types of destinations (urban, cultural site, coastal area, wildlife parks, ecological sites, and small islands) (Manning, 1996).

Issues	Indicators	Suggested measures	Issues	Indicators	Suggested measures
Ecological destruction beach degradation fish stocks depletion	Amount degraded Reduction in catch Extent of erosion caused by tourists Rate of continuing erosion	• % in degraded condition • % of beach eroded • Effort to catch fish • Fish counts for key species • % of surface in eroded state • Visual inspection and photographic record	Human encroachment	Human population in park and surrounding area Activities of people in park and surrounding area	• No. of people within 10 km of boundary • % of park area affected by unauthorized human activity (squatting, wood cutting) • % of surrounding land being used for human purposes such as agriculture (10 km radius)
Overcrowding	Use intensity Length of vehicle line-ups	• Persons per m of accessible beach • No. of visitors (within 10 km boundary) • Ratio of visitors/game animals • Traffic congestion • Length of wait • % of park area affected by unauthorized human activity (wood cutting) • No. of hours spent in vehicle • Cost of entry/lowest average local wage	Ecosystem degradation	Number and mix of species Contented presence of key species in traditionally occupied areas Reproductive success of key species Site degradation Changes in flora Mix and concentration	• Key species count • Count of members of key species/ changes in mix of species • No. of tourist sightings of key species • Areas of species occupation (flora and fauna)/number of road kills of specified species • Primary flora species as a % of total plant cover • No. of outfitters/guides using site • No. of boats using site • % of area negatively affected • Visual inspection and photographic record (local wildlife/biodiversity management offices may provide long-term records for some species)
Violation of social and cultural norms	Languages spoken by locals	• % of community speaking a non-local language	Lack of solitude	Consumer satisfaction	• No. of people at peak period (accessible area only) • Questionnaire on whether solitude objectives met

Lack of safety	Crime levels Types of crime committed Accident levels/ traffic safety Human/animal interaction	• No. and types of crimes reported • Water-related accidents as a % of tourist population • Traffic injuries as a % of population • No. of human–animal contacts reported involving human injury (may be a measure of either more contacts or a change in the level of reporting	Loss of aesthetic qualities/site degradation	Site attraction Restoration costs Levels of pollutants affecting site Measures of behaviour disruptive to site	• Visibility of human presence (e.g. litter counts) • Counts of levels of waste on site • Estimated costs to maintain/restore site per annum • Acidity of precipitation • Traffic vibration (ambient level) • No. of incidents of vandalism reported
Poor species health	Reproductive rate of key species Species diversity Change in mix of animal species	• Monitoring of numbers for animal • groups • Species counts • Key species population counts	Lack of jobs for local population High levels of foreign ownership Currency leakage	Local jobs created through tourism Value of foreign ownership Measures of capital flight	• % of jobs supported by tourism • % of seasonal jobs • % of foreign ownership of tourist establishments • % of exchange leakage from total tourism revenues
Displacement of members of local population	Social impacts	• Average net income of tourists/ average net income of local population • No. of retail establishments/no. of establishments serving local needs (as opposed to tourists)	Health threats	Air pollution measurement	• Air pollution indices (e.g. sulphur dioxide, nitrogen oxide, particulate) • No. of days exceeding specified pollutant standards • Availability of clean water (e.g. can tap water be consumed on site)
	Local satisfaction	• % of local establishments open year-round • No. and type of complaints by locals		Drinking-water quality/ fresh water availability	• Statistics on disease prevalence • Volume of water used by tourists/ volume used by local population on per capita basis
Electricity shortage	Electricity availability	• No. of brown outs • Restriction on use • Changes in cost for electricity use			• Cost to supply water • Cost to supply water/no. of tourists • Estimates of capacity (e.g. volume remaining in reservoir/aquifer)
Sewage disposal	Sewage treatment facilities	• Volume of sewage treated/total volume of sewage • Level of treatment		Type and extent of communicable diseases Noise level	• Records on decibel count of key locations

Table 26.6. A pressure–state–response framework for monitoring tourism impacts (Jamieson, 1997).

Issues	Pressure	State	Response
Visitor impact, natural site degradation	Overcrowding at natural heritage site	Vegetation destruction Physical infrastructure degradation Habitat loss	Zoning to protect fragile areas Limiting access Expenditures on managing conservation
Visitor impact, culture loss and tension	High demand on visiting local religious and cultural events and ceremonies	Local residents do not attend their own festivals Exhibition of hostility by residents towards visitors for intruding Loss of authenticity of cultural/religious events	Allow visitor access to a select number of events Involve residents in determining how much visitation, which events, and how residents could benefit further Develop code of ethics for visitor behaviour

measured. Some broad-based indicators (e.g. habitat fragmentation as an indicator of ecological health) will require specific measures (e.g. species loss as a measure of habitat fragmentation, with certain species selected for monitoring loss criterion). Composite indices (e.g. carrying capacity) are made up of a number of key factors and variables, which at present seem to be site-specific, though a more systematic derivation may be possible in the future (see Manning, 1996). The criteria and guidelines for ecotourism outlined in this chapter should be reflected in the sustainability indicators developed for ecotourism.

As demonstrated in this chapter, there are a variety of indicator models currently in use. The search for sustainability indicators is in its infancy, both with respect to the process and with respect to content of indicator development. Developing good indicators for monitoring and measuring tourism impacts, as well as for monitoring and ensuring the sustainability of ecotourism related resources, remains a critical task in the battle for global sustainability. Debates continue on what constitutes sustainability, what is meant by sustainable development, what should we sustain and why, and how nature is being instrumentally appropriated and controlled through 'ecological modernization', globalization of travel and tourism, and commodification (Eder, 1996). Creating mutually agreed-upon indicators in general, and sustainability-based ecotourism indicators in particular, requires a holistic and integrated planning approach that encompasses all levels of a society from the local to the global. It is crucial that key stakeholders, including local community residents (who stand to be affected by ecotourism in their area) are involved in the decision making process related to developing and applying sustainability indicators. Therefore we have emphasized both the content and the process of indicator development and use in this chapter. Further research using a Delphi technique with an interdisciplinary group of scholars and experts in the field may help to develop a broad 'universal' set of indicators that can be modified by managers and planners to suit their specific ecotourism destination[3].

[3] A Delphi panel of four to five experts in each field including economic, social, cultural, ecological, political, and tourism is suggested. We are in the process of conducting such a study.

Table 26.7. Monitoring strategies and actions (adapted from Jamieson, 1997).

How to monitor, evaluate and respond?	
A Set up a thresholds and monitoring body	A **Assign organizational responsibility** for formulating thresholds and indicators, and for monitoring these. Example: set up a threshold and monitoring committee.
B Set up control systems and mitigation measures, as well as longitudinal database for storing data	B **Develop carrying capacity thresholds and other indicators and measures**. Two distinct sets of indicators are required: performance indicators to measure the results of strategic planning and implementation, and indicators for measuring carrying capacity, use and impacts on the resources used in tourism. These two sets of indicators combined should provide a comprehensive picture of the destination's efforts towards sustainable tourism (for its tourism industry, visitors, natural environment and community) **Develop baseline information** on the indicators developed, as a benchmark against which future results can be measured. Develop alternative scenarios for managing or mitigating impacts and changes in use and tolerance levels, e.g. if use is expected to exceed a threshold in Year X, then alternative actions could be Y Store results in a centralized database system. Longitudinal data (gathered over an extended time frame) is essential for evaluating certain impacts. The lack of noticeable impacts in one year in an area does not necessarily mean that there is no change to the resource; some impacts take a longer time to become evident. Set up a central database system to incorporate base line data and ongoing monitoring results. The community-level database system would contain macro-level data (community and environmental), as well as micro-level data where available from individual organizations
C Conduct integrated evaluation, recognize interdependence of resources and actions	C **Implement indicator based monitoring** with help of the threshold and monitoring body set up in Stage A. Use existing standards, thresholds established and indicator data related to the activity or resource being monitored to conduct an *integrated* evaluation of the impacts of (eco)tourism and make adjustments as required. Also ensure that new standards are established based on evaluation of monitoring data
D Implement management response, with help of previously identified mitigation measures	D **Implement visitor, site and other management actions** (based on evaluation of indicator and threshold information), to ensure proactive management of the carrying capacity of community and environmental resources. These actions should be related to specific objectives set in these areas and to the overall mandate of sustainable tourism. Adjust monitoring strategies and actions as required

References

Allcock, J.B. and Bruner, E.M. (1995) *International Tourism: Identity and Change*. Sage Publications, Thousand Oaks, California.

Azar, C., Holmberg, J. and Lindgren, K. (1996) Socio-ecological indicators for sustainability. *Ecological Economics* 18, 99–112.

The Berlin Declaration (1997) The Berlin Declaration on Biological Diversity and Sustainable Tourism. International Conference of Environment Ministers on Biodiversity and Tourism 6–8 March 1997, Berlin.

Boo, E. (1992) *Wild Lands and Human Needs*. WHN Technical Paper. WWF and USAID, Washington, DC.

Bookbinder, M.P., Dinerstein, E., Arun Rijal, Cauley, H. and Arup Rajouria (1998) Ecotourism's support of biodiversity conservation. *Conservation Biology* 12(6), 1399–1404.

Bossell, H. (1999) *Indicator for Sustainable Development: Theory, Method, and Application.* IISD, Manitoba, Canada.

Britton, S.G. (1982) The political economy of tourism in the Third World. *Annals of Tourism Research* 9(3), 331–358.

Buckley, R. (1999) Tools and indicators for managing tourism in parks. *Annals of Tourism Research* 26(1), 207–210.

Butler, R.W. (1990) Alternative tourism: pious hope or Trojan horse? *Journal of Travel Research* 28(3), 91–96.

Carley, M. (1981) *Social Measurement and Social Indicators: Issues of Policy and Theory.* George Allen & Unwin, London.

Cater, E. (1985) Tourism in the least developed countries. *Annals of Tourism Research* 14(2), 202–226.

Cater, E. (1993) Ecotourism in the third world: problems for sustainable tourism development. *Tourism Management* 14(2), 85–89.

Daly, H.E. and Cobb, J. (1989) *For the Common Good: Redirecting Economy toward Community, the Environment, and a Sustainable Future.* Beacon Press, Boston.

Dann, G.M.S. (1996) *The Language of Tourism: a Sociolinguistic Perspective.* CAB International, Wallingford, UK.

De Kadt, E. (ed.) (1979) *Tourism: Passport to Development?* Oxford University Press, New York.

Diamantis, D. (1997) *The Development of Ecotourism and the Necessity of using Environmental Auditing in its Planning Agenda.* General Technical Report No. NE-232, Northeastern Forest Experiment Station, USDA Forest Service, Radnor, USA, pp. 19–23.

Eagles, P.F.J. (1999) *International Trends in Park Tourism and Ecotourism*, 4th edn. http://www.ahs.uwaterloo.ca/rec/inttrends.pdf

Eder, K. (1996) *The Social Construction of Nature: a Sociology of Ecological Enlightenment.* Sage, London.

Environmental Protection Agency (EPA) (1973) *The Quality of Life Concept.* US Government Printing Office, Washington, DC, pp. 1–10.

Erize, F.J. (1987) The impact of tourism on the Antarctic environment. *Environment International* 13(1), 133–136.

Farrell, B.H. and Runyan, D. (1991) Ecology and tourism. *Annals of Tourism Research* 18(1), 26–40.

Fennell, D. (1999) *Ecotourism: an Introduction.* Routledge, New York.

Getz, D. (1983) Capacity to absorb tourism: concepts and implications for strategic planning. *Annals of Tourism Research* 10(2), 239–263.

Gill, A. and Williams, P. (1994) Managing growth in mountain tourism communities. *Tourism Management* 15(3), 212–220.

Greenwood, D.J. (1989) Culture by the pound: an anthropological perspective on tourism as cultural commodification. In: Smith, V.L. (ed.) *Hosts and Guests: the Anthropology of Tourism.* University of Pennsylvania Press, Philadelphia, pp. 171–186.

Gunn, C.A. (1988) *Vacationscape: Designing Tourist Regions.* Van Nostrand Reinhold, New York.

Gunn, C.A. (1990) The new recreation – tourism alliance. *Journal of Park and Recreation Administration* 8(1), 1–8.

Gunn, C.A. (1994) *Tourism Planning: Basics, Concepts, Cases.* Taylor & Francis, Washington, DC.

Hall, C.M. and McArthur, S. (1998) *Integrated Heritage Management.* Stationery Office, London.

Hammond, A., Adriaanse, A., Rodenburg, E., Bryant, D. and Woodward, R. (1995) *Environmental Indicators: A Systematic Approach to Measuring and Reporting on Environmental Policy Performance in the Context of Sustainable Development.* World Resources Institute, Washington, DC.

Hart, M. (1998) *Indicators of Sustainability.* http://www.subjectmatters.com/indicators

Healy, R.G. (1994) Tourist merchandise as a means of generating local benefits from ecotourism. *Journal of Sustainable Tourism* 2(3), 137–151.

Herremans, I.M. and Cameron, W. (1999) Developing and implementing a company's ecotourism mission statement. *Journal of Sustainable Tourism* 7(1), 48–76.

Holder, J.S. (1988) Pattern and impact of tourism on the environment of the Caribbean. *Tourism Management* 9(2), 119–127.

Inskeep, E. (1991) *Tourism Planning: an Integrated and Sustainable Development Approach.* Van Nostrand Reinhold, New York.

International Development Research Centre (1996) *The Local Agenda 21 Planning Guide: an Introduction to Sustainable Development Planning.* International Council for Local Environmental Initiatives (ICLEI), Toronto, Canada.

Jamal, T. (2000) The social responsibilities of environmental groups in contested destinations. *Tourism Recreation Research* 24(2), 7–18.

Jamal, T. and Getz, D. (1996) Does strategic planning pay? Lessons for destinations from corporate planning experience. *Progress in Tourism and Hospitality Research* 2(1), 59–78.

Jamal, T. and Getz, D. (1999) Community roundtables for tourism-related conflicts: the dialectics of consensus and process structures. *Journal of Sustainable Tourism* 7(34), 356–378.

Jamieson, W. (ed.) (1997) Sustainable Tourism Workbook prepared by the Centre for Environmental Design Research and Outreach, The Faculty of Environmental Design, The University of Calgary.

Kuik, O.A. and Verbruggen, H. (eds) (1991) *In Search of Indicators of Sustainable Development.* Kluwer Academic Publishers, Dordrecht.

Land, K.C. (1975) Social indicator models: an overview. In: Land, K.C. and Spilerman, S. (eds) *Social Indicator.* Russell Sage Foundation, New York.

Lindberg, K. and Hawkins, D. (eds) (1993) *Ecotourism: a Guide for Planners and Managers.* The Ecotourism Society, North Bennington, Vermont.

Liu, Ben-Chieh (1976) *Quality of Life Indicators in US Metropolitan Areas.* Praeger Publishers, New York.

Liverman, D.M., Hanson, M.E., Brown, B.J. and Merideth, R.W. Jr (1988) Global sustainability: towards measurement. *Environmental Management* 12, 133–143.

McIntyre, G. (1993) *Sustainable Tourism Development: Guide for Local Planners.* WTO, Madrid.

Marsh, J. (1993) An index of tourism sustainability. In: Nelson, J., Butler, R. and Wall, G. (eds) *Tourism and Sustainable Development: Monitoring, Planning, Managing.* Department of Geography, University of Waterloo, Canada, pp. 257–258.

Manning, T. (1996) Tourism: where are the limits. *Ecodecision* Spring, 36.

Manning, T. (1999) Indicators of tourism sustainability. *Tourism Management* 20(1), 179–181.

Mathieson, A. and Wall, G. (1982) *Tourism: Economic, Physical and Social Impacts.* Longman, Harlow, UK.

Meadows, D.H., Meadows, D.L., Randers, J. and Nad Behrens, W.W. (1972) *Limits to Growth.* Universal Books, New York.

Miles, I. (1985) *Social Indicators for Human Development.* St Martin's Press, New York.

Mowforth, M. and Munt, I. (1998) *Tourism and Sustainability: New Tourism in the Third World.* Routledge, London.

Mukherjee, R. (1975) *Social Indicators.* Prabhat Press, Meerut.

Murphy, P.E. (1985) Community attitudes to tourism: a comparative analysis. *International Journal of Tourism Management* 2(3), 189–195.

Nash, R. (1967) *Wilderness and the American Mind.* Yale University Press, New Haven, Connecticut.

Nash, R. (ed.) (1968) *The American Environment: Reading in the History of Conservation.* Addison-Wesley Publishing, Reading, Massachusetts.

Nelson, J.G. (1993) Are tourism growth and sustainability objectives compatible? Civil, assessment, informed choice. In: Nelson, J.G., Butler, R. and Wall, G. (eds) *Tourism and Sustainable Development: Monitoring, Planning, Managing.* Department of Geography, University of Waterloo, Canada, pp. 259–268.

Nieto, C.C. (1996) Toward a holistic approach to the ideal of sustainability. *Society for Philosophy and Technology* 2(2), 41–48.

Organization for Economic Cooperation and Development (OECD) (1994) *Environmental Indicators. A Core Set.* OECD, Paris.

Organization for Economic Cooperation and Development (OECD) (1997) *Better Understanding Our Cities: The Role of Urban Indicators.* OECD, Paris.

O'Grady, R. (1990) Acceptable tourism. *Contours (Bangkok)* 4(8), 9–11.

Orams, M.B. (1995) Towards a more desirable form of ecotourism. *Tourism Management* 16(1), 3–8.

Payne, R.J. (1993) Sustainable tourism: suggested indicators and monitoring techniques. In: Nelson, J., Butler, R. and Wall, G. (eds) *Tourism and Sustainable Development: Monitoring, Planning, Managing.* Department of Geography, University of Waterloo, Canada, pp. 249–254.

Peterson, T. (1997) *Sharing the Earth: the Rhetoric of Sustainable Development.* University of South Carolina Press, Columbia.

Place, S.E. (1995) Ecotourism for sustainable development: oxymoron or plausible strategy? *GeoJournal* 35(2), 161–174.

Ross, S. and Wall, G. (1999) Ecotourism: towards congruence between theory and practice. *Tourism Management* 20(1), 123–132.

Rossi, R.J. and Gilmartin, K.J. (1980) *The Handbook of Social Indicators: Source, Characteristics, and Analysis*. STPM Press, New York.

Scace, R.C. (1993) An ecotourism perspective. In: Nelson, J.G., Butler, R. and Wall, G. (eds) *Tourism and Sustainable Development: Monitoring, Planning, Managing*. Department of Geography, University of Waterloo, Canada, pp. 59–81.

Sirakaya, E. (1997) Attitudinal compliance with ecotourism guidelines. *Annals of Tourism Research* 24(4), 919–950.

Sirakaya, E., Sasidharan, V. and Sonmez, S. (1999) Redefining ecotourism: the need for a supply-side view. *Journal of Travel Research* 38(2), 168–172.

Smith, V.L. (ed.) (1977) *Hosts and Guests: the Anthropology of Tourism*. University of Pennsylvania Press, Philadelphia.

Smith, V.L. (ed.) (1989) *Hosts and Guests: the Anthropology of Tourism*, 2nd edn. University of Pennsylvania Press, Philadelphia.

US Department of Commerce (1998) *International Travel and Forecast for the U.S. 1997–2001, Tourism Industries*. International Trade Administration, US Department of Commerce, May.

US Department of Health, Education and Welfare (USHEW) (1969) *Toward a Social Report*. US Government Printing Office, Washington, DC.

Valentine, P.S. (1993) Ecotourism and nature conservation: a definition with some recent developments in Micronesia. *Tourism Management* 14(2), 107–115.

Wallace, G.N. (1993) Visitor management: lessons from Galapagos National Park. In: Lindberg, K. and Hawkins, D.E. (eds) *Ecotourism: a Guide for Planners and Managers*. Ecotourism Society, North Bennington, Vermont, pp. 55–82.

Wallace, G.N. and Pierce, S. (1996) An evaluation of ecotourism in Amazonas, Brazil. *Annals of Tourism Research* 23(4), 843–873.

Wheeler, D. (1995) *California's Ocean Resources: an Agenda for the Future*. Resource Agency, State of California, July.

Wheeller, B. (1993) Sustaining the ego. *Journal of Sustainable Tourism* 1(2), 121–129.

Wheeller, B. (1997) Here we go, here we go, here we go eco. In: Stabler, M.J. (ed.) *Tourism and Sustainability: Principles to Practice*. CAB International, Wallingford, UK, pp. 39–50.

Whelan, T. (1991) Ecotourism and its role in sustainable development. In: Whelan, T. (ed.) *Nature Tourism*. Island Press, Washington, DC, pp. 3–22.

Wight, P.A. (1993) Ecotourism: ethics or eco-sell? *Journal of Travel Research* 31(3), 3–9.

Wilkinson, P.F. (1989) Strategies for tourism in island microstates. *Annals of Tourism Research* 16(2), 153–177.

World Commission on Environment and Development (WCED) (1987) *Our Common Future*. Oxford University Press, Oxford.

World Tourism Organization (WTO) (1995) *Agenda 21 for the Travel and Tourism Industry: Towards Environmentally Sustainable Development*. World Travel and Tourism Council, WTO, and Earth Council, London, Madrid, San José, Costa Rica.

Chapter 27

Rural Development

R.W. Butler
School of Management Studies for the Service Sector, University of Surrey, Guildford, UK

Introduction

The linking of ecotourism and rural development may appear rather paradoxical to some readers, for if there is a single defining characteristic of ecotourism, it is probably its positive relationship with what is thought of as the natural environment. Clearly, not all, if even many, rural areas include much natural environment in the strict sense of the term. The problems and difficulties of defining ecotourism have been fully discussed earlier in this volume and will not be repeated here, however, it is necessary to return briefly to this issue to place the development of ecotourism in rural areas in an appropriate context. The understanding of ecotourism here is taken to include a focus on one or more elements of the natural environment such as wildlife or vegetation, a small rather than a large scale of operation, low rather than high levels of impact (particularly of negative impacts), an emphasis on conservation and sustainability, and a pattern of activity and development which is in harmony with both existing natural and cultural activities and processes in an area. As noted frequently in the literature, such attributes are very similar to those of sustainable tourism (Aronsson, 1994; Burton, 1997; Barkin, 2000).

The problems of defining ecotourism are mirrored in defining the term rural and clarifying what is meant by rural area and rural development (Bramwell and Lane, 1994; Butler *et al.*, 1998). The most simple, but hardly helpful definition might be 'non-urban', but some further elaboration is clearly necessary. To most people, rural implies agriculture, and certainly a settled and modified, normally farmed, environment. Rural rarely, if ever, means an untouched or pristine landscape, thus areas which have never been deliberately modified by mankind are unlikely to be included in such delineation. The Polar regions, high unpopulated mountain ranges, major deserts, uninhabited islands, and even very extensive forests and wetlands are generally excluded from such a definition, at least in part because they have not been permanently settled and modified by humans. Areas such as the foothills of the Rockies, large parts of the Alps, even much of the interior of Australia could be included, however, as these areas are settled and used for agriculture, albeit often in a marginal sense. Large areas of tropical forest are more problematic, for although they may be part of the agricultural landscape in the sense that they are periodically cut and burned, settlement is normally periodic or nomadic and evidence of human modification often relatively fleeting. Small tropical islands tend to have much of their coastal area at least under some form of agriculture and village-based settlement, and thus the term rural is more applicable there. In the context of this chapter, therefore, rural areas are taken to be settled areas which are used primarily for agriculture, in which the pattern of settlement is permanent but may be either village based or dispersed. Under this definition, inevitably little of the land is pristine or free from human modification.

Issues

The working definitions noted above raise some important issues with respect to the links between ecotourism and rural development. Present-day agricultural practices in many parts of the Western world have resulted in very great modification of the environment in which they are undertaken. Most recently the appearance of genetically modified crops and animals has added another dimension to the distancing of agriculture from nature. Much present-day rural visitation and residency for leisure purposes is bound up in nostalgia and the desire to recapture a rural idyll which may never have existed except in art and literature (Dann, 1997; Hopkins, 1998), and already finds itself at odds with agribusiness and a landscape which is often monotonous and bereft of wildlife. Agribusiness and modern large-scale agricultural practices make conventional farm tourism and other traditional rural leisure pursuits increasingly difficult to experience. Farming in many Western countries is under very considerable pressure to increase efficiency and financial returns to the point that the traditional family farm is rapidly disappearing as it becomes difficult, if not impossible, to make a reasonable living solely from a small acreage in most situations, compared with returns from very large, highly mechanized farms. To counteract this, many small-scale farmers need to engage in additional economic activities to remain economically viable. While such a process has meant increasing interest in tourism as one of these options, the maintenance of the original mixed agricultural landscape is under increasing threat. The process of rural change is well documented (Ilbery, 1998) and in general it is a process which does not bode well for ecotourism or most forms of tourism and recreation in rural areas which rely on a traditional farming landscape of considerable diversity.

On the other hand, such a state of affairs does provide an environment in which alternative and supplementary forms of income are particularly welcome (Oppermann, 1998). Thus over the past two or three decades, tourism and recreation have been viewed as potentially positive elements in strategies to preserve the family farm and traditional farming practices in many marginal areas in Western countries in particular (Greffe, 1994). The inclusion of tourism into rural development policies has become extremely common in many strategy and policy documents (Pigram, 1993; Augustin, 1998; Jenkins *et al.*, 1998). The results have not always been successful, it must be stated, in no small part because of a frequent failure to appreciate that tourism of any kind has specific requirements, in both the natural resource and the human resource areas. Locations which cannot produce agricultural output to a satisfactory level are not automatically capable of producing suitable attractions for tourism, and individuals who fail at farming do not automatically succeed in tourism. Nevertheless, many areas which are now marginal for agriculture, because of the trends in that activity noted above, do possess environmental characteristics which may make them suitable for tourism, such as attractive upland scenery, remoteness and quietness, wildlife, and an overall absence of development, along with clean water, woodlands and clean air. Combined with traditional cultural activities and heritage structures and a local population which may be well versed in local nature lore and inherently sociable and friendly to non-locals, potential does exist in many areas for some ecotourism development (see Chapter 20).

Difficulties include the fact that few local communities wish to remain untouched and 'primitive' or 'backward'. While that image may be desired or sought after by visitors, it is one which is often shed as quickly as possible by many local residents. Thus the rural idyll is rapidly changing, and tourism is itself an agent of change and often welcomed as such in many rural areas. Tourism may be seen as one of a very few alternatives to traditional agriculture which allows the local population to remain in their traditional settlements and continue to engage, even if on a reduced scale, in some form of agriculture and other rural activities (Bourke and Luloff, 1996; Sharpley, 1996; Page and Getz, 1997). This can cause some potential conflicts between visitors who wish to see traditional, even mythical rural settings and activities, and local residents who wish to become modernized and abandon what may well be inefficient and uncomfortable ways of life. Ecotourists as much or more than other tourists may be particularly attracted to what they view or expect to view as traditional activities in rural areas and be disappointed if such activities are not present (Ballantine and Eagles, 1994).

As well, many rural activities are incompatible with some of the basic principles of ecotourism. Hunting and fishing have long been staple rural activities, not only for essential food purposes, but also as means of keeping wildlife populations within acceptable limits, and for recreation and pleasure purposes (Hinch, 1998). Trapping and killing of wildlife for skins and fur was a traditional activity in many rural areas and is still practised in parts of northern North America and Russia, and in the southern hemisphere, and is in diametric opposition to the beliefs of many ecotourists and others. A key consideration in rural areas is that such areas are working landscapes, not normally passive settings for nature and most farmers, even traditional family farmers, have scant time for nature given the pressures they are under. The preservation of natural areas and processes will rarely be supported unless it can be shown that by so doing, the rural landowner is also serving his or her own interests. It should not be surprising that support for the establishment of nature reserves, national parks, and other forms of protected environments is often strongest in areas far distant from the proposed areas and often opposed by those in and around such areas. Such proposals are often seen as depriving locals of traditional resources and simply opportunities to meet the desires and preferences of urban and foreign residents at local expense. If ecotourism is to be successful in such situations, then it must yield what is often proposed as one of its key characteristics, economic benefits for local residents (Slee *et al.*, 1997) and, if possible, do so without requiring the sacrifice of other traditional activities and sources of revenue. As will be discussed below, such situations can be found and developed, but are far from automatic or inevitably successful.

Trends

Trends that are apparent in tourism and leisure are both positive and negative for the involvement of ecotourism in rural development, as are trends in agriculture and other rural activities in many areas. As the world's population becomes more urban, more densely settled, more removed from 'natural' conditions and processes, and under more stress, so the potential attraction of non-urban, thinly settled, more natural and less stressful environments increases for at least some segments of society (Cavaco, 1995). Thus the demand for ecotourism and other forms of tourism (from short-term leisure trips to full-time retirement) in rural areas is likely to grow. Similarly, as noted above, as economic pressures grow more severe on marginal agriculture, the search for alternative and supplementary forms of income becomes more significant, and tourism of many forms is increasingly seen as a potentially acceptable addition to the frequently limited range of rural economic activities (Moss and Godde, 2000). As some marginal areas go out of agricultural production, tourism and leisure use may be seen as one of, or even the only, alternative productive economic use of the area (Oppermann, 1998; Saeter, 1998).

One of the more significant and positive trends in rural areas from the point of view of compatibility with ecotourism has been the widespread growth in the market for farm produce and particularly organic and 'natural' agricultural produce. While farm gate sales of products have long been a feature of the rural environment in all parts of the world, in recent decades the deliberate direct marketing of rural produce from the producer has expanded considerably. Marketing of things rural (whether authentic or not) has now become widespread (Hopkins, 1998) and excursions into rural hinterlands for purchasing of rural produce, from food to wine (Hall and Macionis, 1998), and antiques to rural reproductions is now a major form of tourism and recreation. Allied to this trend is the emergence of rural festivals (Butler and Smale, 1991; Janiskee and Drews, 1998), only some of which have well-established roots in local traditions. This trend of increasing popularity of things rural serves to make rural areas popular for a wide range of forms of tourism, of which ecotourism is only one. However, the characteristics of ecotourism noted above make it one of the more compatible forms of tourism with many traditional farm activities and landscapes.

Other trends, however, are less encouraging for the development of ecotourism activities in rural areas. Tourism and recreation uses of rural areas are becoming increasingly varied in number and many of the newer forms of leisure activities which are being pursued in the rural landscape are at the opposite end of the spectrum to ecotourism. These new activities are characterized by being mechanized and dependent on high technology, are often individualistic and competitive, have a high per capita user impact on the environment, are urban related and place little value on the innate naturalness of the environment in which they are practised (Butler, 1998). They include activities such as snowmobiling, mountain biking, all-terrain/off-road vehicle operation, endurance/extreme sports and off-piste downhill skiing and snowboarding. They are, therefore, very different in participant characteristics, motivations, resource requirements and compatibility with other activities, both of a tourism and a traditional rural nature. This is in contrast to the more traditional rural tourism activities such as walking, nature study and observation, sightseeing, visiting historic artefacts and viewing rural operations (Butler, 1998), all of which are compatible with, if not a part of, some forms of ecotourism in rural areas. At a time when ecotourism is being considered as one form of tourism suitable for development in rural areas, those regions are thus experiencing an often rapid growth of participation in the newer forms of tourism and recreation. It is unlikely that these two divergent groups of activities will be able to coexist simultaneously in the same area.

As agriculture is becoming increasingly financially non-viable in some parts of the world, other land uses, generally less compatible with tourism, are competing strongly for the rural land base. In Europe in particular, where distances from metropolitan centres to rural areas are small, land no longer viable for agricultural production is often highly desired for residential or commercial use. The value assigned to the land from such a conversion is much higher than if the land remained in agricultural use or even for low-intensity tourism and recreation. Thus other pressures are exerted on such lands as well as more intensive forms of tourism and recreation. In association with non-agricultural land uses come demands for rural land for transportation routes, storage for a range of products including waste of many kinds, and a range of other land uses not deemed suitable for densely populated areas. In the past, the latter have included military areas and nuclear power stations.

Ironically, as military bases, particularly

in Western Europe and North America are reduced and phased out, some areas which had been used for military training, for example bombing ranges, have become highly suitable for ecotourism. The absence of humans on a permanent basis, and the fact that natural processes have been allowed to continue for many decades, have left many such areas with a unique population of wildlife and vegetation. In one sense they are 'islands' of natural or near-natural landscapes in a sea of otherwise intensely modified landscapes. There often remain problems such as uncleared explosives and relative lack of access, but in most cases the former has to be resolved before the area is released for public access (Butler and Baum, 1999). In Canada, parts of the prairies and southern Ontario have already fallen into such categories, although public access for activities such as ecotourism may face problems with issues of native land claims for such areas and demand for the land for other uses.

The final trend noted here is the general trend towards large-scale commercial farming or agribusiness referred to earlier. Potter (1998) has discussed in some detail the rise of concern over loss of natural values through changing methods of agricultural production and scale of operation and the appearance and implementation of a range of measures designed to conserve natural values in rural areas. He notes increasing concerns from the 1980s onwards as to the ways in which modern farming methods have impacted on natural processes and features in agricultural areas. Larger fields have appeared in the drive for greater economic efficiency through increasing use of mechanization, in turn necessitating removal of hedgerows and trees, increasing drainage and channelization of natural waterways, increasing use of pesticides and fertilizers, and most recently the introduction of genetically modified plants and perhaps animals. To finance the increasing cost of mechanization producers have sought to enlarge farm size and to reduce farm labour, thus encouraging the disappearance of the family farm. Farms have become more specialized and increasingly involved in monoculture. All of these processes result in decreased variety in rural areas, both in terms of agricultural production and the farmed landscape, and more importantly from the point of view of ecotourism, in decreased biodiversity in terms of both vegetation and wildlife. A large part of ecotourism's attraction and the motivation of participants (Eagles, 1992; Ballantine and Eagles, 1994) is the opportunity to view features and elements of the natural or semi-natural landscape. The more varied the landscape and its components, floral and faunal, as well as geologic and aquatic, the more attractive it generally is and the greater the appeal. The appeal of the stereotypical English country landscape, beloved by Christmas card and biscuit tin illustrators, is very much based on its variety, featuring domesticated and wild animals, arable crops and woodland, farm chickens and birds of prey, untamed streams and mill ponds, and peaceful settlements and wild hill scenery. A monocultural landscape of wheat fields with few buildings and no animals or song birds may yield large volumes of low price grain but has little appeal to any form of tourist, least of all an ecotourist.

This is a rather negative picture of rural areas, particularly those in developed 'Western' countries, against which to discuss ecotourism and rural development, yet it is possible to find the successful introduction or expansion of what can be viewed as forms of ecotourism in some rural areas. The following section discusses some examples of ecotourism development in rural areas and endeavours to identify conditions and requirements for the successful introduction and operation of such developments.

Examples

Wheeller (1994) has pointed out very clearly the problems with the cynical marketing of ecotourism and sustainable tourism and the labelling of a wide variety of forms of tourism with these terms in order to penetrate the growing market in this area. Despite these valid concerns, there are examples of what appear to be, at least so far, serious attempts to develop ecotourism opportunities in rural areas in a wide variety of settings. This section discusses a limited number of cases in order to identify characteristics and issues arising which may be of use in the further development of such opportunities in other rural areas.

Fiji

The Fijian islands represent a destination which has experienced a variety of forms of tourism, but which is particularly suitable for the development of ecotourism in a rural context. The natural environment of the South Pacific islands is a major element in the attraction of the region to tourists, as is the local culture of the different island groups. Thus a form of tourism which combines opportunities to experience the undeveloped natural environment and at the same time experience contact with rural Fijians in their home settings and culture has high potential. The distances involved are relatively small and while access is not always easy, compared with long distance interior destinations in other locations, it presents few problems. A range of opportunities exists from one-day trips from conventional tourist resorts to excursions of several days, staying at a number of different villages.

Fiji is one of the few locations which is in the process of developing a policy and strategy specifically for the development of ecotourism in the context of rural villages (Harrison, 1997). It is argued that this specific policy has to be placed in the context of the overall tourism policy for Fiji and reflect the great significance of the land reserve of Fiji as an integral part of the Fijian culture, as well as being a resource for food production and development. The form of ecotourism envisaged in the policy involves small-scale operations, an emphasis on nature and indigenous culture, being locally owned and operated, conserving the natural environment and improving the welfare of local communities (Harrison, 1997, p. 5). Perhaps significantly, it makes the distinction between village-based tourism and ecotourism, a distinction based primarily on the responsibility of tourists and conservation of the human and non-human environments. It recognizes a large number of stakeholders and the need to coordinate current offering and define what is meant by an 'ecotourism product', in part, it is assumed, to avoid the problems noted by Wheeller (1994).

As well as having the goal of improving the welfare of the rural people of Fiji (Harrison, 1997, p. 10), the document also notes the crucial importance of local wishes and aspirations and linkages with traditional arts, crafts and traditions. It calls for a high level of government involvement in training, support and registration of developments, and strong linkages with other elements of tourism in the islands. There is also clear recognition that ecotourism is not a panacea, that not all ecotourism projects are beneficial, that they do result in impacts (some of which may be unforeseen), that they do not result in the influx of vast sums of money, and that not all ecotourism is environmentally friendly.

Since the preparation of the policy, research has been undertaken on ecotourism and village-based tourism in Fiji (Francis, 1998; Reddy, 1998). Reddy (1998) conducted four in-depth case studies of village-based tourism, one of which was an ecotourism development, and two others having minor links with ecotourism. She concluded that the key to the success of the Bouma Forest Park, which linked both the natural and the cultural heritage, was landowner involvement and a commitment to maintaining the integrity of both heritages together. The park, on the island of Tavuni, offers interpretation, forest trails, a spectacular waterfall and, close by, snorkelling, as well as a village visit. The other two projects, a campsite on a small island and a village site were more concerned with catering to general tourists than ecotourists for the most part. She concluded that strong personal or family control and commitment was the key to success, and the greater the number of people and agencies involved, the less successful the project was likely to be (Reddy, 1998, p. 150).

Despite the apparent desire of the government to stimulate and encourage ecotourism ventures as a way to stimulate rural development, Francis (1998) is generally critical of both the government and other major tourism sector bodies for their actual commitment to the concept. Non-participation on committees and boards, allocation of a very small proportion of tourism-related funding to ecotourism promotion and development, and an apparent lack of appreciation of the potential of ecotourism are all cited as reasons for this segment not achieving its full potential (Francis, 1998, pp. 152–153). In reviewing some 68 village-based proposals seeking government support, he concluded 'their ventures appear to be more like what could be called village tourism than ecotourism, involving predominantly village tours, treks to nearby forest areas or the provision of island accommodation' (Francis, 1998, p. 142). He notes, however, that the Ministry for Tourism had titled the list of projects 'Ecotourism Development'. Francis concludes that while ecotourism can make a successful contribution in Fiji, to date it has helped individual Fijian families and entrepreneurs rather than communities and still has a high non-Fijian involvement (Francis, 1998, p. 168). While one would find it hard to disagree with this conclusion, his definition of ecotourism is towards the 'hard' rather than the 'soft' end of the spectrum. Accepting a softer version would allow one to expect that the future for the development of ecotourism in conjunction with the more established village-based tourism would be more successful.

Australia

Australian tourism operates at a very different scale to that of Fiji, but the landscape and wildlife resources of Australia are major tourist attractions, as is the rural heritage and image, particularly of the Outback. Australia is one of very few countries to have a policy or strategy on rural tourism, and The National Rural Tourism Strategy (CDT, 1994b) represents a significant advance in tourism planning. Its definition of rural tourism as 'a multifaceted activity that takes place outside heavily urbanised areas ... characterised by small scale tourism ... in areas where land use is dominated by agricultural pursuits, forestry or natural areas' (CDT, 1994b, p. 3), provides scope for the inclusion of ecotourism. The document, however, does go on to indicate that 'rural tourism can represent to the traveller the essence of country life', which takes much of the emphasis off the natural environment and its attractions. Furthermore, the potential activities listed are much more related to farm-related pursuits and cultural participation. It does include hiking in national parks, rafting and caving, and visits to wildlife reserves, nature-based tours and ecotourism tours (pp. 3–4). Beyond some reference to the National Ecotourism Strategy (CDT, 1994a) in the context of environmental benefits, however, there is nothing specific dealing with ecotourism in the strategy document.

The Rural Tourism Needs Analysis (Morrison, 1995) follows a generally similar line with respect to the limited reference to and role of ecotourism-related projects in rural tourism in Australia. It too places an emphasis on farm tourism, although it does make reference to 'contact with nature and the natural world', heritage, 'traditional' societies and 'traditional' practices (p. 2), and lists among 'optional' elements nature and indigenous peoples and lifestyles (p. 4). Among the case studies examined in this report, however, the majority include references to ecotourism and the natural environment as being significant attractions and elements in the tourism offerings in locations including Katherine (Northern Territory), Cooktown (Queensland) and the Snowy Mountains (New South Wales). Elements include cave tours, wildlife viewing opportunities including crocodiles and birds, guided excursions into the outback, aboriginal tourism visits, and trails through rainforest and desert environments from rural settlement bases. It is clear, however, that such activities are seen as optional additional elements to the conventional farm tourism focus on which rural tourism would be firmly based (Australian Rural Management Services, 1995).

Europe

Interest in ecotourism and green concerns are reasonably well established in many European countries (Becker, 1995), particularly those in the north, and much of the relatively wild and thinly populated land resource is also found in these countries. Much of Scandinavia, upland areas of Scotland, and mountainous regions in Germany, Austria and France, in particular, offer a wide range of natural or semi-natural habitats in highly attractive scenic settings, many of which have long been popular with tourists. Relative freedom of public access in many of these countries, particularly to visitors on foot, has meant that the visitor populations have traditionally used these areas for activities that were low impacting and often involved natural elements of the landscape. In recent years the decline of many upland and marginal peripheral agricultural regions has meant that rural residents have been eager to turn to other forms of income generation, particularly those which could be conducted *in situ* using their local knowledge and expertise. Liedler (1997, p. 64) claimed that for farms in the highest parts of the Austrian Alps, tourism accounted for over 30% of total income, and Rickard (1983) has argued that tourism can help stabilize the rural economic base in declining rural areas in England.

Given the absence of pristine wild land from much of mainland Europe, however, and the widespread nature of agriculture on this continent, conventional 'hard' ecotourism can only be found to any degree in the far north or the highest central mountains. The spread of ski slopes and accompanying technology even on to glacier slopes in the Alps (Zimmermann, 1995) has meant that most of the forms of ecotourism in Europe are of the 'soft' variety (Becker, 1995) and often tied in with forms of adventure tourism, other wilderness activities such as cross-country skiing and hiking, water-based activities such as kayaking, canoeing and sailing, or with conventional farm tourism activities. There is a well-established tradition in Northern and Western Europe of the use of local residences for tourist accommodation, 'bed and breakfast' opportunities abound in the UK and signs indicating 'zimmer frei' (room available) are common in Germany, Austria and Switzerland, as are their equivalents in other countries. There is thus familiarity and acceptance of using local houses, including farms and other premises in rural areas, as the accommodation base for holidays, and the rental of self-catering accommodation, for example, the 'Gites Ruraux', begun in France in the 1950s to stem rural depopulation (Fleischer and Engel, 1997), is equally common and growing rapidly.

Much of the rural-based ecotourism participation is undertaken in parks and protected areas of some kind. Becker (1995, pp. 218–219) notes that 22% of Germany and 8% of France lies within nature parks, with further areas in these countries being under conventional national park protection. All European countries have a variety of protected areas, many of which are actively promoted for tourism in conjunction with neighbouring rural communities. In England and Wales the national parks encompass a large number of rural communities and private land, and rural tourism and ecotourism activities are promoted by bodies such as the Countryside Agency and the National Park Authorities (Parker and Ravenscroft, 2000). In many other countries in Europe and elsewhere, the establishment of national parks and similar reserves was in no small part to stimulate nature-based tourism in an era before ecotourism had been conceived, and to aid rural communities in and adjoining such parks (Butler and Boyd, 2000).

The term ecotourism has not been heavily used in much of Europe to date (see Chapter 10). However, the creation of features such as ecomuseums in France (Dewailly, 1998) and the integration of mountain recreation with rural development in Switzerland and the Alps generally (Becker, 1995), have resulted in increased participation in what are clearly ecotourism activities in many parts of Europe. The development of ecotourism opportunities and participation in such activities is more common in northern Europe than in regions bordering the Mediterranean, which traditionally have not been as environmentally conscious as their northern counterparts. In what may be viewed as a somewhat ironic turn of events, however, rural development involving at least some forms of ecotourism is now taking place in the hinterlands of some of the declining mass tourism resorts in areas such as Spain. Integrated rural planning in areas such as Valencia (Vera and Rippin, 1996) has seen the development not only of wine tourism and more conventional farm tourism, but also trails, natural areas and other opportunities for a more nature-based tourism. The LEADER projects of the European Union (Cavaco, 1995) have been significant in developing such forms of tourism in Portugal and Spain in particular (see Chapter 10).

The Americas

North America can be thought of as the cultural hearth of the concept of ecotourism, and the largest market for its opportunities. Initially most developments took place in the traditional wilderness areas of the north and the west, often within or adjacent to the national and state/provincial parks in the USA and Canada, and the market was heavily internal to North America (Eagles, 1992) (see Chapter 7). Over the past two decades in particular, many ecotourism opportunities have been developed in more clearly rural locations. This reflects the critical need for alternative forms of economic development in rural areas because of similar problems facing family farmers in North America to those discussed above in the context of Europe and elsewhere. The realization of the rapidly growing market for nature-based tourism and the appeal of the name ecotourism have also encouraged this pattern. Weaver (1997) and Weaver and Fennell (1997) have discussed such development opportunities in the context of Saskatchewan, a province that might reasonably be thought initially to have little opportunity for ecotourism, but which has been shown to have considerable potential for targeted development in specific locations, based particularly on wetlands and birdlife.

Central America has long been noted for its ecotourism development, much of its first developments occurring in interior mountainous regions, such as the Monte Verde Cloud Forest in Costa Rica (Fennell and Eagles, 1990). In more recent years there has been widespread development of ecotourism opportunities into many other areas, not only in Costa Rica, but particularly in Belize and in South America (Horwich *et al.*, 1993; Wallace, 1993; Cater, 1996). Some of these newer developments have occurred in rural rather than wild areas, and some doubts have been expressed, both about the suitability of some of the developments and their validity in terms of ecotourism principles (Place, 1998). The rapid growth of ecotourism in parts of Central and South America, particularly in rural areas, raises some concerns over the nature of such growth and subsequent development. If rural residents are to benefit appropriately from such developments, they need to be active participants and have their concerns and desires included in the development and protection of areas.

In Mexico ecotourism potential in rural areas is considerable, and perhaps one of the best examples is in the Sierra Madre of central Mexico, the wintering grounds of the monarch butterfly. Barkin (2000, p. 160) dates interest in this area from publication of details of the migration in *National Geographic* in 1976, and the creation of a biosphere reserve in 1986. The numbers of visitors increased rapidly from 25,000 to over ten times that figure before the end of the century, and he notes the incompatibility of such visitation and development with indigenous rural resident aspirations and traditional activities. The lack of local participation in the designation and management of the reserve, the lack of ability to maximize local economic benefits from the large numbers of ecotourists, and the general lack of integrated planning identified by Barkin (2000, p. 166) show all too clearly that while ecotourism in rural areas can be extremely attractive to tourists, it does not necessarily have the desired economic or other benefits for local rural residents. Similar conflicts between ecotourism development and protection of areas and local rural activities and aspirations have been noted in other areas (Wallace, 1993).

Implications

It seems clear from the above discussion that the form of ecotourism developed in rural areas will vary considerably from area to area, as might be expected. However, in the vast majority of rural areas it will be a form of 'soft' ecotourism (Ziffer, 1989; Acott *et al.*, 1998), where the emphasis is on more passive activities such as viewing and collecting images, than involving intensive physical activity and requiring skill, knowledge and commitment. Long-distance, physically challenging trekking or canoeing opportunities are limited, if not absent, in many rural areas, and the normal substitute is walking along well-established trails with accommodation in formal establishments available close to the footpaths. Exotic wildlife is absent from most rural areas, almost by definition, unless it is contained within wildlife parks or farmed, and thus of only marginal interest to most ecotourists even where it is present. Avifauna is probably the most important element in ecotourism in many rural areas and birdwatching is a major and growing element in most ecotourism offerings in rural areas.

In addition, it is likely that most ecotourism opportunities are or will be offered in conjunction with other forms of tourism, particularly variations of farm tourism. It is likely that in most rural areas ecotourism will not be the primary purpose or motivation of participants but an ancillary activity engaged in for a part of the vacation or holiday period, rather than the predominant or only reason for the visit to that area. Linking ecotourism opportunities with more traditional rural tourism activities such as walking, photography, art, nature study and visiting heritage sites is likely to be the norm. Increasingly, however, there will probably be a range of more contemporary activities offered as well, such as canoeing, rafting, wild camping, mountaineering, caving, fossiking (searching for artefacts and natural deposits primarily for enjoyment, see Jenkins (1992)) and even catch and release fishing, which once may have been regarded as somewhat unsuitable to be linked to ecotourism.

To purists in the ecotourism field such suggestions may be viewed as anathemas, and represent the dilution or 'dumbing down' of ecotourism to cater to popular tastes. In rural areas, however, unlike 'natural' or wilderness areas where there is no local population in need of jobs and income, there is and will continue to be a real need to find alternative forms of economic activity to agriculture. 'Pure' or 'hard' ecotourism opportunities, as noted above, are unlikely to be found in most rural areas and to pretend otherwise is to court economic failure and market rejection (Blamey, 1997). The combination of some limited forms of ecotourism and other generally compatible tourist activities is likely to fit much better into many rural areas, particularly in Western countries, and appeal to a much larger market. In rural areas which are relatively close to major urban centres or established tourist destinations, as is the case in much of Europe and parts of North America, Australia and New Zealand, participation in ecotourism and other forms of tourism is likely to be on a short-term basis, for a few days' duration rather than for the major holiday. Furthermore, a variety of activities is likely to attract participants back for repeat visits more than a single opportunity. Ecotourism, like all forms of tourism, has to achieve economic viability (Slee *et al.*, 1997; Stabler, 1997) as well as meeting environmental and social sustainable criteria in order to survive for any period of time in rural, as in other settings.

The development of such forms of ecotourism in rural areas may imply that ecotourism is entering the mainstream of tourism and becoming simply another part of the overall tourism phenomenon, losing some of its ideological and philosophical focus in the process. It may also imply that ecotourism could become even more polarized in its offerings, rather than being represented by a spectrum of opportunities, with only soft forms appearing in rural areas and hard ecotourism being pushed ever further to the periphery. Such a process is already visible and even in some traditional pure or hard ecotourism areas such as Costa Rica, concerns have already been expressed over the 'softening' and dilution of what were once more exclusive and protected opportunities (Weaver, 1998) (see Chapter 11).

The question has to be raised in this context then, of whether ecotourism in rural areas is little more than a new label on a parcel of fairly traditional leisure activities that have been repackaged and adjusted to complement other, often declining, rural activities such as farming and forestry. It might represent ecotourism being reduced in such areas to a mental construct and an advertising label only. A mental construct in the sense that participants want to participate in ecotourism for the 'feel good' factor described so bitingly by Wheeller (1993) as 'egotourism', and an advertising label because so many public sector agencies are willing to support projects described as ecotourism or sustainable, as noted by Francis (1998) in Fiji. In reality such a state of affairs might not be important if the desired benefits of the developments, such as sustaining rural incomes and employment, regeneration of rural communities, heritage, traditional crafts and culture, and conservation of potentially endangered habitats, species and biodiversity are achieved. Many of those goals fit very well with the ideology and principles of ecotourism (Western, 1993) even if the setting is not an exotic tropical rainforest but an upland farming area in Wales, Canada or New Zealand.

We can argue that village-based trekking into interior woodlands to see birds and plants in Fiji is a legitimate form of ecotourism, because it has a focus on elements such as wildlife and vegetation, is locally based, is small-scale, uses local people and resources, shares in their cultural activities, and has a low impact on the environment. It is logical then to argue that people on individual holidays staying in a bed and breakfast establishment in an English village, visiting the local inn, eating local food, engaging in walking and birdwatching on neighbouring moorland and woodlands are equally engaged in ecotourism. It is more difficult and confusing to decide if the visit is ecotourism when, for example, the tourists in Fiji return after one or two nights, or even only a long day, to a luxury hotel for the rest of their holiday. If the visitors in the English village have driven from their home in a metropolitan centre perhaps only 100 km away for a weekend break, are they really ecotourists or people engaged in leisure and recreation? Such semantic arguments surely miss the point that ecotourism based in any settled area has to be accepted as just one other form of tourism, albeit with certain characteristics and motivations.

Birdwatching is one of the most definitive ecotourism activities, and probably the oldest established form of nature-based tourism. It is practised throughout the world and over the past half century has grown markedly in popularity, moving from a hobby that was often the butt of much humour to a major element in international tourism and domestic recreation and leisure. It now takes many forms, from intensive specialized trips led by experienced guides to Amazonia, Antarctica and wherever interesting or rare bird populations exist, to day trips to municipal sewage works to spot rare migrants. The first type would be regarded clearly as ecotourism of the 'hard' variety, but from first-hand experience this author will vouch that in terms of aspects such as commitment, discomfort, excitement, low impact, and contribution to local economies the latter is equally valid as a form of 'hard' ecotourism. The location is not as important as the activity, the motivation and the participants. Large numbers of birdwatchers participating in intensive birdwatching at a location such as Point Pelee National Park in Canada during migration periods tend to behave and appear more like conventional beach tourists than ecotourists, and are likely to have significant, if unintentional, impacts upon both the birds they are trying to see and the environment they are visiting. It is likely that they would reject such an analogy on the basis that they are committed to the well-being of the birds they are watching, and their environment, and have made major efforts, normally on an individual basis to participate in the activity. They might well regard those on escorted expensive trips where new species are guaranteed as being the 'soft' form of the activity.

Thus, the forms of ecotourism to be found in rural areas should be regarded as much a part of ecotourism as those forms which only occur in remote wilderness. Ecotourism has to be regarded as one extremely varied element in the overall phenomenon known as tourism. When it occurs as part of rural development, its role can be just as crucial in terms of environmental conservation and nature appreciation as when it occurs in remote tropical or polar areas. In terms of fulfilling its role in providing local economic benefits, it is infinitely more successful in a rural setting than in an unpopulated wilderness one.

References

Acott, T.G., La Trobe, H.L. and Howard, S.H. (1998) An evaluation of deep ecotourism and shallow ecotourism. *Journal of Sustainable Tourism* 6(3), 238–253.

Aronsson, L. (1994) Sustainable tourism systems: the example of sustainable rural tourism in Sweden. In: Bramwell, B. and Lane, B. (eds) *Rural Tourism and Sustainable Rural Development.* Channel View Publications, Clevedon, UK, pp. 77–92.

Augustin, M. (1998) National strategies for rural tourism development and sustainability: the Polish experience. *Journal of Sustainable Tourism* 6(3), 191–209.

Australian Rural Management Services (1995) *Rural Tourism Needs Analysis.* ARMS, Parkside, South Australia.

Ballantine, J.L. and Eagles, P.F.J. (1994) Defining the Canadian ecotourist. *Journal of Sustainable Tourism* 2(3), 210–214.

Barkin, D. (2000) The economic impacts of ecotourism: conflicts and solutions in highland Mexico. In: Godde, P.M., Price, M.F. and Zimmermann, F.M. (eds) *Tourism and Development in Mountain Regions.* CAB International, Wallingford, UK.

Becker, C. (1995) Tourism and the environment. In: Montanari, A. and Williams, A.W. (eds) *European Tourism: Regions, Spaces and Restructuring.* John Wiley & Sons, Chichester, pp. 207–220.

Blamey, R.K. (1997) Ecotourism: the search for an operational definition. *Journal of Sustainable Tourism* 5(2), 109–122.

Bourke, L. and Luloff, A.E. (1996) Rural tourism development: are communities in Southwest rural Pennsylvania ready to participate? In: Harrison, L.C. and Husbands, W. (eds) *Practising Responsible Tourism: International Case Studies in Tourism Planning, Policy, and Development.* John Wiley & Sons, Chichester, pp. 277–295.

Bramwell, B. and Lane, B. (eds) (1994) *Rural Tourism and Sustainable Rural Development.* Channel View Publications, Clevedon, UK.

Burton, R. (1997) The sustainability of ecotourism. In: Stabler, M.J. (ed.) *Tourism and Sustainability: Principles to Practice.* CAB International, Wallingford, UK, pp. 357–374.

Butler, R.W. (1998) Rural recreation and tourism. In: Ilbery, B. (ed.) *The Geography of Rural Change.* Longman, Harlow, pp. 211–232.

Butler, R.W. and Baum, T. (1999) The tourism potential of the peace dividend. *Journal of Travel Research* 38(1), 24–29.
Butler, R.W. and Boyd, S.W. (eds) (2000) *Tourism and National Parks*. John Wiley & Sons, London.
Butler, R.W. and Smale, B.J.A. (1991) Geographic perspectives on festivals in Ontario. *Journal of Applied Recreation Research* 16(1), 3–23.
Butler, R.W., Hall, C.M. and Jenkins, J. (eds) (1998) *Tourism and Recreation in Rural Areas*. John Wiley & Sons, Chichester.
Cater, E. (1996) Ecotourism in the Caribbean: a sustainable option for Belize and Dominica? In: Briguglio, L., Butler, R., Harrison, D. and Filho, W. (eds) *Sustainable Tourism in Islands and Small States: Case Studies*. Pinter, London, pp. 122–146.
Cavaco, C. (1995) Rural tourism: the creation of new tourist spaces. In: Montanari, A. and Williams, A.W. (eds) *European Tourism: Regions, Spaces and Restructuring*. John Wiley & Sons, Chichester, pp. 127–150.
Commonwealth Department of Tourism (CDT) (1994a) *National Ecotourism Strategy*. Commonwealth of Australia, Canberra.
Commonwealth Department of Tourism (CDT) (1994b) *National Rural Tourism Strategy*. Commonwealth of Australia, Canberra.
Dann, G.M.S. (1997) The green green grass of home: nature and nurture in rural England. In: Wahab, W. and Pigram, J.J. (eds) *Tourism, Development and Growth: the Challenge of Sustainability*. Routledge, London, pp. 257–276.
Dewailly, J.M. (1998) Images of heritage in rural areas. In: Butler, R.W., Hall, C.M. and Jenkins, J. (eds) *Tourism and Recreation in Rural Areas*. John Wiley & Sons, Chichester, pp. 123–138.
Eagles, P.F.J. (1992) The travel motivations of Canadian ecotourists. *Journal of Travel Research* 3, 3–7.
Fennell, D. and Eagles, P. (1990) Ecotourism in Costa Rica: a conceptual framework. *Journal of Park and Recreation Administration* 8, 23–24.
Fleisher, A. and Engel, J. (1997) Tourism incubators, a support scheme for rural tourism in Israel in World Tourism Organisation. *Rural Tourism: a Solution to Employment, Local Development and Environment*. World Tourism Organization, pp. 82–89.
Francis, J. (1998) Tourism development in Fiji: diversification through ecotourism. MA thesis, University of the South Pacific, Suva.
Greffe, X. (1994) Is rural tourism a lever for economic and social development? In: Bramwell, B. and Lane, B. (eds) *Rural Tourism and Sustainable Rural Development*. Channel View Publications, Clevedon, pp. 22–40.
Hall, C.M. and Macionis, N. (1998) Wine tourism in Australia and New Zealand. In: Butler, R.W., Hall, C.M. and Jenkins, J. (eds) *Tourism and Recreation in Rural Areas*. John Wiley & Sons, Chichester, pp. 197–224.
Harrison, D. (1997) *Ecotourism and Village-based Tourism: a Policy and Strategy for Fiji*. Ministry of Tourism, Transport and Civil Aviation, Suva.
Hinch, T. (1998) Ecotourists and indigenous hosts: diverging views on their relationship with nature. *Current Issues in Sustainable Tourism* 1(1), 120–124.
Hopkins, J. (1998) Commodifying the countryside: marketing myths of rurality. In: Butler, R.W., Hall, C.M. and Jenkins, J. (eds) *Tourism and Recreation in Rural Areas*. John Wiley & Sons, Chichester, pp. 139–156.
Horwich, R.H., Murray, D., Saqui, E., Lyon, J. and Godfrey, D. (1993) Ecotourism and community development: a view from Belize. In: Lindberg, K. and Hawkins, D.E. (eds) *Ecotourism: a Guide for Planners and Managers*. The Ecotourism Society, North Bennington, Vermont, pp. 152–168.
Ilbery, B. (1998) *The Geography of Rural Change*. Longman, Harlow.
Janiskee, R.L. and Drews, P.L. (1998) Rural festivals and community reimaging. In: Butler, R.W., Hall, C.M. and Jenkins, J. (eds) *Tourism and Recreation in Rural Areas*. John Wiley & Sons, Chichester, pp. 157–176.
Jenkins, J. (1992) Fossikers and rockhounds in northern New South Wales. In: Weiler, B. and Hall, C.M. (eds) *Special Interest Tourism*. Belhaven, London, pp. 129–140.
Jenkins, J.M., Hall, C.M. and Troughton, M.J. (1998) The restructuring of rural economies: rural tourism and recreation as a government response. In: Butler, R.W., Hall, C.M. and Jenkins, J. (eds) *Tourism and Recreation in Rural Areas*. John Wiley & Sons, Chichester, pp. 43–68.
Liedler, A. (1997) Tourisme rural et Protection de l'environment. In: World Tourism Organisation (ed.) *Rural Tourism: a Solution to Employment, Local Development and Environment*. World Tourism Organization, Madrid, pp. 63–67.

Morrison, J.B. (1995) *Rural Tourism Needs Analysis.* Australian Rural Management Services Pty Ltd, Parkside, South Australia.

Moss, L.A.G. and Godde, P.M. (2000) Strategy for future mountain tourism. In: Godde, P.M., Price, M.F. and Zimmermann, F.M. (eds) *Tourism and Development in Mountain Regions.* CAB International, Wallingford, pp. 323–338.

Oppermann, M. (1998) Farm tourism in New Zealand. In: Butler, R.W., Hall, C.M. and Jenkins, J. (eds) *Tourism and Recreation in Rural Areas.* John Wiley & Sons, Chichester, pp. 225–254.

Page, S.J. and Getz, D. (eds) (1997) *The Business of Rural Tourism: International Perspectives.* International Thomson Business Press, London.

Parker, G. and Ravenscroft, N. (2000) Tourism, 'National Parks' and private lands. In: Butler, R.W. and Boyd, S.W. (eds) *Tourism and National Parks.* John Wiley & Sons, Chichester, pp. 95–106.

Pigram, J.J. (1993) Planning for tourism in rural areas: bridging the policy implementation gap. In: Pearce, D.G. and Butler, R.W. (eds) *Tourism Research: Critiques and Challenges.* Routledge, London, pp. 156–174.

Place, S.E. (1998) How sustainable is ecotourism in Costa Rica? In: Hall, C.M. and Lew, A.A. (eds) *Sustainable Tourism: a Geographical Perspective.* Longman, Harlow, UK, pp. 107–118.

Potter, C. (1998) Conserving nature: agri-environmental policy development and change. In: Ilbery, B. (ed.) *The Geography of Rural Change.* Longman, Harlow, UK, pp. 85–105.

Reddy, S.W. (1998) The prospects of small-scale village-based tourism in Fiji: problems and opportunities. MA thesis, University of the South Pacific, Suva.

Rickard, R.C. (1983) *The Role of Tourism in the Less Favored Areas of England and Wales.* Agricultural Economics Unit Report 218, University of Exeter, Exeter.

Saeter, J.A. (1998) The significance of tourism and economic development in rural areas: a Norwegian case study. In: Butler, R.W., Hall, C.M. and Jenkins, J. (eds) *Tourism and Recreation in Rural Areas.* John Wiley & Sons, Chichester, pp. 235–246.

Sharpley, R. (1996) *Tourism and Leisure in the Countryside.* Elm Publications, Huntingdon, UK.

Slee, W., Farr, H. and Snowdon, P. (1997) Sustainable tourism and the local economy. In: Stabler, M.J. (ed.) *Tourism and Sustainability: Principles to Practice.* CAB International, Wallingford, UK, pp. 69–88.

Stabler, M.J. (ed.) (1997) *Tourism and Sustainability: Principles to Practice.* CAB International, Wallingford, UK.

Vera, F. and Rippin, R. (1996) Decline of a Mediterranean tourist area and restructuring strategies: the Valencian Region. In: Priestley, G.K., Edwards, J.A. and Coccossis, H. (eds) *Sustainable Tourism? European Experiences.* CAB International, Wallingford, UK, pp. 120–136.

Wallace, G.N. (1993) Visitor management: lessons from Galapagos National Park. In: Lindberg, K. and Hawkins, D.E. (eds) *Ecotourism: a Guide for Planners and Managers.* The Ecotourism Society, North Bennington, Vermont, pp. 55–81.

Weaver, D.B. (1997) A regional framework for planning ecotourism in Saskatchewan. *The Canadian Geographer* 41(3), 281–293.

Weaver, D.B. (1998) *Ecotourism in the Less Developed World.* CAB International, Wallingford.

Weaver, D.B. and Fennell, D.A. (1997) The vacation farm sector in Saskatchewan: a profile of operators. *Tourism Management* 18, 357–365.

Western, D. (1993) Defining ecotourism. In: Lindberg, K. and Hawkins, D.E. (eds) *Ecotourism: a Guide for Planners and Managers.* The Ecotourism Society, North Bennington, Vermont, pp. 7–11.

Wheeller, B. (1993) Sustaining the ego. *Journal of Sustainable Tourism* 1(2), 121–129.

Wheeller, B. (1994) Tourism and the environment. A symbiotic, symbolic or shambolic relationship? In: Seaton, A.V. (ed.) *Tourism. The State of the Art.* John Wiley & Sons, Chichester, pp. 647–655.

Ziffer, K. (1989) *Ecotourism: the Uneasy Alliance.* Conservation International, Washington, DC.

Zimmermann, F.M. (1995) The Alpine region: regional restructuring opportunities and constraints in a fragile environment. In: Montanari, A. and Williams, A.W. (eds) *European Tourism: Regions, Spaces and Restructuring.* John Wiley & Sons, Chichester, pp. 19–40.

Section 6

Planning, Management and Institutions

K.F. Backman
Department of Parks, Recreation and Tourism Management, Clemson University, Clemson, South Carolina, USA

Ecotourism is a field very dependent on effective and efficient planning and policy development at all levels of government, the non-government organization (NGO) sector and business. Thus, it is essential that ecotourism researchers and practitioners are aware of and utilize the best knowledge from the available literature. Whether community or tourism focused, planning involves many actions, participants, fields of knowledge, and levels of decision making and implementation (Branch, 1985). But, according to Gunn (1994), planning for tourism must also consider a number of universal principles. These include a focus on the present, a perspective that goes beyond economic development, the incorporation of all three tourism sectors (business, non-profit organizations and governments), an interactive approach, and integration of three planning scales (community, destination and region). All these principles are equally relevant to ecotourism planning. Furthermore, tourism policy must identify goals and objectives that assist governments in planning the tourism industry (Fennell, 1999). What follows in the next five chapters is current knowledge and insights regarding the best practices in ecotourism policy development, planning and management. The primary institutional players in the field of ecotourism today are also considered.

Chapter 28, titled 'Management Tools and Techniques: an Integrated Approach to Ecotourism Planning' by Sheila Backman, James Petrick and Brett Wright, reviews the complex phenomenon that involves the integration of all ecotourism actions. The chapter provides an overview of many of the integrated conservation and visitor use models that have been discussed in the ecotourism planning and development literature. Relatively simple frameworks that amplify the interrelationships between community development and resource conservation give way to more complex models such as the ecotourism opportunity spectrum, which incorporates concepts from the recreation opportunity spectrum. These provide a conceptual framework from which policy development and optimal decision making can occur. The chapter then focuses on the presentation of an integrated systems model for ecotourism planning developed by the authors. The model starts with ecotourism organizations and their missions, emphasizing the need for coordination between agencies to reduce the fragmentation of the industry. The second major component of the model is the management information system.

Given the scope and amount of information necessary to successfully develop and manage an ecotourism industry, two main elements must be addressed: (i) human systems planning; and (ii) resource systems planning. On the human side knowledge is necessary and usually provided from needs assessments, social and economic impact assessments, attitude surveys and market segmentation studies. Measures relevant to resource systems planning include natural resource inventories, cultural/historical inventories, environmental impact assessments, biodiversity studies and resource sustainability studies. From these two management information system elements, clear and informed ecotourism management decisions can be made which are used to develop measurable objectives for ecotourism planning. Also integrated into the model are elements of ecotourism product and experience that affect image and repeat patronage, and an evaluative/feedback component. In sum, this chapter offers insight into current ideas regarding ecotourism management planning models, and presents an alternative integrated model of decision making that contributes to the planning or management of ecotourism destinations.

Chapter 29, titled 'Policy and Planning' by David Fennell, Ralf Buckley and David Weaver, evaluates the complexity of contemporary ecotourism policy and planning in major world regions. The authors begin by discussing the nature and role of policy development and planning in tourism. Their subsequent analysis shows that Australia's world leadership in this area is related to the interaction and cooperation of the public and private sectors at all levels. In particular, the Australian National Ecotourism Strategy demonstrates a vision for ecological and culturally sustainable ecotourism that serves as a prototype for other countries. Each region of the globe faces its own challenges in regard to policy and planning. For example, the geographic diversity and complexity of the Americas is mirrored in the region's ecotourism policies and plans, as evident in the authors' description of developments in Manitoba, Florida, Guyana and Brazil. In Manitoba, Canada, in development of ecotourism policy, public and private research identifies eight different policy areas as important to the industry's success. These are sustainability, business viability, integrated resource management planning, infrastructure, leadership and cooperation, marketing, aboriginal involvement and awareness and understanding. In the USA, Florida has focused their ecotourism policy around five main components, strategic relationships (between stakeholders), inventory, protection, education and marketing.

In Chapter 30, titled 'Ecotourism-related Organizations', Elizabeth Halpenny discusses the organizations that attempt to minimize the negative impacts and maximize the positive impacts of ecotourism. The chapter begins by reviewing the three categories of organization type, i.e. membership and non-membership NGOs, and public sector or governmental agencies. Halpenny then engages in the difficult task of estimating the number, scope and location of ecotourism organizations. Though these organizations vary in composition, Halpenny demonstrates that they also share many similarities, such as their reasons for *formation* and their *mission*. Form and structure of ecotourism organizations varies between NGOs, member NGOs and government agencies, but within these separate types of organizations, form and structure is similar. Other areas of similarity are in organizational funding, the partnerships formed and the stakeholder groups involved. Regardless of their differences, all indications suggest that ecotourism organizations are going to increase in number and importance worldwide as the tourism industry continues to expand in the next century.

Chapter 31, by Judy Cohen, titled 'Ecotourism in the Inter-sectoral Context' discusses the three primary aspects related to ecotourism and the external arena: (i) the inter-sectoral conflict in ecotourism; (ii) intra- and inter-sectoral coalitions; and (iii) strategies for the intra- and inter-sectoral alliance to achieve ecotourism. In assessing the inter-sectoral conflicts, many industry segments are discussed in relation to their incompatibility in the use of ecosystem resources, e.g. forestry, mining, gas and oil production, manufacturing, etc. The major element of contention here is that these industries do not care whether ecotourism exists in or around their operations. But, for ecotourism to be successful, it is necessary to preserve the pristine environment of an ecotourism area even if these other industries exist there as well. Thus, from the analysis, Cohen identifies two goals for ecotourism to survive. First, the industry can not allow incompatible industries to enter ecotourism areas. Second, the sector must take actions to ensure that ecotourism industry growth does not destroy itself in this growth process. In regard to the coalition, success depends on ecotourism's capacity to provide economic benefits that are distributed equitably to the local population. Also, efforts must be made to ensure compatible industries employ the local population, and that strategies to generate local support for ecotourism are implemented. Suggestions include the empowerment of local people in planning and decision making, making the benefits from tourism clear to all shareholders, ensuring that benefits are equally distributed to local people and recognizing the uniqueness of each ecotourism site. In the chapter's section on intra- and inter-sectoral alliances to achieve sustainable ecotourism the author provides a number of strategies for influencing public policy. The first of these is lobbying of agencies and the petition process. Lawsuits can be used as a means of confrontation. Further, public relations campaigns can be used to either threaten industries or inform the public on damage or potential damage from incompatible industries. A proactive though contentious measure to achieve influence over policy is the formation of alliances with incompatible industries to achieve common goals.

In Chapter 32, titled 'The Place of Ecotourism in Public Policy and Planning', Steven Parker uses worldwide examples to assess how ecotourism operates within the public policy and planning context. Parker examines the numerous governmental programmes, functions and conditions which are external to ecotourism but still have a distinct and significant impact on the ecotourism industry. The first condition is politics and administration. Globally, government ministries such as departments of public works, transportation, environment, immigration, investment and education all set policies impacting ecotourism. One aspect that amplifies the impacts is industry fragmentation, because of its close ties with local units of government and councils of indigenous peoples. A second issue presented is security and the role governments play in the protection of visitors. This is a very important issue to the travelling public particularly in less-developed countries. Numerous examples are provided related to breaks in tourist security and the governmental response to disaster and terrorism in a global context.

Next, Parker describes how government infrastructure policy concerns the planning, finance and construction of ecotourism in a country. Regardless of the size of the project, impacts will occur and subsequent government reaction can lead to success or failure in the ecotourism industry. Two elements related to ecotourism that figure into national policy development are the issues of dependency and leakage. Dependency is a political condition wherein one nation is controlled economically by another. In response, policies need to ensure that countries retain a much greater share of ecotourism profits. Leakage of revenues, on the other hand, requires policies, such as import substitution, subsidization of local labour training, controls over foreign ownership, and transport subsidies, which stop the outflow of revenues generated in the local economies. Parker, like Halpenny, also cites the challenge of convincing former adversaries in incompatible industries to form coalitions.

Fiscal policy, relating to government decisions, includes taxation and public expenditure that affect the ecotourism industry. Areas such as Ecuador's Galapagos Islands, which generate fees from over 80,000 visitors annually, find that this revenue is simply deposited into the country's general treasury. Yet, these destinations need to retain these revenues in order to maintain themselves. Other examples of fiscal policy are the use of various types of taxes such as departure taxes, hotel taxes and custom duties to provide revenues for supporting and preserving areas critical to ecotourism. Finally, the largest issue relating to fiscal policy and foreign countries is the nation's international balance of payments. One measure presented is the 'debt-for-nature swap' process. An example of this is Bolivia, where US$650,000 debt was cancelled in exchange for making a 1.4 million ha addition to the Beni Biological Reserve and thus saving more of this endangered ecosystem.

Financial incentives are closely related to fiscal policy and deal with governmental stimulation and subsidization of an industry. An example at the micro-level is the Costa Rican Tourism Development Incentive Law which offers tax breaks for construction of such facilities as ecolodges. Parker also addresses the maintenance of security in unstable political situations, where policy operates on a macro-level. For ecotourism to continue to grow and to be successful, policies such as those presented in this chapter must be implemented more often and more consistently in the future.

These five chapters approach the subject of planning ecotourism from different directions, perspectives and scales, but all present a similar theme that for this industry to evolve, effective planning is critical regardless of whether the venue is in the USA or Kenya. The only sure means of sustaining vital environments and local populations is through planning, policy development and enforcement.

References

Branch, M.C. (1985) *Comprehensive City Planning: Introduction and Exploration.* American Planning Association, Planners Press, Washington, DC.

Fennell, D.A. (1999) *Ecotourism: an Introduction.* Routledge, London.

Gunn, C.A. (1994) *Tourism Planning: Basic Concepts and Cases*, 3rd edn. Taylor and Francis, New York.

Chapter 28

Management Tools and Techniques: an Integrated Approach to Planning

S. Backman[1], J. Petrick[1] and B.A. Wright[2]

[1]*Department of Parks, Recreation and Tourism Management, Clemson University, Clemson, South Carolina, USA;* [2]*Center for Recreation Resources Policy, George Mason University, Manassas, Virginia, USA*

Introduction

Ecotourism has been described as 'a complex phenomenon, involving integration of many actors including tourists, resident peoples, suppliers, and managers and multiple functions' (Ceballos-Lascuráin, 1993, p. 124). Moreover, these actors, or stakeholders, are engaged in a symbiotic relationship revolving around the idea of tourism as a means of economic development and as a means of promoting conservation of natural resources. If the ecotourism industry is to be developed more fully, and sustained for future generations, it will be necessary to understand the connections between conservation and tourism. In the 1987 report of the World Commission on Environment and Development, *Our Common Future*, it was suggested that problems of the environment and development could be solved if planning and decision making in those two spheres could be linked with cooperation between the tourism industry and government. Therefore the purpose of this chapter is to propose a model of planning, development and management of ecotourism that maximizes the opportunity to solicit input from the various stakeholders groups who could potentially be affected by management decisions.

Lessard *et al.* (1999) described a 'General Framework for Integrated Ecological Assessments' in which they argued that 'Managers are faced more and more with the need to make decisions involving complex cultural, social, economic and environmental issues' (p. 35) and must adapt management strategies and goals to fluid situations. Ecotourism management is an undertaking of such complexity. According to Lessard *et al.* (1999), this involves 'a continuing process of action-based planning, assessment, monitoring, research and adjustment with the objective of improving implementation and achieving desired goals and outcomes' (p. 35).

Fennell and Eagles (1990) described a framework for understanding and integrating conservation and visitor use. They suggested that tour companies, government agencies, remote communities and visitors could all prosper through cooperation and suitable planning. Within their conceptualization (Fig. 28.1), the resource tour (i.e. tourism to see a specific resource) is identified as the central focus of ecotourism,

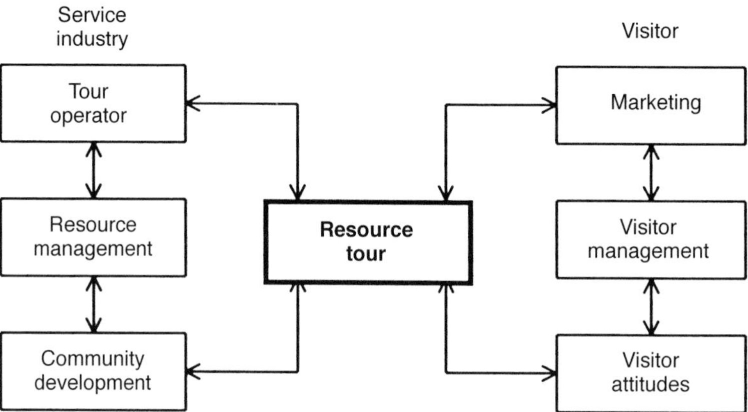

Fig. 28.1. Ecotourism conceptual framework (Fennell and Eagles, 1990).

influenced by both the service industry and the visitor. While some ecotourists seek their own experiences, many avail themselves of the infrastructure and assistance provided by tour operators, guides and/or interpreters. The natural settings that ecotourists desire are typically owned and managed by governments through institutions such as resource management agencies responsible for managing public parks and wildlife refuges. The framework further amplifies the interrelationship between community development and resource conservation.

The visitor component of the model includes marketing, visitor management and visitor attitudes. Marketing is essential to create visitor desire for the destination and can occur at the international, national and local level. The policies created by visitor management ultimately affect the visitors' experience by controlling allowable activities and proscribing others. Finally, the ecotourists' attitudes will be affected by the positive and/or negative experiences during their visit, which may in turn affect their cognitive assessment of their tour. This cognitive assessment will aid in the determination of whether or not to revisit the destination, and will also influence the decisions of others to travel to that destination. Fennell and Eagles (1990, p. 33) conclude that successful ecotourism ventures are dependent upon the 'complicated integration of public policy and private enterprise'.

Another framework for understanding and integrating conservation and visitor use is the ecotourism opportunity spectrum (ECOS) developed by Boyd and Butler (1996). ECOS incorporates concepts from the recreational opportunity spectrum (Clark and Stankey, 1979) and the tourism opportunity spectrum (Butler and Waldbrook, 1991) and was developed to give ecotourism destination management a similar conceptual framework with which to guide decision making. The eight components included in the ECOS framework are: (i) accessibility; (ii) relationship between ecotourism and other resource uses; (iii) attractions offered; (iv) existing tourism infrastructure; (v) level of user skill and knowledge required; (vi) level of social interaction; (vii) degree of acceptance of impacts and control over level of use; and (viii) type of management necessary to ensure long-term sustainability of the destination's resources.

In order to identify the viability of potential ecotourism offerings, the first seven components are assessed on scales ranging from eco-specialist to eco-generalist, with a midpoint of intermediate. An eco-specialist is a tourist requiring minimal infrastructure, having little impact, and generally participating as an individual or in a small group immersed in the local natural and cultural environment. On the other extreme, eco-generalists require a tourism infrastructure in order to be comfortable, have more impact, and usually participate in larger groups. These two poles are comparable to the hard–soft continuum that was described in Chapter 2, where the eco-specialist is the same as the hard ecotourist, and the eco-generalist equates to the soft ecotourist.

With the use of the ECOS, potential opportunities for ecotourism may be identified by management. Further, according to Boyd and Butler (1996, p. 565), if the ECOS can aid tourism marketers in 'attracting and maintaining the desired and appropriate type of ecotourist to a destination, then it could reduce the pressure on the area which a set of undifferentiated users would exert'.

While each of these models is instructive, they do not address the broad nature of ecotourism beyond the ecotour, nor acknowledge the diverse set of organizations and stakeholders that are the hallmark of the ecotourism industry. Therefore, this chapter now proceeds to propose and describe an integrated model for ecotourism planning that acknowledges the multi-organizational structure that constitutes the ecotourism industry and the cooperation and collaboration necessary to make such a structure work efficiently.

An Integrated Systems Model for Ecotourism Planning

Ecotourism organizations and their missions

The ecotourism industry comprises organizations that run the gamut from governmental agencies to non-government organizations (NGOs) and private enterprises (see Chapter 30 for more detail). Stakeholders in this process refer to these organizations, as well as local residents. Also included are interest groups that may benefit from or be affected in some way by the outcome of ecotourism development in a community, even if they themselves are not directly participating in the ecotourism sector (see Chapter 31). This organizational diversity complicates the challenge of ecotourism management, often leading to fragmentation of efforts, a high potential for conflict, and an overall lack of synergy and focus. Even among governmental agencies, the responsibility for managing ecotourism is rarely consolidated under a single department. The management of wildlife, forests, water and other natural resources, for example, is typically assigned to a single agency (e.g. the primary responsibility for managing wildlife is assigned to state wildlife agencies). In contrast, responsibility for developing and managing tourism most often falls under the purview of departments of commerce, economic development, or similar organizations. Coordination between these agencies is often sporadic at best. In many ways, ecotourism suffers from the problem of 'everybody's business is nobody's business'.

Fragmentation results from what Wight (1994, p. 39) termed as the two prevailing views of ecotourism: (i) that public interest in the environment may be used to market a product; and (ii) that this interest may be used to conserve the resources upon which the product is based. But she argues that these views need not be mutually exclusive, and may be complementary. Moreover, ecotourism has suffered historically from fragmentation of efforts among the different strata of organizations whose missions and policies affect the ecotourism industry (i.e. governmental, NGOs and private enterprises). For example, governmental agencies have regulatory power, as well as responsibilities for management of natural resources and economic development. In similar fashion, NGOs have traditionally reflected the more singular orientation of their affiliates in government. Long a stalwart of protection of critical ecosystems through land acquisition, The Nature Conservancy's stated mission, for example, is 'to preserve plants, animals and natural communities that represent the diversity of life on earth by protecting the lands and waters they need to survive'. For their part, private enterprises traditionally have been the 'front line' of suppliers of ecotourism products and services. They benefit economically from the provision of ecotourism goods and services and are the catalyst of economic development within the community. Whereas governmental and non-governmental organizations have been primarily oriented toward management and protection of resources, private businesses historically have been profit-oriented. After all, businesses that do not make a profit do not stay in business long (see Chapter 36).

However, a blurring of the historical distinctions between these organizations is beginning to occur, manifest in programmes related to the concept of ecotourism. An excellent example of this is The Nature Conservancy's Ecotourism Program. This provides 'technical assistance to [their] international partners in order to better harness the potential of ecotourism as a conservation tool that contributes to the long-term protection of biodiversity and the natural resources upon which it is based' (Ecotourism Society, 2000, p. 19). Additionally, private enterprises aspiring to capitalize on societal interest in the environment and recreational travel can no longer fail to recognize their dependency on the natural resource base, nor acknowledge the importance of resource protection and sustainability. Conversely, natural resource management agencies cannot afford to ignore the recreational use of natural and cultural resources. Although arguably still in its infancy, the recent movement by these agencies into the human dimension arena is encouraging, and moves governmental agencies further in line with the common purpose of ecotourism. Many resource management agencies, particularly state wildlife agencies, have begun to develop active human dimensions research programmes, develop positions specifically related to human dimension management, or revise their missions to include the provision of recreation and tourism opportunities. Illustrative of this evolution, in the past decade, is the Virginia Department of Game and Inland Fisheries' initiative to become more aggressive in conducting research with external stakeholders and to hire a 'human dimension specialist' to coordinate these efforts. Further, their mission statement was revised to include the provision of outdoor recreation opportunities

> To manage Virginia's wildlife and inland fish to maintain optimum populations of all species to serve the needs of the Commonwealth; *to provide opportunity for all to enjoy wildlife, inland fish, boating and related outdoor recreation*; to promote safety for persons and property in connection with boating, hunting and fishing.
>
> (Board of Directors, Dept. of Game and Inland Fisheries, 16 May 1990)

Even the longstanding distinction between profit-oriented private enterprise and the non-profit-oriented governmental and non-governmental entities is being eroded (Ziffer, 1989; Fennell, 1999). Many public agencies and NGOs are now actively promoting and delivering ecotourism programmes, tours and events, for a profit. Whether the motivation is economic or stewardship, organizational missions and strategies are beginning to coalesce around the idea of ecotourism, capitalizing on public interest in the environment. Therefore, the need for collaboration, cooperation and synergy among this multitude is as obvious as it will be challenging. In the past, economics have often run counter to protection and preservation interests. But ecotourism, perhaps for the first time, provides a feasible mechanism to align economic incentives with stewardship of the environment. It is at this interface, between the resource systems and human systems, that ecotourism exists and has the potential to flourish.

Management information system

The information necessary to successfully develop and manage a sustainable ecotourism industry is vast. Information needs range from advancing the knowledge of ecology and the tolerance of natural resources to human use, and the market segmentation of ecotourists. Research of interest to the ecotourism industry is regularly conducted by seemingly disparate organizations. For example, resource management agencies typically conduct research on wildlife populations, forests, wilderness and other common resources of interest to ecotourism managers. Inventories of natural resources, as well as cultural and historical resources, are commonly conducted by public agencies, including universities. The environmental impacts of development and the tolerance of resources to increasing human encroachment are assessed and mitigated through environmental impact assessments. Most of this research is undertaken for the purpose of protecting biodiversity, and promoting the preservation and sustainability of our natural and cultural heritage. Unfortunately, even though it has relevance, this research often has not been translated into the context of ecotourism, nor the results disseminated widely.

On the human side, private businesses and tourism-oriented corporations have great knowledge of their customers: their likes, dislikes, motivations and constraints. Chambers of commerce and economic development authorities invest large sums in gathering intelligence on which to base business and development decisions. Attitudinal surveys, needs assessments and social impact assessments have traditionally been the domain of marketers and social scientists, and are only beginning to find their way into natural resource management organizations. The melding of both natural and social science into comprehensive ecotourism research, therefore, is a relatively recent phenomenon. What is needed is a better mechanism for increasing the number of organizations involved in the collaborative conceptualization of ecotourism research, thus increasing the views of more stakeholders in the planning process, and providing better ways of disseminating the results throughout the industry.

The management information system described in our integrated systems model advocates broadening the number and types of organizations involved in ecotourism planning (see Fig. 28.2). This multi-organizational approach to ecotourism planning research has the advantage of being more cost-effective, as well as increasing the efficiency of information dissemination. However, gathering intelligence is only the first step in planning ecotourism projects. Involving stakeholders in the ensuing decision-making process is as important to the success of projects as gathering good intelligence.

Ecotourism management decisions

Although resource systems and human systems planning are depicted in this model as two separate components, this does not suggest that they function independently. Information is transferred between the two. Failure to include information related to stakeholders' preferences and opinions cannot be ignored. Using the information gleaned from the management information system (see above), ecotourism decision makers can begin the process of planning policies, programmes and development of ecotourism experiences. From this, consensus related to plans may emerge. Important actors in the management decision process are the stakeholder groups referred to earlier. Stakeholder participation in the ecotourism management planning process is critical to its success. Stakeholder participation refers to the opportunity for stakeholders to take part in the process of ecotourism development. If governmental bodies are to manage resources for future generations, then it is critical that such organizations gain the support for this approach to ecotourism development from the beginning. Stakeholders such as commercial operators may perceive no harm in attracting unlimited numbers of visitors to the resource each year. Resource planners on the other hand, are likely to espouse another view, considering their mission statement. Resource planners, cognizant of the social and environmental impacts that too many visitors will have on the resource and ultimately contribute to its demise, will have an opinion that may be in opposition to that of some commercial operators. Community residents may have another opinion.

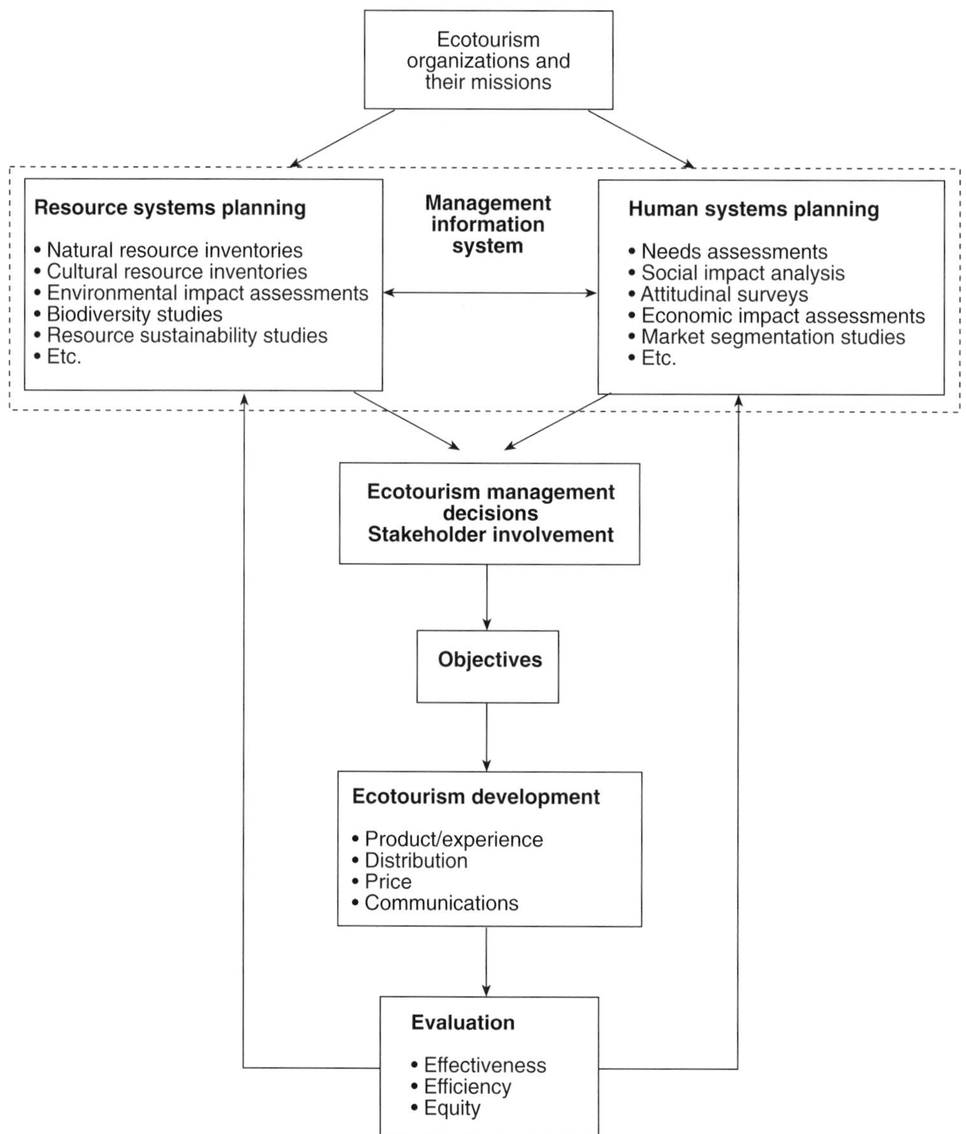

Fig. 28.2. An integrated systems model for ecotourism planning.

Stakeholders have a variety of opportunities or times during which they can participate in the process, including the planning stage, during implementation and evaluation, and in the distribution of benefits. Participation in the planning process includes such tasks as problem identification, formulating alternatives, activity planning, and resource allocation. Participants may also manage or co-manage ecotourism projects, hence stakeholders receive economic, social, cultural and/or other benefits from the project either alone or together.

Paul (1987) suggests these four levels of stakeholder participation: (i) information sharing; (ii) consultation; (iii) decision making; and (iv) initiating action. Information sharing refers to the exchange of information between stakeholders in order to facilitate action. For example, plans to expand a beachfront hotel should be shared with the stakeholders who may be impacted, negatively or positively, by the expansion. Failure to do so will only encourage the development of negative rumours and result in opposition to the project. In the next phase, consultation, stakeholders are not only informed, but also consulted on essential issues during the planning process. Stakeholder groups may possess the type of information needed by an agency to make timely and accurate decisions. The third level, decision making, involves stakeholders in making decisions about the ecotourism project. In these situations, stakeholders are equal partners. When stakeholders take the initiative in terms of actions and decisions related to the ecotourism project, they themselves begin the process.

There are many advantages to incorporating stakeholders in ecotourism projects. One advantage lies with the early warning system that they bring to the table. Stakeholders may provide information that, if suppressed or inaccessible, could otherwise cause conflict. Because stakeholder involvement fosters better planning and decision making, conflicts are brought out earlier and have a greater chance of being resolved. In addition, inclusion of stakeholders would provide opportunities for these participants to become aware of the benefits of the project, and to be more likely to provide support for the project. If stakeholders understand an ecotourism project, they are more likely to become proactive and involved with the project. Their inclusion also provides some aspect of legitimization to the ecotourism project, especially if they represent a diverse array of interests. Additional benefits, such as cost sharing and the protection of cultural norms, may accrue to the ecotourism project.

Ecotourism objectives

Management decisions regarding ecotourism projects and related issues will result in the need to establish specific, measurable project objectives. Just as the missions of the various organizations will vary with respect to its approach to ecotourism management, the objectives formulated for projects will also be somewhat different. Competing objectives result in conflict during implementation. Therefore, care must be taken to be cognizant of the opinions among the various stakeholder groups regarding objectives. Project objectives assist managers in guiding development. Objectives should be couched in terms that are measurable, results-oriented, and time dependent. They should address the concepts of effectiveness, efficiency and equity and form the bases of project evaluation.

Ecotourism development

It is important to consider the development of ecotourism products/experiences in the context of policy. Essential to the development of sustainable ecotourism is consideration of each of the impacts that a management decision may have on the resource's ability to deliver the experience the ecotourist seeks. Several aspects of ecotourism development must be considered together, as follows.

Product/experience

Ecotourism destinations can produce a broad range of experiences. They range from wildlife watching, to mountain biking, river rafting experiences and other activities that incorporate an adventure or cultural component (see Chapter 5). However, these activities are only the means by which ecotourists satisfy their needs. Research has shown that ecotourists are often motivated by the desire to escape or to relax, to see nature, and to participate in nature experiences as a means of socializing with friends (see Chapter 3). If managers concentrate on activities rather than the benefits visitors seek, the experience may be less than satisfactory to the visitor. When experiences do not fulfil ecotourists' needs, then the ecotourist is likely not to return. Failure to manage the benefits associated with the ecotourist's experience can also result in conflict among ecotourists. Mistakes of this nature will ultimately result in product or even destination failure.

Ecotourism managers are also asked to manage interactions between ecotourists and the resource. Vickerman (1988) stated that wildland fauna are particularly impacted by tourist development. Interaction with wildlife is often the critical aspect of the ecotourism experience. To that end, some tour operators provide regular feeding of wildlife such that they will appear and perform at the correct time, thus performing for tourists. But, while this may facilitate the ease with which visitors view wildlife, in most cases, feeding wildlife is frowned upon. These practices may actually do harm to wildlife, thus violating the premise of protecting the resource. Therefore, a common dilemma facing management of ecotourism destinations is how to minimize environmental damage and provide visitors with a memorable tourist experience, while affording local communities benefits from the process. A management tool that can aid in this process is the use of visitor codes. Visitor codes, also termed visitor guidelines or codes of conduct, put in writing principles that a destination would like their patrons to follow. It is believed that with the proper use of well-formulated visitor codes, incorporated within an overall management strategy, the impacts of visitors on natural environments can be reduced.

Several ecotourism destinations, including Belize, Madagascar and Nepal, have already instituted visitor codes. The Countryside Commission in England was one of the initial organizations to employ the use of a code of ethics for visitors more than 20 years ago (Mason, 1994). This code advises visitors to take their refuse home with them, help keep water clean and protect wildlife, plants and trees. The International Association of Antarctic Tour Operators has a code of ethics for visitors to the Antarctic which cites the need to dispose of waste materials properly, and informs visitors not to leave footprints in fragile environments or encroach upon the habitat of seabirds and animals in protected areas. It is hoped that this initiative will decrease the environmental damage caused by shipborne tourism to Antarctica (see Chapter 14).

Ecotourism managers may also utilize temporal and/or spatial zoning as a means of protecting resources and the ecotourism experience. The number of visits may be limited or restricted. For example, backcountry visitation is restricted in areas of Yellowstone and other national parks when the danger of bear–human encounters is high. Man-made structures, such as viewing stands and boardwalks, may be employed to restrict foot traffic in wetlands or other fragile ecosystems to protect the resource from too many ecotourists and their activities.

Distribution of products/experiences

In the past, many environments were protected from ecotourists because they were inaccessible, or due to the high cost of accessing the site. This, however, is no longer the case with most natural areas. The introduction of low cost airfares, new, lightweight backcountry equipment, and other lower participation costs have removed these barriers for many, thus increasing access to the pristine environment. This aspect of ecotourism planning refers to the ecotourists' access to the tourism product/experience. First, the destination must be accessible. How difficult is it for ecotourists to arrive at the destination? What are the destination's hours of operation and seasons? Additionally, decisions related to the use of tour wholesalers, tour brokers, travel agents or other travel intermediaries are made at this time. Selection of partners to distribute ecotourism experiences is critical to the effectiveness of ecotourism development.

Ecotourism destinations must ensure that all members of their distribution network share the same code of ethics for ecotourists. Recently, conflict between tour dive ships and those interested in protecting the resource has occurred. Although dive captains are supposed to instruct their divers not to touch coral reefs, many divers still do so as dive captains watch. Selection of dive ships in which to issue permits becomes an issue of distribution and protection of the ecotourism resource. Other conflicts among stakeholders in marine protection and ecotourism are described in Chapter 17.

Price

Price is one of the four major variables that the ecotourism manager controls. Price is important because it affects 'profitability'. Guided by an organization's objectives and mission statement, ecotourism managers must develop a set of pricing objectives and policies to guide their pricing decisions. These policies should spell out how flexible prices are, at what levels they will be set, and who shall receive discounts. Ecotourists exchange money and non-monetary costs for the benefits they expect to receive from the ecotourism experience. Monetary costs refer to user fees, admission charges, rates, fines, or fees. Monetary costs can also be attributed to the costs associated with travelling to the resource, or with the cost to obtain the equipment necessary to participate. Non-monetary costs include time, opportunity and effort costs.

Not all ecotourism agencies will charge for the experience. The decision to charge or not to charge often depends upon the costs associated with collecting the fees. In some cases it will cost more to collect the fees than the fees collected. Prices are usually charged if ecotourism agencies wish to recover costs or part of the costs from the tourists. Fees can also be charged to motivate the tourist to become aware of the value of the resource. In other instances, fees are charged to motivate resource managers. User fees can also be charged for admission to protected areas, as discussed in Chapters 18 and 23. However, agencies must determine their pricing objectives prior to determining the price to charge. These pricing objectives must be consistent with the agency's objectives. Prices may be charged to generate revenue, or for efficiency, equity or income redistribution, to reduce demand to place it in balance with the area's existing carrying capacity, or to increase the carrying capacity. Efficiency is concerned with getting the most out of a given set of resources. Equity means that the price should be fair. Income redistribution refers to the use of subsidies to assist tourists in the use of the resource.

Ecotourist managers must next decide which proportion of their costs they wish to recover. They can recover all fixed and variable costs, a proportion of fixed or variable costs, only fixed costs or only the variable costs. The key factor for ecotourism agencies to decide in setting the price is to discover what it costs them to deliver the experience. Next they must decide how much of their costs they wish to recover and why. Another consideration in using a price to recover agency costs is the life cycle stage of the experience. New ecotourism destinations may be expensive because access is limited, the costs are high, or because the destination may wish to communicate a value image using price. During the growth stage of the experience the pricing strategy may differ. The price charged to ecotourists could be used to deter certain types of tourists, or to keep visitation levels within manageable parameters. Increasing the price of the experience during peak seasons can also deter some ecotourists for the same reason. Changing the price serves to distribute demand for the experience across other seasons. Prices may be lowered to reflect lower costs, or they may be increased or they may stay the same. During declining popularity ecotourism managers may decide to lower the price so that they cover the minimum costs. The price to charge for the ecotourism experience may be restated as the cost of the experience to stakeholders. What will it cost stakeholders if the ecotourists do not come and what will it cost if the ecotourists do come?

Communications

Ecotourism destinations must develop communications programmes that portray the destination's desired image. The ecotourist destination may include the following components:

1. Advertising
2. Personal selling
3. Public relations
4. Incentives.

Advertising refers to all paid forms of communication from print to broadcast. Personal selling is the act of personal communication between individuals and/or groups. Public relations are those activities the destination engages in as a corporate citizen. For example, ecotourist destinations may sponsor a children's nature photography contest as a means of raising the level of awareness about the ecotourism experience.

Evaluation

Perhaps the most important component of the process is evaluation. Ecotourism managers must evaluate the effectiveness of all of their programmes and policies. Feedback from tourists' evaluation of their experiences can give destination managers the information necessary to change, modify, or delete existing programmes or policies. Problems associated with communication, pricing, products/experiences or distribution can only be addressed if ecotourism managers are aware of the problems. Information needed by ecotourism managers can be obtained from internal and external records held by the agency. Internal records are the data that the organization already have, such as guest records, permits, revenue generated, etc. To answer questions for which the agency does not have the information will require the agency to obtain the data. They may conduct research to answer the questions, or they may use outside sources, government documents for example, to answer the questions. Additionally, ecotourism managers will not be able to assess the agency's progress toward their goals if they are not evaluated.

Ecotourism managers may be interested in the efficiency with which a programme operates. Are the current visitor codes working? On the other hand, ecotourism managers may wish to know about equity. Equity refers to fair allocation of resources, when decisions are made. This attempts to address the issue of balance and who wins and who loses as the consequence of a decision. For example, how does the development of an ecotourism resort affect the residents of the community?

Conclusions

Agencies interested in ecotourism are multifaceted, and as such operate under many conflicting mandates at times. Thus there is likely to be conflict between the stakeholders in the management process. The ideas presented in this chapter offer a process consisting of management tools focused on an integrated approach to decision making. Adoption of this integrated model offers several advantages. Using the model presented in this chapter, assessments conducted at appropriate scope can provide more relevant and cost-effective information for decision making than some limited perspectives. A quality assessment can provide a synthesis of relationships of human and natural resources. Better experience development can be achieved by focusing on the needs and wants of the ecotourist and balancing this with the resources and capabilities of the agency. From human systems planning, stakeholders can learn about the social and economic issues facing ecotourism planners.

References

Boyd, S.W. and Butler, R.W. (1996) Managing ecotourism: an opportunity spectrum approach. *Tourism Management* 17(8), 557–566.

Butler, R.N. and Waldbrook, B.J. (1991) A new planning tool: the tourism opportunity spectrum. *Journal of Tourism Studies* 2(1), 1–14.

Ceballos-Lascuráin, H. (1993) Ecotourism as a worldwide phenomenon. In: Lindberg, K. and Hawkins, D. (eds) *Ecotourism: a Guide for Planners and Managers.* Ecotourism Society, North Bennington, Vermont, pp. 12–14.

Clark, R.N. and Stankey, G.H. (1979) *The Recreation Opportunity Spectrum: a Framework for Planning, Management, and Research.* US Department of Agriculture Forest Service, Pacific Northwest Forest and Range Experiment Station, General Technical Report PNW-98.

The Ecotourism Society (2000) *International Membership Directory.* Ecotourism Society, North Bennington, Vermont.

Fennell, D.A. (1999) *Ecotourism: an Introduction.* Routledge, New York.

Fennell, D.A. and Eagles, P.F.J. (1990) Ecotourism in Costa Rica: a conceptual framework. *Journal of Park and Recreation Administration* 8(1), 23–24.

Lessard, G., Jensen, M., Crespi, M. and Bourgeron, P. (1999) A general framework for integrated ecological assessments. In: Cordell, H.K. and Bergstrom, J.C. (eds) *Integrating Social Sciences with Ecosystem Management.* Sagamore Publishing, Champaign, Illinois.

Mason, P. (1994) A visitor code for the Arctic. *Tourism Management* 15(2), 93–97.

Paul, S. (1987) *Community Participation in Developing Projects: the World Bank Experience.* World Bank, Washington, DC.

Vickerman, S. (1988) Stimulating tourist and economic growth by featuring new wildlife recreation opportunities. *Transactions of the Fifty-Third American Wildlife and Natural Resources Conference.* Wildlife Management Institute, Washington, DC, pp. 414–423.

Virginia Department of Game and Inland Fisheries, Board of Directors' Official Minutes. May 16, 1990.

Wight, P. (1994) Environmentally responsible marketing of tourism. In: Cater, E. and Lowman, G. (eds) *Ecotourism: a Sustainable Option?* John Wiley & Sons, New York, pp. 39–55.

World Commission on Environment and Development (1987) *Our Common Future.* Oxford University Press, Oxford.

Ziffer, K. (1989) *Ecotourism: the Uneasy Alliance*, Working Paper No. 1. Conservation International, Washington, DC.

Chapter 29

Policy and Planning

D.A. Fennell[1], R. Buckley[2] and D.B. Weaver[3]

[1]Department of Recreation and Leisure Studies, Faculty of Physical Education and Recreation, Brock University, St Catherines, Ontario, Canada; [2]International Centre for Ecotourism Research, School for Environmental and Applied Science, Griffith University Gold Coast Campus, Queensland, Australia; [3]Department of Health, Fitness and Recreation Resources, George Mason University, Manassas, Virginia, USA

Introduction

Effective ecotourism management at the macro level is fundamentally related to the existence of appropriate and realistic policy and planning frameworks. The purpose of this chapter is to outline the current situation worldwide with respect to policy and planning within ecotourism itself. The first section defines both of these concepts and indicates their role within the context of tourism. The issue of complexity in the policy context is then addressed. Subsequent sections deal with ecotourism policy and planning in major world regions, including Australia, Asia, the Americas, Europe and Africa. Given space restrictions, this chapter does not include each of the hundreds of jurisdictions worldwide that engage in tourism and ecotourism policy and planning. Instead, a case study approach is adopted to demonstrate the variation among destinations, best practice, and general regional patterns. These case studies include both national and subnational examples.

Nature and Role of Policy and Planning in Tourism

'Policy' can simply be defined as a course of action that is adopted and pursued by a given stakeholder, and especially one in government. Policy provides the broad guidelines that are intended to shape the development of a particular sector or sectors in a way presumed by the relevant authority to be desirable. 'Planning' is the process by which policy is implemented, and a 'plan' is a document that articulates this intended process, usually after having been the focus of public consultation and political debate (Wilkinson, 1997). While plans are usually well defined, policy can be situated along a continuum from that which is formal and highly articulated within some kind of official document, to that which is informal, unwritten, and based on implied convention or consensus (Boothroyd, 1995).

In the tourism sector, the existence of policy and planning is associated with the potential for market failure, which describes the inability to attain equilibrium between demand and supply over the long term at the macro (i.e. destination) level. This can occur because of the unwillingness of individual businesses to undertake or contribute toward the management or marketing of the broader destination in which they are situated, since such investments will benefit their business competitors as well as themselves. (See the related discussion on 'social traps' in Chapter 41.) Given the relative weakness and fragmentation of the tourism sector as a whole, it is also unlikely that tourism-related industry associations will fulfil these essential marketing and management functions, the neglect of which leads to product deterioration and diminishing demand (Williams and Shaw, 1998). The fact that macro-level market failure does *not* occur in most destinations is due to the intervention of government as the vehicle through which destination management and marketing are invested and undertaken in the interests (theoretically) of all individual stakeholders. Accordingly, it is government that normally formulates and approves tourism-related policy and planning at the macro level, toward the fostering of a sustainable tourism sector. The main government engine driving the tourism policy and planning process at a country level is commonly referred to as a national tourism organization (NTO).

Complexity of the Policy and Planning Context

The clear positioning of tourism policy and planning as a government prerogative, however, does not mean that implementation is therefore a simple or taken-for-granted task. First, there is the problem of achieving compromise among all the stakeholders who constitute the tourism sector within a particular planning jurisdiction. Beyond this internal context, tourism policy can be pursued concurrently and in an often contradictory way by authorities at a local, regional, national and international level, each of which assiduously maintains and often attempts to expand its own sphere of influence. Similarly, just as tourism does not exist independently of competing resource sectors (see Chapter 31), tourism policy and planning is necessarily affected by the policy and planning that occurs within those other spheres, and by the public policy that governs the country, province or municipality as a whole. The most articulate tourism plan, for example, will be effectively derailed if the national government decides suddenly to impose strict visa requirements and dissuasive entry fees on all foreign arrivals in response to a perceived problem with illegal immigration (see Chapter 32). The tourism sector and their NTOs, in response to such threats, have often failed to demonstrate effective lobbying clout, thereby placing themselves in the disadvantageous position of having to react and adapt to, rather than influence, those external forces (Goeldner *et al.*, 2000). Adding to this complexity are the overlapping life expectancies of tourism and other plans; overall policy, for example, might change dramatically as a result of a national election, even though the tourism sector for that country may be only half way through its own 5-year plan.

Clearly, the framework within which policies and plans are actually articulated and implemented is immensely complex, leading Ackoff (1979) to use the term 'mess' to describe the real nature of the policy context. It is for this reason that policy statements are often deliberately worded in a vague and non-specific fashion, and that many tourism plans do not progress substantively beyond their release to the media. For a subsector such as ecotourism, the situation is even more perilous. In the first instance, it is highly unlikely that a nested hierarchy of tourism destinations (e.g. municipal, state, national) will all recognize the importance of ecotourism within the overall tourism sector, and even if they do, that they will all agree on its definition. Some, furthermore, may choose to articulate a discrete ecotourism plan, while others may imbed ecotourism within the overall tourism plan.

Ecotourism policy makers and planners must, in addition, contend not only with potentially incompatible external sectors such as the forestry and agriculture industries, but also with tourism activities that may interfere negatively with ecotourism, such as hunting and mass beach-based tourism. With few exceptions, and even in ecotourism strongholds such as Costa Rica and Kenya, it is often the interests of these more conventional tourism sectors that prevail. Accordingly, it is even less likely that an *ecotourism* plan will be carried through to full fruition, especially at a national scale where the 'mess' is evident in its full glory. More promising is the policy and planning framework that attends the scale of an individual protected area, since in such situations the array of potentially conflicting stakeholding sectors and jurisdictions is more curtailed.

This critical assessment is in no way intended to imply that the process of ecotourism-related planning and policy formulation should be abandoned. For one thing, such issues attend policy and planning within *any* sector. To the contrary, these processes at the very least provide an opportunity for ecotourism stakeholders to consult, interact, negotiate and hopefully emerge with some kind of consensus as to how the sector currently situates, how it should subsequently evolve, and what concerted action therefore needs to be taken. Allowance is usually made for periodic review and reassessment to take into account changes in the internal and external environment. If undertaken properly, plans and policy statements can exercise a significant amount of influence within and beyond the ecotourism sector, even if only some or none of this material is actually adopted as official policy.

Australia

Australia is the first 'region' to be presented, given its status as a world leader in ecotourism policy and planning. In reality, there is no indication that the mainstream tourism industry as a whole has necessarily moved further towards ecotourism in Australia than in other countries. However, where Australia arguably leads the world is in the explicit adoption of ecotourism principles, in name as well as in concept, in government planning and policies and in associated industry initiatives. Public-sector examples include the *National Ecotourism Strategy* (Australia, Commonwealth Department of Tourism, 1994) by the federal government of the day, ecotourism plans and strategies by several state governments, notably Queensland (see below), and a range of ecotourism plans, policies and principles prepared or adopted in piecemeal fashion by local governments throughout the country. High profile private-sector initiatives related to policy and planning include the very active Ecotourism Association of Australia and its National Ecotourism Accreditation Program (see Chapter 37), which has been endorsed by various government agencies. Other relevant initiatives include the National Nature and Ecotour Guide Certification Program, and references to ecotourism principles in the Strategic Plan produced by Australia's peak tourism industry association, Tourism Council Australia's *Our Heritage Our Business*.

One of the reasons that this progress has been possible may well be that there is no historical tradition of competing terminology, and few organized lobby groups to impede the institutionalization of ecotourism. It was therefore possible for a relatively small number of individuals dedicated to the promotion of ecotourism, in government as well as industry, conservation groups and research organizations, to generate relatively rapid understanding and acceptance of the term. However, there are still many tour companies which advertise 'ecotours' that at best fall under the category of 'nature-based' (Buckley and Araujo, 1997). On the other hand, the State Government of Western Australia has accurately described its tourism strategy for parks and public lands as a nature-based tourism strategy, recognizing that not all of the tourism activities concerned would qualify as ecotourism.

National Ecotourism Strategy

Australia's National Ecotourism Strategy was produced by the federal government tourism portfolio in the mid 1990s with the vision that Australia should have 'an ecologically and culturally sustainable ecotourism industry that will set an international example for environmental quality and cultural authenticity'. Its aims were to identify issues, develop a national framework, and formulate policies and programmes. It followed on from a National Tourism Strategy released in 1992, and a National Strategy for Ecologically Sustainable Development, also formulated in 1992. The Strategy incorporated comments from 149 submissions and 103 critiques of a draft report. It was coupled with small government funding programmes intended to encourage ecologically sustainable tourism.

The Strategy argues that ecotourism is a management philosophy, a market segment, and an integral part of the tourism industry. It summarizes stakeholders, definitions, impacts, issues and means of implementation. In concert with many policies and plans, it is comprehensive rather than detailed, addressing important issues such as integrated regional planning, infrastructure creep, impact monitoring, operator accreditation, public liability insurance, and economic instruments. The Strategy identifies ecotourism-related priorities, including accreditation, market research, energy and waste minimization, minimal-impact infrastructure, education and baseline studies and monitoring. Chapter 5 of the National Ecotourism Strategy for example, listed issues, objectives and actions under the following headings:

- Ecological sustainability
- Integrated regional planning
- Natural resource management
- Regulation
- Infrastructure
- Impact monitoring
- Marketing
- Industry standards and accreditation
- Ecotourism accreditation
- Involvement of indigenous Australians
- Viability
- Equity considerations.

Each of these, in turn, is associated with various sub-issues. Under the heading of 'impact monitoring', for example, the Strategy listed issues such as:

- sustainable levels of usage in different environments,
- which activities relate to which environmental impacts, and
- how to avoid and control degradation.

It specified, as Objective 6 of the Strategy, to 'undertake further study of the impacts of ecotourism to improve the information base for planning and decision making'. And it suggested seven actions, namely:

- develop the information base of natural and cultural attractions;
- investigate relevant indicators to monitor the environmental, social and cultural impacts of tourism;
- undertake ecological baseline studies and investigate the limits of acceptable change for ecotourism destinations;
- initiate long-term monitoring of the impacts of current tourism and recreation activities within, and adjacent to, protected areas and fragile ecosystems and review past experience;
- investigate means by which the ecotourism industry could contribute to research and monitoring of ecotourism impacts;
- investigate the economic and social significance of ecotourism;
- promote and facilitate the wide dissemination of ecotourism data.

Under the heading of industry standards and accreditation, the Strategy lists five actions, namely:

- develop and promote industry standards for ecotourism as a basis for industry self-regulation;
- investigate methods to identify and give recognition to ecotourism operators who establish and adhere to high standards;
- examine options for developing a national ecotourism accreditation system;

- develop environmental accreditation modules to encourage the adoption of best practice in ecotourism;
- explore the use of a logo for marketing of ecotourism products.

As of 2000, the National Ecotourism Strategy was apparently no longer official policy of Australia's federal government, which was of a different political persuasion from that which originally adopted the document. It was a very influential document, however, and clearly ahead of its time. The federal tourism portfolio has in fact followed up with actions or more detailed studies on each of the priorities identified, with one glaring exception: environmental baseline studies and impact monitoring remain almost completely neglected, except for independent initiatives by academic ecologists and protected area agencies.

Queensland

The maturity of Australia with respect to ecotourism policy and planning is indicated by the active involvement in ecotourism at the state and regional level. The *Queensland Ecotourism Plan* is a state government publication produced by the Queensland Department of Tourism, Sport and Youth. A draft plan was circulated for public comment in 1995, and a final version published in 1997. Its vision is for 'ecotourism in Queensland to be ecologically, commercially, culturally and socially sustainable'. Ecotourism, moreover, is expounded as a model for other forms of environmentally responsible tourism.

The Queensland Ecotourism Plan defines ecotourism as 'nature-based tourism that involves education and interpretation of the natural environment and is managed to be ecologically sustainable'. It recognizes that this 'involves an appropriate return to the local community and long-term conservation of the resource'. The Plan acknowledges three broad styles of ecotourism, which it refers to as self-reliant, small-group and popular. The last of these is essentially the mass nature-based tourism or 'soft ecotourism' that is described in Chapter 2.

The Plan contains numerous suggestions for practical implementation, but since many of these are within the domain of other government agencies and little funding has been provided, few have yet been adopted. However, the environment branch of the state tourism promotion agency, Tourism Queensland, has subsequently done a great deal to promote and enhance ecotourism in Queensland. For example, it has been a strong supporter of the National Ecotourism Accreditation Program, has carried out market surveys and studies of ecotourists, liaises closely with land management agencies, and publishes a very useful newsletter, *Ecotrends* (also available on the Internet).

Differences between the draft and final versions of the Queensland Ecotourism Plan provide an interesting illustration of the way in which such documents evolve. Both versions contain tables listing visitor characteristics, biogeographical regions, and responsibilities of various levels of government, for example, but the draft is more detailed. Similarly, both versions noted that the Commonwealth (federal) Government is responsible for 'protection of outstanding universal values of World Heritage properties', and that it does this through 'regulations to prohibit activities which might place World Heritage values at risk'. In the draft this was in the main text (Table 6.1); in the final version it was in an appendix (A 3.1).

The draft version contained six detailed tables of strategies, listing policy areas, desired outcomes, existing situations, recommended actions and agencies involved. The six major strategy areas were:

- environmental protection
- product development
- infrastructure development
- marketing and promotion
- local community involvement
- planning and management.

In the final version, these were expressed as Action Plans, listing actions, responsibilities and time frames. Environmental protection and management planning were linked, and so were product development and marketing and promotion. Infrastructure development was unchanged, but community involvement was divided into two categories, local and general community development.

South Pacific

Despite its high profile as an ecotourism destination, New Zealand is embryonic with respect to the development of its ecotourism policy and planning (see Chapter 9). Ironically, Fiji is more advanced in this respect, despite that country's concentration on mass beach-based tourism, having prepared an official ecotourism plan in 1997 (see Chapter 9). Among the remaining South Pacific destinations, Samoa is notable for its apparent pursuit of an ecotourism-dominated tourism policy that is evident in the formulation of a National Ecotourism Programme and a Samoan Ecotourism Network (Weaver, 1998).

Asia

Historically, tourism in Asia has been focused on cultural and architectural attractions, ancient and modern, and in some enclaves (e.g. Singapore, Hong Kong, Bangkok) on shopping, nightlife and gambling. In recent years, however, specialist tour operators in a number of Asian nations have begun to offer adventure and nature-based tours, often linked with cultural attractions, and some marketed under the term ecotourism (see Chapter 8). Adventure tourism opportunities have been promoted through international outdoor sport competitions, and by magazines such as *Action Asia*.

Ecotourism planning and policy in Asian nations has most commonly been initiated by non-government organizations rather than national government tourism agencies. The Worldwide Fund for Nature, for example, initiated work on a National Ecotourism Plan for Malaysia, adopted by the Malaysian Government in 1997. In 1999 the World Conservation Union, after several years of planning, held a national workshop on a National Ecotourism Strategy for Vietnam, in conjunction with the Vietnam National Administration for Tourism and other relevant government agencies including both the investments and environment portfolios. The workshop was co-funded by multilateral and bilateral aid agencies. While the workshop was very successful, it is worth noting that advertising by the Vietnamese government and government-run tourism agencies at a large international tourism trade show in Hong Kong, held at the same time, did not reflect an ecotourism theme. It remains to be seen whether this will change in future years. The Vietnam workshop produced a proceedings volume (Luong *et al.*, 1999), but has not yet yielded any official government strategy on ecotourism.

Thailand

According to an article in one of Thailand's regional newspapers, the *Pattaya Mail* (www.pattayamail.com/306/features), the Thailand Institute for Scientific and Technological Research (TISTR) produced an 'action plan to facilitate a national ecotourism strategy' some time in the mid 1990s. The publication list on the TISTR web site (www.tistr.or.th), however, makes no mention of such a strategy. Other sites, including that of the Tourism Authority of Thailand (TAT) (www.tat.or.th) mention various promotional campaigns and directories for nature and adventure tourism, such as the 'Amazing Thailand' Campaign and the Thailand Travel and Adventure Guide (www.thailand-travelsearch.com). The 'Amazing Thailand' site includes 'eco-adventure' company listings, but no policies. There are also references in magazines such as *Action Asia*, to a Green Hotels project, TAT environmental awards, a Thai Ecotourism Tour Operators Association, and a Thai Ecotourism and Adventure Travel Association. It is not clear whether these are functioning organizations or just good ideas. They do not appear to maintain web sites. In any event, there does not seem to be any coordinated national ecotourism strategy that is actually used as a basis for government policy or action in Thailand. Government publications from the mid-1990s, however, do state that TAT in 1996/97 was pushing for the formulation and adoption of a national ecotourism policy (TAT, nd).

Bhutan

The Kingdom of Bhutan, a small country in the eastern Himalayas, apparently does not have a written ecotourism strategy by that name, but its management of foreign visitors and attractions effectively means that its entire national tourism industry is arguably a form of upmarket ecotourism. Hence, this may be described as an example of *de facto* ecotourism policy. Bhutan has been predominantly Buddhist since the 7th century, and still has a very rich natural environment that supports over 700 bird species and larger mammals such as snow leopard and golden langur. Visitors are limited to about 5000 per year in total (5361 in 1999) and must travel on preplanned itineraries and stay in tourist hotels (www.kingdomofbhutan.com).

The Americas

The Americas (North, Central and South) comprise over 50 countries and dependencies which, combined, contain a tremendous variety of social and ecological conditions. This unbroken chain of land is the most geographically extensive in the world, ranging from about 80°N in Canada, to 55°S in Chile. It is also the most ecologically rich, containing almost all of the major world ecoregions found on Earth. These range from icecap and tundra provinces in the northern portions of Canada, to rainforest and rainforest altitudinal zones in Central and South America.

Such heterogeneity has meant that, until recently, there was virtually no unified look at ecotourism in the Americas. Edwards *et al.* (1998a, b) changed this through their comprehensive overview of ecotourism policy in this broad region, which examined each country and dependency (and separate states and provinces in the USA and Canada) on the basis of definition, policy development, and other related policy issues. These authors found that in Anglo-America, for example, Canada has developed written policy at the national level, whereas the USA has not. In the Latin American countries (LAC), 25 nations/dependencies had developed ecotourism policy documents (Box 29.1). The authors also discovered differences in the nature of policy among US states and provinces, and the policy developed in LAC nations. For example, more of the US and Canadian ecotourism contact persons were found in marketing-related positions than their LAC counterparts. In addition, more of the LAC contact people (19%) were housed directly in ecotourism or environment positions, whereas only 6% of US and Canadian contacts were in similar positions. This finding suggests that LAC countries have governmental structures that are more ecotourism specific in their orientation in comparison to those in Anglo-America.

More specific analysis of ecotourism policy in the Americas, beyond the

Box 29.1. Country/dependency demonstrating policies in ecotourism (Edwards *et al.*, 1998a).

Antigua and Barbuda	Grenada
Belize	Guatemala
Bolivia	Guyana
Brazil	Honduras
Canada	Mexico
Chile	Nicaragua
Colombia	Paraguay
Costa Rica	Peru
Cuba	Saba
Curaçao	St Lucia
Dominican Republic	Suriname
Ecuador	Trinidad and Tobago
El Salvador	

Edwards *et al.* study, demands a focus on the policies that have been developed by individual countries, states/provinces, or dependencies. While some documents have been in print for up to 5 years (in the case of Brazil), others are still evolving. Consequently, this section is very applied in response to the dearth of theoretical and conceptual literature on the subject. Four case study destinations (the province of Manitoba in Canada, the state of Florida, USA, Brazil and the Republic of Guyana) will be featured to represent policy development initiatives in this broad region.

Manitoba, Canada

In Canada, provinces and territories are responsible for recreation and tourism planning. In such a large and diverse country, this gives political jurisdictions the needed flexibility to develop policies that are reflective of their unique social and ecological milieux. In response to the increased demand for ecotourism in Canada, policy documents have emerged over the past few years from several provinces and territories. The mid-western province of Manitoba is a representative example of ecotourism policy development in Canada. Travel Manitoba is the agency responsible for tourism in the province, and also for the discussion paper on ecotourism (Keszi, 1997). This document has the following policy goal: 'to contribute to Manitoba's economic and environmental well-being by promoting the development of an ecotourism industry that is domestically viable, internationally competitive, and sensitive to the surrounding ecological, cultural and economic environments'. To accomplish this end, objectives have been established for eight different policy areas: sustainability, business viability, integrated resource management planning, infrastructure, leadership and cooperation, marketing, aboriginal involvement, and awareness and understanding. As an illustration, the policy directives for sustainability have been articulated as follows:

Sustainability: Adopting a supply-driven approach to tourism development requires that activities be limited according to the ability of an area to deliver a sustainable tourism experience.

Policy 1.1: Ecotourism activities shall be planned and developed in a manner that respects the economic integrity, and the ecological, cultural and visitor carrying capacity of the host area.

Policy 1.2: Monitoring efforts shall be undertaken to ensure that ecotourism development is being carried out in a manner that is respective of the environment and people.

Policy 1.3: The economic involvement of local residents in ecotourism development activities shall be encouraged.

Policy 1.4: The incorporation of conservation efforts into ecotourism development activities shall be promoted.

A key policy issue in Manitoba relates to incidental or secondary activities (see Fennell, 1999). Joe Keszi, of Travel Manitoba (personal communication, 4 November 1999), suggests that more consumptive activities, such as fishing, are often central to the ecotourism experience (in addition to wildlife viewing and interpretation). He suggests that one of the only ways to sell ecotourism to the northern outfitters is to include a more well-rounded package of ecotourism, adventure tourism and cultural tourism. Tourists from abroad have come to expect this type of diversified experience as part of the overall 'ecotourism' package. A similar issue relates to the principle of sustainability and its applicability to ecotourism. It appears that in Manitoba, like many other Canadian provinces, consumptive tourism products such as fishing are considered ecotourism if they are deemed sustainable. However, as the literature on tourism sustainability suggests, any form of tourism has the potential to be sustainable, which is not to say that it is also ecotourism.

Florida, USA

In Florida, ecotourism policy has been developed through a cooperative effort of many public, private and not-for-profit enterprises, including all levels of government, commercial enterprises, conservation organizations, historical and archaeological groups, museums, the tourism industry, tourism commissions and councils, and operators. Florida's ecotourism policy, developed by the Ecotourism/Heritage Tourism Advisory Committee (1997), is a blueprint for the state's future development of ecotourism and heritage tourism. It identifies goals, strategies and recommendations to protect and promote the natural, coastal, historical and cultural assets of Florida, with the purpose of linking these to commercial tourism in Florida.

Florida's policy is based on five main components: strategic relationships (between stakeholders, as identified above), inventory (of natural and cultural resources), protection (of the natural and cultural heritage of the state), education (of those involved in the industry), and marketing (recommendations on how to overcome the fragmented state of ecotourism marketing in Florida). Each of these components is further broken down into a series of goals and strategies. For example, the education component has three goals, the first of which is as follows:

Goal 1. Develop local and regional training and credentialing/certification programs for ecotourism and heritage tourism providers.

Strategies.

1.1 Support and encourage the creation of guidelines to support local and or regional and state incentive-based credentialing/certification programs developed for ecotourism and heritage tourism providers.

1.2 Develop incentives for those participating in the credentialing/certification programs.

1.3 Develop a system to recognize and approve credentialing/certification programs.

Like many other political jurisdictions in North America, Florida has developed a liberal definition of ecotourism. In Florida ecotourism is defined as: 'Responsible travel to natural areas which conserves the environment and sustains the well-being of local people while providing a quality experience that connects the visitor to nature' (Ecotourism/Heritage Tourism Advisory Committee, 1997). Included in this definition are nature-based tours; managed access to sanctuaries; wildlife viewing; visitation to natural areas such as beaches, forests, lakes and greenways; Native American Reservations; and outdoor recreational activities such as hiking, canoeing, snorkelling, horseback riding, boating, diving and fishing. As such, virtually any type of outdoor activity is ecotourism. There is an inherent danger in this 'shotgun' approach to ecotourism, as it assumes that more consumptive forms of outdoor recreation, like fishing, are synonymous with other non-consumptive, low-impact forms of ecotourism like birding and nature appreciation (see Fennell, 2000). However, given the massive extent to which conventional tourism has infiltrated the state of Florida, such a broad approach may be appropriate.

Brazil

As the world's fifth largest country by size, Brazil is one of the most biologically diverse regions on Earth. The Amazon River basin, as Brazil's ecological hub, contains some 20% of the world's fresh water and the world's most extensive rainforest system. Occupying some 42% of the Brazilian landmass, the Amazon region contains 2500 different kinds of fish, 50,000 higher plant species, millions of insect species, and over 1000 tributaries (Taylor, 1996). This ecological wealth has made Brazil a key destination for the ecotourist. Weaver (1998), for example, writes that Brazil, along with Costa Rica, is a top ecotour venue, listed by two widely distributed eco-guide publications. The value of ecotourism to Brazil is underscored through the development of a national ecotourism policy (Grupo de Trabalho Interministerial MICT/MMA, 1994). This document, like Manitoba, addresses many of the key social, ecological, and economic issues related to ecotourism development. The document defines ecotourism as: 'A segment of tourism that uses, in a sustainable way, natural and cultural patrimony, encouraging its conservation, which seeks the formation of environmental awareness by way of environmental interpretation, promoting the well-being of the populations involved'.

The main objectives of the policy document are to: (i) make ecotourism activities compatible with the conservation of natural areas; (ii) strengthen inter-institutional cooperation; (iii) make possible the effective participation of all relevant segments of the sector; (iv) promote and stimulate the capacity of human resources for ecotourism; (v) promote, stimulate and provide incentives for the creation and improvement of infrastructure for ecotourism; and (vi) promote the use of ecotourism as a vehicle for environmental education. These main objectives are to be achieved through a number of prioritized actions and strategies (see Table 29.1), involving non-government organizations (NGOs), government, industry, and other interested stakeholder groups. The strategies for objective four, Quality control of ecotourism products, include the following.

- Inspect ecotourism services and operations.
- Establish and develop processes and methodologies to evaluate the impacts of ecotourism on the environment.
- Propose ways to involve ecotourists in the monitoring and carrying out of inventories and research in visited natural areas.
- Identify reference models for ecotourism services and operations.
- Encourage the creation of a self-regulatory system in the private sector, with the participation of consumers.
- Foster and develop research directed at ecotourism quality control.

The concept of self-regulation in the private sector is one that will be questioned in Brazil, and other countries (Gunningham and Grabowsky, 1998). While ethical conduct may exist in a market culture environment, it is often not intrinsic, but rather extrinsic in terms of avoiding punishment (domestic laws) or seeking rewards (good ethics relates to a good reputation) (Malloy and Fennell, 1998).

Guyana

Located on the north coast of South America, Guyana is a small nation with tremendous biodiversity in environments that range from tropical rainforest (the northern reaches of the Amazon basin) to dry, barren lands. Historically Guyana has had trouble attracting tourists, which is a consequence of the lingering image of the Jonestown Massacre of 1978. Other contributing factors include few human resources, a legacy of anti-tourism sentiment within the government, malaria, inadequate financing, and poor international and national transport networks (www.excite.com/travel/countries/guyana/?page=overview). However, given the natural environment and greater political stability, Guyana is now making a concerted effort to develop its ecotourism industry. The national plan for ecotourism development in Guyana (Republic of Guyana, 1997) has defined ecotourism as follows: 'a form of travel for pleasure that has a low impact on the natural and cultural environment, gives the visitor a better understanding of the unique qualities of the place being visited, contributes to the well-being of local Guyanese, and promotes conservation of Guyana's resources'.

Guyana's ecotourism policy is based on 24 ecotourism policy subjects (see Box 29.2). Each subject is further broken down into a policy statement, details related to

Table 29.1. Brazil's guidelines for a National Ecotourism Policy (Diretrizes para uma Politica Nacional Ecoturismo, 1994).

Action	Strategy
Regulating ecotourism	Provide the ecotourism segment with its own legal structure, which is in harmony with federal, state and municipal spheres, and follows adequate parameters and criteria
Strengthening inter-institutional interaction	Promote the articulation and the exchange of information and experience between government agencies and private sector entities
Formation and training of human resources	Encourage the formation and training of personnel to perform the diverse functions relevant to ecotourism
Quality control of ecotourism products	Promote the development of methodologies, models and systems to accompany, evaluate and perfect ecotourism, linking the public and private sectors
Information management	Carry out a search for information at a national and international level, aiming at the formation of a database and obtaining indicators for ecotourism development
Incentives for ecotourism development	Promote and stimulate the creation of adequate incentives for the improvement of technology and service, the amplification of existing infrastructure and the implementation of ecotourism enterprises
Implementation and suitability of infrastructure	Promote the development of technology and the introduction of infrastructure in priority ecotourism destinations
Tourist awareness and information	Promote the activities that are inherent in ecotourism to the tourist, and orient tourists to behave appropriately in visited areas
Community participation	Seek to employ the communities located in potential and existing ecotourism destinations, encouraging them to identify ecotourism as a viable economic alternative.

the particular subject (supporting community enterprises), responsible parties (e.g. Tourism Association of Guyana), consulting parties (e.g. Local Development Councils), time frame for implementation, cost implications and benefits. Each of these is outlined below, using accreditation as an example:

Ecotourism Policy Subject:
Accreditation

Details:
Accreditation shall fall within the established legal definition of ecotourism.
Applicants shall demonstrate that they substantially comply.
Monitoring shall be done on an annual basis or more frequently if non-compliance issues are at issue.

Responsible Parties:
Ministry of Trade, Tourism and Industry
Tourism Advisory Board

Consulting Parties:
Tourism Association of Guyana
Private Sector Commission of Guyana

Time Frame for Implementation:
1998 establish compliance model
1999 communicate model to operators to apply for accreditation
2000 begin accreditation and monitoring process

Cost Implications:
Training of one additional staff member as field monitor for Department of Tourism.
Initial training and establishment of accreditation procedures = US$9500.
Annual salary = US$20,000

Benefits:
International travel agents will be assured of a reasonably consistent range of ecotourism products and services in Guyana. This will reassure international visitors concerned with quality and safety.

Guyana is clearly at an early stage in the development of ecotourism. A search of the Specialty Travel Index found no advertisements for ecotourism in Guyana. While there appear to be some significant constraints facing the industry (as stated

Box 29.2. Ecotourism policy subjects for Guyana (Guyana Ministry of Trade, Tourism and Industry, 1997).

1. Definition of ecotourism
2. Assuring local involvement and benefits
3. Contribution to conservation
4. Respecting rights of indigenous peoples
5. Integration with parks and protected areas
6. Integration with other land uses
7. Waste management
8. Energy and water conservation
9. Facility design standards
10. Preservation of architectural heritage
11. Life safety and security standards
12. Hotel green management
13. Monitoring environmental impacts
14. Training/education standards
15. Guidelines for marketing ecotourism
16. Financial support and incentives
17. Public awareness
18. Destination promotion
19. Accreditation
20. Supporting institutional framework
21. Measuring success
22. Creating regional networks
23. Environmental education
24. Transportation links

above), Guyana is advantaged by the fact that it has a detailed definition and policy that will aid in the ecotourism development process. Guyana can follow a logical progression of: (i) definition; (ii) policy; (iii) practice; and (iv) implications, whereas other countries have had to deal with the implications of an ecotourism industry that has already been established long before the articulation of appropriate definitions and policies.

Europe

Unlike the Americas, no comprehensive study has yet been undertaken with respect to European ecotourism policy. However, from surveys of the relevant tourism authorities, and examination of the pertinent academic and government literature, regional trends can be discerned. In the first instance, it soon becomes clear that explicit ecotourism policies are exceptional rather than normative in the European context. Since the collapse of the Soviet bloc in the early 1990s, eastern European countries such as Romania and Albania have given prominence to ecotourism in their tourism policies, although Hall and Kinnaird (1994) suggests that such adoptions are more rhetoric than substance.

In most European countries, ecotourism policy, such as it is, tends to be implicit and indirect. In many cases, this *de facto* recognition entails references to the importance of 'nature-based' and 'rural' tourism within the overall national tourism strategy. Hence, even if the term 'ecotourism' itself is not used, these oblique references may fulfil the first fundamental criterion of the sector, which is the focus on the natural environment as an attraction base (see Chapter 1). The third criterion, sustainability, is almost universally apparent to the extent that virtually every European tourism policy purports an adherence to the principles of environmental and socio-cultural responsibility. A survey of European NTOs, conducted in the early 1990s, supports this contention. Results revealed the 'promotion of environmental tourism' to be an important policy consideration in many countries, but especially Belgium, Italy and Portugal. Overall, this item rated a mean score of 3.1 out of 5 among all responding NTOs, behind 'attraction of high spending tourists' (4.7), 'improving the quality of the tourism product' (4.2), 'reducing seasonality' (4.0), and 'spreading the benefits of tourism beyond areas of concentration' (3.6) (Akehurst *et al.*, 1993). All of these policy objectives are compatible with ecotourism, and suggest tacit recognition of the latter even in situations where no explicit mention is made. The case of Switzerland is illustrative. The NTO representative has indicated (personal communication, 1995) that the term 'ecotourism' was not used in national policy because of the confusion surrounding the term. However, sustainability was identified as a prime directive of all Swiss tourism development, and the importance of the natural environment as the core component of the national tourism product was emphasized. Other 'green' policies that provide similar recognition have been put forward by Poland, Slovenia and Slovakia (Hall, 1998).

Declaration of adherence, be it explicit or tacit, does not equate with implementation, and it is useful to consider the extent to which different types of jurisdiction are most likely to translate policy into action. In general, tourism policy at the national level is strongest among the more prescriptive, centralized (or unitary) states, including France, Portugal and Greece, and among microstates such as San Marino and Liechtenstein, which are also unitary but function in effect at the level of a single-tier municipality. Tourism policy, in contrast, is weakest in decentralized federal states such as Germany and Belgium, and in regional states such as the UK (Akehurst *et al.*, 1993). A larger dynamic related to the latter group is the gradual concession of power at the state level in favour of sub- and super-state structures. The devolution to 'regional' or 'national' authorities is already evident in the UK, where Scotland, England and Wales are largely left to their own volition (Airey and Butler, 1999), and in the autonomous regions of Spain, where nature-based tourism policies are present and actively being implemented (Valenzuela, 1998).

Multilateral policy structures, and especially those associated with the European Union, are also becoming increasingly relevant to ecotourism (see Chapter 10). For example, the 1985 European Commission (EC) policy framework for tourism, and the 1991 EC Action Plan for tourism, both emphasize environmental protection (i.e. sustainability), the promotion of rural tourism, support for small business and the diversification of regional economies (Williams and Shaw, 1998). Intriguingly, some multilateral programmes, such as LEADER (liaison entre actions de développement de l'économie rurale), are directed specifically at the regional rather than country level (Jenkins *et al.*, 1998), thereby indicating the emergence of a sub-state/super-state framework that will increasingly supersede country-level tourism policy and planning. Other ecotourism-related, super-state policy initiatives are directed toward the politically emancipated countries of Eastern Europe. The Phare programme, for example, promotes sustainable rural development and the establishment of natural/cultural heritage trails in Macedonia, the Czech Republic and Slovakia.

Africa

Ecotourism-related policy in Africa is most evident within the 'Safari corridor' that extends from Kenya to South Africa (see Chapter 16). We say 'ecotourism-related' because this sector tends to be embedded within a broader tourism policy framework that emphasizes an amalgam of wildlife resources, protected areas, and community participation and empowerment (Weaver, 1998; Honey, 1999). Explicit ecotourism policy is rare, although the South African government did declare 1996 to be the 'Year of Eco-Tourism' as part of its strategic product positioning. While the more common allusions to wildlife and community participation seem compatible with ecotourism, a careful analysis of policy in this region reveals several serious contradictions. Several countries have explicitly embraced big game hunting as an acceptable mode of wildlife-based tourism, the best known example being Zimbabwe's CAMPFIRE programme (Derman, 1995). Calling into question the apparent commitment to community-level engagement is the tendency for tourism planning and policy to be highly centralized at the state level, although South Africa may be exceptional given its regional/federal system of government.

Conclusions

As stated in the introduction, space restrictions preclude an inclusive examination of global ecotourism policy and planning. Hence, the intent of this chapter has been to identify relevant patterns on a regional basis, and to provide a feel for these patterns through the presentation of specific material from a variety of relevant case studies. Australia stands out for the relative sophistication of its ecotourism policy and planning, while Asian engagement is more tentative and more likely to involve NGOs. In North America, ecotourism policy tends to have a liberal connotation, with jurisdictions such as Manitoba and Florida including consumptive activities such as fishing under this rubric. In contrast, ecotourism in Latin America tends to be defined more narrowly, and is more independent administratively and in practice. Further, Canada and the USA show less leadership in ecotourism from the national level, and more from regional governments, whereas LAC countries seem to rely more on national directives for the administration of ecotourism. Accreditation, regulation and/or monitoring of operators are themes that are consistent with all case studies in the Americas. In Europe, ecotourism policy is seldom explicit, but usually embedded within broader tourism policies through references to nature-based tourism, rural tourism and sustainability. State-level structures are also giving way to sub-state and super-state frameworks as the drivers of tourism policy, suggesting a development that may occur in other regions when they reach a similar level of geopolitical integration. Finally, African ecotourism policies, similarly, are oblique, being situated within a broader framework of wildlife, protected areas and community participation.

Whether policies are direct or indirect, actual implementation in all regions is impeded by the reality that ecotourism is relatively new, and hence relatively weak in terms of influence in comparison to more established segments of the tourism industry and, just as importantly, other industries such as agriculture and forestry. Where ecotourism is disadvantaged in this way, it may indeed be logical to embed the sector within a broader policy framework, as is apparently the case in Europe and Africa, or within a broader definitional framework, as in North America.

Acknowledgement

Thanks to Steve Edwards of *Conservation International* for providing information for the Americas section of this chapter.

References

Ackoff, R. (1979) The future of operational research is past. *Operational Research Journal* 30(1), 93–104.

Airey, D. and Butler, R. (1999) Tourism at the regional level. In: Keller, P. and Bieger, T. (eds) *Future-Oriented Tourism Policy: a Contribution to the Strategic Development of Places.* AIEST, St-Gall, Switzerland, pp. 71–90.

Akehurst, G., Bland, N. and Nevin, M. (1993) Tourism policies in the European Community member states. *International Journal of Hospitality Management* 12(1), 33–66.

Australia, Commonwealth Department of Tourism (1994) *National Ecotourism Strategy.* Australian Government Publishing Service, Canberra.

Boothroyd, P. (1995) Policy assessment. In: Vanclay, F. and Bronstein, D.A. (eds) *Environmental and Social Impact Assessment.* John Wiley & Sons, New York, pp. 83–128.

Buckley, R.C. and Araujo, G. (1997) Green advertising by tourism operators on Australia's Gold Coast. *Ambio* 26, 190–191.

Derman, B. (1995) Environmental NGOs, dispossession, and the state: the ideology and praxis of African nature and development. *Human Ecology* 23, 199–215.

Ecotourism/Heritage Tourism Advisory Committee (1997) Recommendations on the statewide plan to protect and promote the natural, coastal, historical, cultural, and commercial assets of Florida.

Edwards, S.N., McLaughlin, W.J. and Ham, S.H. (1998a) *Comparative Study of Ecotourism Policy in the Americas – 1998. Vol. II Latin America and the Caribbean.* Organization of American States.

Edwards, S.N., McLaughlin, W.J. and Ham, S.H. (1998b) *Comparative Study of Ecotourism Policy in the Americas – 1998. Vol. III USA and Canada.* Organization of American States.

Fennell, D.A. (1999) *Ecotourism: an Introduction.* Routledge, London.

Fennell, D.A. (2000) What's in a name: conceptualising natural resource-based tourism. *Tourism Recreation Research* 25(1), 97–100.

Goeldner, C., Ritchie, J. and McIntosh, R. (2000) *Tourism: Principles, Practices, Philosophies*, 8th edn. John Wiley & Sons, Chichester, UK.

Grupo de Trabalho Interministerial MICT/MMA (1994) *Diretrizes para uma Política Nacional de Ecotourismo/Coordenaçao de Silvio Magalhaes Barros II e Denise Hamú M de La Penha.* Embratur, Brasília.

Gunningham, N. and Grabowsky, P. (1998) *Smart Regulation: Designing Environmental Policy.* Clarendon Press, Oxford.

Guyana Ministry of Trade, Tourism and Industry (1997) *National Plan for Ecotourism Development.* Recommended Support Policies, Republic of Guyana.

Hall, D.R. (1998) Central and Eastern Europe: tourism, development and transformation. In: Williams, A.M. and Shaw, G. (eds) *Tourism and Economic Development: European Experiences.* John Wiley & Sons, Chichester, UK, pp. 345–373.

Hall, D.R. and Kinnaird, V. (1994) Ecotourism in Eastern Europe. In: Cater, E. and Lowman, G. (eds) *Ecotourism: a Sustainable Option?* John Wiley & Sons, Chichester, UK, pp. 111–136.

Honey, M. (1999) *Ecotourism and Sustainable Development: Who Owns Paradise?* Island Press, Washington, DC.

Jenkins, J.M., Hall, C.M. and Troughton, M. (1998) The restructuring of rural economies: rural tourism and recreation as a government response. In: Butler, R., Hall, C.M. and Jenkins, J. (eds) *Tourism and Recreation in Rural Areas.* John Wiley & Sons, Chichester, UK, pp. 43–67.

Keszi, J. (1997) Discussion document: ecotourism policy in Manitoba. Prepared for Department of Industry, Trade and Tourism. Tourism Development Branch, Winnipeg, Manitoba.

Luong, P.T., Koeman, A., Thi Lam, N., Cuong, N.D.H. and Cam, H.D. (1999) *Proceedings of a National Workshop on a National Ecotourism Strategy.* IUCN and Vietnam National Administration for Tourism, Hanoi.

Malloy, D.C. and Fennell, D.A. (1998) Ecotourism and ethics: moral development and organizational cultures. *Journal of Travel Research* 36, 47–56.

Queensland, Department of Tourism Small Business and Industry (1997) *Queensland Ecotourism Plan.* QD TSBI, Brisbane.

Republic of Guyana (1997) *National Plan for Ecotourism Development*, chapter 11.0: *Recommended Support Policies.* Ministry of Trade, Tourism and Industry.

TAT (nd) *The Approach to Ecotourism in Thailand*, brochure. Tourism Authority of Thailand, Bangkok.

Taylor, E. (1996) *Brazil.* Thomas Allen & Son, Markham, Ontario.

Valenzuela, M. (1998) Spain: from the phenomenon of mass tourism to the search for a more diversified model. In: Williams, A.M. and Shaw, G. (eds) *Tourism and Economic Development: European Experiences.* John Wiley & Sons, Chichester, UK, pp. 43–74.

Weaver, D.B. (1998) *Ecotourism in the Less Developed World.* CAB International, Wallingford, UK.

Wilkinson, P. (1997) *Tourism Policy and Planning: Case Studies from the Commonwealth Caribbean.* Cognizant Communications, New York.

Williams, A.M. and Shaw, G. (1998) Tourism policies in a changing economic environment. In: Williams, A.M. and Shaw, G. (eds) *Tourism and Economic Development: European Experiences.* John Wiley & Sons, Chichester, UK, pp. 375–391.

Chapter 30

Ecotourism-related Organizations

E.A. Halpenny

Nature Tourism Solutions, Almonte, Ontario, Canada

Introduction

Following from the previous chapters in this section, a third mechanism that is used to minimize the negative impacts and maximize the positive impacts of ecotourism is the ecotourism-related organization. Ecotourism organizations are administrative or functional structures that are concerned with ecotourism. For this chapter they are sorted into three categories: (i) membership non-government organizations (NGOs); (ii) non-membership NGOs; and (iii) public sector or governmental agencies. Ecotourism organizations, found throughout the world, play important roles ranging from grass-roots advocacy to international policy making.

This chapter will examine ecotourism organizations, first through an investigation of their distribution, number and level at which they operate. Relevant characteristics will also be outlined, including their structure, stakeholders, mission, funding sources and history. Perhaps most importantly, the role that ecotourism organizations play in making ecotourism a more sustainable tool for conservation and community development, will also be explored.

Much of the information herein is drawn from past correspondence and research materials collected at The Ecotourism Society (TES), the main international membership-based ecotourism NGO, at which the author was the Projects Director for 4 years. Much of this chapter is informed by an informal survey of ecotourism organizations recently undertaken by TES. Also useful was a study performed in 1996 by the same organization, which examined the feasibility of establishing an international network of ecotourism associations coordinated by TES (Gruin, 1996). While the study determined that such a network was financially not feasible, the data collected from each association constitutes a valuable research resource.

The subject of ecotourism organizations, however, remains somewhat ambiguous, leading to a tentative summary here. Readers should be aware of this, and examine the information provided with the understanding that much more research on this subject must take place. One of the greatest barriers to gathering information on this subject is the lack of time most ecotourism NGOs, and some ecotourism-related public sector agencies have to document and answer questions about their organizations. This is especially true in developing counties. Another challenge is the overlap that can occur with other chapters in this book, especially when discussing the role of government organizations in ecotourism. Additional information on this topic can be found in the other chapters in this section.

Location, Scope and Numbers

The essential mandate of an ecotourism organization is to minimize the negative impacts and maximize the positive impacts of ecotourism. The number of organizations striving to accomplish this on a daily basis throughout the world is growing rapidly along with the phenomenon of ecotourism itself. Most countries that host ecotourism activity have their own unique set of organizations dedicated to this task. However, the composition of this structure is determined in large part by the resources that are available. For this reason, developed countries generally host a greater number of these organizations, although many of their actual field programmes are run in the developing countries, which are the primary destinations for most ecotourism activity.

It is impossible to estimate the number of ecotourism organizations worldwide because of the wide-ranging nature of these organizations. However, Table 30.1 provides a list of ecotourism associations (membership NGOs who deal regularly with ecotourism issues) that have corresponded with TES in the past 5 years. Not all associations actively corresponded with TES in 1999, therefore their validity as active ecotourism membership organizations is not verified. The list illustrates the varied scale at which ecotourism associations can operate, a pattern that also holds true for public sector-related ecotourism organizations.

Ecotourism organizations can be found at the grass roots or local level, addressing concerns of local stakeholders. They can also appear at state, regional, national and international levels. Each has a set of constituents that they serve. Below, the levels at which the various types of ecotourism organizations operate are described. Table 30.2 provides examples.

International organizations

In the international arena, many different organizations address ecotourism-related issues. In the past the World Tourism Organization (WTO) chose to ignore ecotourism, citing the fact that it is difficult to measure statistically due to its ambivalent definition. As well, the WTO has cited its limited resources as a rationale for focusing on mainstream tourism concerns. This neglect, however, is changing as preparations gain momentum for the year 2002, the Year of Ecotourism. WTO recently hosted a meeting on the subject of ecotourism in Costa Rica, and signed a letter of understanding with TES in preparation for 2002.

The United Nations Development Program (UNDP) is another international government organization that deals with ecotourism, largely through its international development assistance programme. UNDP channels aid to countries in need of development assistance. Often this aid is used to develop ecotourism enterprises as part of a larger regional development scheme. For example, UNDP channelled financial assistance from the Global Environment Facility to the South Pacific Regional Environment Programme (SPREP), whose objectives include making community-based conservation areas viable through the promotion of income-generating activities such as ecotourism. SPREP embarked on a series of product development, training and capacity-building initiatives to promote appropriate and sustainable tourism activities in various Pacific islands. An example of an initiative supported by SPREP is found in Markira Province, Solomon Islands. The project has helped several communities run their own tours, working in partnership with overseas tour companies who send tourists. The community operates the tours and sets the prices. They provide the guides, cooks and cultural performers, decide how many tourists should visit and what rules they should follow (SPREP, unpublished).

WTO and UNDP are just two examples of how international government eco-

Table 30.1. Ecotourism associations.

Ecotourism association	Country	Level/arena
Albanian Ecotourism Association	Albania	National
Alaska Wilderness Recreation and Tourism Association (AWRTA)	USA	State
Armenian Ecotourism Association	Armenia	National
Asociacion Ecuatoriana De Ecoturismo (ASEC)	Ecuador	National
Belize Ecotourism Association	Belize	National
Bolivian Ecotourism Association	Bolivia	National
Eco Brazil – Associacao Brasileira de Ecoturismo	Brazil	National
Ecotourism Association of Australia (EAA)	Australia	National
The Ecotourism Association of Papua New Guinea	PNG	National
The Ecotourism Society (TES)	USA	International
The Ecotourism Society of Kenya (ESOK)	Kenya	National
Ecotourism Society of Saskatchewan (ESS)	Canada	State
Ecuador Ecotourism Association	Ecuador	National
Estonia Ecotourism Association	Estonia	National
Fiji Ecotourism Association (FEA)	Fiji	National
Fundacion Ecoturismo Argentina	Argentina	National
Georgia Greens Movement	Georgia	National
Hawaii Ecotourism Association (HEA)	Hawaii	State
Honduras Ecotourism Association	Honduras	National
Japan Ecotourism Society (JES)	Japan	National
La Societe Duventor	Canada	Regional
Mexican Association of Adventure Tourism and Ecotourism (AMTAVE)	Mexico	National
Morovo Lagoon Ecotourism Association	Solomon Islands	Local
Partners in Responsible Tourism (PIRT)	USA	National
South Carolina Nature-Based Tourism Association	USA	State
Swedish Ecotourism Association	Sweden	National
Toledo Ecotourism Association	Belize	Local
Virginia EcoTourism Association (VETA)	USA	Regional
Zanzibar Ecotourism Association	Tanzania	Local

Table 30.2. Different arenas and types.

Level/arena	Type	Examples
International	Government	UNDP; WTO
	Membership NGO	TES; Tourism Concern
	Non-member NGO	CI; TNC; IUCN
National	Government	KWS and Kenya Ministry of Tourism; Fiji Ministry of Tourism and Transport and University of the South Pacific; Germany's BMZ and GTZ
	Membership NGO	ESOK; FEA; EAA
	Non-member NGO	INDECON
Regional, state and local	Government	Queensland Tourism (Environment Department); Tourism Saskatchewan
	Membership NGO	ESS
	Non-member NGO	Redberry Pelican Project

tourism-related organizations can play a role in making ecotourism a tool for sustainable development. At the international level NGOs also play a role. Tourism Concern, a UK-based NGO dedicated to ensuring tourism is a just and sustainable form of business, has worked for many years to make tourism more sustainable. Although Tourism Concern has never worked solely on ecotourism issues, it recognizes the importance of the phenomenon, especially for people living in developing countries. The international advocacy group distributes a collection of fact sheets on ecotourism, stressing education among consumers, the media and governments.

The US-based TES is dedicated solely to ensuring that ecotourism is a viable tool for biodiversity conservation and community development. It plays many roles, including education of professionals and consumers, research, policy formation, and facilitating the establishment of global professional networks. TES, given its limited annual budget of less than US$500,000, lacks sufficient resources to truly address the needs of all its constituents, and has therefore chosen to focus annually on different world regions and specific issues. For example, Southeast Asia was a region of focus in 1999, and international development policy as it relates to ecotourism was one of its foci.

Non-member NGOs also play a significant role in the international arena. Much like the example of UNDP given earlier, international NGOs can act as channels for financial assistance to developing countries. The Nature Conservancy (TNC), and Conservation International (CI), are two US-based conservation NGOs that have their own ecotourism departments, and play this role. The actions of these departments are moulded by the conservation agendas set forth by their parent organizations. They are assigned to work in regions of the world where the organization's biodiversity conservation mandate dictates. The efforts of TNC and CI focus largely on infrastructure and skills building, thereby generating the capacity to run successful ecotourism enterprises within local communities. A third international NGO, the World Conservation Union, generally focuses more on developing visitor and resource management strategies and tools for ecotourism activities inside protected areas, distributing them to park management professionals.

Interestingly, TNC is a membership organization, but it has been placed in the non-membership category since its membership activities have little to do with the organization's international ecotourism efforts. The one exception to this is the organization's travel programmes for its members. There are hundreds of NGOs and universities in developed countries that offer travel programmes to their members as part of their membership benefits package. When the travel programmes feature ecotourism destinations, the package is generally intended for members in search of stimulating, learning vacations in biodiverse or culturally rich regions. TNC successfully supports many of its ecotourism enterprises development projects in developing countries by sending their membership travel programmes to them.

National organizations

Further examples of ecotourism organizations can be found at the national level. Government plays an important role in the national arena. Government-related ecotourism organizations active at this level generally come from four areas: parks management agencies, universities, tourism ministries, and environment or natural resource ministries. In Kenya, much of the government-related ecotourism activity at the national level is performed by the Kenya Wildlife Service (KWS), a quasi-governmental organization whose mandate is the management of wildlife in the country. The KWS's role as Kenya's protected areas manager makes it a leading national force in guiding ecotourism development in the country. The KWS must work with industry associations such as the Kenyan Association of Tour Operators and the Ministry of Tourism to make ecotourism a success.

From Fiji comes another example of organizations working at the national level on ecotourism issues. There, the Ministry of Tourism and Transport recently completed and released *Ecotourism and Village-Based Tourism: a Policy and Strategy for Fiji* (Sawailau and Malani, *c.* 1998). The policy was a collaborative effort, drawing ideas from stakeholders throughout the country. The University of the South Pacific was also an active government-related contributor, with the participation of Dr David Harrison as the policy's editor.

A final example of a national-level government-related ecotourism organization comes from Germany and the efforts of the German Federal Ministry for Economic Cooperation and Development (BMZ). Very different from the examples from Fiji and Kenya, the BMZ has been investigating the value of ecotourism through its Working Group on Eco-Tourism, publishing the book *Ecotourism as an Instrument of Nature Conservation? Possibilities of Increasing the Attractiveness of Conservation Measures* in 1995. More recently the BMZ has been investing in ecotourism projects in developing countries through the government-owned Deutche Gesellschaft fur Technische Zusammenarbeit (German Technical Co-operation, GTZ). Like the development assistance programmes described earlier, GTZ's objective is to 'strengthen the performance capability of people and organizations in developing counties, as well as their capacity for self-help. To this end, technical, economic and organizational skills are transferred and measured improving the conditions for application' (GTZ, 1997). Outlined in Table 30.3 are some of the projects that have been funded through GTZ in recent years.

Also at the national level, member NGO ecotourism organizations play an important role in making ecotourism a tool for sustainable development. Returning to Kenya, a national ecotourism association, the Ecotourism Society of Kenya (ESOK), played an organizational role in the groundbreaking conference 'Ecotourism at a Crossroads'. Developed by KWS and TES, the 1997 conference represented one of the first times in East Africa where the failure of tourism management in parks and the decline of the mainstream tourism industry in Kenya was openly discussed. After a few growing pains, ESOK has again emerged, with donor funding support to take an active lead on Kenya's national stage, organizing the establishment of a national accreditation programme for ecotourism.

The Fiji Ecotourism Association (FEA) is another example of a national membership ecotourism NGO. Established in 1995, the organization experienced a period of dormancy when over-committed board members could not devote enough time to the fledgling organization. The FEA has recently experienced a renewal, with the release of the Ecotourism and Village-Based Policy. The FEA is hosting a conference in June 2000 to stimulate the progress of ecotourism development in the country.

Perhaps the most successful national ecotourism membership organization has been the Ecotourism Association of Australia (EAA). The association boasts a wide membership including industry, government, academic and NGO representation. The organization hosts an annual conference at which the quality and financial health of the nation's ecotourism industry is assessed. The association currently manages the national ecotourism accreditation programme (NEAP), and has also recently launched a new nature guide accreditation programme.

ESOK, FEA and EAA all play an active role at the national level, giving their membership a voice in national decision making. There are other ecotourism NGOs who do not have memberships, but also take an active role on the national stage. An example of this is the Indonesian Ecotourism Network (INDECON), a network and information source on ecotourism for Indonesia. Through fostering dialogue among ecotourism stakeholders, building the capacity in-country for ecotourism planning and management, and encouraging model ecotourism developments INDECON is attempting to educate Indonesian communities, government officials and entrepreneurs about ecotourism.

Table 30.3. Select tourism in projects executed by GTZ on behalf of the German Federal Ministry of Economic Cooperation and Development (BMZ) (Haep, 1999).

Project title	Country	Relation to tourism	Project duration	German contribution (DM, million)
Promotion of Sustainable Development through Tourism	Central America	1,3,6	1997–2002	5.00
Promotion of Small Enterprises	Benin	4	1997–2000	2.75
Prorenda – Promotion of Small Enterprises in Pernambuco	Brazil	4	1997–2001	3.00
Promotion of the Economy and Employment	Bulgaria	4,6	1998–	3.80
Environmental Management in the IX. Region	Chile	3,5	1997–2000	3.00
Conservation of the Tai National Park	Ivory Coast	3,5	1997–2000	2.75
Support of the National Environment Agency	Gambia	3,5,6	1993–1999	2.50
Know-How-Transfer for Waste Water Management	Jamaica	3	1996–1999	3.50
Conservation of Petra	Jordan	3,5	1996–2000	3.00
Promotion of vocational training	Cape Verde	2	1996–2000	3.00
Integrated Conservation in East Congo	Dem. Rep. of Congo	3,5	1996–2001	16.40
Parks and Wildlife	Malawi	3,5	1996–2002	3.40
Advisory to the Ministry of Development	Macedonia	3,5,6	1997–	2.87
Combating Desertification	Namibia	5,6	1993–1999	9.5
Promotion of Tourism	Palestine	1,3,6	1996–2000	4.60
Promotion of small and medium-sized enterprises	Peru	4	1997–2000	4.80
People and Parks	South Africa	3,5	1996–1999	4.20
Communal Wildlife Management	Tanzania	3,5	1998–2000	2.5

1 Tourism as main activity; 2 Tourism as part of vocational training; 3 Mitigating negative impacts of tourism; 4 Tourism in enterprise promotion; 5 Tourism to achieve conservation objectives; 6 Tourism Policy Development.

Sub-national organizations

Below the national level are found regional, state and local arenas for action. Ecotourism organizations play a role at each of these levels. These arenas have been combined to simplify Table 30.2, as most of the characteristics and challenges that ecotourism faces at these levels are similar.

Queensland Tourism and Tourism Saskatchewan are two examples of state-level public sector ecotourism organizations. Both have displayed agendas with a strong focus on ecotourism in the last 5 years. In Australia, Queensland Tourism's environment division publishes a quarterly newsletter titled *EcoTrends* informing industry, NGOs, universities and the public sector about ecotourism-related events, accreditation recipients, department research and policy. Queensland's tourism department regularly promotes the development of quality tourism products through its support of research on small business development, ecotourism markets, visitor and park management, and through the nation-wide NEAP.

In Canada, the province of Saskatchewan's tourism department, Tourism Saskatchewan, also supports the development of the local ecotourism industry. This effort was launched with a province-wide conference on ecotourism early in 1998 at which 21 objectives were announced (Ecotourism Task Force Recommends, 1998) (Box 30.1). With the support of volunteers, NGOs, community groups and other government agencies, 16 of the objectives have been achieved. The department's effectiveness did however suffer a setback when a key policy maker left his position. However, Tourism Saskatchewan's commitment now appears to be re-established, as they work towards a second major conference scheduled for later in 2000 (J. Hnatiuk, Bennington, Vermont, 2000, personal communication).

Tourism Saskatchewan is working with another key player in the province, the Ecotourism Society of Saskatchewan (ESS). ESS is an example of a state or local-level membership NGO. Aside from helping to organize an ecotourism conference, ESS is actively working to set up a provincial accreditation programme for ecotourism products. The chief goal of this volunteer organization is to ensure that a standard of quality for ecotourism products is established in the province. Finally, a Saskatchewan organization that works on ecotourism-related issues at the municipal level is the non-member NGO Redberry Pelican Project. The Redberry Pelican Project Foundation's mission is 'conservation through research, education and tourism'. Initiated as a model project for using tourism to encourage local efforts to conserve a colony of white pelicans within a national migratory bird sanctuary at Redberry Lake, the foundation has grown as an example of ecotourism excellence for the whole of Canada. Operating without a stable funding base, the foundation relies on visitor revenue, and partnership agreements and contractual arrangements that include monitoring the sanctuary and leading the development of standards and accreditation for ecotourism. The existence of the Redberry Pelican Project has led to the protection of a unique ecosystem, substantial economic benefit to local communities, and a leadership role for the province of Saskatchewan in the development of ecotourism in Canada (Redberry Pelican Project, 1999).

In summary, ecotourism organizations consist of non-member NGOs, member NGOs and public sector organizations that operate at local, regional and state levels, as well as in national and international arenas. While varying in composition and in other characteristics, they do share many similarities, as described below.

Ecotourism Organization Characteristics

As the number of ecotourism organizations increases, certain shared characteristics can be identified among them. These characteristics have been documented in the recent survey of ecotourism organizations by TES, and is based on the author's own knowledge of individual organizations.

Formation

There are three common reasons or watershed moments which act as catalysts for the formation of ecotourism organizations. One of the most prevalent is the occurrence of major conferences on ecotourism or related issues. The preparation for the conference and the actual event itself help stakeholders identify each other, communicate, and identify common goals for establishing an ecotourism organization. Certainly this may be truer for ecotourism associations such as TES, which was established at the 1990 International Symposium: Ecotourism and Resource Conservation, held in Miami, Florida. However, conferences may also play a role in the decision of universities to redirect curriculum or form centres of study on ecotourism, such as the Centre for Ecotourism at the University of Pretoria. Conferences may also inspire policy makers to emphasize ecotourism in government agency mandates. The latter instance occurred in Kenya following the 'Ecotourism at a Crossroads' conference.

Box 30.1. Saskatchewan Ecotourism Task Force Recommendations: announced at the first Saskatchewan Ecotourism Conference, January 1998.

1. An accreditation programme be implemented for Saskatchewan's ecotourism products and services.
2. Elder's Guidelines should be established by the Federation of Saskatchewan Indian Nations and implemented by local bands and applied to aboriginal product development and marketing.
3. Interpretive services be required for all accredited ecotour packages, and that an inventory of such services be produced.
4. Establish a programme to promote the visitor's responsibilities as an ecotourist.
5. Develop a new 'outfitter' licensing category, to be established and implemented to recognize the less consumptive nature of ecotourism-related businesses.
6. Upgrade northern fishing camps to ecolodge standards, in recognition of the overwhelming perception of northern Saskatchewan as the main ecotourism attribute of the Province. Facilitating investments by the aboriginal community and developing a management mentoring programme will be key to this process.
7. Support the development of culturally-unique services and attractions, such as vacation farms, overnight stays in teepees, dog-sledding and fowl suppers.
8. Ensure that tourism interests be directly represented in all land-use planning forums, and that regulatory authorities be required to utilize an interdisciplinary approach to all future land-use planning.
9. Ensure that co-operative relationships with other land users be confirmed by Statements of Mutual Recognition and Respect, formalizing a consultative process regarding future land-use plans.
10. Establish a process where visitors contribute a portion of their expenditures to a fund which has, as its objective, the preservation and enhancement of the natural environment.
11. Encourage local employment and suppliers through integration into the accreditation programme.
12. Establish ecolodge facilities that model sustainable systems and technologies by a non-profit organization (such as the Saskatchewan Research Council) in each of the Province's main ecozones.
13. Amend Crown-land lease policies and prepare guidelines to allow for the establishment of ecotourism-related businesses. Also, it needs to be determined whether a new tax assessment policy more conducive to multiple land-use and habitat preservation could be developed for land currently zoned as agricultural.
14. Ensure that the Tourism Industry Association of Canada's Guiding Principles on Ecotourism be adopted by all stakeholders.
15. Establish a three-part programme to develop the ecotourism industry in Saskatchewan, including support for:
    - infrastructure projects which demonstrate innovative and environmentally-friendly technology,
    - assessment of the changes to environments that result from ecotourism activities, including base-line studies, monitoring and environmental audits,
    - integration of ecotourism input into regional planning and development.
16. Support the efforts of ecotourism industry representatives to work with financial institutions to establish a lending programme which would match a guaranteed/insured deposit base with targeted lending to ecotourism businesses.
17. Ensure that Saskatchewan's ecotourism industry strongly supports the introduction of sustainable tourism courses at the province's universities and technical institutions.
18. Formulate a strategy to accelerate market development of Saskatchewan's sustainable tourism products and services.
19. Encourage Saskatchewan ecotourism stakeholders to initiate discussions with neighbouring jurisdictions for the purpose of developing integrated planning by eco-zone.
20. Encourage the development of sustainable tourism policies through continuation of the task force process, making adjustments as required, with the objective of securing broad industry acceptance of these recommendations and determining the process associated with their implementation.
21. Establish an accountability process (who is accountable, what for and to whom) which could involve the formation of a body to oversee policies and programmes designed to further develop sustainability in the industry.

(After: The Ecotourism Task Force Recommends, 1998)

Some ecotourism organizations arise in response to catastrophic events. An example of this comes from Alaska. The Alaska Wilderness Recreation and Travel Association (AWRTA) was established from a coalition of small tourism businesses and community members who depend on the continued existence of pristine wilderness for their livelihoods. Wilderness areas along the Alaska coast were damaged by the *Exxon Valdez* oil spill in 1989. Following clean-up efforts, AWRTA was formed in 1992.

In some instances, other types of organization evolve into ecotourism organizations, changing their mission to address ecotourism-related goals. An example of this is the recent development of formalized guide training programmes in Canadian universities and colleges. Addressing a need in the ecotourism and adventure travel industries, public sector agencies such as schools are modifying their curriculum to produce a labour force tailored for the rapidly expanding nature-based tourism sector.

Reason for formation: mission

Each of the above examples of ecotourism organization formation display underlying reasons for establishing such an institution. In general there are three themes that tend to characterize the reason for formation, and are reflected in the present day missions of each organization. These are:

1. Stewardship of resources (both cultural and natural).
2. Addressing community, conservation and business interests (including promotion, enterprise development, networking).
3. Collecting and disseminating information about ecotourism.

Table 30.4 outlines the missions of selected ecotourism associations. Each of the missions fits at least one of the above reasons for formation. Sources of funding are discussed later in this chapter.

Form and structure

The form and structure of ecotourism organizations varies between non-member NGOs, member NGOs and government agencies (see Table 30.5 for examples). Large, non-member NGOs and government agencies, such as Conservation International's ecotourism department and Queensland Tourism's environment division, are subject to the organizational structures of their parent organizations. They generally have a limited number of staff, interns and contract employees, all on the main organization's payroll. They are influenced heavily by the policies of the parent organization, and are generally reliant to some degree for funding from the parent organization. Smaller, non-member NGOs, such as INDECON, are characterized by a small staff, with a few additional volunteers. Funding for these organizations is even less stable than for the larger, non-member NGOs and government agencies. All non-member NGOs must report to a board of directors on issues such as annual budgets and organizational strategies and planning.

Membership-based NGO ecotourism organizations are slightly different. They too must report to a board, which is almost always composed of volunteers, on financial management and strategic planning issues. However, they must also report to membership, usually in the form of a newsletter or regular meetings. For example, the EAA has a general meeting on a yearly basis at its annual conference, and publishes a periodic newsletter, *Ecotourism News* (EAA, 1996b). The EAA has a committee of up to nine members, elected by the membership, which includes an executive comprising a President, Vice President, Secretary and Treasurer. Committee members are assigned specific projects for the year, e.g. revising the NEAP programme. The EAA also has one full-time office manager (EAA, 1999a). The largest ecotourism membership NGO, TES, has a staff of six, who report to a board of directors, nominated by members and elected by directors, and a board of advisors, nominated by

Table 30.4. Ecotourism association formation, mission and funding sources.

Association	Date formed	Mission	Ecotourism definition	How was it formed?	Current funding sources
Alaska Wilderness Recreation and Tourism Association (AWRTA)	1992	To support stewardship of the wild in Alaska and the development of healthy, diverse travel businesses and communities by linking business, community and conservation interests		Following an oil spill in Prince William Sound in 1989, those involved in small tourism businesses and recreationists who worked and played on the Sound helped with the clean up and formed a coalition to support those whose businesses depend on pristine wilderness	Membership, foundation grants
AMTAVE (Mexican Association of Adventure Tourism and Ecotourism)	1994	To promote and protect the ecotourism and adventure tourism sites in Mexico, contributing to sustainable development of each region involving actively the local communities	IUCN definition	Members had common worry of promoting adventure and eco-tourism, and protecting nature	Membership and tourists' fairs, and adventure sports championships
Ecotourism Society of Saskatchewan	1998/99	To support and promote the development of a Saskatchewan ecotourism industry and the protection and perpetuation of host natural ecosystems and cultures	Enlightening nature travel experience that contributes to conservation of the ecosystem and to the culture and economic resources of the host communities	The organization evolved from the Watchable Wildlife Assn., which was perceived to be too narrow in its focus	Membership

Ecotourism Society, The (TES)	1991	To unite the conservation community with travel professionals to make ecotourism a tool for sustainable development and conservation worldwide	Responsible travel to natural areas that conserves the environment and sustains the well-being of local people	Nature tourism business, academics and conservation NGOs perceived a need for an international network/assn. to address growing ecotourism field	Membership, book sales, foundation grants, training and education contracts
Fiji Ecotourism Association	1995		Ecotourism is a form of nature-based tourism which involves responsible travel to relatively undeveloped areas to foster an appreciation of nature and local cultures, while conserving the physical and social environment,respecting the aspirations and traditions or those who are visited, and improving the welfare of local communities	Objective was originally to consolidate all those in the private sector who were interested in ecotourism in terms of ecological, cultural, historical, nature-based and adventure-based tourism	
Georgia Greens Movement	1997	Develop ecotourism	Ecology and healthy tourism		Membership
Green Tourism Promotion (India)	1997	To provide information, training and marketing support to ecotourism organizations and professionals committed to working for sustainable environment and safe-guarding the interests of local people	Ecotourism is responsible travel to natural areas that conserves the environment and sustains the well-being of local people	Started as an independent voluntary initiative, to provide marketing support to ecotourism destinations (in India) and to facilitate networking, exchange of information and training programmes for effective management of ecotourism operations. There was a lack of centralized information available on ecotourism products, etc.	Membership, sale of goods and services

Continued

Table 30.4. *Continued*

Association	Date formed	Mission	Ecotourism definition	How was it formed?	Current funding sources
Hawaii Ecotourism Association	1995		Nature- and culture-based tourism that is ecologically sustainable and supports the well-being of local communities	Following a state-sponsored Hawaii Conference on Ecotourism in 1994. Small businesses that provided ecotourism experience needed to form an organization	Membership, some government grants
INDECON (Center for Indonesian Ecotourism Research, Training and Promotion)	1995	To develop and promote ecotourism based on scientific research, training and promotion	Ecotourism is responsible travel to the protected natural areas, as well as to unprotected natural areas, which conserves the environment (natural and cultural) and improves the welfare of local people)	The network initiated by the Institute for Indonesia Tourism Studies, Bina Swdaya Tours and Conservation International Indonesia Program	Foundations and corporate grants, sale of goods and services, and training and education programmes
Japan Ecotourism Society	1998	To provide an open forum for people who work in ecotourism so they can exchange and share information amongst themselves, to introduce advanced case studies and to provide opportunities for human resource training for the purpose of developing ecotourism appropriate to respective areas		A study group formed in 1990 composed of travel industry and researchers joined with journalists to study ecotourism	Membership and sale of goods and services

members and elected by advisors and directors. The boards are designed to represent the organization's diverse membership, based on regional and professional categories. TES also relies on volunteer interns for administrative and research assistance in the office, official 'research associates' who are called upon for their academic expertise, and occasional voluntary support from select members when projects dictate the need.

Most membership NGOs have open access to membership. There are, however, some exceptions, such as the self-proclaimed membership organization, the Ecotourism Society of Sri Lanka, whose board and membership is the same. The organization acts more like a 'think-tank', and plays a role similar to INDECON in Indonesia (C. Gurung, Columbo, Sri Lanka, 1999, personal communication). It is difficult to differentiate the purpose and structure of the two organizations, thus making the classification of membership NGO less useful in this instance.

Other membership NGOs such as the Namibia Community-based Tourism Association (NACOBTA) focus on grass-roots empowerment. The organization was formed to close the gap between community-based tourism and private sector tourism in Namibia, achieving greater well-being for rural communities through economic development and empowerment. NACOBTA has a management committee of seven elected community members who oversee the work of a programme manager and support a staff of seven permanent positions (Schalken, 1999).

Funding, partnerships and stakeholders

Funding can be derived from a variety of avenues for ecotourism organizations. Public sector ecotourism organizations, such as protected area and resource management agencies or tourism ministries, generally receive their funds from taxpayers via the government. However, in developing countries they may also receive assistance from donor agencies funded by groups such as the European Union. The latter was the chief funding agency for the 'Ecotourism at a Crossroads' conference in Kenya in 1997. Another example is the Inter-American

Table 30.5. Form and structure of select ecotourism organizations.

Type	Organization	Membership	Decision making	Salaried staff
Non-member NGO	Conservation International Ecotourism Department	Not applicable (NA)	Dictated in part by parent organization	10+
Government Agency	Queensland Tourism Environment Department	NA	Dictated in part by parent organization	5+
Non-member NGO	Indonesia Ecotourism Network	NA	Staff	3+
Membership NGO	Ecotourism Association of Australia	Open 435+ members	Committee with executive, staff	1
Membership NGO	The Ecotourism Society	Open 1500+ members	Boards, staff	6+
Membership NGO	Ecotourism Society of Sri Lanka	Restricted	Board/membership	0
Membership NGO	Namibia Community-Based Tourism Association	Open 42+ members	Management committee, staff	

Development Bank, which is scheduled to loan Brazil US$250 million to develop the Brazilian Amazon as an ecotourism destination.

Membership NGOs have access to a unique source of funding, i.e. membership revenue. Some ecotourism NGOs such as TES base their fee structures on benefit-related membership levels, while other NGOs base their fees on ability to pay. Generally, the higher the fee, the greater the benefits that are expected. Non-member and member NGOs share several funding sources. Both can receive funds from governments. For example, the Canadian Tourism Commission, Canada's tourism development and promotion agency, helped to fund the country's 'Ecotourism Product Club' (P. Kingsmill, Bennington, Vermont, 1999, personal communication). The product club concept is designed to help small- and medium-sized businesses to pool their efforts to build networks they could not otherwise afford on their own. They usually include cooperative ventures and partnerships that build new packages, fund research projects on customer needs, and assist in product development and marketing strategies (CTC, 1999).

Funding for NGOs can also come from private foundations. For example, in 1998 the MacArther Foundation granted the Indonesia Ecotourism Network US$75,000 to 'strengthen links among Indonesian non-profit groups, the government, and the tourism industry and to develop ecotourism guidelines for Indonesia' (New Grants, 1999, p. 19). A third source of funding for ecotourism-oriented NGOs is donor agencies. Development assistance can be used to fund the activities of NGOs in developing countries for a time. In 1998 NACOBTA received funding from the Swedish government's International Development Agency. It also received funding from an international conservation NGO, the World Wide Fund for Nature (Schalken, 1999).

Another source of funding for ecotourism-NGOs consists of donations from corporations. An example of this comes from southern Belize. The Toledo Institute for Development and Environment (TIDE) received assistance in its efforts to use tourism as a tool for community development and biodiversity conservation in the Punta Gorda region. International NGO TNC stepped in to assist with expertise and money, and introduced TIDE to corporate partner Orvis, a leading sport fishing equipment retailer. Tapping into the tremendous potential for fly fishing in the region, Orvis donated equipment and training to local fishermen, providing them with an alternative and lucrative livelihood as catch and release fishing guides. Orvis' and TNC's combined travel programmes supply a guaranteed stream of customers largely from the USA to the Punta Gorda region.

The example of Orvis, TIDE and TNC provides a good case study on the importance of partnerships for ecotourism organizations. Partnerships can involve all stakeholders involved in, or affected by ecotourism, including tourism entrepreneurs, NGOs, community members, the media and government agencies. Another example comes from TES's launch of a consumer education campaign in early 2000. TES approached its larger wholesale operator and lodge members to support a new consumer education campaign through a renewal of their Supporting level membership (US$1000). When combined with a grant from a private foundation, these funds were then used to complete a pilot study documenting the contributions the ecotourism industry was making towards conservation and local communities. The data collected from this effort were then used to inform a larger campaign to help consumers to understand that the tourism businesses they chose to patronize would 'make a difference' to conservation efforts. The contributions by TES industry members funded an upgrade of the consumer section of the TES web site (www.ecotourism.org/choice). Further sponsorship by TES industry members, as well as Continental Airlines in the form of advertorial purchases in a special issue of a leading US-based nature magazine, and an information pamphlet designed to launch the campaign, helped to make the project a success. The travel industry and travel media proved to be invaluable partners in the effort to educate consumers about choosing travel products that meet the principles of ecotourism.

The type of partnerships that the organization will engage in is directly related to the organization's mission and its constituents. It is the ecotourism organization's constituency that is the last subject of discussion in this section. The constituents or stakeholders related to an ecotourism organization again vary with the type of organization. Public sector ecotourism organizations have the widest constituency to serve. Communities, other public sector organizations, industry, NGOs, educational institutions, the media, neighbouring countries or regions can all be impacted by the actions of governmental ecotourism organizations. Many of these are discussed in other chapters within this book.

Membership NGOs' direct constituency is their membership. From the TES survey of ecotourism associations, membership seems to fall into the following categories:

1. Industry association
2. Consulting association
3. Multi-sectoral association.

The Thai Ecotourism & Adventure Travel Association may be an example of an ecotourism membership NGO that is dominated by industry members. Little evidence could be found to indicate that the association does more than promote ecotourism and adventure travel business. A consulting association classification characterizes the previously mentioned Ecotourism Society of Sri Lanka. A multi-sectoral association, which hosts a rich assortment of professionals in its ranks, is the category to which most ecotourism associations belong. For example FEA, TES and EAA all claim industry, community, governmental and NGO representation among their memberships.

Funding sources and constituencies are the entities that an ecotourism organization must appeal to in its daily efforts to minimize negative impacts associated with ecotourism, and maximize positive impacts. Its decision making structure, administrative form and available resources, including human, financial and intellectual capital influence the effectiveness of an ecotourism organization's actions.

The Role of Ecotourism Organizations

There are many roles that ecotourism organizations play in their efforts to make ecotourism a viable tool for sustainable development. These roles include:

1. Research
2. Regulation
3. Development of standards (guidelines, codes of practice and certification)
4. Industry/enterprise development, including promotion
5. Lobbying and advocacy
6. Education of consumers and professions (including the industry and other professionals)
7. Policy development and implementation
8. Fundraising for communities and biodiversity conservation
9. Development of management tools and strategies.

Ecotourism organizations have played a major role in researching the viability of ecotourism as a sustainable development tool. Universities provide funding and facilities for this research, and also function as a forum for related debate. Government agencies also play a major role in this effort. A good recent example of this is the Canadian Tourism Commission's efforts to document best practice within the Canadian ecotourism industry in the publication *Catalogue of Exemplary Practices in Adventure Travel and Ecotourism* (Wight, 1999), which will serve as a learning tool for other ecotourism entrepreneurs. Member and non-member NGOs play a lesser role in research, largely due to their limited budgets. Nevertheless, valuable research is accomplished by these organizations, often tailored to meet the very practical needs of the constituents, such as the recent documentation of the ecolodge industry's financial and economic status (Sanders and Halpenny, 2000).

Government ecotourism organizations

play the main role in regulation of development. For example, public sector agencies, such as Australia's Great Barrier Reef Marine Park Authority, have a great impact on ecotourism activities through regulatory oversight and enforcement of visitation to protected areas. NGOs can also play a role, for example through the development of guiding certification programmes, ensuring safety and quality through regulations for ecotourism guiding.

A precursor to certification is the establishment of basic standards or guidelines. All ecotourism organizations have played a role in developing codes of conduct for ecotourists. Industry membership NGOs also often set their own codes of practice, guidelines tailored for their own activities or setting. Table 30.6 outlines the activities of select ecotourism associations regarding the establishment of codes of conduct (for tourists), codes of practice (for industry) and ecotourism certification programmes. Government agencies have also played a role in establishing certification programmes in their own countries. Tourism Saskatchewan is currently working with ESS to accomplish this. Australia and Costa Rica's national governments have successfully completed the development of national ecotourism certification or accreditation programmes.

A fourth important role played by ecotourism organizations is the development of an ecotourism industry. Ecotourism businesses are often small and isolated, and lack the financial capital to reach the marketplace effectively, or to network with potential partners. The example given earlier in this chapter of the Canadian Ecotourism Product Club is one solution provided by a public sector ecotourism organization. Another common problem is that ecotourism entrepreneurs often lack the skills necessary to achieve a financially successful business (see Chapter 36). Non-member NGOs such as Conservation International and governmental agencies such as UNDP and GTZ try to address this issue through technical assistance programmes. Non-member NGOs pursue similar goals by hosting annual conferences at which the latest

Table 30.6. Ecotourism associations' development of guidelines and certification programmes.

Organization	Code of conduct	Code of practice	Guide certification	Certification programme status
AWRTA (Alaska Wilderness Recreation and Tourism Association)	Yes	Yes	Yes	No
AMTAVE (Mexican Association of Adventure Tourism and Ecotourism)	None	Yes		Planning/pilot phase
EAA (Ecotourism Association of Australia)		Yes	Yes	Complete/in operation
TES (The Ecotourism Society)	None	Yes	None	None
ESOK (The Ecotourism Society of Kenya)				Planning/pilot phase
ESS (Ecotourism Society of Saskatchewan)	Yes	Yes		Planning/pilot phase
FEA (Fiji Ecotourism Association)				Planning
Georgia Greens Movement	Yes	No		No
HEA (Hawaii Ecotourism Association)	No	Yes	Yes	No
JES (Japan Ecotourism Society)	No	No		Planning
Swedish Ecotourism Association				Planning
VETA (Virginia EcoTourism Association)			Yes	Planning

best practice examples and trends can be shared, or through the publication of books and newsletters. Membership NGOs also play an active advocacy role. Small ecotourism operators, as was described in the formation of AWRTA, are given a greater voice through their membership association. The successful campaign led by TES in 1993 to increase park fees in Costa Rica is a results-oriented example of what membership NGOs can do to maximize the positive impacts of ecotourism (TES, 2000).

Ecotourism organizations also play a role in education; for example education of consumers by NGOs such as TES and Tourism Concern. Education of professionals is a role that all types of ecotourism organizations have addressed. As already discussed, university and technical assistance programmes additionally increase the pool of qualified labour available to the ecotourism industry (see Chapter 40). Also of great importance is the education of professionals working in other sectors of the economy, including resource managers, media and policy makers. The education of international policy makers through the dissemination of educational publications on sustainable nature tourism by the UK's Department of International Development and Germany's BMZ at the United Nations Committee on Sustainable Development (UNCSD) meeting in April 1999, is an excellent example of government agencies taking a leadership role in educating their cohort about the potential of ecotourism. The UNCSD April 1999 meeting was an important United Nations event at which most national governments debated the nature and viability of sustainable tourism. Both publications (DFID, *c.* 1999; Steck *et al.*, 1999) are written by government agencies for other policy makers, facilitating an easy understanding of the subject matter.

This influence on policy is yet another important role played by ecotourism organizations. Again, all three types of ecotourism organization are involved in this. An example comes from TES's efforts in 1999 to focus on the subject of ecotourism development policy. TES's decision to focus on the subject was based on its awareness that an increased amount of financing was being made available from the international development community for ecotourism projects in developing countries. In order to encourage the wise investment of these funds, TES invited the donor community (e.g. The World Bank, InterAmerican Development Bank) to meet with NGOs, government representatives from developing countries and ecotourism entrepreneurs. The meeting was designed to help each group understand the needs and challenges of each other. It also highlighted success stories in ecotourism investment. The overall intention was to begin to influence the way policy makers at different donor institutions make decisions about investing in ecotourism.

Fundraising for community or environmental causes is another role that ecotourism organizations play. At the international level non-member NGOs such as Conservation International devote significant time fundraising for their own programme, which in turn provides funding for projects in biodiverse developing countries. Other, more locally based NGOs, may raise money for a mooring buoy programme or park management.

A final role for ecotourism organizations is the development and refinement of management tools and strategies for ecotourism activities. Certainly government agencies devote resources to developing methods for minimizing the impacts of visitors on communities and natural areas visited. NGOs also play a role in this, as seen in ESS's efforts to work with Tourism Saskatchewan and other provincial agencies to implement the objectives outlined earlier in this chapter. An example of an international non-member NGO's efforts to manage visitors to parks and protected areas can be seen in the publication *Tourism at World Heritage Cultural Sites: the Site Manager's Handbook* (ICOMOS, 1993) published by the International Council on Monuments and Sites.

Conclusion

Ecotourism organizations play an important role in maximizing the positive impacts and minimizing the negative impacts associated with ecotourism. The effectiveness of ecotourism organizations is influenced by several factors, including access to adequate resources, the context in which they operate, their internal structure, and their official missions.

Regardless of these factors, ecotourism organizations are bound to increase in number in the coming years as ecotourism expands worldwide. Additional effort should therefore be made to document the strengths and weaknesses of established ecotourism organizations, using these lessons to improve the operations of current organizations, and lessen the difficulties experienced by new entries. The roles played by ecotourism organizations, ranging from regulation to fundraising, will continue to be critical in addressing the challenges associated with making ecotourism a true tool for sustainable development.

References

Anon (1998) Ecotourism task force recommends. Proceedings of the First Saskatchewan Ecotourism Conference, 12–13 January 1998.

Anon (1999) New Grants. *The Chronicle of Philanthropy* 25 March, 19.

CTC (1999) $1.5 Million SME Program Seeks Bids. *Communique* November 3(2), 1, 3.

DFID (*c.* 1999) *Changing the Nature of Tourism: Developing an Agenda for Action.* Department for International Development, London.

EAA (1999a) *Ecotourism Association of Australia Draft Business Plan, 2000–2001.* Ecotourism Association of Australia, Brisbane.

EAA (1999b) *Ecotourism News*, Newsletter of the Ecotourism Association of Australia. Spring.

Gruin, M. (1996) *Recommendation and Strategy Plan for the Establishment and Implementation of an International Partners Program for The Ecotourism Society*, Review Draft. Greenpoint Ltd, Hershey, Pennsylvania.

GTZ (1997) *Tourism and Conservation of Biological Resources: Ecotourism in Germany's Technical Cooperation with Developing and Transforming Countries.* Deutsche Gesellschaft für Techische Zusammenarbeit (GTZ), Germany.

Haep, R. (1999) Tourism Projects of German Development Cooperation on behalf of the Federal Ministry of Economic Cooperation and Development (BMZ). An unpublished paper prepared for the Ecotourism Development Policy Meeting, Washington, DC, 24 September 1999.

ICOMOS (1993) *Tourism at World Heritage Cultural Sites: the Site Manager's Handbook.* The ICOMOS International Committee on Cultural Tourism.

Redberry Pelican Project (1999) *The Redberry Pelican Project (Canada) Foundation.* Redberry Pelican Project, Hafford, Saskatchewan.

Sanders, E. and Halpenny, E.A. (2000) *The Business of Ecolodges: a Survey of Ecolodge Economic and Finance.* The International Ecotourism Society, Burlington, Vermont.

Sawailau, S. and Malani, M. (*c.* 1998) In: Harrison, D. (ed.) *Ecotourism and Village-Based Tourism: a Policy and Strategy for Fiji.* Ministry of Tourism and Transportation, Suva, Fiji.

Schalken, W. (1999) Where are the wild ones? The involvement of indigenous communities in tourism in Namibia. *Cultural Survival Quarterly: World Report on the Rights of Indigenous Peoples and Ethnic Minorities* 23(2), 40–42.

Steck, B., Strasdas, W. and Gustedt, E. (1999) *Tourism in Technical Co-operation: a Guide to the Conception, Planning and Implementation of Project-Accompanying Measures in Regional Rural Development and Nature Conservation.* Deutsche Gesellschaft für Technische Zusammenarbeit (GTZ), Germany.

TES (2000) TES Initiatives. *The Ecotourism Explorer.* The Ecotourism Society web site http:www.ecotourism.org/inits.html

Weaver, D.B. (1998) *Ecotourism in the Less Developed World.* CAB International, Wallingford, UK.

Wight (1999) *Catalogue of Exemplary Practices in Adventure Travel and Ecotourism.* Canadian Tourism Commission, Ottawa, 86pp.

Chapter 31

Ecotourism in the Inter-sectoral Context

J. Cohen
Marketing Department, Rider University, Lawrenceville, New Jersey, USA

Introduction

The basic concept underlying ecotourism is that tourism generates revenue that is used to conserve the ecology of an area that attracts the tourists in the first place. The concept of ecotourism has been expanded in some cases to include not only ecological preservation but also cultural preservation. Cultural issues will be considered in this paper, but the focus will be on the more restrictive definition favoured by Brandon and Margolius (1996), i.e. that the underlying goal of ecotourism is to preserve the ecological environment (henceforth referred to simply as 'the environment'). However, more than just the ecological environment must be of concern if ecotourism is to succeed. As Brandon and Margolius (1996) point out, two of the requirements for successful ecotourism are that there must be an economic incentive for ecological preservation and that ecotourism should provide economic benefits to the local population. Therefore, when discussing ecotourism in an inter-sectoral context, both ecological and economic issues will be considered.

This chapter will discuss the three primary aspects related to ecotourism in the inter-sectoral context, namely: (i) the inter-sectoral conflict in ecotourism (i.e. compatible versus incompatible industry); (ii) intra- and inter-sectoral coalitions; and (iii) strategies for the intra- and inter-sectoral alliance to achieve sustainable ecotourism.

The Inter-sectoral Conflict in Ecotourism: Compatible versus Incompatible Industries

No industry operates in isolation. However, ecotourism is particularly sensitive to the presence of other industries. The long-term viability of ecotourism depends on a pristine environment. Therefore, there is an inherent incentive for the ecotourism industry to preserve the environment. This incentive is lacking in many other industries, notably manufacturing/resource extraction (henceforth noted as MRE) industries. Some resource extraction industries are inherently unsustainable, such as the mining and petroleum industries. Once the resource is depleted in a certain location, a new location must be found. Other resource extraction industries are potentially sustainable, e.g. timber and fishing. However, the success of the industry does not necessarily depend on sustainability in any one area. If one location is depleted, another location can often be found. While one can argue that eventually the world's resources will be depleted, this is not a very convincing argument to an individual company that is concerned with its short-term profit picture. For resource extraction industries, profits only depend on availability of that resource to sell. Unless the company has decided to position itself as an ecologically responsible company, there is no inherent incentive to protect the environment (although there may be external factors such as legal requirements, discussed further below).

Even when the product itself is harvested in a sustainable manner, the rest of the environment might be damaged. For example, a timber farm may be sustainable but the process of felling trees destroys the flora and fauna in the area. For example, record timber harvesting is damaging salmon streams in some areas (Van Dyk and Gardner, 1990). The process used in a sustainable industry may also create visual and noise pollution that is hostile not only to local wildlife but also to tourists.

For manufacturing industries, the inherent incentive to protect the environment may be even lower than for resource extraction industries. First, manufacturing industries can source their raw materials from anywhere in the world. If an area's resources have been depleted, suppliers from other areas are usually available. Second, the method of manufacturing, even if it is ecologically harmful, does not affect the quality of the product (unless the manufacturer is positioning the product as a green product) (Cohen and Richardson, 1995). In fact, the manufacturer might be able to make the product more competitive by lowering costs, and therefore the price, by engaging in ecologically damaging activities such as dumping chemicals.

MRE industries that damage the environment are obviously incompatible with ecotourism. Ecotourism differs from MRE industries in that: (i) the customer must be brought to the ecotourism site (= *in situ* consumption); and (ii) a relatively pristine environment is the primary product attribute that is being marketed. Therefore, there is an inherent incentive for the tourism industry to preserve the pristine environment of the ecotourism site. However, from the MRE perspective, the presence of tourists does not interfere with their operations, as long as those tourists and the tourism industry allow them to operate. Therefore, conflict is inevitable between ecotourism and incompatible industries.

What industries are incompatible with ecotourism? Many industries qualify, but those that have been at the forefront of battles over the environment include mining, the timber industry and ranching. Mining is inherently an unsustainable industry, as well as often unsightly. The timber industry vies for woodlands that are likely to be attractive to a variety of flora and wildlife, and thus also to tourists. Ranchers alter or destroy local habitat to provide grazing. They oppose the reintroduction of animals such as wolves which were previously exterminated because of attacks on livestock (Herrick, 1995). Petroleum extraction disrupts wildlife and the potential oil spills are deadly. While the Alaska pipeline does not seem to have caused the ecological destruction many feared it would, environmentalists say that only half of the planned development has taken place and that destruction has been minimized because of the pressure they have put on oil companies (Lamb, 1987).

While the types of incompatible industries discussed above are easily identified as adversaries of ecotourism, a less obvious industry that is incompatible with ecotourism may be ecotourism itself. '[E]cotourism can't possibly become a growth engine for the tourism industry and stay true to its environmental roots' (Terhune, 1997). The problem is that when there is a tourist attraction that is very desirable, each hotelier (i.e. someone who provides lodging) has an incentive to offer more space in order to increase sales (*The Economist*, 1991). But the biggest threat to the environment is the development of lodging for tourists (Terhune, 1997). Lodging destroys habitat, as does the infrastructure needed to support it. In the Caribbean, effluent from hotels is destroying barrier coral reefs (*The Economist*, 1991).

However, lodging is not the only problem. Even if tourists lived in tents, the environment would be affected. In Nepal, firewood is being used up to provide heat for tourists and to cook their food; deforestation is the result (*The Economist*, 1991). On the Gulf coast of Florida, divers and snorkellers are given the chance to interact with manatees, which they often chase and harass (Terhune, 1997). Islands are particularly fragile environments. '"Tourism is the most destructive thing on any island," according to J. Alan Gumbs, chairman of the Anguilla Tourist Board,' (Shattuck, 1997). The Galapagos Islands, which have had a large increase in tourists, have experienced trail erosion and litter. Other problems involve animals such as turtles, which mistake plastic bags for jellyfish and die as a result. Tourists feed the animals to the extent that the animals become dependent on these handouts and when unavailable, cannot feed themselves (Roe *et al.*, 1997). Additional negative impacts of wildlife tourism are reviewed by Roe *et al.* (1997). These include stress on animals, wildlife mortality, soil and vegetation damage, disturbance of wildlife, air pollution caused by vehicles, habitat disturbance due to infrastructure development, removal of natural attractions (such as molluscs) by souvenir collectors, introduction of diseases, and litter.

When the environment at an ecotourism site becomes degraded, tourists can simply find a new destination. A study of repeat visitors to the Florida Keys, USA, which has experienced steady development, showed that satisfaction with water quality, amount of wildlife they could view, and the state of the parks and protected areas decreased over a 5-year period (Terhune, 1997). In this way, uncontrolled ecotourism is similar to strip mining; an area is developed until the resources are depleted, and then abandoned. The difference is that in mining, the mining company has other areas to mine. In the case of ecotourism, the tourist has other areas to visit but the tourism industry in the abandoned site is not easily relocated. Certainly small, local tourism marketers cannot easily relocate. Even chains incur a large cost to shut down in one area and rebuild elsewhere.

Based on this analysis of incompatible industries, it is clear that the ecotourism industry must have two goals in order to survive. First, the ecotourism industry must prohibit incompatible industries from sharing a potential or actual ecotourism site. This may include preventing incompatible industries from entering ecotourism sites (e.g. protected areas) or restricting the activities of, or expelling completely, incompatible industries currently operating at a potential ecotourism site. Second, the ecotourism industry must take actions to ensure that it is not destroyed by its own growth. This raises the question as to who can actually fight the battle against incompatible industries. In the next section, the importance of coalition building is discussed.

The Intra- and Inter-sectoral Coalition in Ecotourism

Because of the small-scale, fragmented nature of the ecotourism industry, individual ecotourism marketers will by themselves not be economically or politically powerful enough to battle against incompatible industries, which may be extremely large and well organized. Therefore, alliances must be formed in order to develop a successful megamarketing strategy. Megamarketing has been defined by Kotler (1986) as 'the strategically coordinated application of economic, psychological, political and public relations skills to gain the cooperation of a number of parties in order to enter and/or operate in a market' (p. 118). In this section, the megamarketing strategy for forming alliances will be discussed.

As with any marketing activity, there must be value in exchange. Each member of the ecotourism coalition must have something of value to offer the coalition and must receive something of value in return. Cohen and Richardson (1995) identify: (i) potential coalition members; (ii) resources that they are able to contribute to the coalition; and (iii) the goal of each potential member (i.e. the benefit which each member can expect to receive from achieving the goal of sustainable ecotourism). They also suggest strategies to promote the coalition to each potential member (Table 31.1). Two interrelated target coalition member groups that merit particularly close inspection are the local population and compatible industries. Closer analysis of these groups will show that they are interrelated.

As discussed at the beginning of this chapter, ecotourism must provide economic benefits to the local population. This is especially true in developing and less-developed countries, where the population is relatively immobile. Often this has not been the case (Roe *et al.*, 1997). Also, Cater (1992) found, in countries such as Belize, that as much as 90% of tourism development was foreign owned and provided little benefit to local residents. If ecotourism does not provide an adequate economic base creating new employment

Table 31.1. Coalition building in ecotourism stakeholder groups (adapted from Cohen and Richardson, 1995).

Target coalition member	Resources	Goals	Strategies
Members with economic goals			
As yet uninvolved members of the ecotourism industry	Political clout; perhaps money	Sustain industry	Stay mutually informed; convince of self-interest; power in numbers
Manufacturers of ecotourism-related equipment/supplies	Money; organizing customers	Keep customer base	Identify appropriate companies; tie into customer base
Compatible industries	Economic and political clout	Economic sustainability	Stay mutually informed; convince of self-interest, power in numbers
Governmental and international bodies concerned with development	Official power; legal acumen; information; possibly money	Economic sustainability	Make case that ecotourism is economically viable
Public policy officials	Political power	Get re-elected; career goals within agency	Lobby, testify, convince to take action
Members with economic and ecological goals			
Local population	Political clout	High quality of life (including clean environment and healthy economy; perhaps protecting ancestral lands)	Inform of potential threats to the environment; convince and help develop economically; get involved politically
Members with ecological goals			
Tourists	Money; potential public support	Have site to visit	Keep informed; if they have clout, suggest writing to politicians
Governmental and supragovt'l bodies concerned with ecology	Official power; legal acumen; information; possibly money	Ecological sustainability	Make case that area is special
Ecological groups	Expertise; information; lobbying relationships; grass-roots ties in area	Ecological sustainability	Choose appropriate group, depending on site

opportunities, the poor and landless people will be forced to continue to engage in cattle ranching, slash-and-burn agriculture, and the killing of animals for food (Hively, 1990). Instead of these practices, ecotourism could provide jobs such as managing or working in small restaurants, driving tourists from their lodgings to and/or through the ecotourism site, or engaging in sustainable agriculture. For example, in Guatemala, palm fronds are sustainably harvested for the floral industry in a 2 million acres reserve, which is protected in part by money donated by The Nature Conservancy (Hively, 1990).

Another example of successful ecotourism in Guatemala is Eco-Escuela, which was opened in 1993 by Conservation International. Tourists come to study Spanish and local ecology (Tanner, 1998). The local population, instead of cutting down the rainforest for farms and firewood, are now engaged in supporting the school. Local people are employed as teachers. Families provide room and board. One woman, who owns one of the three telephones in town, charges US$1 to make credit card calls and at night turns her courtyard into a cantina. Other people sell food, do laundry, act as tour guides, and rent motorized canoes. One woman buys bedding in bulk to sell to host families. Eco-Escuela was turned over to the local community in 1996 and is now run by a cooperative of teachers, host families and administrators.

These examples are in sharp contrast to some other ecotourism sites. For example, in the Arctic, while local tour operators are present, there are also non-local companies that employ transient workers and repatriate their profits (Price, 1994). There is a higher probability that the profits of ecotourism will benefit the local population when growth is limited by public policy. The potential for large profit not only attracts large, non-local companies who then repatriate profits, but also ultimately destroys ecotourism. St Martin/St Maarten, in The Netherlands Antilles, was an idyllic island until the introduction of casino and hotel development. Now, the island experiences chronic pollution and social problems. This contrasts with the island of Anguilla, which still has unspoiled beaches and clean waters. To date, they have not overbuilt, and only cruise ships with a maximum capacity of 200 people are allowed to dock there (Shattuck, 1997).

Compatible industries must also employ the local population. For example, in addition to establishing Eco-Escuela in Guatemala, Conservation International assisted local people to start other businesses in the region, including those that manufacture pot-pourri, spices and cosmetic oils using non-endangered jungle plants. This gives the local population further incentive to preserve the rainforest (Tanner, 1998). In Algarve, Portugal, the tourist season is July and August. The rest of the year, the local population engages in farming and fishing (Lambert, 1999).

Sometimes, however, industries that seem completely compatible are in fact compatible with respect to resource use but not with respect to time use. Some fragile ecological areas, such as deserts, mountains and arctic areas, have climatic extremes at certain times of year, so tourism is limited to those time periods when the climate is more temperate. However, these times are also the prime agricultural, herding or hunting seasons. In these areas, traditional, compatible industries may be neglected and local people may start to depend on outside sources of income, whereas they were once self-sufficient. For example, in Nepal, male farmers can make more money from being a porter for a few weeks than working their fields for a whole year, so they leave their wives in charge of their farms. In some areas this has led to a decline in agricultural output, resulting in the need to import rice and other basic foods. Similarly, in arctic and desert societies, the prime hunting and gathering seasons are also the tourist seasons. People do not have time to collect their own food, and therefore depend on imports. If the tourists stop coming, the local people may eventually forget their traditional ways to gather food along with other aspects of their cultural heritage (Price, 1994).

While it is true that tourists can be fickle, and uncontrolled growth of ecotourism can be self-destructive, one should not overly romanticize traditional agriculture-based economies. Self-sufficiency does occur during good years, but the usual result is famine when output is lean. For most areas, the important issue then is economic diversity. Ecotourism can be only one component of a diverse economy. Sometimes the added presence of tourists can promote diversity. For example, in Ladakh in Kashmir, the presence of tourists has led to a wider range of vegetables being grown and thus more diversification in diet (Price, 1994).

Economic diversification is also important in industrialized areas. Van Dyk and Gardner (1990) point out that the timber industry, which is usually incompatible with ecotourism, is constantly warning that tourism is an unstable industry. In reality, it is the timber industry that has been experiencing steady job loss. Van Dyk and Gardner (1990) suggest that compatible industries for ecologically sensitive tourism include the electronics, aerospace, software and health-care industries.

Ecotourism needs to offer more than the promise of jobs for local people. Brandon and Margolius (1996) note that there is an implicit assumption among ecotourism advocates that ecotourism offers a competitive advantage over other potential income-earning opportunities. This, however, is not necessarily the case. Ecotourism cannot simply substitute one form of undesirable employment for another. In a dialogue hosted by the UN Commission on Sustainable Development (M2 Presswire, 1999), a trade union representative pointed out that workers in the tourism industry are often employed in substandard conditions. Furthermore, quality-of-life issues must be addressed. In developing countries, the designation of areas as protected has prevented local people from using the land for traditional uses such as gathering wood and using local plants for food and medicine (Leader-Williams *et al.*, 1990). Do people in these societies really want to replace their traditional lifestyles by supporting themselves with jobs as servants to foreigners? Even if they have jobs that do not result in direct contact with tourists, tourists can behave in ways which are unacceptable to local populations; for example, by using non-traditional hunting weapons (Talarico, 1989).

Ultimately, power must be given to the local population regarding the fate of their area. Without their cooperation, no coalition of outsiders will succeed in establishing ecotourism into an area. Brandon (1993) suggests several strategies to generate local support for ecotourism. These include:

- empowering local people in planning and decision making;
- making clear that the benefits from tourism come from protecting the environment;
- distributing benefits to many local people, not just a few;
- working with local community leaders; and
- recognizing that each site needs its own specific strategy.

Public policy officials

Coalition members with economic goals need to be convinced, if they are not already, that 'small is beautiful'. The only way for ecotourism to succeed in the long run is to enforce limits to growth. Because self-regulation is difficult in this area, public policy ultimately needs to be employed (see Chapter 32). Although public policy is not a panacea, one can argue that public policy officials are ultimately the most important group in this segment of the coalition. Specific strategies to draw public policy makers into the coalition are described further below.

Among those potential coalition groups whose goal is ecological, tourists are perhaps the most difficult to both recruit and maintain. Some tourists may have a long-term relationship with and love for a certain ecological area. For example, some people have a special relationship with the

New Jersey, USA, seashore because they have been visiting it since they were children. However, this is not the case for many ecotourists, who have a world full of potential locations to visit. If one area becomes unpleasant, they can find another. The best way to make this type of ecotourist feel loyal to an area is to ensure that their experiences meet their expectation, i.e. before the ecotourist area begins to degrade. When ecotourists do feel loyal enough to a specific area to fight for it, they can be useful allies. They are a well-educated group, generally with at least 4 years of further education (Terhune, 1997; source: Ecotourism Society). As such, they can be articulate spokespersons.

Environmental groups

Ecological groups should be chosen for the coalition based on their area of primary interest. For example, ecotourism areas by seashores would be the bailiwick of groups such as the American Littoral Society. Different ecological groups can work with each other to form an even stronger coalition (Hertzog, 1985). Governmental and supra-governmental groups concerned with ecology are extremely important as public policy makers. As with public policy makers concerned with economic goals, they are an important part of the coalition but also must be an important target for mega-marketing strategies to bring into the coalition.

Strategies for the Intra- and Inter-sectoral Alliance to Achieve Sustainable Ecotourism

A variety of strategies have been used to try to prohibit or dislodge incompatible industries from an area to ensure the viability of an ecotourism site. One of the most common types of strategy is to influence public policy. Traditional strategies for influencing public policy are discussed in this section. Additionally, groups concerned about the ecology of an area have been creating other, innovative strategies, as examined below.

Influencing Public Policy

Lobbying

One venerable method for attempting to influence public policy is lobbying. For example, in the USA a coalition of several ecological and Native American groups is trying to make Hell's Canyon in Hell's Canyon National Recreation Area much less accessible to tourists (Bernton, 1998). While the US Forest Service originally rejected the plan presented to them by this group, lobbyists in Washington pressured them to review and develop a new plan. Lobbying was also one of many traditional methods used to try to prevent MacMillan Bloedel, a giant timber company, from cutting the ancient rainforest in Clayoquot Sound, in western Canada (Bossin, 1999).

Petitions

On a more grass-roots level, petitioning public policy makers can be effective. For example, a coalition of environmental groups was able to stop development in the Willmore Wilderness area in Canada by collecting 65,000 names on a petition and presenting it to the government (Hertzog, 1985). Petitions may be even more effective than lobbying, because they let lawmakers know that the public in general (which includes voters) is concerned about development, rather than just a few special interest groups.

The process of collecting names for a petition can be efficient and effective with modern marketing methods. The important thing to remember is that people are more likely to engage in a target behaviour if it is easy and quick to do. Door-to-door or other types of personal canvassing are one option, although it means the potential petitioner may be interrupted at an inconvenient time. Other methods may be more viable. Petitioners can be mailed postcards that they simply need to sign, stamp and post to their local public policy maker. Even more efficient and cost-effective is the use of the Internet. Petitioners can be given a form, through email, to send to their local public policy maker. There is no need for a stamp or even a mailbox.

Lawsuits

When laws exist to protect the whole or one aspect of an ecological area, the coalition can use these laws to prevent incompatible industries from operating in the area. For example, in the USA the Endangered Species Act can be used if operations such as logging will destroy habitat or disturb the nesting sites of certain species. The most famous case is perhaps that of the spotted owl in north-west USA. Unfortunately, sometimes the wording of a law is not clear. For example, a 1975 US congressional act was supposed to protect Hell's Canyon National Recreation Area. The act says that management of the area should be compatible with protecting wilderness, fish, wildlife and its unique ecosystem. However, the act also recognizes 'ranching, grazing, farming, timber harvesting and the occupation of homes and lands' (Bernton, 1998), so the intention of the lawmakers is not clear and it could be difficult to invoke this act in a lawsuit.

Robert Reich, former US Secretary of Labor, feels that lawsuits are becoming more important than enacting laws to try to regulate industry (Reich, 1999). However, he points out that lawsuits are not as efficient as laws. Judges do not have the expertise in the issues that they deal with; in contrast, regulatory agencies specialize in an area. Nonetheless, the age of big government is over. Furthermore, Reich says, the close ties between politicians and large companies lower the probability that politicians will pass regulations that companies do not want. Therefore, fewer laws will be enacted and the courtroom will be the arena in which to curb activities of businesses that are harmful to a variety of stakeholder groups.

Public relations campaigns

Public relations campaigns can be used, or the threat of them can be used, to try to prevent incompatible industries from damaging ecological areas. The above-mentioned coalition of ecological groups and Native Americans used many strategies to try to save the ancient rainforests of Clayoquot Sound from logging. One strategy was to threaten Pacific Bell with a public relations campaign that would inform its customers that its telephone books were made from 1000-year-old trees (Bossin, 1999).

Civil disobedience

The coalition fighting to save Clayoquot Sound also engaged in civil disobedience. Ten thousand protesters tried blockading the road into Clayoquot. The result was 900 arrests (Bossin, 1999). While such acts of civil disobedience are not likely to eliminate incompatible industries from the target area, they can certainly alert potential coalition members to the issues at stake. However, civil disobedience may also alienate some potential coalition members who feel that the behaviour is too extreme.

In all of the above strategies, members of the ecotourism coalition take an adversarial stance toward incompatible industries. The goal is to subdue or conquer. However, there are other strategies that are more likely to result in a win–win situation for all parties. These strategies may not all be embraced by all members of the ecotourism coalition, however.

Quid pro quo

Sometimes an agreement can be made with an incompatible industry to allow it to operate or expand in one area in exchange for keeping another area pristine. For example, the Irvine Company, which owns Fashion Island Mall, in Newport, California, wanted to expand its mall but was being opposed by a group called Stop Polluting Our Newport (SPON). SPON withdrew its opposition in return for the Irvine Company agreeing to turn some of the rest of its 400 acres of Newport land into a park. The environmental groups involved want to keep the area as undeveloped as possible, with just trails and an education centre (Landman, 1988).

The quid pro quo approach is not always acceptable to some members of ecotourism coalitions. For example, ecological groups are protesting a plan by the Natural Resources Ministry in Ontario, Canada, to use publicly owned parks and protected areas for forestry, mining, tourism, fishing and hunting. The provincial government has declared that mining companies will be allowed to operate in such areas if a substantial potential for mining activity can be demonstrated. The government will then change the boundaries of the park and create a new protected area with the mining company which has the same size and equivalent ecological value (McAndrew, 1999).

Similarly, in Moosehead Lake (Maine, USA), a plan has been developed by paper companies, a local advisory group and the State Planning Office to deal with the rapid development that has taken place in the past decade. The plan would maintain 160 km of the eastern shore in a pristine condition, while allowing development on the western shore. Paper companies, which own half of the land on the 370 km shoreline, would be able to sell development rights for land they own on the eastern side to those on the western shore. The plan is being opposed by local companies such as real estate developers, who say the law is too restrictive (Chamberlain, 1990).

Forming alliances with incompatible industries

Bossin (1999) describes an unlikely alliance in Canada between the giant timber company MacMillan Bloedel, ecological groups such as Greenpeace and Sierra Club, and Clayoquot Native Americans regarding logging rights in the ancient rainforest of Clayoquot Sound. After a 20-year battle which included traditional adversarial approaches, MacMillan Bloedel is joining with Native Americans to form a company, 'Lissaak', which means 'respect' in the local Native American language. MacMillan Bloedel is asking the Canadian government to transfer its cutting rights to this company. The company will decide which timber can be cut in a way that would make the forest ecologically sustainable. The timber will have a new ecocertification, the 'green stamp', which will be actively marketed by Greenpeace.

Even ranchers can be part of the solution rather than the problem. Jack Turnell, a rancher in Wyoming, has changed his operating practices to avoid the traditional damage associated with ranching due to overgrazing, erosion of stream banks and depletion of water supplies. Instead, practices include rotating pastures, minimizing the use of fertilizers and pesticides, and crossing cattle with breeds that do not congregate around water. As a result, rivers and streams are lined with plants which provide habitat for a wide variety of wildlife, including antelope, deer, moose, elk, bear and mountain lions (Kenworthy, 1992). By making the incompatible industry part of the alliance, all parties benefit. Those concerned with the ecological environment will have a protected area that is suitable for ecotourism. The incompatible industry will be able to operate in the disputed area, although in a controlled manner. This still offers more value to the incompatible industry than spending millions of dollars to fight opposition groups, as MacMillan Bloedel had done over the last 20 years (Bossin, 1999). Ecologically supportive practices can themselves be more profitable than traditional practices. Rancher Turnell's profit has increased because better quality grasses make his cattle grow larger (Kenworthy, 1992).

Not all incompatible industries are potential partners for ecotourism. For example, except for the traditional low-technology method of panning for gold, it would be hard to conceive of a mining process which does not create visual and noise pollution and which does not damage the ecological environment.

Private ownership of ecotourism site

An ecological site can be secured for ecotourism use through private sector ownership. This is most likely to occur in developing countries where governments do not have the funds to manage a site (Hively, 1990; Roe *et al.*, 1997). For example, two tourist safari companies have developed projects with local communities in northern Tanzania. The marketers offer low volume, low impact ecotourism. The only infrastructure developed consists of access tracks and campsites. Local communities receive annual payments and other fees. Villagers are allowed to use the areas for seasonal grazing. Incompatible industries are prohibited on the contracted lands, including agriculture, permanent settlements, charcoal production, hunting and live bird capture (Dorobo Tours and Safaris and Oliver's Camp, 1996). If total ownership is not a viable option, members of the ecotourism coalition can financially help governments to manage a site. For example, The Nature Conservancy donates money to Latin American and Caribbean countries to help protect parks (Hively, 1990).

Private ownership of ecotourism sites is certainly a viable option (see Chapter 19). However, agreements must be carefully drawn. Contracts should stipulate the level of development allowed. Contracts should also ensure that the site retains its biodiversity. For example, predator animals should not be killed in order to provide more prey animals for tourists to photograph or hunt, as the US government once did in Yellowstone (Chase, 1987). Once the natural ecological order is disrupted, a ripple effect occurs and the ecological system is further upset.

Summary and Conclusions

For ecotourism to succeed, it must operate in an area free from industries that disturb the ecological environment that attracts tourists. These incompatible industries include many manufacturing and resource extraction industries. However, incompatible industries also include more developed forms of tourism. In order to ensure that such incompatible industries are restricted from the ecotourism site, coalitions must be formed with other interested parties who have ecological and/or economic interests in an ecologically pristine area. Coalitions are necessary to achieve the political and economic power necessary to deal with incompatible industries, many of whom include large, powerful corporations. Coalitions should be created by using a marketing strategy which offers something of value (economic or ecological) to each coalition member. In return, the coalition receives resources (including financial, expert, grass-roots ties, lobbying power, etc.).

The ecotourism coalition has many strategies to choose from when trying to prohibit incompatible industries from operating in an ecotourism site. In the past, most strategies were adversarial. Today, some coalitions are recognizing that incompatible industries may sometimes be made to be compatible with ecotourism, and are forming alliances with their former adversaries. Given the broad spectrum of ideologies in the ecotourism coalition, however, 'selling' the idea of working with a former adversary may be as hard as converting the adversary to an ally. Herein may lie the greatest challenge to ecotourists.

References

Bernton, H. (1998) Competing visions collide over Hells Canyon. *Sunrise* 20 December 20, D1.

Bossin, B. (1999) Lissaak. *The Globe and Mail* 22 May, D2.

Brandon, K. (1993) Basic steps toward encouraging local participation in nature tourism projects. In: Lindberg, K. and Hawkins, D.E. (eds) *Ecotourism: a Guide for Planners and Managers.* The Ecotourism Society, North Bennington, Vermont, pp. 134–151.

Brandon, K. and Margolius, R. (1996) Structuring ecotourism success: framework for analysis. Paper presented at *The Ecotourism equation: Measuring the Impacts, International Society of Tropical Foresters Yale University, 12–14 April 1996, New Haven, Connecticut, USA.* http://www.Ecotourism.org/tessrsfr.html

Cater, E. (1992) Profits from paradise. *Geographical Magazine* 64(3), 15–21.

Chamberlain, T. (1990) A battle for the heart and soul of Maine. *The Boston Globe* 26 October, 44.

Chase, A. (1987) *Playing God in Yellowstone.* Harcourt, Brace, Jovanovich, Orlando, Florida.

Cohen, J. and Richardson, J. (1995) Nature tourism vs. incompatible industries: megamarketing the ecological environment to ensure the economic future of nature tourism. *Journal of Travel and Tourism Marketing* 4(2), 107–116.

Dorobo Tours and Safaris and Oliver's Camp (1996) Potential models for community-based conservation among pastoral communities adjacent to protected areas in Northern Tanzania. In: Leader-Williams, N., Kayera, J.A. and Overton, G.L. (eds) *Community-Based Conservation in Tanzania.* World Conservation Union (IUCN), Gland, Switzerland, pp. 101–107.

The Economist (1991) How to strangle your children. 23 May, 15–17.

Herrick, T. (1995) Lobo plan has ranchers up in arms: wildlife agents want to reintroduce wolves. *Houston Chronicle* 20 August, 1.

Hertzog, S. (1985) Fight brews on land use in Alberta. *The Globe and Mail* 5 January, 8.

Hively, S. (1990) The tropics – imperiled plots. *The Plain Dealer Cleveland,* 5 October.

Kenworthy, T. (1992) The lesson of the black-footed ferret: grazing and conservation compatible, preaches Wyoming Cattleman. *The Washington Post* 2 August, a10.

Kotler, P. (1986) Megamarketing. *Harvard Business Review* 64 (March/April), 117–124.

Lamb, D. (1987) Concerns linger even as Alaska Pipeline becomes routine. *Los Angeles Times* (11/24), 16.

Lambert, S. (1999) Found: the secret Algarve. *Associated Newspapers Ltd* 22 May, 58.

Landman, R. (1988) Regional park may come true: Upper Newport Bay land being considered again. *The Orange Country Register* 28 January, 1.

Leader-Williams, N., Harrison, J. and Green, M.J.B. (1990) Designing protected areas to conserve natural resources. *Science Progress* 74, 189–204.

M2 Presswire (1999) UN: promotion of sustainable practices in tourism industry focus of Sustainable Development Commission. *M2 Communications* 20 April.

McAndrew, B. (1999) Green group sees red over mining in parks – says province duped them in land-use accord. *The Toronto Star* 23 May, 1.

Price, M. (1994) A fragile balance. *Geographical Magazine* 66(9), 32–33.

Reich, R.B. (1999) Regulation is out, litigation is in. *Newsweek* 11 February, 15A.

Roe, D., Leader-Wiliams, N. and Dalal-Clayton, B. (1997) *Take Only Photographs, Leave only Footprints: the Environmental Impacts of Wildlife Tourism.* IIED Wildlife and Development Series No. 10. http://www.Ecotourism.org/tessrsfr.html

Shattuck, H. (1997) Caribbean in transition: what price tourism? Islands are taking diverse approaches. *Houston Chronicle* 18 May, 1.

Talarico, D. (1989) *Wildlife Viewing Tourism on the Yukon North Slope: Considerations Relevant to Aklavik, Northwest Territories.* Report to the Fish and Wildlife Branch, Department of Renewable Resources, Government of Yukon, Canada.

Tanner, J. (1998) Where ecotourism may save a jungle. *Business Week* 16 January, 38J.

Terhune, C. (1997) Rise of ecotourism sparks fears of more runaway development. *Wall Street Journal: Florida Journal* 12 March, Florida.

Van Dyk, R. and Gardner, M. (1990) A skeptical view of timber industry's concern for jobs. *The Seattle Times* 12 September, A11.

Chapter 32

The Place of Ecotourism in Public Policy and Planning

S. Parker
Department of Political Science, University of Nevada, Las Vegas, Nevada, USA

Introduction

Unlike most of the other entries in this volume, this chapter is not designed to discuss some specific aspect of ecotourism or protected-area management. Instead, its function is to focus on the public policy and planning context within which these two operate. In so doing, numerous governmental programmes, functions and conditions which are external to ecotourism but that have some significant effect upon it, are examined. How ecotourism's impact on this sector differs from the impact on mass or traditional tourism is also assessed.

Ecotourism is an activity carried on within a set of social and physical conditions that are heavily influenced by governmental decision making. Those that will be discussed below include politics, security, infrastructure, dependency, fiscal policy, and financial incentives and measures designed to deal with problems of dependency. The latter element is included because most of this article's analysis deals with ecotourism in the world's less-developed countries (LDCs). While many different LDC policy areas will be analysed, the one that is perhaps the most significant will be avoided: the connection between ecotourism and rural development. That subject is excluded here since it is presented elsewhere in this encyclopedia (see Chapter 27).

Politics and Administration

The way in which a government organizes for managing its ecotourism and protected-area activities has a substantial impact on how these functions can be carried out. Typically they are vested in a Department of Agriculture or Forestry, as in Ecuador, Indonesia and Tanzania. When submerged in this way, within a larger whole, their effectiveness is inevitably compromised. In Costa Rica, for example, responsibility for national parks is given to the Ministry of Natural Resources, Energy and Mines. The placement of a conservation unit so unambiguously within an organization whose main goal is mineral exploitation lessens that unit's organizational strength.

Ecotourism and protected area management are also affected by the policies and operations of numerous other government agencies whose primary missions deal with different functions (Richter, 1980) and whose main loyalties are thus found elsewhere. Departments of public works, transportation, environment, immigration and customs, cultural affairs, investment, education and planning all set policies that impact ecotourism. This fragmentation, in turn, makes leadership and policy coordination extremely difficult.

In addition to these multipurpose ministries are those that are in direct competition with ecotourism: agriculture, fisheries, forestry, mining and energy. In most of the developing world these agencies serve as governmental advocates for countries' primary industries, and because of this fact they normally enjoy more political power in national capitals (Hall, 1994). Agencies in charge of tourism and protected area management tend, in contrast, to be politically weak and less effective in advancing their agendas (Wells and Brandon, 1993). This is partially because their constituencies are weak (Greffe, 1994; McNeely, 1995) and less likely to be supported by the local population.

Similarly, the industry fragmentation that is so widespread in traditional tourism (Elliott, 1983) is amplified in the case of ecotourism because of its close ties with local units of government and councils of indigenous peoples. The need for coordination in such decentralized situations has given rise to a call for the creation of advisory and coordinating committees in major ecotourism countries (Boo, 1990). Such an approach has been tried in Australia where, for example, a Consultative Committee brings together major stakeholders from both the public and private sectors to advise the government on policy relating to the Great Barrier Reef. It has been able to do this in an effective manner because its members include representatives from conservation organizations, fishery and agricultural associations, councils of Aboriginal elders, marine park tourism operators and several levels of government.

Ordinarily, however, it is extremely difficult for such bodies to demonstrate initiative and operate efficiently. As de Kadt (1992) has pointed out, officials working in public agencies that exist to serve commercial clientele groups such as fisheries can expect few rewards for cooperating with outsiders interested in tourism policy. The would-be collaborators have different training, different loyalties, different interpretations of the public interest and different diagnoses of the resource problem, in addition to having different masters to serve. A further example of how such dynamics can create critical problems concerns the relationship between tourism and immigration agencies. The latter have border protection and visitor regulation as their primary *raisons d'être* and have therefore frequently been unwilling to facilitate tourist movement. We might hypothesize that the intensity of this problem is ordinarily related to such factors as a country's level of political stability, its history of xenophobia and its need for foreign exchange earnings.

Frequently, the identities of these 'different masters' are determined by a country's own political dynamics and can have an enormous impact on the industry. In Thailand, for example, the practice of allocating ministries to different political parties has meant a very inconsistent application of tourism policy (Elliott, 1987). When the Tourism Authority is managed by one political party, and the Natural Resources ministry by another, each can pursue its own goals, ignoring calls for cooperation, with no fear of recrimination. Such a situation is exacerbated when the stronger departments like Fisheries and Agriculture are given to one of the dominant political parties in a coalition while the weaker Tourism or Environment portfolios are entrusted to a minor party.

In many LDCs, the military is another governmental institution becoming increasingly involved with ecotourism and resource management (D'Souza, 1995). Sometimes the reason for this cooperation rests on the simple fact that the military can provide a source of readily available manpower. In Jordan, for example, special units of army conscripts are used to assist in conservation programmes while Venezuela relies on its armed forces to assist with various programmes designed to combat deforestation. Other policy explanations have to do with the availability of skills and the access to needed technology. A case in point is Ecuador, which uses military aircraft to transport tourists 640 km from the coast to the Galapagos Islands. It also relies on naval patrol craft to stem the tide of unlicensed boats bringing sightseers to the area. Mexico utilizes Naval personnel in a similar way to regulate dive tourism at its Revillagigedos Islands 320 km from shore in the Pacific. Applications such as this are primarily related to a country's efforts to control smuggling or establish a claim of sovereignty over distant possessions, but they have a secondary impact on ecotourism since they become a means to regulate it.

Because wildlife conservation is such an integral part of resource management, another military application involves the use of anti-poaching patrols (Pleumarom, 1994). Tanzania relies heavily on the army for this purpose and in Nepal the military provides a guard force for The Royal Chitwan National Park. Perhaps the most intense use of the armed forces for this purpose occurs in Kenya where poachers from neighbouring Somalia cross the border to hunt elephants. Because movement across the frontier is so easy, Kenyan park rangers are frequently assisted by regular army personnel who plan strategic actions against these bands much as they would against a small invading force whose purpose was political rather than criminal and commercial.

Security

Security is one of the most fundamental roles for the military. This is a policy and planning area of central importance because a government unable to protect its visitors will quickly lose them. With traditional tourism, security ordinarily entails responsibilities like protecting people from pickpockets or burglars. In many LDCs, however, the nature of the task is much broader because it can involve issues of political instability, violence and insurrection. A nation in the throes of civil strife is a dangerous place, and tourism is extremely vulnerable to such disruptions. For example, ethnic violence in Sri Lanka in the 1980s halved that country's number of tourist arrivals, while Guatemala's civil war cut tourism from half a million visitors in 1979 to less than 200,000 in 1984 (Lindberg, 1991). Furthermore, it must be remembered that ecotourism venues are usually far afield and likely to be located in difficult terrain. For these reasons, they are harder to defend and secure than urban areas or resorts. By the very nature of their activity, it is ecotourists, rather than any other type, who are most likely to come into contact with armed dissidents.

Many destination countries have only recently achieved independence and they tend to be geographically defined by international boundaries that were drawn generations ago by occupying colonial powers. For this reason they often include numerous ethnic and religious groups that have been traditional enemies but have never had to coexist before under a single regime administered by just one ethnic group. Such situations make these countries especially vulnerable to the problems of secession and civil war, due to these same rivalries. Traditional tribal enemies, the Hutus and Tutsis killed tens of thousands of people as they fought for dominance in Rwanda during the early 1990s. In the process, places like the Parc National des Volcans were decimated. Habitat and infrastructure were destroyed; the flow of tourists dried up and the number of tour operators in that particular park declined from fifteen to one between 1990 and 1993 (Shackley, 1995).

A similar decline occurred in the wake of Fiji's 2000 military coup, an event sparked by a power struggle between native Fijians and those of Indian descent. Not only did visitation fall by 90% in the months immediately after the coup, but the availability of investment funds evaporated as well. Two coup attempts in Dominica in 1981 had the same predictable consequences (Weaver, 1991), and even after the civil war in Mozambique had come to an end in 1992, the government's decision to authorize a large ecotourism project instead of a pulp mill was based on lingering postwar considerations (Massinga, 1996).

The 1999 kidnapping and murder of eight Westerners on a gorilla-watching expedition in Uganda is merely one of the best-known illustrations of the problem of political violence. In that case an ethnic civil war in neighbouring Rwanda spilled over into Uganda and targeted British and American tourists because their governments, along with Uganda, had given aid to the new Rwandan regime. One of the rebels' goals was to punish Uganda by disrupting its tourism industry. Their logic held that if tourists feared for their lives, then they would not visit the country. This, in turn, would cause commercial hardship and a consequent loss of tax revenues for the hated government. All of these things did, in fact, happen because ecotourism could be targeted so easily.

Similarly, the terrorist bombings of US embassies in Kenya and Tanzania in 1998 had a devastating effect on the wildlife tourism industry in those two countries. The incidents stemmed partially from the inability of local officials to guarantee the safety, not of tourists, but of the embassies themselves. In this case tourists specifically were not targeted, but the effect on the industry was the same. Thus, again we can see clearly how issues and policies that are only collateral to ecotourism can have a major impact upon this sector.

In what was only a slight variation on this theme, Ecuadorian fishermen stormed the Charles Darwin Research Station in the Galapagos Islands in 1995. Wielding clubs and machetes, they blockaded the administrative headquarters to protest a government fishing moratorium. It was the vulnerability of nature tourism that ultimately caused the government and the world to pay attention to their demands. Similarly, when Palestinian extremists attacked a major tourism site in Luxor, Egypt in 1996, their goal was to use the event to publicize their grievances against the Egyptian regime. Such grievances were already well known, but the notoriety of the event put them back near the top of the international agenda.

The ability to protect visitors from physical violence is a *sine qua non* for destination governments, but the maintenance of political stability (the absence of unpredictable changes in government) is almost as important. Widely regarded as one of the most stable LDCs in the Western hemisphere, Costa Rica has been able to offer investors a sense of certainty and security that has been a major factor in its rise to pre-eminence as an ecotourism destination (Weaver, 1994). In contrast, neighbouring Guatemala and Nicaragua have been deemed much less stable and thus developmental funds have not been as readily available. It is not a coincidence that both are still recovering from recent civil wars.

Infrastructure

Government infrastructure policy concerns the planning, financing and construction of projects ranging from airports and utilities to roadways and housing. Such public works are of vital importance to all types of tourism, but perhaps less so to ecotourism than to other types. This is the case because mass tourism utilizes much greater amounts of natural energy sources, and recycling and low-impact activities require less capital construction, to mention just a few of the reasons. Of course it is true that ecotourism must rely on certain types of standard infrastructure to provide services like visitor transportation and access, yet ordinarily the scale required is far more modest.

Together, the projects that deliver such services comprise a significant part of the physical context within which the industry must operate. However, government planners tend to be considerably less sympathetic to ecotourism than to the traditional variety because of considerations relating to the economies of scale (Inskeep, 1987). Concentrating visitors in large resort areas like Bali produces a much more efficient use of capital resources. According to Jenkins (1982), considerations of scale predispose decision makers in the direction of mass tourism. Frequently 'bigger is better' to investors and policy makers because it may mean the ability to distribute fixed costs over a larger number of units and visitors.

Whether the infrastructure decision concerns a large or a small-scale project, impacts will follow. To begin with, consider the case of Shark Bay in Western Australia (Dowling, 1991). There, a 1985 decision to pave the access road made it much easier for visitors to reach the area and thereby helped to generate a 15-fold increase in tourism over the next 6 years. Infrastructure can thus be a major factor in transforming ecotourism into mass tourism, as is happening on Mexico's Yucatan Peninsula, where a super highway from Cancún to Playa del Carmen has made virtually all of the region's marine and cultural sites available to thousands of visitors daily. Policy relating to such construction can thus be a major factor in propelling a destination from one stage to the next in Butler's (1980) model of a tourist area cycle.

Such a progression is not inevitable, however, as is indicated by a contrasting example. Costa Rica's Monteverde Cloud Forest Preserve illustrates how policy and planning can be used to ensure that the scale of infrastructure and the scale of tourism are compatible. There the community specifically decided to keep the improvements simple and not pave the 40 km dirt access road from the main highway, The Interamericana (Honey, 1994). They reasoned that a paved road would generate more tourists but less per-person spending. If the Preserve could be reached in 30 min instead of the usual 2.5 h then many erstwhile overnight visitors would become merely day-users instead. Because they would not have to stay in local lodges, eat in area restaurants, etc., they would generate less income for residents who would, in turn, have a reduced monetary interest in conservation. This monetizing of environmental benefits is, of course, part of the basic logic of ecotourism, and these contrasting cases illustrate how infrastructure decision making can either support it or operate as an impediment.

While such commitments are frequently made as part of an overall tourism plan, they may also be reached as the means to other ends, but then generate 'spillover effects' for the industry. Consider, for example, the opening of the Cairns International Airport in 1984. It was part of a broader Australian plan for the development and economic diversification of northern Queensland. Today that infrastructure project is the major link that brings nearly 2 million visitors annually to the Great Barrier Reef. With visitation doubling since 1990, park officials there have had to generate new strategies for the licensing of operators and for the management of the World Heritage Area itself. Most of these impacts can ultimately be traced back to the airport decision.

A similar scale of change was initiated in the Galapagos Islands after the Ecuadorean government decided to construct two airports there (Kenchington, 1989). With some 80,000 visitors now arriving each year, the problems of overcrowding at sites visited by Charles Darwin are all too well known. Many of them began with infrastructure decisions that relied on economies of scale and thus needed more and more tourists to make them affordable.

While there certainly are a number of negative consequences that can flow from such commitments, the indisputable fact is that these public works projects are an essential part of the policy and planning context within which ecotourism operates. To glimpse the consequences of their absence or inadequacy one need look no further than the case of Boracay in the Philippines (Smith, 1992). Initially tourism there was small scale and cottage based, but it produced enormous environmental damage due to deficiencies in the water and sanitation facilities. Today a similar threat hangs over the island of Bonaire in The Netherlands Antilles (Roberts and Hawkins, 1994), because there is no wastewater treatment plant to process the effluent generated by the 65,000 divers and other guests who visit there annually.

Dependency

Ecotourism also figures in a nation's policies to combat the problems of dependency and leakage. As Mathews and Richter (1991) have pointed out, dependency is a political condition in which the economic well-being of one nation is controlled by an agenda established elsewhere. With regard to tourism, in general, this means that the pace of change and development is set by financial choices made not in Samoa, but in Los Angeles or Tokyo, because that is where the investment funds, the expertise and the tourists are to be found. Since so many tourism assets around the world are not owned locally, most profits are not reinvested locally and this is one of the major dimensions of leakage, the loss of earnings back to the metropolitan centre (Pleumaron, 1994). According to both Lindberg (1991) and Boo (1990), less than half of all tourism monies spent in the developing world remain there. They are instead repatriated to the countries from which the investments and the tourists originated.

There are numerous policies that a destination government can adopt in order to mitigate this problem and the stimulation of ecotourism is, itself, one of them. This is because the small-scale nature of many of these enterprises, together with their emphasis on local funding sources and the provision of employment for locals (Hawkins *et al.*, 1995), means that they can retain a far greater share of the profit within the country than can mass tourism operations.

Another cause of leakage is the need of hoteliers and others to purchase tourist goods (food, recreational equipment, furnishings, clothing, etc.) from abroad. A policy to combat this problem, known as import substitution, focuses both on finding similar goods locally and on creating the conditions under which native entrepreneurs will increase their production. Lindberg and Huber (1993) have examined several ways in which governments can accomplish such goals. They include improving linkages with local sources of transportation via subsidies; encouraging the development of local food sources through training and research; broadening the use of local labour and materials in the construction of lodges and stimulating local handicrafts through such mechanisms as the creation of cooperatives. Because of the differences in scale and in the levels of comfort demanded by clients, these techniques have so far proven to be more effective in the ecotourism context than in that of traditional or mass tourism.

Perhaps the most significant type of dependency involves airline service (Britton, 1982). This particular form of vulnerability affects ecotourism and traditional tourism almost equally and it cannot be mitigated by such targeted policies as those mentioned above. The reliance of many South Pacific islands like Yap and Palau on Continental Airlines' Air Micronesia or the dependence of Caribbean destinations such as the Turks and Caicos Islands on American Airlines provide appropriate examples. The future of such places is almost totally in the hands of external decision makers. If they are unable to recruit additional air carriers, local policy makers have few other options available than the extremely expensive one of operating and subsidizing their own airlines as do destinations like Papua New Guinea, the Cayman Islands and The Netherlands Antilles. Of course, given the technological requirements that go with such an investment, the decision to follow this policy then leaves the nation subject to further problems of leakage since both the planes and their spare parts normally have to be purchased abroad.

Another dimension of dependency concerns the foreign ownership of land, and most destination countries have taken some form of action to redress this problem. One difficulty, however, is that there have proven to be too many conditions or loopholes in public policy by which such laws can be circumvented. Thailand (Elliott, 1983) and the Philippines (Smith, 1992) illustrate a common approach. Both of these countries regulate foreign ownership, but outside investors can neutralize their laws by simply establishing partnerships with native entrepreneurs. Similarly, Cater (1994) has shown that while Belize limits the amount of land that can be held by any one foreign individual, in practice the policy's cumulative effect has been to simply allow a multitude of small-scale foreign holdings, such as those focused around ecolodges.

An additional area of public policy designed to mitigate dependency, deals with the subject of expatriate labour and the goal of providing preferred employment opportunities for the native population (Mathieson and Wall, 1982). Here again, ecotourism can be useful to policy makers because its scale and attenuated need for technology generate less reliance on outsiders than does mass tourism. However, this is not always an accurate assessment because needed technicians and other specialists may not be available locally. Unfortunately, host countries frequently pay too little attention to providing the training programmes that will allow their own nationals to qualify for such jobs. Because of the near inevitability of such problems, most governments allow for exemptions when local skills are unavailable. When expatriates do have to be hired, the relevant public-policy issues are concerned with the levels at which they are taxed; whether their earnings can be remitted out of the country, and if so, at what percentage. As in Costa Rica, labour policy can also provide disincentives to the recruitment of foreign nationals, through the pricing structure applied to work permits.

Fiscal Policy

Fiscal policy, the set of government decisions relating to taxation and public expenditure, is a major force affecting tourism and ecotourism. Perhaps the main reason for this is the fact that decision makers, like those in Honduras, see it as a way of earning needed foreign exchange (Stonich *et al.*, 1995). Many governments use ecotourism as a revenue source by levying taxes on goods and services and by diverting user fees away from the national parks at which they are collected. Ordinarily these fees are simply deposited in the general treasury, as is done with payments collected from the 80,000 annual visitors to Ecuador's Galapagos Islands (Honey, 1994). Of course, if fiscal policy allowed the funds to be used for improvements in the parks where they were collected, then habitat management would be greatly improved (Lindberg, 1991). However, when the funds are simply siphoned off for other purposes, then fiscal policy degrades conservation policy, as it did until recently in national parks in the USA (Parker, 1998).

Other taxation options available to host governments include hotel taxes, airport departure taxes, customs duties that are levied on imports used for tourism purposes and income taxes on companies and individuals working in the industry, as well as the proceeds from taxes assessed on tourism properties (Inskeep, 1991). Several of these tools affect ecotourism less than they do mass tourism. For example, since ecolodge owners usually rely more heavily on local farmers for the purchase of foodstuffs, they are less subject to the effect of food import duties. The same is true with regard to the overall impact of property taxes since the scale of ecotourism construction is ordinarily much more modest. Another fiscally relevant difference between these two segments of the industry is that ecotourism operators pay a far greater proportion of their fees and taxes to local villages than do their counterparts in mass tourism. The fragmented and hinterland-based nature of the industry means that ordinarily it will be less at the mercy of the central government's tax collectors.

Undoubtedly, the most important connection between fiscal policy and tourism/ecotourism concerns the seemingly universal predilection of governments to use these two as means for increasing their foreign exchange earnings (de Kadt, 1976). Such a policy can create substantial social and ecological problems (Stonich *et al.*, 1995), and a sizeable literature exists on this subject. For example, Ceballos-Lascuráin

(1991) has written on how the Canary Islands' fiscal policy, designed to maximize the economic return from parks, has led to a rash of 'questionable development'. Large hotels and golf courses bring in more tourists and thereby generate more tax revenue and foreign exchange, but in the process they degrade the nearby protected areas. In a related study, Stephenson (1993) examined the habitat disturbance and general environmental degradation that has occurred on Madagascar since that island's government decided to authorize primate viewing as a means of generating foreign currency.

One other dimension of fiscal policy dealing with a nation's international balance of payments is the relatively new and innovative financing mechanism known as the 'debt-for-nature swap' (Mieczkowski, 1990; Visser and Mendoza, 1994). Once a host government has sanctioned its usage, this practice reduces the amount of debt owed to foreign lenders in exchange for that government's creation of protected areas. Here, instead of fiscal policy impeding conservation and ecotourism, it actually works to their benefit. Rather than being sacrificed to concerns of international finance, environmental protection is enhanced through a government's pursuit of its monetary goals. For example, Bolivian officials worked with the group Conservation International and secured the cancellation of a US$650,000 debt in exchange for making a 1.4 million ha addition to The Beni Biological Reserve (Cartwright, 1989). This arrangement both improved the government's international fiscal position and provided legal protection for the addition to the reserve. Not all such sites are opened for ecotourism, but for those venues that are, it is fiscal policy rather than environmental policy that operates as the key factor.

Financial Incentives

Closely related to such fiscal policy considerations are those dealing with governmental stimulation and subsidization of the industry. These activities are usually implemented via tax law and they have tended to apply far more to mass- than to ecotourism. This is so because governments view such aid as a kind of fiscal investment; one that will yield a higher return if applied to larger projects. Hannigan (1994) has argued that there is a very great danger in such scenarios, since the financial returns expected may not be congruent with sustainability goals. Using Ireland as an example, he shows how the governmental setting of economic targets can lead to the sacrifice of environmental objectives.

Ecolodges and similar facilities do continue to be desirable investments in situations where funds are limited (Sherman and Dixon, 1991), but because government can earn higher returns and create more jobs with larger projects, this sector suffers by comparison. In Costa Rica, for example, The Tourism Development Incentive Law offers tax breaks for construction, but in order to qualify for participation a project must have a minimum of 20 rooms (Hill, 1990). Such a requirement frequently excludes ecolodges and related facilities, as well as the participation of locals, who have far more limited access to investment funds.

Nonetheless, there are exceptions such as the occasional use of 'micro-loans' and 'incubator' assistance. Both of these mechanisms were relied upon, for example, in the early 1990s in Belize to help small businesses during the establishment of the Community Baboon Reserve (Horwich *et al.*, 1993). In this case, incentives were provided by both public and non-government organization entities. Governments can also bundle small projects together to reach the lending thresholds required by international banks (Hawkins *et al.*, 1995).

On Anguilla, the government uses tax incentives to encourage limited-scale development and the construction of cottage-style accommodations in guest houses and small hotels (Wilkinson, 1989). Small investors on the island of Dominica benefit from The Hotel Aids Ordinance and can import hotel equipment and building materials duty free. Once the facility is in operation, all investors are exempt from income tax on their earnings for 12 years (Boo, 1990). A similar tax moratorium on the Island Territory of Bonaire has helped to stimulate the growth of marine ecotourism in The Netherlands Antilles (Pieters and Gevers, 1995). Other targeted incentives in use in the Caribbean include investment tax credits, cash grants, leasing of property and the provision of subsidized utilities (Lindberg, 1991). Kenya offers tax rebates and the duty-free import of equipment (Olindo, 1991). Similarly, the Brazilian government has used tax incentives to generate ecolodge investment in the Amazon Basin (Ruschmann, 1992).

While all such devices find application in alternative tourism settings, their broadest employment is in the service of mass tourism because it yields a more substantial return to the granting agency, the government. Australia, for example, provides incentives such as low interest loans, accelerated depreciation and a programme for lending money to private investors (Hall, 1995). However, these are applied with much greater frequency to mass- than to ecotourism operations.

An additional reason why this is so has to do with the concept of mobility of capital. Tourism destinations around the world are in competition with each other to generate more investment. A tax advantage offered by one island microstate will have to be matched, or bettered, by another or large investors will take their funds elsewhere. Generally speaking, because of the size of their projects, ecotourism entrepreneurs do not enjoy the same kind of leverage. When the entrepreneurs are local villagers and the ecotourism is community based, the difficulty is frequently magnified (Drumm, 1998). For example, Costa Rica's Tortuguero National Park has become very popular with nature travellers in search of opportunities to observe nesting sea turtles. This trend has generated numerous investment opportunities. However, the pace of development has proven to be too rapid to allow local villagers the time needed to accumulate sufficient capital of their own to invest in the construction of locally owned tourism facilities (Place, 1991). Since no government programme exists to help them in such an endeavour, they have benefited only minimally from the area's popularity.

Conclusion

The purpose of this chapter has been to summarize and assess the numerous ways in which government action affects protected area management and the ecotourism industry. While we have seen that there is an enormous impact, there also appears to be a range in the specificity of this impact. In some instances, such as the maintenance of security in unstable political situations, policy operates on a *macro-level*, since its manifest purpose relates to conditions in a nation's entire political system. In others, such as financial incentive programmes, the goals are much more tightly drawn, and might therefore be referred to as *micro-level* policies.

Using these two examples, we will close by suggesting that the relationship between ecotourism and public policy may be very usefully conceptualized as a series of impact points arrayed along a continuum of perspectives. Such a continuum is presented in Fig. 32.1. The macro perspective relates to concerns such as domestic security or the promises made by national political parties. However, these are matters that will attract the attention of participants from most major economic sectors and social groups. In such a context the voice of ecotourism is usually an attenuated one, diluted by the fact that it is just one among many. On the other hand, when the issue is micro in nature and relates to a particular infrastructure decision like wastewater treatment or ferry service for a small island, then the industry is in a better position to generate input and create true partnerships.

Policy processes in much of the world are conditioned by the fact that participating groups ordinarily have only a small stake in major system-wide decisions and

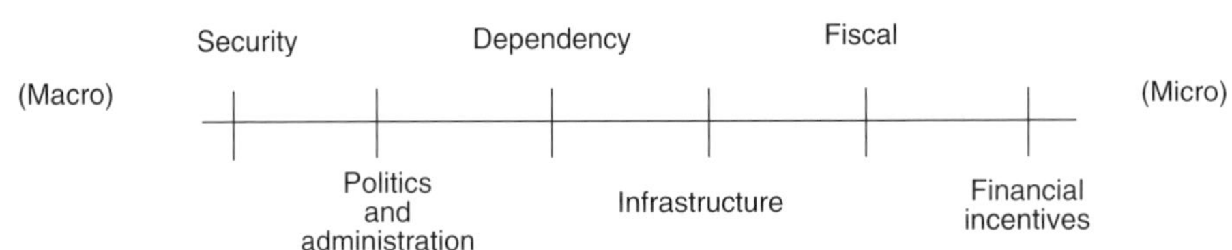

Fig. 32.1. The range of public impact on ecotourism.

by the related fact that their potential to change the outcome of them is minimal. For this reason their involvement tends to be episodic and dilatory. Conversely, there are other issues that affect only a few groups, but because their resolution is so vital to stakeholder well-being they tend to participate with high levels of intensity. Public policy relating to ecotourism illustrates this dynamic quite well, and therefore members of the industry have generally experienced their greatest efforts and successes on the micro end of the continuum with issues like financial incentives, transportation decisions and very specific environmental management practices. Future, longitudinal studies would do well to examine the long-term accuracy and applicability of this observation, in order to ascertain whether it is something that is generally true or is merely related to the dynamics of this industry at one particular stage in its development.

All of the policy areas discussed in this chapter have been external to the industry itself, but some have had impacts of a much more immediate nature than others. Thus, in Fig. 32.1, the six policy types are represented along a continuum arranging them from left to right, based on the extent to which their impact is indirect (macro) or direct (micro).

References

Boo, E. (1990) *Ecotourism: the Potentials and Pitfalls.* World Wildlife Fund, Washington, DC.

Britton, S.G. (1982) The political economy of tourism in the third world. *Annals of Tourism Research* 9(3), 331–358.

Butler, R.W. (1980) The concept of a tourist area cycle of evolution: implications for management of resources. *Canadian Geographer* 24(1), 5–12.

Cartwright, J. (1989) Conserving nature, decreasing debt. *Third World Quarterly* 114–127.

Cater, E. (1994) Ecotourism in the third world: problems and prospects for sustainability. In: Cater, E. and Lowman, G. (eds) *Ecotourism: a Sustainable Option?* John Wiley & Sons, New York, pp. 69–86.

Ceballos-Lascuráin, H. (1991) Tourism, ecotourism and protected areas. *Parks* 2(3), 31–35.

de Kadt, E. (1976) *Tourism: Passport To Development?* Oxford University Press, New York.

de Kadt, E. (1992) Making the alternative sustainable: lessons from development for tourism. In: Smith, V.L. and Eadington, W.R. (eds) *Tourism Alternatives: Potentials and Problems in the Development of Tourism.* University of Pennsylvania Press, Philadelphia, pp. 47–75.

Dowling, R. (1991) Tourism and the natural environment. *Tourism Recreation Research* 16(2), 44–48.

Drumm, A. (1998) New approaches to community-based ecotourism management. In: Lindberg, K., Wood, M. and Engeldrum, D. (eds) *Ecotourism: a Guide for Planners and Managers.* The Ecotourism Society, North Bennington, Vermont, pp. 197–213.

D'Souza, E. (1995) Redefining national security: the military and protected areas. In: McNeeley, J. (ed.) *Expanding Partnerships in Conservation.* Island Press, Washington, DC, pp. 156–165.

Elliott, J. (1983) Politics, power and tourism in Thailand. *Annals of Tourism Research* 10(3), 377–393.

Elliott, J. (1987) Government management of tourism: a Thai case study. *Tourism Management*, 223–232.

Greffe, X. (1994) Is rural tourism a lever for economic and social development? *Journal of Sustainable Tourism* 2(1), 22–40.

Hall, C.M. (1994) *Tourism and Politics: Policy, Power and Place.* John Wiley & Sons, New York.

Hall, C.M. (1995) *Introduction to Tourism in Australia.* Longman, Sydney.

Hannigan, K. (1994) National policy, European structural funds and sustainable tourism: the case of Ireland. *Journal of Sustainable Tourism* 2(4), 179–192.

Hawkins, D., Wood, M.E. and Bittman, S. (1995) *The Ecolodge Sourcebook.* The Ecotourism Society, North Bennington, Vermont.

Hill, C. (1990) The paradox of tourism in Costa Rica. *Cultural Survival Quarterly* 14(1), 14–19.

Honey, M. (1994) Paying the price of ecotourism: two pioneer biological reserves face the challenge brought by a recent boom in tourism. *Americas* 46(6), 40–47.

Horwich, R., Murray, D., Saqui, E., Lyon, J. and Godrey, D. (1993) Ecotourism and community development: a view from Belize. In: Lindberg, K. and Hawkins, D. (eds) *Ecotourism: a Guide for Planners and Managers.* The Ecotourism Society, North Bennington, Vermont, pp. 152–168.

Inskeep, E. (1987) Environmental planning for tourism. *Annals of Tourism Research* 14(1), 118–135.

Inskeep, E. (1991) *Tourism Planning: an Integrated Sustainable Development Approach.* Van Nostrand Reinhold, New York.

Jenkins, C.L. (1982) The effects of scale in tourism projects in developing countries. *Annals of Tourism Research* 9(2), 229–249.

Kenchington, R. (1989) Tourism in the Galapagos Islands: the dilemma of conservation. *Environmental Conservation* 16(3), 227–236.

Lindberg, K. (1991) *Policies for Maximizing Nature Tourism's Ecological and Economic Benefits.* World Resources Institute, Washington, DC.

Lindberg, K. and Huber, R. (1993) Economic issues in ecotourism management. In: Lindberg, K. and Hawkins, D. (eds) *Ecotourism: a Guide for Planners and Managers.* The Ecotourism Society, North Bennington, Vermont, pp. 82–115.

Massinga, A. (1996) Between the devil and the deep blue sea: development dilemmas in Mozambique. *The Ecologist* 16(2), 73–75.

Mathews, H. and Richter, L. (1991) Political science and tourism. *Annals of Tourism Research* 18(1), 120–135.

Mathieson, A. and Wall, G. (1982) *Tourism: Economic, Physical and Social Impacts.* Longman, New York.

McNeely, J. (1995) *Expanding Partnerships in Conservation.* Island Press, Washington, DC.

Mieczkowski, Z. (1990) *World Trends in Tourism and Recreation.* Peter Lang, New York.

Olindo, P. (1991) The old man of nature tourism: Kenya. In: Whelan, T. (ed.) *Nature Tourism: Managing for the Environment.* Island Press, Washington, DC, pp. 23–38.

Parker, S. (1998) Concessions policy in America's national parks. *National Social Science Journal* 10(1), 114–121.

Pieters, R. and Gevers, D. (1995) A framework for tourism development on fragile island destinations: the case of Bonaire. In: Conlin, M.V. and Baum, T. (eds) *Island Tourism: Management Principles and Practice.* John Wiley & Sons, New York, pp. 123–132.

Place, S. (1991) Nature tourism and rural development in Tortuguero. *Annals of Tourism Research* 18(1), 186–201.

Pleumarom, A. (1994) The political economy of tourism. *The Ecologist* 24(4), 142–148.

Richter, L. (1980) The political uses of tourism: a Philippine case study. *The Journal of Developing Areas* 237–257.

Roberts, C. and Hawkins, J. (1994) *Report on the Status of Bonaire's Coral Reefs.* University of the Virgin Islands, St Thomas.

Ruschmann, D. (1992) Ecological tourism in Brazil. *Tourism Management* 13(1), 125–128.

Shackley, M. (1995) The future of gorilla tourism in Rwanda. *Journal of Sustainable Tourism* 3(2), 61–72.

Sherman, P.B. and Dixon, J.A. (1991) The economics of nature tourism: determining if it pays. In: Whelan, T. (ed.) *Nature Tourism: Managing for the Environment.* Island Press, Washington, DC, pp. 89–131.

Smith, V. (1992) Boracay, Philippines: a case study in 'alternative' tourism. In: Smith, V.L. and

Eadington, W.R. (eds) *Tourism Alternatives: Potentials and Problems in the Development of Tourism.* University of Pennsylvania Press, Philadelphia, pp. 135–157.

Stephenson, P. (1993) The impacts of tourism on nature reserves in Madagascar. *Environmental Conservation* 20(3), 262–265.

Stonich, S., Sorenson, J. and Hundt, A. (1995) Ethnicity, class and gender in tourism development: the case of the Bay Islands, Honduras. *Journal of Sustainable Tourism* 3(1), 1–28.

Visser, D. and Mendoza, G. (1994) Debt for nature swaps in Latin America. *Journal of Forestry* 92(6), 13–16.

Weaver, D. (1991) Alternative to mass tourism in Dominica. *Annals of Tourism Research* 18(3), 414–432.

Weaver, D. (1994) Ecotourism in the Caribbean Basin. In: Cater, E. and Lowman, G. (eds) *Ecotourism: a Sustainable Option?* John Wiley & Sons, New York, pp. 159–176.

Wells, M. and Brandon, K. (1993) The principles and practice of buffer zones and local participation in biodiversity conservation. *Ambio* 22(2–3), 157–162.

Wilkinson, P.R. (1989) Strategies for tourism in island microstates. *Annals of Tourism Research* 16(1), 153–177.

Section 7

The Business of Ecotourism

B. McKercher

Department of Hotel and Tourism Management, Hong Kong Polytechnic University, Hung Hom, Kowloon, Hong Kong, China

---

Ecotourism and business? The two words seem to be an oxymoron. Many people view ecotourism as an ideologically or philosophically pure construct that should not be sullied by sordid business considerations. Tourism business is often regarded as being interested only in profit and of showing no respect for host communities or environments. How can ecotourism and business be linked? What many people do not realize is that net positive benefits from ecotourism can only accrue if it is a self-supporting activity, which means that it must be commercially viable. Cultural, social and economic benefits will only occur when ecotourism generates sufficient profit. Non-viable activities represent a net drain on scarce resources and expose host communities to exploitation. Likewise, ecotourism is most likely to achieve its ecologically sustainable mandate when it is also financially sustainable. It is only a powerful force for the conservation of natural areas when it provides an attractive economic alternative to more resource-consumptive activities. Similarly, profitable businesses can afford the costs associated with being environmentally responsible, when non-profitable businesses will struggle, allocating their limited resources to core activities needed to keep the business afloat.

Indeed, as you read this Encyclopedia, consider that virtually everything discussed is predicated on the usually unstated assumption of a financially viable ecotourism sector. Yet, while being at the heart of ecotourism, business considerations have been the subject of relatively little discussion in either the academic or mainstream literature. These five chapters introduce the reader to some of the business issues affecting ecotourism. The chapters discuss generic business as well as marketing topics, as well as discussing issues relating to specific subsectors of this activity.

Three chapters (Chapters 33, 34 and 35) examine ecotourism operations on the ground. In Chapter 33, John Gardner offers a personal account of his experiences in developing an ecotourism accommodation facility in Bermuda. He argues effectively that 'accommodation and facilities set the tone for the guest's experience'. Gardner describes the range of accommodation types available, from the vernacular to converted historic buildings, contemporary styles, and portable accommodation. He also reviews a number of practical development issues including site use, facility location so as not to hinder the quality of the experience and the use of alternative power, water and waste management. This chapter concludes with a fascinating personal account of how he developed Daniel's Head in Bermuda. Gardner discusses how he had to design the facility to fit the ambience and character of the site, how he has laid out the units and also how he physically developed the superstructure.

In contrast, Bryan Higgins talks about non-facility dependent ecotourism in his chapter on ecotourism operators (Chapter 34). The ecotourism operator, the organization that takes people on tours of ecotourism sites, represents the mainstream activity of most ecotourism businesses. However, while there are many different styles and types of ecotourism operations available, Higgins points out that relatively little information is available on this sector. Apart from describing the diversity of products, he also focuses on two key areas of ecotourism: ecotourism networks and types of ecotour operator organization. Ecotourism involves introducing people to natural or near natural areas. To do so, though, a complex set of relationships must exist to link the prospective tourist with the operator. Inbound, outbound and local tour operators all have a role to play in successful ecotourism. Likewise, there are many different types of operators, ranging from conservation associations that run ecotourism as a fundraising activity, to community groups, universities and commercial tour operators. Each has a slightly different focus and slightly different operational ethos.

Betty Weiler and Sam Ham (Chapter 35) look at the operation of ecotourism on the ground in their chapter on tour guides and interpretation. This chapter examines the roles of tour guides, the principles of interpretation and then looks at the combined role of the two in the ecotourism experience. The tourist often regards the guide's main role as that of an interpreter who can make a site come alive. From an operational perspective, though, the guide's role is far more complex, also ensuring that any tour runs smoothly and considering the safety and comfort of clients. Interpretation is the focus of this chapter, however. The two authors provide a comprehensive introduction to interpretation and suggest that it is different from teaching. Good interpretation must be enjoyable, relevant, well organized, must provide an opportunity to learn and should also be themed, rather than topic based. The authors conclude their chapter by looking at the role of the guide and interpretation in providing a quality experience.

A number of generic business and marketing issues are discussed by McKercher in Chapter 36. He reports that many businesses are only marginally viable, which has led some people to question the true size of the commercial ecotourism market. The chapter illustrates that ecotourism operators face the same types of business challenges as all other small businesses: a lack of business planning, a lack of marketing expertise, a shortage of start up and working capital and unrealistic expectations on entry. He also shows that the ecotourism sector has been bedevilled by phantom demand, that is, participation rates reported by state and national tourism organizations do not translate into demand for commercial products. Indeed, one of the hard lessons learned by ecotourism operators is that visiting a national park does not automatically make someone an ecotourist. The chapter then proceeds to look at ecotourism from a marketing perspective and argues that operators must consider their product from the perspective of how it satisfies the needs of potential users and that to succeed, all those involved in ecotourism must adopt a strategic marketing focus.

Issaverdis (Chapter 37) pulls this entire section together with his chapter looking at the interrelated issues of benchmarking, accreditation, best practice and audits. As a sign of a maturing sector, many operators are now realizing that they must set standards that will enhance the performance of the industry as a whole. Any sector is only as strong as its weakest member. Creating target standards should benefit all operators. Issaverdis illustrates how each of the four elements identified above is critical to the upskilling of this sector. Benchmarks set performance targets; importantly, though, they are dynamic and provide a foundation for a continuous learning process. Accreditation recognizes those organizations that have achieved a certain standard of excellence. Further, the establishment of accreditation processes and the assumed marketing benefits that accrue from them will motivate the industry to improve its standards. Best practice, defined as the optimal approach to management, is useful in showing operators how to improve their performance. Auditing is a confirmation process that corroborates the above processes. Cumulatively, these five chapters provide the ecotourism stakeholder with a comprehensive and well-articulated exposure to the hard business aspects of the sector that are so vital to its success, yet often overlooked or deliberately ignored.

Chapter 33
Accommodations

J. Gardner
Cooper and Gardner Architects, Hamilton, Bermuda

Introduction

Imagine that a certain holiday was one of the top ten best experiences of your life. Many factors would have contributed to this: the location, season, weather, service, cost, value, accommodations, activities and your state of mind; even who you travelled with or met. Perhaps, also, it was your expectations and the extent to which they were met or even exceeded. If any one factor fell short, your experience may still be memorable but it would not be the best that it could have been.

The recent surge in popularity of ecotourism has much to do with the search for a richer holiday experience by the guest. It also has much to do with the efforts of hospitality business people to differentiate their product in a world market that is awash with choices. By definition, ecotourism is connected directly to the natural and cultural environment of the destination. It enables the traveller to better appreciate a new environment, to experience a sense of adventure, to relax and perhaps to learn. This is life enrichment and sometimes it is escapism. It is also good business when done well.

Every culture expresses itself and reveals its past through the creative actions of its people, which may be recreational as well as utilitarian. The activity of creating shelter is one of the most ancient of the human race's activities. It is therefore without any surprise that the built form of any culture reflects its deepest roots as well as its current direction. A tourism product that can sensitively recognize this, work within it or complement it, will guarantee a unique product that is attractive to the travelling customer. The resulting facilities will enhance the visitor experience and have the potential to define its success.

The accommodations and facilities of an ecotourist operation set the tone for the guest experience. They are the structure that enables the business to prosper and guests to enjoy themselves. They are an integral part of the backdrop and stage set for the total combined experience. It is essential that the facilities are well considered in all respects and that they inform, and are informed by, the operations of the property, and are harmonious with the culture and natural landscape of their environment. The purpose of this chapter is to outline general concepts in the area of ecotourism accommodations, such as the practical issues associated with their development, and to illustrate these concepts with a case study from Daniel's Head, in Bermuda.

General Concepts on Ecotourism Accommodations

Building type opportunities

Ecotourism facilities are in a uniquely advantageous position since they are inherently more diverse than traditional hospitality products. A wide range of options is available when folding cultural activities with contemporary, vernacular or historical building forms (or variations of them) and applying them to the activities that are promoted in each facility (see Table 33.1). The design and construction of buildings, however, is not and should not be a casual decision. The need to decide how a new structure may be built for practical purposes, or how an existing structure can be renovated, is also an opportunity to communicate a message. This message can then be determined through the design and finish of this structure to support the goals and objectives of the operations and financial plan.

A property that chooses to affix the label of being an ecotourism facility invariably conjures up the image of an alternative, rustic, low density or low impact facility. This is understandable given the beginnings of ecotourism operations as tents, platforms, vernacular huts and similar lower density, less visible and relatively isolated facilities. This is also exciting in that the guest is able to have access, in comfort, to a lifestyle that would normally be unattainable without discomfort, risk or a time commitment they cannot afford. However, the ecotourism property is not necessarily restricted to this type. The 'eco' in ecotourism can be the 'back to nature', 'get away from it all' or 'adventure' style of experience. It can also have a greater affiliation with the local culture and can be tied to the educational aspects of environmentally responsible living. There is no reason why an ecotourism property cannot exist just as easily as a contemporary structure in a suburban or urban environment provided it is carefully located in a site or building of special affiliation to that culture through its vernacular architecture or iconographic status. It would, in addition, need to be accompanied by an operations programme of appropriate educational, cultural or environmentally focused activities.

An exciting and memorable facility also has the option of being sited in renovated or newly built facilities. The former may be more challenging to adapt but could also be more economical. In any event, existing buildings come with a history and a narrative of their past. It enriches a visitor's experience to be aware of and appreciate

Table 33.1. Potential ecotourism building types by category.

Vernacular building types	Historical building types	Contemporary structures	Portable and low-impact structures
Indigenous structures Grass huts, mud structures, caves, elevated halls, house boats, reed platforms and buildings, yurts, tree platforms, ice houses, teepees, cliff dwellings, stick houses	*Developed vernacular* Colonial architecture, residentially derived styles, commercially developed styles Military architecture, colonial architecture, ecclesiastical architecture, monuments, industrial building, palaces and great homes	*Prefabricated structures* Masonry, glass fibre, reinforced concrete, rigid tents, inflatable structures *Traditional tourism* Cottage colonies, inns, guest houses, homes	Rigid tents, collapsible tents, elevated huts, inflatable structures, vehicles, jungle hammocks

the events, the people and the modifications that have engaged their destination. It can be argued that the ecotourist is invariably an interested tourist. The facilities have to be presented intelligently, and they need to be available to provide the layers of interest that can enhance the vacation experience. If these can be made available at the guests' choosing then so much the better, for they will be able to relax and be stimulated when they wish.

In all instances it is critical to avoid being casual in the process of choosing spatial allocations, materials and furnishings. These all have the ability to inform the guest on the culture and activities associated with the property. They can set a standard for comfort, or not, that can then be coordinated with the rate structures and expectations of the guest.

A traditional hotel invariably consists of one building with racked rooms above a ground floor assemblage of public spaces. A variation on this theme is a central facilities structure with individual accommodation units in the adjacent grounds. Motels and cottage colonies are polar opposites in the quality of this type of accommodation. The ecotourism product comes into its own most readily with this model of development. There is an inherent sense of an individual identity, and a small scale that appeals to the message of a quality vacation experience.

The field is, however, wide open in terms of considering the possible accommodation types. This is particularly applicable when the accommodation is the passport to access new experiences. Mountain huts for adventure tourists are without any central facilities and are waypoints in a journey. A vehicle, like a boat or bus, may become the base for high-end luxury or communal living. Each contributes in its own way to providing a completely unique experience. A low impact tent may permit building approvals where a permanent structure would not normally be allowed. A mobile tent may permit a facility to have limited life span and be viable for a limited time measured in weeks or just a few years. This could allow an operation to exist where land ownership or long-term commitments are not available.

A lightly furnished room can set the tone for a sustainable lifestyle. This may be in keeping with a message on the limited nature of resources. Furnishings do not need to be cheap because they can evoke the quality of an austere oriental, simple living or high modern aesthetic but cast in the culture of the place in which the property is sited. Colour, as opposed to clutter, can also enrich both the interior and exterior. Architectural and design solutions may also be eschewed in favour of landscape options in which the materials and horticulture can define the atmosphere. In the appropriate climate this has the added benefit of a constantly changing appearance. Small furnishings and fixtures, for example, can be something as simple as a mirror edged in sea glass. In this instance it can evoke childhood memories, can be made by local artisans, or on the property. It can demonstrate the recycling of materials (especially those that are discarded) and provides a unique accessory to a room's decor. It can also be made available for sale, thus providing additional revenue. The same could apply with wind chimes that have the added benefit of offering an additional (auditory) sensory experience.

Development practicalities

The ideal of an ecotourism development concept is to marry a vision of the experience with environmental sustainability and the constraints of a successful business venture, in such a way that the enterprise can function pragmatically. A traditional hotel must equate the return on investment of an initial development or renovation with the cost of that investment. Land or property acquisition and development costs become the greatest initial expense against which future revenue must be equated. Theory holds that the greater the investment then the better the product could be and the higher the fee that can be charged. This becomes a more marginal proposition in developed markets where competition is greater and costs are higher. The ecotourism product can circumvent this by offering other advantages to the developer and different standards to the guest. Box 33.1 lists various development strategies that need to be taken into account to maximize the probability of product success.

A useful question is 'How much does it really take for someone to be happy?' Less is sometimes more and a richness of experience can be offered in preference to opulence. Comfort cannot be compromised, but optimal comfort does not have to be bought by offering all modern conveniences. A delicate balance between rates and standards can be attained in which costs and, thus, rates can be lowered and the amenities reduced, while taking maximum advantage of the natural environment, providing a wide range of activities and taking the moral high ground of operating a responsible and environmentally sustainable facility.

There are also fundamental infrastructure requirements that face all guest facilities no matter what scale they happen to be. Guests need to eat and drink and they also require, almost always, the consumption of packaged products. Support services include cleaning and maintenance and the management of all human, commercial and, sometimes, livestock waste. Power and potable water demands are quite often difficult demands to meet. The former is often restricted by location, capacity and storage constraints. The latter is similarly affected by environmental collection, storage, purification and, sometimes, distribution constraints. Communications is, interestingly, most easily addressed given recent advances in technology. However, this is paralleled by increased expectations on the part of guests.

The greater the volume of people, the more sensitive and sophisticated the management of these issues needs to be. Certain thresholds in the number of guests can require (under certain building codes or even sensible adherence to moral standards) the importation of sophisticated and expensive systems. For example, sewage treatment plants may be required for facilities of over, say, 100 beds. The good news is that there is an array of exciting and alternative methods of addressing infrastructural issues. The bad news is that they are too often not as cost effective as traditional systems, especially at a large scale.

It can be presumed that the majority of guests are likely to come from Westernized cultures. The daily services to which these people are accustomed are currently managed within their own technologically sophisticated cultural/industrial economies. Interestingly, this level of consumption is not sustainable on a global basis if it were to be applied to the world population. It is equally likely that this level of consump-

Box 33.1. Development strategies for ecotourism accommodations.

- Siting to maximize natural amenities in lieu of newly built amenities
- Maximum interfacing with the natural environment
- Downsizing of central facilities offset by operational phasing and efficiencies
- Facilities and activities that minimize the emphasis on individual accommodations
- Communal living options
- Services and facilities provided by local development partners
- Self-sufficiency of the occupant to minimize operational overheads and central facilities
- Use of low impact and lower cost structures
- Use of indigenous building types that capture the local 'sense of place'
- Use of prefabricated building types
- Operations to de-emphasize site requirements
- An emphasis on landscaping

tion will not be sustainable in many areas that are candidates for ecotourism at least in the traditional, unspoiled, naturalistic and undeveloped environment. Properties in these locations therefore have the opportunity to convert this constraint into a genuine marketing tool by demonstrating to their visitors how comfortable they can be and, possibly, how much happier they can be with a well-managed and more sustainable, alternative lifestyle. In contrast to this, there is also no inherent reason why an ecotourism operation cannot operate in a highly developed culture where the use of alternative and more sustainable services and utilities is not required. In these instances the property can act as an educational tool for the visiting guest, and also as a prototype for adjacent communities.

Construction means and methods also provide opportunities for utilizing environmentally friendly systems, some of which are listed in Table 33.2. Quite often, the indigenous building systems have developed out of natural solutions using readily available materials. The benefits of considering building in the same manner as the existing culture are considerable: readily available first construction and maintenance skills, aesthetic consistency with the vernacular and culture, and the support and cooperation of the local community. It is also likely that in the very long term the materials used can be returned to the environment without generating pollution in the event that the facility is reinvented, relocated or abandoned.

If local materials are not used then other factors can come into play. Prefabricated components can offer educational opportunities and technological advances for the local population. If designed properly they can be installed so that they may be removed in the future without damaging the environment. It is also possible that the finishes and furnishings can be procured from locally sustainable sources or from recycled products. These provide very powerful marketing opportunities for the operator. They will be able to demonstrate to their clients, to local authorities and the hospitality industry that its product is an asset to its community in all respects and that it sets an example for its guests to demonstrate a better way of living.

A Case Study: Daniel's Head, Bermuda

Daniel's Head is a new environmentally focused tourism facility developed by Destination Villages with an opening date in the spring of 1999. Its conception originated in Stanley Selengut's pioneering facilities on St John in the Virgin Islands (see below). The development model is a village of individual tent cottages that are customized to suit the environment in which they are located. They are arranged on a 5 ha peninsula, an abandoned Canadian Military Base, and operate in consort with the remaining built structures. The project is a paradigm of the factors and pressures that influence the development of an environmental facility. The intriguing difference is that the site is in a sophisticated tourism destination on an affluent and densely populated suburban island. Bermuda is not a traditional ecotourism scenario.

Table 33.2. Alternative infrastructure options for ecotourism accommodations.

Power	Water	Waste management
Passive and active solar, wind, water, geothermal, tidal, building siting to take advantage of prevailing climate benefits	Non-potable flushing, grey water collection, reverse osmosis supply, roof collection, sewage treatment recycling and reduction in consumption	Composting toilets, biological treatment, greenhouse filtration, recycling, glass and aluminium crushing, commercial links to waste as raw material (glass, aluminium, paper and cardboard)

Inception

In 1995 the Government of Bermuda had four military bases returned to the island. These represented almost 10% of the total land area of the 54 km^2 island. The population in this limited area is approximately 70,000 permanent residents. Bermuda also hosts about 550,000 visitors a year and was voted the fourth most favourite tropical island destination in the world in a 1999 Conde Nast traveller poll. The then government of Bermuda applied a rigorous land use analysis and zoning exercise through a base utilization committee and assigned the development to a quasi-autonomous non-governmental organization (QUANGO), a purpose-formed company called The Bermuda Land Development Company (BLDC).

One of the four parcels of land is a peninsula of exceptional beauty known as 'Daniel's Head'. In the early 20th century this property had been part of a farm, as was much of Bermuda. The British Admiralty converted this into an electronic listening station during the Second World War and flattened half the site, a small hill, into a circular promontory at about 5 m above sea level. After the War and when the British Admiralty reduced its operations significantly, this base was operated by the Canadian military as a Cold War listening post. Additional buildings and barracks were constructed during this tenure.

The site was zoned for tourism use with a small beach segment held out as a national park to address local community concerns. The BLDC sought proposals from interested developers. Approximately six were received, all proposing housing, traditional cottage colonies and a proposal for a park. They were all found unsatisfactory, and an alternative tourism-related proposal was sought. After some research the BLDC decided to contact Stanley Selengut who had recently received a number of awards for his low impact and environmentally focused properties on the island of St John in the US Virgin Islands: 'Maho Bay', 'Harmony' and 'Concordia'. Selengut teamed up with a development partner, who eventually assumed control over the project and made a proposal. This proposal was accepted, and the project proceeded.

The St John, Virgin Islands properties as precedents

Destination Villages is a new company that is building four hospitality facilities worldwide in scenic areas. Each facility consists of central facilities and a 'Village' of private accommodation units. There is a minimum of 75 and a maximum of 150 of these units in each development. The Daniel's Head development utilizes the existing buildings for the central facilities and is constructing 120 accommodation units all as new structures. These are 5 m × 5 m tent cottages and they are an evolutionary design from the St John properties adapted to the Bermuda climate, building regulations and economic circumstances.

'Maho Bay' is Stanley Selengut's original development and is located (as are all the St John properties) on a steep hill. The initial development consisted of low impact wooden platforms on which tents could be pitched and connecting walkways inspired by the US National Park Service requirements in delicate natural areas. In time the tents became simple frame structures with suspended fabric for weather protection. They continued to develop until after 15 years, in the mid-1990s, they were fairly sophisticated rigid frame tents. These are characterized by having natural ventilation through window screens and perimeter shelves. The floor is constructed from wood planks or a recycled plastic and wood board. The plan is 5 m × 5 m square with one corner open as a terrace. They can accommodate two people comfortably and another two, less comfortably, on a futon-styled couch. The decor is spartan and reflects the wood structure and fabric finish. The advantage is the security of a fixed structure with the amenity of a tent's intimate relationship to the natural environment. An LPG camping stove and a portable ice cooler allow the guests to have a self-sufficient holiday if they wish. The central facilities do provide a self-serving open-air restaurant with stunning views to the west and the bay below. There is also a small convenience store, laundry, activities desk, beach concession, maintenance area and offices. The facility is powered by mains electricity and the guests use communal bathhouses which are connected to a central sewage treatment plant.

'Maho Bay' has been exceptionally well marketed as an environmental resort. Its primary strength is its low impact design and secure camping atmosphere. The quality of sunlight and the shadows of leaves on the walls and roofs of the tents are a welcome alternative to the traditional hotel room. The hillside location also offers the effect of staying in a tree house. A variety of environmental initiatives have been initiated and in the context of a traditional hotel these are fairly significant: water savers are installed in the bathhouses, low flush urinals, composting and a limited organic garden. Other initiatives have not yet taken hold in a truly significant way. Some of the glass is recycled and melted into souvenirs but this is a small operation and some of the aluminum cans are similarly melted into accessories. A solar oven and a solar icemaker were tried, but the large scale of the property and the damage by hurricanes prevented these from becoming operational staples.

A second development, 'Harmony', has been built on the hill above Maho Bay. A different approach has been taken since these are traditional solid frame buildings. A variety of recycled building products and alternative energy systems are used. 'Concordia', on the other side of the island, started as a second condominium style 'Harmony', but has evolved into a different facility with ten independent 'Ecotents' beside this structure. These have been built in two phases. The first phase is a series of six 5 m × 5 m tents with a variety of split-levels and loft bunks. The individual tent facilities are improved over 'Maho' in that they have expanded kitchen facilities and private toilets and showers in different configurations. This evolution is not arbitrary. The 'Concordia' site does not have the benefit of a central dining infrastructure and the nearest village is not only too far to walk but a long enough drive in a rented car to be inconvenient as the only place to eat.

It had also been found at 'Harmony' and at the original 'Concordia' that the guests preferred the experience of being in the fabric-clad structures connected by the raised wooden walkways. Generally the aesthetic is the same as the other facilities and the interior remains undecorated as a direct expression of its construction. The second phase of the tents has refined this by presenting the wood structure as a more rationalized system and exploring a generally tidier level of detailing. There remains a sense of adventure and tranquillity which is directly associated with the guests' experience of occupying a different, but comfortable, building type. This is enhanced by the requirements for a form of environmental self-sufficiency.

Daniel's Head

Daniel's Head develops the approach taken in St John by providing a facility that combines the Central Facilities and scale of 'Maho' with the increased sophistication of the 'Concordia' Eco tents. Importantly, Daniel's Head provides the entire facility as a single development in time. The St John facilities were built incrementally on a virtual cash flow basis. Daniel's Head has required a different level of planning particularly since it was placed under the constraints of a firm budget which amounted to approximately 50% of the cost of a traditional hotel development.

Site design

The existing buildings are sited on a sloped site on the south of the property. The tents are generally arrayed throughout the site. At first glance the density may be felt to be rather too even and saturated throughout the property. Given more land, it is generally preferable to concentrate development in a focused area. The nature of this site as a part of a suburban island meant that the site itself and the adjacent land were insufficient to do this. The siting strategy was therefore intended to place a premium on site landscaping and reforestation to the maximum that the budget would allow. The interstitial space between the tents would become a dense green zone. This would take time to grow so the tents are sited to the perimeter of the property wherever possible. This reduces density generally by effectively expanding the usable site area. It also, and importantly, permits the amenity of the tents to be derived from the best attributes of the site: the ocean and shoreline. Some of the tents are directly over the water. This is an environmentally risky move in that it appears to be at variance with normal requirements not to intrude on natural areas. In this instance this is mitigated by the overall plan to reforest the site, and by the ability to remove the tents with minimal trace. Furthermore, the economic viability of the project requires an attractive product in the early years.

This is an emerging form of ecotourism in which the concept leverages its viability by gently intruding into, and perhaps even improving, a natural environment so that it may create an economic base for the local community. This in turn assists a desired industry to occur and grow, creates employment and permits the site to be restored over time without being an added burden on the tax base. It also permits the developer to have sufficient security to make further investments.

Tent design

The tents are 5 m × 5 m squares with internal bathrooms alternating on different sides. They have a verandah on one side and a covered entry opposite. They are designed with white roofs in keeping with the Bermuda architectural vernacular. The white roofs also keep the tents cool, and look attractive as a neutral palette against the vibrant colours of the landscape and the water. The walls are a neutral tone to reflect as much heat as possible without being white, which will show dirt easily. The tents over the water are a light blue colour. In the existing trees they are clad in green. The intent has been to gain a maximum cost benefit from a repetitive building form but to still provide variety of experience by varying the finish and the orientation, and through site placement.

The tents are placed on a platform of six or nine piles depending on their height and the ground condition. One important factor was the building code that required the tents' structure be able to withstand hurricane force winds at 110 mph while the fabric has to withstand an 80-mph wind loading. Important technical factors like this can have a significant impact on the cost of a project. It is therefore useful to resolve these as early as possible in the design/cost process.

The structure of the tents was originally conceived as a wood frame built from laminated 2 × 4s. This then developed to a 10 cm × 10 cm wood structure. Faced with budget issues, a prefabricated tubular pipe structure was considered and this ultimately developed to become a prefabricated aluminum frame. A result of the structure change was to add an interior liner of fabric attached with an industrial grade Velcro. This provides a superior finish, assists with cleaning, provides an insulating layer in the wall and conceals all the services. The ceiling is formed in an upside-down tray, which is a traditional Bermuda architecture configuration. This provides an extra sense of height. The flat of the tray is formed out of mesh (which is not a standard detail) and so the roof void is accessible to act as a cooling plenum in the summer via vents in the gables. These are closed in the winter, thereby enabling the roof to act as a passive heat source. In both instances a central ceiling fan can distribute the air in an upward or downward direction, as required.

The tents are connected to the mains electrical supply. It was originally hoped to have all the tents off the grid, but further research indicated that this would have required a major centralized solar plant. Since there was insufficient space or first cost budget for this configuration, it was decided to limit the number of tents to five for demonstration purposes as a practical compromise. Almost one-third have a remote integrated solar hot water heater to reduce electrical costs, which are exceptionally high in Bermuda. These are sited away from current shaded areas.

The interior of the tents is designed to be comfortable in a bright, casual, cottage holiday style. Furnishings are simple, with a queen-size bed as standard, a pull out or trundle sofa and various tables and chair options. A hammock is provided as a standard amenity and the verandah is structured so that the hammock can be hung from it. At Daniel's Head it is also intended that this interior acts as the conceptual starting point for the conversion of 16 rooms in an existing barracks. Fifteen of the tents over the water have a window in the floor to view the water and shoreline below. This has become an opportunity to coordinate with the furnishings and so these windows are aligned to match glass-topped coffee tables above. The result is a visible amenity in a location where it can be appreciated, where it will not be covered by a carpet and where it is not unsafe for guests to walk over if it becomes slippery. This level of integration, between environment, structure and interior design, results in a seamless product that contributes to the guests' satisfaction.

The Daniel's Head tent is the result of an extended development process that goes back to an evolving prototype which has been adapted to suit another climate, market, budget and construction method. The result is a tent that is appropriate to its new site yet still retains the original fundamentals of its environmental considerations, size and use. In a different climate and culture it would evolve as a different form. This is occurring in Destination Villages' Hawaii site where, for example, the climate permits the windows to be left as screens only.

Central facilities

The Central Facilities at Daniel's Head are of a different aesthetic in that they are in a conversion of a single existing structure. The final form and programme is the condensing of various elements that were originally planned to be distributed through four buildings. This was reduced by budget constraints. This demonstrates that an environmentally focused hospitality operation can successfully compromise usual standards in the interests of utilizing fewer resources and funds. In this instance the compromises would be in a blending of lounge and bar functions, smaller offices and a simplification of the food service amenities.

Environmental overview

At the initial planning stages the regulatory authorities were sceptical that Daniel's Head would be an ecotourism facility. It is not located in an isolated area of profound environmental value such as a rainforest. This scepticism was reasonable given the classic perceptions about the appropriate location of ecotourist properties. Similarly, the accommodations were different from the rustic structure, lodge or hut that might be envisaged. This scepticism extended to the activities which would interest the guests, since there were no wide-open tracts of land to appreciate. The ultimate realization of the government of Bermuda, and hence the approval, was that Daniel's Head had the potential to be a leadership project in environmental restoration. It could also act as an incubator for environmental education as well as adding another dimension to the traditional tourism product that was available.

The accommodations and facilities of Daniel's Head enable a damaged site to be restored in a business environment. Its open spaces are the incredible views over the water even though its land plan is quite dense. Its accommodations are comfortable enough to provide a reasonable alternative to the traditional products that are available, yet simple enough to appeal to the luxury camper. Its design adheres to the culture of its location and this can be enhanced through the direct integration of the local population in the facilities' activities. Finally, and perhaps most importantly, the alternative energy, composting, sustainable design and sustainable operations programmes can provide an example to the local community. Table 33.3 provides a synopsis of the environmentally appropriate practices that characterize this facility.

Summary

Ecotourism accommodations offer a new dimension and invigoration to a local tourism product. They have the ability to provide a different and wider range of accommodations and will challenge the traditional properties to revisit their own requirements and standards. They will also provide an educational resource for their community to improve not only their visitors' experience but also their own quality of life. There is tremendous latitude in the developer's ability to seek an appropriate idiom for each facility. A good grounding in the local culture and an emphasis on sustainable design, construction and operations will contribute immensely to the honesty and success of the enterprise.

Table 33.3. Daniel's Head: environmental attributes.

Land use	Eco-tents	Services/ infrastructure	Flora	Fauna
Rejuvenating a military base Reusing existing buildings Minimize site works Environmental construction plan	Efficient prefabricated structures Low impact design Removable new construction Windows on three sides Natural ventilation for cooling Adherence to vernacular architecture Maximize site environmental appreciation Recycled and local materials specification	Pedestrian site Raised walkways Mulch pathways No air-conditioning Solar photovoltaic system Solar hot-water heating Composting toilets Minimized electrical coverage Minimal site lighting Roof water collection and site cisterns	Site rejuvenation through landscaping Transplanted trees Endemic planting philosophy Composting	Bermuda tropic bird nesting programme Endemic lizard rejuvenation programme Marine life custodial programme

Chapter 34

Tour Operators

B.R. Higgins
Department of Geography and Planning, State University of New York at Plattsburgh, Plattsburgh, New York, USA

Tour operators occupy a critical role in the tourism industry, given their role as intermediaries that design, organize, package, market and operate vacation and other tours (Morrison, 1996). The purpose of this chapter is to discuss the role and status of tour operators within the ecotourism sector. A preliminary step for understanding ecotourism-related operators is to review research about tour operators in general within the tourism literature. Remarkably, even though tour operators have become major actors within the global travel industry, only a few studies have investigated their character and most of these are somewhat dated (Touche Ross, 1975; Britton, 1978; Ascher, 1985; Sheldon, 1986; Delaney-Smith, 1987; Urry, 1990; Vellas and Becherel, 1995; Ioannides, 1998). Focusing on three of the more recent studies, Urry (1990) indirectly analysed tour operators and the expansion of the package tour industry, with special attention to Europe and the UK. He argued that specialization in the tourism industry, which would include the establishment of ecotour operators, had changed tour production from a concentrated, mass-market complex into a more segmented, post-Fordist economic system.

With a similar European focus, Vellas and Becherel (1995) include a section on tour operators in their review of international tourism from an economic perspective. Their coverage of tour operators provides detailed information and an excellent analysis of the rise and fall of the major operators in selected European countries, but almost nothing about tour operators in other world regions. Ioannides' (1998) important study takes a critical look at tour operators as it sketches the evolution of the package tour industry, identifies the contemporary activities of tour operators and reports on a pilot survey of US specialist tour operators that includes ecotours. Ioannides observes that tour operators are key manipulators of tourist origin–destination flows who have displayed little loyalty to specific destinations. Overall, given the limited consideration of tour operators within the tourism literature, comparisons of ecotourism-related operators with the tourism industry as a whole will be limited.

Ecotour Operators in the Literature

Surprisingly, a literature review on the more specialized topic of ecotourism-related operators reveals more publications than on tour operators in general as well as more attention to recent developments in this field. Currently, two rather distinct types of publications are available in regard to ecotourism-related operators. One kind of publication offers a guide on how to start and manage an ecotourism business. This practical genre addresses a variety of 'nuts and bolts' business issues including organization of the company, product development, finance, marketing and other essential business development topics. Currently, four ecotourism-related business guides of high quality are available: a guide for developing a business in adventure tourism with a Canadian focus (Cloutier, 1998); two guides for establishing nature-based tourism businesses from an Australian perspective (Beeton, 1998; McKercher, 1998); and a step-by-step manual for developing nature and culture-based tourism operations from a US publisher (Patterson, 1997). Together these books identify and clarify the key business components that are essential for an ecotourism-related business.

Another set of publications has taken a more detached approach to investigating the character of ecotour operations. Since relatively little is known about ecotour operators, the research design in these publications usually includes a survey of operators. They will now each be briefly sketched according to the world region they investigate. The earliest of these operator surveys, and the only one focused on a destination country, examined five businesses that specialized in nature travel and their related development needs in Ecuador (Wilson, 1987). Next, five separate studies have explored the character of North American-based ecotour operators. First, an early survey of 32 operators who specialized in nature tourism by Ingram and Durst (1989) noted that just three firms served over 1000 clients each during 1986. In the 1990s, independent surveys of North American-based ecotour operators by Rymer (1992), Yee (1992), Higgins (1996) and Lew (1998) expanded this knowledge base and demonstrated the dynamic character of ecotour operators in this world region. Together they indicated that the number and size of ecotour operations had been growing significantly during the 1990s and that market share within the USA was concentrated. In fact, the five largest operators served a total of 50,000 travellers in 1994 and 35 of the 82 firms responding served more than 1000 clients, thus capturing 90% of the market (Higgins, 1996). In addition, almost all of the US operators conducted their ecotours outside the USA and, even though most had trips to more than one country or world region, very few had itineraries in all of the major world regions (Higgins, 1996; Lew, 1998).

In the southern hemisphere, Weiler (1993) critically examined the environmental commitment of 27 nature-based tour operators in Australia, while McKercher and Robbins (1998) and McKercher (1998) profiled and analysed Australian nature-based business. McKercher (1998), citing a study by Cotteril, estimated that Australia had 600 ecotourism operators, without mentioning their inbound or outbound orientation, which are typically small businesses of four or fewer staff. He also noted that national surveys estimate that Australian inbound operators serve about 70% of all inbound tourists and 90% of the visitors from Asia.

Finally, Holden (1996) profiled UK outbound tour operators who were 'environmentally friendly'. He reported on the limits of their environmental commitment and noted that the orientation of their destinations was Europe with 65%, Southeast Asia 51%, South America, Central America and Africa at 46% and Australasia 31%. Together, these studies provide an important baseline profile of ecotour operators in Australia, Ecuador, the UK and North America. At the same time, they implicitly demonstrate the strong national segmentation among ecotour operators and researchers. However, since each of the studies defined its tour operator population in a different manner, utilized a unique set of survey questions and was primarily exploratory in design, it is not a simple matter to summarize their results. It is also noteworthy that, for the most part, these studies gave limited attention to each other's surveys or the patterns of international organization among ecotour operators.

Ecotour Product Development and Evaluation

During the past two decades a diversity of new tourism products has appeared in the global marketplace. In addition, specific products have become associated with new tourism themes, some of which are shown in Box 34.1 (Zurick, 1995; Mowforth and Munt, 1998). It should be noted that this list of new tourism themes is meant to illustrate the trend in specialization and not to serve as a definitive inventory. Ecotourism has become one of these new tourism themes, even though its definition and character continue to generate substantial debate (Blamey, 1995; Finucane and Dowling, 1995; Weaver, 1998b) (see Chapters 1 and 2). Given the wide scope of these new tourism themes as well as the variable criteria used to identify ecotourism, the relationship between ecotour operators and new tourism products will be examined in more detail.

First, it is important to note the diversity, potential overlap and tension that may be found within these new tourism themes. Since each of these tourism segments has its own set of stakeholders, interest groups and marketing connections, operators usually consider the relation of their products to this broader marketing context. (See Middleton and Hawkins (1998) for a detailed examination of the importance of market segmentation for sustainable tourism.) Thus, operators consider not only specific activities, such as kayaking, whale-watching or wine tasting, but also integrate their product with broader themes. For ecotour operators this growth of new tourism products and themes has consequently provided both opportunities and challenges. The increasing diversity of specialized interests, growing number of clients and segmentation of the market offer substantial opportunities for savvy businesses. In fact, most ecotours include a variety of activities and integrate more than one of these tourism themes, thereby contributing to the sorts of tourism hybrids cited by Weaver in Chapter 5.

At the same time, it has become a challenge to assure that the integration of diverse activities does not alienate a particular market segment or muddle the overall character of a trip. Unfortunately, with small ecotour operators little time or money is available to consider such issues in preparing a trip (McKercher, 1998). Alternatively, large ecotour operators may expend considerable resources over the course of a year to assess and plan a new trip. In such cases their planning may involve a preliminary assessment of market demand, visiting potential venues, determining costs, and negotiating contracts for new ecotour experiences. However, systematic research has not yet identified how operators conduct this process.

Despite the lack of empirical research, all the ecotour business publications stress the importance of ecotour product development and offer suggestions for itinerary planning, themes and pricing (Patterson, 1997; Beeton, 1998; Cloutier, 1998; McKercher, 1998). According to Patterson (1997), pricing analysis should include calculating variable costs per person, fixed costs per person, overhead and marketing costs per person, commissions, profit, discounts and refunds and credits. Cloutier (1998) goes further to recommend plotting

Box 34.1. New tourism themes.

- Adventure tourism
- Cultural tourism
- Ecotourism
- Farm and rural tourism
- Festival tourism
- Food tourism
- Environmental tourism
- Heritage tourism
- Nature tourism
- Scientific tourism
- Soft tourism
- Sustainable tourism
- Wine tourism

a product-positioning map that compares the price and quality of a new product with the position of competitor's products in the area. He also discusses how to calculate the break-even point for a tour as well as establish a tour's price using a bottom-up approach.

As discussed by Weiler and Ham in Chapter 35 of this Encyclopedia, ecotourism research has only recently begun to appreciate and evaluate interpretation. Clearly this is likewise true about the assessment of ecotours in general. Little is known about how ecotour operators conceptualize their tours or how this compares with the assessments of ecotourists. One innovative example of ecotour assessment was the Green Evaluation project that was conducted by The Ecotourism Society and the Ecuadorian Ecotourism Association during 1996 and 1997 (Wood, 1998). In this very original programme, ecotourists were asked to critically evaluate their ecotour experience based upon key criteria of ecotourism (Sirakaya, 1997). While reputable consumer evaluation organizations have been established in numerous countries (e.g. Consumer Reports in the USA) and engineering criteria are available for quality control with industrial production (e.g. ISO guidelines), the evaluation of ecotours has not yet been developed in such a systematic manner. Two important exceptions are the excellent work of Blake and Becher (1997) in designing and compiling sustainable tourism ratings for selected ecotourism operations in Costa Rica, and the Ecotourism Association of Australia product accreditation system (see Chapter 37).

The development of routes, itineraries and products for independent ecotourists is also an important issue. Since independent travellers comprise the largest segment of the ecotourism market, it is important to consider how they select operators and products. One of the few studies to critically explore how independent tour routes, products and operators develop, and what impacts they have on local communities, is Zurick's (1995) analysis of independent tourists and adventure travel. He observed that the profusion of travel books for independent travellers create distinct but parallel routes to see non-Western countries. Thus, while independent travellers may scorn fully packaged tours offered by outbound operators, the routes, itineraries, and local tour operators they select from alternative travel sources similarly embody a prearranged and commercialized experience. Since this selection of destinations, operators and accommodations has a profound impact on operators in destination countries, more research should examine this distinct process for independent ecotourists.

Even though the experience offered by ecotour operators is clearly a crucial aspect of ecotourism, the production of ecotours has received much less attention within the literature than other topics such as protected areas, environmental impacts, biodiversity or economic valuation (Higgins, 1996; Eagles and Higgins, 1998). An examination of ecotourism-related economic studies indicates the development of a substantial literature, as demonstrated in the excellent review by Lindberg (1998). The scope of economic studies has included a number of public sector issues such as setting fees for protected areas (Laarman and Gregersen, 1996) and estimating economic impacts (Dixon and Sherman, 1990). It has also included more theoretical subjects, such as assessing demand curves (Crouch, 1995) and the contingent evaluation of natural resources (Navrud and Mungatana, 1994). Overall though, the economic studies have had little to say about the supply side of the ecotourism industry or the character, organization, strategies, or geographical distribution of ecotour operators. In fact, this blind spot within ecotourism is part of a broader gap in our knowledge about the supply side of tourism in general. As Ioannides and Debbage (1998) have noted, the tourist industry has not received the same attention that manufacturing and other economic sectors have been given in the economics, tourism or geography literature. Consequently, our knowledge of ecotour production as well as mainstream tour production is still limited.

Several features make it a challenge to

conduct research about the production of ecotours. First, governments have published little if any secondary information regarding ecotour operators, ecotourists, or ecotour products. An important exception is the Bureau of Tourism Research in Australia where considerable high quality research about nature and backpacker tourists was conducted during the 1990s (Blamey, 1995; Haigh, 1995; Buchanan and Rossetto, 1997; Blamey and Hatch, 1998). Second, much of the relevant information concerning ecotour operators and products is either proprietary or confidential in nature. Since it is unusual for tour operators to disclose sales figures, costs, financing, profit margins, characteristics of clients, marketing strategies, business networks or other crucial business information, it has been difficult for researchers to analyse patterns within this sector or among destination communities. In addition, the operation of tours is a largely unregulated activity that involves significant risks and has recently resulted in numerous bankruptcies (Waters, 1998, p. 149). Furthermore, these operations are typically small businesses with very limited time and energy. Thus, ecotour operators simply may not have time for an interview or choose to be deliberately vague when discussing their operations.

Finally, several definitions of ecotourism have been used in different contexts, and a variety of related tourism themes, as shown in Box 34.1, may also be used by ecotour operators (Blamey, 1995; Zurick, 1995; Weaver, 1998a, b). With the increasing specialization and importance of market segmentation, tour operators have frequently aligned themselves within one or more of these broader notions and related market segments as they identify an aesthetic for their tours (Mowforth and Munt, 1998). Trekking and adventure tours, such as the type that are offered in the hill country of northern Thailand (Cohen, 1989; Dearden, 1989), are examples of this alignment. In this multi-layered context, the use and meaning of 'ecotourism' by either an operator or researcher becomes an issue that requires critical analysis. Since critical assessments may alienate operators who are aligned with other speciality notions or embarrass operators who have used the term ecotour primarily as a marketing ploy (Wight, 1993; Campbell, 1999), developing a thoughtful research strategy has become increasingly important. For a variety of reasons then, researchers have encountered new challenges when collecting primary data or requesting unpublished documents through personal contacts. Given the special challenges that may be encountered while attempting to collect evidence from tour operators, such as refusals, non-response and evasiveness, researchers should carefully consider the methodology of studies to assure that the observations and conclusions do not exceed the quality of the evidence that has been collected.

Ecotour Operator Networks and Organization

International market and product distribution networks

The most common framework to conceptualize ecotour operators and production is to distinguish between outbound, inbound and local ecotourism operators. In this rudimentary framework, outbound ecotour operators are based primarily in the industrialized countries, inbound operators are found in international gateway cities of destination countries and local operators are located in the rural service-centres of destination countries (Ashton and Ashton, 1993; Higgins, 1996; Wood, 1998; Honey, 1999). In select cases, the supply of ecotours has been conceptualized to also include travel agents/retailers, who play a significant role in select world regions such as Australia (Ziffer, 1989; McKercher, 1998). Even though this geographic scheme has frequently been used to distinguish types of ecotour operators, the networks of ecotour operators and organization of ecotour production have seldom been explicit topics of discussion in the literature. To address this oversight the chapter will map some of the key networks among ecotour operators and consider the significance of these organizational niches for ecotourism as a whole.

A number of issues make it important to go beyond a focus on individual ecotourism operators and their internal organization to also examine relations between operators and related support organizations. First, as Tremblay (1998) has argued in detail, the transaction costs and organizational diversity of tourism have limited the ability of research that focuses primarily on firms and ownership to understand the tourism industry. As an alternative he presented a network approach to the economic organization of tourism that he believes offers more robust potential to address the importance of alliances, cooperative ventures, and partnerships in tourism. Since the relations among distinct tour operator businesses and within distinct market niches are clearly crucial for ecotourism, this chapter will sketch some of the distinct networks in this economic environment. (For a detailed study of the geography of tourism businesses and the natural environment in a particular region see Higgins and Holmes (1999).)

Vertical operator networks

To appreciate the role of connections and networks for ecotourism, consider a typical outbound ecotour operation. Such businesses usually have purchasing agreements with international airlines as well as one or more inbound tour operators which are usually responsible for the ground transportation, food and accommodations in the destination country. The inbound operator may be an independent business, a subsidiary company that is wholly owned by the outbound ecotour operator or a joint venture involving a local business and the outbound operator. Its role is to provide a package of goods and services, usually offered by smaller, independent operators who lack the necessary distribution resources to enter the international marketplace, which the outbound operator can link with airfare to provide an easy to purchase product for the tourist. Although empirical studies have seldom systematically mapped these production alliances, such ecotourism business connections are clearly pivotal for ecotour operators of all kinds and sizes.

Under this schema, a series of complementary yet separate tourism businesses organize themselves to create an integrated package. Such integrated networks usually include air travel, ground transportation, accommodations, food and guiding, as depicted in Fig. 34.1. They frequently involve ecotour operators from industrialized countries making connections with national operators located in the international gateway cities of destination countries. The national inbound operator then networks with operators in different areas to provide a diversity of itineraries and products. These networks are the basis of most outbound tourism (Ioannides, 1998; Tremblay, 1998). Establishing such a network allows operators to negotiate price, availability, quality and other features that are crucial to their ability to market and manage their operations. Such networks may be implemented in a variety of ways, including annual contracts, exclusive offering agreements, block purchase or reservation of services and minority investment schemes.

However, the growing accessibility of fax, Internet access, and computer reservation systems has had a significant impact on tourism product distribution (Mason and Milne, 1998; Milne, 1998; Milne and Gill, 1998). Changing technology has modified the character of ecotourist market demand, since it has become easier for ecotourists from industrialized countries to independently arrange their own ecotours by contacting inbound and local ecotour operators in destination countries directly. As a result, a variety of new operator networks are emerging, enabling the local, destination operator to sell his or her product directly to the international consumer.

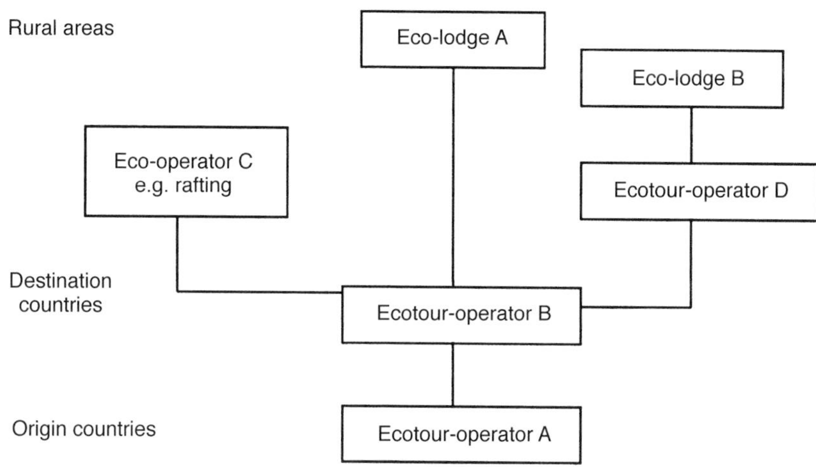

Fig. 34.1. Vertical operator network.

Ecotourism associations

Another popular form of operator collaboration is the formation of a trade or industry association (discussed in more detail in Chapter 30). Forming a speciality tourism organization usually involves creating a hybrid organization that has a strong focus on the needs and perspectives of tour operators, but also includes other tourism stakeholders and interests. These groups are frequently formed at the national level to organize and highlight the needs of this special tourism segment. Examples include the Australian Ecotourism Association, Tourism Watch based in Germany, Indonesian Ecotourism Network, the Brazilian Ecotourism Society, E-Travel Canada, the International Ecotourism Society based in the USA and the Ecuadorian Ecotourism Association. These groups typically include not only ecotour operators but environmental non-government organizations (NGOs) as well as professionals and academic researchers (see Fig. 34.2). A key component of these networks is to build a broad-based organization that may share business services, provide branding, improve marketing and advocate policy alternatives for sustainable tourism.

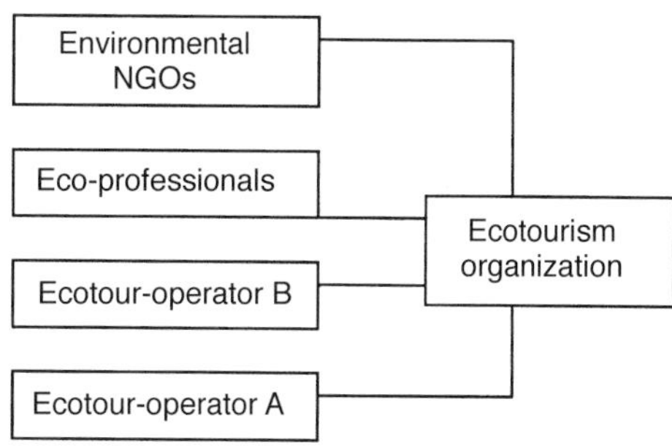

Fig. 34.2. Speciality tourism organization.

In select cases these speciality groups have also formed at a sub-national level as with the various state ecotourism groups in the USA (e.g. Alaska, Florida and Hawaii) or at the world-regional level as with the Pacific Asia Travel Association. These organizations usually provide client referrals and marketing for members as well as developing plans and projects to serve their broader interests. Also, since governments have frequently been reticent to evaluate or regulate tourism, these organizations have sometimes provided an alternative method to deal with the negative impacts of tourism as well as advocate sustainable tourism guidelines.

Community-based operator

Another type of organization that has been growing in both popularity and significance is the community-based ecotour operator, as illustrated in Fig. 34.3. A number of features are special about this type of network. First, they are usually based within indigenous communities. This introduces important cultural dimensions, both in terms of ecotourist appreciation and impact as well as business management and community development. In addition, they are typically established within and designed to serve a particular village or group of communities. This distinguishes them from the majority of ecotour operations that are usually established for control by an individual or family and with less explicit connections to a particular place or community. Of course, the systematic issues of indigenous development and community-based enterprises are not unique to tourism and have been discussed in many other contexts.

A wide variety of studies has addressed community-based ecotourism operators. In fact, research in this sub-field has been growing so rapidly that this section should only be considered prefatory in coverage. First, Zeppel (1998) has written an excellent overview of sustainable tourism and indigenous peoples with examples from many world regions. Another insightful overview, with examples from around the world, is the special issue of *Cultural Survival* (Wood, 1999). This impressive edition included 14 separate pieces with a special focus on protecting indigenous culture and land through ecotourism. In addition, Honey's (1999) critical review of ecotourism directs substantial attention to locally controlled development and community-based operators both in her systematic analysis and in the seven nations she discusses in detail. Besides these internationally focused works, many studies of particular community-based tourism operations are also available. During the 1990s one of the hot spots for indigenous and community-based tourism was clearly Ecuador. Insightful work on different aspects of community tourism initiatives in Ecuador include work by Rogers (1996), Schaller (1996), Smith (1996a, b), Wesche (1996), Drumm (1998) and Wood (1998). In fact, Schaller's work even included an interactive web site where visitors can plan and manage a locally controlled ecotourism project in the Ecuadorian Amazon (http://www.eduweb.com/amazon.html). Chapter 22 of this volume, by Hinch, also refers to community control issues in the context of indigenous communities.

Other studies of community-based operators in South America include an analysis of Lake Titicaca's campesino-controlled tourism in Bolivia (Healy and Zorn, 1988, 1994). In Central America a guide to community-based ecotourism in Belize (Belize Ministry of Tourism and Environment, 1994) and an analysis of select community sites in Belize by Horwich *et al.* (1993) are good examples from this region. In North America, tourism operations at Acoma have been very effectively developed by Native Americans and serve as a model for tribal tourism in this region (Smith, 1996). In Africa, a detailed description and analysis of four hypothetical approaches to community involvement in Namibia was developed by Ashley and Garland (1994) and the problems of integrating Maasai tradition with tourism were analysed by Berger (1996). Another region that has become a hot spot of community-based tourism alternatives is Australia–New Zealand. The Northern Territory of Australia alone had 52 Aboriginal tours listed with its tourism office in 1996 (Zeppel, 1998). For extensive references and more analysis

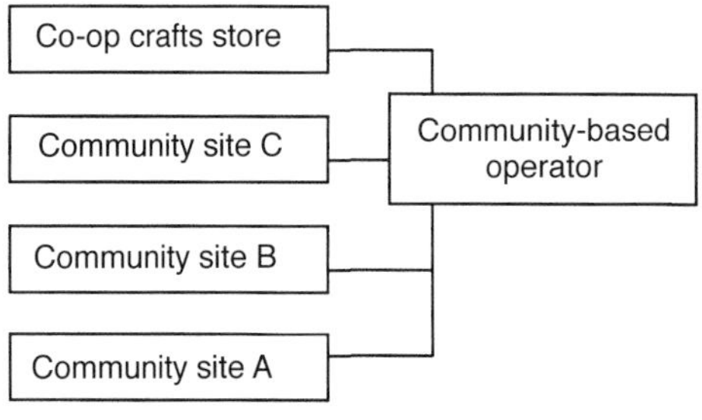

Fig. 34.3. Community-based operator.

of these initiatives see Zeppel's (1998) detailed consideration of both Australia and New Zealand. Overall then, the number of community-based ecotour operators is small. However, in comparison to the total number of operators or their gross revenue, the high growth rate, wide international distribution, and the abundance of research on community-based operators suggests that these organizations will be increasingly important in the future.

Non-profit environmental organizations offering travel

As the ecotourism industry continued to grow during the 1990s one of its major sources of support was the growth of non-profit environmental organizations (see Chapter 30). The tremendous growth in the number of organizations, their membership and the amount of global funding for environmental NGOs has influenced ecotour operators in numerous ways. First, the significant funds that these organizations receive and their collaboration with bilateral and multilateral development agencies has had a significant impact on the scope of international conservation initiatives (Princen and Finger, 1994; Honey, 1999). In addition, many of these organizations have developed travel programmes for their members. In some cases the ecotours in the travel programme are run exclusively by the NGO but in most cases the travel programmes sign contracts for nature trips with for-profit operators. Even though it is difficult to estimate the size of this market, research in 1994 indicated that 11 of 83 or 13% of the US-based ecotour operators studied were non-profit organizations who cumulatively served over 20,000 clients or 17% of the total market (Higgins, 1996). This sector is the focus of an increasingly large body of literature (Weiler and Richins, 1995; Higgins, 1996; McLaren, 1998; Mowforth and Munt, 1998; Wood, 1998; Honey, 1999).

University travel groups

Another recent development in non-profit travel organizations has been the growth of university travel programmes. Alumni and educational travel programmes offer travel alternatives that usually include trips highlighting nature, conservation and local communities. It should be noted, though, that while university travel programmes are extensive within the USA, at the present time they are much less common in other parts of the world. Within the USA this subsector has been developing for some time. In 1985, for example, a for-profit business named Travel Learning Conferences was incorporated that organizes annual conferences to provide connections between non-profit institutions and for-profit tour operators. The institutional survey for their 1997 meeting indicated that the 200 institutions participating at this meeting sold trips to more than 37,000 passengers in the previous year, primarily to foreign destinations (TLC, 1998). It is also important that for-profit tour operators that are contracted by these non-profit institutions conduct almost all of this foreign travel. Together, the growing number of ecotourists from environmental NGOs and university travel programmes has created a distinctive and substantial ecotour operator niche as shown in Fig. 34.4.

Ecoresort complex

The recent construction of inclusive ecolodges on private nature reserves with quality guides and eco-packaging has created alternative ecotourism products. A number of forces have been fuelling this changing environment, where upscale lodges become more environmentally sensitive, at the same time as the appreciation of indigenous design and heritage are incorporated more extensively. First, progressive architects, landscape architects, land planners and developers have been working to improve the sustainability of lodges and parks for some time. In the

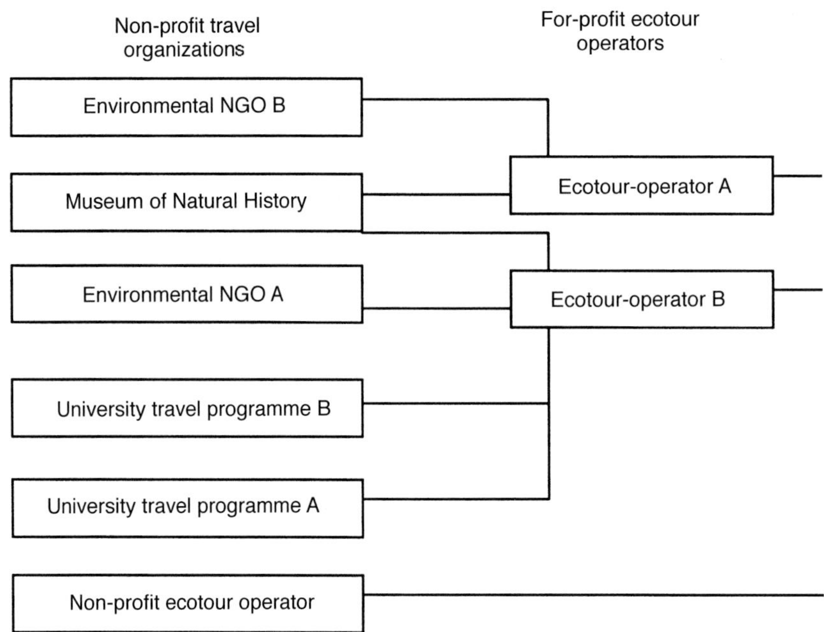

Fig. 34.4. Non-profit travel organizations.

1990s this was highlighted for ecotourism with the publication of works on the topic of ecolodge design, environmental landscape layout and park planning (NPS, 1992; Andersen, 1993; Hawkins *et al.*, 1995; Ceballos-Lascuráin, 1996).

Second, resort developers have been struggling to address the mounting environmental criticism of large-scale projects and establish positive alternatives (Ayala, 1995, 1996). The hybrid alternative that has appeared is termed an ecoresort complex as shown in Box 34.2. This is fundamentally different from the other networks since it seeks to integrate a wide variety of tour components with corporate management into one site or complex. While this has been a well-known strategy of international resorts for many years, only recently have developers attempted to integrate mega-project elements with increased sensitivity to the environment and the experience of place. Of course, designing and building such elaborate systems requires significant planning, development and management experience as well as large amounts of capital. As a result, bilateral and multilateral development agencies have recently funded a variety of such alternatives and large transnational corporations have initiated such products as a multi-layered approach to attract an upscale clientele (Mowforth and Munt, 1998; Honey, 1999). To understand these initiatives by tour operators and accommodation enterprises see the examples and analysis of Middleton and Hawkins (1998) as well as Hawkins *et al.*, (1995).

Box 34.2. Ecoresort complex.

Recreation activities
Nature reserve
Accommodation
Food
Entertainment
Interpretation

Future Research, Planning and Policy

Changing patterns in the global supply and demand of tourism have produced many impacts during the past two decades. One of the more visible trends has been the growth of ecotourism as a specialized form of global tourism. This development has many dimensions, as discussed in the various chapters of this Encyclopedia. One of the key components in this transformation has been the formation and growth of thousands of ecotourism-related tour operators functioning within every world region. Even though our current knowledge of these actors is often limited, a diverse and vibrant group of researchers, consultants and operators has made important contributions to the growing literature on this topic. Yet, significant topics have yet to be systematically addressed. For example, what contributions have ecotourism operators made to environmental conservation and biodiversity preservation? How do the community impacts of ecotourism-related operators compare to mainstream tour operations? What strategic interventions would leverage the most benefit for a large number of small ecotour operators? What is the best method to evaluate and maintain the quality of ecotours? What are the key gender and labour issues for ecotour operators? As ecotourism research, planning and policy studies address these and other questions, the economic geography of ecotourism-related operators will hopefully become clear and relations between the human and natural environment improved.

References

Andersen, D. (1993) A window to the natural world: the design of ecotourism facilities. In: Lindberg, K., Wood, M. and Engeldrum, D. (eds) *Ecotourism: a Guide for Planners and Managers*, Vol. 2, The Ecotourism Society, North Bennington, Vermont, pp. 116–133.

Ascher, F. (1985) Transnational tour operator corporations. In: *Tourism: Transnational Corporations and Cultural Identities*. UNESCO, Paris, 57–69.

Ashley, C. and Garland, E. (1994) *Promoting Community-Based Tourism Development: Why, What, and How?* Research Discussion Paper Number 4, Directorate of Environmental Affairs, Windhoek, Namibia.

Ashton, R. and Ashton, P. (1993) *An Introduction to Sustainable Tourism (Ecotourism) in Central America: Paseo Pantera Ecotourism Program*. Wildlife Conservation International, Gainesville, Florida.

Ayala, H. (1995) Ecoresort: a 'green' masterplan for the international resort industry. *International Journal of Hospitality Management* 3–4, 351–374.

Ayala, H. (1996) Resort ecotourism: a master plan for experience management. *Cornell Hotel and Restaurant Administration Quarterly* 37(5), 54–61.

Beeton, S. (1998) *Ecotourism: a Practical Guide for Rural Communities*. Landlinks Press, Collingwood, Australia.

Belize Ministry of Tourism and Environment and Belize Enterprise for Sustained Technology (1994) *Guide to Community-Based Ecotourism in Belize*. Belize.

Berger, D. (1996) The challenge of integrating Maasai tradition with tourism. In: Price, M. (ed.) *People and Tourism in Fragile Environments*. John Wiley & Sons, Chichester, UK, pp. 175–198.

Blake, B. and Becher, A. (1997) *The New Key to Costa Rica*. Ulysses Press, Berkeley, California.

Blamey, R. (1995) *The Nature of Ecotourism*, Occasional Paper No. 21. Bureau of Tourism Research, Canberra.

Blamey, R. and Hatch, D. (1998) *Profiles and Motivations of Nature-Based Tourists Visiting Australia*, Occasional Paper No. 25. Bureau of Tourism Research, Canberra.

Britton, R. (1978) International tourism and indigenous development objectives: a study with special reference to the West Indies. PhD thesis, University of Minnesota, Minneapolis, Minnesota, USA.

Buchanan, I. and Rossetto, A. (1997) *With my Swag Upon my Shoulder: a Comprehensive Study of International Backpackers to Australia*, Occasional Paper No. 24. Bureau of Tourism Research, Canberra.

Campbell, L. (1999) Making the best of a bad situation: promoting ecotourism in rural Costa Rica, presentation at The Society for Human Ecology Conference, Montreal, May 1999.

Ceballos-Lascuráin, H. (1996) *Tourism, Ecotourism and Protected Areas.* World Conservation Union, Gland, Switzerland.

Cloutier, R. (1998) *The Business of Adventure.* Bhudak Consultants, Kamloops, British Columbia.

Cohen, E. (1989) 'Primitive and remote': hill tribe trekking in Thailand. *Annals of Tourism Research* 16, 30–61.

Crouch, G.I. (1995) A meta-analysis of tourism demand. *Annals of Tourism Research* 22, 103–118.

Dearden, P. (1989) Tourism in developing societies: some observations on trekking in the highlands of north Thailand. *World Leisure and Recreation* 31(4), 40–47.

Delaney-Smith, P. (1987) The tour operator: new and maturing business. In: *The Travel and Tourism Industry: Strategies for the Future.* Pergamon Press, New York, pp. 94–106.

Dixon, J. and Sherman, P. (1990) *Economics of Protected Areas: a New Look at Benefits and Costs.* Island Press, Washington, DC.

Drumm, A. (1998) New approaches to community-based ecotourism management. In: Lindberg, K., Wood, M. and Engeldrum, D. (eds) *Ecotourism: a Guide for Planners and Managers.* The Ecotourism Society, North Bennington, Vermont, pp. 197–214.

Eagles, P. and Higgins, B. (1998) Ecotourism market and industry structure. In: Lindberg, K., Wood, M. and Engeldrum, D. (eds) *Ecotourism: a Guide for Planners and Managers.* The Ecotourism Society, North Bennington, Vermont, pp. 11–43.

Finucane, S. and Dowling, R. (1995) The perceptions of ecotourism operators in Western Australia. *Tourism Recreation Research* 20(1), 14–21.

Haigh, R. (1995) *Backpackers in Australia*, Occasional Paper No. 20. Bureau of Tourism Research, Canberra.

Hawkins, D., Wood, M. and Bittman, S. (eds) (1995) *The Ecolodge Sourcebook for Planners and Developers.* The Ecotourism Society, North Bennington, Vermont.

Healy, K. and Zorn, E. (1988) Lake Titicaca's campesino-controlled tourism. In: Annis, S. and Hakim, P. (eds) *Direct to the Poor: Grassroots Development in Latin America.* Lynne Rienner Publishers, Boulder, Colorado, pp. 45–72.

Healy, K. and Zorn, E. (1994) Taquiles's homespun tourism. In: Kleymeyer, C. (ed.) *Cultural Expression and Grassroots Development: Cases from Latin America and the Caribbean.* Lynne Rienner Publishers, Boulder, Colorado, pp. 135–147.

Higgins, B. (1996) The global structure of the nature tourism industry: ecotourists, tour operators, and local business. *Journal of Travel Research* 35(2), 11–18.

Higgins, B. and Holmes, T. (1999) *Tourism Business, Community and Environment in the Adirondack Park.* Wildlife Conservation Society, New York.

Holden, A. (1996) A profile of U.K. outbound 'environmental friendly' tour operators. *Tourism Management* 17(1), 60–64.

Honey, M. (1999) *Ecotourism and Sustainable Development.* Island Press, Washington, DC.

Horwich, R., Murray, D., Sacqui, J., Lyon, J. and Godfrey, D. (1993) Ecotourism and community development: a view from Belize. In: Lindberg, K. and Hawkins, D. (eds) *Ecotourism: a Guide for Planners and Managers.* The Ecotourism Society, North Bennington, Vermont, pp. 152–168.

Ingram, D. and Durst, P. (1989) Nature-oriented tour operators: travel to developing countries. *Journal of Travel Research* 28, 11–15.

Ioannides, D. (1998) Tour operators: the gatekeepers of tourism. In: Ioannides, D. and Debbage, K. (eds) *The Economic Geography of the Tourist Industry: a Supply-side Analysis.* Routledge, London, pp. 139–158.

Ioannides, D. and Debbage, K. (eds) (1998) *The Economic Geography of the Tourist Industry: a Supply-side Analysis.* Routledge, London.

Laarman, J.G. and Gregersen, H.M. (1996) Pricing policy in nature-based tourism. *Tourism Management* 17(4), 247–254.

Lew, A. (1998) The Asia-Pacific ecotourism industry: putting sustainable tourism into practice. In: Hall, C. and Lew, A. (eds) *Sustainable Tourism: a Geographical Perspective.* Longman, Harlow, UK, pp. 92–106.

Lindberg, K. (1998) Economic aspects of ecotourism. In: Lindberg, K., Wood, M. and Engeldrum, D. (eds) *Ecotourism: a Guide for Planners and Managers*, Vol. 2. The Ecotourism Society, North Bennington, Vermont, pp. 87–115.

McKercher, B. (1998) *The Business of Nature-Based Tourism*. Hospitality Press, Melbourne.

McKercher, B. and Robbins, B. (1998) Business development issues affecting nature-based tourism operators in Australia. *Journal of Sustainable Tourism* 6(2), 173–188.

McLaren, D. (1998) *Rethinking Tourism and Ecotravel*. Kumarian Press, West Hartford, Connecticut.

Mason, D. and Milne, S. (1998) *Putting Paradise On-Line*, Case Study No. 1998–06. University of Wellington, Victoria, New Zealand, pp. 1–18.

Middleton, V. and Hawkins, R. (1998) *Sustainable Tourism: a Marketing Perspective*. Butterworth Heinemann, Oxford.

Milne, S. (1998) Tourism and sustainable development: exploring the global-local nexus. In: Hall, C. and Lew, A. (eds) *Sustainable Tourism: a Geographical Perspective*. Longman, Harlow, UK, pp. 35–48.

Milne, S. and Gill, K. (1998) Distribution technologies and destination development. In: Ioannides, D. and Debbage, K. (eds) *The Economic Geography of the Tourist Industry: a Supply-side Analysis*. Routledge, London, pp. 123–138.

Morrison, A. (1996) *Hospitality and Travel Marketing*, 2nd edn. Delmar Publications, New York.

Mowforth, M. and Munt, I. (1998) The industry: lies, damned lies and sustainability. In: Mowforth, M. and Munt, I. (eds) *Tourism and Sustainability: New Tourism in the Third World*. Routledge, London, pp. 188–236.

Navrud, S. and Mungatana, E.D. (1994) Environmental valuation in developing countries: the recreational value of wildlife viewing. *Ecological Economics* 11, 135–151.

National Park Service (1992) *Sustainable Design: a Collaborative National Park Service Initiative*. US Government Printing Office, Denver, Colorado.

Patterson, C. (1997) *The Business of Ecotourism: the Complete Guide for Nature and Culture-Based Tourism Operations*. Explorer's Guide Publishing, Rhinelander, Wisconsin.

Princen, T. and Finger, M. (1994) *Environmental NGOs in World Politics*. Routledge, New York.

Rogers, M. (1996) Beyond authenticity: conservation, tourism, and politics of representation in the Ecuadorian Amazon. *Identities* 3(1–2), 73–125.

Rymer, T. (1992) Growth of U.S. ecotourism and its future in the 1990's. *Florida International University Hospitality Review* 10(1), 1–10.

Schaller, D. (1996) Indigenous ecotourism and sustainable development: The case of Rio Blanco, Ecuador. Masters thesis, The University of Minnesota, Department of Geography, Minneapolis, USA.

Sheldon, P.J. (1986) The tour operator industry: an analysis. *Annals of Tourism Research* 13, 349–365.

Sirakaya, E. (1997) Attitudinal compliance with ecotourism guidelines. *Annals of Tourism Research* 24(4), 919–950.

Smith, R. (1996a) *Crisis Under the Canopy: Tourism and Other Problems Facing the Present Day Huaorani*. Abya-Yala, Quito, Ecuador.

Smith, R. (1996b) *Manual de Ecoturismo Para Guias y Communidades Indigenas de la Amazonia Equatoriana*. Abya-Yala, Quito, Ecuador.

Smith, V. (1996) The four Hs of tribal tourism: Acoma – a pueblo case study. *Progress in Tourism and Hospitality Research* 2, 295–306.

TLC (1998) *Non-profits in Travel – Manual for the Twelfth Annual Conference*. Travel Learning Conferences Inc., Ronan, Montana.

Touche Ross and Company (1975) *Tour Wholesale Industry Study*. Touche Ross, New York.

Tremblay, P. (1998) The economic organization of tourism. *Annals of Tourism Research* 25(4), 837–859.

Urry, J. (1990) The changing economics of the tourist industry. In: *The Tourist Gaze: Leisure and Travel in Contemporary Societies*. Sage Publications, London, pp. 40–65.

Vellas, F. and Becherel, L. (1995) Selling and marketing the tourism product. In: *International Tourism: an Economic Perspective*. St Martin's Press, New York, pp. 163–193.

Waters, S. (1998) *Travel Industry Yearbook*. Child and Waters, New Hampshire.

Weaver, D. (1998a) *Ecotourism in the Less Developed World*. CAB International, Wallingford, UK.

Weaver, D. (1998b) *Nature-Based Tourism in Australia and Beyond: a Preliminary Investigation*. CRC for Sustainable Tourism Pty Ltd, Griffith University, Queensland, Australia.

Weiler, B. (1993) Nature-based tour operators: are they environmentally friendly or are they faking it? *Tourism Recreation Research* 18(1), 55–60.

Weiler, B. and Richins, H. (1995) Extreme, extravagant and elite: a profile of ecotourists on Earthwatch expeditions. *Tourism Recreation Research* 20(1), 29–36.

Wesche, R. (1996) Developed country environmentalism and indigenous community controlled ecotourism in the Ecuadorian Amazon. *Geographische Zeitschrift* 84(3/4), 157–168.

Wight, P. (1993) Ecotourism: ethics or eco-sell? *Journal of Travel Research* 31, 3–9.

Wilson, M. (1987) *Nature Oriented Tourism in Ecuador: Assessment of Industry Structure and Development Needs.* Southeastern Center for Forest Economics Research, Research Triangle Park, North Carolina.

Wood, M. (1998) New directions in the ecotourism industry. In: Lindberg, K., Wood, M. and Engeldrum, D. (eds) *Ecotourism: a Guide for Planners and Managers*, Vol. 2. The Ecotourism Society, North Bennington, Vermont, pp. 45–62.

Wood, M. (1999) Ecotourism, sustainable development and cultural survival: protecting indigenous culture and land through ecotourism, guest editor. *Cultural Survival* 23(2), 25–59.

Yee, J. (1992) *Ecotourism Market Survey: a Survey of North American Ecotourism Tour Operators.* Pacific Asia Travel Association, San Francisco, California.

Zeppel, H. (1998) Land and culture: sustainable tourism and indigenous peoples. In: Hall, C. and Lew, A. (eds) *Sustainable Tourism: a Geographical Perspective.* Longman, Harlow, UK, pp. 92–106.

Ziffer, K. (1989) *Ecotourism: the Uneasy Alliance.* The Ecotourism Society, North Bennington, Vermont.

Zurick, D. (1995) *Errant Journeys: Adventure Travel in a Modern Age.* University of Texas Press, Austin.

Chapter 35

Tour Guides and Interpretation

B. Weiler[1] and S.H. Ham[2]
[1]*Department of Management, Monash University, Berwick Campus, Narre Warren, Victoria, Australia;* [2]*College of Forestry, Wildlife and Range Sciences, Department of Resource Recreation and Tourism, University of Idaho, Moscow, Idaho, USA*

Introduction

The role of the tour guide in ecotourism, and in particular the guide's use of interpretation as an indispensable tool for achieving the goals of ecotourism, is the focus of this chapter. Many definitions of ecotourism found in the literature (see Chapter 1) acknowledge that education and/or interpretation is a key element of ecotourism. In other words, the ecotourism experience is meant to engender an intellectual, emotional and even spiritual connection between people and places as much as it does a physical experience with land and water. The key to establishing these links between people and places is interpretation. Originally defined by Tilden (1957), interpretation is an educational activity aimed at revealing meanings and relationships to people about the places they visit and the things they see and do there. The premise of this chapter is that interpretation lies at the heart and soul of what ecotourism is, and what ecotour guides can and should be doing.

The chapter begins with a discussion of the multiple roles of the tour guide, and provides an argument as to why interpretation should be regarded as a critical and indispensable element of what ecotour guides do. This is followed by a section that outlines the principles of interpretation as they apply to guided ecotours, and illustrates their potential to deliver visitor satisfaction while at the same time achieving the aims of ecotourism. The authors then examine the importance of quality interpretive guiding in tourist decision making and thus the importance of providing information about the interpretation to be offered when marketing guided tours. This is followed by an analysis of selected research findings that link quality guiding, and in particular the use of interpretation by guides, to the quality of the ecotourist experience. Finally, the chapter reviews and critiques current practices and issues in tour guide recruitment, professional development (including training and qualifications), best practice standards, and recognition and reward schemes. While the last two sections are based largely on Australian experience and evidence, they are illustrative of trends and issues that have global relevance.

Special emphasis is placed on face-to-face interpretation delivered by tour guides in both developed and developing countries. This includes interpretation on land or on water, as part of guided walks, guided tours using non-motorized forms of travel (e.g. canoe, raft, mountain bike or horseback), and vehicle-based tours (e.g.

bus, four-wheel drive, riverboat or sea-going vessels). It should be noted that much of what is discussed applies to any tour guide wishing to provide a quality visitor experience, whether as part of an ecotourism operation or as part of an adventure, cultural or other nature-based tour, attraction or resort programme. Tour guides often have the freedom and thus the important opportunity to practise the principles of ecotourism and interpretation regardless of where or for whom they work, and irrespective of whether the product is labelled and marketed as an ecotour or as some other type of nature-based product.

This chapter does not concern itself directly with non-personal or 'static' interpretation such as printed materials, signs, exhibits, self-guided walks, pre-recorded tour commentaries on cassettes or videos, virtual tours, and other electronic media. Many of these interpretive media can be effective in enhancing visitors' understanding and appreciation of the environments being visited and the various natural and cultural phenomena experienced. They are important elements of the ecotourism experience, and many of the principles and issues discussed in this chapter are applicable to these static interpretation media. Further discussion on the use of static interpretive media may be found in Zehr *et al.* (1990), Trapp *et al.* (1991), Ham (1992) and Sorrell (1996).

Interpretation: Just One of the Many Roles of an Ecotour Guide?

The tour guide's 'role' has been the subject of scholarly discussion and analysis for just over a decade. Arguably the main conceptual framework used to dissect and analyse the various roles and functions of the tour guide has been Cohen's model (1985). This model acknowledges both the traditional 'pathfinding' role and the more recent 'mentoring' role of all tour guides, and uses these to produce a 2 × 2 matrix of tour guide roles. The authors feel that the main value of this model is the recognition that guides have accountabilities both within the group (i.e. to facilitate learning and enjoyment of individual clients, and to nurture and manage interaction between clients) and outside the group (i.e. to facilitate and mediate interaction between clients and host communities). Weiler and Davis (1993) added a third dimension to the model for nature-based tour guides: interaction with the natural environment itself. However, there has been almost no published research, with the exception of Haig (1997), testing or applying these frameworks.

A number of studies have investigated the roles and impacts of both mainstream tour guides as well as specialist guides, including ecotour guides (Holloway, 1981; Weiler and Richins, 1990; Weiler *et al.*, 1992). The overwhelming finding of these studies has been that the tour guide is expected to play a large number of different roles, in response to the expectations and demands of various stakeholders in the ecotourism experience.

From a business perspective, it is useful to frame a discussion of the roles of the tour guide in the context of the various stakeholders to which guides are accountable. Whether the tour is a brief, 20-min guided walk in the rainforest, or an extended four-wheel drive tour covering thousands of kilometres and a range of ecosystems, the guide has many masters. Key among these are the tour operator, the client, the host community, land managers, and the tourism industry itself, each of which is discussed in turn in the subsections that follow.

The tour operator

Guides may be employed on a permanent, casual or freelance basis. Regardless of employment status, the guide is presumably accountable to deliver what is specified in his or her job description. Traditionally, however, much of what the tour operator expects of the guide is determined by informal understanding rather than formal documentation. Increasingly, operators are able and willing to articulate what they perceive to be the duties and, to a lesser extent, the outcomes and performance indicators of their guides. From the tour operator's perspective, for example, a guide's duties often include:

- ensuring the safety, health and comfort of clients;
- providing courteous and quality customer service;
- responding to the needs and expectations of visitors from other cultures and those with special needs due to age, a disability or special interests;
- managing interactions within client groups;
- delivering the tour cost-effectively;
- providing high quality, informative and entertaining commentary;
- meeting the legal and moral obligations and expectations of protected area managers, host communities and clients.

The development of industry-wide competency or occupational standards, with input from operators and guides, has been a relatively recent (1990s) phenomenon that has helped to formalize and standardize employers' expectations, at least in developed countries like Australia and Canada (CTHRC, 1996; Tourism Training Australia, 1999). However, the conditions under which tour guides must work, the remuneration associated with satisfactory (or better) performance, and the extent to which guides are held accountable for what they do on the job varies widely. Indeed, for many small tour operators, the guide is also the owner/operator, so is accountable only to him/herself. Given the immediacy of the financial, marketing and administrative demands of operating a business (see Chapter 36), insufficient attention is likely to be given to at least some of the above duties.

At the other extreme, companies organizing tours to remote areas or in countries other than where they are based may never actually observe the tour guide's performance on the job, relying totally on the guide's own assessment of the tour and, to some extent, client feedback. The delivery of tour commentary is assumed to be part of every guide's job, but the extent to which the guide is expected to practise principles of effective interpretation varies widely. Many operators have little or no idea what interpretation is (Weiler *et al.*, 1992; Weiler, 1993b), and no idea whether their clients' understanding and appreciation of nature and culture are enhanced as a result of their guided tour experience.

The visitor

The terms 'visitor', 'ecotourist' and 'client' are used here to refer to tour participants, regardless of whether they are local residents, domestic tourists or international visitors, and irrespective of who pays or whether an overnight stay is involved.

Visitor satisfaction is, many would argue, the ultimate measure of success in any tourism business (Ryan, 1995). Visitor satisfaction is a complex variable, influenced to some extent but not entirely by expectations and on-site perceptions, for which it is often very difficult to obtain valid measures (Blamey and Hatch, 1996; Childress and Crompton, 1997; Ryan, 1998). For example, a visitor may indicate that his/her expectations were met, but is this satisfaction? Meeting visitor expectations is usually necessary, but may not be sufficient, to producing satisfied clients who will become repeat customers for the company and/or recommend the tour to others, both of which are extremely important in the ecotourism industry (see Mancini, 1990; Blamey and Hatch, 1996).

Unfortunately, our understanding of what contributes to visitor satisfaction or dissatisfaction, particularly with respect to the performance of the tour guide, is still very poor. Some research, however, has indeed demonstrated a link between the quality of guiding and tourist satisfaction (see Lopez, 1981; Geva and Goldman, 1991). Hughes' (1991) study of a boat tour on Palm Island (Australia) found that the ability of the guide to effectively interact with the group, provide a commentary of interest and ensure smooth running of the tour, emerged as the most important components of the guide's role. Similarly, Geva and Goldman (1991) in a study of 15 guided tours from Israel to Europe and the USA found that the guide's conduct (relations with tour participants) and expertise were the most important of 15 tour attributes in determining client satisfaction. Finally, another Australian study on guided tours in Tasmania (Forestry Tasmania, 1994) found that clients placed considerable value on the guide's interpretation of the environment and land management. In addition, a growing body of qualitative evidence, based on the practical experiences of businesses and organizations, suggests that providing quality interpretive services makes real business sense. According to Conservation International, a Washington, DC-based environmental nongovenment organization, high quality interpretation 'can also improve business by increasing the quality of guests' experience, increasing repeat visitation and occupancy rates, providing unique marketing opportunities and allowing hotels to charge higher rates' (Sweeting *et al.*, 1999, p. 27).

In recent years, and as reflected in Chapter 3, a number of authors (e.g. Weiler and Richins, 1995; Blamey and Hatch, 1996; Wight, 1996a, b) have argued that ecotourists are not a homogeneous group. Some of these authors have suggested a number of proposed dimensions or continua along which ecotourists may vary. These have included, for example:

- preferred type of natural setting;
- desire for physical challenge/degree of self-reliance;
- desire for intellectual challenge/motivation to learn something new;
- desire to be directly and actively involved with the natural environment;
- desire to be environmentally responsible (ranging from minimizing impacts to actively engaging in enhancing or restoring visited environments).

Some authors (Weiler and Richins, 1995) have assumed that hard-core ecotourists are at the extreme end of each of these continua (i.e. that if they want physical challenge, they also want intellectual challenge, a highly participatory experience, and an environmentally responsible one; see Chapter 2). However, in the absence of data to support this notion, it is more likely that a vast range of market segments exists, each with its own unique set of motivations and expectations. In summary, while there is some evidence that clients expect and appreciate quality interpretation, this may vary depending on individual ecotourists, their motivations generally, and their expectations of each particular tour. Much more research is needed to determine the relationship between tour guides' competencies as interpreters, their on-the-job performance, and client satisfaction (Blamey and Hatch, 1996).

The host community

In many situations, the tour guide acts as a mediator between visitors and hosts (Cohen, 1985). In the context of ecotourism, this is most apparent on tours that visit indigenous sites or communities and tours to foreign countries. In some of these cases, the operator is legally required to consult and gain approval from indigenous communities, and the guide may have contractual obligations spelled out verbally or in writing. Even where this is not a legal requirement, the ethics and essence of ecotourism are that tours are conducted in ways which minimize negative sociocultural impacts and contribute in a positive way toward the host community (see, for example, Nash, 1996). The tour guide's role in doing so is to provide accurate and culturally appropriate interpretation of the site or resource in a way that enhances visitors' understanding and appreciation of indigenous culture, history, contemporary lifestyles, values and issues (see for example, Harvey and Hoare, 1995). According to recent experiences in Ecuador and elsewhere (see Sweeting *et al.*, 1999), ecotourists are willing to pay more for such 'culturally sensitive' tours, especially when the community itself governs the format and density of use. The tour guide also plays an important role in monitoring and modelling appropriate cross-cultural behaviour so that visitors impact in a positive way economically, culturally, socially and environmentally (see Chapters 23–28 for elaboration of these issues).

There is some debate as to the whether a local guide (i.e. one who is native to the destination country or region or even local community being visited) is 'better' than a foreign guide. Clearly, the training and employment of local guides helps to ensure that more of the economic benefits are felt by the host community, an important consideration particularly for tourism in developing countries and regions. Moreover, local guides are more likely to understand the protocols and sensitivities associated with visiting and experiencing cultural sites and communities within the host country. On the other hand, guides with cultural and/or socio-demographic backgrounds similar to their clients are more likely to understand visitors' expectations, fears and likely faux pas. A guide who has never travelled or lived outside his/her home community may have little understanding of appropriate customer relations and the Western values and attitudes about the environment that many ecotourists hold. In reality, good training can overcome the shortcomings of either. The goals of ecotourism are probably best met by having a local and foreign guide working together, a policy larger tour companies are beginning to embrace in some countries.

Land managers

Protected-area managers increasing rely on commercial tour operators to deliver interpretation within parks and on other protected lands both in developed and developing countries. Policies that support and promote privatization of guiding and interpretation via commercial licences, concessions and outsourcing of guided activities within parks have been put in place in many countries without a clear understanding of whether commercial tours can and do meet the objectives of land management agencies. For the visitor whose experience within protected areas is just one component of a guided tour, the role of the tour guide is critical. According to many writers (e.g. MacKinnon *et al.*, 1986; Sweeting *et al.*, 1999), it is the application of interpretive principles that will ensure visitors gain an understanding and appreciation of the parks they are visiting. Land managers are also increasingly dependent on tour guides and operators to monitor their own and their clients' impacts on the natural environment, and to articulate and model minimal impact practices for their clients (DNRE, 1999).

The tourism industry

Other players in the tourism industry may have expectations of the operator and/or the tour guide and have little or no understanding of what 'interpretation' is or what benefits quality interpretive services might engender for individual companies and the tourism industry as a whole. In Australia, this has become obvious in the development of some state tourism policies and plans, and in the early developmental stages of the National Ecotourism Accreditation Program (NEAP)[1] (see Chapters 9 and 37), where some operators and other industry representatives use the terms 'education' and 'interpretation' interchangeably. In some situations, tour wholesalers and travel agents with no understanding of interpretation suggest and even dictate the content, method and language in which the tour content should be communicated. In some countries and specific destinations such as the Galapagos, tour guiding is regulated and controlled by government or industry bodies, and guides must complete specified training, obtain qualifications or demonstrate particular competencies in order to be licensed as tour guides (Britton and Clarke, 1987). To date, these schemes

[1] From January 2000 'education' has been removed from the interpretation section of the accreditation document.

have revolved largely around legal, health and safety issues, and have tended to underestimate the importance of interpretation in law enforcement, in client safety and satisfaction, and in the overall viability of the guided tour sector of the tourism industry (see Chapter 37).

From the perspective of all stakeholders, it is apparent that tour guides play a pivotal role in ecotourism and are critical to meeting the needs and expectations of operators, clients, host communities, land managers, and the wider tourism industry in both developed and developing countries. The foregoing discussion provides a backdrop against which the guide's current role as interpreter, facilitating understanding and appreciation of natural and cultural phenomena can be critically examined. Clearly, interpretation is not just one of the many roles that an ecotour guide plays; when it is done well, it is the distinguishing feature of 'best practice' in guiding. The remainder of this chapter focuses on the interpretive role of tour guides.

Principles of Interpretation as Applied to Ecotourism

The premise of this chapter is that the tour guide can and should play a key role in facilitating clients' understanding and appreciation of natural and cultural phenomena. It has also been argued that the use of interpretation, and more particularly the application of interpretive principles by the tour guide, is an essential element of ecotour guiding. But what are these 'principles' of interpretation, and why are they integral to guided ecotourism experiences? Although this question has been addressed in detail elsewhere (see Ham, 1992), a brief summary of these principles is needed here in order to critique current guiding activities and identify issues. There are at least five such principles:

1. Interpretation is not teaching or 'instruction' in the academic sense.
2. Interpretation must be enjoyable for visitors.
3. Interpretation must be relevant for visitors.
4. Interpretation must be well organized so that visitors can easily follow it.
5. Interpretation should have a theme, not just a topic.

Interpretation is not teaching or 'instruction' in the academic sense

Although interpretation involves the transfer of information about places and cultures from guides to tour clients, guides are not teachers in the sense that visitors must master or remember all the information. Indeed, research shows (see Thorndyke, 1977; Ham, 1983; Beck and Cable, 1998) that visitors may forget much of the supportive factual information obtained through a commentary yet still internalize a deep overriding notion of the importance of it all. But research also shows that tour clients will pay attention to a commentary only if they choose to. Since they are not accountable to master the information, the only motivation they have to pay attention is that it promises to be a rewarding expenditure of their time. Ecotourists are therefore a voluntary audience. Ham (1992) termed this type of audience a 'non-captive' audience because, unlike students in a classroom (the classic 'captive' audience), tour clients are not held prisoner by an external reward system involving grades and qualifications. The best tour guides know this, and they work hard to capture and maintain their audiences' attention. Students in search of a grade will try to pay attention to boring teachers. But visitors in search only of a rewarding experience will not make the same effort with a boring guide. Unlike some classroom teachers, guides never have the option of being boring or pedantic. If they are, they will, quite simply, lose their audiences and ultimately their jobs.

Interpretation must be enjoyable for visitors

A simple yet under-appreciated idea among some guides is that interpretation aimed at a non-captive audience like ecotourists must be enjoyable and fun for them (Sweeting *et al.*, 1999). Although entertainment is not interpretation's main goal, it must certainly be considered one of its essential qualities. Visitors who join guided tours are pleasure-seekers. Even 'ecotourists', who are sometimes (and probably inaccurately) cast as serious students of nature and culture, seek fun and entertainment from their tour experiences. Of course, what constitutes 'fun' will vary among different types of visitors, and successful guides pay close attention to these differences and idiosyncrasies. Among almost all groups, however, one aspect of fun is informality, for example, involving visitors in a playful competition, a role play, or simply concentrating on using common, everyday language and a conversational tone. Some studies have even demonstrated that the more an interpretive medium reminds an audience of academia, the less interesting and provocative it becomes (Washburne and Wagar, 1972; Ham and Shew, 1979). Experiences in both developed and developing countries throughout the world suggest that effective interpretation may have many qualities, but a common one everywhere is that it is fun for its audiences (Ham and Sutherland, 1992; Ham *et al.*, 1993; Sweeting *et al.*, 1999).

Interpretation must be relevant for visitors

Beyond the sheer entertainment value of a guide's commentary, tour clients pay special attention to those things the guide can relate to their interests and personalities. Simply put, people pay attention to what they can understand and what they care about (Tilden, 1957; Ham, 1992). When a guide makes what he or she is saying or showing understandable to a group, visitors pay attention, because the new information is meaningful to them. Experienced guides make their commentaries meaningful by using common language and by employing analogies, metaphors and other methods of bridging the unfamiliar world of the tour route, content and environment to the things already known and familiar to the group (Ham, 1992; Sweeting *et al.*, 1999). Similarly, when commentaries focus visitors' attention on things they already care about, an attentive audience is almost guaranteed (see Moray, 1959; Cherry, 1966; Ham, 1983). Ham (1992) terms this type of communication 'personal' since it connects what is being described or displayed to something personally important or significant to visitors (their well-being, their loved ones, their deepest values, strongest principles or convictions, or things of profound symbolic importance). According to Ham (1992), when a guide has made her or his commentary both meaningful and personal, she or he has made it highly relevant.

Interpretation must be well organized so that visitors can easily follow it

As pleasure-seekers, visitors on guided tours (a non-captive audience) switch attention at will whenever what they are hearing, seeing or doing, fails to gratify them. Often, attention switching has to do with how much effort a visitor thinks will be required to follow a presentation. As Ham (1992) and Beck and Cable (1998) have discussed, a well-organized presentation will require less effort to follow than the same presentation which lacks organizational structure. As Miller (1956) demonstrated nearly a half century ago, humans can manage more information with less effort if it is organized into no more than 5–9 categories or units. Within this 5–9 range, Miller's research found no relationship between a person's level of educational attainment or intelligence and how many categories of information he or she could follow. Rather, it is the amount of experience or prior interest we have with the subject at hand (irrespective of our IQ) that determines how many categories of information we can effectively manage and keep sorted out (see Solso, 1979). Any group of tour clients is likely to include a cross-section of people who have more or less experience with, and prior interest in, the subjects a guide will discuss. Accordingly, Ham (1992) argues that a tour guide is more likely to sustain the attention of the whole tour group by delivering the interpretive commentary organized around five or fewer main ideas, regardless of the type of tour.

Interpretation should have a theme, not just a topic

A central defining principle of ecotourism is its focus on conservation and perpetuation of the natural and cultural values inherent in the land and water. How ecotourism achieves this lofty goal is often debated, but widespread agreement seems to exist in the simple fact that what a guide says to his or her clients can influence how they think and behave with respect to the places they visit. In other words, the messages that a guide imparts to a group of visitors, relative to the protected values of a place, may in large part determine what they will think, feel and do both in the short term (on-site) and possibly even in the long term, once they have returned home.

According to Ham and Krumpe (1996) and others (see, for example, Orams, 1997; Knapp and Barrie, 1998), one of the most important things guides can do is to facilitate a bonding between their clients and the places they lead them to, thus connecting people and places in powerful ways that nurture respect and caring about those places. Guides do this by communicating strong themes to their tour clients (Lewis, 1980; Ham, 1992; Ham and Krumpe, 1996; Beck and Cable, 1998; Sweeting *et al.*, 1999, and others). Themes are messages, factual but compelling statements about a place or a thing. Skilful tour guides, according to most contemporary writers, practise thematic interpretation by imparting compelling messages to their clients about the places they visit.

Themes are whole ideas, morals to the story, an overriding message that a visitor takes home (Lewis, 1980; Ham, 1992). They are contrasted to *topics* in the sense that a topic is merely the subject matter of presentation (geology, religion, plants, wildlife, etc.) whereas a theme is a specific message about that topic. Drawing on research conducted by Thorndyke (1977), Ham (1992) concluded that visitors will forget isolated facts from a guided tour, even fascinating, mind-boggling facts, but they will remember and possibly internalize the bigger ideas, or themes, of the tour. These themes, according to Ham and Krumpe (1996) are tantamount to 'beliefs' which social psychologists have demonstrated to be the building blocks of attitudes and related behaviours, including attitudes and behaviours about conservation. Thus, guides who concentrate on imparting strong themes in their tours do far more to further ecotourism's aim of facilitating conservation than guides who concentrate on the relatively simple task of saying interesting things about a topic in entertaining ways. Guides who are armed with strong themes, and who then creatively package their presentations in entertaining, relevant and organized tours, are the ones who make the real difference in helping ecotourism to achieve its loftiest of goals, conservation.

Tour Guides and Interpretation as Factors in Tourist Decision-making

So far, this chapter has examined the multiple roles of an ecotour guide, and the centrality of interpretation in what ecotour guides do. The chapter now turns to the business implications of quality tour guiding and interpretation from a number of perspectives. One of these perspectives deals with how the quality of guiding and interpretation affects the visitors' decision-making behaviour and therefore tour product marketing.

The tourist decision-making process is in fact not a single decision but a series of decisions, beginning with a decision to travel or not to travel. Other pre-trip decisions can include choice of destination, season and dates of travel, budget, choice of travelling partner(s), mode(s) of transport, accommodation, and selection of other tourism products such as packaged tours and guided ecotours. For each decision, there are many factors internal and external to the individual, the latter including destination and product attributes that can influence the tourist's decision (Sonmez and Graefe, 1998). For example, school holidays may be a major factor affecting season and dates of travel, with some having to travel at these times and others avoiding them.

Most tourist decisions are made without consideration of tour products, let alone the tour guide. Only if the consumer decides to pre-select and pre-book a package tour or a guided ecotour will he or she consider the quality of the tour guide. In such cases, most consumers are concerned mainly with choosing a tour company they can count on to provide an experience that is safe, reliable and convenient, and offers value for money (Fay, 1992). Such pre-trip decisions are often made on the advice of a family member or friend; other influences include travel agents, tour company brochures, and web sites. Any consideration of the guide's ability would probably be limited to issues such as the guide's practical skills (driving, cooking, navigation), language competency, and knowledge of the destination/area which might be perceived to affect the safety and well-being of the customer. It is likely that only hard-core ecotourists would make the effort to examine the qualifications, knowledge or skills of the guide for such pre-trip bookings, and in many cases, there may be no way of determining who the actual individual guide is going to be.

Once en route or at the destination, the tourist then makes a series of decisions, and it is usually only at this stage that visitors may give greater attention to the guide's competencies. For example, the visitor may or may not choose to join a guided tour. If the choice is to join a tour, then a particular company as well as a particular tour product may be important considerations that determine choice, usually based on one or a combination of the following: word-of-mouth (especially other travellers), tour brochures, and telephone calls or visits to the tour operator's office front. At this decision making stage, there is some evidence that tourists, and ecotourists in particular, do examine carefully any information about the qualifications, knowledge, skills and abilities of the guide. A study by Weiler *et al.* (1992) found that the qualifications of the guide were promoted in 18% of brochures and promotional materials for guided ecotour products. Again, certain types of ecotourists may be more likely to look for product differentiation with respect to the quality of the guide, and more research is needed to identify the needs and expectations of these market segments.

The Tour Guide and Interpretation as Integral Elements of the Ecotourist Experience

Compared to what we know about tourist decision making, the evidence of a relationship between quality tour guiding and interpretation and the visitor experience is more revealing. In a study of 295 ecotourists on 23 very diverse guided day tours, Weiler and Crabtree (1998) found that visitors were surprisingly perceptive about the knowledge and skills of their guides and able to provide detail on what they liked and disliked about their guide. The study revealed that although clients were largely complimentary about most aspects of the guides' performance, their most common criticisms focused on the lack of or incorrectness of conservation themes imparted by the guide. Given that these were all tours at the 'soft' end of the ecotourism spectrum (i.e. short tours of 1 day or less, with limited or no physical challenge and limited active involvement by the visitor), it is intriguing to note the respondents' interest in, and desire for, interpretation and minimal impact messages.

Additionally, Weiler and Crabtree's study found that despite the guides' strong performance on most evaluative criteria dealing with site knowledge, tour management and interpersonal communication skills, they performed the poorest on indicators pertaining to interpretation methods and conservation themes. These include: (i) delivering organized and thematic interpretation (e.g. evidence of a theme, sequencing, introduction and conclusion); and (ii) providing messages on ecologically sustainable practices and behaviours (e.g. monitoring group behaviour and communicating minimal impact themes, both on-site and post-tour).

These results suggest that guided ecotours may be falling short of their potential to deliver a quality visitor experience while imparting strong conservation themes. There may be a number of reasons for this. Firstly, the operator and guide often have very limited information about clients, particularly for day tours, where there may be little more than a name on a list prior to departure. Research on the interpretive needs and expectations of clients of particular ecotour products is virtually non-existent, as individual operators usually lack the time, money and expertise to conduct such research.

Secondly, there is often a mix of clientele on any particular tour, particularly day tours. As mentioned earlier, ecotourists vary widely in their interests, motivations and expectations, some of which directly impact on the way interpretive services might be offered on a tour. Visitors may range from wanting to be passive listeners and observers to highly active and involved tour participants. Some may want only to take away a basic message (theme) from the tour, while others may want to be able to recall facts and acquire conservation knowledge and skills that they can apply back home.

If tour products are highly targeted, the guide can make some assumptions and develop the content and style of the tour to match the targeted market. For example, if the tour product is targeting hard-core ecotourists looking for physical and intellectual challenge and willing to become actively involved, a product along the lines of the experiences offered by Earthwatch International (Weiler *et al.*, 1993) but underscored by high-quality interpretation would be appropriate. However, more often than not there are wide variations within a single tour group and certainly there are variations on different tours, requiring significant adaptability and resourcefulness between one tour and the next. Guides who are versatile and able to apply the principles of interpretation outlined earlier are clearly at an advantage in being able to provide all members of their tour group with a quality experience. This is the third and perhaps most significant reason why guided tours often fall short of delivering a quality visitor experience. Guides frequently lack the knowledge and/or skills to apply interpretive principles to the design and delivery of their guided tours.

In summary, improving the tour guide's potential to provide a quality visitor experience hinges on three factors. First, better target market research is needed in order to deliver an interpretive experience that meets the expectations and 'needs' of tour clients and effectively apply the principle of product–market match (see Chapter 36). Second, the guide must harness the principles of interpretation as a way of meeting the needs and expectations of all tour group members, and in most cases this means better guide training in interpretation. As we have argued elsewhere (Ham and Weiler, 1999), more research is needed in countries other than Australia to ascertain what may be a range of other training needs, but the need for interpretation and minimal impact training is clear. According to research and conventional thinking, guides who succeed in bringing these qualities to their approach will produce more satisfied clients and contribute in significant ways to the expressed goals of ecotourism. Third, operators and protected area managers must ensure that the tour itinerary and operating conditions are conducive to excellent interpretation, and this means protected area manager and tour operator education about interpretation and product development.

Current Practices and Issues in Ecotour Guiding

In Australia, one of the most ecotourism-advanced countries, the tour operations industry currently has no industrial awards, i.e. there are no industry-wide legal specifications regarding the rights and obligations, qualifications, working conditions and rates of pay, for either tour operators or guides. Some of the larger tour companies have enterprise agreements which specify such conditions, but generally speaking, the criteria by which operators recruit, select, retain and remunerate guides are largely undocumented. Equally important is the fact that there is no government or regulatory body requiring an individual to be qualified or licensed in order to work as a guide.

Operators appear to place a heavy emphasis on hiring guides who have practical skills and experience, who are willing to work in difficult working conditions (e.g. unsociable days and hours in remote locations), and who are multi-skilled, e.g. fluency in a second language. Since most tour operators in Australia and elsewhere have only a rudimentary understanding of interpretation, it is unlikely that they are recruiting guides for their expertise as interpreters and/or increasing their pay if they upgrade their interpretive skills or qualifications. And generally speaking, tour operators are unlikely to see any reason to change what they already consider to be an acceptable level of tour guide performance, based on what they perceive to be satisfied tour clients and sustained demand for their products.

For practising tour guides, there is also little incentive for individuals to actively upgrade their interpretive skills or qualifications or improve the quality of their tours. To date there has been no evidence that interpretive qualifications or skills lead to preferential treatment either in gaining employment or in career advancement. Thus, despite the evidence that interpretation is key to delivering a quality guided ecotourism experience, neither tour operators nor guides are likely to initiate dramatic change in how much time or financial resources they allocate to improving the interpretive competence of their guides. It then falls to the wider tourism industry, educational institutions, protected area managers and other government bodies to find ways to, first, raise the awareness of what is 'good practice' in ecotour guiding, and second, provide incentives to improve the standard of guiding practice throughout the industry. As discussed in Chapter 37, the pursuit of excellence in the tourism industry has led to a number of ways of measuring and rewarding quality. The usefulness and relevance of these strategies for improving tour guide performance is discussed in this section.

The earliest and perhaps weakest forms of quality control are professional associations and codes of conduct. There are many professional tour guide associations, including at least four in Australia alone, but most are focused on city guides or guides working for inbound tour operators. The Ecotourism Association of Australia (EAA) now has a membership category specifically for ecotour guides, and a regular column in its newsletter devoted to guide news and issues, which is helping to raise the level of awareness of guides and guiding. As for codes of conduct, numerous behavioural guidelines have been developed for nature-based tourism, but they tend to be targeted at either the operator or the visitor. The EAA has both, but neither has any influence beyond raising awareness as to what is sound environmental practice when operating tours in natural areas.

Competency standards or occupational standards are another form of benchmarking, and can be very useful if linked to employee recruitment and selection processes, rates of pay, the development of training curricula, and qualification schemes. In Australia, this was not the case for ecotour guides or tour guides of any kind as of early 2000. Although considerable work including industry consultation has gone into the development and subsequent revision of the tour guide standards, the vast majority of tour operators and guides are unfamiliar with the tour guide competency standards. Almost any training provider can offer a tour guide training programme, and while a number of training providers have worked hard to align their guide training programmes to these standards, few incentives exist for guides to select competency-based courses over those not aligned with the standards. The vast majority of the tourism industry, and certainly the travelling public, have no understanding of tour guide qualifications and competencies. This is unfortunate, as many aspects of the tour guide's job are well-articulated in the tour guide competency standards, including those addressing interpretation (see Tourism Training Australia's web site: www.tourismtraining.com.au).

In Australia, this situation is poised for change, largely due to industry-based accreditation and certification schemes such as the NEAP, which has been in operation since 1997, and the National Ecotour Guide Certification Program (NEGCP), which was officially launched in late 2000. To be an accredited tour product under NEAP, for example, a tour operator must demonstrate that the product meets the education and interpretation criteria spelled out in the accreditation document. Once the NEGCP is fully implemented, it is proposed that a minimum percentage of a tour operator's guides will need to be NEGCP-certified in order to gain and/or maintain NEAP accreditation. The certification criteria are closely aligned with the tour guide competency standards mentioned earlier.

The most powerful and controversial aspect of the ecotour guide certification scheme is its potential links to the licensing and/or issuing of permits to tour operators by protected area managers. In Australia, protected areas are managed largely by state land management agencies such as Western Australia's Department of Conservation and Land Management, Queensland's National Parks and Wildlife Service, and Parks Victoria. Currently, tour operators who are NEAP-accredited have special privileges with some of these protected area management agencies such as Parks Victoria, which grants extended licences and permission to operate in wilderness areas not accessible by non-accredited operators. If access to these protected areas were limited to NEAP-accredited operators using only NEGCP-certified guides or others who can demonstrate that they meet the agency's criteria through some other means, then both operators and guides would be more likely to see the value of meeting the standards set by the guide certification programme, including interpretive competency.

Another important step in encouraging professionalism in ecotour guiding is to link certification to existing education and training programmes. In the USA, interpreters are certified via the National Association for Interpretation (NAI), which specifies a university undergraduate degree as a minimum educational requirement or, alternatively, a 2-year technical degree and 5000 h of documented experience (NAI web site, 1999). In Australia, there appears to be a preference for recognizing educational qualifications in such a way that students graduating from such programmes are neither privileged, nor excluded from certification. Relevant issues that still need to be resolved include the stipulation of a minimum threshold of industry experience, the requirement for at least some assessment of competency in the workplace rather than entirely in the classroom, the need for regular re-certification and the requirements associated with renewal, and the appropriateness of levels of certification to reflect levels of competence (either through study or through work experience) (see Chapter 40).

Finally, there are some very successful tourism award schemes that recognize excellence in both ecotour guiding (e.g. the Golden Guide Award in Western Australia) and interpretation (e.g. the Interpretation Australia Association and NAI in the USA). These are truly 'best practice' awards that complement the minimum standards approach of the other initiatives, while simultaneously serving to motivate and educate guides as well as to regulate the practice of guiding.

Conclusion

Best-practice ecotour guiding need not be confined to the ecotourism industry. Many freelance guides move between operators who vary widely in their management philosophies and target markets. These guides have the opportunity to apply ecotourism and interpretation principles to tours ranging widely in group size, tour length, subject matter and location, and to reach clients who vary greatly in their understanding of and commitment to the principles of ecotourism. There is widespread agreement that guides play a pivotal role not only in the quality of the ecotourist's experience, but in facilitating the conservation goals of ecotourism. Sven Olof Lindblad, owner of Lindblad Expeditions, a New York-based ecotourism cruise company, proclaimed in a recent fundraising campaign for the Galapagos Islands that 'it will be the passion and insistence of the traveler that will ultimately save the world's special places' (see Ham and O'Brien, 1998). Lindblad's prophecy strikes at the heart and soul of interpretation's central role in quality ecotour guiding. Saving the world's 'special places' remains both the premise and promise of ecotourism. Interpretation, creatively packaged and powerfully delivered, lies at the heart of both.

References

Beck, L. and Cable, T. (1998) *Interpretation for the 21st Century.* Sagamore Publishing, Champaign, Illinois.

Blamey, R. and Hatch, D. (1996) *Profiles and Motivations of Nature-based Tourists Visiting Australia,* Occasional Paper Number 25. Bureau of Tourism Research, Canberra.

Britton, S. and Clarke, W.C. (eds) (1987) *Ambiguous Alternative: Tourism in Small Developing Countries.* University of the South Pacific, Suva, Fiji.

Canadian Tourism Human Resource Council (CTHRC) (1996) *National Occupational Standards for the Canadian Tourism Industry.* Heritage Interpreter, Ottawa.

Cherry, C. (1966) *On Human Communication.* Massachusetts Institute of Technology, Cambridge, Massachusetts.

Childress, R.D. and Crompton, J.L. (1997) A comparison of alternative direct and discrepancy approaches to measuring quality of performance at a festival. *Journal of Travel Research* 35(Fall), 43–57.

Cohen, E. (1985) The tourist guide: the origins, structure and dynamics of a role. *Annals of Tourism Research* 12(1), 5–29.

Department of Natural Resources and Environment (DNRE) (Victoria) (1999) *Best Practice in Park Interpretation and Education.* A Report to the ANZECC Working Group on National and Protected Area Management Benchmarking and Best Practice Program, Melbourne.

Fay, B. (1992) *Essentials of Tour Management.* Prentice Hall, Englewood Cliffs, New Jersey.

Forestry Tasmania (1994) *Guided Nature-Based Tourism in Tasmania's Forests: Trends, Constraints and Implications.* Forestry Tasmania, Hobart.

Geva, A. and Goldman, A. (1991) Satisfaction measurement in guided tours. *Annals of Tourism Research* 18(2), 177–185.

Haig, I. (1997) Viewing nature: can the guide make the difference? Unpublished Master of Arts thesis, Griffith University, Brisbane.

Ham, S. (1983) Cognitive psychology and interpretation: synthesis and application. *Journal of Interpretation* 8(1), 11–27.

Ham, S. (1992) *Environmental Interpretation: a Practical Guide for People with Big Ideas and Small Budgets.* Fulcrum/North American Press, Golden, Colorado.

Ham, S. and Krumpe, E. (1996) Identifying audiences and messages for nonformal environmental education: a theoretical framework for interpreters. *Journal of Interpretation Research* 1(1), 11–23.

Ham, S. and O'Brien, T. (1998) Unpublished fundraising campaign for the Galapagos Conservation Fund. Special Expeditions, Inc., New York.

Ham, S. and Shew, R. (1979) A comparison of visitors' and interpreters' assessments of conducted interpretive activities. *Journal of Interpretation* 4(2), 39–44.

Ham, S. and Sutherland, D. (1992) Crossing borders and rethinking a craft interpretation in developing countries. In: Machlis, G. and Field, D. (eds) *On Interpretation Sociology for Interpreters of Natural and Cultural History*, revised edn. Oregon State University Press, Corvallis, Oregon, pp. 251–274.

Ham, S. and Weiler, B. (1999) Training ecotour guides in Central America: lessons learned and implications for regional guide training and certification. Paper presented to the Central American Regional Conference on Cleaner Technology and Environmental Management for the Tourism Industry, Guatemala City, Guatemala, 1–3 September.

Ham, S., Sutherland, D. and Meganck, R. (1993) Applying environmental interpretation in protected areas in developing countries. *Environmental Conservation* 20(3), 232–242.

Harvey, J. and Hoare, A. (1995) Benefits of ecotourism in local communities. In: HaySmith, L. and Harvey, J. (eds) *Ecotourism and Nature Conservation in Central America.* Wildlife Conservation Society, Gainesville, Florida, Chapter 6.

Holloway, J.C. (1981) The guided tour: a sociological approach. *Annals of Tourism Research* 8(3), 377–401.

Hughes, K. (1991) Tourist satisfaction: a guided 'cultural' tour in North Queensland. *Australian Psychologist* 26(3), 166–171.

Knapp, D. and Barrie, E. (1998) Ecology versus issue interpretation: the analysis of two different messages. *Journal of Interpretation Research* 3(1), 21–38.

Lewis, W. (1980) *Interpreting for Park Visitors.* Eastern National Park and Monument Association, Eastern Acorn Press, Philadelphia.

Lopez, E. (1981) The effect of tour leaders' training on travellers' satisfaction with tour quality. *Journal of Travel Research* 1(4), 23–26.

MacKinnon, J., MacKinnon, K., Child, G. and Thorsell, J. (1986) *Managing Protected Areas in the Tropics.* IUCN/World Conservation Union, Cambridge.

Mancini, M. (1990) *Conducting Tours: a Practical Guide.* South-Western Publishing Co., Cincinnati, Ohio.

Miller, G. (1956) The magical number seven, plus or minus two: some limits on our capacity for processing information. *Psychological Review* 63(2), 81–97.

Moray, N. (1959) Attention in dichotic listening: effective cues and the influence of instructions. *Quarterly Journal of Experimental Psychology* 11(1), 56–60.

Nash, D. (1996) *Anthropology of Tourism.* Pergamon Press, Oxford.

National Association for Interpretation (NAI) (1999) http://www.interpnet.com/

Orams, M. (1997) The effectiveness of environmental education: can we turn tourists into 'greenies?' *Progress in Tourism and Hospitality Research* 3, 295–306.

Ryan, C. (1995) *Researching Tourist Satisfaction: Issues, Concepts, Problems.* Routledge, London.

Ryan, C. (1998) The travel career ladder: an appraisal. *Annals of Tourism Research* 25(4), 936–957.

Solso, R. (1979) *Cognitive Psychology.* Harcourt Brace Jovanovich, New York.

Sonmez, F.S. and Graefe, R.A. (1998) Terrorism risk of foreign tourism decisions. *Annals of Tourism Research* 25(1), 112–145.

Sorrell, B. (1996) *Exhibit Labels: an Interpretive Approach.* AltaMira Press, Walnut Creek, California.

Sweeting, J., Bruner, A. and Rosenfeld, A. (1999) *The Green Host Effect: an Integrated Approach to Sustainable Tourism and Resort Development.* Conservation International, Washington, DC.

Thorndyke, P. (1977) Cognitive structures in comprehension and memory of narrative discourse. *Cognitive Psychology* 9(1), 77–110.

Tilden, F. (1957) *Interpreting Our Heritage.* University of North Carolina Press, Chapel Hill, North Carolina.

Tourism Training Australia (1999) http://www.tourismtraining.com.au

Trapp, S., Gross, M. and Zimmerman, R. (1991) *Signs, Trails and Wayside Exhibits: Connecting People and Places.* University of Wisconsin, Stevens Point Press, Inc., Stevens Point, Wisconsin.

Washburne, R. and Wagar, J. (1972) Evaluating visitor response to exhibit content. *Curator* 15(3), 248–254.

Weiler, B. (1993a) The tour leader: that 'special' ingredient in special interest tourism. In: Veal, T., Jonson, P. and Cushman, G. (eds) *Proceedings of Leisure and Tourism: Social and Environmental Change.* World Congress of the World Leisure and Recreation Association, UTS Sydney, pp. 687–691.

Weiler, B. (1993b) Nature-based tour operators: are they environmentally friendly or are they faking it? *Tourism Recreation Research* 18(1), 55–60.

Weiler, B. and Crabtree, A. (1998) Assessing ecotour guide performance: Findings from the field. Presented at the 1998 Heritage Interpretation International Conference, Sydney.

Weiler, B. and Davis, D. (1993) An exploratory investigation into the roles of the nature-based tour leader. *International Journal of Tourism Management* 14(2), 91–98.

Weiler, B. and Richins, H. (1990) Escort or expert? Entertainer or enabler? The role of the resource person on educational tours. In: O'Rourke, B. (ed.) *The Global Classroom Proceedings.* Department of Continuing Education, University of Canterbury, New Zealand, pp. 84–94.

Weiler, B. and Richins, H. (1995) Extreme, extravagant and elite: a profile of ecotourists on Earthwatch expeditions. *Tourism Recreation Research* 20(1), 29–36.

Weiler, B., Davis, D. and Johnson, T. (1992) Roles of the tour leader in environmentally responsible tourism. In: Weiler, B. (ed.) *Ecotourism* 1991 Conference Papers. Bureau of Tourism Research, Canberra, pp. 228–233.

Weiler, B., Richins, H. and Markwell, K. (1993) Barriers to travel: a study of participants and non-participants on Earthwatch Australia expeditions. *Building a Research Base in Tourism. Proceedings of the National Conference on Tourism Research.* Bureau of Tourism Research, Canberra, pp. 151–160.

Wight, P. (1996a) North American ecotourism markets: motivations, preferences and destinations. *Journal of Travel Research* Summer, 3–10.

Wight, P. (1996b) North American ecotourists: market profile and trip characteristics. *Journal of Travel Research* Spring, 2–10.

Zehr, J., Gross, M. and Zimmerman, R. (1990) *Creating Environmental Publications: a Guide to Writing and Designing for Interpreters and Environmental Educators.* University of Wisconsin, Stevens Point Press, Inc., Stevens Point, Wisconsin.

Chapter 36

The Business of Ecotourism

B. McKercher

Department of Hotel and Tourism Management, Hong Kong Polytechnic University, Hung Hom, Kowloon, Hong Kong, China

Introduction: Ecotourism the Growth Industry?

For years, we have been reading that ecotourism is one of the fastest growing tourism segments. Annual growth rate estimates have varied from 10 to 30% (Wight, 1990; Lindberg, 1991), or about two to five times the background growth rate for tourism in general. The World Tourism Organization (WTO) has even gone as far as suggesting that ecotourism accounts for 20% of world travel (WTO, 1998). At the same time, a number of articles have been written in the academic and public media espousing the great potential of ecotourism. It is so appealing that a number of countries have developed formal policies to capitalize on the apparent demand (Dowling, 1998). Articles such as 'Ecotourism: the rising star of Australian tourism' (Dowling, 1996), 'Eco-profitable' (Oliver, 1994) and papers presenting regional overviews (Dowling, 1998; Edmonds and Leposky, 1998) imply that the commercial potential of ecotourism is almost unlimited.

At an operational level, though, the interest in ecotourism does not appear to have translated into the emergence of a vibrant, profitable business sector. With the exception of the accommodation sector and some ecotourism businesses with mass market appeal, most ecotourism enterprises are finding it difficult to identify and reach viable market segments. The failure to translate apparent demand into profitable business ventures is driven by two flawed assumptions made by government agencies, academics and many potential operators. The first is the misinterpretation of statistics that confuse visitation to natural areas with demand for commercial ecotourism products (Blamey, 1996a, b). Most people can consume an ecotourism experience without the need to purchase a product from a commercial operator. The real market demand for ecotourism activities is, therefore, only a small fraction of the apparent market interest (McKercher, 1998a). The second reason is based more on ideological hope than commercial reality. Ecotourism is recognized widely as an ecologically, morally and ethically preferred form of tourism that, if done correctly, optimizes social, cultural and ecological benefits, while providing the tourist with an uplifting experience (Wight, 1993; Dowling, 1998; Malloy and Fennell, 1998). Further, it is recognized that ecotourism can only be sustainable if it is profitable for ecotourism operators (McKercher, 1998a; Tisdell, 1998). Thus, it is felt that because this activity is so morally, ecologically and ethically sound, a mass market *must* exist for such activities.

What little empirical research that has been conducted examining the profitability of this sector (Tourism Canada, 1988; McKercher and Davidson, 1995; Tisdell, 1995; Cotterill, 1996; Wild, 1996) suggests that many businesses are under-performing. Indeed, the challenges of running a nature-based tourism business are similar to the generic types of challenges faced by all small businesses which, typically, have a 90% plus failure rate (McKercher and Robbins, 1998). These issues include: the failure to identify real market opportunities, lack of business planning, under-capitalization, little or no marketing skills and poor financial management skills. Much of this is explained by the fact that many people apparently enter this sector out of their love for certain non-commercial recreation activities (Bransgrove, 1992). Their motives are driven by a desire to play, rather than to develop a quality tourism product. Many new operators often find it difficult to adjust from the low pressure situation of leading friends on tour to the greater expectations and demands of fee-paying clients.

This chapter examines some of the business issues affecting the delivery of ecotourism products. It begins with a brief overview of the common business planning and business management issues affecting eco-businesses, as identified by Australian ecotourism operators. It is acknowledged that it is impossible to examine all the business issues in depth in an overview chapter such as this. Instead, the bulk of the chapter will focus on the three most important issues affecting the viability of the commercial ecotourism sector: the market, the product and the nexus between the two.

Overview of Business Planning and Marketing Issues in Ecotourism

It is surprising that many people entering the ecotourism sector have a remarkable lack of business skills and, of more concern, do not appear to have considered their lack of skills as a problem. Most will bring strong operational skills, based on their many years of participation in their respected activity. Many people choose to enter the field for lifestyle reasons, thinking that they can get paid to play. Few have ever run a business before and few seem concerned about business issues. But unless they learn about running a business, the lifestyle dream can turn into a nightmare; poverty is not an attractive lifestyle for an aspiring ecotourism operator. The following discussion summarizes the key findings of a study conducted of the business planning and marketing needs of the ecotourism sector as identified by successful Australian ecotourism businesses (McKercher and Robbins, 1998).

Few prepare business plans

Entering business with a clear business plan, is or should be one of the first tasks completed by prospective operators. Business plans serve two main purposes. The first is that they force prospective operators to work through the fine detail of their proposed venture prior to its establishment. By following a logical and rigorous process, operators can test their idea to see if a real business opportunity exists. The goal of the exercise prior to operation becomes one of demonstrating that a true market opportunity exists or aborting a non-viable idea prior to start-up. While a business plan will not guarantee success, it has been shown to reduce failure by 60% or more. The second key role played by business plans begins once the business starts. Business plans outline a desired growth path for the enterprise and, importantly, establish benchmarks by which to monitor the performance of the business. Operators can then adjust their business plans, or alter their operations if the business does not seem to match the projections.

Most new entrants do not prepare business plans. As a result, they fail to identify clear goals and objectives for their business or, alternatively, identify unrealistic goals. Many do not know if a viable market opportunity exists for their idea and do not have a realistic assessment of the costs involved in starting and incubating a business. Typically, they assume unrealistically high sales and profit figures and budget unrealistically low cost figures. The result is that many businesses are under-capitalized and because revenue is not what was hoped for, cannot thrive. Successful operators appreciate that it takes many years for a business to establish itself. To succeed, businesses must have both a plan to pursue and the resources required to survive during those fragile early years.

Financial management skills lacking

Financial management is a significant issue for most new operators. It is not surprising considering that many come from an operational background and have never had to manage an organization's finances before. Slower than expected growth, coupled with higher than anticipated costs cause many operators to face financial crises in the formative years of the business. A lack of resources will affect all areas of a business, from marketing and advertising to equipment purchases and the ability to hire staff. A serious miscalculation of the real costs of running a business can effectively strangle growth for many years. In addition, many operators do not understand the concept of cashflow. Businesses run on cash, just as a car runs on petrol. No matter how expensive a car is, if it does not have any petrol in the tank, it won't work. The same applies to a business. Once a business runs out of cash, it ceases to function. Few new operators seem to appreciate the importance of cash, believing that the annual budgeting exercise is all the financial planning that is required.

Lack of marketing knowledge

In a similar manner, there is a general lack of marketing knowledge among many new entrants. A request for marketing help often belies a deeper problem in the business, namely the failure to identify a commercially viable business opportunity right from the start. Regardless, many ecotourism operators do not understand what marketing is or how marketing works. Given that most have limited resources, one would think that they would adopt a targeted marketing approach, allocating those scarce resources in such a way that they optimize the chance to reach their desired markets. Instead, many adopt what is known in the trade as a 'shotgun' marketing approach, squandering their resources and hoping that by chance they will find a desired client. This occurs usually as a function of not knowing who the target market is, not understanding its habits and buying power and not having a clear idea of how to reach that market.

Many operators are also quite ignorant of how the tourism distribution system works and also appear wary of having to pay between a 10 and 35% commission to someone else to sell their product. A complex system of retail travel agents, wholesale tour operators and inbound tour operators has evolved over the past 150 years to act as the conduit between the provider of the product and the marketplace. This system works on a commission basis. If anyone is interested in gaining international distribution for their product, or of even gaining broad national distribution, they must work through the travel trade. Failure to cost products so that they are attractive to the travel trade, coupled with a failure to take advantage of these networks, constrains the ability of most businesses to achieve wide market penetration of their products.

Price is an interesting issue for this sector, for many operators do not understand what they are really selling and, therefore, are reluctant to charge what the product is worth. At one level, price is an accounting concept: you must charge enough to cover your costs and to provide you with an acceptable profit. At another level, though, price is an important marketing concept, for the price charged reflects the value of the product. Many operators adopt a cost-plus approach to pricing by adding up the component costs of the trip (food, guide, vehicle, accommodation, etc.) and then adding a profit component. These operators invariably under-price their product. In reality, the true worth of a product and, therefore, the price that can be charged comes not from the component costs, but from the value-adding made by the operator, often brought about by the ease of access provided by the product and the quality of the personalized service offered. The more unique the product being offered, the higher the price that can be charged. Ironically, charging a low price, especially if all around you are charging a higher price for the same product, may be counter-productive. Consumers may compare your product with the competition and wonder what it is you are leaving out, not what un-needed extras they are adding. After all, a US$14,000 car cannot compare with a US$45,000 car, even if both will take you from point A to point B.

Ecotourism is all about selling experiences. Good operators have come to realize that good ecotourism practices begin with good customer service. Staff must be both professional and approachable. Further, staff must engender a sense of confidence, especially if the experience involves, or is perceived to involve some risk. Effective communication skills are a vital ingredient of good customer satisfaction.

Do you have the right personal skills?

Finally, not everyone is cut out to be in the ecotourism game. Operators must possess a number of personal attributes. Prospective operators must have patience, determination, drive, enthusiasm, a love of hard work, be willing to work long hours, genuinely enjoy meeting and being with new people, have a friendly disposition and be able to act in a clear, concise and professional manner. Moreover, stress and the risk of personal burn-out in this sector are high, especially during those critical early years of the business. New operators are faced with long hours and often low rewards. In most cases, the business will become all consuming, becoming an extension of the operator and not just his or her job. To survive, operators must, therefore, ensure that they have a healthy balance between their work and non-work lives.

Are viable business opportunities identified?

The lessons identified above are valuable for anyone planning on entering the sector commercially as a tour operator, accommodation provider or attraction owner. They highlight that most ecotourism businesses fail to identify a viable business opportunity. Should they be lucky enough to do so, many operators do not know how to capitalize on that opportunity, either because of a failure to plan or for a lack of marketing skills. In this author's opinion, the single greatest problem facing this sector is the failure of prospective operators to identify a commercially viable market and to deliver products that satisfy the needs, wants and desires of that market. Instead, they try to push inappropriate products on to an undifferentiated market that is really not very interested in what they are trying to sell. The key to any successful ecotourism business is predicated on developing a deep knowledge of who the market is and why it wants to purchase an operator's product (see Chapter 3 by Wight). It is only from this knowledge that an operator can develop products that meet the needs of the market and can plan to position them attractively for the right markets. The rest of this chapter examines the product market nexus.

The Illusory Commercial Ecotourism Market

How large is the market for commercial ecotourism products? No one knows for sure, but it is fair to say it is only a small fraction of the total number of people who have been labelled as 'ecotourists'. Indeed, phantom demand seems to be driving much of the unrealistic expectations of the sector. On the surface, ecotourism seems to be a booming activity, which would imply the existence of substantial business opportunities. The Australian Bureau of Tourism Research, for example, suggests that up to 50% of the more than 4 million international tourists to that country engaged in what could loosely be described as an ecotourism experience during their visit (Blamey, 1996b). Similarly, the aforementioned comment by WTO suggests that ecotourists now account for 20% of world travel.

But these figures are misleading, for most of these 'ecotourists' are labelled thus based on a visit they made to a protected area sometime during their trip. But, as Acott *et al.* (1998) argue, visiting an ecotourism site does not make one an ecotourist. Whether or not you agree with their assertion, it is true that visiting an ecotourism site certainly does not make one a prospective commercial ecotourism client. A deeper examination of Australian tourism statistics reveals that only 5% of international visitors indicated that experiencing outdoor or nature-based activities particularly influenced their decision to travel to Australia (Blamey, 1996a). It is likely that an even smaller percentage purchased ecotourism goods and services. In reality, the commercial ecotourism market is a true niche market; small, specialized and discrete. Successful Australian nature-based tourism operators contacted in relation to another study used the terms 'tiny', 'micro' or 'minute' to describe the market (McKercher and Robbins, 1998).

Why few people are commercial ecotourists

One factor leading to over-estimation of the size of the market is the unclear context in which the term 'ecotourism' is used. To a large extent, the legitimacy of ecotourism rests primarily in its ability to achieve sustainable land uses rather than as a product category (Lawrence *et al.*, 1997). Using this supply-side definition, any visitor to land managed along sustainable ecotourism principles can be labelled an ecotourist, regardless of the activity the person pursued. In addition, a lack of clarity over the use of the term in a tourism context has further confused the issue. There is no doubt that the term 'ecotourism' has been misused by tourism marketers and tourism promotion agencies alike, to the extent that almost any form of non-urban tourism can be labelled as ecotourism (Wight, 1993; Wheeler, 1995; Lindberg and McKercher, 1997). But even when used legitimately, ecotourism can describe simultaneously destinations, experiences and products (parks, attractions, accommodation and tours). Many ecotourism destinations and experiences can be consumed at little or no cost, especially when one remembers that most land use management agencies have a legislative obligation to provide recreational opportunities at minimal cost to the public. Further ecotourism destinations may not provide business opportunities for commercial ecotourism operators, especially if the ecotourism attraction is a protected area. Many national parks, for example, preclude the provision of commercial activities inside park boundaries.

The result is a disparity between the number of people who visit ecotourism areas and the actual number of 'real' ecotourists. At an operational level, the commercial realities of running ecotourism businesses further limit the number of people who are willing or able to pay for services. Why would anyone purchase a commercial product if they could participate in exactly the same experience at little or no cost? The answer is few would, unless the ecotourism product offered something of added value that the consumer could not otherwise get. Many recreational ecotourists are philosophically opposed to what they see as the commercial exploitation of protected areas and eschew commercial operators. Many former operators have made the fatal mistake of looking at participation rates in conservation/recreational clubs and associations (birdwatching, bushwalking, etc.) and assumed that a ready market was available to them. What they failed to appreciate is that one of the reasons why people join these organizations is to gain inexpensive access to high quality experiences. Many of these people are exactly the *wrong* market for the commercial ecotourism sector for they have substitute products that provide an experience of equal or greater quality at a fraction of the cost.

The most successful ecotourism businesses fill the need-satisfaction gap that the traveller would otherwise not have. For example, ecotourism operators that visit the Great Barrier Reef in far north Queensland, Australia, provide their clients with access that most would not be able to provide themselves. In a similar manner, fly-in wilderness fishing lodges in the Canadian Shield area of North America provide access that would otherwise not be available to all but the hardiest wilderness traveller. Many marginally viable businesses feel their unique selling point is the quality of the information they impart to the consumer, when the consumer clearly feels this is not essential to their being able to enjoy the experience.

Finally, it is expensive to operate an ecotourism business, and these expenses must be passed on to the clients in the form of relatively high prices. This financial reality further limits the market by pricing ecotourism as an exclusive activity. Ecotourism businesses rarely enjoy the economies of scale of mass-tourism ventures, and as such, have relatively high fixed costs per client. In addition, these ventures often must overcome small group sizes, seasonality issues, higher per capita marketing costs, in-built volatility in the marketplace and a whole host of other factors that drive up their costs and consequently their prices. Space precludes a detailed discussion of this issue. Readers requiring more information on costing and pricing issues of commercial operators are advised to refer to *The Business of Nature-Based Tourism* written by McKercher (1998a).

Profiling the Commercial Ecotourism Market

As discussed in Chapter 3, commercial ecotourists are affluent, well travelled, well educated, independent, have a high disposable income, eschew normal packaged tours aimed at the mass market and seek an alternative vacation experience (Eagles, 1992; Sorensen, 1993; Weiler *et al.*, 1993; Pearce and Wilson, 1996; Wight, 1996; Meric and Hunt, 1998). Their age can vary, with two dominant age groups evident: older experienced travellers who are prepared to purchase up-market products and younger travellers who are seeking innovative experiences (Pearce and Wilson, 1996). Most are women and many travel on their own (McKercher and Davidson, 1995; Meric and Hunt, 1998). They are very active and belong to environmental organizations more than the general population as a whole. Their involvement in environmental and public interest group organizations means that not only do they have a keen interest in environmental issues, they also tend to have a greater knowledge of environmental issues. They choose ecotourism experiences because they want to be active, meet people with similar interests, learn new skills and optimize their time use.

Women are the dominant client group

That fact that most ecotourists are women is something that many in the industry seem to recognize, but at the same time seem to overlook in their product design and promotional activities. It has been estimated that women constitute up to 75% of the nature-based and cultural tourism market (Bond, 1997; Meric and Hunt, 1998). This market is affluent: women adventure travellers in the USA, for example, represent a US$55 billion retail sales market (Bond, 1997). Most of these women do not want to travel in women-only travel parties. While, in private conversation, operators acknowledge that women constitute their dominant market, few operators appear to target their promotional activities, or tailor their trips specifically to suit the demands of women (McKercher and Davidson, 1995). Women and men respond differently to promotional activities and read different media. Moreover, women participate in nature-based tourism trips for a variety of reasons that operators do not seem to appreciate fully. Research into the nature-based tourism industry in the Australian state of Victoria showed that women purchased commercial nature-based tourism products for the following reasons: ease and convenience of participation, safety and security provided by a guided trip; and the opportunity to travel with others, especially if the woman was travelling by herself. The industry, on the other hand, thought that women purchased their products because of the trip attributes, such as route selection, equipment provision and the area being visited (McKercher and Davidson, 1995). Dissonance between this market's real needs and the operator's perception of these needs may explain low participation rates in some businesses.

Different types of ecotourists with different needs

It is a mistake to assume that the nature-based tourism market is a single, unified market. Instead, different types of people are attracted to different activities (ATC, nd; Pearce and Wilson, 1996; Weaver *et al.*, 1996; Wight, 1996a, b; Meric and Hunt, 1998). This overview chapter does not permit a detailed examination of each submarket; however, it is important to appreciate that a continuum of ecotourists exists that can be defined based on the traveller's commitment to ecotourism and the centrality of an ecotourism experience in their vacation choice (Wight, 1996a; Acott *et al.*, 1998; Meric and Hunt, 1998). These different market segments have been labelled alternatively as 'specialist' and 'generalist' ecotourists, or have been placed on a continuum from 'hard core' to 'casual' ecotourists, or deep to shallow (Acott *et al.*, 1998). Dowling and Charters (1999), for example, have developed a product market segment matrix for Queensland ecotourists (see Chapter 2 by Orams).

Specialist or hard core ecotourists are committed ecotourists who seek the experience because it provides them with an intense, nature-based tourism activity. They are likely to participate in longer trips, be knowledgeable about the experience offered and be experienced travellers. The nature-based tourism experience is the main purpose of the trip for these people. Specialist nature-based tourists are prepared to make sacrifices to participate in their tourism experience. They are prepared to travel long distances to participate in the activity and do not want nor do they expect luxurious accommodation, food and night life (Lang *et al.*, 1996). This market is quite small and is seeking exclusive activities.

'Generalist' or 'casual' ecotourists, on the other hand, participate in ecotourism as one of many activities they will do while on holiday. Their decision to participate is convenience based. They are likely to take short duration trips, have limited knowledge about the area being visited, book at the last minute and may only have a casual interest in the activity. These people are less likely to travel large distances explicitly for the nature-based tourism experience, but may be prodigious consumers if an experience is easy to consume in close proximity to their current destination. Depending on the activity, this market is large, and in fact, more closely approximates the mass tourism market, especially for easy to consume discretionary ecotourism experiences provided in mainstream destinations.

Market moving to soft ecotourism

As the popularity of ecotourism grows, the market is shifting more toward the casual or soft end. This shift presents a number of opportunities and challenges for ecotourism providers, as soft ecotourists have substantially different needs, wants and desires from those of hard ecotourists.

Anyone entering the 'hard' ecotourism market must appreciate that the absolute market for their activities is small. To succeed, operators must provide truly unique experiences that offer exceptional quality experiences. These businesses will most likely be small scale, with high prices being charged. Location is a critical factor in their success, for the area must be recognized as an important ecotourism destination. It will be difficult to run a successful ecotourism venture if the market considers the location as not having the necessary qualities it wants in an eco-product. One of the factors mitigating against the success of eco-resorts in Finland, for example, is that it is not perceived to have high quality environmental assets needed to sustain such businesses (Bjork, 1997).

Conversely, recreation, fun, activity and ease of access seem to drive demand for soft ecotourism experiences. In these cases, the thrill of the experience plays a stronger role in attracting visitors than the skill or desired learning outcomes. Products that provide an enjoyable, recreational and, dare I say, an escapist and commodified, outdoor-oriented tourism experience are desired by this market. While this assertion may be an anathema to committed, hard-core ecotourists, it none the less represents the reality of the desires of a market that consumes an eco-experience as one of many experiences they might consume during a holiday. Further, to be able to compete, such experiences must be seen as being attractive alternatives to other recreational experiences. Emanuel de Kadt (1979) said some 20 years ago 'the normal tourist is not to be confused with an anthropologist or other researcher. Tourists are pleasure seekers, temporarily unemployed and above all, consumers; they are taking their trip to get away from everyday cares'. To a large extent, this maxim still holds true today for the vast majority of tourists who may be interested in an eco-experience.

Ecotourism 'Products' Provide Personal Needs Satisfaction

What is it that ecotourists buy when they purchase a commercial product? What does anyone buy when he or she buys a holiday? Do you buy airfare, accommodation and transfers or are you really buying something else? You may pay for airfare, accommodation and transfers, but in reality you are buying something that will provide you with an intrinsic or personal benefit. Far too many operators look at their product from the perspective of what they are selling to the tourist and not from the perspective of what the tourist is buying. As such, they tend to focus on the tangible elements that make up the product. In fact, these features are simply elements that facilitate consumption of an experience.

What the ecotourist, and indeed any consumer purchases when he or she buys a product is personal needs satisfaction. Products are defined as 'anything that can be offered to a market for attention, acquisition, use or consumption that might satisfy a need or a want.' (Kotler and Turner, 1989, p. 435). The key words are 'satisfying a need or want'. It is difficult to do, but operators must always think of their product from the perspective of the person who is buying the good or service. When consumers buy nature-based tourism products, they are buying a range of personal benefits (Hawkins, 1994). Increasingly, tourists are buying products that satisfy their lifestyle preferences, with the quest for adventure or the desire to see nature becoming increasingly important aspects of many people's lifestyles (Donoho, 1996).

The key questions to consider when examining any ecotourism product are 'what needs are being satisfied by purchasing this product?' and 'why would anyone prefer my product over someone else's?' These questions are so basic that many people fail to ask them, let alone answer them effectively. Is the decision to establish an ecotourism business in a certain location based on the ability to satisfy customer needs effectively, or is made simply because that is where the operator lives?

Are tour schedules, itineraries and components of a trip selected based on customer needs or on the idea of providing a 'good package'? More importantly, do you know what it is that the consumer is really buying?

Conceptualization of a 'product'

Products exist on three levels: core, tangible and augmented (Kotler and Turner, 1989). The core product represents the most fundamental level of any product and answers the question of what needs, wants or desires are being satisfied by buying the product. Someone purchasing a hotel room is buying the need satisfaction of sleep. The bed, colour TV and decor are the tangible components that facilitate this need satisfaction. Someone buying a computer may be purchasing the need satisfaction of being able to communicate with the world via the Internet and email. The components that go into the computer, the RAM and ROM and the like, only serve to facilitate this need satisfaction. The same is true with any ecotourism product.

It is only once the core benefit has been identified that the rest of the product can be tailored to deliver the need satisfaction. The tangible aspect of a product represents the elements that are assembled to facilitate need satisfaction. In the ecotourism industry, these are usually features like accommodation, transport, food, guides and the like. There is a challenge in developing the tangible component of the product when the good being sold is experiential or ethereal in nature, for the consumer cannot inspect the product and assess its value (Bharadwaj *et al.*, 1993). Augmented product elements are those additional features and services that make the product complete. Augmented products for the tourism industry might include easy reservation services, credit or cancellation policies, the provision of travel insurance, the inclusion of wildlife manuals while on tour or the sale of film.

First and foremost, the operator must define which core personal benefits the product will provide. Is it learning, escape, understanding, fun, status, a personal challenge or something else? Once the benefit is decided, then and only then, can the operator consider how to transform them into something tangible that the consumer can purchase. Most operators focus on the tangible products without really understanding what it is that they are selling, or more importantly, what it is that their clients are buying. As a former operator, this is understandable, for good tour operators spend weeks and months developing their 'product' until they feel it is right. Tour operators must research routes, accommodation houses, check added benefits, food, prices, reservation procedures and the like. This is a time consuming and all encompassing task. But, unless they determine what benefit they are providing to the client, it may also be a futile task.

Belatedly, a number of operators have come to realize that success in the nature-based tourism business comes not from a sales approach of pushing products on to clients, but from a marketing approach of delivering products that satisfy the clients' needs, wants and desires. As a result, many operators have had to modify their products substantially over the years. One operator related to me, for example, that his clients were upper-middle-class residents, and as such they were not happy being provided with a barbecue every night. They would rough it but really appreciated some attempt to lift catering standards. A tag-along four-wheel drive operator (four-wheel drive tours where the clients bring their own vehicles and 'tag-along' with the guide) expected his clients to be satisfied with simple guiding and support services, but soon realized that they also wanted meals and equipment (McKercher and Robbins, 1998).

Many operators do not fully appreciate the core need that their product is satisfying. Indeed, it may have less to do with the eco-experience and may have more to do with the ease with which the experience can be consumed. The appeal of purchasing a commercial package lies not in the provision of access to an area that everyone can access, nor is it necessarily the quality of the interpretation, unless the product is targeted at hard ecotourists. Instead, the appeal is that the consumer is offered a one-stop shopping experience. With one phone call, an extremely busy professional woman can overcome all the problems associated with negotiating departure dates, researching the trip route, booking all accommodation, food and services, obtaining the equipment needed, overcoming safety and security concerns and, in many cases, finding suitable travel partners. The opportunity/cost savings provided by a relatively low-involvement purchasing decision enables clients to overcome real or perceived time barriers that would have otherwise blocked their participation.

This is not to suggest for an instant that the tangible product is not important in the successful delivery of tourism products. In fact the opposite is true. If the tangible product is substandard or deficient in any way, clients will not be satisfied because their core needs are not being met. The tangible product becomes a reflection of the core product and, as such, should be developed to satisfy fully core needs. Further, tangible products have a quality- or value-adding level that makes the product as a whole more valuable than the sum of the individual component parts. They have distinctive features that distinguish them from other similar products. They have some styling that usually reflects the personality of the owner/operator. Further, as a separate entity, they can be branded. Finally, tangible products have some level of packaging (such as a brochure) that can be used to make the product attractive.

The Product–Market Nexus: the Most Important Business Consideration

The most important business factor influencing the success or failure of ecotourism businesses is the ability of the operator to match the needs of the consumer with the product being offered (McKercher, 1998b). Alternatively, some astute operators can succeed by matching the benefits produced by the product with the desires of target markets. The failure to do either is a recipe for failure. The logic is simple. What would you rather do, provide a service people want or try to push something people do not want on to an unsupportive marketplace?

Clearly, the former is preferred, but in far too many instances, the latter is what is delivered. It is true that all great businesses begin with great ideas. It is equally true that not all great ideas lead to viable business prospects. The difference between a good idea and a business opportunity lies in the ability to determine if a market is interested in a product, if sufficient numbers of people are willing to pay for it, whether or not the product can be delivered in a cost-effective manner and whether or not alternative products satisfying the same needs of the same markets already exist.

Four basic tenets

The product–market nexus is predicated on four basic tenets that should drive all businesses:

1. Any successful business delivers products that satisfy the needs, wants and desires of its clients.
2. This objective can only be achieved if a business knows who its clients are and understands their motives.
3. Successful businesses must be able to target their products at market segments that are large enough to be profitable, have growth potential, are interested in the products on offer, have unique characteristics that identify them as discrete segments, are easy to access in a cost-effective manner and are willing to pay the price being asked.
4. The products must be seen as being preferred products by the target market.

These four tenets need little elaboration. The first law is predicated on a basic understanding of why products exist. The second tenet recognizes the assumption of understanding the motives that drive the decision making process of the market. It is only through a deep understanding of what influences a market that anyone can deliver products to satisfy their needs. The third tenet identifies the features of attractive markets. They must be sufficiently large to support the business, yet at the same time share common characteristics that enable the operator to identify potential consumers, isolate them from the broader marketplace and reach them in a cost-effective manner. The fourth recognizes that competition is intense. Successful products must somehow offer something to their target audience that makes them preferred over the noise in the marketplace.

In essence, these rules argue that successful businesses in general, and successful small businesses in particular, must develop specialized products targeted tightly at defined markets. Most operators are faced with a number of challenges. They have limited resources that must be allocated effectively and efficiently. Further, they are forced to compete in a crowded marketplace where, in some cases, operators have literally dozens of potential competitors.

Understanding the premise behind the product–market nexus, also allows the operator to understand the basic concept of strategic marketing management (Aaker, 1995). Strategic marketing argues that no organization, not even the largest global companies, can be all things to all consumers. In order to survive, businesses must make a number of strategic decisions about what they want to be, and who they want to target. They must further differentiate their product offerings in a way that is meaningful to the consumer in order to gain a preferred place in the consumer's mind.

Think strategically

Strategic marketing is all about defining an organization's product–market nexus by defining:

- the products it chooses to offer;
- the products it chooses not to offer;
- the markets it chooses to target;
- the markets it chooses not to target;
- the competitors it chooses to compete with;
- the competitors it chooses to avoid.

The act of defining the products offered and the markets targeted dictates the types of business the operator will compete against. To succeed, one must know exactly what products it is offering at what markets and who its competitors are. It is only in this manner that one can develop business and marketing strategies to position the product as the preferred choice for its target markets. By the same token, it is equally important to know what the business does not do, which markets it does not target and which businesses it does not choose to compete with. This is especially true for small businesses with limited budgets that must allocate scant resources wisely. Too many tourism businesses try to do too many things. As a result, they lose their focus, and in doing so forget what their core product is and who their core markets are. Moreover, the features of the product usually have to be compromised in order to broaden the appeal of the product to as wide a market as possible. As a result, the quality and uniqueness of the product is diminished. A true recipe for failure is to assume that any business must compete against all other tourism businesses worldwide.

Conclusion

The product–market nexus is the most critical issue affecting the viability of the commercial ecotourism sector. Without clearly identifying markets and understanding them thoroughly, and without then developing products that will be regarded as the preferred choice for these markets, no ecotourism business can thrive. The product–market nexus forms the foundation from which a business plan evolves and is the central feature in the development of successful marketing tactics. Ecotourism is like all other small businesses; the likelihood of success is slim. However, the failure rate and under-performance of many existing businesses can be reduced dramatically if prospective operators see their venture for what it is. It is a business whose success will depend on how well it is planned, financed, managed and marketed as a business. It is not a lifestyle, nor is it a means of getting paid to play. The future development and, indeed, the future existence of commercial ecotourism relies on the development of appropriate business skills within this sector.

References

Aaker, D. (1995) *Strategic Marketing Management*, 4th edn. Wiley, Brisbane.

Acott, T.G., la Trobe, H.L. and Howard, S.H. (1998) An evaluation of deep ecotourism and shallow ecotourism. *Journal of Sustainable Tourism* (6)3, 238–254.

ATC (nd) *Asia, Europe, Japan Market Segmentation Studies, Executive Summary*. Australian Tourist Commission, Sydney.

Bharadwaj, S.G., Varadarajan, P.R. and Fahy, J. (1993) Sustainable competitive advantage in service industries: a conceptual model and research propositions. *Journal of Marketing* 57(4), 83–99.

Bjork, P. (1997) Marketing of Finnish eco-resorts. *Journal of Vacation Marketing* 3(4), 303–313.

Blamey, R. (1996a) Profiling the ecotourism market. In: Richins, H., Richardson, J. and Crabtree, A. (eds) *Taking the Next Steps*. Ecotourism Association of Australia, Brisbane, pp. 1–9.

Blamey, R. (1996b) *The Elusive Market Profile: Operationalising Ecotourism*. Bureau of Tourism Research, Canberra.

Bond, M. (1997) Women travellers: a new growth market. *What new in Pacific Asia?* PATA, Singapore: http://www.hotel-online.com/Neo/New..../PataWomenTravellers_Nov1997.htm

Bransgrove, C. (1992) *Tour Operators Business Guidebook*. Small Business Development Corporation, Victorian Tourism Commission, Melbourne.

Cotterill, D. (1996) Developing a sustainable ecotourism business. In: Richins, H., Richardson, J. and Crabtree, A. (eds) *Taking the Next Steps*. Ecotourism Association of Australia, Brisbane, pp. 135–140.

de Kadt, E. (1979) *Tourism: Passport to Development*. Oxford University Press, New York.

Doncho, R. (1996) Broadening their horizons. *Sales and Marketing Management* 148(3), 126–128.

Dowling, R. (1996) Ecotourism: the rising star of Australian tourism. Paper presented at Strategic Alliances: Ecotourism Partnerships in Practice. Ecotourism Association of Australia, Brisbane.

Dowling, R. (1998) Ecotourism in Southeast Asia: appropriate tourism or environmental appropriation? *Asia-Pacific Journal of Tourism Research*, 12 pp. http://www.hotel-online.com/Neo/Trends/AsiaPacificJournal/July98_Ecotourism SoutheastAsia.html

Dowling, R. and Charters, T. (1999) Ecotourism in Queensland. In: Molloy, J. and Davies, J. (eds) *Tourism and Hospitality: Delighting the Senses 1999* Part One *Proceedings of the Ninth Australian Tourism and Hospitality Research Conference*. CAUTHE Bureau of Tourism Research, Canberra, pp. 262–275.

Eagles, P. (1992) The motivations of Canadian ecotourists. In: Harper, G. and Weiler, B. (eds) *Ecotourism*. Bureau of Tourism Research, Canberra, pp. 12–20.

Edmonds, J. and Leposky, G. (1998) Ecotourism and sustainable tourism development in Southeast Asia. *Asia-Pacific Journal of Tourism Research* 6 pp. http://www.hotel-online.com/Neo/Trends/AsiaPacificJournal/July98_EcotourismSustainable.html

Hawkins, D.E. (1994) Ecotourism: opportunities for developing countries. In: Theobald, W.F. (ed.) *Global Tourism: the Next Decade*. Butterworth Heinemann, Melbourne, pp. 261–273.

Kobler, P. and Turner, R.E. (1989) *Marketing Management*, Canadian 6th edn. Prentice Hall, Scarborough.

Lang, C.T., O'Leary, J.T. and Morrison, A.M. (1996) Trip driven attributes of Australian outbound nature travellers. In: Prosser, G. (ed.) *Tourism and Hospitality Research: Australian and International Perspectives*. Bureau of Tourism Research, Canberra, pp. 361–377.

Lawrence, T.B., Wickins, D. and Philips, N. (1997) Managing legitimacy in ecotourism. *Tourism Management* 18(5), 307–316.

Lindberg, K. (1991) *Policies for Maximising Nature Tourism's Ecological and Economic Benefits.* World Resources Institute, Washington, DC.
Lindberg, K. and McKercher, B. (1997) Ecotourism: a critical overview. *Pacific Tourism Review* 1(1), 65–79.
Malloy, D.C. and Fennell, D.A. (1998) Ecotourism and ethics: moral development and organizational cultures. *Journal of Travel Research* 36(Spring), 47–56.
McKercher, B. (1998a) *The Business of Nature-based Tourism.* Hospitality Press, Melbourne.
McKercher, B. (1998b) Matching products with markets: essential to survival and a source of competitive advantage. In: *Proceedings of Leading Outdoor Organisations: 1998 ORCA Conference.* Outdoor Recreation Council of Australia, Sydney, Australia, pp. 179–192.
McKercher, B. and Davidson, P. (1995) Women and commercial adventure tourism: does the industry understand its dominant market? In: Faulkner, B., Fagence, M., Davidson, M. and Craig-Smith, S. (eds) *Tourism Research and Education in Australia.* Bureau of Tourism Research, Canberra, pp. 129–140.
McKercher, B. and Robbins, B. (1998) Business development issues affecting nature-based tourism operators in Australia. *Journal of Sustainable Tourism* 6(2) 173–188.
Meric, H.J. and Hunt, J. (1998) Ecotourists' motivational and demographic characteristics: a case of North Caroline travellers. *Journal of Travel Research* 36(Spring), 57–61.
Oliver, S. (1994) Eco-profitable. *Forbes* 153(13), 110.
Pearce, D. and Wilson, P.M. (1996) Wildlife viewing tourists in New Zealand. *Journal of Travel Research* Fall, 19–26.
Sorensen, L. (1993) The special interest travel market. *Cornell Hotel and Restaurant Administration Quarterly* 34(3), 24–28.
Tisdell, C.V. (1995) Investment in ecotourism: assessing its economics. *Tourism Economics* 1, 375–387.
Tisdell, C. (1998) Ecotourism: aspects of its sustainability and comparability with conservation, social and other objectives. *Australian Journal of Hospitality Management* 5(2), 11–22.
Tourism Canada (1988) *Adventure Travel in Western Canada.* MacLaren Plansearch, Tourism Canada, Ottawa.
Weaver, D., Glenn, C. and Rounds, R. (1996) Private ecotourism operations in Manitoba, Canada. *Journal of Sustainable Tourism* 4(3), 135–146.
Weiler, B., Richins, H. and Markwell, K. (1993) Barriers to travel: a study of participants and non-participants in Earthwatch Australia programs. In: Hooper, P. (ed.) *Building a Research Base in Tourism, Proceedings of the National Conference on Tourism Research.* Bureau of Tourism Research, Canberra, pp. 151–161.
Wheeler, M. (1995) Tourism marketing ethics: an introduction. *International Marketing Review* 12(4), 38–49.
Wight, P. (1993) Ecotourism: ethics or eco-sell. *Journal of Travel Research* 31(3), 3–9.
Wight, P. (1996a) North American ecotourists: market profile and trip characteristics. *Journal of Travel Research* Spring, 2–10.
Wight, P. (1996b) North American ecotourism markets: motivations, preferences and destinations. *Journal of Travel Research* Summer, 3–10.
Wild, C. (1996) The North American ecotourism industry. In: Richins, H., Richardson, J. and Crabtree, A. (eds) *Taking the Next Steps.* Ecotourism Association of Australia, Brisbane, pp. 93–96.
World Tourism Organization (WTO) (1998) Ecotourism now one-fifth of market. *World Tourism Organisation News* 1, 6.

Chapter 37

The Pursuit of Excellence: Benchmarking, Accreditation, Best Practice and Auditing

J.-P. Issaverdis
Victoria University, Footscray Park Campus, Melbourne, Australia

Introduction

Benchmarking, accreditation, best practice and auditing for the ecotourism sector need to be better understood. This chapter discusses each of these elements involved in the delivery of quality ecotourism experiences, explains the relationships between them and reveals how the use of quality processes can enhance service delivery, business viability and environmental sustainability. The chapter is written within the framework of the Australian experience of achieving measurable standards and identifying the common challenges facing operators, administrators, researchers, public land managers and educators. Australia is, arguably, the world leader in this area. The content however, is relevant to ecotourism destinations in other countries as well.

Ecotourism is defined broadly in Australia as:

> nature-based tourism that involves interpretation and education, and is managed to be ecologically sustainable [and] recognises that the 'natural environment' includes cultural components and that 'ecologically sustainable' involves an appropriate return to the local community and long-term conservation of the resource.
>
> (Commonwealth Department of Tourism, 1994)

The criterion of environmental sustainability, in particular, requires the establishment of effective benchmarking, accreditation, best practice and auditing procedures.

In Australia, most ecotourism businesses are small, owner-operator or family-run businesses. Such businesses have limited resources and available time. Those involved in ecotourism are primarily motivated by their enthusiasm for being in the natural environment and introducing others to such settings. Ecotourism activities are usually based on relatively undisturbed natural resources such as national parks, natural reserves and wilderness areas, with few operating on privately owned land. Within the sector there are many that feel the establishment of standards is important, but are challenged by the question of how to improve standards. In general, true ecotourism operators are disturbed by the increasing number of nature-based operators who are simply adopting the term ecotourism as a marketing opportunity, and are not committed to the delivery of a high quality ecotourism product to the consumer. For this reason benchmarks, accreditation, best practice and auditing programmes can help ensure the long-term commercial and environmental sustainability of the sector.

The major benefit for operators who subject themselves to the processes of benchmarking, best practice, auditing and accreditation is often not the outcome itself but rather the value of closely scrutinizing the operation. The process can assist in the identification of areas requiring improvement and the implementation of measures which will enhance the value of the experience for consumers, improve environmentally sustainable practices and increase yield for the business.

Figure 37.1 demonstrates the interrelationship between the four elements. Benchmarking, accreditation and best practice exist independently as well as being interlinked. Auditing is the necessary component to ensure the other concepts are valid and that reliable measures of improving performance are defined. The centre of the model represents the ideal situation where an ecotourism business has established operational benchmarks, has become accredited, is performing at the level of best practice, is conducting regular internal audits and has undergone an external audit to verify its level of performance.

Key Issues for Consideration

Before discussing benchmarking, accreditation, auditing and best practice in greater detail, the reader must be made aware of a number of key issues affecting the sector and the successful implementation of any programme related to these four criteria. These include:

- *The cost of delivering programmes.* The costs of delivering industry-led benchmarking, accreditation, best practice and auditing programmes must be recognized. Documentation, staff, assessment, database and marketing collateral must be established to market and deliver programmes effectively, in a way which offers value, integrity and validity for the sector.
- *The need to offer tangible inducements.* Many operators expect tangible benefits, which extend beyond the simple pursuit of standards for the greater good of the industry. Inducements (real or implied) might include preferred or discounted access to government marketing pro-

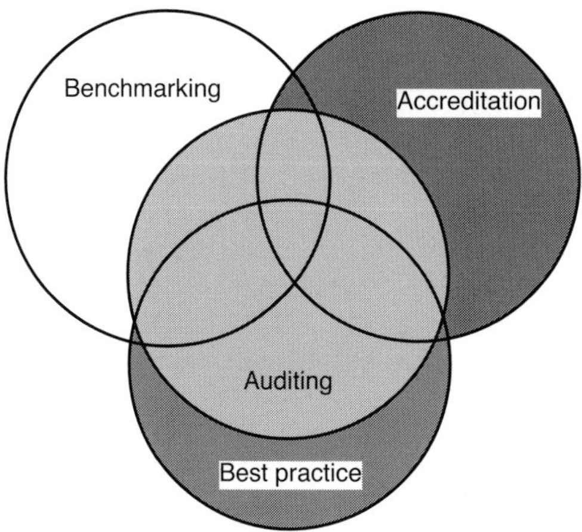

Fig. 37.1. Benchmarking, accreditation, best practice and auditing systems model.

grammes, special access to restricted sections in national parks, licence or permit extensions, branding and marketing benefits.

- *The need to provide impartial standards.* Auditing procedures must be established which are low cost, consistent, non-threatening and can be delivered in remote areas. Issues to be considered include the selection of auditors, training and skilling, and ensuring consistent application of assessment values.
- *The potential benefits of codes of conduct.* Codes can play a part in the overall continuum of developing industry professionalism and can be a useful means of introducing standards to sectors of the tourism industry. A successful example of an ecotourism code currently in practice is the *Guidance for Visitors to the Antarctic* code of conduct established by the International Association of Antarctic Tour Operators. The Code covers issues such as the protection of Antarctic wildlife, respect for protected areas, respect for scientific research, safety and the conservation of Antarctica's pristine environment. Because of the uniqueness of the destination, the close monitoring of visitor experiences and the limited numbers of visitors and tours to the Antarctic each year, the Code has been perceived positively as exerting a significant impact on operator performance.
- *The potential weaknesses of codes of conduct.* Moworth and Munt (1998) suggest that industry-based codes of conduct 'attempt to influence attitudes and modify behaviour' and are mainly voluntary. Problems associated with codes include difficulty of monitoring, and abuse for short-term marketing advantage. Some operators also object to the establishment of codes, for they see them as a covert attempt to regulate the industry.
- *The role of the consumer in improving performance.* Establishing benchmarks, becoming accredited, achieving best practice and conducting regular audits can only be approached within the context of delivering a better product to the consumer. One approach to performance improvement is simply asking customers what they want and implementing reasonable suggestions. Regular communication with the customer is essential and can be achieved through customer surveys, comment sheets, feedback forms and face-to-face discussions.

Benchmarking

There is a difference between benchmarking and best practice. Codling (1998) suggests that 'Benchmarking is often considered as a tool to enable systematic comparison of the performance of an organization against that of others'. In a more traditional manufacturing sense, Büyüközkan and Maire (1998) define a benchmark as 'a point of reference from which measurements and comparisons of any sort may be made'.

Confusion arises when the terms are mixed in the literature. Pearce *et al.* (1998) suggest that companies can use the benchmarking process to determine best practices and standards of performance by comparing characteristics and business practices of similar organizations. Voss *et al.* (1997) suggest that 'benchmarking can be defined as the search for industry best practices that lead to superior performance' and Povey (1998) states that 'benchmarking is the art of gathering information about external best practice'. It is important therefore that a clear distinction is made between benchmarking and best practice.

Thomas and Neill, cited in Pigram (1998), suggest that 'Benchmarking is a continuous learning process designed to compare products, services, and practices with reference to external competitors and then implement procedures to upgrade performance to match or surpass these'. This captures the essence of benchmarking, namely a continuous process of gathering information both internally and externally and the use of this information to improve the performance of the operation.

The benchmarking process

For tourism businesses to implement benchmarks effectively, the following process needs to occur:

- there must be a proper selection of the target features and practices to be emulated;
- there must be careful selection of the organizations chosen to compare performance; and
- there must be an effective monitoring and feedback system to ensure outcomes are improved (Pigram, 1998).

Tourism operators often fail to recognize the value of spending the money and making the time available to experience other tourism product as a means of benchmarking their own operation. Experiencing other tours, accommodation or attractions is often the simplest method of gaining information in view of the time and dollar constraints facing many small operators. Another simple and effective means of comparing performance is the establishment of networks with other operators offering the same, or similar, product. Discussing performance and possible improvements with one's peers does require a degree of openness and confidence. Some operators may find this threatening. Yet, the process may prove to be invaluable as an inexpensive means of measuring and improving performance over time.

Publications produced by government agencies or industry organizations are another effective source of measuring and improving performance. These publications are usually relatively inexpensive and can be a valuable resource for assessing the implementation of water and energy conservation measures, waste minimization, recycling, building materials and future directions. The Tourism Council of Australia (TCA), for example, has recently released two publications: *Being Green is your Business* (1999) and *Being Green Keeps You Out of the Red* (1998), which offer guidelines to tourism operators on improving environmental performance and the commercial benefit to be gained. Economic benchmarks suggested by TCA in their publications for the accommodation and attraction sector include:

Energy
- expenditure on energy/turnover;
- kWh per guest night (or per visitor);
- MJ per guest night.

Water
- kilolitres supplied per guest night (or per visitor);
- kilolitres supplied per meal cover;
- kilolitres supplied per ha of grounds;
- kilolitres supplied/turnover. (Tourism Council Australia, 1998.)

It should be recognized that many of these 'innovative practices' are becoming reasonably affordable and are being accepted as standard practice across the ecotourism sector which, given the criterion of sustainability, is expected to demonstrate such innovations. Numerous sources of information now exist, including publications and the Internet, which outline simple processes which can be implemented in the area of business management, interpretation, water and energy conservation, waste minimization, recycling, appropriate building materials and future directions. In addition to these resources, it is important that the sector continues to seek improved performance standards and establish 'new' benchmarks.

Accreditation

The concept of accreditation is gaining increased acceptance by tourism industry managers, as a means of enhancing standards (McKercher, 1998; Fennell, 1999). A number of factors are providing impetus for the process. These include the increasing expectation of standards and awareness of service quality by consumers, the increasing expectation by travel intermediaries that the product will be safe and environmentally responsible, an increased industry awareness of sustainable business practice and a growing interest in research.

The term 'accreditation' has wide currency across service and manufacturing industries. Historically the tourism industry is familiar with accommodation rating systems such as 'star' or 'crown' systems. I define tourism accreditation as 'programmes that provide a means of establishing the extent to which a business offering tourism experiences meets industry nominated standards. The programme encourages the delivery of consistently high quality products and promotes continuous improvement'.

Key considerations for accreditation programmes include:

- the establishment and continuous improvement of industry standards;
- providing a focus on industry-nominated standards, rather than government-nominated standards;
- establishing a continuum of measures to improve professionalism, standards and quality of product delivery;
- applying the term to individuals, industry organizations, tourism businesses and ecotourism products;
- recognizing that accreditation should not be compulsory, but rather encourage participation through marketing opportunities and incentives (Issaverdis, 1998).

Accreditation and environmental sustainability

The Australian Government report on ecologically sustainable development (1991) suggests that properly managed tourism can minimize negative environmental impacts and lead to long-term benefits for the environment. The report recommends that 'An important move to both assist the tourism industry and conserve Australia's biodiversity would be to establish a national representative system of protected areas together with nationally consistent management standards and practices' (Commonwealth of Australia, 1991). Where it is supported by the industry and properly administered, tourism accreditation offers the prospect of providing nationally consistent standards and environmental management practices. This may assist the long-term protection of environmental assets.

The inclusion of environmentally sustainable elements within the accreditation process can serve alongside the more established licensing and permit systems operated by many government environmental conservation agencies. The Australian National Ecotourism Strategy identifies that the diversity, and lack of standards in environmental management and use by the tourism industry has:

> prompted some ecotourism operators and natural resource managers to call for accreditation systems for accommodation, tour guides, and field operations that would identify their products in the marketplace, enhance the desirability of products and minimise the impact on the natural environment.

The strategy outlined accreditation systems and environmental Codes of Practice operating internationally and concluded that 'Accreditation systems involve formal acknowledgment of adherence to agreed standards' (Commonwealth Department of Tourism, 1994). An Australian accreditation programme is being implemented, as discussed in the case study.

Self-regulation

A key aim of self-regulation by industry is to avoid excessive government interference. For this strategy to be successful, the ecotourism sector must prove that it is capable of handling its own affairs to the satisfaction of relevant stakeholders, including consumers, entrepreneurs and government. Accreditation systems can provide a basis for industry self-regulation and, if marketed successfully, provide consumers with useful information to assist in the selection of ecotourism products. It appears that the more successful industry-led accreditation schemes are self-funding. Success will also depend on the degree to which industry standards meet the expectations of both consumers and natural-resource managers and provide tangible benefits to operators, such as a marketing advantage over competitors. If the long-term objective is to ensure that all operators are accredited, a challenge for programme managers is to demonstrate sustainable advantage.

Public land management: licensing, permits and tourism accreditation

It is widely acknowledged that because environmental tourism relies on natural and cultural resources in a fundamental way, pressures and problems are created for sustainable management. Effective environmental management can lead to integration between protection and resource use. It is also important that land management agencies recognize the commercial needs of tourism operators. Appropriately licensed and accredited nature-based or ecotourism operators may assist in bringing about both sustainable use and long-term protection.

Licensing and the issuing of permits lie within a broader legislative framework. They provide a means to control tourism operator numbers and their activities. In Australia, government appears to be taking a more active role in setting standards, a trend that is perceived as a threat by certain operators. Examples include public land permits, vehicle licences, driving licences and workplace health and safety permits. Proponents of accreditation programmes argue that accredited tourism operators are better placed to control the behaviour and impact of tour groups. This is regarded as preferable to leaving individuals and groups to their own devices and unsupervised. Public land managers often acknowledge that accredited tourism operators are an extension of the provision of government conservation and land management services. In this context they may assist in the process of environmental management and the protection of natural heritage areas.

An opportunity exists for government conservation and land management agencies to support accredited tourism businesses by providing them with tangible benefits. These may include extensions on the tenure of commercial tour permits or licences, cooperative marketing ventures between operators and land management agencies, and greater access by commercial tour operators to restricted areas of national parks. Such incentives may prompt tourism businesses to secure accreditation thereby continuing the enhancement of professional industry standards and environmental management practices. Australian government agencies have acknowledged the role tourism industry accreditation can play in the achievement of sustainable tourism practices (Issaverdis, 1998).

Accreditation case study: National Ecotourism Accreditation Program

The National Ecotourism Accreditation Program (NEAP) was launched in Australia in 1996. Ecotourism operators had expressed their desire to differentiate the nature-based tourism and ecotourism product. The programme focuses on the environmental management philosophies and practices of ecotourism businesses (Ecotourism Association of Australia, 1996).

Ecotourism accreditation aims to encourage businesses to benchmark their product against nominated industry standards, to provide a high standard of interpretation, to encourage the provision of high quality ecotourism experiences and to strive for best practice environmental management. Designed as a self-assessment programme, NEAP is co-managed by the Ecotourism Association of Australia (EAA) and the Australian Tourism Operators Network (ATON). Operators complete the self-assessment programme and submit the application for independent assessment and verification. The applications are then forwarded to an independent panel for approval and, if successful, the ecotourism product is accredited for a period of 3 years subject to an annual renewal (Issaverdis, 1998).

Ecotourism accreditation covers various product sectors including accommodation, attractions and tours. NEAP may be distinguished from other business accreditation programmes by its focus on individual products rather than accreditation of the whole organization. This approach particularly suits those organizations that provide a range of tourism experiences and are able to accredit specific ecotourism product. The concept is a close parallel of the programme accreditation undertaken in the outdoor recreation industry (Bassin *et al.*, 1992; Gass and Williamson, 1995).

The programme is designed as a standalone programme. It is, however, compatible with the completion of business management accreditation through other accreditation programmes such as the ATON Tourism Accreditation Program (1996). Completion of both programmes by a business would ensure coverage of both environmental considerations for ecotourism products and the business management issues necessary to operate tourism businesses effectively.

Two levels of accreditation are available, namely *Ecotourism Accreditation* and *Advanced Ecotourism Accreditation*. The accreditation logo differentiates the product's level of accreditation (see Fig. 37.2). The difference between the two levels reflects the commitment and operational practices of the business to sustainable tourism practices and environmental interpretation. It should be noted that a third level of accreditation focuses on nature-based tourism operators and is less rigorous than ecotourism accreditation. The intention of this level is to recognize nature-based operators that have a commitment to environmentally sustainable practices. This level forms part of the revised NEAP that was launched in 2000.

At the time of writing, a total of 184 products have been accredited, representing the product of 79 ecotourism businesses at Ecotourism Accreditation or Ecotourism Advanced Accreditation level (see Table 37.1). Accredited products may be further broken down into sectors, showing the number of accredited tours, attractions and accommodation (see Table 37.2). Cotterill (1996) has suggested that the ecotourism sector in Australia comprises approximately 600 operators, thereby providing a sizeable potential market for the programme. Despite the lack of tangible benefits currently available for accredited operators nationally, including limited government marketing initiatives apart from those offered by Tourism Queensland (Dowling and Charters, 1999), the NEAP take-up rate by ecotourism businesses has approximately doubled each year.

Fig. 37.2. NEAP logos.

Consumer and industry benefits

NEAP provides industry and consumers with an assurance that an accredited ecotourism product is backed by a commitment to developing best practice environmental management and the provision of quality ecotourism experiences. Accreditation offers consumers and industry a branding to identify ecotourism product (EAA, 1996).

Operator benefits

NEAP provides existing ecotourism businesses with criteria to measure sustainable practices and interpretation performance. It also assists new operators to develop genuine ecotourism product by providing operational guidelines and information to

Table 37.1. NEAP accredited level product breakdown as at June 1999 (n = 79 accredited businesses) (NEAP, Brisbane, 1999, personal communication).

Level of accreditation	Total number of accredited products	% of total accredited product
Advanced accredited products	123	66.8
Accredited products	61	33.2
Total	184	100

Table 37.2. NEAP accredited product sectors as at June 1999 (n = 184 accredited products) (NEAP, Brisbane, 1999, personal communication).

Activity type	Totals of activity type	% of total accredited product
Tours	153	83.2
Attractions	6	3.2
Accommodation	25	13.6
Total	184	100

make nature-based businesses more sustainable (EAA, 1996). The branding allows operators to seize promotional opportunities available to recognized ecotourism product. The programme offers operators the opportunity to improve performance on an ongoing basis and to establish best practice.

Ecotourism accreditation: what should it include?

Approval is needed from State and/or national industry and government organizations for accreditation programmes. Programmes must be easily accessible to operators and the process of accreditation for the operator must be relatively simple. An independent assessment must be in place to ensure the integrity of the accreditation process. Ideally, an on-site audit should be conducted, but this may be too costly depending on location (see Auditing section for further discussion). Upon approval, the operator should receive access to a nationally recognized brand that clearly distinguishes it as an accredited product and is easily recognized by consumers and by the industry.

Ecotourism accreditation programmes should require businesses to include information on:

- business operations and management;
- legal compliance;
- marketing plans and practices;
- customer service;
- human resource management;
- risk management and emergency procedures;
- environmentally sustainable practices;
- interpretation and education;
- cultural and local community involvement (Issaverdis, 1998).

Consumer research

It is important to acknowledge that little consumer research has been done in the area of tourism accreditation. A study by Nielsen *et al.* (1995) sampled nature-based tourist attitudes towards ecotourism and ecotourism accreditation. The results showed strong support for an ecotourism accreditation programme that accurately identified operators committed to the principles of ecotourism.

A further consumer-focused study, which encompassed accreditation for the adventure tourism sector, addressed the issue of risk management and general business practices, rather than environmental management (Bergin and Jago, 1999). The study showed that consumers are supportive of accreditation, but would not necessarily opt to purchase accredited product over non-accredited product. This may be attributed to a lack of awareness on the part of consumers. It recommended the conduct of further research to consider accreditation awareness and the actual and perceived need for accreditation by consumers.

The future of ecotourism accreditation in Australia

For ecotourism accreditation to have a long-term future, consumers must see a value in purchasing accredited product. Ecotourism operators must see the flow-on effect of improved yield and comprehensive industry support by both government marketing agencies and public land managers.

Accreditation is steadily developing support through national and state government marketing organizations such as the Australian Tourist Commission and Tourism Queensland (Dowling and Charters, 1999). Public land managers such as Parks Victoria and the Department of Conservation and Land Management Western Australia (Field and Shea, 1995) are viewing accreditation as an additional means to ensure effective management of tour operators on public land. Parks Victoria is offering extended tenure on public land permits to accredited operators. Operators are showing increased willingness to participate in the programme as awareness is raised of the potential benefits.

Best Practice

Achieving best practice is a goal for most industries, including tourism. Best practice may be distinguished from benchmarking in that it identifies those practices that are considered the most effective and efficient at the time. Best practice is a management approach to operations and customer service, which demands the highest standard of performance at all times.

Pigram (1998) suggests that best practice is the optimal approach to operations management 'relative to levels of performance in comparable firms and operations'. Australia's NEAP defines best practice as 'a condition, which is considered to be of the highest quality, excellence, or standing. A highly desirable and advantageous state which has been created and managed in a way for others to reflect on' (EAA, 1996).

Several principles are commonly associated with businesses that achieve best practice. These include a commitment to change and continuous improvement, retaining a highly skilled workforce, having a team-based management structure, adopting innovative technology, focusing on customer needs, ensuring superior communication processes, using performance measurement systems, and benchmarking. A key component for best practice ecotourism operators is the 'integration of environmental management into all operations of the business' (Pigram, 1998). Outstanding economic and environmentally sustainable practices and highly effective interpretation of the environment distinguish those best-practice organizations from the rest.

Implementing best practice in ecotourism

Many businesses focus on achieving best practice in a single area of the organization. This may be the financial or business management aspect, the operational components or the environmental management requirements. To be considered a 'best practice organization' businesses must adopt a philosophy of continuous improvement in all areas of the operation including business, operational, environmental, risk management, interpretation, marketing and service standards. It is acknowledged that this can often be a complex and time-consuming process for most ecotourism businesses.

Management must seek out new and cleaner technologies and apply these to ensure that resources such as water and energy are conserved and recycling practices implemented (including grey- and blackwater reuse and recovery systems). Ecotourism businesses, by definition, should be seen to be leading the tourism industry in adopting environmentally sustainable practices. Other key areas of the operation include environmental interpretation provided to consumers, ongoing education of consumers and local communities, and training and up-skilling of staff and active involvement with the local community.

What is the optimal method to achieve best practice? Operators may choose to invest in the services of consultants, refer to industry experts, read industry guides or documents, or simply network with other operators. In seeking improved environmental practices worldwide, there are a greater number of options for operators to choose from that will enable environmental, operational, business and service objectives to be met. An example is the design of accommodation facilities with effective, passive solar building design, which will reduce energy costs, improve environmental performance and ensure guest comfort (Commonwealth Department of Tourism, 1995).

Cost-benefit analysis

Ensuring that businesses conduct an effective cost-benefit analysis is a requirement for achieving best practice. It is important that a long-term approach is taken and that all costs are considered including the environmental costs and time costs. Examples of different cost-benefit approaches may include:

1. *Is it a good idea and can I afford it?* This is often an approach taken by small business. Decisions are usually based upon little formal analysis, but rather personal beliefs regarding the benefits and availability of funds. This approach may result in decisions that do not effectively reduce costs and environmental impacts.
2. *Purchase cost comparison.* A simple cost comparison approach in which purchasing price is reviewed. This approach is fairly naive as it focuses purely on cash flow and limited capital, again a reality for most small businesses. The issues of ongoing operating cost, maintenance cost, replacement costs, servicing requirements and possible environmental impacts are often not considered. This approach is extremely risky and is not recommended.
3. *Payback Period Comparison.* This approach considers the time required to repay the extra capital costs of the equipment or installation. This approach focuses purely on the financial cost aspects and tends to ignore the savings generated and broad, long-term benefits of investing in positive environmental practices.
4. *Life cycle cost comparison.* This approach considers all of the costs and benefits for each option. Various levels of analysis may be used, from sophisticated computer modelling to actual case study comparison. This approach is inclusive of all factors affecting the long-term success of the investment and is recommended (Commonwealth Department of Tourism, 1995).

NEAP innovative best practice

Australia's NEAP encourages operators, as part of their application for accreditation, to nominate innovative best practice in categories including impact assessment, interpretation and education, land use and location, conservation initiatives, natural area management and working with local communities. Examples of possible best practice are provided in the programme as a guide to encourage improvement (EAA, 1996).

Analysis of 'real-life' examples of best practice, as nominated by operators who have completed the NEAP programme between 1997 and 1999, suggests that many ecotourism operators are still relatively unsophisticated in their understanding of industry best practice. The nominations are reflective of what Australian ecotourism operators believe best practice to be. The operators' nomination of best practice is based upon their knowledge and experience gained through research, training, networking, visitation and experiencing other ecotourism product. Many of the nominations for best practice would be considered as minimum expectations for ecotourism businesses. This presents the question as to whether the level of information and type of education provided by the industry is appropriate, or whether the expectations of the world's best practice are unrealistic for most ecotourism businesses. It is suggested that the answer lies somewhere in-between.

What is needed to achieve ecotourism best practice?

Based on the personal experience of the author, the following elements are likely to be found in businesses that are striving for ecotourism best practice:

- flexible management structure encouraging all staff to share the goal of and take responsibility for achieving operational excellence;
- business and operational management systems or strategies, which encourage organizational excellence;
- staff experienced and skilled in environmental management techniques, good problem-solving abilities and willing to take responsibility for the day-to-day performance of the organization;
- consumers are encouraged to participate in improving environmental sustainability and contribute to the ongoing improvement of the ecotourism operation;
- positive presentation in the marketplace, confident of its marketing competitiveness;
- regular review of performance, open to outside scrutiny by making results available to industry experts and through external audits;
- a leader in its field actively contributing to industry issues through conferences, workshops, local tourism and environmental organizations.

The marketing benefits of achieving best practice

Businesses intentionally strive to achieve best practice because it brings opportunity and value. Recognition of achieving best practice among peers and industry will lead to increased profile and greater marketing opportunities. Government tourism organizations will seek to profile those businesses with travel journalists, travel agent and wholesaler familiarizations and overseas trade missions. Operators will gain greater exposure through industry conferences, educational institutions and other training and development programmes. Provided this exposure is used effectively in marketing opportunities to consumers, it can result in making consumer choices easier when selecting one ecotourism product over another.

In 1999, the Canadian Tourist Commission published a *Catalogue of Best Practices in Adventure Travel and Ecotourism* that has useful, practical information on business management, product and delivery, customer service and relations, training and human resources development, resource protection and sustainability, social and community aspects, packaging, marketing and promotion, and product development. The catalogue provides a useful means to identify best practices by industry, and to profile operators performing at this level. The intent is to identify and market excellence in Canadian nature-based and ecotourism product, and to distribute the information nationally and internationally to media, travel industry, government and industry tourism organizations (Wight, 1999).

Auditing

Auditing is the process a business undergoes to identify and confirm benchmarks, to provide accreditation with reliability and validity, and measure and verify best practice. Appropriate internal and/or external audits can provide ecotourism businesses with valuable information regarding performance levels and can offer tangible means to improve standards across the organization. The audit must be based upon a sound set of guidelines and clear measurement criteria must be established. The actual audit process is based on questions about the various operational and business management practices of an organization. A constructive report based upon the audit, which can be used to improve performance, should form part of the complete process.

Internal and external audits

Internal audits may be conducted by individuals within the organization, provided those individuals objectively measure performance levels against pre-determined criteria. Ideally the person should have undergone some form of auditor training to ensure reliability, objectivity and accuracy in the auditing process. External audits, such as Green Globe, are considered to be preferable as they are less prone to bias, tend to be more objective, and are often more constructive for the operator. External audits may be conducted by trained auditors, industry experts, other operators (peer review) or even consumer-based audits such as customer reviews, surveys and group feedback sessions.

External audits are critical in ensuring that development programmes such as industry-led accreditation programmes do in fact measure operational performance and that operators are delivering the service that they promise. The long-term credibility of such programmes is dependent on effective external auditing being part of the process. Ideally, the audit should be conducted as part of the initial application stage. This may not be feasible, however, given cost, time and geographic location. It is recommended that an external audit is conducted within 12 months of accreditation and at various ongoing stages in the product's lifetime, for example every 2–3 years.

Auditing considerations

Cost is a real factor in establishing viable external auditing programmes, from the perspective of both the operator and the managing organization such as NEAP. Ecotourism businesses often have to bear the cost of an external audit by paying for the service themselves. The expertise, time and transportation of the auditor and the additional cost of hosting the auditor during the audit process may be costly given the remote location of many ecotourism businesses. This fact alone prevents many small businesses from seeking external audits.

From the programme manager's perspective, the cost of building in auditing costs to programmes such as those discussed above, may make the programme too costly for operators to consider. In addition the cost of auditor training, ensuring consistent delivery of assessment standards, selection of appropriate auditors and ensuring confidentiality are challenging factors. Auditors must have empathy with the operators. They must be familiar with the various challenges facing ecotourism businesses, such as low yield, seasonality, the lack of business management training and the high cost of recyclable materials. Auditors must approach the audit with a positive attitude to ensure that the operator accepts the outcomes of the audit.

Training providers as auditors

The Australian accreditation process has considered linking auditing to traditional training providers such as training colleges or universities. Strong views exist both supporting and opposing this consideration. In support of the concept is the fact that training providers are often based regionally and therefore travel costs for the auditor can be minimized. The fact that professional trainers are skilled communicators, understand the process of developing skills in others and should be objective in their assessment lend credibility to the concept. The use of training providers would also fulfil the desirable outcome of removing the auditing responsibility from the programme's managing organization.

The opposing view is that training providers are often narrow in their base of expertise, often have little practical or operational experience and may see a commercial advantage in conducting negative audits to pressure operators into seeking formalized training. The matter is still unresolved. By utilizing carefully selected training providers who have operational experience and are already based in the local area, then developing their auditor skills through training, an effective auditing programme may be developed. It could be considered the ideal to establish an independent audit organization although, as previously discussed, the issue of cost becomes a major factor. Such organizations need to be commercially viable and therefore profit based.

Environmental auditing

Most ecotourism businesses are more familiar with the concept of environmental audits, as opposed to business audits. Environmental auditing is described by Pigram (1998) as 'a process whereby operations of an organization are monitored to determine whether they are in compliance with regulatory requirements and environmental policies and standards'. This implies the need for businesses to be familiar with relevant environmental legislation and the industry standards set in programmes such as NEAP or Green Globe (World Travel and Tourism Council, 1997). The audits can be conducted through internal or external review. The person conducting the audit, the purpose of the audit and for whom, as well as the broad operational and business environment also need to be considered (Moforth and Munt, 1998).

Environmental audits as management tools

Environmental audits are a management tool that can provide a regular and objective evaluation of the environmental performance of the organization. It is important that the process is systematic and addresses all components of the operation (Goodall, 1995). Environmental audits have similar objectives to environmental impact assessments and should be seen as complementary processes in achieving sustainable tourism practices among ecotourism businesses. The audit confirms compliance with environmental management planning regulations and ensures that the organization is maintaining environmental standards over time through relevant management and operational procedures.

The importance of environmental auditing programmes in establishing benchmarks and achieving best practice is well documented. Environmental auditing programmes:

> increase the overall level of environmental awareness of the industry, assist tourism management to improve environmental standards through 'benchmarking' against proven performance, identify opportunities to reinforce positive environmental interactions and accelerate the achievement of best practice environmental management in the industry.
>
> (Pigram, 1998)

Operators adopting environmental audits may achieve the following tangible, commercial benefits:

- more efficient use of resources and waste minimization will result in cost savings;
- environmental problems may be identified before they become liabilities;
- positive environmental practice benchmarks may be established;
- an organization's corporate image may be improved;
- various marketing advantages may be realized;
- investors, regulators, customers and the community will have increased confidence in the product;
- higher quality employees, who are better motivated, may be recruited (Goodall, 1995).

The value of audits is undisputed. They form an integral part of benchmarking, accreditation and best-practice ecotourism management. The challenge is to establish effective audit programmes which are not cost prohibitive, can be easily implemented, and deliver acceptable outcomes for operators, which can be continuously reviewed. Audits must be perceived by the sector as an essential tool to improve performance, not some form of 'test' or criticism.

Conclusion

It has been argued that one can only be as green as one can afford to be. Is it going to take an environmental or operational disaster to 'force' ecotourism businesses to undergo business improvement processes? The ecotourism sector would be well advised to demonstrate a greater commitment to continuous raising of standards, if it wishes to maintain its marketing advantage as a sector which achieves sustainability and offers high-quality products.

The pursuit of establishing benchmarks, supporting industry-led accreditation programmes, striving for best practice and implementing internal and external auditing processes must become an integral part of the ongoing business practices of all ecotourism operators. Wearing (1995) suggests the ecotourism sector needs to prepare effectively for a growth in professionalism, with increased demand for ecotourism product, an increasing number of ecotourism operators and more informed tourists. As discussed in the chapter, a commitment to the various processes intended to improve performance must come with a commitment to underpinning the integrity of these processes through auditing. Consideration should also be given to linking with other types of business accreditation and award programmes, education and staff development programmes, such as formalized training for managers, guides and other employees.

Threats to the success of industry-led improvement programmes exist in the form of industry complacency, operators perceiving the process to be too hard, consumers purchasing ecotourism product based on price alone, and lack of tangible benefits for operators from government and industry organizations. It is vital that the benefits of purchasing accredited and best practice product are communicated to consumers.

Recognizing the 'best' ecotourism practice

Perhaps it is time for the ecotourism sector to make a positive move towards recognizing outstanding environmental, business and operational performance. Awards have not been discussed in the chapter but are an area of growing interest. An award provides recognition of the 'best' operators in a category. There must be a tangible value in encouraging operators to go through the rigorous process of award submission and assessment such as a marketing advantage over competitors. The process must be inclusive, open to scrutiny and open to all comers. The award process should also be linked to national or international programmes to further enhance credibility and increase the value to operators. The value to operators as well as consumers, and ultimately to sustainable environmental practices is the fundamental purpose of all improvement programmes.

Adopting improved ecotourism business practices

The growing awareness and understanding of the many challenges facing the sector in defining and improving benchmarks, implementing accreditation programmes, establishing best practice and developing operational audit programmes which have value to both consumers and the sector is clearly highlighted. The ultimate goal is that ecotourism businesses must perceive the advantage in achieving audited benchmarks, accreditation and best practice, or at least be convinced of the disadvantage of not pursuing this approach.

The advantages must clearly relate to the economic benefits to the business. The goal of these programmes is not to have all ecotourism businesses adopting the same approach to business management, but rather to encourage all businesses to do better in all areas of their operation, particularly environmental practice (Pigram, 1998). Industry development programmes must allow for the diverse, dynamic nature of the ecotourism sector and must allow well-performing ecotourism operations opportunities for further innovation.

Acknowledgement

Gratitude is expressed to the National Ecotourism Accreditation Program for supplying the NEAP Accreditation product figures and operator Best Practice material, and Victoria University in Melbourne, Australia.

References

Australian Tourism Operators Network (1996) *ATON Tourism Accreditation Program – Tour Operators and Attractions.* Australian Tourism Operators Network, Melbourne.

Bassin, Z., Breault, M., Fleming, J., Foell, S., Neufield, J. and Priest, S. (1992) AEE organizational membership preference for program accreditation. *Journal of Experiential Education* 15(2), 21–27.

Bergin, S. and Jago, L.K. (1999) Accreditation of adventure tour operators: the consumer perspective. In: Molloy, J. and Davies, J. (eds) *Delighting the Senses, Proceedings of the Ninth Australian Tourism and Hospitality Research Conference*, CAUTHE, Adelaide, pp. 305–316.

Büyüközkan, G. and Maire, J.L. (1998) Benchmarking process formalization and a case study. *Benchmarking for Quality Management and Technology* 5(2), 101–125.

Codling, B. (1998) Benchmarking: a model for successful implementation of the conclusions of benchmarking studies. *Benchmarking for Quality Management and Technology* 5(3), 158–164.

Commonwealth Department of Tourism (1994) *National Ecotourism Strategy.* CdoT, Canberra.

Commonwealth Department of Tourism (1995) *Best Practice Ecotourism: a Guide to Energy and Waste Minimisation.* CdoT, Canberra.

Commonwealth of Australia (1991) *Ecologically Sustainable Development Working Groups – Final Report Tourism.* CdoT, Canberra, pp. 14–36.

Cotterill, D. (1996) Developing and sustainable ecotourism business. In: Richins, H., Richardson, J. and Crabtree, A. (eds) *Ecotourism and Nature-Based Tourism: Taking the Next Steps. Proceedings of the Ecotourism Association of Australia National Conference.* EAA, Brisbane, pp. 135–140.

Dowling, R. and Charters, T. (1999) Ecotourism in Queensland. In: Molloy, J. and Davies, J. (eds) *Delighting the Senses: Proceedings of the Ninth Australian Tourism and Hospitality Research Conference.* CAUTHE, Adelaide, pp. 262–275.

Ecotourism Association of Australia (EAA) (1996) *National Ecotourism Accreditation Program.* EAA/ATON, Brisbane.

Fennell, D.A. (1999) *Ecotourism: an Introduction.* Routledge, New York.

Field, G. and Shea, S. (1995) Tour accreditation – beyond licences and workshops. In: Richins, H., Richardson, J. and Crabtree, A. (eds) *Ecotourism and Nature-Based Tourism: Taking the Next Steps. Proceedings of the Ecotourism Association of Australia National Conference.* EAA, Brisbane, pp. 211–213.

Gass, M. and Williamson, J. (1995) Accreditation for Adventure Programs. *JOPERD* January, 22–27.

Goodall, B. (1995) Environmental auditing: a tool for assessing the environmental performance of tourism firms. *Geographical Journal* 161(1), 29–37.

Issaverdis, J. (1998) Tourism industry accreditation – a comparative critique of developments in Australia. MBus thesis, Victoria University, Melbourne, Australia.

McKercher, B. (1998) *The Business of Nature-based Tourism.* Hospitality Press, Melbourne.

Moforth, M. and Munt, I. (1998) *Tourism and Sustainability: New Tourism in the Third World.* Routledge, London.

Nielsen, N., Birtles, A. and Sofield, T. (1995) Ecotourism accreditation: shouldn't the tourists have a say? In: Richins, H., Richardson, J. and Crabtree, A. (eds) *Ecotourism and Nature-Based Tourism: Taking the Next Steps. Proceedings of the Ecotourism Association of Australia National Conference.* EAA, Brisbane, pp. 235–242.

Pearce, P.L., Morrison, A.M. and Rutledge, J.L. (1998) *Tourism: Bridges across Continents.* McGraw-Hill, Sydney.

Pigram, J.J. (1998) Best practice environmental management and the tourism industry. In: Cooper, C. and Wanhill, S. (eds) *Tourism Development: Environmental and Community Issues.* John Wiley & Sons, Chichester, UK, pp. 117–127.

Povey, B. (1998) The development of a best practice business process improvement methodology. *Benchmarking for Quality Management and Technology* 5(1), 27–44.

Tourism Council Australia (1998) *Being Green Keeps you Out of the Red.* TCA, Sydney.

Tourism Council Australia (1999) *Being Green is your Business.* TCA, Sydney.

Voss, C., Ahlstrom, P. and Blackmon, K. (1997) Benchmarking and operational performance: some empirical results. *International Journal of Operations and Production Management* 17(10), 1046–1059.

Wearing, S. (1995) Professionalism and accreditation of ecotourism. *Leisure and Recreation* 37(4), 31–36.

Wight, P. (1999) *Catalogue of Exemplary Practices in Adventure Travel and Ecotourism.* Canadian Tourism Commission, Ottawa.

World Travel and Tourism Council (1997) *Green Globe.* World Travel and Tourism Council, London.

Section 8

Methodologies, Research and Resources

D.B. Weaver

Department of Health, Fitness and Recreation Resources, George Mason University, Manassas, Virginia, USA

Ecotourism, as stated several times throughout this volume, is still in its infancy as a focus of both research and practice. The purpose of this final collection of chapters is to assess how far ecotourism has evolved during the past 15 years with respect to the means by which knowledge in this area is being generated and disseminated, and with regard to the knowledge that is still required. Good research is the key to the development of a reliable knowledge base upon which sound management decisions can be made, and therefore the findings of Backman and Morais in Chapter 38 give cause for concern. In an analysis of a major refereed journal that showcases much of the leading edge ecotourism research, they show that most articles use some kind of quantitative approach to the collection of data. However, the great majority of these articles do not progress beyond simple frequencies and distributions, or beyond 'exploratory' studies. Statistical techniques that demonstrate cause-and-effect relationships or facilitate categorization and analysis, such as cluster and factor analysis, or analysis of variance, are notably inconspicuous. Furthermore, few attempts were evident to test or propose general theories, or even to engage in comparative case study analysis. The articles that employed qualitative research are frequently based on perfunctory methods providing no confidence that the results can be extended beyond each particular case in point. In short, ecotourism (like tourism in general, it should be added) has yet to demonstrate the same rigour in the application of methodology that characterizes some of the more mature social sciences. Until it does so, the reliability of its underlying database will be a matter of concern.

In Chapter 39, Eagles pursues the theme of information needs and sources within the ecotourism sector, again pointing out the need for a higher level of knowledge and expertise across an array of areas as the sector becomes increasingly complex and competitive. Product managers, for example, must provide quality interpretation in order to ensure visitor satisfaction, and this means that the standard and knowledge of guides must be constantly improved. Yet, relatively little information is currently available, and that which is available may not be accessible to or known by the relevant practitioners. The same can be said for market segmentation research, much of which is controlled by private companies. The situation is improving in some areas. For example, Eagles cites a recently published book (McKercher, 1998) that details the practical business aspects of ecotourism and compiles a great deal of useful generic information that hitherto had been scattered among hundreds of separate sources. It is to be hoped that similar works will appear in such critical areas as interpretation, marketing, impact management, and so on. Much good information is already available in these areas, though in a highly fragmented form, and often in disciplines not directly related to ecotourism or tourism in general.

Eagles rightly states that good ecotourism management will require the kind of broad and rigorous training that can only be provided by the tertiary education sector. In Chapter 40, Lipscombe and Thwaites take stock of the situation with respect to ecotourism training and education, focusing on Australia because of the leadership demonstrated by that country in this field. They report that the ecotourism-trained graduate must possess a variety of skills, attitudes and knowledge in order to meet the needs of the sector, including evaluative and analytical skills that require familiarity with various research techniques. An increasing number of programmes have been established in Australian universities and community colleges to produce such graduate outcomes, and this is a very positive sign of the sector's growing maturity. From virtually no relevant programmes in 1994, there were about 75 on offer in Australia by 2000, only a small number of which, however, actually include the word ecotourism in the course name. Despite this progress, a major outstanding issue is the lack of interaction between the tertiary sector and the industry that will eventually absorb most of the ecotourism graduates. Industry tends to favour 'real life' experience, while graduates complain that industry does not value the analytical and theoretical skills that are obtained in a university environment. The promotion of a more formal dialogue between the education and private sector is, therefore, an imperative if each is to benefit from exposure to the other.

The final chapter in this section reflects on the research needs that are required in the ecotourism sector. Like several of the other contributors to this section, Fennell points out and laments the lack of rigour, sound empirical data, and theory-building within the field of ecotourism, though acknowledges that this is largely explained by its infancy. According to Fennell, the areas that should be recognized and pursued as priority research foci in ecotourism are related to ethics, values, attitudes and impacts. This is because of the special onus that ecotourism places on issues such as learning, interaction with sensitive environments, and the imperative of sustainability. Hence, even the more utilitarian aspects of ecotourism, such as the business components described by Eagles, are underpinned by these deeper considerations. Operators need to balance environmental sustainability with their own financial viability, while ecotourists need to balance the former with their desire for satisfying experiences. For example, ecotourism operators must be aware of 'social traps' wherein the environment is degraded because of a tendency to consider the individual rather than collective welfare. Fennell also discusses values and ethics as dimensions that underlie the types of experience that are sought by ecotourists and offered by practitioners, codes of ethics being one example of a related practical outcome. In sum, all of these often neglected dimensions must be engaged by all participants in the sector if ecotourism is to fulfil its mandate as a 'responsible' form of tourism.

Reference

McKercher, B. (1998) *The Business of Nature-based Tourism.* Hospitality Press, Elsternwick, Victoria, Australia.

Chapter 38

Methodological Approaches Used in the Literature

K.F. Backman[1] and D.B. Morais[2]

[1] *Department of Parks, Recreation and Tourism Management, Clemson University, Clemson, South Carolina, USA;* [2] *School of Hotel, Restaurant and Recreation Management, Pennsylvania State University, University Park, Pennsylvania, USA*

The reviewing of past research efforts facilitates an improvement and understanding of research and reveals the philosophical, conceptual, substantive and technical problems of research in a field as broadly defined as ecotourism (Wells and Picou, 1981; Reid and Andereck, 1989; Malhotra *et al.*, 1999). This process can be particularly beneficial to a developing field like ecotourism, where there is a limited knowledge base regarding research practices and techniques employed.

As research in areas such as tourism and marketing show (Dann *et al.*, 1988; Reid and Andereck, 1989; Malhotra *et al.*, 1999), a starting point to gain this knowledge is through assessing academic journals. These publications constitute an indicator of the direction a field has taken and provide an index of the level of research proficiency achieved by the field. Thus, in the examination of the research techniques used, it is possible to assess the methodological sophistication of current ecotourism research efforts and compile an inventory of the popularity of various techniques utilized.

The purpose of this chapter is to summarize the current state of research in ecotourism by reviewing the breadth and popularity of research techniques used in the field of ecotourism. This was accomplished by reviewing the primary thrust of articles published in the *Journal of Sustainable Tourism* (*JST*) between the years 1994 and1999, along with a review of some technical reports and a sample of articles from other journals such as the *Annals of Tourism Research* (*ATR*) and the *Journal of Travel Research* (*JTR*) during the same 6-year time period. The chapter also presents a brief review of a sample of the different academic fields as they applied their methods to the study of ecotourism. Additionally, a cross-classification of various research techniques using ecotourism research was done and observations on the application of these research techniques to address methodological issues in ecotourism are reported. Finally, the chapter provides a discussion of the limitations of this study and some direction ecotourism research could take as we move further into the 21st century.

Method

The main data analysed for this chapter were the research methodologies used in academic journals. The use of academic journals as a sample source for this study was deemed appropriate because one of the steps in the scientific process is the communication of research findings to a wider audience, and particularly to one's academic peers. Also, academic journals, though sometimes prolonging the delay in publication because of the peer-review and printing processes, are still generally considered the primary source for 'cutting edge' research in a field (Babbie, 1995; Witt, 1995). The major criticism against academic journals has been the limited accessibility and applicability of many of their ideas to practitioners in professional settings. However, for the purposes of this chapter, the important component to consider is the identification of the 'cutting edge' methods being used in ecotourism research, not their potential application by professionals. The reason that the three journals listed above were selected over other journals publishing tourism research was their major focus on tourism sustainability, which is a core criterion of ecotourism. Although a large proportion of the articles in *JST*, *ATR* and *JTR* therefore deal directly or indirectly with ecotourism, this is not the sole focus of these journals. The total number of articles examined over this 6-year period was 147. The research methods were differentiated first between the quantitative and qualitative methods used, then evaluated for the type of statistical or analytical method employed. The third evaluation used in this study was geographic, in terms of the location of the research studies as well as the institutions with which the authors are affiliated. Because this is only a descriptive assessment of what research methods have been used in the ecotourism literature, no in-depth analysis of the data was attempted.

A Sample of Disciplinary Fields Studied in Ecotourism

Economic studies

Economic benefits of tourism are often the driving force for implementing ecotourism. As a consequence, there has been an abundance of studies of the economic impact of ecotourism development. The large majority of those studies use an input–output economic model (see Chapter 23). A problem that becomes apparent upon analysis of the economic impacts of ecotourism is an evident increase in leakage due to control of the industry by foreign investors. Two main factors limit the economic benefits of the local community: (i) all-inclusive lodging is centred on multinational chain hotels blocking local ownership from direct profits; and (ii) high paying jobs are not offered to locals, which limits the impact through employment. For example, some studies of national parks in Kenya revealed that national parks' visitor revenues were used to maintain the park ranger system, and were not used to improve the quality of life of the local residents or preserve the natural environment (Dieke, 1991).

Ecological studies

Many studies in the literature have focused on the impact of ecotourism in the environment, and have used a more diverse array of methods than is evident in the economics literature. In Kenya, studies revealed that daily observation of the cheetah inhibited them from hunting and gave them undue difficulties to survive and mate. Other studies in Central America examined the management policies necessary to control the negative impact on coral reef and fish populations that has been registered in Belize and other popular marine scuba-diving destinations. For example, researchers have discovered that rays in a marine park in Belize have been suffering from skin burns (from tourist handling), and changes in behaviour and feeding habits (tourists feeding them) (Lindberg *et al.*, 1996). In Antarctica, observation methods borrowed from biological sciences were employed to examine the influence of the impact of ice-cruises on the wildlife and flora of the most popular landing areas of the South Shetland Islands (Aeero and Aguirre, 1994).

Social-psychological studies

A few studies in the ecotourism literature attempted to evaluate social-psychological aspects of ecotourist behaviour. Initial studies of motives were carried out in areas as diverse as Australia, Canada and the USA (Eagles, 1992; Wight, 1996; Blamey and Braithwaite, 1997). Another approach is to segment the ecotourism market, as was done in South Carolina, USA, using psychographic characteristics of visitors. For example in one study in the south-eastern region of the USA, ecotourists were found to be highly involved in their ecotourist behaviour and considered themselves to be opinion leaders in the south-eastern region of the USA (Jamrozy *et al.*, 1996).

Types of Methods Used in the Ecotourism Literature

The quality of qualitative and quantitative methods

The research design of a study generally begins with the selection of two things: first a topic and second a paradigm (Creswell, 1994). Paradigms are used in social science to help understand phenomena. They also advance assumptions about the world being studied, how the science should be conducted. As Creswell (1994) states, paradigms generally encompass both theories and methods, and the two most widely used in the literature are the qualitative and quantitative paradigms. In this chapter, a qualitative study is defined as being consistent with the assumptions of a qualitative paradigm. It includes an enquiry process of understanding an ecotourism problem by building a complex, holistic picture formed with words, reporting the detailed views of informants and conducting this research in a natural setting (Creswell, 1994).

As opposed to the qualitative process, the quantitative study is consistent with the quantitative paradigm, and is an enquiry into ecotourism problems based on testing a theory composed of variables, measured with numbers and then analysed with appropriate statistical procedures (Creswell, 1994). The results of these statistical procedures help to determine whether the predictive generalizations of the theory hold true in the ecotourism context (Creswell, 1994; Babbie, 1995).

The question then comes to mind, why select one paradigm over another? The reasons for selection relate to the researchers' views of the world, their training and experience in research, the researchers' psychological attributes and the attitudes and nature of the problem being researched. In the qualitative study method two approaches to the unit of observation can be used. The *emic* approach emphasizes the importance of collecting data in the form of verbatim texts from the informants in order to preserve the original meaning of the information (Pelto and Pelto, 1978). The second approach is the *etic* which studies human behaviour as the classification of body motions in the terms of the effect these emotions have on the environment (Pelto and Pelto, 1978). Thus, some *emic*-focused study methods are participant observation, key-informant interviewing, collection of life histories and structured interviews and surveys. Some *etic*-focused study methods are measurement of social interaction, proxemics and videotape research, content analysis of folktales and other literature, archives records and technical equipment in fieldwork.

The qualitative study approaches that have been used in tourism research in general and to some lesser extent ecotourism research are methods focused around areas such as managerial perspectives, national, regional and municipal perspectives, industry perspectives and impact assessment. Particular data collection methods used in tourism research include surveys (pre-trip, en route and post-trip), delphi technique methods, model building, multidimensional scaling, marketing assessment (of communication, advertising, conversion studies, etc.), forecasting tourist demand, and industry structure (Ritchie and Goeldner, 1994; Smith, 1995).

Finally, the thrust of the debate between which research study method (qualitative or quantitative) is most appropriate is moot, for what that decision should be guided by is the research problem identified. The researchers in ecotourism may be guided by the concept that if the problem is concentrated to a few subjects and you need a great deal of information on them a qualitative research method would be most appropriate. If, on the other hand, the problem relates to a large number of subjects, but only requires information on a few variables then probably a quantitative method might be performed.

There are several criteria that may be used to assess the quality of the methodology used in a study. An important quality of the data obtained is its neutrality. The concept of neutrality is related to the fact that both qualitative and quantitative methods use some degree of subjectivity in the process of data collection. To minimize subjectivity of data collection, and thereby attain more valid findings, authors attempt to document their judgements with evidence. They might also use multiple observers and seek consistency among their observations and interpretation of the data obtained (Newman and Benz, 1998). In the ecotourism literature, few authors report explicit attempts to minimize the subjectivity of their data collection and interpretation process. The problem of subjectivity, in contrast, was prevalent in case studies where data was obtained from selected secondary sources and their interpretation was based on the experience of the authors.

A challenge applicable mainly to studies adopting a qualitative methodology is that sometimes the data obtained may accurately describe an occurrence, but not capture the true essence of the phenomenon. Also the data may not detect if the occurrence is frequent or sporadic, and therefore atypical of the phenomenon. To minimize these challenges, authors need to conduct their study with prolonged engagement on-site to detect trends and abnormal events, and they need to conduct persistent observation to determine the frequency of events (Newman and Benz, 1998). In the ecotourism literature, researchers frequently conduct participant observations consisting of only a few visits of short duration to a destination. Many then fail to recognize the limitations of their data and assume to have captured the essence of the culture they are visiting.

Often in ecotourism research, authors become very attached to the destination they are researching, and tend to collect data and interpret them based on their subjective beliefs. To compensate for this tendency researchers may debrief their findings with other experts, the authors may go back to their original subjects and check if their interpretation was accurate, or the authors may obtain data from multiple sources and compare the consistency of the findings (i.e. triangulation) (Newman and Benz, 1998). The ecotourism literature has been poor in reporting methods of reducing researcher bias. Only a limited number of studies used the triangulation of qualitative and quantitative methods to gain a more accurate insight into the phenomenon being studied.

Qualitative methods

A significant portion of the literature employs qualitative methods to examine various ecotourism phenomena. The use of qualitative research methods can arguably be very useful in this field due to the lack of knowledge and need for understanding the meaning of phenomena that do not necessarily obey the theories developed under the scope of mass tourism. However, many of the qualitative studies encountered in the ecotourism literature are characterized by poor methodology. For example, many authors use personal notes from informal interviews as the main source of data for their discussion. Others engage in participant observation during a reduced number of visits to a destination, and use the limited insight from that data collection and their personal experience to reach broad conclusions and recommendations. Frequently, regardless of the data collection methods used, authors do not give an explanation of how they conducted the analysis of the information they collected. Finally, only a very limited number of studies (e.g. Getz and Jamal, 1994) used some type of triangulation to improve their accuracy.

On the other hand, there are a small number of studies that have used qualitative methodologies to their best capability. Very often, the qualitative methods were used in parallel with quantitative ones, which typically strengthened the study. A good example of a study incorporating both quantitative and qualitative methodologies was the Banff-Bow Valley Study (Ritchie, 1998). This study used focus-groups and on-site surveys to gather information about the impact of tourism in an ecologically sensitive destination. Other studies have incorporated structured interviews or participant observation to bring depth to more superficial but extensive findings obtained through on-site surveys and analysis of secondary demographic data (Wall, 1993; Barron and Prideaux, 1998).

Case studies

The qualitative method most frequently used in ecotourism literature has been the case study (see Table 38.1). This methodology is also used very frequently in the general tourism literature (Ryan, 1995). Case studies have been mainly used to describe the evolution of several variables supposedly due to the implementation of specific tourism developments. Many articles consist of the analysis of the environment under the perspective of the researcher. The discussion is based on personal experience, or based on secondary data collected from local, regional or national tourism-governing agencies. Despite the value of this type of report, they lack generalizability and therefore have limited importance to the development of the body of knowledge addressing ecotourism. Furthermore, these idiosyncratic studies seldom use knowledge from previous empirical findings, and often do not consider the results from similar case studies conducted elsewhere. In addition, these studies are often undertaken by individuals associated in some form with the governing agencies of the study area, which could potentially compromise the credibility of the document. Therefore it would be advisable that researchers try to strengthen their contribution to the ecotourism literature by seeking collaborative efforts with academics or other independent researchers that are clearly removed from pressures of local tourism agencies and community. An example of a study that incorporated the collaboration of independent researchers is Palacio and McCool's (1997) analysis of the ecotourist segments in Belize.

Interviews

Other methods used frequently to collect data are structured and unstructured interviews (see Table 38.1). These interviews are conducted either with tourists, with local residents, or with travel agents. In-depth interviews are not used as frequently as either structured or informal interviews. Burton (1998) used in-depth interviews of ecotourism operators to assess their strategies to cope with tourism growth in Australia. Focus groups are a special form of interviewing that benefit from the interaction of various subjects with the help of a moderator and are useful to help identify specific research topics to concentrate on and to suggest questions and issues important to the subject population. For example Hobson and Mak (1995) used focus groups to gain insight into the characteristics and motivations of tourists who participated in a tour focused on the culture and community of Hong Kong.

Observation methods

Another method used frequently in qualitative studies has been participant observation. This method, however, has not been used to its full potential. Participant observation involves prolonged contact with the study subjects either informing them of the study or hiding it from the subjects. As an example, Thomlison and Getz (1996) participated in a number of prolonged

Table 38.1. Data collection methods used in the ecotourism literature.

Methodology	Data collection	Frequency	Percentage
Qualitative	Case study	21	14.3
	Structured interviews	9	6.1
	Informal interviews	6	4.1
	Participant observation	4	2.7
	Content analysis	4	2.7
	Focus groups	2	1.4
	In-depth interviews	2	1.4
	Other forms of observation	2	1.4
	Analysis of images	1	0.7
Quantitative	On-site surveys	16	10.9
	Mail-out surveys	15	10.2
	Secondary data	15	10.2
	Phone surveys	5	3.4
	Mechanical/systematic observation	3	2.0
Qualitative and quantitative		3	2.0
Conceptual		39	26.5
Total		147	100

Some articles used multiple methods of data collection.

ecotours in Central America to gain in-depth, first-hand knowledge of the working environment of ecotourism companies. Other forms of observation methods used by ecotourism researchers have been mechanical forms of observation, and methods borrowed from the biological sciences to measure impact on the environment. A mechanical observation device (automatic person and bicycle counter unit) was used by Cope *et al.* (1998) to assess the use level of a long-distance cycle route.

Analysis of qualitative data

Only rarely did studies report the specific techniques used to analyse the data obtained from qualitative methods. Some of the techniques reported in the ecotourism literature were coding (1.3%), sorting (1.3%), imagery and semiotics (1.3%), looking for dominant themes (2.6%), and phenomenological interpretation (1.2%) (see Table 38.2).

Quantitative methods

Studies employing quantitative methods in the field of ecotourism can be classified into two broad categories. First, there are studies that examine essentially demographic data through the use of descriptive statistics, typically with the objective of characterizing the economic and social environment of a host population or target market. Second, more comprehensive studies examine complex relationships and cause-and-effect associations between variables describing the hosts or markets and their behaviour. These latter studies often employ factor-cluster segmentation procedures, or other simpler forms of clustering procedures. They typically include different statistical tests for examining the differences between the groups obtained. The tests observed were chi-squared, *t*-test, and analysis of variance. In addition, some studies used multiple regression to model the variables explaining certain behavioural patterns.

There was an apparent difference between authors that used more sophisticated statistical analysis, and those who limited their data analysis to description. That is, it seems that some authors tested principles and relationships of mass tourism in the ecotourism field, and for that purpose they used more sophisticated methods. On the other hand, other authors

Table 38.2. Data analysis techniques used in the ecotourism literature.

Methodology	Data analysis	Frequency	Percentage
Quantitative	Descriptive statistics	43	55.9
	Chi-squared	4	5.2
	T-tests	4	5.2
	ANOVA	3	3.9
	MANOVA	1	1.3
	Discriminant analysis	3	3.9
	Regression	3	3.9
	Factor analysis	4	5.2
	Cluster analysis	5	6.5
Qualitative	Coding	1	1.3
	Sorting	1	1.3
	Imagery + semiotics	1	1.3
	Triangulation	1	1.3
	Dominant themes	2	2.6
	Phenomenological interpretation	1	1.2
Total		77	100

Conceptual articles and articles without specified methods of data analysis were not included in this table; some articles used multiple types of techniques.

seemed to be analysing the ecotourism phenomenon as a completely new field of research and therefore had to start by describing the context.

Results of Examination of Ecotourism Literature

Method of data collection

The most common method of data collection used in the sampled articles was the on-site survey, which represented 10.9% of methods (see Table 38.1). This method has been attributed with the advantage of providing a short turn-around on the data collection process. The next two most commonly used methods were mail-out surveys and secondary data analysis at 10.2%. The most commonly used quantitative method used for data collection was structured interviews at 6.1%. The second most popular method used was informal interviews at 4.1%, followed by participant observation and content analysis each at 2.7%.

When including all types of articles addressing ecotourism in the literature, the two most frequently used methodologies were conceptual articles at 26.5% and case studies at 14.3% (see Table 38.1). These findings tend to support Ross and Wall's (1999) argument that in the research and practice of the ecotourism field, the one missing element has been the absence of ecotourism theory and the operationalization of that theory. The articles reviewed for this chapter suggest that some effort is being made to move toward the development of more theoretical grounding in ecotourism research but that, at the point this body of research is today, the field has further to go to get to the level Ross and Wall (1999) are suggesting the field should reach.

Ecotourism data analysis methods

The results of an evaluation of the types of data analysis used in the ecotourism literature suggested that for quantitative studies the most frequently used method of analysts was the reporting of descriptive statistics, at 55.8% of the total (see Table 38.2). However, this summary reporting, in most instances, was all the analysis needed and was appropriate for the purposes of the studies conducted and reported in the journal. Other types of statistical methods used besides descriptives were cluster analysis 6.5%, and factor analysis, chi-squared analysis, and *t*-tests at 5.2% each. As for qualitative data analysis methods used in the literature, these were used much less often, but are certainly of no less importance. Identification of dominant themes in the data collected were used in 2.6% of the studies, while coding, sorting, imagery and semiotics, triangulation and phenomenological interpretation were all used at an equal frequency of 1.3% (see Table 38.2).

In summary, the previous discussion suggests that the vast majority of the studies using a quantitative methodology in the ecotourism field used descriptive statistics to present their findings. This does not mean quantitative research methodology is superior to qualitative but more likely means that quantitative methods are quicker to complete and easier to interpret, especially if conducted at such a rudimentary level. In fact there is a lot to gain by incorporating both methodologies in one study. Only three (2.0%) of the studies examined incorporated both methodological approaches (see Table 38.1), but it was evident that these studies presented strengths and provided in depth and rigorous analysis and discussion of the findings (Hobson and Mak, 1995; Barron and Prideaux, 1998; Ritchie, 1998).

Geographic Distribution of Research and Authors

Location of destinations

The third aspect of research methodology in the ecotourism literature considered in this analysis includes the destination where the data was collected. This aspect of the research process is important to the assessment of which regions of the world are taking the lead in adopting ecotourism as a viable industry and where most new knowledge regarding the field is originating. In this study, the USA was the most popular location at 13.8% (see Table 38.3), followed by Canada at 11.7%, Australia at 10.6%, and New Zealand, UK, Indonesia and Antarctica at 5.2% each. These countries were followed by many other countries on every continent. This shows that ecotourism research is a global concern and the likelihood is that ecotourism will continue to grow in importance as we move further into this century.

Location of ecotourism researchers

The study characteristic analysed with regard to the ecotourism literature is not so much a true methodology component but rather relates to the origin of who is doing ecotourism research. Thus, we assessed the literature in search of the country of origin of the authors of the papers on ecotourism, that is, the location of the author's academic or employer affiliation as reported in the published articles. It was found that authors from the UK, at 22.7%, were most represented (see Table 38.4). This group was followed by authors from Australia 20.3%, the USA at 18.8% and Canada at 17.9% of all authors publishing in this journal. Again, many other countries from around the globe were represented but at a much lower frequency of publication than the first four mentioned. Finally, as Figs 38.1 and 38.2 reveal, the number one geographic region in which ecotourism has been researched is the Americas, and this is the number one geographic region of origin for ecotourism researchers as well. However, the figures from Oceania, though significantly lower, are remarkable in their own right given the relatively small population of this region. This shows that despite the global nature of the ecotourism field much of the current research is concentrated in the more developed countries.

Conclusions

With regard to the development of a research agenda, a field such as ecotourism that is in its relative infancy requires assessment of its research agenda and

Table 38.3. Destinations studied in the ecotourism literature.

Destination			
Continent	Country	Frequency	Percentage
America	USA	13	13.8
	Canada	11	11.7
	Costa Rica	3	3.1
	Belize	2	2.1
	Honduras	2	2.1
	Ecuador	1	1.1
	Jamaica	1	1.1
	Dominica	1	1.1
	Cayman Islands	1	1.1
	Brazil	1	1.1
Sub-total		36	38.3
Oceania	Australia	10	10.6
	New Zealand	5	5.2
Sub-total		15	15.8
Europe	UK	5	5.2
	Ireland	2	2.1
	Austria	2	2.1
	Spain	2	2.1
	Poland	1	1.1
	Sweden	1	1.1
	Greece	1	1.1
	Switzerland	1	1.1
	Romania	1	1.1
	Germany	1	1.1
Sub-total		17	18.1
Asia	Indonesia	5	5.2
	Thailand	3	3.1
	Pacific islands (Fiji, Tahiti, Hawaii, Tonga)	2	2.1
	Malaysia	1	1.1
	Hong Kong	1	1.1
	Maldives	1	1.1
	India	1	1.1
	China	1	1.1
Sub-total		15	15.9
Antarctica		5	5.3
Africa	Zambia	1	1.1
	Tanzania	1	1.1
	Uganda	1	1.1
	Rwanda	1	1.1
	Ghana	1	1.1
	Kenya	1	1.1
Sub-total		6	6.6
Grand total		94	100

Conceptual articles were not considered in this table.

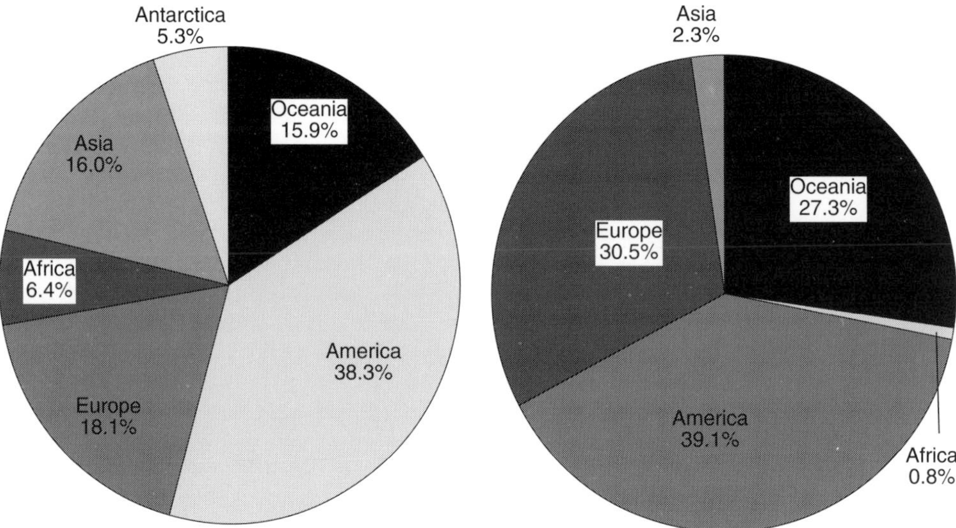

Fig. 38.1. Percentage of continents included as ecotourism research sites.

Fig. 38.2. Percentage of authors publishing ecotourism research origin.

methods in current use to see in which direction it needs to go. In the evaluation conducted for this chapter, ecotourism research has shown little common direction and has appeared scattered across the spectrum of research methodologies. This is not necessarily a negative situation for ecotourism research but rather reflects the field's stage of development. In reviewing the research being carried out currently in ecotourism, the quality level of research was found to be excellent in most cases. The selection of research methodology appropriateness was also found to be excellent in the current body of literature available in ecotourism. As was presented previously, there has been some over-representation of certain research methods in the literature compared with others, such as the use of case studies. While case studies can be a viable way for approaching certain problems or research questions, they tend not to provide a great contribution to the knowledge in the ecotourism field beyond describing yet another location and its local population. These case analyses are needed. However, for the field and its knowledge base to evolve, more rigorous research efforts are required as well. Only generalizable findings from sound research can provide a clearer understanding of what ecotourism is, who ecotourists are, and how they differ from other tourists (if they do). So, where does ecotourism research go from here?

The findings in this chapter need to be qualified by identifying certain limitations in the assessment process. Specifically, the findings presented are based primarily on the research published in just one tourism journal, and second, the evaluation is not without a certain amount of subjective bias. Therefore, to truly understand the extent of the body of literature published in ecotourism today, it would be necessary to compile and analyse a database that reviewed all the refereed journal material published globally, which is beyond the scope of this chapter.

In assessing the state of ecotourism research methods over the past 6 years, two trends become apparent. First is that the number of research studies being conducted and published is growing at an extremely high rate, whereas in the early 1990s only two or three articles a year were published on ecotourism. Currently, six or eight articles a year are being published in

Table 38.4. Location of ecotourism authors.

Origin of authors			
Continent	Country	Frequency	Percentage
America	USA	24	18.8
	Canada	23	17.9
	Belize	2	1.5
	Argentina	1	0.8
Sub-total		50	39.0
Oceania	Australia	26	20.3
	New Zealand	9	7.0
Sub-total		35	27.3
Europe	UK	29	22.7
	Switzerland	2	1.5
	Austria	2	1.5
	Ireland	1	0.8
	Spain	1	0.8
	France	1	0.8
	Sweden	1	0.8
	Greece	1	0.8
	Denmark	1	0.8
Sub-total		39	30.5
Asia	Indonesia	1	0.8
	Hong Kong	1	0.8
	Israel	1	0.8
Sub-total		3	2.4
Africa	Uganda	1	0.8
Grand total		128	100

Articles with authors from multiple nationalities were considered as one observation for each of the different nationalities; articles with multiple authors from one nationality were considered as only one observation for the respective country of origin.

each of the major journals. The second trend is the move from articles being exploratory, case study-oriented and conceptual in nature to more of a focus on application of traditional tourism principles applied or tested in the context of the ecotourism field. But, as for a clear movement from a particular research approach or statistical approach to another, there are no trends that are currently apparent.

One process that may prove useful in helping to guide ecotourism researchers in the future is presented by Smith (1995) in which he states tourism researchers face a number of challenges in trying to understand the industry called tourism. These challenges seem equally appropriate for ecotourism researchers and may provide direction for their research:

1. The lack of credible measurements for describing the size and impact of tourism.
2. Great diversity in the industry, with some analysts questioning whether tourism is a single industry or group of related industries (see, e.g. Leiper, 1993).
3. Spatial and regional complexities.
4. A high degree of fragmentation (Smith, 1995, p. 14).

To briefly address these challenges from an ecotourism perspective it would be appropriate to say that ecotourism still lacks credibility in the eyes of many decision makers (with a few exceptions as presented in earlier chapters). In addition, ecotourism researchers have far to go in estimating the scope and breadth of the field. The second challenge as to whether ecotourism is a single industry or a group of related industries needs researching due to the fact that the field still has difficulty accepting or supporting any unified definition of what ecotourism is (Sirakaya *et al.*, 1999). Moreover, the field has difficulty in deciding what activities and services constitute an ecotourism experience. Certainly, when you look at research published on ecotourism a point that immediately becomes apparent is spatial and regional complexities in this research. As presented previously, studies were completed on every continent, with the majority being conducted in the Americas and Australia. Still, more fundamentally, researchers need to decide at what level ecotourism should be studied; i.e. nationally, state/province or locally, or all of the above. Finally, fragmentation not only exists in the current body of research methods used but also among the sectors involved in ecotourism. Whether from the perspective of natural resources, economic impacts, social or cultural implications, planning and policy, or administration, all have tended to minimize coordination and cooperation related to product development and marketing.

In the investigation of research in ecotourism in the published literature, one great void that can be identified is the lack of longitudinal research studies. Most or at least 95% of research published, regardless of study method used, was a cross-sectional study of a one-time analysis of a problem or issue. Rather than trying to understand the causal process of these problems over an extended period, conclusions are drawn and recommendations made on the basis of observations made from one point in time. Studies are needed which use each of these three longitudinal types of research especially if researchers are going to truly understand and be able to explain a phenomenon such as ecotourism. These methods are: (i) trend studies that examine changes within the host populations and visitor populations over time (years); (ii) cohort studies which examine subpopulations such as visitor segments as they change over time; and (iii) panel studies which examine a set of travellers to ecotourism destinations over a period of years. Then, through the coordination of the current and longitudinal studies, ecotourism researchers will begin to more fully understand this complex industry and its market segments.

Thus, it would appear to be most appropriate to focus future research around an agenda such as this rather than around whether it is preferable to use qualitative as opposed to quantitative research methods. The only certainty the future has with regard to ecotourism research is that to better understand this phenomenon the field should take advantage of new technology such as geographical information systems, global positioning units, and sophisticated and path analytical modelling. Their use as research tools will become more important to help understand what is the most effective and efficient development of ecotourism.

References

Aeero, J.M. and Aguirre, E.A. (1994) A monitoring research plan for tourism in Antarctica. *Annals of Tourism Research* 21(2), 295–302.

Babbie, E. (1995). *The Practice of Social Research*, 7th edn. Wadsworth Publishing Company, Belmont, California, USA.

Barron, P. and Prideaux, B. (1998) Hospitality education in Tanzania: is there a mood to develop environmental awareness? *Journal of Sustainable Tourism* 6(3), 224–237.

Blamey, R.K. and Braithwaite, V.A. (1997) A social values segmentation of the potential ecotourism market. *Journal of Sustainable Tourism* 5(1), 29–45.

Burton, R. (1998) Maintaining the quality of ecotourism: two Australian marine environments. *Journal of Sustainable Tourism* 6(2), 117–142.

Cope, A.M., Doxford, D. and Hill, T. (1998) Monitoring tourism on the UK's first long-distance cycle route. *Journal of Sustainable Tourism* 6(9), 210–223.

Creswell, J.W. (1994) *Research Design: Qualitative and Quantitative Approaches.* Sage Publications, Thousand Oaks, California.

Dann, G., Nash, D. and Pearce, P. (1988) Methodology in tourism research. *Annals of Tourism Research* 15(1), 1–28.

Dieke, P.V.E. (1991) Policies for tourism development in Kenya. *Annals of Tourism Research* 18(2), 269–294.

Eagles, P.F. (1992) The travel motivations of Canadian ecotourists. *Journal of Travel Research* 32(2), 3–7.

Getz, D. and Jamal, T.B. (1994) The environment-community symbiosis: a case study for collaborative tourism planning. *Journal of Sustainable Tourism* 2(9), 152–173.

Hobson, J.S.P. and Mak, B. (1995) Home visit and community-based tourism: Hong Kong's family insight. *Journal of Sustainable Tourism* 3(4), 179–190.

Jamrozy, U., Backman, S.J. and Backman, K.F. (1996) An investigation into the relationship of involvement and opinion leadership in nature-based tourism. *Annals of Tourism Research* 23(4), 908–924.

Leiper, N. (1993) Industrial entropy in tourism systems. *Annals of Tourism Research* 20, 221–226.

Lindberg, K., Enriquez, J. and Sproule, K. (1996) Ecotourism questioned: case studies from Belize. *Annals of Tourism Research* 23(3), 243–562.

Malhotra, N.K., Peterson, M. and Kleiser, D.B. (1999) Marketing research: a state-of-the-art review and directions for the twenty-first century. *Journal of the Academy of Marketing Science* 27, 160–183.

Newman, I. and Benz, C.R. (1998) *Qualitative-Quantitative Research Methodology.* Southern Illinois University Press, Carbondale, Illinois.

Palacio, V. and McCool, S.F. (1997) Identifying ecotourism in Belize through benefit segmentation: a preliminary analysis. *Journal of Sustainable Tourism* 5(3), 234–243.

Pelto, P.J. and Pelto, G.H. (1978) *Anthropological Research: the Structure of Inquiry*, 2nd edn. Cambridge University Press, New York.

Reid, L.J. and Andereck, K.L. (1989) Statistical analyses used in tourism research. *Journal of Travel Research* 28, 21–24.

Ritchie, J.R. (1998) Managing the human presence in ecologically sensitive tourism destinations: insights from the Banff-Bow Valley Study. *Journal of Sustainable Tourism* 6(4), 293–313.

Ritchie, J.R. and Goeldner, C.R. (1994) *Travel, Tourism and Hospitality: a Handbook for Managers and Researchers*, 2nd edn. John Wiley & Sons, New York.

Ross, S. and Wall, G.C. (1999) Ecotourism: towards congruence between theory and practice. *Tourism Management* 20, 123–132.

Ryan, C. (1995) *Researching Tourist Satisfaction: Issues, Concepts, Problems.* Routledge, London.

Sirakaya, E., Sasidharan, V. and Sonmez, S.C. (1999) Redefining ecotourism: the need for a supply-side view. *Journal of Travel Research* 38, 168–172.

Smith, S.L.J. (1995) *Tourism Analysis: a Handbook*, 2nd edn. Longman Group Limited, Harlow, UK.

Thomlinson, E. and Getz, D. (1996) The question of scale in ecotourism: case study of two small ecotour operators in the Mundo Maya Region of Central America. *Journal of Sustainable Tourism* 4(4), 183–200.

Wall, G. (1993) International collaboration in the search for sustainable tourism in Bali, Indonesia. *Journal of Sustainable Tourism* 1(1), 38–47.

Wells, R. and Picou, J. (1981) *American Sociology: Theoretical and Methodological Structure.* University Press of America, Washington, DC.

Wight, P.A. (1996) North American ecotourism markets: motivations, preferences, and destinations. *Journal of Travel Research* 35(1), 3–10.

Witt, P.A. (1995) Writing for publication, rationale, process and pitfalls. *Journal of Park and Recreation Administration* 13(1), 10–17.

Chapter 39

Information Sources for Planning and Management

P.F.J. Eagles

Department of Recreation and Leisure Studies, University of Waterloo, Waterloo, Ontario, Canada

Introduction

All decisions are dependent upon information. The better the information available to the planner and manager, the better the chance for a good decision. Ecotourism, like the broader field of tourism, is a challenging activity to plan and manage. Ecotourism functions in an information-rich environment. The markets are often global, with worldwide information needs and sources. The destinations are sensitive and use of these sites requires special knowledge sets. The private operators are often small with few efficiencies of scale for a global market. Some of the main elements of the ecotourism activity are large, such as national parks and national tourism agencies, but are frequently naive about ecotourism management. The social and cultural implications of tourism in remote areas are large and important. This complexity requires decision makers to be cognizant of the requirement for careful consideration of the many facets of the activity.

Ecotourists demand high levels of accurate and competently communicated information. Much of the information required by planners and managers can also be used in ecotourist information and interpretation programmes. Ecotourism planning and management require information from a wide variety of fields: ecology, park management, marketing, social impacts, cultural management, finance, law, guide training and personnel management. Ecotourism is built upon the notion that individuals are travelling, and spending their hard-earned money and time, to learn and experience nature. These travellers are looking for experiences, knowledge, personal satisfaction and social contact. Those that service such needs must provide a rich ecological, social and cultural environment around the tourist experience. So, both the tourist experience and the management of the production system that provides this experience are based upon a strong, information-rich environment.

This chapter proposes a model to describe the planning and management of the ecotourism experience (Fig. 39.1). This model provides a synthetic pattern for a discussion of the components of ecotourism, and is a framework for the description of the information sources that are necessary for effective planning and management. The central component of ecotourism is the *ecotourist experience* by a traveller. This experience is influenced by four important managerial activities. The quality and presentation of the natural environment is critical to the experience, and therefore *environmental management* of that environment is key. The travel experience is heavily influenced by the *business management* elements of all those involved in producing the experience. Poor business management means that ecotourism will fail. Ecotourism typically involves travel to remote places, often with local populations that are heavily influenced by the tourism. The *social and cultural management* policies and procedures are vital elements of ecotourism. Ecotourism only exists if individual people find out about the destination, discover attractive activities, detect that the site and activities will provide useful personal benefits, determine suitable levels of safety and comfort, find suitable transport, and then decide to travel. *Marketing* is therefore a critical precursor to the ecotourist experience. This chapter discusses the information sources and needs for ecotourism using these categories as a framework.

The literature on ecotourism is large and growing rapidly. The Ecotourism Society, in cooperation with Parks Canada and the University of Waterloo, publishes an ecotourism bibliography of English language publications at frequent intervals (Eagles and Nilsen, 2001). The Ecotourism Society is constantly publishing highly relevant material in the field and the publications can be reviewed through a web site and purchased from the Society (The Ecotourism Society, 1999).

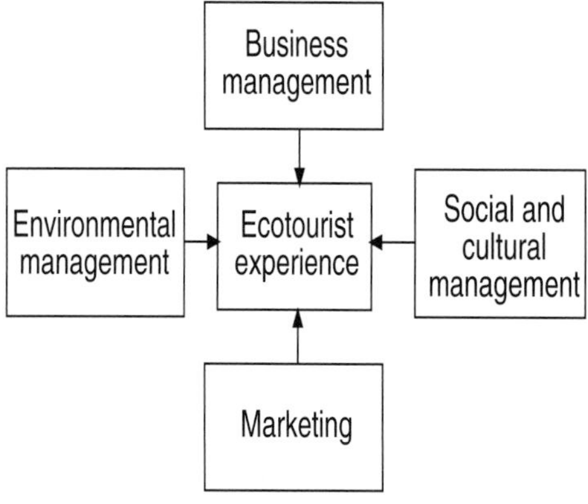

Fig. 39.1. Model of the ecotourism experience.

Higgins (1996) found that the ecotourism industry in the USA grew dramatically over the previous two decades. This growth, which was almost certainly mirrored in other industrial nations, will continue into the foreseeable future. This growth will increase the demand for suitable information to serve all aspects of the expanding ecotourism industry.

Describing the Ecotourist

Before one can discuss numbers, flows and impacts of ecotourism, one must have an operational definition of ecotourism and the ecotourist. Elsewhere in this book are discussions of the definitions of ecotourism and ecotourists. Here we only briefly introduce the topic, for the purpose of understanding information needs.

Before one describes ecotourists one needs to have a definition that allows for measurement (Ballantine and Eagles, 1994). Blamey (1995, 1997) developed a minimalist approach to the definition of ecotourism. He concluded that for market research purposes the following definitions are most useful:

> An ecotourism experience is one in which an individual travels to what he or she considers to be a relatively undisturbed natural area that is more than 40 km from home, the primary intention being to study, admire, or appreciate scenery and its wild plants and animals, as well as any existing cultural manifestations (both past and present) found in these areas.
>
> An ecotourist is anyone reporting to have undertaken at least one ecotourism experience in a specified region during a specified period of time (for example, during a stay in Australia).

Importantly, Blamey was successful in providing definitions of ecotourism and ecotourists that can be used operationally in social and market surveys. It is critical that such definitions are operational when volume and trend measurements are required.

Blamey and Braithwaite (1997) reported that an understanding of social values assists with the comprehension of the ideals that a person has about their world, their country and their community. Those ideals impact on many issues, such as the proper method of payment for park management, income taxes versus user fees, for example. They make the telling point that people's social values of community affect the structure and form of an ecotourism experience.

The largest market study of ecotourism undertaken so far was done for British Columbia and Alberta in Canada in 1995 (HLA and ARA, 1995a, b). For this study, the term ecotourism was used and was defined broadly and simply as 'nature, adventure and cultural experiences in the countryside' (HLA and ARA, 1995a, p. ES-1). The study found a very large ecotourism market in Canada and the USA. In the seven metropolitan areas studied, Seattle, San Francisco, Los Angeles, Dallas, Chicago, Toronto and Winnipeg, a market of 13.2 million potential ecotravellers was found, a high percentage of the area's population. This was much larger than anticipated, and showed a substantial market in North America. Wight (1996a, b, 1998) provided further analysis of this data and described the market profile and trip characteristics of potential North American ecotourists. Unfortunately, the excellent market studies from Australia and Canada are much too rare. More such studies are needed in all industrial countries, and need to be repeated at a frequent rate.

Visitor Experience

All ecotourism is based on a personal interaction between a traveller and a natural environment. This intimate personal experience has several key elements that must be well understood and managed. The Office of National Tourism (1999a) of Australia produced guidelines to assist and direct potential ecotourists. These include tips on selecting a tour operator, on minimum impact behaviour in sensitive sites, and on cultural sensitivity. The visitor experience is composed of several elements: visitor satisfaction, guides and guide training, accommodation, food, transport, as well as site and trip information.

Visitor satisfaction

The visitor's satisfaction with an experience is based upon a mixture of expectations and experiences. It is critical that the ecotour manager attempts, as much as possible, to understand the existing visitor knowledge level and expectations. The development of a visitor's background knowledge and expectations typically occurs before the actual on-site experience. It is important, based upon an understanding of the visitor's knowledge level and expectations, for the manager to design programmes to further influence the visitor's expectations in the desired direction. It is then necessary to operate the programme to provide the desired experiences.

The interactions among the visitor, the natural environment, the ecotour provider, and the other tourists provide the site basis for the visitor experience. All of those involved in the design and provision of the visitor experience must have sufficient information on each of the elements of the experience so as to design a suitable and satisfying programme. This information is often difficult to obtain, but sources are available. The managers should ensure the provision of detailed information for the needs of management, and for the needs of the visitors.

The natural environment knowledge of the planner, manager and operator comes from two primary sources. The first is formal, academic preparation of the decision maker in ecology and resource management. Abundant and relevant information is available in universities all over the world in these fields. The second source is primary on-site experience with the natural environment of the local area. It is best if this on-site experience builds upon the theory and background obtained by formal academic training. It is naive to think that technical knowledge can be obtained accurately and comprehensively without attendance at an academic institution (see Chapter 40). Guides, and others, that are self-trained may know the local flora and fauna intimately, but usually have little concept of the fundamental properties of ecology, geology, biology, meteorology, limnology, fisheries, marine biology, genetics, ornithology, mammalogy, forestry, conservation biology, landscape ecology, or park management. The lack of theoretical knowledge means that any information provided will be shallow in scope, without an understanding of background principles. Additionally, ecotourists are a highly educated population and only a well-educated guide can deal with a full range of information needs of such a population (see Chapter 3).

For the ecotour operator to provide a fulfilling ecotourist experience, it is important that this operator knows more about the site features and about the experience than does the visitor. However, in many places in the world, especially those in very remote sites and those involving high cost, the ecotourists are often very knowledgeable about the natural environment. With high levels of knowledge come high expectations of further learning, widespread background in the field and the need for carefully designed visitor programmes. It is critical that the guide, when one occurs, has sufficient knowledge in the subject area and of techniques of interpretation.

Guides and guide training

The guide is a critically important part of the ecotour. In the case of non-personal ecotours, the guide may be a guide book, a park brochure or a specialized publication on special site characteristics. These types of non-personal interpretation sources are vitally important because the independent traveller may rely heavily on one source of information, such as a guide book. The huge demand for such information has led to an industry of guide books, of which the *Lonely Planet* guides have become particularly successful.

There is a wide variety of types of personal guides available. They may be specially trained guides just for ecotourism, or they might be someone whom the ecotourist just happens to encounter on their trip. The latter may be problematic because the information might not be accurate, or might be purposely misleading due to some ulterior motive of the provider. I know of some people who asked for directions in Venezuela. The directions they received from a local boy took them right into the hands of a robber. It behoves the international and national tourism bodies to ensure that sufficient accurate information is available in order to warn tourists of the pitfalls, as well as provide correct directions. However, government and private tourist companies are sometimes loath to provide accurate and potentially alarming information on problems and dangers due to a concern that this will scare away potential clients.

Specially trained guides are sometimes needed. In the cases of sensitive sites, crime-ridden communities, difficult travel situations, and highly specialized conditions, some site managers require all tourists to be accompanied by a trained guide. The types and amounts of training required for guides is often under debate, though over time, as the ecotourism industry matures, the guides are expected to gain higher levels of formal training. As more information becomes available, consumers expect that their ecotour operators have staff that keep up to date with that information.

Those responsible for hiring guides and interpreters usually require training in the subject matter of environmental studies, and in the techniques of interpretation. It is much easier to teach a formally trained biologist, for example, techniques for public speaking, safety and interpretation, than it is to train an interpreter the complexity of biology and ecology. The lack of accreditation for guides in most countries leads to much confusion for the potential ecotourist. From the point of view of the potential consumer of a service, there are information, personality, cost and safety concerns. It is very difficult for the ecotourist to evaluate the effectiveness of a guide before the purchase decision. And in many cases the ecotour operator provides low levels of information to the consumer on the guide's characteristics. In mature tourism markets, as in most European countries, the guides are all accredited and licenced. This leads to a much higher level of quality across the guiding fraternity and to more quality assurance by the consumer. A typical example is the 'Blue Badge Guide' training system in place across much of the UK (British Tourist Authority, 1999; The Guild of Registered Tourist Guides, 1999).

It is critical that the ecotour operator, whether public or private, provides accurate and sufficiently detailed information on the guide so as to enable the potential client to make an informed decision and to form appropriate levels of expectation.

Accommodation

A critical element of any trip is the accommodation. It must have suitable amenities, pricing and location. Recently, with the advent of the ecolodge, the accommodation facilities have become ecotour attractions in themselves, with abundant natural environmental elements around the rooms, a good library, some minor scientific supplies, and suitable opportunities for on-site interpretation.

It is crucial for all those involved in ecotourism to fully assess the suitability of accommodation for ecotourists. This can be done using the normal information sources of brochures, familiarization tours and discussions with other operators. Good references are available on many aspects of ecolodge development. *The Ecolodge Sourcebook* contains information on site selection, finance, planning, design, alternative energy applications, conservation education, guidelines and an impressive set of resources, including a variety of architectural plans for ecolodges (Hawkins *et al.*, 1995). Recently, the advent of the World Wide Web has allowed ecolodges to provide a global view of their facilities to their potential clients. This availability of information means that all those involved in ecotourism can get site information in a speedy and comprehensive fashion. It also means that the ecotourist can be very well informed of sites well before the site visit occurs.

Food

Food is an important, but standard part of any travel experience. It is important that all people involved in ecotourism understand their clients' food preferences. It is doubtful that ecotourists' desires and needs for food are much different from any other tourists. However, ecotourists are probably more likely than the average tourist to request locally grown food, and to be concerned that the food production is environmentally suitable.

Transport

Transport to and from the primary ecotourism site typically consists of standard forms such as aeroplanes, boats, buses and private cars. The biggest difference with ecotourism is the specialized forms of transport that occur on-site. Over the last 50 years many specialized transport vehicles have been developed just for ecotourism. These include the safari vehicles, of several types, developed and used throughout Africa, the snowmobiles developed in Canada for snow-based movement, the specialized submarines built in Vancouver for moderate depth ocean viewing and used extensively in marine parks in the tropics, and the sport utility vehicles that are now utilized all over the world. All of this transport requires a thorough knowledge of the capability, functionality and cost of the vehicles. In many private ecotour companies the vehicles are the biggest single expenditure unit and the most time-consuming part of the business operation. Information on such equipment can be obtained at trade shows, from other operators, from equipment dealers, and from the Internet.

Site and trip information

The ecotourists want to know about the site they are considering visiting. Site information includes: the general ecological features of the site, special site features to be expected, special or sensitive species to be encountered, available tourism services, prices and expected behaviours. They also want to know key information about the trip. This trip information includes: the cost, the length, the start time, the end time, the types of accommodation, the types of transport, the expected weather, potential dangers, required clothing, safety issues, special expectations of them by the tour providers, special equipment needs, the physical difficulty of the trip, and cultural expectations.

It is best that all of this information is provided to the ecotourists well before the trip. This enables proper preparation and the development of suitable expectations. The information can be provided in written format. This is inexpensive and effective, if it is read. In high volume settings specialized videos are available. These can be sent to the person's home before the visit, and can be repeated in the visitor centre or on the bus on the way to the site. In some cases the company offers special meetings or training sessions. These are usually done for long-distance, high-cost trips. Increasingly the Internet allows for the relatively easy, inexpensive and fast transfer of written, visual and video information to the traveller's home well before the trip.

Marketing

Information for marketing of ecotourism can be categorized into: understanding the client, attracting the ecotourist, setting prices and programme evaluation.

Understanding the client

All ecotourism is dependent upon the personal desires of individual people. It is critical that those involved in ecotourism understand these desires and how they are derived. Several key questions need answers:

- Why does this person choose to travel, rather than save their money or spend it on another service or a good?
- What personal attitude set predetermines a desire to take this trip?
- What type of experience is being sought?
- What memories, products or experiences are looked for?
- How much experience with ecotravel does the person have?
- What is the person's financial capability?
- What is their physical capability?
- What language do they speak and read?
- What information sources do they utilize?

A solid ecotourism marketing effort is based on a good understanding of the answers to these questions. Such answers are relatively difficult to obtain due to a lack of publicly available research in this area, but a few studies provide useful information.

Eagles and Higgins' (1998) case studies of the ecotourism market in Kenya and Costa Rica revealed principles that can be useful elsewhere. The development of a solid, sustainable ecotourism industry is dependent upon several key factors. Important ecological sites must be protected within a set of national parks and reserves that are well managed and available for tourism use. Most of the ecotourism visitation in these two countries involves the national reserves, but private reserves and ecolodges play a role through the provision of specialized programmes and services. The long-term success of park tourism requires cooperation between the public and private sectors. An ecotourism industry will only survive if the quality of the natural environment is maintained.

Protected sites best survive politically when a mobilized constituency, including the tourism industry, argues for their existence.

Forestry Tasmania (1994) found that many private ecotourism operators misunderstood the desires and intentions of their clients. For example, most ecotourists placed higher emphasis on the quality and content of the environmental interpretation than did the operators providing the service. This study revealed that even when in constant contact with their clients, the ecotour operators were unable to properly assess a key part of their clients' needs. This reinforces the need for independent and thorough evaluation of ecotourism programmes, and especially the component of understanding the ecotourist. Wild (1998) described the ecotourism marketing experience in Canada. Silverberg *et al.* (1996) found that psychographic research techniques were useful in differentiating segments within the nature-based tourism market. The information on the lifestyle characteristics of the tourists was derived from a consumer survey. Statistical analysis of the research data found at least six group types within their sample. Carefully designed and implemented consumer surveys are important primary sources of information on ecotourists.

Attracting the ecotourist

There is an abundant literature describing the characteristics of ecotourists, and more is becoming available on a continuous basis. Eagles (1992) provided an early study of the social and attraction motivations of Canadian ecotourists. Ballantine and Eagles (1994) described Canadian ecotourists visiting Kenya. Crossley and Lee (1994) found that ecotourists differed significantly from mass tourists in several ways: age, education, income, occupation, trip duration, number in tour group, trip partner, trip season, type of lodging, trip planning, and percentage of tour cost spent for transport. They found no difference in gender or total trip cost per day. Weiler and Richins (1995) studied the ecotourists who participated in Earthwatch Expeditions in Australia. McCawley and Teaff (1995) described the characteristics and environmental attitudes of divers on the coral reefs of the Florida Keys in the USA. Obua and Harding (1996) studied the visitor characteristics and travel attitudes of the visitors to Kibale National Park in Uganda. Silverberg *et al.* (1996) investigated the lifestyle characteristics of tourists in an area of the USA. They concluded that psychographics can be used to differentiate segments of the nature-based tourism market. Wight (1996a, b) provided a thorough and sophisticated summary of North American ecotourists. Woods and Moscardo (1998) described Australian, Japanese and Taiwanese ecotourists, one of the first papers in English to discuss the emerging ecotourist populations in Japan and Taiwan.

These studies, and many others like them, provide a reasonably thorough description of the characteristics of ecotourists in the countries studied. These data provide a good basis for the development of a market analysis and marketing strategy aimed at attracting the ecotourist to a particular destination or product. For the operator in the field, one limitation in accessing the information available is its presence in unpublished government reports, and in academic libraries. Furthermore, there is very little market information available, in English, on the ecotourists of many countries, such as Germany, Japan and Italy.

Setting prices

Surprisingly, there is little available on ecotourism pricing policies. J. Laarman and H. Gregerson (unpublished) prepared a summary of the policy and administrative aspects of nature-based tourism pricing. Silverberg *et al.* (1996) found that there are subtle differences in various sectors of ecotourism, each of which require a different pricing approach. Much more work needs to be undertaken on setting prices for ecotourism.

Programme evaluation

All programmes can be improved with formal, frequent evaluation. Since marketing is such a key function in the ecotourism business, it is critical that such programmes are frequently evaluated. It is important to check if the tourists are arriving at the site properly prepared psychologically, physically and with proper clothes and equipment. Did all aspects of the food, transport, and guiding function properly? What made the clients choose one destination or company over others? Many of these, and other important business elements, are best evaluated by the marketing section of the business or government agency. This arm must maintain a certain level of independence from the rest of the operation, so that it can function as an independent source of critical information for the decision makers.

Forestry Tasmania (1994) undertook an elegant and practical evaluation of the professionally guided tours taking place in the state forests of Tasmania. This study found fundamental weaknesses in the private industry's ability to earn profit and to respond to the market demand. The findings in this useful report reveal the importance of a carefully structured evaluation of ecotourism operations and programmes.

Business Management

Business management in ecotourism can be seen as composed of several fields: finance, law, licences and permits, accreditation, liability and insurance, staff training, and personnel management. The most compact source of business management information on ecotourism is McKercher's (1998) book. This one source provides a solid background to all major aspects of business management for ecotourism. The value of a formal business education, as found in a business school or a leisure studies school with a business programme, is high and should not be underestimated. There is a lot of naivety in tourism. Too many feel that they can be successful in a tourism business, with only good intentions and hard work. The high failure rates and turnover rates reveal that good intentions can never replace solid educational preparation.

The Canadian Tourism Commission (CTC) (1995) produced a summary of the potential for ecotourism and adventure tourism in Canada, revealing important business need in Canada. They undertook a road show of adventure and ecotourism consultation across the country in 1997. This effort produced a useful set of documents describing the business opinion of the industry and outlining policy needs (Villemaire and Murray, 1997). This document provides a useful lens into the business concerns of the many small ecotourism businesses in that country. The Department of Environmental Affairs and Tourism in South Africa (1996) produced a status and summary statement on ecotourism in that country. Given the importance of ecotourism to the national economy in South Africa, national government policies are very important in encouraging and guiding the business development in this rapidly growing industry. Lindberg *et al.* (1998) edited one of the most useful books available on ecotourism management, entitled: *Ecotourism: a Guide for Planners and Managers*. This book has an international focus and provides solid descriptions of many aspects of ecotourism business management.

Finance

All operations, big and small, government and private, require finance. This chapter and this book are not the appropriate places to fully discuss the details of business finance. Typically, one needs the assistance of financial experts when designing and implementing a business plan. This can be obtained by using specialized finance consultants or, in simpler situations, banks and loan companies may have staff that can assist. All predictions of tourism volume, cash flow and expenses should be conservative, in order to err on the positive.

The Australian tourism bodies frequently produce useful publications in ecotourism management. An excellent way to reduce financial costs and to reduce the ecological footprint of an operation is to adopt policies to minimize chemical use, energy use and water use. Excellent guidelines on environmentally sensitive tourism design and management are available from the Australian Department of Tourism (Commonwealth Department of Tourism, 1995).

Law

All business operations must function within the laws of the country of business. In ecotourism, the international aspect of the travel means that a complicated, multijurisdictional approach to the legal aspects of business must be adopted. The inbound operator may be based in one country, the transport company in a second country, the outbound operator in a third country, and the ecotourists from many others. This creates challenging contractual, monetary and operational issues.

It is critical that the base of operations of any business is in a country with a fair and operational rule of law. The police must be honest and the judiciary independent and honest. Without these key factors any property ownership, contract, licence or agreement will be exposed to the whims of corruption and bribery. Many ecotour operations have failed when corrupt government officials, police or local operators engaged in fraudulent activities. It is imperative that any ecotour business becomes familiar with the legal situation in their area of operation.

Licences and permits

All ecotour operators, both public and private, require licences and permits in order to operate. These can include: access permits for parks, vehicle safety permits, food handling permits, firearm permits, business operations licences, land-use development approvals, building permits, tax collection licences, proof of incorporation, workplace health and safety approval, proof of insurance, employee training certification, bus driver licences, and many others. The licences can be issued by the national, provincial, regional or local governments. Park and protected area agencies often require special permits before sites can be used. All licences and permits must be comprehensively handled by competent staff and kept up to date.

Accreditation

Ecotourists look for independent indicators of product quality, and accreditation is an excellent way of indicating the achievement of such quality. Accreditation is a formal process for the determination of product and service quality. In a competitive market, such as tourism, accredited operators and sites have a competitive advantage. Given the difficulty faced by potential tourists in their ability to assess the travel product before purchase, independent accreditation is seen as a very useful sign of quality. Given this market pressure, the more mature the tourism markets, the higher the levels of accreditation.

The best example of accreditation in ecotourism in the world today is found in Australia (see Chapter 37). The national ecotourism accreditation scheme is functioning well, is widely accepted and has become a model for other countries. The programme has helped to raise the standards of ecotourism across the country. It allows competent operators to clearly reveal to others their high level of business management. It provides a framework for continuous improvement in the industry. Up-to-date information on this programme can be found on the web site of the Office of National Tourism (1999b).

The Green Globe programme of tourism accreditation is gaining wider acceptance globally. Some of the larger operators are adopting the ISO 9000 series of programme operation standards to improve their operations and to prove to their clients that they are serious about quality. In all cases, it is important for tourism operators to see some form of independent assessment of quality, such as accreditation. Once this is achieved, this must be communicated to potential clients.

Liability and insurance

All people providing a service to others in return for financial remuneration have a duty of care to their clients. Typically, this duty of care involves providing a level of service and safety as would be normally expected in similar circumstances in the area of operation. However, tourists from distant locales may have service and safety expectations quite different from that normally occurring in the local area of operation. All operators, both public and private, should make themselves aware of the standard of care normally expected in their area. They then must put operational procedures into place to ensure that all parts of their operation provide this standard of care. It is important that sufficient levels of insurance coverage are maintained. It is important that the legal structure of the country of operation is sufficiently mature to allow insurance to function at internationally acceptable levels. Mature ecotourist operations provide information to their clients on the levels of security, safety and insurance coverage that they hold. Experienced ecotourists demand such information before making a travel decision.

Staff training

Ecotourism is an information-rich, highly personal activity. Ecotourists generally have high service, safety and information expectations. Those who have travelled widely know very well what it means to be serviced by personnel with appropriate levels of ecological, service and interpretation experience. Therefore, all staff involved in ecotourism, from the safari driver to the booking agent, require high levels of service training. Typically, this is achieved by local or regional colleges, or in some countries by industry training bodies.

Personnel management

After the features of the natural environment that attracts the ecotourists, the qualities of the personnel are the most important component of any ecotourism business. It is critically important to attract, properly reward and retain good personnel. All ecotourism businesses need to be very concerned about professional levels of personnel management. Unfortunately, small size and seasonality may limit the abilities of many ecotourism operators to attract and retain well-qualified staff. Some countries, such as Kenya and Australia, have specialized college programmes for ecotourism training. The outcomes of such training are obvious in these countries, with higher than average service quality, more skilled information provision and better operations overall.

Social and Cultural Management

The policies for and management of the social and cultural elements of ecotourism are important for long-term success. Such management can be seen as being composed of several areas: community relations, cultural impact, and local economic impact.

Community relations

Most ecotourism occurs in beautiful, remote locales. In most of those locales local people are affected by the tourism activities. Generally, the local residents are used to having a resource-based economic system, based on the exploitation of the forests, the fisheries, or the agricultural potential of the area. In very remote locales the local people may not be in a market economy, and are used to trading between each other for needed goods and services. Tourism is quite different. Those who exploit natural resources see the environment as a source of physical products, and frequently do not understand an economic system that sees the environment as a source of experiences.

There are many potential sources of conflict between the ecotourism industry and the local people. The goal interference between those who want to physically utilize nature and those who want to spiritually experience nature can be the basis of substantial conflict. The influx of foreign money, new ideas and new power structures may leave the locals in a state of concern. Therefore, the ecotour manager, whether he or she works for government or a private company, must be astute in developing positive community relations. Every community has its own needs, but experiences elsewhere in the world show some general principles worth considering (Ashley and Roe, 1998). These principles are outline below.

In most cases, it is very important that the ecotourism provides obvious and appreciated local community benefit. It takes intelligence gathering to develop an understanding of what type of benefit is appropriate. And the benefit must be sufficiently visible so that the local community sees it occurring and understands where it is coming from. For example, the provision of jobs for local people is an obvious benefit, and one that is usually appreciated and valued. People who work in the ecotourism business naturally become supportive of that industry and support it in community decision processes.

In many parts of the world there are consultants available who specialize in community relations and community development. For example, throughout eastern and southern Africa, there is a cadre of people with expertise in the development of community relations plans for the application of development programmes. In both Kenya and Tanzania the national park agencies have specialized staff whose role is that of community relations. These people work with local political bodies, develop personal relationships with politicians, serve as a sounding board for complaints, and importantly develop a list of community needs. Ruaha National Park is in southern Tanzania. When the roof blew off the local government council building in a storm, the Ruaha community development officer quickly found the funds to build a new roof. The symbolism of the local town council meeting each week under a park-provided roof is strong evidence of the tourism benefit flowing to the town. In eastern Kenya, the Kenya Wildlife Service has built schools for local communities, and in one case provided a cement mixer to a village. In all three examples the agencies made sure that signs on the structures showed, to all who could read, that they were provided with funds derived from the nearby park agencies, with the money ultimately coming from tourism income.

Some private companies and many government agencies have programmes to provide specialized training to talented young people. This can take the form of small scholarships, or can be substantial grants for education at colleges or universities. Utali College in Nairobi, Kenya, is a fine example of a college designed specifically for the training of people for all aspects of the tourism industry. This college provides training for all types of jobs in tourism, such as room cleaners, cooks, waiters, safari drivers, guides and hotel managers.

Cultural impact

Tourism has value-changing impact on all involved. Given that ecotourism has an inherent goal of personal education, it has the potential for higher levels of personal change for the travellers and for the local people than many other forms of travel. Many ecotourist destinations have been substantially changed by the tourism industry. It is a hotly debated topic whether these changes are positive or negative, and in most cases they are probably both, depending on the view of the observer (Travis, 1982).

All involved in the ecotourism industry must be aware of the cultural impact inherent in this activity. This requires a certain humility, recognizing that people's ideas, values and behaviours will be changed forever due to tourism. Tourism planners and managers may have to hire or otherwise involve local expertise in cultural management. For long-term programme success positive cultural management policies will ensure a supportive local community.

Local economic impact

Positive local economic impact is one of the strongest arguments made by governments and local people when they support ecotourism development. The financial gain can come from three sources: (i) the spending of the tourists; (ii) the investment of the private sector; and (iii) the investment of the public sector. Typically, the public sector investment, in the form of special protection regimes, such as the creation of national parks and wildlife reserves, comes first. This is followed by small tourist flows. As the private sector recognizes a potential market, its investment of money and expertise often results in much larger tourist flows.

Many studies have looked at the economic impact of ecotourism (Butler and Hvenegaard, 1988; Hvenegaard, 1992; Filion *et al.*, 1994; Lindberg and Enriquez, 1994; Driml and Common, 1995; Kangas *et al.*, 1995). These studies are typically available in university libraries. However, economic impact from tourism is not always obvious. This is especially true when many other economic activities are also occurring in the same area. In these latter cases, ecotourism managers have a responsibility to document the local economic impact, and then ensure that this impact is communicated to the appropriate people.

There are many ways to approach the measurement of local community impact. The provincial economic impact model developed for the Federal Provincial Parks Council in Canada is a user-friendly computer model for the calculation of the economic impact of a park (Stanley *et al.*, 1999). The model has been specifically designed to be useful to a manager who has no special training in economics, but wants a competent estimate of the economic impact of a park or a cultural event. New South Wales National Park and Wildlife Service spearheaded an effort to undertake local economic impact studies in Australia. Examples include the economic impact of Coolah Tops National Park (Conner and Christiansen, 1998) and Montague Island Nature Reserve (Christiansen and Conner, 1999). These Australian and Canadian approaches are useful examples of the approaches that can be effectively used for the calculation of the economic impact of parks and the associated ecotourism. The Canadian approach calculates provincial-level impact, the Australian model calculates community-level impact.

Environmental Management

The effective management of the natural environment and its presentation for ecotourists are key components of ecotourism. Environment management is composed of several fields: site sensitivity and ecological impact, environmental protection and sustainable operations, and park management.

Site sensitivity and ecological impact

Almost by definition, sites that have high ecotourism potential are ecologically sensitive. Therefore, all activities must be very carefully planned to ensure maximum positive ecological benefits and minimum negative ecological impacts. Many governments have laws and policies requiring the creation of an environmental impact study of proposed developments in ecologically sensitive areas. Adherence to such laws or policies requires special studies undertaken by highly trained specialists. Large government agencies, such as park agencies, often have such staff within their own organizations. Corporations typically hire consultants to undertake the work. Small corporations must rely on their own expertise and that gleaned from government staff and private experts. After the impact studies are complete, site design and construction must be carried out in a manner that follows the recommendations from the impact studies. Facility and programme operation must adhere to developed environmental standards.

The field of environmental management surrounding ecotourism is too large to discuss within this particular chapter. However, such management is heavily data dependent. Large amounts of current and relevant data must be collected on the environmental features of the subject area. These data must be intelligently analysed in order to produce useful information for planning and management purposes.

The SEACAM programme in eastern Africa produced useful guidelines for the assessment of the impact of tourism facilities in coastal areas (Grange and Odendaal, 1999). This book describes in detail the information needs, the procedures, and many applied examples of coastal tourism development. The book is a good combination of theory and practice in planning for tourism in sensitive environmental and cultural areas. The book is designed for coastal tourism, but has general utility in the environmental assessment of tourism.

Environmental protection and sustainable operations

Virtually none of the Earth's surface goes without some form of human use. The issue is not one of use, it is one of type and amount of use. Ecotourism is inherently protective because the ecotourists want to see the environment sensitively managed so that it is available to provide experiences with nature.

The Australian national government, private ecotourism operators and research groups are the best sources of environmentally positive operational guidelines and procedures for ecotourism. The Commonwealth Department of Tourism (1995) of Australia produced a useful set of guidelines aimed at minimizing the use of energy, water and supplies in ecotourism projects. Boele (1996) outlined in detail energy efficiency measures in tourism. The Tourism Council of Australia (TCA and CRC Tourism, 1998) published a guide to best-practice approach for developing environmental sustainability and maintaining a viable business. More recently Basche (1999) wrote a manual explaining the general principles of sustainable tourism as adopted by the Tourism Council of Australia. The principles are explained under the following headings: use resources sustainably, reduce over-consumption and waste, maintain diversity, integrate tourism into planning, support local economics, involve local people and indigenous communities, consult stakeholders and the public, train staff, market tourism responsibly, and undertake research. This book provides understandable directions for sensitive and sustainable tourism. Details are provided on seven specific activities: road transport, marine tours, camping tours, bushwalking tours, horse-riding, raft and kayak tours, light aircraft and helicopters. This is the only publication of its type available.

Park management

A substantial amount, and possibly a majority, of the world's ecotourism occurs in national parks and other forms of protected areas. This cultural institution now contains the best pieces of wild nature that still exist on Earth. This high ecological value, the cultural heritage connections within society, the sense of increasing scarcity, and the increasing demand for use makes parks and protected areas key areas for ecotourism.

Many parks contain very large-scale tourism activities, and have considerable expertise in the field. It behoves those outside park management to become familiar with the legal, policy and political structure surrounding park management in their area. It behoves those within park management to become familiar with the needs, constraints and operations of the private ecotourism operators who work locally. Tourism operations can become a source of major conflict in park management. However, properly designed and sensitively handled tourism park relationships can become a major, positive force in parks and in local communities.

Summary

Ecotourism is a complicated and sophisticated enterprise. Good management requires a flow of useful and accurate information from many fields. It is critical that managers put themselves into a position so that they get the information they require when it is most needed. It is necessary that the managers sufficiently understand the sources and methodologies used to collect the information, so its accuracy and utility can be properly evaluated.

Managers must ensure that they get all information, not just that which is positive. It is easy to develop a system where the sources for a manager provide only good news, leaving out the difficult bits. This may be easy on the psyche, but only for a short period. If a problem is developing with a local community, with a staff person or with the cash flow, the manager must become aware at the earliest possible opportunity.

It behoves the government tourism and statistical agencies to provide a continuous, relevant and accurate flow of data on tourism. Whenever possible, the data should be interpreted so that the widest range of users can see their utility (CTC, 1995, 1997; DEAT, 1996). It is important for the government and private users of such data to demand their collection, and then to support the collection agencies politically. There are a large number of demands made on governments, and if tourism information flows are to be maintained they must be cultivated, just as one would water and fertilize a garden.

The private sector often collects expensive and proprietary data on tourism. These are used for their own business interests, naturally enough. However, the private sector is often negligent in not making such data available to others in the industry when they are no longer of direct utility to the collecting company. Such rich datasets should be made available to university researchers, to park agencies and to the general tourism industry at the earliest feasible date.

Ecotourists require high levels of information. Much of the information required for competent planning and management can also be useful in the design and delivery of interpretive services to ecotourists. Those operations with higher levels of suitable information stand a much better chance of making more suitable decisions in environmental, cultural, social, economic and business areas of operation. The use of information is dependent on the education and abilities of the managers who are making decisions. There is no substitute for a well-educated planning and management staff.

Acknowledgements

Special thanks to Stephanie Yuill, David Weaver and Anne Ross for many insightful comments on an earlier draft of this chapter.

References

Ashley, C. and Roe, D. (1998) *Enhancing Community Involvement in Wildlife Tourism: Issues and Challenges*, Wildlife and Development Series No 11. International Institute for Environment and Development, London.

Ballantine, J.L. and Eagles, P.F.J. (1994) Defining Canadian ecotourists. *Journal of Sustainable Tourism* 2(1), 1–5.

Basche, C. (1999) *Being Green is your Business*. Tourism Council Australia, Woolloomooloo, New South Wales.

Blamey, R.K. (1995) *The Nature of Ecotourism*. Bureau of Tourism Research, Canberra.

Blamey, R.K. (1997) Ecotourism: the search for an operational definition. *Journal of Sustainable Tourism* 5(2), 109–131.

Blamey, R.K. and Braithwaite, V.A. (1997) A social values segmentation of the potential ecotourism market. *Journal of Sustainable Tourism* 5(1), 29–45.

Boele, N. (1996) *Tourism Switched On: Sustainable Energy Technologies for the Australian Tourism Industry*. Tourism Council Australia, Barton, Australia.

British Tourist Authority (1999) Travel Trade. http://www.bta.org.uk/traveltrade/bluebadge.htm

Butler, J.R. and Hvenegaard, G.T. (1988) The economic values of bird watching associated with Point Pelee National Park, Canada and their contribution to adjacent communities. Unpublished paper presented at The Second Symposium on Social Science in Resource Management. University of Illinois, Urbana, Illinois.

Canadian Tourism Commission (CTC) (1995) *Adventure Travel in Canada: an Overview of Product, Market and Business Potential*. Canadian Tourism Commission, Ottawa, Ontario.

Canadian Tourism Commission (1997) *Adventure Travel and Ecotourism: the Challenge Ahead*. Canadian Tourism Commission, Ottawa, Ontario.

Christiansen, G. and Conner, N. (1999) *The Contribution of Montague Island Nature Reserve to Regional Economic Development*. NPWS Environment Economics Series, New South Wales National Parks and Wildlife Service, Hurstville, New South Wales.

Commonwealth Department of Tourism (1995) *Best Practice Ecotourism: a Guide to Energy and Waste Minimisation*. Commonwealth Department of Tourism, Canberra.

Conner, N. and Christiansen, G. (1998) *The Contribution of Coolah Tops National Park to Regional Economic Development*. NPWS Environment Economics Series, New South Wales National Parks and Wildlife Service, Hurstville, New South Wales.

Crossley, J. and Lee, B. (1994) Characteristics of ecotourists and mass tourists. *Visions in Leisure and Business* 13(2), 4–12.

Department of Environmental Affairs and Tourism (DEAT) (1996) *The Development and Promotion of Tourism in South Africa: White Paper*. Government of South Africa, Cape Town.

Driml, S. and Common, M. (1995) Economic and financial benefits of tourism in major protected areas. *Australian Journal of Environmental Management* 2(2), 19–39.

Eagles, P.F.J. (1992) The travel motivations of Canadian ecotourists. *Journal of Travel Research* 31(2), 3–7.

Eagles, P.F.J. and Higgins, B.R. (1998) Ecotourism market and industry structure. In: Lindberg, K., Epler Wood, M. and Engeldrum, D. (eds) *Ecotourism: a Guide for Planners and Managers*, 2nd edn. The Ecotourism Society, North Bennington, Vermont, pp. 11–43.

Eagles, P.F.J. and Nilsen, P. (2001) *The Ecotourism Bibliography*, 5th edn. The International Ecotourism Society, Burlington, Vermont (in press)

The Ecotourism Society (1999) *TES Bookstore*, http://www.ecotourism.org/books.html

Filion, F.L., Foley, J.P. and Jacquemot, A.J. (1994) The economics of global ecotourism. In: Munasinghe, M. and McNeely, J. (eds) *Protected Area Economics and Policy: Linking Conservation and Sustainable Development*. The World Bank, Washington, DC, pp. 235–252.

Forestry Tasmania (1994) *Guided Nature-based Tourism in Tasmania's Forests*. Forestry Tasmania, Hobart, Tasmania.

Grange, N. and Odendaal, F. (1999) *Guidelines for the Environmental Assessment of Coastal Tourism*. The Secretariat for Eastern African Coastal Area Management (SEACAM), Maputo, Mozambique.

The Guild of Registered Tourist Guides (1999) *The Professional Association of Blue Badge Guides within the British Isles*. http://www.blue-badge.org.uk/guild/frameset.html

Hawkins, D.E., Epler Wood, M. and Bittman, S. (1995) *The Ecolodge Sourcebook*. The Ecotourism Society, North Bennington, Vermont.

Higgins, B.R. (1996) The global structure of the nature tourism industry: ecotourists, tour operators and local business. *Journal of Travel Research* 35(2), 11–18.

HLA Consultants and ARA Consulting Group (1995a) *Ecotourism – Nature/Adventure/Culture: Alberta and British Columbia Market Demand Assessment: Main Report*, 1st edn. Ministry of Small Business, Tourism and Culture, Victoria, British Columbia.

HLA Consultants and ARA Consulting Group (1995b) *Ecotourism – Nature/Adventure/Culture: Alberta and British Columbia Market Demand Assessment: Ecotourism Market Literature Review*, 1st edn. Ministry of Small Business, Tourism and Culture, Victoria, British Columbia.

Hvenegaard, G.T. (1992) Marketing and economic benefits of ecotourism at Point Pelee National Park, Canada. Unpublished paper presented at The First World Congress on Tourism and the Environment, Wetlands Management Field Seminar, Belize City, Belize.

Kangas, P., Shave, M. and Shave, P. (1995) Economics of an ecotourism operation in Belize. *Environmental Management* 19(5), 669–673.

Lindberg, K. and Enriquez, J. (1994) *An Analysis of Ecotourism's Economic Contribution to Conservation and Development in Belize.* World Wildlife Fund, Washington, DC.

Lindberg, K., Epler Wood, M. and Engeldrum, D. (1998) *Ecotourism: a Guide for Planners and Managers*, Vol. 2, 1st edn. The Ecotourism Society, North Bennington, Vermont.

McCawley, R. and Teaff, J.D. (1995) Characteristics and environmental attitudes of coral reef divers in the Florida Keys. In: *Linking Tourism, the Environment, and Sustainability*, Vol. Gen. Tech. Rep. INT-GTR-323. US Department of Agriculture, Forest Service, Intermountain Research Station, Ogden, Utah, pp. 40–46.

McKercher, B. (1998) *The Business of Nature-based Tourism.* Hospitality Press, Elsternwick, Victoria, Australia.

Obua, J. and Harding, D.M. (1996) Visitor characteristics and attitudes towards Kibale National Park, Uganda. *Tourism Management* 17(7), 495–505.

Office of National Tourism (1999a) *Ecotips for Travellers to Australia.* http://www.tourism.gov.au/ecotour/ecotip.html

Office of National Tourism (1999b) *National Ecotourism Accreditation Program (NEAP).* http://www.tourism.gov.au/ecotour/ecoacred.html

Silverberg, K.E., Backman, S.J. and Backman, K.F. (1996) A primary investigation into the psychographics of nature-based travelers to the south eastern United States. *Journal of Travel Research* 35(2), 19–28.

Stanley, D., Smeltzer, S. and Perron, L. (1999) *Provincial Economic Impact Model: Instruction Manual.* Department of Canadian Heritage, Hull, Quebec.

Tourism Council Australia (TCA) and CRC Tourism (1998) *Being Green Keeps You Out of the Red.* Tourism Council Australia, Woolloomooloo, New South Wales.

Travis, A. (1982) Managing the environmental and cultural impacts of tourism and leisure development. *Tourism Management* 3(4), 256–262.

Villemaire, A. and Murray, J. (1997) *Adventure Travel and Ecotourism Implementation Workshops: Summary Report.* Canadian Tourism Commission, Ottawa, Ontario.

Weiler, B. and Richins, H. (1995) Extreme, extravagant and elite: a profile of ecotourists on Earthwatch expeditions. *Tourism Recreation Research* 20(1), 29–36.

Wight, P. (1996a) North American ecotourists: market profile and trip characteristics. *Journal of Travel Research* 34(4), 2–10.

Wight, P. (1996b) North American ecotourism markets: motivations, preferences, and destinations. *Journal of Travel Research* 34(5), 3–10.

Wight, P. (1998) Appealing and marketing to the North American ecotourist. In: Johnston, M., Twynam, G.D. and Haider, W. (eds) *Shaping Tomorrow's North: the Role of Tourism and Recreation.* Lakehead University, Thunder Bay, Ontario, pp. 75–97.

Wild, C. (1998) Lessons from developing and marketing ecotourism: a view from Canada. Paper presented at the Ecotourism Association of Australia annual meeting, Margaret River, Australia.

Woods, B. and Moscardo, G. (1998) Understanding Australian, Japanese, and Taiwanese ecotourists in the Pacific Rim region. *Pacific Tourism Review* 1, 329–339.

Chapter 40

Education and Training

N. Lipscombe and R. Thwaites
School of Environmental and Information Science, Charles Sturt University, Albury, New South Wales, Australia

Introduction

This chapter discusses formal education and training programmes developed to service the ecotourism industry. It will endeavour to differentiate ecotourism from other forms of tourism as a base on which to justify the need for and the rapid development of formal education programmes. Particular reference is made to Australia, as Australia was the first country to develop specific ecotourism degree programmes. The authors also refer to their own experiences in developing and teaching in Australia's first ecotourism degree programme. The chapter is structured under five key issues now facing educational institutions, graduates and the ecotourism industry. Since ecotourism education is still a relatively new phenomenon, the chapter seeks to raise questions that need addressing, rather than provide definitive answers to specific issues. As such, the discussion leads to the identification of issues that have arisen during the short history of ecotourism education in Australia that pertain to educational sectors elsewhere. For the authors, and for all the educators who have a vested interest in delivering industry relevant education and seeing graduates into the industry of their training/education, the benefit of such an approach lies in having the questions/concerns/issues exposed for future discussion and debate.

The 'Greening' of Tourism

Recent decades have seen a growing awareness of environmental issues and interest in natural and cultural heritage conservation by the world's communities. This has had flow-on and growth effects in the tourism industry with commercial operators, worldwide, finding it economically viable to offer tour experiences which focus on the natural and cultural values of an area and/or which are provided in a natural setting (Buckley and Pannell, 1990). In addition to the trend toward the use of natural areas, the growing emphasis on ecologically sustainable practices with respect to activities and developments is also evident.

The concept of ecotourism, which has evolved at the same time, has a number of principles which set it apart from other types of tourism, which in turn, create different educational challenges when compared with mainstream tourism programmes. These principles include:

- *Natural area focus*: ecotourism relies on the use of natural areas in general, protected areas, and/or places with special biological, ecological or cultural interest.
- *Education and interpretation*: ecotourism should include components of education and interpretation of natural and cultural aspects of a place. Visitors should learn about and develop a respect for the culture of the places they visit, and develop an understanding of nature and natural processes of that place and, through this process, for other places and conservation in general.
- *Ecologically sustainable management*: ecotourism is managed to avoid or minimize negative impacts and to confer benefits on host communities and environments, for present and future generations.
- *Contribution to conservation*: ecotourism must benefit conservation; the benefit can be a net benefit: changed community norms through education and, consequently, changed political and social priorities. Ecotourism must be low impact, or at least well managed.
- *Benefits to local communities*: ecotourism should generate economic, cultural and social benefits for local people. This may be in the form of increased employment and entrepreneurial opportunities or, equally, it may be by way of strengthening specific cultural traits or values. At the very least, ecotourism should have net benefits on local social and economic development.
- *Cultural content*: ecotourism recognizes that it is possible to identify and apply management approaches that reduce the stresses on communities and maximize the flow-on of benefits to them.

National and international figures suggest that ecotourism is one of the fastest growing tourism sectors (Reingold, 1993 cited in Blamey, 1995; Wight, 1996), although some people question this status (Blamey, 1995a, b; Blamey and Hatch, 1998). In Australia, for example, the Office of National Tourism reported a rise of 45% in the number of international visitors visiting indigenous sites and attractions in one year from 1995 to 1996 (ONT, 1998). Recent research undertaken by the Australian state of Queensland's Tourist and Travel Corporation (1998) showed the importance of ecotourism to this region (which encompasses the Great Barrier Reef and World Heritage listed rainforest). The study of 780 tourists revealed that 27.6% of respondents could be defined as 'definite' ecotourists, and another 29% as 'probable' ecotourists. The survey showed further that over half of the respondents had an underlying disposition towards nature and learning as part of their vacation, but only about a quarter actually demonstrate this in planning and undertaking vacations. On the other hand, Blamey (1995a, b) and Blamey and Hatch (1998) found little evidence to support the claim that ecotourism in Australia is growing at a rate greater than that of inbound tourism as a whole. They found, however, that participation in certain activities such as white-water rafting, outback safari tours and visits to Aboriginal sites have grown faster than visitor arrivals. While they do not dispute that the ecotourism sector is growing, is it growing faster than inbound tourism as a whole?

It appears evident that ecotourism is providing increasing opportunities for employment, regardless of whether it is growing at the same pace or faster than tourism in general. Further, trend studies and forecasts (TFC, 1999) suggest it will continue to provide employment opportunities into the future. However, as identified in Box 40.1, a number of questions arise about the nature of the employment provided, which further raises issues relating to the type of ecotourism education and training required.

Cotterill (1996) estimated that Australia had some 600 ecotourism operators employing 6500 full-time, part-time and casual employees or around 4500 full-time equivalent employees (about 1% of total tourism employment). He offered a conservative estimate of the industry payroll to be AU$115 million, with turnover about

Box 40.1. Industry sector and employment growth.

Is the perceived growth in ecotourism resulting in increasing employment opportunities?
Is ecotourism really growing, or is there just increased recognition of the place of certain nature-based operations within the overall market?
Is there a growing demand for qualified employees?
What jobs are available, and what skills, experience and training are employers looking for in applicants to fill these jobs?
What are the needs of the ecotourism industry of prospective employees?
Does the ecotourism industry offer a 'spectrum' of opportunities, seeking employees with a range of skills and experience?
Is there a growing demand for graduates in ecotourism?
Do ecotourism operators need to employ university graduates?
Where do university and TAFE-trained graduates fit into an ecotourism job 'spectrum'?

AU$250 million. His profile of the ecotourism industry suggested that the average operator employs about ten staff, or six full-time equivalents, but that most operators have only four or fewer employees. McKercher (1998) described the nature-based tourism sector as typified by businesses which are

> run by owner-operators who have few or no full-time staff other than family members. Most have no formal business or tourism training. Many of the businesses are marginal, and many owners are forced to seek a second income to keep them operational. Too many operators say they are in the tourism game as a 'lifestyle' choice. Distressingly, the drop-out rate of failed businesses is extremely high.
>
> (McKercher, 1998, p. 2) (see also Chapter 36)

Why an Ecotourism Education?

The need for formal education programmes was recognized in the Australian National Ecotourism Strategy (Allcock *et al.*, 1994, p. 41) as a means of ensuring that high quality industry standards are met. The strategy called for the development of 'environmental education modules to encourage the adoption of best practice in ecotourism (industry, conservation groups, tertiary education institutions)'. The Queensland Ecotourism Plan (Department of Tourism, Small Business and Industry, 1997, p. 30) was more specific, referring directly to undergraduate and postgraduate degree courses which include ecotourism components but which are also productive and active research programmes aimed at the delivery of quality ecotourism products. It states

> Ecotourism operators and their employees require specialised training in areas such as ecology, environmental education, environmental and resource management, communication and business skills. Training should be competency based, tailored to the particular requirements of the industry and provided in a culturally appropriate manner.

Such outcomes, clearly, are desirable not just in Queensland and the rest of Australia, but in other ecotourism destinations as well.

In developing a bachelor of applied science degree in ecotourism, Charles Sturt University, located in southern New South Wales, consulted with various industry and government agency representatives on course structure and content. A range of knowledge, skills and attitude requirements were identified which would need to be developed in students to ensure that on graduation they could contribute to the development of the ecotourism industry (N. Lipscombe, unpublished). These are summarized in Box 40.2.

Box 40.2. Educational requirements identified by Charles Sturt University, industry and government representatives.

Knowledge
Contemporary philosophy and ethics towards ecotourism
Basic ecological and geomorphological principles pertaining to Australia
Australian wildlife and vegetation dynamics and interrelationships
Environmental impact occurrence and management relating to ecotourism
Ecologically sustainable development and environmental management principles
Cultural heritage and cultural heritage management principles
Business management theory
Ecotourism business practices
Communication and interpretation theory relating to natural and cultural heritage
Leadership theory

Skills
Ability to apply philosophical and ethical practice in ecotourism management
Skills in communicating the dynamics, interrelationships and management of natural and cultural heritage
Skills in business management and ecotourism business practices
Skills in the application of leadership theory
Skills in recognizing, evaluating and resolving tourism-related environmental, social and cultural impacts
Ability to implement ecologically sustainable development principles pertaining to ecotourism

Attitudes
An appreciation of ecologically sustainable development principles
An appreciation of ethical business practice
An awareness of the importance of environment and heritage management to the ecotourism industry
An appreciation of ethical ecotourism operations
The encouragement of an ethical profile of ecotourism to the public through environmentally responsible ecotourism operations

For the full potential of ecotourism to be reached it is vital that there is a highly skilled workforce with the capacity to market, interpret and deliver ecologically, culturally, socially and financially sustainable products (A. Crabtree, unpublished). A credible tourism industry is dependent on training and education that provides these specialized skills, especially at the tourist/operator interface: the guide. At this level there needs to be particular emphasis on training that includes a significant proportion of environmental and cultural content, combined with training in interpretation (see Chapter 35). At the 'operator' level there needs to be a balance of this against business management, accounting and marketing components. At all levels, training should aim to enhance visitors' experiences and lead to more sustainable environmental practices; to apply the principles on which ecotourism is based.

Extent and Nature of Ecotourism Education

Tertiary and post-secondary educational institutions in Australia and in other parts of the developed world are currently in a state of change as federal and state government policies move education towards 'fee for service' arrangements. As education budgets are diminished, institutions are subject to organizational and financial constraints which, while limiting their ability in the short term to respond to changing industry training needs, has forced the identification of niche markets which, on the basis of predicted growth, can expect high student demand.

It was on the basis of an expected increasing market in ecotourism employment opportunities that universities and post-secondary colleges (Colleges of Technical and Further Education (TAFE) in Australia, Community Colleges in the USA) have in recent times developed courses designed to provide qualified and trained graduates to fill positions within the ecotourism sector. In the years leading up to the development of the first ecotourism course in Australia, tourism researchers were beginning to recognize the need for combining tourism management with recreation or environmental studies (Bowden, 1991; Richins *et al.*, 1995). Bowden (1991) stressed that all people involved in the travel and tourism industry, including educators, needed to respond to the growing market demand for environmentally friendly tourism. Bowden also speculated that ethical practices within the industry would only be adopted if all people in the industry received some environmental education.

Before 1994, there were very few courses being presented offering students information specifically related to ecotourism. Today there are a large number of courses offered at different levels, dealing with ecotourism, cultural tourism, and interpretation of natural and cultural heritage. In the Ecotourism Education Directory compiled by the Australian government (CDOT, 1996), a total of 75 ecotourism courses are listed. Twenty-eight of these are university courses, ranging from graduate certificates offered part-time over 1 year, to Bachelor degrees of 3 years' full-time study. Honours degrees, graduate diplomas and Masters degrees are also listed. As well as these, postgraduate research degrees are listed at ten different universities. The 47 ecotourism courses listed which are not university based are very diverse, ranging from 1-day seminars to 1-year full-time TAFE certificates.

Of the 75 courses listed, eight include ecotourism in the course name. Five of these are offered at universities, including Bachelor Applied Science (Ecotourism), Bachelor of Science (Ecotourism), Bachelor of Technology (Ecotourism), and Graduate Certificate (Ecotourism). The other three include an Introduction to Ecotourism Certificate, the Ecotourism Planning and Management Training Program, and a Certificate in Ecotourism Operations. A further nine courses have a strong emphasis on ecotourism, sustainable or environmental tourism, including two at universities (Bachelor Applied Science (Environmental Management and Tourism), and Graduate Certificate (Natural and Cultural Heritage Interpretation)). Seven are offered outside universities (certificates in tour guiding, environmental and cultural tourism, and short courses in eco-awareness and ecologically sustainable diving).

Twenty-four of the 75 courses focus specifically on interpretation and guiding in the natural and cultural environment. Many of these are training programmes targeting specific groups, such as Discovery Rangers, and Aboriginal communities, offering courses from 1 day only to 1 full year. Five of the listed courses are university based, offering Bachelor, Grad. Dip. and Masters degrees in outdoor education, Masters of social and environmental education, and Masters of tourism interpretation. A further 13 courses seem to be focused in the area of general tourism, with no strong emphasis on ecotourism or interpretation and guiding in the natural or cultural environment, while 21 of the listed courses seemed to only indirectly relate to tourism and ecotourism. These include university degrees in environmental science and environmental management, nature conservation, protected area management, coastal resource management, recreation planning and management, and environmental education. The non-university courses covered areas such as resource management, landcare, conservation, landscape management and reef biology.

While 17 of the courses listed focus specifically on ecotourism, environmental or sustainable tourism, and 24 on interpretation and guiding in the natural and cultural environment, 34 (almost half) seem to have little specific reference to ecotourism, relating to tourism more generally or environmental management. Other existing courses in areas such as tourism, tourism marketing, tourism and hospitality, and tourism management are not listed in this directory of ecotourism courses. However, in recent years, ecotourism, environmental tourism, nature-based tourism and sustainable tourism have become more commonly available minor or major study options for students undertaking traditional tourism, hospitality and marketing degrees.

Since the publication of the Ecotourism Education Directory in 1996 (CDOT, 1996) the number of courses which are either fully or fractionally devoted to ecotourism study in the post-secondary and tertiary sectors appears to have grown rapidly. Hence, it would be difficult to determine with any accuracy the number of students involved and the number of graduates who have found their way into the industry. What is evident is that courses in the ecotourism, nature-based tourism and sustainable tourism fields range from 1-day intensive workshops and seminars to 3 years of full-time study with the full range giving certification. What is also evident is that ecotourism subjects are becoming a commonly available option as elective or minor streams for a variety of environmental/tourism courses.

While the number of students undertaking ecotourism or related studies across Australia has increased rapidly since 1994, the fact that courses are so differentiated in terms of depth, breadth and duration of study poses a dilemma for the professional body and the industry. This dilemma stems from an unclear understanding of the nature of the education acquired by the diversity of graduates from the different courses and institutions. In relation to the issue of ecotourism education, Box 40.3 shows the issues that need to be addressed.

The University Sector

In general, the degree courses offered by the 36 Australian universities are studied full-time and normally take 3 years to complete by full-time mode and 6 years by part-time mode (depending on the amount of credit awarded for previous tertiary or post-secondary study). Provision is made for part-time, including evening study, and some universities have well-established systems of external studies in the same courses as those offered full-time. Courses aim to develop knowledge, skills and attitudes through subjects that incorporate theoretical, conceptual and practical elements.

All courses are internally developed, assessed and reviewed. However, because the system in place requires an assessment of the need and demand for the course initially, as well as the development of up to 24 subjects, the course development process can take in excess of 1 year before being offered for study. Universities, therefore, find it difficult to react quickly to changes in vocational markets. In order to be able to react more quickly to changes in the market, universities, in recent years, have made a number of changes. These include:

Box 40.3. Ecotourism education.

Are tourism operators aware that university training exists in ecotourism, and are they aware of the skills which graduates might possess?
Are tourism operators aware of the different skills that TAFE and university trained graduates might possess, and the different ways in which these skills could be put to use?
How can a stronger link be made between educational institutions and the industry to ensure that they are providing the education and training needed, that the industry has access to a steady supply of quality graduates, and that graduates are aware of the opportunities existing in the marketplace for them?

- allowing students to study individual subjects at a set fee without being enrolled in a course;
- developing niche market minors (four subjects) as an option for students enrolled in established degrees;
- providing industry training workshops;
- offering summer schools to niche markets; and
- providing non-accredited short courses to industry and professional groups.

The TAFE Sector

TAFE colleges in Australia (similar in purpose and function to US community colleges) are a significant provider of industry training. Delivered through a network of some 215 major colleges (91 metropolitan and 124 non-metropolitan colleges) initial skills in vocational and preparatory courses reach in excess of 1 million students each year (Noordhoorn, 1990, p. 27). Most of the courses are aimed at providing vocational education. TAFE has been and is the most accessible tertiary provider, both in geographical and educational terms. In addition, it has strong links with the labour market and is directly affected by changes in labour market conditions. As far as leisure/recreation/tourism is concerned, TAFE has been primarily a provider of programmes and courses to the public. More recently, however, courses to train personnel in outdoor tour guiding, resource and environmental management, and outdoor education have emerged in response to a demand from the tourism and ecotourism industry. Tourism training and education, and related courses, largely consist of single-entry post-secondary courses (pre-employment courses, certificates, advanced certificates and associate diplomas from 6 weeks' to 2 years' duration) which provide access, through articulation, into a university programme.

The perception and knowledge that the industry has regarding the diversity of education and training opportunities available, and its suitability for meeting employment requirements, need to be discussed by addressing certain questions (Box 40.4).

Current ecotourism training is actually characterized by both top-heavy and entry-level courses. Career paths have not been developed in this industry and the number of management-level jobs is restricted, unless you start your own business. At the other extreme, entry-level training is frequently deemed inappropriate given that the industry demands a great deal of sophistication and wealth of experience from both guides and operators. Respondents to a recent survey of ecotourism graduates (R. Thwaites, unpublished) on training needs suggested that lifetime experiences and practical skills were often considered more important than formal education or qualifications. Ecotourism employers are often far more concerned about appropriate licences and industry-based skills to ensure 'duty of care', and workplace health and safety requirements and practical experience, than any evidence of environmental or cultural content knowledge or interpretation skills picked up through a formal training course (no matter how ecotourism specific). Multi-skilling is a necessity, not a luxury for employment in the ecotourism industry; most operators are small businesses and multiple-functions and cross-industry competencies are the norm. A guide may need

Box 40.4. Diversity of education opportunities.

Are university graduates competing for the same jobs with graduates from 6-week short courses and TAFE certificate courses?
What are employers looking for, in terms of formal education and training, for ecotourism positions?
In the eyes of ecotourism operators and employers is there a difference between university and TAFE qualifications?
Which certification carries with it a competitive advantage in terms of employment?

to hold appropriate activity-specific skills, qualifications or licences (e.g. language skills, coxswain's certificate, bus driver's licence, swift water rescue) as well as the more routine skills of a guide. An owner-operator will have to juggle business-related functions and skills (marketing, accounting, etc.) with the operational aspects of running an ecotour.

Because of the rather eclectic range of skills that ecotourism employment demands, the student choosing an appropriate stand-alone training course is in a difficult, if not impossible position because not all stand-alone courses provide all the skills and knowledge required by employers. Opportunities, however, do exist for upskilling and continued development. A range of alternative training providers (e.g. adult education organizations, continuing education programmes at university, private providers, non-profit organizations and others) offer a variety of programmes that supply content on specific knowledge areas, environments or cultural groups. Industry-based qualifications are also available. Training not geared specifically to ecotourism, but to a number of related fields (outdoor education, parks and wildlife management, etc.) can be adapted to specific needs. Potential students and trainees are well advised to do careful research and to approach training providers directly to ensure that the education and training offered serves, or can be adapted to serve, their particular needs. There is, however, now a burgeoning number of more relevant courses and education packages that offer both entry-level training and upskilling and are good examples of 'training designed by industry for the industry' (Haase, 1995, p. 163).

The Industry/Academia Link

Within Australia, as in other parts of the world, the need to establish closer links between 'business' and 'academia' is well recognized. The Business/Higher Education Round Table established in 1990, has several goals of which the establishment of such links is deemed to be of primary importance. In this context an implicit goal is to enhance the relevance of the education provided to the needs of the business community.

Educational institutions are quick to adopt mechanisms to involve employers and professional associations in course design and delivery. However, this practice clearly continues to be an issue for those in the industry who have 'achieved' in their business development without tertiary qualifications and who regard business experience as being more relevant than a tertiary education. In relation to the content of tertiary courses, educational institutions could certainly be doing more to make students aware of the requirements of the industry to which they are aiming beyond those attributes provided by their course. In addition, educational institutions could often do more to provide the support to facilitate work experience programmes and give more emphasis to the development of business management and personal communication skills. At the industry level, tertiary educational institutions are well placed (and is a growing expectation) to be offering seminars, training workshops, consulting services, short courses, conference involvement, facilitation, committee involvement, ongoing liaison and industry-specific research: all mechanisms for fostering links and involvement with the industry. While opportunities in regard to these activities are yet to be fully realized, there is a need also for closer links to be driven by the industry.

Formal education and training for tour guides and tour operators is still relatively new to institutional programmes. While the tourism industry has well-established education and training programmes and awards for the management and hospitality/travel consultant sectors, tour guide and tour operator standards remain outdated with a focus on service-related functions. Tour guide standards have recently undergone a much needed review (EAA, 1998) reflecting the dramatic changes that have occurred in the tourism industry in the last decade, with the market becoming increasingly sophisticated, increasingly knowledgeable on environmental and cultural issues and demanding more authentic experiences.

The Ecotourism Association of Australia, while encouraging educational institutions to develop programmes of study to service the ecotourism industry (by being represented on course accreditation boards), have instituted an industry accreditation programme (NEAP) in an endeavour to increase the service delivery standards (see Chapter 37). The programme 'is designed to provide a range of benefits for ecotourism businesses, potential ecotourism clients, natural areas where ecotourism operations occur, natural area managers and local communities where ecotourism could or does occur' (EAA, 1996, p. 3). The 'self-assessed' programme is comprehensive in identifying the accreditation criteria and bringing industry closer to its professional association. However, it places very little emphasis on the need for operators and/or guides to have attained a recommended level of education and training, and does not differentiate in any way the value of the numerous levels of education and training available. A clear and purposeful link between education and industry, and between education/industry and the professional/industry association would seem therefore, to be a mutually beneficial issue to discuss. Box 40.5 shows the questions which could form the basis of such discussion.

An important issue not addressed to date in forums that combine educators and business operators, is the need for both groups to be realistic about their expectations of new graduates. While students do mature during a 2- or 3-year period in an education/training institution, and while fieldwork/placements provide some experiential learning, it should not be expected that the graduate who emerges from a course is a fully fledged professional. As in any profession, the industry must recognize that the experiences gathered during early years of employment, complementing previous education and training, are a necessary aspect of professional development for graduates. As Prosser (1990, p. 5) points out, 'the potential for misunderstanding on both sides therefore, contributes to the importance of developing effective mechanisms for liaison between the parties. It also illustrates the importance of institutions tracking the destination and career path of graduates to provide performance indicators'.

Similarly, there has long existed a perceived schism between academic learning (regardless of the level) and the learning that is gained by actually performing the multiple tasks required of any position within the workplace. In response to expressed concerns regarding the lack of correspondence between theory and practice, educators have devised models of, and methods to provide, experiential learning. Definitions and explanations of what comprises experiential learning are diverse but they appear to have in common a concern that students are actively involved and are able to apply relevant learning to seek

Box 40.5. Education/industry/professional association links.

How can educational and training institutions make employers more aware of the benefits of employing graduates?
How can educational/training institutions establish and develop flexible and responsive training systems that involves the industry, the professional association?
Do educators and employers need to develop a relationship which enables a more detailed understanding of each other's functional attributes and requirements?
Are 'continuing education' courses an essential part of career development?
How can courses better reflect industry requirements?
Should the professional body be doing more to bridge the gap between educators and the industry?
If the industry and the professional association is serious about ecotourism accreditation (NEAP), is there a need to include the levels of education and training (qualifications) desirable for operators and guides?

flexible solutions to immediate practical problems. Relevant objectives of experiential learning may include: developing an understanding of the interactive nature of the relationship between theoretical concepts and practice; consideration of the perceptions and professional assumptions of an occupational group; flexibility in responding to situations; and awareness of the demands of professional practice (Weil and McGill, 1990). From an industry and institution's perspective, this is best gained by being actively involved in the industry through industry placements during the training course. The understanding of the theoretical and practical relationship in course development and employment preparation for both educational institutions and the industry forms the basis of another issue from which further questions could be posed (Box 40.6).

Now that there are education and training institutions providing ecotourism-specific courses with 'ecotourism' as part of the certification, for some institutions there is a growing concern for qualification flexibility and portability. Because graduates do not always end up in the industry for which their qualification is aimed, are institutions providing a disservice to graduates by including ecotourism in the name of the qualification? While there has been a great deal of debate among academics and certain sectors of the industry over the use of the term 'ecotourism' and its role in describing and marketing certain types of tourism activities, is it a useful 'marketing' tool for graduates to have 'ecotourism' as a part of the title of their qualification? The experience of graduates from Charles Sturt University might suggest that for a small number of jobs, and employers, it may be useful, but for a large group of employers, and greater number of jobs, it has little meaning or importance, or may even elicit negative responses. A further question for discussion therefore could be: What does it mean to the industry to have graduates trained specifically in 'ecotourism'?

Conclusion

Some of the issues raised in this chapter have arisen from responses to questionnaires and discussions with students and graduates of the Bachelor Applied Science degree in Ecotourism at Charles Sturt University, Australia, a world pioneer in this sphere. The first intake of students into this 3-year course began in 1995, and in early 1999, all past and present students were surveyed to gauge the progress of the course, and the expectations and experiences of the students. While this course was developed with the assistance of industry and government representatives and offers a diverse range of theoretical and practical subjects, the survey raised some interesting questions about the course, and the suitability of such a qualification in the ecotourism marketplace. Students identified one of their strongest likes of the course as its practical orientation, and the opportunities provided to gain industry experience. However, they also described unmet expectations related to a similar range of practical issues associ-

Box 40.6. Graduate employment.

What are the expectations of graduates in relation to the course they have studied and its ability to prepare them for a role in the industry?
What role should industry be playing in the training and education of those preparing to enter the industry? (Work experience during study and ongoing professional development after employment?)
How can this role be further facilitated?
Is it possible to develop career paths for graduates?
How can the relationship between theory and practice be better understood: a closer working relationship between educational institutions and the industry?

ated with the connection between the university, the industry and the professional association. A perception that a university education was not valued by the industry, that its relevance was questioned by the industry, and the difficulty graduates face meeting job selection criteria based on TAFE (non-university training institutions) training curricula, were real issues for graduates. These individuals also cited the lack of awareness within the industry of university courses, and of the skills possessed by university graduates. The experience of university graduates seeking employment in a competitive market, therefore, raises questions about the role of education in the ecotourism industry, and the relationship between education institutions and the industry.

Educational institutions have been quick to react to the notion of rapid growth in this sector of the tourism industry often without any real understanding of the nature of the industry and its capacity and desire to absorb graduates from the plethora of courses now available. This chapter has presented a number of issues and many more questions about education and training which seriously need to be addressed if educational programmes are to remain viable and the industry credible. We have not endeavoured to resolve the issues or provide the answers to the questions. Rather, by expressing what are very real concerns for educational providers, the professional association and the industry, we hope that in some future education-specific forum the process of discussion, debate and resolution will take place.

References

Allcock, A., Jones, B., Lane, S. and Grant, J. (1994) *National Ecotourism Strategy*. Australian Government Publishing Service, Canberra.

Blamey, R.K. (1995a) Profiling the ecotourism market. In: The Ecotourism Association of Australia, National Conference, *Taking the Next Steps*, Alice Springs, November.

Blamey, R.K. (1995b) *The Nature of Ecotourism*, BTR Occasional Paper No. 21. Bureau of Tourism Research, Canberra.

Blamey, R. and Hatch, D. (1998) *Profiles and Motivations of Nature-based Tourists Visiting Australia*, BTR Occasional Paper No. 25. Bureau of Tourism Research, Canberra.

Bowden, D. (1991) Ecotourism – the implications for the environmental movement for tourism. In: Ward, B., Wells, J. and Kennedy, M. (eds) *Tourism Education in Australia and New Zealand*, Conference Papers. BTR, Canberra, pp. 198–202.

Buckley, R. and Pannell, J. (1990) Environmental impacts of tourism and recreation in national parks and conservation reserves. *Journal of Tourism Studies* 1(1), 24–32.

Commonwealth Department of Tourism (CDOT) (1996) *Directory of Ecotourism Education*. Commonwealth Department of Tourism, Canberra.

Cotterill, D. (1996) Developing a sustainable ecotourism business. In: Richins, H., Richardson, J. and Crabtree, A. (eds) *Taking the Next Steps*. Ecotourism Association of Australia, Brisbane.

Department of Tourism, Small Business and Industry (1997) *Queensland Ecotourism Plan*. QGP, Brisbane.

Ecotourism Association of Australia (EAA) (1996) *National Ecotourism Accreditation Program*. Commonwealth of Australia, Canberra.

Ecotourism Association of Australia (EAA) (1998) *Australian Ecotourism Guide 1998/99*. Ecotourism Association of Australia, Brisbane.

Haase, C. (1995) Tourism training – so? In: Richins, H., Richardson, J. and Crabtree, A. (eds) *Taking the Next Steps*. Ecotourism Association of Australia, Brisbane.

McKercher, R.D. (1998) *The Business of Nature-based Tourism*. Hospitality Press, Elsternwick, Victoria.

Noordhoorn, U. (1990) The role of TAFE in recreation education and training. In: Prosser, G. (ed.) *The Challenge for the Future*. Department of the Arts, Sport, the Environment, Tourism and Territories, AGPS, Canberra.

Office of National Tourism (ONT) (1998) *Aboriginal and Torres Strait Islander tourism.* Tourism Facts No. 11. Office of National Tourism, Canberra.

Prosser, G. (ed.) (1990) *Recreation Education and Employment: the Challenge of the Future.* Department of the Arts, Sport, The Environment, Tourism and Territories, AGP, Canberra.

Queensland Tourist and Travel Corporation (1998) *Ecotrends,* September.

Reingold, L. (1993) Identifying the elusive ecotourist. In: *Going Green,* a supplement to *Tour and Travel News.* 25 October, 36–39. Cited in Blamey, R.K. (1995) *The Nature of Ecotourism,* BTR Occasional Paper No. 21. Bureau of Tourism Research, Canberra.

Richins, H., Richardson, J. and Crabtree, A. (1995) Proceedings of the Ecotourism Association of Australia National Conference, *Taking the Next Steps,* Alice-Springs.

Tourism Forecasting Council (TFC) (1999) *Forecast: the Ninth Report of the Tourism Forecasting Council,* Vol. 5, No. 1, August, Tourism Forecasting Council, Canberra.

Weil, S. and McGill, I. (1990) A framework for making sense of experiential learning. In: Weil, S. and McGill, I. (eds) *Making Sense of Experiential Learning.* The Society for Research into Higher Education and Open University Press, Bury St Edmunds, UK, pp. 3–24.

Wight, P.A. (1996) North American ecotourism markets: motivations, preferences and destinations. *Journal of Travel Research* 35(1), 3–10.

Chapter 41
Areas and Needs in Ecotourism Research

D.A. Fennell
Department of Recreation and Leisure Studies, Faculty of Physical Education and Recreation, Brock University, St Catharines, Ontario, Canada

Research Needs: Beyond the Basics

In 1997 I had the opportunity to attend a large international ecotourism conference, which attracted many researchers, governmental representatives, and industry executives from around the world. Knowing that this would probably be the only major tourism meeting that I might be attending over the course of the year, I had high expectations. However, as the conference progressed, I felt more and more jaded by the fact that the conference did not seem to living up to its billing as ecotourism's 'way to the future'. Sessions were not well attended nor were they well conceived, and unfortunately, as the conference wore on, my patience wore out. Yes, I was treated to a good display of slides on new and old ecotourism projects, trips, and developments from around the world. Missing, however, was any hint of theory building or development, empiricism, or new conceptualizations that should form the basis of a conference and the evolution of a field of study.

The field of ecotourism has a good excuse for this lack of academic rigour: infancy. Having been around for only 15 years or so is reason enough to warrant a period of grace to enable the field to organize or develop a body of knowledge. While some may argue that the research that currently exists is consistent with the progression of the tourism field in general, others, as I do, suggest the need for less descriptive research using many of the same old methods, principles and ideas. Are we becoming stale? Perhaps. Do we need to broaden our horizons in order to generate some new perspectives? Yes. Let me briefly outline some observations regarding the state of ecotourism research before venturing into a discussion of potential areas of research, which is the topic of the next section.

1. Ecotourism has a shortage of sound empirical data. This sentiment is mirrored by Lindberg *et al.* (1996), who suggest that 'little quantitative analysis of ecotourism's success in achieving conservation and development objectives has been reported' (p. 544). A good example of new research that makes a significant empirical contribution to the field is the work of Sirakaya and McLellan (1998) on tour operator compliance with ecotourism principles. Another is Bookbinder *et al.* (1998), whose empirical research near Royal Chitwan National Park in Nepal established that the economic benefits of ecotourism are not as great as the community assumed.

2. Closely linked to the first point is the sense that the field would greatly benefit from research methods and analyses that are more sophisticated in nature (e.g. confirmatory factor analyses on, for example, tourism impact scales). In particular there is a dearth of experimental (pre- and post-test analyses), quasi experimental, and comparative research in the literature. Consequently, most of the research to date is exploratory and descriptive.

3. Researchers are quick to note in their studies that ecotourism is the fastest growing sector of the largest industry in the world; we see this time and time again in the literature. Possibly as a result of its relative importance as an income earner, ecotourism research has been given a high priority for funding in some countries. Not so in other countries, however, where the sector continues to be misunderstood and misrepresented by both government and industry (perhaps because they are either unaware or unmoved by the supposed growth of ecotourism). In the latter case, researchers must continue to demonstrate the need for funding in order to carry out ecotourism research. This demands better linkages with government and industry, and the employment of solid research methodologies in securing these monies, as well as in demonstrating the alleged growth rates of ecotourism.

4. Until recently there has been what might be termed an 'insularity' in the focus of ecotourism research which, if continued, will do very little to further the field. As such, strong growth within the field will be achieved by diversifying the theoretical repertoire that researchers have conventionally relied upon (as suggested earlier). Because a strong theoretical foundation is virtually non-existent in ecotourism research, researchers will be forced to draw upon other fields. The long-established disciplines of psychology, sociology, anthropology, economics, ecology and geography will provide answers to some of ecotourism's most pressing questions, as they have been doing for tourism in general. In exploring these, researchers may gain recognition both within the field and in other disciplines, by testing the work of others.

In general, based on the points set out above (the need to build theory, empiricism and conceptual frameworks in ecotourism research), the suggestions that follow may help to stimulate more discussion on research methods and techniques. While some theories happen to be older, the feeling is that rather than simply ignore these perspectives, we need to, in the words of Einstein, 'think in a radically old way'.

Areas of Research to Explore

This section concentrates on specific areas of research, principally from outside the realm of tourism studies. The reader will note the absence of a discussion on ecotourism definitions. While it is true that a universally accepted definition does not exist, there is some consensus on the principles that ought to be included in a definition of ecotourism. These include, but are not limited to, the fact that it is a part of a broader nature-based or natural resource-based tourism; it contributes to conservation or preservation of natural areas; it provides benefits to local people; it facilitates learning; it should be sustainable and ethically based, and it should be effectively managed. Not coincidentally, many of the theoretical perspectives outlined below either directly or indirectly relate to some of the principles set forth in a comprehensive definition of ecotourism, especially sustainability, education and responsibility (see Chapter 1). In doing so, they, in my opinion, relate to many of the most important underlying issues confronting the ecotourism industry. As such, the discussion on ethics, values, attitudes, impacts and carrying capacity, for example, have not only social and economic implications, but also ecological ones. Finally, while many case studies have been included in the chapter, constraints on length preclude in-depth analysis of any one area.

Social traps

The social trap theory posits that the choices open to an individual depend on the system of which one is a part. If these choices are unacceptable or limited, they may require the individual to act on decisions that are self-defeating for the person or for society as a whole (Platt, 1973). Individual behaviours (i.e. resource consumption) are rewarding in the short run (to the individual), but lead to negative outcomes for the collective – the commons – in the long run (Edney, 1980) (see also Hardin's (1968) tragedy of the commons). For example, a chronic problem in the commercial fishing industry is the depletion of fish stocks. While fishermen often understand the importance of viable populations, their immediate need is to sustain themselves economically year after year, despite the dwindling supply of fish. The same holds true with tourism in that the structure of the industry often does not work to the advantage of the individual worker, especially in developing world economies due to multinational domination of resources (airlines, hotels, food, etc.). In cases where local people have no other employment alternatives, they must submit to the policies and procedures of those who control the industry. Much like the case of overfishing, ecotour operators must often compete for a share of a finite natural-resource base. Too many operators accommodating too many tourists can become somewhat self-defeating. (The Galapagos Islands is a good example of the social trap phenomenon, whereby the decisions by policy makers to periodically increase visitation undermine initial attempts to control the impacts of tourism in this sensitive region. The lure of tourists and associated tourist spending is too powerful.) According to Platt, morality and greed are not the central problem in this phenomenon, but rather the arrangement in time of costs and rewards of those involved.

While self-enhancement and greed are not central to Platt's social trap theory, Edney (1980) explores a number of theories which may be used to examine problems of resource allocation within society. He underscores the philosophy of Hobbes in suggesting that a person's basic drives are built on self-interest, egotism and competition. People in an open-market system are ruled by competition, and these aggressive tendencies are thought by some to be biologically driven (see Ardrey, 1970). Similarly, Dawkins' (1976) selfish gene theory posits that social behaviours are a function of biological selection. Selfishness is thought to be necessary and basic to one's survival. Conversely, universal love and welfare of the group 'are concepts which simply do not make sense' (p. 2). The influence of competition and selfishness are so invasive, according to Dawkins, that associations of mutual benefit within the community are often unstable because stakeholders continually strive to obtain more out of the association than they put in. Sharing and openness are concepts that have been described as critical to the success of community development. One wonders, however, whether ecotourism can be used as a vehicle to secure social and economic equity within a community, as many authors are wont to illustrate (Sproule, 1996). Is, therefore, the desire to secure economic well-being too strong for the individual ecotourism operator, or is the philosophy and spirit of ecotourism persuasive enough to counter individuality?

As much of the social psychological literature has demonstrated, there are setting and situational factors that strongly influence the behaviour of individuals, which may in part explain tendencies that some individuals have towards competition and greed, beyond the biological perspective.

Values

One of the very timely conclusions of the work of Edney is the realization that the examination of *values* should prove to be a more fruitful way of exploring conflict that exists within human systems. The rationale for this is that effective decision making can only occur if we understand the underlying principles – equity, freedom of choice, competition – that help shape the fabric of the system in question. Madrigal (1995) suggests that because of the link to a person's cognitive structure, values may be instrumental in predicting human behaviour. As defined by Schwartz (cited in Oishi *et al.*, 1998), values are 'desirable, transitional goals, varying in importance, that serve as guiding principles in people's lives' (p. 1177).

Although Oishi *et al.* (1998) illustrate that the study of values has been overshadowed throughout the 1970s, 1980s, and part of the 1990s by attitudes, attributions, social cognition and group processes, a few pioneering studies have emerged. One of the first was based on the work of Mitchell (1983), who undertook a comprehensive study of values in American society. His research culminated in the development of the values and lifestyles typology (VALS), which comprises four main groups, subdivided into nine lifestyles, each outlining a unique way of life within American society on the basis of values, beliefs and drives. A second typology that was developed to examine people's values is the list of values (LOV) scale, initiated by Kahle (1983). The nine values included in this scale are self-respect, security, warm relationships with others, sense of accomplishment, self-fulfilment, sense of belonging, being well respected, fun and enjoyment in life, and excitement. Kahle feels that this scale, while being more parsimonious than VALS, also has the advantage of having greater predictive utility. Finally, Schwartz (1994) developed a scale of values (SVS) on the basis of ten value types: power, achievement, hedonism, stimulation, self-direction, universalism, benevolence, tradition, conformity, and security. These values are organized in a circular structure with individuals pursuing adjacent values (e.g. power and achievement) over opposite or conflicting values (e.g. power and benevolence). For example, individuals who are high on universalism are also high on benevolence and self-direction. They tend to pursue goals related to equality and protection of the natural world.

Tourism researchers have only recently begun to investigate the role of values and, as outlined by Madrigal (1995), most of this research has related to segmenting the travel market. Madrigal and Kahle (1994) used the LOV to establish a value-system segmentation of tourists visiting Scandinavia. They examined whether vacation activity importance ratings differed across segments comprising tourists who were grouped on the basis of their selection of personal values. However, Madrigal's (1995) more recent work, also using the LOV scale, demonstrates the importance of values in exploring personality type and travel style. In addition, while tourism researchers have developed tourist typologies, these have generally not been hinged upon a values-based philosophy. As such, given the importance of values as a potential for understanding human nature, it is worthwhile to consider the merits of the VALS, LOV and SVS typologies in explaining and predicting the behaviour of ecotourism stakeholders.

Machiavellianism

The work of Machiavelli spawned a social-personality theory that examines how people are manipulated, and the tactics which are employed by those who are influential versus those who are not. Christie and Geis (1970) developed a Machiavellianism scale for the purpose of differentiating people on the basis of power, influence and control. In general, the scale is organized around the following three substantive areas: tactics, views of human nature and morality; and contrast the High Mach (those who are self-interested and who feel the end justifies the means) against the Low Mach (those who promote common good).

Tactics
High Mach: 'A white lie is often a good thing'
Low Mach: 'If something is morally right, compromise is out of the question'
Views of Human Nature
High Mach: 'Most people don't really know what's best for them'

Low Mach: 'Barnum was wrong when he said a sucker is born every minute'
Morality
High Mach: 'Deceit in conduct of war is praiseworthy and honourable'
Low Mach: 'It is better to be humble and honest than important and dishonest'

People fall along a continuum of standards of behaviour based on their responses to the scale. High Machs have relative standards of conduct ('Never tell anyone the real reason you did something, unless it's useful to do so'), while Low Machs are said to have absolute standards ('Honesty is always the best policy'). In addition, High Machs flourish under the following conditions: (i) when interaction is face-to-face; (ii) when rules and guidelines are minimal; and (iii) when emotional arousal is high, contributing to poor task performance for Low Machs. In essence, the High Mach demonstrates a cool detachment, which enables them to keep an emotional distance from other people and situations.

In related research, Hegarty and Sims (1979) found that High Machs behaved significantly less ethically than Low Machs, while Singhapakdi and Vitell (1990) reported that High Mach marketers tended to be more reluctant to punish unethical behaviour. In another study by these authors (Vitell and Singhapakdi, 1991), it was concluded that 'Since Machiavellians tend to value their personal interests but not their clients' interests, the implication for management is to make the self-interest of these individuals intertwined with the clients' interests' (p. 67). This statement is certainly illuminated in the following example related to the environmental ethics of tour operators. Masterton (1992) found that almost all tour operators interviewed agreed that the abuse of the planet is a bad thing, in theory. In practice, however, most operators did not want to discuss their environmental responsibilities, and one went so far as to say the following: 'I have to make a living, and if people want to go to polluted, or overcrowded, or disgustingly commercial tourist traps, then they're going to go. So why shouldn't I get the commission for the trip' (Masterton, 1992, p. 18).

One of the most significant problems facing the ecotourism industry is the practice of eco-opportunism: tourism companies using the ecotourism label to sell their various products; products that may not actually be ecotourism (however defined). The notion that people will quite naturally take care of themselves before the needs of others, and manipulate them to meet their own ends, is indeed a very intriguing element of the tourism operator–tourist relationship which certainly merits further consideration. In addition, the general theory of marketing ethics proposed by Hunt and Vitell (1986) may have utility in providing further guidance in the analysis of decision making in marketing situations involving ethical issues.

Ethics

An area of research that has recently (since the late 1980s) started to attract some attention in tourism and ecotourism studies is ethics (see Wheeler (1994) for an overview of the emergence of ethics in tourism and hospitality). While some of this work is empirically based (see Hall, 1989; Whitney, 1989; Wheeler, 1992), most is descriptive. Codes of ethics, in particular, have been a focus of tourism researchers, along with their use in attempting to curtail sociological and ecological impacts. Tourism researchers may benefit from the general, empirically based literature on codes of ethics in business and marketing where, for example, Singhapakdi and Vitell (1990) and Laczniak and Murphy (1985) write that enforcement is a key component of the employment of a code of ethics within an organization. What seems to be most important about the viability of the code is that top level executives must be seen to support the document in order for others to follow suit (see also Brooks, 1989). While some of these issues have been addressed in the tourism literature (see Blangy and Nielson, 1993), they have remained largely untested.

The concept of organizational culture (OC) is one that is strongly linked to the development of codes of ethics, in business and tourism. OC is defined by Denison (1996, p. 6) as:

> the deep structure of organizations, which is rooted in the values, beliefs, and assumptions held by organizational members. Meaning is established through socialization to a variety of identity groups that converge in the workplace. Interaction reproduces a symbolic world that gives culture both great stability and a certain precarious and fragile nature rooted in the dependence of the system on individual cognition and action.

Codes of ethics, therefore, may be the tacit, outward expression of the organizational culture of a firm. The work of Schein (1985) stands out as prominent in this area, through his creation of the peeled onion conceptualization of OC. Malloy and Fennell (1998) use this multi-layered model to identify the basis of OC in ecotourism businesses. The model and its application to ecotourism is explained as follows:

> Artifacts (layer 1) encompass the physical characteristics of the organization, such as its reward and punishment structure, the advertisements and brochures it produces, the tours it promotes, and the codes of ethics it stipulates to employees and to clientele. The second layer, behaviors, consists of the actual behaviors that can be observed by employees, clientele, and the public at large ... The third layer includes the organization's values. These values can be internalized by employees in a strong clan-like culture. Or they may be less strongly held ... The core of the model – basic assumptions – consists of the very basic ontological realities of the organization. This core describes the metavalues (i.e., unquestioned tacit beliefs) that have come about as a result of the ability of these concepts to traditionally assist in the resolution of organizational dilemmas.
>
> (Malloy and Fennell, p. 48)

Malloy and Fennell argue that the need to address the ethical conduct of ecotourism firms from an organizational perspective becomes increasingly more important as the sector matures and prospers into the 21st century. As such, the sector may demand further exploration in the future in the same way that the general business literature has attempted to do. However, while the principles inherent within codes of ethics and corporate values may be easily conceived, researchers have warned that they may be quite difficult to measure and study in organizations (Payne, 1980).

In related work, Fennell and Malloy (1999) undertook a study to analyse a number of different tourism operators (ecotourism, adventure, fishing, cruise line and golf) to determine ethical differences among these groups. They used the multidimensional ethics scale (MES) first developed by Reidenbach and Robin (1988, 1990), which employs the use of an eight-item scale based on the philosophies of justice, deontology and relativism. Respondents (in this case tour operators) were asked to respond to three tourism scenarios – economic, social and ecological ones – using the scale. The authors found that ecotourism operators were in fact more ethical than the other operators included in the study, on the basis of the responses to the scenarios. In addition, ecotourism operators were found to: (i) use a corporate code of ethics more often than other tourism operators; (ii) have higher levels of education; and (iii) have smaller organizational sizes than other operators. All three of these measures have been found to relate to higher ethical standards in business practice. The MES is just one ethics scale that has been used in the business and marketing literature. Other scales which hold potential in future ecotourism research include the ethical behavior scale, which Fraedrich (1993) used in gauging the ethical behaviour of retail managers. This scale, like the MES, uses a number of moral philosophy perspectives – rule and act utilitarianism, rule and act deontology, and egoism – to examine business ethics. Finally, the defining issues test developed by Rest (1979) has been used to measure moral reasoning of various stakeholders (usually in a business context), and may therefore lend itself nicely to examining the moral stances of operators, local people, government officials and tourists.

All of the above-mentioned scales are designed to test moral viewpoints on the basis of theoretical perspectives. While these scales are some of the best in terms of the generation of empirical data, Fennell (2000a) has suggested that the tourism industry could benefit from the creation of centres of applied ethics, in the same way centres of business ethics and environmental ethics have benefited those fields. These tourism centres would be responsible for the collection of both empirical and descriptive data, and provide guidance to operators, for example, on the day-to-day issues related to their operations. Tourist input would also be encouraged in terms of the evaluation of the ethical behaviour of ecotour operators (see Appendix). Precedents for this type of tourist involvement have already been set by the national accreditation programmes of Australia and the Canadian Tourism Commission. In both cases tourists are encouraged to provide feedback on ecotour operators. Surveys, such as the example in the Appendix, could easily be employed as a means by which to gather information on both good (ethical) and bad (unethical) operators, and in doing so address the 'responsibility' principle found in so many definitions of ecotourism. Furthermore, such information, along with the aforementioned scales could be used to further classify operations on the basis of their ethical and organizational behaviour and, in doing so, provide a clear perspective on the mission and intent of so-called ecotourism businesses.

Benefits

Research on the benefits of participation in leisure activities has attracted the attention of a number of leisure researchers over the past decade. Benefit, as defined by Driver *et al.* (1991), refers to any improvement in one's condition (or a group or society) that is viewed to be advantageous, and such benefits may be categorized as social, psychological, spiritual, emotional, physiological and environmental. Driver and his colleagues set out to identify a series of human needs that could be gratified by participation in leisure pursuits. Through a number of adjustments, these authors developed the recreation experience preference (REP) scales. As of the early 1990s, over 40 REP scales had been developed to measure the benefits of leisure experiences. One of the key features of the REP, over other leisure scales, is its focus on outdoor recreation environments, hence its applicability to ecotourism. This can be seen in the following examples of REP scales and the preference domains into which the scales are grouped.

- *Enjoy nature:* scenery; general nature experience
- *Share similar values:* be with friends; be with people having similar values
- *Outdoor learning:* general learning; exploration; learn about nature
- *Escape physical stressors:* Tranquility/solitude; privacy; escape crowds; escape noise
- *Achievement:* seeking excitement; social recognition; skill development
- *Risk reduction:* risk moderation; risk prevention.

Many of these domains have been discussed in the ecotourism literature, and especially those related to the enjoyment of nature and outdoor learning, which correspond to the ecotourism literature on different types of ecotourists (see Laarman and Durst, 1987; Lindberg, 1991). Furthermore, there is evidence to support the fact that some dedicated ecotourists and adventure tourists appear to prefer the company of like-minded or similarly skilled individuals in their experiences ('Share similar values'). For example, Ewert (1985) found that beginner climbers were more extrinsically motivated to participate in mountain climbing activity, whereas experienced climbers did so for intrinsic reasons.

Other domains of the REP, i.e. 'Achievement/stimulation' and 'Risk reduction', may also provide a theoretical basis from which to analyse the differences that exist between adventure tourists and ecotourists. Although there is a strong intuitive sense that ecotourists are different from adventure tourists in terms of motivation and benefits sought (see Dyess, 1997), there is really only anecdotal evidence explaining how they differ. While the literature on adventure pursuits focuses on risk, skill and competence, Fennell (2000b) suggests that natural resource-based tourists, such as adventure tourists and ecotourists, may be differentiated on the basis of consumptive values, impact, focus of learning and reliance on technical skills.

Finally, the 'Seeking excitement' experience preference may help to further differentiate adventure tourists and ecotourists. Zuckerman's (1979) sensation seeking scale (SSS) has been employed to identify a personality trait defined as a 'need for varied, novel, and complex sensations and experiences and the willingness to take physical and social risks for the sake of such experience' (p. 10). While the scale has recently been applied to groups such as international travellers (Fontaine, 1994), it can be argued that a modified form of the SSS may be appropriate for individuals currently engaging in a number of types of adventure and ecotourism holidaying (both nationally and internationally). This would allow for an assessment of whether the type of excursion they actually experienced is consistent with their need for sensation. For example, results of the SSS might indicate whether an individual is more suited toward an adventure holiday, an ecotourism-based holiday, or a cultural holiday, or a combination of the three. Further analysis of individual responses to questionnaires would then allow researchers to determine what the above categories are perceived to be by holiday travellers. Form V of the SSS contains 40 questions in a paired comparison format. Examples of some of these comparisons are as follows:

A. I like to explore a strange city or section of town by myself, even if it means getting lost.
B. I prefer a guide when I am in a place I don't know well.

A. I would like to take up the sport of water-skiing.
B. I would not like to take up water-skiing.

A. I prefer the surface of the water to the depths.
B. I would like to go scuba-diving.

Research on benefits, therefore, in the manner outlined above, would help further our understanding of different educational benefits that appeal to ecotourists and other types of tourists.

Attitudes

The development of a more progressive environmentalism in the 1970s led to the belief that global ecological problems derive from a dominant social paradigm (DSP) that influences the way we do business, develop policy, and approach science and technology. Dunlap and Van Liere (1978) suggested that a new environmental value was taking hold in society that opposed the largely anthropocentric ways of the DSP. This new environmental paradigm (NEP), an ecocentric way of thinking, was described as being more supportive of a holistic, integrative view of humanity, and is one that places humanity into the context of nature and not above nature. Dunlap and Van Liere developed a NEP scale, consisting of 12 items, designed to gauge how accepting the public was of the NEP (Table 41.1). Their initial study found that the scale was reliable, valid and unidimensional; however, subsequent research determined that the scale is in fact multidimensional (see Albrecht *et al.*, 1982; Geller and Lasley, 1985). Geller and Lasley (1985), for example, in their study of farmers and urbanites confirmed a three-factor model on the basis of nine of the 12 items from the original scale. The three dimensions included 'Balance of nature' (items 1,

Table 41.1. The new environmental paradigm scale.

1. The balance of nature is very delicate and easily upset
2. When humans interfere with nature it often produces disastrous consequences
3. Humans must live in harmony with nature to survive
4. Mankind is severely abusing the environment
5. We are approaching the number of people the Earth can support
6. The Earth is like a spaceship with only limited room and resources
7. There are limits to growth beyond which our industrialized society cannot expand
8. To maintain a healthy economy we will have to develop a 'steady-state' economy where industrial growth is controlled
9. Mankind was created to rule over nature
10. Humans have the right to modify the natural environment to suit their needs
11. Plants and animals exist primarily to be used by humans
12. Humans need not adapt to their natural environment because they can remake it to suit their needs

2, 3, 4), 'Limits to growth' (items 6, 7, 8), and 'Man over nature' (items 9, 10).

While historically the NEP has been subject to some debate, researchers have shown renewed interest in the scale. In a study of visitors to Biscayne Bay National Park, USA, Jurowski *et al.* (1995) found that it was the younger visitors to the park that maintained more ecocentric attitudes. This group favoured allocating park resources to protecting the park environment, while the older and more anthropocentric clientele, in contrast, favoured efforts to develop the park environment. The authors concluded by suggesting that park managers may wish to include alternative management practices for the two groups. In related research on attitudes, Jackson (1986) found evidence that individuals who prefer what may be termed appreciative outdoor pursuits (e.g. hiking) maintain stronger pro-environmental attitudes than those who are more consumptive in their outdoor recreational pursuits (e.g. hunting and trail biking). Also, the work of Kellert (1985) suggests the existence of a typology of attitudes toward animals and the natural environment. These range from positive attitudes (e.g. ecologistic and naturalistic), to negative attitudes where there is a primary concern for the material value of animals, the control of animals, or the avoidance of animals. He developed a number of scales to measure the attitude types, and found that committed birdwatchers were among the most knowledgeable of a number of animal-oriented groups.

Attitude research will help to further differentiate between ecotourists and non-ecotourists, and soft and hard path ecotourists. In regard to the continuum of ecotourists, the results of attitude-based analyses may be used for the purpose of structuring different programmes and experiences for these disparate groups. The assumption to be made is that different types of ecotourists want different types of experiences.

Impacts

Mayur (1996) wrote that environmentalists often focus too strongly on the symptoms of environmental deterioration, rather than on their underlying causes. He likened this to administering cough medicine to a tuberculosis patient. While the cough may abate, the untended disease may yet kill. This is certainly also true of the tourism industry, which by many accounts is one of the worst representations of unfettered capitalism. Our predisposition as tourism researchers is to focus on the impacts of the sector, and this has left us with little means by which to mitigate and control such impacts. This is most emphatically stated by Meadows *et al.* (1972, p. xi):

> It is the predicament of mankind that man can perceive the problematique yet despite his considerable knowledge and skills, he does not understand the origins, significance and interrelationships of its many components and is thus unable to devise effective responses. This failure occurs in large part because we continue to examine single items in the problematique without understanding that the whole is more than the sum of the parts, and that change in one element means change in the others.

Some researchers (see Dowling, 1993; McKercher, 1993) cite a fundamental lack of sound models and data to aid in the continuing struggle to overcome the impact dilemma (impacts are often identified but not controlled). Now, at the turn of the century, it is painfully apparent that there is still an abundance of unresolved issues that confront the tourism industry, some which may prevail despite our best remedial efforts. The recent development of tourism impact scales may help in our attempts to better understand the pressure that tourism exerts on various regions. The 35-item tourism impact scale developed by Ap and Crompton (1998), for example (based on a number of domains, including society and culture, economics, crowding and congestion, and environment) is an example of such a measure. A predecessor to this scale is the tourism impact attitude scale which Lankford (1994) developed and implemented in Oregon and Washington, USA, for the purpose of gauging attitudes and perceptions toward tourism and rural development.

Impacts will continue to play a significant role in determining what is/is not acceptable in terms of appropriate and responsible tourism development. This holds true for the tourism industry as a whole but also ecotourism where the impact scales, as discussed above, may be utilized to examine the perceptions of developers, local inhabitants, and tourists. In particular, ecotourists may be contrasted against other types of tourists in order to gain an understanding of the pressure that each exerts in different environments of a destination.

Carrying capacity and norms

A closely related concept to impacts is carrying capacity. Since the mid-1960s, researchers in outdoor recreation have looked closely at issues related to numbers of participants in outdoor settings and their effects on the natural world (see Lucas, 1964; Wagar, 1964). An assumption was made that by controlling numbers to these settings, many of the effects would simply disappear. (Carrying capacity can be defined as the amount of use a particular area can absorb over time before there is an unacceptable impact to either other users or the resource base.) In one of the earliest studies on the concept, Wagar (1964, p. i) wrote that:

> The study ... was initiated with view that the carrying capacity of recreation lands could be determined primarily in terms of ecology and the deterioration of the areas. However, it soon became obvious that the resource-oriented point of view must be augmented by consideration of human values.

From this study and others, researchers have learned not only that biological environments are dynamic in reference to carrying capacity, but so too are human values, needs, benefits, expectations and levels of satisfaction. The setting of specific numerical limits in outdoor settings, therefore, is often not the best course of action in controlling the effects of outdoor recreational use. As such, according to Lindberg *et al.* (1997), the focus has shifted away from 'How many is too many?' to one of 'What are the desired conditions?'. In response to the shortcomings of carrying capacity, a number of preformed planning and management frameworks have been developed with the purpose of balancing biological and social components of outdoor recreation settings, experience, and use. These include the recreation opportunity spectrum (ROS), limits of acceptable change (LAC), visitor impact management (VIM) and the visitor activity management process (VAMP). All are USA-developed except VAMP, which was developed for Parks Canada (an excellent overview of these models can be found in Payne and Graham, 1993). In general, these models have been virtually untouched by tourism researchers. Exceptions include the work of Butler and Waldbrook (1991) who adapted the ROS into a tourism opportunity spectrum model, and Harroun (1994), who discussed VIM and LAC as models appropriate for analysing ecological impacts in developing countries. In addition, Dowling (1993) developed his own tourism-specific model, the environmentally based tourism planning framework, which links tourism development and environmental conservation.

The value of ROS, LAC, VAMP and other preformed planning and management models is that they lend an element of sophistication to the management of people in protected areas, such as national parks. As such, there is merit in using these models or re-fitting or creating new models, as Dowling, and Butler and Waldbrook have done, in order to better plan and manage the tourism industry. A constraint to their use is the fact that both expertise and resources are needed to implement these models, and given their comprehensiveness, the models may have the potential of intimidating individuals and/or agencies lacking social and ecological planning backgrounds.

Finally, and briefly, a strong theme in outdoor recreation research over the past few years is the adoption of encounter norms or crowding norms as a means by which to objectively and systematically determine levels of use in outdoor settings. As outlined by Lewis *et al.* (1996), encounter norms (there is a difference between personal norms and group norms) are viewed as 'visitors' individual or shared beliefs about appropriate use levels and social situations' (p. 144). Acceptable levels of other users and user groups may therefore differ within and between groups. For example, canoeists may feel more crowded, even though they encounter fewer other parties during the day, than white-water rafters. The assumption is that canoeists want a much different experience from rafters, and expectations of the setting will probably differ between the two groups (see Roggenbuck *et al.*, 1991; Shelby and Vaske, 1991; Heywood, 1996).

Animal–human interactions

By nature, ecotourism is an activity that involves an interaction between people and nature (plants and animals). At times this interaction can be antagonistic. For example, Burger *et al.* (1995) reported that different bird species have different levels of tolerance, and are not consistent in their responses to ecotourists. In this study the researchers found that birds have distinct behavioural patterns at different times of the year (breeding, migration, feeding and so on). In general the authors felt that managers need to consider the use of separate management techniques for different species in various settings. It was suggested that the following needed to be considered in understanding human–bird interactions:

1. *Response distance.* The distance between the bird and the intruder at which the bird makes some visible or measurable response.
2. *Flushing distance.* The distance at which the bird actually leaves the site where it is nesting or feeding.
3. *Approach distance.* The distance to which one can approach a bird, head-on, without disturbing it.
4. *Tolerance distance.* The distance to which one can approach a bird without disturbing it, but in reference to passing by the bird tangentially.

Blane and Jackson (1994) further acknowledge the importance of strict environmental monitoring and management in order to safeguard the ecotourism industry and wildlife. On the basis of some 370 observations of ecotourism boat–whale interactions, they discovered a series of different whale behaviour patterns. These included avoidance behaviour (increasing their speed and surfacing less often), interactions between whales and boats (investigating the boats), pre-disturbance behaviour (in 75% of cases, whales resumed their pre-disturbance behaviour), location (whales reacted differently in different settings), and boat variables (boat type, speed, range and angle of approach). While other research supports the fact that killer whales do not seem to be affected by regulated boat traffic (see Obee, 1998), there does appear to be the need to further document animal–human interactions among different species, and in different settings. It is the environment and the level and type of use which may determine the response patterns of animals like whales.

Another area of research on human–wildlife encounters that has potential relates to the concept of habituation. This doctrine, which has long been a topic of interest for psychologists, involves learning *how not* to respond to a stimulus. When an individual is exposed to some type of stimulus, there is often an immediate and vigorous response. However, after repeated and sustained exposure to the stimulus, responses typically lessen and may disappear altogether. Habituation may provide the needed explanation as to why animals in zoos remain indifferent to the overtures of visitors on a day-to-day basis, and also explain the lethargy demonstrated by a pride of lions at the arrival of several vehicles. The extent to which ecotourism affects these species, in various settings, should also be further explored in an attempt to ameliorate any negative effects. Variables such as frequency of interaction, numbers of tourists, time, spatial patterns and sensitivities with respect to predator and prey relationships could be considered in the decisions of whether to include a species on an ecotourism itinerary or not.

As a final note, funding bodies, at least in Canada, have consistently shown a preference for financing the work of biologists and other natural scientists in natural areas over social research. Perhaps in our efforts to understand and manage ecosystems we have spent too much time and money radio-collaring bears and snakes, when we ought to be radio-collaring tourists. The oft-quoted adage that 'bear management is 90% people management' should be taken seriously. It means that the real problem in parks and protected areas is tourists, not bears. Park management has recognized the importance of the social sciences in managing natural areas over the past 15 years, but there is still a great deal of work required. This includes obtaining an understanding of the spatial and temporal movement of tourists (and different tourism groups), and the pressure that they exert on different regions of a destination, as stated earlier.

Consequently, there is a moral issue at hand which ought to be considered in deciding what qualifies as acceptable human–wildlife interactions. As ecotourism continues to expand in how it is interpreted and practised, ecotourism policy makers and leaders will be forced to take a closer look at current definitions of the term. Activities such as fishing qualify as ecotourism under some broad definitions of ecotourism (see Holland *et al.*, 1998). However, Fennell (2001) argues that such pursuits, which intentionally harm and physically and emotionally stress animals or resources, are more anthropocentric in nature and lie outside the realm of acceptable ecotourism practice.

Conclusion

There is a myriad of research theories, techniques and approaches that have been developed in the social and natural sciences which, if pursued, will add value to ecotourism investigations. The intriguing thing about the field of ecotourism is that it bridges the gap between these social and natural science realms. While this is exciting, it also presents itself as a challenge to find common ground and to bring meaning and direction to this new field of study. While advocates of ecotourism argue that it is a more responsible form of tourism, the theories and philosophies stated above may be used to empirically demonstrate whether a particular 'ecotourism' product or activity can be considered responsible (whatever 'responsible' means). It should further be noted that some of the points made in this paper are open to debate, and have been raised to stimulate much-needed discussion on where ecotourism research perhaps should go in the future. Hopefully researchers will continue to address many of the most pressing issues that confront the ecotourism industry and find commonalities between applied and theoretical research, using natural and social investigative means.

References

Albrecht, D., Bultena, G., Hoiberg, E. and Nowak, P. (1982) The new environmental paradigm scale. *Journal of Environmental Education* 13, 39–43.

Ap, J. and Crompton, J.L. (1998) Developing and testing a tourism impact scale. *Journal of Travel Research* 37, 120–130.

Ardrey, R. (1970) *The Social Contract.* Dell, New York.

Blane, J. and Jackson, R. (1994) The impact of ecotourism boats on the St Lawrence beluga whales. *Environmental Conservation* 21(3), 267–269.

Blangy, S. and Nielson, T. (1993) Ecotourism and minimum impact policy. *Annals of Tourism Research* 20(2), 357–360.

Bookbinder, M.P., Arun, E. and Arup, H. (1998) Ecotourism's support of biodiversity conservation. *Conservation Biology* 12(6), 1399–1404.

Brooks, L.J. (1989) Corporate codes of ethics. *Journal of Business Ethics* 8, 117–129.

Burger, J., Gochfeld, M. and Niles, L.J. (1995) Ecotourism and birds in coastal New Jersey: contrasting responses to birds, tourists, and managers. *Environmental Conservation* 22(1), 56–65.

Butler, R.W. and Waldbrook, L.A. (1991) A new planning tool: the tourism opportunity spectrum. *The Journal of Tourism Studies* 2(1), 2–14.

Christie, R. and Geis, F.L. (eds) (1970) *Studies in Machiavellianism.* Academic Press, New York.

Dawkins, R. (1976) *The Selfish Gene.* Oxford University Press, New York.

Denison, D.R. (1996) What is the difference between organizational culture and organizational climate? A native's point of view on a decade of paradigm wars. *Academy of Management Review* 21, 619–654.

Dowling, R. (1993) An environmentally based planning model for regional tourism development. *Journal of Sustainable Tourism* 1(1), 17–37.

Driver, B.L., Tinsley, H.E.A. and Manfredo, M.J. (1991) The paragraphs about leisure and recreation experience preference scales: results from two inventories designed to assess the breadth of the perceived psychological benefits of leisure. In: Driver, B.L., Brown, P.J. and Peterson, G.L. (eds) *Benefits of Leisure.* Venture, State College, Pennsylvania.

Dunlap, R.E. and Van Liere, K. (1978) The 'new environmental paradigm'. *Journal of Environmental Education* 9(4), 10–19.

Dyess, R. (1997) Adventure travel or ecotourism? *Adventure Travel Business* April, 2.

Edney, J. (1980) The commons problem. *American Psychologist* 35, 131–150.

Ewert, A. (1985) Why people climb: the relationship of participant motives and experience level to mountaineering. *Journal of Leisure Research* 17(3), 241–250.

Fennell, D.A. (2000a) Tourism and applied ethics. *Tourism Recreation Research* 25(1), 59–69.

Fennell, D.A. (2000b) What's in a name? Conceptualising natural resource-based tourism. *Tourism Recreation Research* 25(1), 97–100.

Fennell, D.A. (2001) Ecotourism on trial: the case of billfish angling as ecotourism. *Journal of Sustainable Tourism* 8, 341–345.

Fennell, D.A. and Malloy, D.C. (1999) Measuring the ethical nature of tourism operators. *Annals of Tourism Research* 26(4), 929–943.

Fontaine, G. (1994) Presence seeking and sensation seeking as motives for international travel. *Psychological Reports* 75, 1583–1586.

Fraedrich, J.P. (1993) The ethical behaviour of retail managers. *Journal of Business Ethics* 12, 207–218.

Geller, J.M. and Lasley, P. (1985) The new environmental paradigm scale: a reexamination. *Journal of Environmental Education* 17(1), 9–12.

Hall, S. (1989) Ethics in hospitality: how to draw your line. *Lodging* Sept., 59–61.

Hardin, G. (1968) The tragedy of the commons. *Science* 162, 1243–1248.

Harroun, L.A. (1994) *Potential Frameworks for Analysis of Ecological Impacts of Tourism in Developing Countries.* WWF, Washington, DC.

Hegarty, W.H. and Sims, H.P. (1979) Organizational philosophy, policies, and objectives related to unethical decision behavior: a laboratory experiment. *Journal of Applied Psychology* 64(3), 331–338.

Heywood, J.L. (1996) Conventions, emerging norms, and outdoor recreation. *Leisure Sciences* 18, 355–363.

Holland, S.M., Ditton, R.B. and Graefe, A.R. (1998) An ecotourism perspective on billfish fisheries. *Journal of Sustainable Tourism* 6(2), 97–116.

Hunt, S.D. and Vitell, S. (1986) A general theory of marketing ethics. *Journal of Macromarketing* 6(1), 5–16.

Jackson, L.E. (1986) Outdoor recreation participation and attitudes to the environment. *Leisure Studies* 5, 1–23.

Jurowski, C., Muzaffer, U., Williams, D.R. and Noe, F.P. (1995) An examination of preferences and evaluations of visitors based on environmental attitudes: Biscayne Bay National Park. *Journal of Sustainable Tourism* 3(2), 73–86.

Kahle, L.R. (1983) Dialectical tensions in the theory of social values. In: Kahle, L.R. (ed.) *Social Values and Social Change.* Praeger, New York.

Kellert, S.R. (1985) Birdwatching in American society. *Leisure Sciences* 7(3), 343–360.

Laarman, J.G. and Durst, P.B. (1987) Nature travel and tropical forests, FPEI Working Paper Series, Southeastern Center for Forest Economics Research, North Carolina State University, North Carolina.

Laczniak, G.R. and Murphy, P.E. (1985) Incorporating marketing ethics into the organization. In: Laczniak, G.R. and Murphy, P.E. (eds) *Marketing Ethics: Guidelines for Managers.* Lexington Books, Lexington, Massachusetts, pp. 97–105.

Lankford, S.V. (1994) Attitudes and perceptions toward tourism and rural regional development. *Journal of Travel Research* 32, 35–43.

Lewis, M.S., Lime, D.W. and Anderson, D.H. (1996) Paddle canoeists' encounter norms in Minnesota's Boundary Waters Canoe Area Wilderness. *Leisure Sciences* 18, 143–160.

Lindberg, K. (1991) *Policies for Maximising Nature Tourism's Ecological and Economic Benefits.* World Resources Institute, Washington, DC.

Lindberg, K., Enriquez, J. and Sproule, K. (1996) Ecotourism questioned: case studies from Belize. *Annals of Tourism Research* 23(3), 543–562.

Lindberg, K., McCool, S. and Stankey, G. (1997) Rethinking carrying capacity. *Annals of Tourism Research* 24(2), 461–464.

Lucas, R.C. (1964) Wilderness perception and use: the example of the Boundary Waters Canoe Area. *Natural Resources Journal* 3(3), 394–411.

Madrigal, R. (1995) Personal values, traveller personality type, and leisure travel style. *Journal of Leisure Research* 27(2), 125–142.

Madrigal, R. and Kahle, L.R. (1994) Predicting vacation activity preferences on the basis of value-system segmentation. *Journal of Travel Research* 32(3), 22–28.

Malloy, D.C. and Fennell, D.A. (1998) Ecotourism and ethics: moral development and organizational cultures. *Journal of Travel Research* 36, 47–56.

Masterton, A.M. (1992) Environmental ethics. *Island Destinations* (a supplement to *Tour and Travel News*) November, 16–18.

Mayur, R. (1996) *Earth, Man and Future.* International Institute for Sustainable Future, Mumbai, India.

McKercher, B. (1993) Some fundamental truths about tourism: understanding tourism's social and environmental impacts. *Journal of Sustainable Tourism* 1(1), 6–16.

Meadows, D.H., Meadows, D.L., Randers, J. and Behrens, W.W. (1972) *The Limits to Growth.* New American Library, New York.

Mitchell, A. (1983) *The Nine American Lifestyles.* MacMillan, New York.

Obee, B. (1998) Ecotourism boom: how much can wildlife take? *Beautiful British Columbia* 40(1), 6–17.

Oishi, S., Schimmack, U., Diener, E. and Suh, E.M. (1998) The measurement of values and individualism-collectivism. *Personality and Social Psychology Bulletin* 24(11), 1177–1189.

Payne, R.J. and Graham, R. (1993) Visitor planning and management in parks and protected areas. In:

Dearden, P. and Rollins, R. (eds) *Parks and Protected Areas in Canada.* Oxford University Press, Toronto, pp. 185–210.

Payne, S.L. (1980) Organizational ethics and antecedents to social control processes. *Academy of Management Review* 5(3), 409–414.

Platt, J. (1973) Social traps. *American Psychologist* 28, 641–651.

Reidenbach, R.E. and Robin, D.P. (1988) Some initial steps toward improving the measurement of ethical evaluations of marketing activities. *Journal of Business Ethics* 7, 871–879.

Reidenbach, R.E. and Robin, D.P. (1990) Toward the development of a multidimensional scale for improving evaluations of business ethics. *Journal of Business Ethics* 9, 639–653.

Rest, J.R. (1979) *Development in Judging Moral Issues.* University of Minnesota Press, Minneapolis.

Roggenbuck, J.W., Williams, D.R., Bange, S.P. and Dean, D.J. (1991) River float trip encounter norms: questioning the use of the social norms concept. *Journal of Leisure Research* 23(2), 133–153.

Schein, E.H. (1985) *Organizational Culture and Leadership.* Jossey-Bass, San Francisco.

Schwartz, S.H. (1994) Are there universal aspects in the structure and contents of human values? *Journal of Social Issues* 50, 19–45.

Shelby, B. and Vaske, J.J. (1991) Using normative data to develop evaluative standards for resource management: a comment on three recent papers. *Journal of Leisure Research* 23(2), 173–187.

Singhapakdi, A. and Vitell, S.J. (1990) Marketing ethics: factors influencing perceptions of ethical problems and alternatives. *Journal of Macromarketing* 12, 4–18.

Sirakaya, E. and McLellan, R.W. (1998) Modelling tour operators' voluntary compliance with ecotourism principles: a behavioral approach. *Journal of Travel Research* 36, 42–55.

Sproule, K. (1996) Community-based ecotourism development: identifying partners in the process. Paper presented at the Ecotourism Equation: Measuring the Impacts Conference. School of Forestry and Environmental Studies. 12–14 April.

Vitell, S.J. and Singhapakdi, A. (1991) Factors influencing the perceived importance of stakeholder groups in situations involving ethical issues. *Business and Professional Ethics Journal* 10(3), 53–72.

Wagar, J.A. (1964) The carrying capacity of wildlands for recreation. Society of American Foresters, *Forest Service Monograph* 7, 23.

Wheeler, M. (1992) Applying ethics to the tourism industry. *Business Ethics – a European Review* 1(4), 227–235.

Wheeler, M. (1994) The emergence of ethics in tourism and hospitality. *Progress in Tourism, Recreation, and Hospitality Management* 6, 46–56.

Whitney, D.L. (1989) The ethical orientations of hotel managers and hospitality students: implications for industry, education and youthful careers. *Hospitality Education and Research Journal* 13(3), 187–192.

Zuckerman, M. (1979) *Sensation Seeking: Beyond the Optimal Level of Arousal.* Erlbaum, Hillsdale, New Jersey.

Appendix: Evaluation of Ecotour Operator Ethics

This evaluation form includes many of the most frequently cited aspects and concerns related to an ecotour experience. Please rate your ecotour operator's ethical behaviour on how well they satisfied each of the criteria below, from a 'Superior', 'Adequate' or 'Inferior' perspective.

Operator: ____________________ Tour dates: ____________________

Tour description: ______________________________________

Evaluation criteria	**Operator's ethical behaviour**			
	Superior	**Adequate**	**Inferior**	**N/A**
General ethics:				
Respect shown to animals				
Respect shown to plants				
Operator's treatment of financial matters related to the tour				
Respect shown to local people				
Operator following a code of ethics				
Respect shown to ecotourists				
Local people:				
Operator employing local/indigenous people as guides or front-line personnel				
Operator efforts to employ local/ indigenous people in middle or upper management positions				
Local people were in control of the decision making in relation to ecotourism in their community				
Local people owned and ran the ecotour operation				
Environmental education:				
The interpretation of the natural world				

Evaluation criteria	Operator's ethical behaviour			
	Superior	**Adequate**	**Inferior**	**N/A**
Knowledge of plant and animal life				
Level of pre-trip information provided to the client				
Level of on-site information provided to the client				
Operator/guide had formal education in ecotourism or ecotourism-related fields				
Operator professionalism:				
Operator/guide was accredited or certified				
Knowledge of land use and park policies				
Knowledge of the site or region itself (e.g. location, history, weather)				
A low guide-to-participant ratio				
Language proficiency of guides				
Degree of safety of the experience				
Understanding of the importance of minimizing tourism impacts				
Contribution to conservation:				
Monetary (e.g. park fees, support of local nature groups)				
Physical (e.g. removal of garbage, tree planting)				
Accommodation and transportation:				
Reliance on 'green' hotels				
Use of hotels that are locally owned				
Use of hotels that are built with local materials				
Use of hotels that have a low carrying capacity				
Use of vans/buses that use cleaner petroleum				
Over reliance of vehicles at ecotour sites				

Glossary

(**Bold** terms within definitions are listed elsewhere as glossary main entries)

ACE tourism: a hybrid form of tourism that combines Adventure, Cultural and Ecotourism; recognizes that many tourism products, such as **trekking**, combine a variety of experiences, attractions and motivations, and therefore cannot be neatly placed within a single category.

Accreditation: a quality control mechanism that formally recognizes businesses or products that meet nominated industry standards, usually associated with sectoral **best practice**; the National Ecotourism Accreditation Program (NEAP) of Australia is a world leader in ecotourism accreditation, but most countries are far behind, indicating the relative immaturity of this sector overall.

Adaptancy platform: a view of tourism popular during the 1980s which put forward options such as **sustainable tourism** and ecotourism that were perceived to be more benign and appropriate than **mass tourism** for most kinds of destinations; it was a logical continuation of the **cautionary platform**, and adhered to a similar set of ideological assumptions about tourism and development.

Adventure tourism: usually a form of **nature-based tourism** that incorporates an element of risk, higher levels of physical exertion, and the need for specialized skills; often hybridizes with ecotourism and other forms of tourism, as in **ACE tourism**.

Advocacy platform: a capitalist perspective, dominant prior to the 1970s, in which tourism is perceived as a great benefit for virtually any kind of community; more tourism equates with more benefits, and hence **mass tourism** is the best possible scenario.

Alternative tourism: tourism that is deliberately fostered as a more appropriate small-scale, community-controlled option to **mass tourism** in environmentally or socio-culturally sensitive destinations; ecotourism was originally conceived as an environmentally based form of alternative tourism during the era of the **adaptancy platform**.

Auditing: a process that a business undergoes to identify and confirm benchmarks, provide **accreditation** with reliability and validity, and measure and verify **best practice**; to be credible, auditing must be carried out by a third party, though businesses can and should conduct their own internal audits as a prelude to the latter.

Benchmarking: a quality control mechanism in which the performance of an operation in specified areas is evaluated through comparison with similar operations, and usually those that adhere to sector **best practice**; the implication is that the sector constantly improves as businesses strive to meet and exceed existing best practice standards.

Best practice: a set of operational standards that are considered to be the most effective and efficient within a sector with respect to certain desired outcomes, such as environmental sustainability and effective interpretation in the case of ecotourism; **accreditation** of businesses and products is usually based on adherence to best practice standards, which are commonly recognized through awards.

Buffer zone: an area, usually surrounding a high order **protected area**, that acts as a cushion between the latter and external spaces that are unprotected; buffer zones usually have the status of a low order multi-use protected area, and often provide accommodations and other services and facilities for ecotourists.

Carrying capacity: the amount of tourism-related activity that a site or destination can sustainably accommodate; often measured in terms of visitor numbers or visitor-nights over a given period of time, or by the number of available accommodation units; management techniques such as **site hardening** can be employed to raise a site's carrying capacity.

Cautionary platform: a left-leaning 'anti-tourism' perspective that emerged in the late 1960s as a reaction to the perceived excesses of unrestricted **mass tourism**, especially within the Third World.

Code of ethics: a list of recommended practices intended to effect environmentally and/or socio-culturally sustainable behaviour and outcomes within the targeted group, such as the International Association of Antarctic Tour Operators (IAATO); often criticized for being vague, voluntary and based on a system of self-regulation, codes of ethics are considered to be a weak type of quality control mechanism.

Constant capital rule: an outcome of **sustainable development**, wherein one generation leaves the same amount of 'capital' to the next generation that it inherited from the previous generation; there is considerable disagreement about the extent to which man-made and semi-natural capital can and should be substituted for natural capital.

Consumptive tourism: commonly refers to hunting and fishing, which extract or 'consume' resources from the natural environment; the term is contentious, since it can be argued that all forms of tourism have both a consumptive and non-consumptive element; the common tendency to equate consumptive with unsustainable is also unwarranted.

Demonstration effect: process whereby local people, through direct exposure to tourists, begin to imitate their behaviour, often resulting in the modification or shunning of their own culture; usually associated with mass tourism, though the intrusion of ecotourists into the **host community** may induce a similar effect.

Dominant social paradigm: a view of man's technological and scientific supremacy that has dominated Western society for the past 300 years, and is commonly regarded as the underlying cause of many contemporary environmental and social problems, including unsustainable **mass tourism**; accordingly, is now being challenged by the **new environmental paradigm**.

Ecolodge: a specialized form of accommodation that caters specifically to ecotourists; usually a small, upmarket facility located in or near a **protected area** or wilderness setting; ecolodges are a high profile form of ecotourism accommodation, but account for only a small proportion of all ecotourist visitor-nights.

Ecoresort complex: a hybrid form of accommodation/attraction that incorporates features of a conventional resort and an **ecolodge**; that is, these are responses to changing market trends that tend to be large, managed in an environmentally sustainable way, and designed to provide a variety of recreational opportunities, including ecotourism.

Ecotourism: a form of tourism that is increasingly understood to be: (i) based primarily on nature-based attractions; (ii) learning-centred; and (iii) conducted in a way that makes every reasonable attempt to be environmentally, socio-culturally and economically sustainable.

Ecotourism opportunity spectrum (ECOS): a model, incorporating elements of the recreational opportunity spectrum (ROS) and the tourism opportunity spectrum (TOS), that identifies the viability of potential ecotourism sites by assessing these against various criteria, and rating them on a scale ranging from an eco-specialist to an eco-generalist orientation.

Ecotourism organization: a membership-based, **non-government organization** that is focused on the promotion and enhancement of ecotourism within a particular jurisdiction; these occur at a global (The International Ecotourism Society), national (e.g. Ecotourism Association of Australia) and sub-national scale.

Endemism: the degree to which the flora and fauna of a place occur naturally only in that place; a high level of endemism is associated with isolated locations such as islands and **rainforest** valleys; the implication for ecotourism is that those wishing to see these species in their natural environment can only do so by visiting the place where these endemic species are found.

Hard ecotourism: ecotourism that tends toward longer, specialized trips by small groups within a wilderness or semi-wilderness setting mediated by minimal services; also called active, deep or eco-specialist ecotourism, this constitutes only a very small portion of the total ecotourism sector.

Host community: a group of people in a small-scale destination, usually permanent residents, who are thought to have a common interest and bond in maintaining a high quality of life for themselves; support by the host or local community is now widely considered crucial for tourism or ecotourism in particular to be successful, and this is often achieved through community control and involvement in tourism.

Iconic attraction: an attraction that symbolizes and dominates a destination; iconic ecotourism attractions include the Great Barrier Reef of Australia and Kruger National Park of South Africa.

Indigenous people: a term that is open to some interpretation, but generally referring to the original inhabitants of an area; they often form a dominant and increasingly assertive population group in peripheral and relatively undisturbed areas, and hence are growing in importance as a stakeholder in the ecotourism sector.

Input–output (IO) analysis: a class of statistical techniques that are used to calculate the economic impacts of tourism within a destination; several empirical studies have focused on ecotourism.

Interpretation: in ecotourism, the process whereby the significance and meanings of natural and associated cultural phenomena are revealed to visitors, usually with the intent of providing a satisfying learning experience while at the same time inducing and encouraging more sustainable behaviour among those experiencing this interpretation.

Inter-sectoral realm: other sectors, especially agriculture, fishing, forestry, and various conventional forms of tourism, that use the same resource base as ecotourism; the relationship between these sectors and ecotourism can range from conflicting to complementary; the inter-sectoral realm is part of the broader external environment surrounding ecotourism, which also includes government.

IUCN classification system: a system put forward by the World Conservation Union (IUCN) that divides **protected areas** into seven basic categories, according to the types of human intervention that are permitted in each; this system eliminates the confusion arising from each jurisdiction establishing its own protocol, and as a result has become the international standard for protected area classification.

Knowledge-based platform: a tourism perspective appearing in the early 1990s that attempted to transcend the ideological biases of earlier platforms by emphasizing the importance of applying sound scientific techniques to obtain knowledge about tourism; attempted to divorce the association between scale and value judgements about tourism.

Limits of acceptable change (LAC): a land management philosophy that identifies specific indicators of environmental quality and tourism impacts, and defines thresholds within which the conservation goals of a **protected area** are met.

Marine protected area (MPA): an area of sea or ocean that has been designated as a **protected area**; the global MPA system is regarded as being in an incipient state of development compared with the terrestrial system.

Market segmentation: the division of a larger consumer market into smaller, homogeneous groups on the basis of geographical, socio-demographic, psychographic and/or behavioural characteristics, in order to better effect target marketing and management strategies; ecotourists are commonly segmented into **hard ecotourism** and **soft ecotourism** sub-groups.

Mass tourism: large-scale tourism, typically associated with 3S (sea, sand, sun) resorts and characteristics such as transnational ownership, high leakage effects, seasonality, and package tours; ecotourism can conceivably be a form of mass tourism under the logic of the **knowledge-based platform**, whereas previously the two were considered mutually exclusive.

Multiplier effect: the amount of ongoing indirect and induced economic activity that is generated within a destination as a result of direct tourist expenditures; ecotourism is commonly assumed to generate a high multiplier effect because of the involvement of local communities.

National park: often used synonymously with **protected area**, and used by various jurisdictions as a formal designation to describe a range of protected area arrangements; the term is most effectively employed, however, as the name for an IUCN category II protected area, that is, a highly protected space that is managed to accommodate a sustainable level of visitation; this is the most important type of protected area from an ecotourism perspective.

Nature-based tourism: any form of tourism that relies primarily on the natural environment for its attractions and/or settings; incorporates ecotourism as well as substantial portions of **adventure tourism** and **3S tourism**, neither of which are necessarily sustainable or learning-centred.

New environmental paradigm (NEP): a biocentric or ecocentric way of thinking that advocates a holistic, integrative view of humanity, and places people within and not above nature; this view, also called the green paradigm, is challenging the **dominant social paradigm** that places man above nature; ecotourism can be seen as one manifestation of the NEP.

Non-government organization (NGO): an association or body that is not tied to a government, and usually operates in relation to a specific sector; membership-based **ecotourism organizations**, for example, focus on ecotourism, while environmental NGOs such as Conservation International are also involved in this sector as an adjunct to their broader conservation mandate.

Plan: a document that articulates the planning process through which a given **policy** is implemented; a few countries, such as Australia, have well-articulated and well-supported ecotourism plans, as do some sub-national jurisdictions such as the states of Australia.

Policy: a course of action, usually by government, that provides the broad guidelines for shaping the development of a particular sector or sectors in a way deemed desirable; some countries such as Australia have a focused ecotourism policy, though in most cases ecotourism is indirectly affected by broader government policies.

Protected area: a designated portion of land or water (i.e. **marine protected areas**) to which regulations and restrictions have been applied, thereby affording a given degree of protection against on-site activities that threaten the environmental integrity of the area; protected areas are usually described as being either public or private, and are most commonly categorized according the **IUCN classification system**.

Qualitative paradigm: a theoretical and methodological framework for research in ecotourism and other areas that builds a complex and holistic knowledge base through the analysis of words; sometimes said to derive detailed information about a small sample of subjects.

Quantitative paradigm: a theoretical and methodological framework for research in ecotourism and other areas that measures phenomena with numbers, and analyses these with appropriate statistical techniques to derive predictive generalizations; sometimes said to derive limited information about a large number of informants.

Rainforest: a closed canopy, layered forest that results in tropical, subtropical or humid temperate environments from high levels of precipitation; these are considered one of the most attractive settings for ecotourism, although deforestation and degradation are steadily reducing the amount and quality of the world's rainforests.

Restoration ecotourism: ecotourism that focuses on the rehabilitation or reconstruction of degraded environments; provides an incentive for such efforts, and offers opportunities for volunteer participation.

Safari tourism: refers to a wildlife-viewing expedition, usually undertaken by small groups in the **savannah** regions of sub-Saharan Africa by way of four-wheel drive or other motorized transport; in the past, also referred to as a hunting expedition.

Savannah: an area of subtropical open woodland, shrubs and grass; as an ecotourism venue, these occur mainly in southern and eastern Africa, where the presence of numerous big game animals has attracted a thriving **safari tourism** sector in countries such as Kenya and South Africa.

Site hardening: the implementation of site modifications, such as the paving of a hiking trail or the construction of a sewage treatment system, that increase the **carrying capacity** of that site to receive visitors.

Soft ecotourism: **ecotourism** that tends toward shorter, multi-purpose trips within well-serviced areas frequented by large numbers of soft ecotourists; also called passive, shallow, popular or eco-generalist ecotourism, this accounts for most ecotourism activity.

Sustainability indicators: variables that provide information about the extent to which a particular destination is environmentally, socio-culturally and/or economically sustainable; the identification of appropriate indicators and their critical thresholds is a major challenge for operationalizing the concept of **sustainable tourism**, and ecotourism specifically.

Sustainable development: development carried out in such a way as to meet the needs of the present without compromising the ability of future generations to meet their own needs; an elusive and complex concept popularized in 1987 by the Brundtland Report, and since used as an underlying principle and objective within many sectors, including tourism and ecotourism; the **constant capital rule** is an example of the underlying complexity of this concept.

Sustainable tourism: as a direct follow-up to the concept of **sustainable development**, sustainable tourism is tourism that meets the needs of the present generation without compromising the ability of future generations to meet their own needs; more commonly perceived as tourism that does not negatively impact the environment, economy, culture and society of a particular destination; ecotourism is a form of sustainable tourism.

3S tourism: sea, sand and sun tourism, usually equated with **mass tourism** in a coastal resort setting; ecotourism complements 3S tourism in destinations such as Costa Rica, Kenya and Australia, and overlaps with 3S tourism in activities such as scuba-diving.

Tour operator: a critical intermediary in the tourism system that is responsible for the design, organization, packaging, marketing and operation of vacation and other tours at the outbound, inbound or local level; these are becoming increasingly important as ecotourism becomes larger and more formalized.

Trekking: a form of **ACE tourism** that incorporates elements of long-distance walking and exposure to local cultures and the natural environment; commonly associated with the Himalayas and northern Thailand.

Whale-watching: a specialized form of marine-based ecotourism that also includes other cetaceans such as dolphins.

Wilderness: a subjective concept, but generally regarded in Western culture as a relatively extensive area of mainly undisturbed natural environment; considered to be an important venue for **hard ecotourism**.

Zoning: a management technique, commonly used within protected areas, whereby certain areas are designated to accommodate specific kinds of tourism and other activity; in part, these designations are based on inherent **carrying capacity**, though once designated, measures such as **site hardening** may be implemented to raise the area's carrying capacity.

Index